CSA Standards Update Service

C22.1-06
January 2006

Title: *Canadian Electrical Code, Part I*
Pagination: 605 pages (xxxi preliminary and 574 text)

Automatic notifications about any updates to this publication are available online.

To register for e-mail notifications, and/or to download any existing updates in PDF, enter the Online Store at **www.ShopCSA.ca** and click on **My Account** on the navigation bar. Then, click on **E-mail Services** and click on **CSA Standards Update Service**.

The **List ID** for this document is **2017350**. Please enter this **List ID** in the appropriate field to sign up for updates to this publication.

CSA Standard

C22.1-06
Canadian Electrical Code, Part I

Safety Standard for Electrical Installations

(Twentieth Edition)

CANADIAN STANDARDS ASSOCIATION

®Registered trade-mark of Canadian Standards Association

- *The* Canadian Electrical Code, Part I, *is a voluntary code for adoption and enforcement by regulatory authorities.*
- *The* Canadian Electrical Code, Part I, *meets the fundamental safety principles of International Standard IEC 60364, Electrical Installations of Buildings.*
- *Consult with local authorities regarding regulations that adopt and/or amend this Code.*

Published in January 2006 by Canadian Standards Association
A not-for-profit private sector organization
5060 Spectrum Way, Suite 100, Mississauga, Ontario, Canada L4W 5N6
1-800-463-6727 • 416-747-4044

Visit our Online Store at www.ShopCSA.ca

ISBN 1-55436-023-4

Technical Editor: Rick Gilmour

Contents

Committee on *Canadian Electrical Code, Part I*
(Membership lists as of August 2005)

A.Z. Tsisserev *(Chair)*	City of Vancouver, British Columbia
G. Lobay *(Vice-Chair)*	CANMET, Ottawa, Ontario
R.C. Gilmour *(Project Manager)*	CSA, Mississauga, Ontario

Representing **Provincial Electrical Inspection Authorities**

T. Collins	Government Service Centre, Cornerbrook, Newfoundland
J. Einarson	Department of Community Services, Whitehorse, Yukon
T. Kitson	Department of Community and Cultural Affairs, Charlottetown, Prince Edward Island
R. Leduc	Department of Municipal Affairs, Edmonton, Alberta
D.R.A. MacLeod	Department of Environment and Labour, Halifax, Nova Scotia
R. Marion	Government of Northwest Territories, Yellowknife, Northwest Territories
R. May	B.C. Safety Authority, New Westminster, British Columbia
W.G. McMullan	Manitoba Hydro, Winnipeg, Manitoba
G. Montminy	Régie du bâtiment du Québec, Québec, Québec
T. Olechna	Electrical Safety Authority, Mississauga, Ontario
S. Paulsen	Department of Public Safety, Fredericton, New Brunswick
L. Radom	SaskPower, Regina, Saskatchewan
E. Zebedee	Government of Nunavut Community and Government Services, Iqaluit, Nunavut

Representing **Municipal Electrical Inspection authorities**

M.S. Anderson	City of Winnipeg, Manitoba
M.D. Gardener	City of Calgary, Alberta
A.Z. Tsisserev	City of Vancouver, British Columbia

Representing **Committee on Use of Electricity in Mines**

G. Lobay	CANMET, Ottawa, Ontario

Representing **Electro-Federation Canada**

P. Desilets	Leviton Manufacturing of Canada Limited, Pointe-Claire, Québec
J. Neu *(Associate)*	Electro-Federation Canada, Mississauga, Ontario
B.F. O'Connell	Tyco Thermal Controls (Canada) Ltd., Trenton, Ontario
K.L. Rodel	Hubbell Canada LP, Pickering, Ontario
B. Savaria	Eaton Electrical Canada Operations, Burlington, Ontario
M. Smith	Rockwell Automation Canada Inc., Cambridge, Ontario

Representing **Forest Products Association of Canada**

T. Branch	Weyerhaeuser Dryden Operations, Dryden, Ontario

Representing **CSA Consumer Network**

D.H. Dunsire	Winnipeg, Manitoba

Representing **Railways**

B.A. Biglow	Edmonton, Alberta

Representing **Education**

P. McDonald	Northern Alberta Institute of Technology, Edmonton, Alberta
T. Simmons	British Columbia Institute of Technology, Burnaby, British Columbia

Representing Communication Industry

C.B. Chan	MTS Communication Inc., Winnipeg, Manitoba
J. Poulin	Bell Canada, Longueuil, Québec

Representing Labour

V. Clendenning	International Brotherhood of Electrical Workers, Winnipeg, Manitoba

Representing Canadian Electrical Contractors Association

P. Liberatore	Corporation of Master Electricians of Québec, Montréal, Québec

Representing Public Works Canada

A. Sutherland	Public Works and Government Services Canada, Hull, Québec

Representing Division of Building Research NRC

P. Rizcallah	National Research Council Canada, Ottawa, Ontario

Representing Canadian Electricity Association

R.A. Burpee	Saint John Energy, Saint John, New Brunswick
J. Cote	Hydro-Québec, Montréal, Québec
F.L. Kaempffer	British Columbia Hydro, Burnaby, British Columbia
H. Sam *(Associate)*	Canadian Electricity Association, Montréal, Québec

Representing National Elevator & Escalator Association

D. McColl	Otis Canada, Inc., Mississauga, Ontario

Representing Fire Insurers

G. Currie	Portage la Prairie Mutual Insurance Company, Portage la Prairie, Manitoba

Representing Canadian Association of Petroleum Producers

V.G. Rowe	Marex Canada Limited, Westbank, British Columbia

Representing Underwriters' Laboratories of Canada

N. Breton *(Associate)*	Underwriters' Laboratories of Canada, Scarborough, Ontario

Representing Bahamas

C. Bartlett *(Associate)*	The Grand Bahama Development Company Ltd., Grand Bahama Island, Bahamas
D. King *(Associate)*	Ministry of Works, Nassau, Bahamas

Representing International Association of Electrical Inspectors

J. Carpenter *(Associate)*	International Association of Electrical Inspectors, Richardson, Texas, USA
S.W. Douglas	Electrical Safety Authority, Cambridge, Ontario

Representing Underwriters Laboratories Inc.

T. Lichtenstein *(Associate)*	Underwriters Laboratories Inc., Northbrook, Illinois, USA

Representing National Electrical Code Committees

M.W. Earley *(Associate)*	National Fire Protection Association, Quincy, Massachusetts, USA

Representing National Electrical Manufacturers Association

J.T. Pauley *(Associate)*	Square D Company, Lexington, Kentucky, USA

Representing Mexico

M. Jimenez *(Associate)*	ANCE, Mexico
G. Merodio *(Associate)*	Condumex, Mexico

Representing Canadian Home Builders Association

D. Johnston *(Associate)*	Canadian Home Builders Association, Ottawa, Ontario

***Representing* National Electrical Contractors Association**

H.B. Stauffer *(Associate)*	National Electrical Contractors Association, Bethesda, Maryland, USA

Ex Officio Members

D.E. Clements	Nova Scotia Power Inc., Halifax, Nova Scotia
T.W. Odell	General Motors of Canada Limited, Oshawa, Ontario

Former Members

In addition to the members of the Committee, the following former members made valuable contributions to the development of this Code:

S. Bond	Toronto, Ontario
J.D. Chaplow	General Motors of Canada Limited, Oshawa, Ontario
R. Chauhan	National Research Council Canada, Ottawa, Ontario
D.A. Coleman	Ministry of Community, Aboriginal and Women's Services, Electrical and Elevating Devices Safety Branch, New Westminster, British Columbia
S.J. Coles	Eaton Electrical Canada Operations, Burlington, Ontario
R.E. Edwards	Alcan Cable, Mississauga, Ontario
E. Gabryl	Municipal Electric Association, Toronto, Ontario
J.R. Layden	Government Service Centre, St. John's, Newfoundland
E. Marinoff	Office of the Fire Marshal of Ontario, Toronto, Ontario
J. Peters	Department of Community Services and Attorney General, Charlottetown, Prince Edward Island
J.-L. Robert	Régie du bâtiment du Québec, Québec, Québec
S. St-Antoine	Hydro-Québec, Montréal, Québec

Regulatory Authority Committee

S. Paulsen *(Chair)*	Department of Public Safety, Fredericton, New Brunswick
T. Olechna *(Vice-Chair)*	Electrical Safety Authority, Mississauga, Ontario
M.S. Anderson *(Associate)*	City of Winnipeg, Manitoba
D. Clements *(Associate)*	Nova Scotia Power Inc., Halifax, Nova Scotia
T. Collins	Government Service Centre, Cornerbrook, Newfoundland
J. Einarson	Department of Community Services, Whitehorse, Yukon
M.D. Gardener *(Associate)*	City of Calgary, Calgary, Alberta
T. Kitson	Department of Community and Cultural Affairs, Charlottetown, Prince Edward Island
R. Leduc	Department of Municipal Affairs, Edmonton, Alberta
D.R.A. MacLeod	Department of Environment and Labour, Halifax, Nova Scotia
R. Marion	Department of Public Works and Services, Yellowknife, Northwest Territories
R. May	British Columbia Safety Authority, New Westminster, British Columbia
W.G. McMullan	Manitoba Hydro, Winnipeg, Manitoba
G. Montminy	Régie du bâtiment du Québec, Québec, Québec
L. Radom	SaskPower, Regina, Saskatchewan
A.Z. Tsisserev *(Associate)*	City of Vancouver, British Columbia
E. Zebedee	Government of Nunavut Community and Government Services, Iqaluit, Nunavut
R.C. Gilmour *(Project Manager)*	CSA, Mississauga, Ontario

Executive Committee

A.Z. Tsisserev *(Chair)*	City of Vancouver, British Columbia
G. Lobay *(Vice-Chair)*	CANMET, Ottawa, Ontario
R. Burpee	Saint John Energy, Saint John, New Brunswick
R. Leduc	Alberta Municipal Affairs, Edmonton, Alberta
P. Liberatore	Corporation of Master Electricians of Québec, Montréal, Québec
S. Paulsen	Department of Public Safety, Fredericton, New Brunswick
T. Simmons	British Columbia Institute of Technology, Burnaby, British Columbia
M. Smith	Rockwell Automation Canada Inc., Cambridge, Ontario
R.C. Gilmour *(Project Manager)*	CSA, Mississauga, Ontario

National Building Code/Canadian Electrical Code Liaison Committee

A.Z. Tsisserev *(Chair)*	City of Vancouver, British Columbia
P. Rizcallah *(Vice-Chair)*	National Research Council Canada, Ottawa, Ontario
M.S. Anderson	City of Winnipeg, Manitoba
G. Dupont-Laneuville	Régie du bâtiment du Québec, Montréal, Québec
T. Fazzari	Mohawk College, Stoney Creek, Ontario
R.A. Nelson	CSA, Mississauga, Ontario
S. Paulsen	Department of Public Safety, Fredericton, New Brunswick
R.C. Gilmour *(Project Manager)*	CSA, Mississauga, Ontario

Section Subcommittees

Section 0 — Object, scope, and definitions

G. Lobay *(Chair)*	CANMET, Ottawa, Ontario
D.H. Dunsire	Winnipeg, Manitoba
M.D. Gardener	City of Calgary, Alberta
R.C. Gilmour	CSA, Mississauga, Ontario
D. Heron	Electrical Safety Authority, Worthington, Ontario *(Representing International Association of Electrical Inspectors)*
N. Mancini	CSA, Toronto, Ontario
G.A. Moberg	Gloucester, Ontario

Section 2 — General Rules

E. Zebedee *(Chair)*	Government of Nunavut Community and Government Services, Iqaluit, Nunavut
D.H. Dunsire	Winnipeg, Manitoba
B. Haydon	CSA, Mississauga, Ontario
N. Mancini	CSA, Toronto, Ontario
D.G. Morlidge	Fluor Canada Ltd., Calgary, Alberta
T. Olechna	Electrical Safety Authority, Mississauga, Ontario
D. Roberts	Schneider Canada Inc., Mississauga, Ontario

Section 4 — Conductors

S. Paulsen *(Chair)*	Department of Public Safety, Fredericton, New Brunswick
L. Asselin	Laval, Québec *(Representing International Association of Electrical Inspectors)*
G.R. Beer	Jay Electric Ltd., Brampton, Ontario

G. Brunt	F.C. O'Neill Scriven & Associates, Halifax, Nova Scotia
T. Edwards	Alcan Cable, Atlanta, Georgia, USA
B. Haydon	CSA, Mississauga, Ontario
N. Mancini	CSA, Toronto, Ontario
G. Montminy	Régie du bâtiment du Québec, Québec, Québec
R.T. Neal	Toronto, Ontario
R.A. Nelson	CSA, Mississauga, Ontario
D.S. Reith	Nexans Canada Inc., Markham, Ontario
A.Z. Tsisserev	City of Vancouver, British Columbia

Section 6 — Services and service equipment

T. Olechna *(Chair)*	Electrical Safety Authority, Mississauga, Ontario
R.A. Burpee	Saint John Energy, Saint John, New Brunswick
D.H. Dunsire	Winnipeg, Manitoba
J. Gamble	C. Gamble Electric (1982) Ltd., Winnipeg, Manitoba
M.D. Gardener	City of Calgary, Alberta
B. Haydon	CSA, Mississauga, Ontario
D. Letcher	Don Letcher (E.S.C.O.) Enterprises, Sherwood Park, Alberta *(Representing International Association of Electrical Inspectors)*
P. Liberatore	Corporation of Master Electricians of Québec, Montréal, Québec
N. Mancini	CSA, Toronto, Ontario
W.G. McMullan	Manitoba Hydro, Winnipeg, Manitoba
E.J. Power	Stanhope, Prince Edward Island
V. Yu	Code Instructor Association of B.C., Burnaby, British Columbia

Section 8 — Circuit loading and demand factors

D.E. Clements *(Chair)*	Nova Scotia Power Inc., Halifax, Nova Scotia
S. Douglas *(Vice-Chair)*	Electrical Safety Authority, Cambridge, Ontario *(Representing International Association of Electrical Inspectors)*
Y. Boodram	Schneider Canada Inc., Toronto, Ontario
R.C. Gilmour	CSA, Mississauga, Ontario
N. Mancini	CSA, Toronto, Ontario
R. Moberg	DeBray Solutions, North Gower, Ontario
G. Montminy	Régie du bâtiment du Québec, Québec, Québec
D. Singh	Scarborough, Ontario
J.E. White	J.E.C. White Consulting, Burlington, Ontario
V. Yu	Code Instructor Association of B.C., Burnaby, British Columbia

Section 10 — Grounding and bonding

R. Leduc *(Chair)*	Department of Municipal Affairs, Edmonton, Alberta
K.D. McLennan *(Vice-Chair)*	Islay, Alberta
K. Almon	Dartmouth, Nova Scotia
S. Bygrave	Michelin North America (Canada) Inc., New Glasgow, Nova Scotia
E. Court	Court Consulting Ltd., Calgary, Alberta
J. Courteau	Alcan Cable, Ville St-Laurent, Québec
J. Fotheringham	International Association of Electrical Inspectors, Winnipeg, Manitoba
M.D. Gardener	City of Calgary, Alberta
B. Haydon	CSA, Mississauga, Ontario

N. Mancini	CSA, Toronto, Ontario
D.G. Morlidge	Fluor Canada Inc., Alberta
T. Olechna	Electrical Safety Authority, Mississauga, Ontario
I. Simpson	Ground-It.Com Consulting Ltd., North Vancouver, British Columbia

Section 12 — Wiring methods

M.D. Gardener *(Chair)*	City of Calgary, Alberta
S. Douglas *(Vice-Chair)*	Electrical Safety Authority, Cambridge, Ontario
D.J. Andrews	D.J.A. Engineering, Calgary, Alberta
L. Baker	Arcon Electric Ltd., London, Ontario *(Representing International Association of Electrical Inspectors)*
C.W. Beile	Allied Tube and Conduit Corporation, Wheaton, Illinois, USA
J.B. Biollo	Biollo Agency Ltd., Leduc, Alberta
G. Currie	Portage la Prairie Mutual Insurance Company, Portage la Prairie, Manitoba
D.H. Dunsire	Winnipeg, Manitoba
B. Haydon	CSA, Mississauga, Ontario
N. Mancini	CSA, Toronto, Ontario
D.T. Mansfield	Calgary, Alberta
G. McCue	Harlock-Schultz Electric Ltd., Guelph, Ontario
K.D. McLennan	Islay, Alberta
S. Paulsen	Department of Public Safety, Fredericton, New Brunswick
L. Radom	SaskPower, Regina, Saskatchewan
P. Schmaltz	Reggin Technical Services Ltd., Calgary, Alberta

Section 14 — Protection and control

B. Savaria *(Chair)*	Eaton Electrical Canada Operations, Burlington, Ontario
T. Branch	Weyerhaeuser Dryden Operations, Dryden, Ontario
S. Bygrave	Michelin North America (Canada) Inc., New Glasgow, Nova Scotia
S. Davies	Segment Engineering Inc., Calgary, Alberta
G.T. Gingara	Associated Engineering (Sask) Ltd., Saskatoon, Saskatchewan
D. Heron	Electrical Safety Authority, Worthington, Ontario
N. Mancini	CSA, Toronto, Ontario
G. Montminy	Régie du bâtiment du Québec, Québec, Québec
K.E. Morris	Morris Electric Ltd., Edmonton, Alberta
T. Pope	CSA, Mississauga, Ontario
M. Pullan	Bright Electric Limited, Mississauga, Ontario
W.C. Rossmann	Jacobs Canada Inc., Calgary, Alberta
N. Scott	Orillia, Ontario *(Representing International Association of Electrical Inspectors)*
D. Singh	Scarborough, Ontario
C. Thwaites	Ferraz Shawmut Canada Inc., Toronto, Ontario

Section 16 — Class 1 and Class 2 circuits

T. Simmons *(Chair)*	British Columbia Institute of Technology, Burnaby, British Columbia
J.B. Biollo	Biollo Agency Ltd., Leduc, Alberta *(Representing International Association of Electrical Inspectors)*
R.M. Leighton	Burlington, Ontario

| N. Mancini | CSA, Toronto, Ontario |

Section 18 — Hazardous locations

V.G. Rowe *(Chair)*	Marex Canada Limited, Westbank, British Columbia
G. Lobay *(Vice-Chair)*	CANMET, Ottawa, Ontario
J.A. Bossert	Hazloc Inc., Portland, Ontario
D. Clements	Nova Scotia Power Inc., Halifax, Nova Scotia
M. Cole	Hubbell Canada LP, Pickering, Ontario
J.H. Dymond	Hammonds Plains, Nova Scotia
R. Leduc	Department of Municipal Affairs, Edmonton, Alberta
K.D. McLennan	Islay, Alberta
T. Olechna	Electrical Safety Authority, Mississauga, Ontario
T. Pope	CSA, Mississauga, Ontario
R. Robertson	Shell Canada Limited, Calgary, Alberta
J.W. Rogers	Human Resources and Skills Development Canada, Sydney, Nova Scotia
W.M. Shao	WmShao Consultant, Edmonton, Alberta

Section 20 — Flammable liquid and gas dispensing and service stations, garages, bulk storage plants, finishing processes, and aircraft hangars

G. Lobay *(Chair)*	CANMET, Ottawa, Ontario
D.E. Clements	Nova Scotia Power Inc., Halifax, Nova Scotia *(Representing International Association of Electrical Inspectors)*
J.G. Demers	Marex Canada Ltd., Calgary, Alberta
D.H. Dunsire	Winnipeg, Manitoba
W. Mayko	EPCOR, Edmonton, Alberta
S. Misyk	The Inspection Group Inc., Edmonton, Alberta
G. Montminy	Régie du bâtiment du Québec, Québec, Québec
T. Pope	CSA, Mississauga, Ontario
E.J. Power	Stanhope, Prince Edward Island
V.G. Rowe	Marex Canada Limited, Westbank, British Columbia
W.R. Sutherland	Electrical Safety Authority, London, Ontario

Section 22 — Locations in which corrosive liquids, vapours, or excessive moisture are likely to be present

S. Paulsen *(Chair)*	Department of Public Safety, Fredericton, New Brunswick
L. Baker	Arcon Electrical Ltd., London, Ontario *(Representing International Association of Electrical Inspectors)*
G. Currie	Portage la Prairie Mutual Insurance Company, Portage la Prairie, Manitoba
D.H. Dunsire	Winnipeg, Manitoba
G.T. Gingara	AMEC America Limited, Saskatoon, Saskatchewan
N. Mancini	CSA, Toronto, Ontario
D. Wilson	Manitoba Hydro, Winnipeg, Manitoba

Section 24 — Patient care areas

T. Olechna *(Chair)*	Electrical Safety Authority, Mississauga, Ontario
A.Z. Tsisserev *(Vice-Chair)*	City of Vancouver, British Columbia
M.S. Anderson	City of Winnipeg, Manitoba
A.M. Dolan	University of Toronto, Toronto, Ontario

H. Dowhan	Stantec Consulting Ltd., Edmonton, Alberta
P.M. Gelinas	Hôpital du Sacré-Cœur de Montréal, Montréal, Québec
D. Letcher	Don Letcher (E.S.C.O.) Enterprises, Sherwood Park, Alberta
	(Representing International Association of Electrical Inspectors)
P.E. Paasche	University of New Brunswick, Fredericton, New Brunswick
D. Roberts	Schneider Canada Inc., Mississauga, Ontario
G.A. Schidowka	CSA, Mississauga, Ontario
W. Woodley	Markham, Ontario
T. Woolhouse	Ellard-Wilson Engineering Limited, Markham, Ontario

Section 26 — Installation of electrical equipment

R. Leduc *(Chair)*	Department of Municipal Affairs, Edmonton, Alberta
P. Desilets	Leviton Manufacturing of Canada Limited, Pointe-Claire, Québec
D.H. Dunsire	Winnipeg, Manitoba
M. Earley	National Fire Protection Association, Quincy, Massachusetts, USA
R.C. Gilmour	CSA, Mississauga, Ontario
T.L. Harman	University of Houston, Clear Lake, Houston, Texas, USA
P. Liberatore	Corporation of Master Electricians of Québec, Montréal, Québec
N. Mancini	CSA, Toronto, Ontario
W. Mayko	EPCOR, Edmonton, Alberta
W.G. McMullan	Manitoba Hydro, Winnipeg, Manitoba
R.A. Nelson	CSA, Mississauga, Ontario
T. Olechna	Electrical Safety Authority, Mississauga, Ontario
S. Paulsen	Department of Public Safety, Fredericton, New Brunswick
B. Savaria	Eaton Electrical Canada Operations, Burlington, Ontario
A.Z. Tsisserev	City of Vancouver, British Columbia

Section 28 — Motors and generators

M. Smith *(Chair)*	Rockwell Automation, Cambridge, Ontario
M.S. Anderson	City of Winnipeg, Manitoba
T. Branch	Weyerhaeuser Dryden Operations, Dryden, Ontario
D.E. Clements	Nova Scotia Power Inc., Halifax, Nova Scotia
	(Representing International Association of Electrical Inspectors)
S. Davies	Segment Engineering Inc., Calgary, Alberta
P. Desilets	Leviton Manufacturing of Canada Limited, Point-Claire, Québec
E.J. Friesen	E.J. Friesen and Associates Inc., Calgary, Alberta
M. Henville	CSA, Mississauga, Ontario
N. Mancini	CSA, Toronto, Ontario
B. Mead	National Refrigeration & Air Conditioning Products Inc., Brantford, Ontario
G.A. Moberg	Hudson, Florida, USA
R.A. Nelson	CSA, Mississauga, Ontario
L. Silecky	Ferraz Shawmut Canada Inc., Toronto, Ontario
	(Representing International Association of Electrical Inspectors)
D. Singh	Scarborough, Ontario
W. Somerville	Calgary, Alberta

Section 30 — Installation of lighting equipment

P. Desilets *(Chair)*	Leviton Manufacturing of Canada Ltd., Pointe-Claire, Québec
D.E. Clements *(Vice-Chair)*	Nova Scotia Power Inc., Halifax, Nova Scotia
G. Brunt	F.C. O'Neill Scriven & Associates, Halifax, Nova Scotia
J.A. Davidson	Manitoba Hydro, Virden, Manitoba *(Representing International Association of Electrical Inspectors)*
D. Hulford	CSA, Mississauga, Ontario
A. Milne	21st Olympiad Sales, Agincourt, Ontario
G. Montminy	Régie du bâtiment du Québec, Québec, Québec
T. Olechna	Electrical Safety Authority, Mississauga, Ontario
D. Rittenhouse	Maple Ridge, British Columbia

Section 32 — Fire alarm systems and fire pumps

M.S. Anderson *(Chair)*	City of Winnipeg, Manitoba
R. Florio	Tyco Thermal Controls — Pyrotenax, Toronto, Ontario
D. Gendebien	TornaTech Inc., St-Laurent, Québec
T. Pope	CSA, Mississauga, Ontario
P. Rizcallah	National Research Council Canada, Ottawa, Ontario
V. Rochon	Rochon Engineering Inc., Concord, Ontario
S.W. Smith	Electrical Safety Authority, Mississauga, Ontario
R. Stewart	Electrical Safety Authority, Toronto, Ontario *(Representing International Association of Electrical Inspectors)*
A.Z. Tsisserev	City of Vancouver, British Columbia
D. Weber	Vipond Systems Group, Mississauga, Ontario

Section 34 — Signs and outline lighting

L. Radom *(Chair)*	SaskPower, Regina, Saskatchewan
L. Catton	Acme Neon Signs (Windsor) Limited, Windsor, Ontario
K. Devine	Electra Sign, Winnipeg, Manitoba
D.H. Dunsire	Winnipeg, Manitoba
T. Elantis	Allanson International Inc., Toronto, Ontario
M. Golly	Alberta Municipal Affairs, Edmonton, Alberta
D. Hulford	CSA, Mississauga, Ontario
C. Mak	Teksign Inc., Mississauga, Ontario
G. Montminy	Régie du bâtiment du Québec, Québec, Québec
E.J. Power	Stanhope, Prince Edward Island
S. Scarrow	PRO SIGN, Saskatoon, Saskatchewan
D. Slowski	Direct Electronic Inc., Lisle, Ontario

Section 36 — High-voltage installations

F.L. Kaempffer *(Chair)*	British Columbia Hydro, Burnaby, British Columbia
J. Butts	Electrical Safety Authority, Alliston, Ontario *(Representing International Association of Electrical Inspectors)*
J. Côté	Hydro Québec, Montréal, Québec
J.M. Gallagher	Bayer Corporation, Bayton, Texas, USA
M.D. Gardener	City of Calgary, Alberta
R.B. Hamilton	Calgary, Alberta
W.H. Khella	W.H. Khella Enterprises, Mississauga, Ontario

A.C. Lawrence	Scarborough, Ontario
N. Mancini	CSA, Toronto, Ontario
G. Montminy	Régie du bâtiment du Québec, Québec, Québec
T. Olechna	Electrical Safety Authority, Mississauga, Ontario
T. Pope	CSA, Mississauga, Ontario
A.N. Sunley	Voltech Engineering Ltd., Calgary, Alberta

Section 38 — Elevators, dumbwaiters, material lifts, escalators, moving walks, lifts for persons with physical disabilities, and similar equipment

D. McColl *(Chair)*	Otis Canada Inc., Mississauga, Ontario
D. Balmer	Accessibility Equipment Manufacturer's Association, Brampton, Ontario
B. Blackaby	Otis Elevator Company, Farmington, Connecticut, USA
A.D. Brown	KONE Inc., Toronto, Ontario
A. Byram	Department of Labour, Fredericton, New Brunswick
R.E. Droste	Avon, Connecticut, USA
R. Hadaller	Technical Standards and Safety Authority, Toronto, Ontario
R. Kennedy	Department of Labour, Halifax, Nova Scotia
R. MacKenzie	CSA, Toronto, Ontario
S. Mercier	Régie du bâtiment du Québec, Montréal, Québec
M. Pedram	Thyssenkrupp Elevator, Toronto, Ontario
A. Rehman	Schindler Elevator Corporation, Scarborough, Ontario
M. Sterguic	Technical Standards and Safety Authority, Mississauga, Ontario

Section 40 — Electric cranes and hoists

S. Douglas *(Chair)*	Electrical Safety Authority, Cambridge, Ontario
B.A. Biglow	Edmonton, Alberta
N. Mancini	CSA, Toronto, Ontario
L. McQuerry	Demag Cranes & Components Corp., Cleveland, Ohio, USA
W.R. Sutherland	Electrical Safety Authority, London, Ontario *(Representing International Association of Electrical Inspectors)*
L.G. Uruski	Department of Labour, Winnipeg, Manitoba

Section 42 — Electric welders

P. Liberatore *(Chair)*	Corporation of Master Electricians of Québec, Montréal, Québec
N. Mancini	CSA, Toronto, Ontario
L. Silecky	Ferras Shawmut, Toronto, Ontario *(Representing International Association of Electrical Inspectors)*

Section 44 — Theatre installations

L. Radom *(Chair)*	SaskPower, Regina, Saskatchewan
W. Gillard	Saskatchewan Centre of the Arts, Regina, Saskatchewan
T. Olechna	Electrical Safety Authority, Mississauga, Ontario
R. Ouellette	Electrical Inspection Edmundston Region, Edmundston, New Brunswick *(Representing International Association of Electrical Inspectors)*
M. Perreault	Canadian Broadcasting Corporation, Montréal, Québec
J. Ritenburg	Ritenburg and Associates Ltd., Regina, Saskatchewan
G. Rose	Leviton Manufacturing of Canada, Toronto, Ontario
M. Wilson	CSA, Mississauga, Ontario

Section 46 — Emergency systems, unit equipment, and exit signs

A.Z. Tsisserev *(Chair)*	City of Vancouver, British Columbia
M.S. Anderson	City of Winnipeg, Manitoba
R.M. Bartholomew	Electric Power Equipment (1986) Ltd., Vancouver, British Columbia
S. Bygrave	Michelin North America (Canada) Inc., New Glasgow, Nova Scotia
T. Fazzari	Mohawk College, Stoney Creek, Ontario
B. McAllister	City of Camrose, Alberta
R.A. Nelson	CSA, Mississauga, Ontario
T. Pope	CSA, Mississauga, Ontario
P. Rizcallah	National Research Council Canada, Ottawa, Ontario
V. Rochon	Rochon Engineering Inc., Concord, Ontario
R. Sutherland	Electrical Safety Authority, London, Ontario *(Representing International Association of Electrical Inspectors)*

Section 48 — Motion picture studios, projection rooms, film exchanges including film-vaults, and storehouses for pyroxylin plastic and nitrocellulose X-ray and photographic film

L. Radom *(Chair)*	SaskPower, Regina, Saskatchewan
M. Perreault	Canadian Broadcasting Corporation, Montréal, Québec
R. Stewart	Electrical Safety Authority, Mississauga, Ontario *(Representing International Association of Electrical Inspectors)*
M. Wilson	CSA, Mississauga, Ontario

Section 50 — Solar photovoltaic systems

M.S. Anderson *(Chair)*	City of Winnipeg, Manitoba
T. Simmons *(Vice-Chair)*	British Columbia Institute of Technology, Burnaby, British Columbia
K.S. Brightwell	Electrical Safety Authority, Belleville, Ontario
P.M. Cusack	S.A. Armstrong Limited, Toronto, Ontario
P. Drewes	Sol Source Engineering, Newmarket, Ontario
D. Egles	Soltek Solar Energy Ltd., Victoria, British Columbia
G. Howell	Howell-Mayhew Engineering Incorporated, Edmonton, Alberta
D. Hulford	CSA, Mississauga, Ontario
S. Martel	Natural Resources Canada, Varennes, Québec
C.R. Price	Quail Engineering, Winfield, British Columbia
E. Smiley	British Columbia Institute of Technology, Burnaby, British Columbia
D. Turcotte	Natural Resources Canada, Varennes, Québec

Section 52 — Diagnostic imaging installations

D.R.A. MacLeod *(Chair)*	Department of Environment and Labour, Halifax, Nova Scotia
M.B. Raber *(Vice-Chair)*	Winnipeg, Manitoba
J. Einarson	Department of Community Services, Whitehorse, Yukon
E. Carlson	CSA, Toronto, Ontario *(Representing International Association of Electrical Inspectors)*
M. Wilson	CSA, Mississauga, Ontario

Section 54 — Community antenna distribution and radio and television installations

J. Poulin *(Chair)*	Bell Canada, Longueuil, Québec
E. Chantigny	General Electric Canada, Pointe-Claire, Québec
B. Nameh	Rogers Cable Systems Ltd., Don Mills, Ontario

P. Olders   Terra Communications, Inc., Scarborough, Ontario
(Representing International Association of Electrical Inspectors)

L. Radom   SaskPower, Regina, Saskatchewan

G. Tubrett   CSA, Mississauga, Ontario

Section 56 — Optical fiber cables

C.B. Chan *(Chair)*   MTS Communications Inc., Winnipeg, Manitoba

S. Finnagan   Algonquin College, Ottawa, Ontario

B. Haydon   CSA, Mississauga, Ontario

P. Olders   Terra Communications, Inc., Scarborough, Ontario
(Representing International Association of Electrical Inspectors)

J. Poulin   Bell Canada, Longueuil, Québec

V.G. Rowe   Marex Canada Limited, Westbank, British Columbia

A.Z. Tsisserev   City of Vancouver, British Columbia

Section 60 — Electrical communication systems

J. Poulin *(Chair)*   Bell Canada, Longueuil, Québec

D.J. Andrews   D.J.A. Engineering, Calgary, Alberta

C.B. Chan   MTS Communications Inc., Winnipeg, Manitoba

E. Chantigny   General Electric Canada, Pointe-Claire, Québec

P. Desilets   Leviton Manufacturing of Canada Limited, Pointe-Claire, Québec

S. Finnagan   Algonquin College, Ottawa, Ontario

E.S. Guevara   Industry Canada, Ottawa, Ontario

D. Schultz   TELUS Communications (B.C.) Inc., Edmonton, Alberta

R. Smith   Aliant Telecom, Moncton, New Brunswick

A.Z. Tsisserev   City of Vancouver, British Columbia

M. Wilson   CSA, Mississauga, Ontario

Section 62 — Fixed electric space and surface heating systems

V.G. Rowe *(Chair)*   Marex Canada Limited, Westbank, British Columbia

J. Turner *(Vice-Chair)*   Swansea Consulting, Toronto, Ontario

J. Adam   Syncrude Canada Limited, Fort McMurray, Alberta

R. Barth   Thermon Manufacturing Company, San Marcos, Texas, USA

P. Desilets   Leviton Manufacturing of Canada Limited, Pointe-Claire, Québec

T.S. Driscoll   Shell Canada, Calgary, Alberta

W.E. Hanthorn   Tyco Thermal Controls (Canada) Ltd., Trenton, Ontario

T. Pope   CSA, Mississauga, Ontario

R. Stromer   Imperial Oil Resources Limited, Calgary, Alberta

S. Tetreault   Shell Canada Ltd., Montréal, Québec

J. Thomson   Electrical Safety Authority, Corunna, Ontario
(Representing International Association of Electrical Inspectors)

Section 66 — Amusement parks, midways, carnivals, film and TV sets, TV remote broadcasting locations, and travelling shows

G. Montminy *(Chair)*   Régie du bâtiment du Québec, Québec, Québec

D. Burke   Victoria, British Columbia

T. Olechna   Electrical Safety Authority, Mississauga, Ontario

A. Paquette   Régie du bâtiment du Québec, Montréal, Québec

M. Perreault   Canadian Broadcasting Corporation, Montréal, Québec

L. Radom   SaskPower, Regina, Saskatchewan
A. Wanuch   Robertson Electric Wholesale, Toronto, Ontario
W. White   City of Vancouver, British Columbia

Section 68 — Pools, tubs, and spas
S.W. Douglas *(Chair)*   Electrical Safety Authority, Cambridge, Ontario
     (Representing International Association of Electrical Inspectors)

T. Bartoffy   CSA, Mississauga, Ontario
D. Letcher   Don Letcher (E.S.C.O.) Enterprises, Sherwood Park, Alberta
     (Representing International Association of Electrical Inspectors)

T. Minna   EPI Electrical Contractors, Brampton, Ontario
G. Montminy   Régie du bâtiment du Québec, Québec, Québec
T. Olechna   Electrical Safety Authority, Mississauga, Ontario
L.B. Ross   Newmarket, Ontario
D.K. Stuebing   Solon Enterprises Ltd., Peace River, Alberta
K. Tomihiro   Pool and Hot Tub Council of Canada, Markham, Ontario

Section 70 — Electrical requirements for factory-built relocatable structures and non-relocatable structures
R. May *(Chair)*   British Columbia Safety Authority (BCSA), New Westminster, British Columbia

M.S. Anderson   City of Winnipeg, Manitoba
J. Einarson   Department of Community Services, Whitehorse, Yukon
K. Maynard   Canadian Manufactured Housing Institute, Ottawa, Ontario
R. Morin   Economical Insurance Group, Oshawa, Ontario
     (Representing International Association of Inspectors)

Section 72 — Mobile home and recreational vehicle parks
R. May *(Chair)*   British Columbia Safety Authority (BCSA), New Westminster, British Columbia

M.S. Anderson   City of Winnipeg, Manitoba
J. Baker   OPCA, Embro, Ontario
W. Donald   Winnipeg, Manitoba
J. Einarson   Department of Community Services, Whitehorse, Yukon
D. Letcher   Don Letcher (E.S.C.O.) Enterprises, Sherwood Park, Alberta
     (Representing International Association of Electrical Inspectors)

G. Montminy   Régie du bâtiment du Québec, Québec, Québec
T. Olechna   Electrical Safety Authority, Mississauga, Ontario

Section 74 — Airport installations
R. May *(Chair)*   British Columbia Safety Authority (BCSA), New Westminster, British Columbia
D. Henry *(Vice-Chair)*   Department of National Defence, Westwin, Manitoba
E.J. Alf   Transport Canada, Ottawa, Ontario
G.W. Bradbury   B.T.E. Engineering Technology Services, St. Petersburg, Florida, USA
     (Representing International Association of Electrical Inspectors)

D.H. Dunsire   Winnipeg, Manitoba
G.T. Gingara   AMEC Americas Limited, Saskatoon, Saskatchewan
R. Kowalik   Alberta Transportation and Utilities, Edmonton, Alberta
N. Mancini   CSA, Toronto, Ontario

Section 76 — Temporary wiring

W.G. McMullan *(Chair)*	Manitoba Hydro, Winnipeg, Manitoba
S. Douglas *(Vice-Chair)*	Electrical Safety Authority, Cambridge, Ontario *(Representing International Association of Electrical Inspectors)*
B. Doan	Summer Electric London Ltd., Komoka, Ontario
N. Mancini	CSA, Toronto, Ontario
B. O'Donnell	AC Powerline Construction, Pickering, Ontario
T. Olechna	Electrical Safety Authority, Mississauga, Ontario

Section 78 — Marinas, yacht clubs, marine wharves, structures, and fishing harbours

A. Sutherland *(Chair)*	Department of Public Works and Government Services Canada, Hull, Québec
T. Branch	Department of Public Safety, Bathurst, New Brunswick
A. Donaldson	Ontario Marinas Operators' Association, Penetanguishene, Ontario
T.A. Fekete	Scarborough, Ontario
K. McCormick	Electrical Safety Authority, Cobourg, Ontario
K.L. Rodel	Hubbell Canada LP, Pickering, Ontario
M. Vollmer	Burlington, Ontario

Section 80 — Cathodic protection

T. Simmons *(Chair)*	British Columbia Institute of Technology, Burnaby, British Columbia
E. Court	Court Consulting Ltd., Calgary, Alberta
S.J. Croall	Manitoba Hydro, Winnipeg, Manitoba
J.G. Demers	Marex Canada Ltd., Calgary, Alberta
N. Mancini	CSA, Toronto, Ontario
R.J. Maynard	Aurora Environmental Consulting Ltd., Calgary, Alberta
W.G. McMullan	Manitoba Hydro, Winnipeg, Manitoba
D. Schill	SaskPower, Yorkton, Saskatchewan *(Representing International Association of Electrical Inspectors)*
R. Stromer	Imperial Oil Resources Ltd., Calgary, Alberta
A.Z. Tsisserev	City of Vancouver, British Columbia
R.G. Wakelin	Correng Consulting Services, Markham, Ontario

Section 82 — Closed-loop and pre-closed-loop power distribution

C.B. Chan *(Chair)*	MTS Communications Inc., Winnipeg, Manitoba
G.N. Bowling	Ottawa, Ontario
D. Juden	C.C.G., Ottawa, Ontario
N. Mancini	CSA, Toronto, Ontario
D. Pilon	SaskPower Electrical Inspectors, Prince Alberta, Saskatchewan *(Representing International Association of Electrical Inspectors)*

Section 84 — Interconnection of electric power production sources

F.L. Kaempffer *(Chair)*	British Columbia Hydro, Burnaby, British Columbia
M.S. Anderson	City of Winnipeg, Manitoba
R.A. Burpee	Saint John Energy, Saint John, New Brunswick
D. Desrosiers	Direction Principale, Montréal, Québec
D.H. Dunsire	Winnipeg, Manitoba
D. Heron	Electrical Safety Authority, Worthington, Ontario *(Representing International Association of Electrical Inspectors)*

A. Mak	EPCOR Distribution and Transmission Inc., Edmonton, Alberta
T. Pope	CSA, Mississauga, Ontario
V.G. Rowe	Marex Canada Limited, Westbank, British Columbia
T. Simmons	British Columbia Institute of Technology, Burnaby, British Columbia

Section 86 — Electric vehicle charging systems

T.W. Odell *(Chair)*	General Motors of Canada, Oshawa, Ontario
R. Field	Norvik Technologies Inc., Mississauga, Ontario
D. Hulford	CSA, Mississauga, Ontario
C. Keyes	Kinetrics Inc., Toronto, Ontario
S. Lines	Natural Resources Canada, Ottawa, Ontario
N. Mancini	CSA, Toronto, Ontario
A.Z. Tsisserev	City of Vancouver, British Columbia

Appendix C
A.Z. Tsisserev *(Chair)*   City of Vancouver, British Columbia

Appendix D
S. Paulsen *(Chair)*   Department of Public Safety, Fredericton, New Brunswick

Appendix E
V.G. Rowe *(Chair)*   Marex Canada Limited, Westbank, British Columbia

Appendix F
V.G. Rowe *(Chair)*   Marex Canada Limited, Westbank, British Columbia

Appendix G
A.Z. Tsisserev *(Chair)*   City of Vancouver, British Columbia

Appendix H
V.G. Rowe *(Chair)*   Marex Canada Limited, Westbank, British Columbia

Appendix J
V.G. Rowe *(Chair)*   Marex Canada Limited, Westbank, British Columbia

Δ

Preface

This twentieth edition of the *Canadian Electrical Code, Part I,* was approved by the Committee on the *Canadian Electrical Code, Part I,* and by the Regulatory Authority Committee at their June 2005 meetings in Kelowna, British Columbia. This twentieth edition supersedes the previous editions, published in 2002, 1998, 1994, 1990, 1986, 1982, 1978, 1975, 1972, 1969, 1966, 1962, 1958, 1953, 1947, 1939, 1935, 1930, and 1927.

Sections 0 to 16 and 26 are considered general sections, and the other sections supplement or amend the general sections.

Various requirements were revised as a result of the continuing efforts toward harmonization. In addition, there are significant changes to Sections 2, 10, 18 (particularly in Appendices B and J), 26, and 32. Sections 74 and 84 have been revised considerably to reflect new technology and industry practices.

A new Subsection was added to Section 22 to include the requirements for Sewage Lift and Treatment Plants. A new Appendix I has been added listing the interpretations that have been approved since the last edition, in the cases where the Rule in question has not been clarified yet. A new Appendix K has been added giving the fundamental safety principles contained in IEC 60634-1.

General arrangement

The Code is divided into numbered Sections, each covering some main division of the work. The Sections are divided into numbered Rules, with captions for easy reference, as follows:

(a) **Numbering system** — With the exception of Section 38, even numbers have been used throughout to identify Sections and Rules. Rule numbers consist of the Section number separated by a hyphen from the 3- or 4-digit figure. The intention in general is that odd numbers may be used for new Rules required by interim revisions. Due to the introduction of some new Rules and the deletion of some existing Rules during the revision of each edition, the Rule numbers for any particular requirement are not always the same in successive editions.

(b) **Subdivision of Rules** — Rules are subdivided in the manner illustrated by Rules 8-204 and 8-206, and the subdivisions are identified as follows:

00-000	Rule
(1)	Subrule
(a)	Item
(i)	Item
(A)	Item

(c) **Reference to other Rules, etc.** — Where reference is made to two or more Rules, the first and last Rules mentioned are included in the reference. Where reference is made to a Subrule or Item in the same Rule, only the Subrule number and/or Item letter and the word "Subrule" or "Item" need be mentioned. If the reference is to another Rule or Section, then the Rule number and the word "Rule" shall be stated (e.g., "Rule 10-200(3)" and not "Subrule (3) of Rule 10-200").

The principal changes that have been made between the 2002 edition of the *Canadian Electrical Code, Part I,* and this new edition published in 2006 are marked in the text of the Code by the symbol delta (Δ) in the margin. Where revisions to or deletions from the text have caused existing Rules to be renumbered, only the first renumbered Rule in the sequence is marked. Users of the Code are advised that the change markers in the text are not intended to be all-inclusive and are provided as a convenience only; such markers cannot constitute a comprehensive guide to the reorganization or revision of the Code. Care must therefore be taken not to rely on the change markers to determine the current requirements of the Code. As always, users of the Code must consider the entire Code and any local amendments.

Acknowledgement

Acknowledgement is made for the use of material contained in the *National Electrical Code.*

The history and operation of the *Canadian Electrical Code, Part I*

The preliminary work in preparing the *Canadian Electrical Code* was begun in 1920 when a special committee, appointed by the main Committee of the Canadian Engineering Standards Association, recommended that this project be undertaken. A third meeting of this Committee was held in June 1927 with representatives from

Nova Scotia, Québec, Ontario, Manitoba, Saskatchewan, and British Columbia attending. At this meeting, the revised draft, which had been discussed at the previous two meetings, was formally approved and a resolution was made that it be printed as Part I of the *Canadian Electrical Code.*

The present Committee on *CE Code, Part I,* is composed of 41 members, with representation from inspection authorities, industry, utilities, and allied interests. The main Committee meets once a year and deals with reports that have been submitted by the 42 Section Subcommittees, which work under the jurisdiction of the main Committee. Suggestions for changes to the Code may be made by any member of the Committee or anyone outside the Committee as outlined in Clause C6.

January 2006

Notes:

(1) *Although the intended primary application of this Standard is stated in its Scope, it is important to note that it remains the responsibility of the users of the Standard to judge its suitability for their particular purpose.*

(2) *CSA Standards are subject to periodic review, and suggestions for their improvement will be referred to the appropriate committee.*

(3) *All enquiries regarding this Standard, including requests for interpretation, should be addressed to Canadian Standards Association, 5060 Spectrum Way, Suite 100, Mississauga, Ontario, Canada L4W 5N6.*

Requests for interpretation will also be accepted by the Committee (see Clause C9). They should be worded in such a manner as to permit a specific "yes" or "no" answer based on the literal text of the requirement concerned.

Interpretations are published in CSA's periodical Info Update, *which is available on the CSA Web site at www.csa.ca.*

Metric units

Symbols and conversion factors for SI units

Recognized symbols for SI units have been used in the *Canadian Electrical Code, Part I*. For the convenience of the user, these symbols and the units they represent have been listed in the following table; the table also gives a multiplying factor that may be used to convert the SI unit to the previously used unit.

Symbol	SI unit	Multiplying factor for conversion to previously used unit	Previously used unit
A	ampere(s)	1	ampere(s)
cm^3	cubic centimetre(s)	0.061	cubic inch(es)
°(s)	degree(s) (angle)	1	degree(s) (angle)
°C rise	degree(s) Celsius	1.8	degree(s) Fahrenheit
°C temperature	degree(s) Celsius	1.8 plus 32	degree(s) Fahrenheit
h	hour(s)	1	hour(s) (time)
Hz	hertz	1	cycles per second
J	joule(s)	0.7376	foot-pound(s)
kg	kilogram(s)	2.205	pound(s)
kJ	kilojoule(s)	737.6	foot-pound(s)
km	kilometre	0.621	mile(s)
kPa	kilopascal(s)	0.295	inch(es) of mercury
		0.334	feet of water
		0.145	pound(s) per square inch (psi)
kW	kilowatt	3415.179	BTU/h
lx	lux	0.093	foot-candle(s)
L	litre	0.220	gallon(s)
m	metre(s)	3.281	feet
m^2	square metre(s)	10.764	square feet
m^3	cubic metre(s)	35.315	cubic feet
MHz	megahertz	1	megacycles per second
min	minute(s)	1	minute(s)
mL	millilitre(s)	0.061	cubic inch(es)
mm	millimetre(s)	0.03937	inch(es)
mm^2	square millimetre(s)	0.00155	square inch(es)
Ω	ohm(s)	1	ohm(s)
Pa	pascal(s)	0.000295	inch(es) of mercury
		0.000334	feet of water
		0.000145	pounds per square inch (psi)
V	volt(s)	1	volt(s)
W	watt(s)	1	watt(s)
µF	microfarad(s)	1	microfarad(s)

Conduit sizes

In the previous edition of the Code, changes to the identification of the size of conduit were made throughout the Code. The metric trade designator was given first, with the value in inches following in parentheses. In this edition, only the metric trade designator is used.

Conduit trade sizes

1998 edition	2002 edition	2006 edition
3/8 (12)	12 (3/8)	12
1/2 (16)	16 (1/2)	16
3/4 (21)	21 (3/4)	21
1 (27)	27 (1)	27
1-1/4 (35)	35 (1-1/4)	35
1-1/2 (41)	41 (1-1/2)	41
2 (53)	53 (2)	53
2-1/2 (63)	63 (2-1/2)	63
3 (78)	78 (3)	78
3-1/2 (91)	91 (3-1/2)	91
4 (103)	103 (4)	103
5 (129)	129 (5)	129
6 (155)	155 (6)	155

Reference publications

This Standard refers to the following publications, and the year dates shown indicate the latest editions available at the time the Standard was approved:

CSA (Canadian Standards Association)

B44-04, *Safety Code for Elevators and Escalators*

CAN/CSA-B44.1-04/ASME-A17.5-2004, *Elevator and Escalator Electrical Equipment*

B52-05, *Mechanical Refrigeration Code*

CAN/CSA-B72-M87 (R2003), *Installation Code for Lightning Protection Systems*

CAN/CSA-B137.1-02, *Polyethylene Pipe, Tubing, and Fittings for Cold-Water Pressure Services*

CAN/CSA-B149.1-05, *Natural Gas and Propane Installation Code*

CAN/CSA-B149.2-05, *Propane Storage and Handling Code*

CAN/CSA-B355-00, *Lifts for Persons with Physical Disabilities*

CAN/CSA-B613-00, *Private Residence Lifts for Persons with Physical Disabilities*

CAN/CSA-C22.2 No. 0-M91 (R2001), *General Requirements — Canadian Electrical Code, Part II*

C22.2 No. 1-04, *Audio, Video, and Similar Electronic Equipment*

C22.2 No. 3-M1988 (R2004), *Electrical Features of Fuel-Burning Equipment*

CAN/CSA-C22.2 No. 14-95 (R2001), *Industrial Control Equipment*

C22.2 No. 22-M1986 (R2004), *Electrical Equipment for Flammable and Combustible Fuel Dispensers*

C22.2 No. 25-1966 (R2004), *Enclosures for Use in Class II Groups E, F, and G Hazardous Locations*

C22.2 No. 30-M1986 (R2003), *Explosion-Proof Enclosures for Use in Class I Hazardous Locations*

C22.2 No. 33-M1984 (R2004), *Construction and Test of Electric Cranes and Hoists*

C22.2 No. 41-M1987 (R2004), *Grounding and Bonding Equipment*

C22.2 No. 59-M1987 (R2004), *Attachment Plugs, Receptacles, and Similar Wiring Devices for Use in Hazardous Locations: Class I, Groups A, B, C, and D; Class II, Groups G, in Coal or Coke Dust, and in Gaseous Mines*

C22.2 No. 59.1-M1987 (R2001), *Fuses (Both Plug and Cartridge-Enclosed Types)*

C22.2 No. 82-1969 (R2004), *Tubular Support Members and Associated Fittings for Domestic and Commercial Service Masts*

CAN/CSA-C22.2 No. 106-M92 (R2001), *HRC Fuses*

C22.2 No. 107.1-01, *General Use Power Supplies*

C22.2 No. 111-00, *General-Use Snap Switches*

CAN/CSA-C22.2 No. 114-M90 (R2005), *Diagnostic Imaging and Radiation Therapy Equipment*

C22.2 No. 124-04, *Mineral-Insulated Cable*

C22.2 No. 125-M1984 (R2004), *Electromedical Equipment*

C22.2 No. 126.1-02, *Metal Cable Tray Systems*

CAN/CSA-C22.2 No. 130-03, *Requirements for Electrical Resistance Heating Cables and Heating Device Sets*

CAN/CSA-C22.2 No. 130.1-M90 (R2001), *Heat-Tracing Cable Systems for Use in Industrial Locations (withdrawn)*

C22.2 No. 137-M1981 (R2004), *Electric Luminaires For Use in Hazardous Locations*

C22.2 No. 141-02, *Unit Equipment for Emergency Lighting*

C22.2 No. 145-M1986 (R2004), *Motors and Generators for Use in Hazardous Locations*

C22.2 No. 152-M1984 (R2001), *Combustible Gas Detection Instruments*

CAN/CSA-C22.2 No. 157-92 (R2002), *Intrinsically Safe and Non-Incendive Equipment for Use in Hazardous Locations*

C22.2 No. 174-M1984 (R2003), *Cables and Cable Glands for Use in Hazardous Locations*

C22.2 No. 178-1978 (R2001), *Automatic Transfer Switches*

C22.2 No. 211.0-03, *General Requirements and Methods of Testing for Nonmetallic Conduit*

C22.2 No. 213-M1987 (R2004), *Non-incendive Electrical Equipment for Use in Class I, Division 2 Hazardous Locations*

CAN/CSA-C22.2 No. 248 series (R2005), *Low-Voltage Fuses*

CAN/CSA-C22.2 No. 601 series, *Medical Electrical Equipment*

CAN/CSA-C22.2 No. 60529:05, *Degrees of Protection Provided by Enclosures (IP Code)*

CAN/CSA-C22.3 No. 1-01, *Overhead Systems*

C22.3 No. 7-94 (R2004), *Underground Systems*

C83-96 (R2005), *Communication and Power Line Hardware*

CAN3-C235-83 (R2000), *Preferred Voltage Levels for AC Systems, 0 to 50 000 V*

CAN/CSA-C282-00, *Emergency Electrical Power Supply for Buildings*

CAN/CSA-C1264-99 (R2004), *Ceramic Pressurized Hollow Insulators for High-Voltage Switchgear and Controlgear*

CAN/CSA-C50052-99 (R2003), *Cast Aluminium Alloy Enclosures for Gas-Filled High-Voltage Switchgear and Controlgear*

CAN/CSA-C50064-99 (R2003), *Wrought Aluminium and Aluminium Alloy Enclosures for Gas-Filled High-Voltage Switchgear and Controlgear*

CAN/CSA-C50068-99 (R2003), *Wrought Steel Enclosures for Gas-Filled High-Voltage Switchgear and Controlgear*

CAN/CSA-C50069-99 (R2003), *Welded Composite Enclosures of Cast and Wrought Aluminium Alloys for Gas-Filled High-Voltage Switchgear and Controlgear*

CAN/CSA-C50089-99 (R2003), *Cast Resin Partitions for Metal-Enclosed Gas-Filled High-Voltage Switchgear and Controlgear*

CAN/CSA-E60079-0-02, *Electrical Apparatus for Explosive Gas Atmospheres — Part 0: General Requirements*

CAN/CSA-E60079-1-02, *Electrical Apparatus for Explosive Gas Atmospheres — Part 1: Flameproof Enclosures "d"*

CAN/CSA-E60079-2-02, *Electrical Apparatus for Explosive Gas Atmospheres — Part 2: Electrical Apparatus — Type of Protection "p"*

CAN/CSA-E60079-5-02, *Electrical Apparatus for Explosive Gas Atmospheres — Part 5: Powder Filling "q"*

CAN/CSA-E60079-6-02, *Electrical Apparatus for Explosive Gas Atmospheres — Part 6: Oil-Immersion "o"*

CAN/CSA-E60079-7-03, *Electrical Apparatus for Explosive Gas Atmospheres — Part 7: Increased Safety "e"*

CAN/CSA-E60079-11-02, *Electrical Apparatus for Explosive Gas Atmospheres — Part 11: Intrinsic Safety "i"*

CAN/CSA-E60079-15-02, *Electrical Apparatus for Explosive Gas Atmospheres — Part 15: Electrical Apparatus with Type of Protection "n"*

CAN/CSA-E79-18-95 (R2004), *Electrical Apparatus for Explosive Gas Atmospheres — Part 18: Encapsulation "m"*

CAN/CSA-E61241-1-1-02, *Electrical Apparatus for Use in the Presence of Combustible Dust — Part 1-1: Electrical Apparatus Protected by Enclosures and Surface Temperature Limitations — Specification for Apparatus*

CAN/CSA-M421-00, *Use of Electricity in Mines*

CAN/CSA-S413-94 (R2000), *Parking Structures*

CAN/CSA-Z32-04, *Electrical Safety and Essential Electrical Systems in Health Care Facilities*

CAN/CSA-Z240 MH Series-92 (R2001), *Mobile Homes*

CAN/CSA-Z240 RV Series-99 (R2004), *Recreational Vehicles*

CAN/CSA-Z241 Series-03, *Park Model Trailers*

CAN/CSA-Z267-00, *Safety Code for Amusement Rides and Devices*

CAN/CSA-Z662-03, *Oil and Gas Pipeline Systems*

ANSI/API (American National Standards Institute/American Petroleum Institute)
505 (1997), *Recommended Practice for Classification of Locations for Electrical Installations at Petroleum Facilities Classified as Class I, Zone 0, and Zone 2*

ANSI/ASME (American National Standards Institute/American Society of Mechanical Engineers)
B1.20.1-1983 (R2001), *Pipe Threads, General Purpose (Inch)*
Z535.4-2002, *Product Safety Signs and Labels*

ANSI/IEEE (American National Standards Institute/Institute of Electrical and Electronics Engineers)
487-2000, *Recommended Practice for the Protection of Wire-Line Communication Facilities Serving Supply Locations*

ANSI/NEMA (American National Standards Institute/National Electrical Manufacturers Association)
WD 6-2002, *Wiring Devices — Dimensional Requirements*

API (American Petroleum Institute)
500 (1997), *Recommended Practice for Classification of Locations for Electrical Installations of Petroleum Facilities Classified as Class I, Division 1 and Division 2*
2216 (2003), *Ignition Risk of Hydrocarbon Liquids and Vapors by Hot Surfaces in the Open Air*

CEA (Canadian Electricity Association)
249 D541-1989, *Simplified Rules for Grounding Customer-Owned High Voltage Substations*

IEC (International Electrotechnical Commission)
60079-1A:2003, *Electrical apparatus for explosive gas atmospheres — Part 1: Flameproof enclosures "d"*
60079-4:1975, *Electrical apparatus for explosive gas atmospheres — Part 4: Method of test for ignition temperature*

60079-4A:1970, *Electrical apparatus for explosive gas atmospheres — Part 4: Method of test for ignition temperature — First supplement*

60079-10:2002, *Electrical apparatus for explosive gas atmospheres — Part 10: Classification of hazardous areas*

60079-12:1978, *Electrical apparatus for explosive gas atmospheres — Part 12: Classification of mixtures of gases of vapours with air according to their maximum experimental safe gaps and minimum igniting currents*

60079-13:1982, *Electrical apparatus for explosive gas atmospheres — Part 13: Construction and use of rooms or buildings protected by pressurization*

60079-14:2002, *Electrical apparatus for explosive gas atmospheres — Part 14: Electrical installations in hazardous areas (other than mines)*

60079-16:1990, *Electrical apparatus for explosive gas atmospheres — Part 16: Artificial ventilation for the protection of analyser(s) houses*

60079-17:2002, *Electrical apparatus for explosive gas atmospheres — Part 17: Inspection and maintenance of electrical installations in hazardous areas (other than mines)*

60079-19:1993, *Electrical apparatus for explosive gas atmospheres — Part 19: Repair and overhaul for apparatus used in explosive atmospheres (other than mines or explosives)*

60079-20:1996, *Electrical apparatus for explosive gas atmospheres — Part 20: Data for flammable gases and vapours, relating to the use of electrical apparatus*

60079-25:2003, *Electrical apparatus for explosive gas atmospheres — Part 25: Intrinsically safe systems*

60079-26:2004, *Electrical apparatus for explosive gas atmospheres — Part 26: Construction, test and marking of Group II Zone 0 electrical apparatus*

60364-1:2001, *Electrical installations of buildings — Part 1: Fundamental principles, assessment of general characteristics, definitions*

60364-5-51:2005, *Electrical installations of buildings — Part 5-51: Selection and erection of electrical equipment — Common rules*

60529:2001, *Degrees of protection provided by enclosures (IP Code)*

60721, *Classification of Environmental Conditions*

60781:1989, *Application Guide for Calculation of Short Circuit Currents in Low-Voltage Radial Systems*

61241-1-1:1999, *Electrical apparatus for use in the presence of combustible dust — Part 1-1: Electrical apparatus protected by enclosures and surface temperature limitation — Specification for apparatus*

61241-1-2:1999, *Electrical apparatus for use in the presence of combustible dust — Part 1-2: Electrical apparatus protected by enclosures and surface temperature limitation — Selection, installation and maintenance*

61241-2-1:1994, *Electrical apparatus for use in the presence of combustible dust — Part 2: Test methods — Section 1: Methods for determining the minimum ignition temperatures of dust*

61241-2-2:1993, *Electrical apparatus for use in the presence of combustible dust — Part 2: Test methods — Section 2: Method for determining the electrical resistivity of dust in layers*

61241-2-3:1994, *Electrical apparatus for use in the presence of combustible dust — Part 2: Test methods — Section 3: Method for determining minimum ignition energy of dust/air mixtures*

61241-3:1997, *Electrical apparatus for use in the presence of combustible dust — Part 3: Classification of areas where combustible dusts are or may be present*

61241-4:2001, *Electrical apparatus for use in the presence of combustible dust — Part 4: Type of protection "pD"*

IEEE (Institute of Electrical and Electronics Engineers)

80-2000, *Guide for Safety in AC Substation Grounding*

484-2002, *Recommended Practice for Installation Design and Implementation of Vented Lead-Acid Batteries for Stationary Applications*

835-1994, *Standard Power Cable Ampacity Tables*

837-1989, *Standard for Qualifying Permanent Connections Used in Substation Grounding*

1584-2002, *Guide for Performing Arc-Flash Hazard Calculations*

C62.41-1991, *Recommended Practice on Surge Voltages in Low-Voltage AC Power Circuits*

Institute of Petroleum

Model Code of Safe Practice — Part 15: Area Classification of Locations for Petroleum Installations

ISA (Instrument Society of America)

RP 12.6-2003, *Wiring Practices for Hazardous (Classified) Locations — Instrumentation — Part 1: Intrinsic Safety*

ISO (International Organization for Standardization)

965:1998, *General Purpose Metric Screw Threads — Tolerances*

NEMA (National Electrical Manufacturers Association)
VE 1-2002, *Metal Cable Tray Systems*

NFPA (National Fire Protection Association)
No. 20-2003, *Standard for the Installation of Stationary Fire Pumps for Fire Protection*
No. 51A-2001, *Standard for Acetylene Cylinder Charging Plants*
No. 70-2005, *National Electrical Code*
No. 70E-2004, *Electrical Safety in the Workplace*
No. 72-2002, *National Fire Alarm Code*
No. 91-2004, *Standard for Exhaust Systems for Air Conveying of Vapors, Gases, Mists, and Noncombustible Particulate Solids*
No. 96-2004, *Standard for Ventilation Control and Fire Protection of Commercial Cooking Operations*
No. 496-2003, *Standard for Purged and Pressurized Enclosures for Electrical Equipment*
No. 497-2004, *Classification of Flammable Liquids, Gases, or Vapors and of Hazardous (Classified) Locations for Electrical Installations in Chemical Process Areas*
No. 505-2006, *Fire Safety Standard for Powered Industrial Trucks Including Type Designations, Areas of Use, Conversions, Maintenance, and Operation*
No. 655-2001, *Standard for Prevention of Sulfur Fires and Explosions*
No. 820-2003, *Standard for Fire Protection in Wastewater Treatment and Collection Facilities*

NRCC (National Research Council Canada)
National Building Code of Canada, 2005
National Farm Building Code of Canada, 1995
National Fire Code of Canada, 2005

UL (Underwriters Laboratories Inc.)
1449-1996, *Standard for Safety for Transient Voltage Surge Suppressors*

ULC (Underwriters' Laboratories of Canada)
CAN/ULC-S524-2001, *Installation of Fire Alarm Systems*
CAN/ULC-S531-2002, *Smoke Alarms*

Other publications
Natural Resources Canada, *Atlas of Canada*, 2006

Section 0 — Object, scope, and definitions
(See Appendix G)

Δ

Object (see Appendix B)

The object of this Code is to establish safety standards for the installation and maintenance of electrical equipment. In its preparation, consideration has been given to the prevention of fire and shock hazards, as well as proper maintenance and operation.

Compliance with the requirements of this Code and proper maintenance will ensure an essentially safe installation. Safe installations may also be ensured by compliance with the objective-based fundamental safety principles of IEC 60364-1 (see Appendix K). Compliance with these objective-based installation criteria by industrial and similar users may be achieved through the implementation of specific quality management programs or equivalent programs acceptable to the authorities adopting and enforcing this Code.

Wiring installations that do not make provision for the increasing use of electricity may be overloaded in the future, resulting in a hazardous condition. It is recommended that the initial installation have sufficient wiring capacity and that there be some provision made for wiring changes that might be required as a result of future load growth.

This Code is not intended as a design specification nor as an instruction manual for untrained persons.

The requirements in this Code address the fundamental principles of protection for safety contained in Section 131 of IEC 60364-1, which encompasses protection against electric shock, thermal effects, overcurrent, fault currents, and overvoltage.

Scope

This Code covers all electrical work and electrical equipment operating or intended to operate at all voltages in electrical installations for buildings, structures, and premises, including factory-built relocatable and non-relocatable structures, and self-propelled marine vessels stationary for periods exceeding five months and connected to a shore supply of electricity continuously or from time to time, with the following exceptions:

Δ
(a) installations or equipment employed by an electric, communication, or community antenna distribution system utility in the exercise of its function as a utility, as recognized by the regulatory authority having jurisdiction, and located outdoors or in buildings or sections of buildings used for that purpose;

(b) equipment and facilities that are used in the operation of an electric railway and are supplied exclusively from circuits that supply the motive power;

(c) installations or equipment used for railway signalling and railway communication purposes, and located outdoors or in buildings or sections of buildings used exclusively for such installations;

(d) aircraft; and

(e) electrical systems in ships that are regulated under Transport Canada.

For mines and quarry applications, see also CAN/CSA-M421.

This Code and any standards referenced herein do not make or imply any assurance or guarantee by the authority adopting this Code with respect to life expectancy, durability, or operating performance of equipment and materials referenced herein.

Definitions

For the purpose of correct interpretation, certain terms have been defined and where such terms or their derivatives appear throughout this Code they shall be understood to have the following meanings. The ordinary or dictionary meaning of terms shall be used for terms not specifically defined in this Code.

Acceptable — acceptable to the authority enforcing this Code.

Accessible (as applied to equipment) — admitting close approach because the equipment is not guarded by locked doors, elevation, or other effective means.

Accessible (as applied to wiring methods) —
(a) not permanently closed in by the structure or finish of the building; and
(b) capable of being removed without disturbing the building structure or finish.

Accredited certification organization — an organization that has been accredited by the Standards Council of Canada, in accordance with specific criteria, procedures, and requirements, to operate, on a continuing basis, a certification program for electrical equipment.

Alive or **live** — electrically connected to a source of voltage difference, or electrically charged to have a voltage different from that of the earth; the term may be used in place of the term "current-carrying", where the intent is clear, to avoid repetition of the longer term.

Aluminum-sheathed cable — a cable consisting of one or more conductors of approved type assembled into a core and covered with a liquid- and gas-tight sheath of aluminum or aluminum alloy.

Ampacity — the current-carrying capacity of electric conductors expressed in amperes.

Approved (as applied to electrical equipment) —
(a)   such equipment has been certified by a certification organization accredited by the Standards Council of Canada in accordance with the requirements of
    (i)   CSA standards; or
    (ii)   other recognized documents, where such CSA standards do not exist or are not applicable; or
(b)   such equipment conforms to the requirements of the regulatory authority.

Authorized person — a qualified person who, in his or her duties or occupation, is obliged to approach or handle electrical equipment; or, a person who, having been warned of the hazards involved, has been instructed or authorized to do so by someone having authority to give the instruction or authorization.

Auxiliary gutter — a raceway consisting of a sheet metal enclosure used to supplement the wiring space of electrical equipment and to enclose interconnecting conductors.

AWG — the American (or Brown and Sharpe) Wire Gauge as applied to non-ferrous conductors and non-ferrous sheet metal.

Bathroom — a room containing bathing or showering facilities and that may also contain a wash basin(s) and/or water closet(s).

Bonding — a low impedance path obtained by permanently joining all non-current-carrying metal parts to ensure electrical continuity and having the capacity to conduct safely any current likely to be imposed on it.

Bonding conductor — a conductor that connects the non-current-carrying parts of electrical equipment, raceways, or enclosures to the service equipment or system grounding conductor.

Box connector — see **Connector**.

Branch circuit — see **Circuit**.

Building — a structure that stands alone or that is cut off from adjoining structures by firewalls, unpierced or with openings, protected by approved fire doors.

Bus — a conductor that serves as a common connection for the corresponding conductors of two or more circuits.

Busway — a raceway consisting of metal troughing (including elbows, tees, and crosses, in addition to straight runs) containing conductors that are supported on insulators.

Cabinet — an enclosure of adequate mechanical strength, composed entirely of non-combustible and absorption-resistant material, designed either for surface or flush mounting, and provided with a frame, mat, or trim, in which swinging doors are hung.

Cable tray — a raceway consisting of troughing and fittings formed and constructed so that insulated conductors and cables may be readily installed or removed after the cable tray has been completely installed, without injury either to conductors or their covering.

> **Ladder cable tray** — a prefabricated structure consisting of two longitudinal side rails connected by individual transverse members, with openings exceeding 50 mm in a longitudinal direction.

> **Nonventilated cable tray** — a prefabricated structure without openings within the integral or separate longitudinal side rails.

> **Ventilated cable tray** — a prefabricated structure consisting of a ventilated bottom within integral longitudinal side rails, with no openings exceeding 50 mm in a longitudinal direction.

Cell — one of the hollow spaces, suitable for use as a raceway, of a cellular metal or cellular concrete floor, the axis of the cell being parallel to the longitudinal axis of the floor members.

Cellular floor — an assembly of cellular metal or cellular concrete floor members, consisting of units with hollow spaces (cells) suitable for use as raceways and, in some cases, non-cellular units.

Circuit (see Appendix B) —

Branch circuit — that portion of the wiring installation between the final overcurrent device protecting the circuit and the outlet(s).

Communication circuit — a circuit that is part of a communication system.

Control circuit — the circuit that carries the electric signals directing the performance of a control device, but that does not carry the power that the device controls.

Extra-low-voltage power circuit — a circuit, such as a valve operator and similar circuits, that is neither a remote control circuit nor a signal circuit, but that operates at not more than 30 V and that is supplied from a transformer or other device restricted in its rated output to 1000 V•A and approved for the purpose, but in which the current is not limited in accordance with the requirements for a Class 2 circuit.

Low-energy power circuit — a circuit other than a remote control or signal circuit that has the power supply limited in accordance with the requirements for Class 2 remote control circuits.

Multi-wire branch circuit — a branch circuit consisting of two or more ungrounded conductors having a voltage difference between them and an identified grounded conductor having equal voltage between it and each ungrounded conductor, with this grounded conductor connected to the neutral conductor.

Non-incendive circuit — a circuit in which any spark or thermal effect that may occur under normal operating conditions or due to opening, shorting, or grounding of field wiring is incapable of causing an ignition of the prescribed flammable gas or vapour.

Remote control circuit — any electrical circuit that controls any other circuit through a relay or an equivalent device.

Signal circuit — any electrical circuit, other than a communication circuit, that supplies energy to a device that gives a recognizable signal, such as circuits for doorbells, buzzers, code-calling systems, signal lights, etc.

Circuit breaker — a device designed to open and close a circuit by non-automatic means and to open the circuit automatically on a predetermined overcurrent without damage to itself when properly applied within its ratings.

Instantaneous-trip circuit breaker — a circuit breaker designed to trip only under short-circuit conditions.

Communication circuit — see **Circuit**.

Communication system — see **System**.

Community antenna distribution system — see **System**.

Concealed — rendered permanently inaccessible by the structure or finish of the building.

Conductor — a wire or cable, or other form of metal, installed for the purpose of conveying electric current from one piece of electrical equipment to another or to ground.

Conduit — a raceway of circular cross-section, other than electrical metallic tubing and electrical non-metallic tubing, into which it is intended that conductors be drawn.

Flexible metal conduit — a metal conduit that may be easily bent without the use of tools.

Liquid-tight flexible conduit —
(a)   a flexible metal conduit having an outer liquid-tight jacket; or
(b)   a flexible liquid-tight non-metallic conduit.

Rigid conduit — a rigid conduit of metal or a non-metallic material.

Rigid HFT conduit — a rigid non-metallic conduit of halogen-free thermoplastic.

Rigid metal conduit — a rigid conduit of metal made to the same dimensions as standard pipe and suitable for threading with standard pipe threads.

Rigid non-metallic conduit — a rigid conduit of non-metallic material that is not permitted to be threaded.

Rigid PVC conduit — a rigid non-metallic conduit of unplasticized polyvinyl chloride.

Rigid RTRC conduit Type AG — a rigid non-metallic conduit of reinforced thermoset material suitable for direct burial or encasement in concrete and for exposed or concealed work.

Rigid RTRC conduit Type BG — a rigid non-metallic conduit of reinforced thermoset material suitable for direct burial or encasement in concrete.

Rigid Type DB2/ES2 PVC conduit — a rigid non-metallic conduit of PVC for direct burial or encasement in concrete or masonry.

Rigid Type EB1 PVC conduit — a rigid non-metallic conduit of PVC for encasement in concrete or masonry.

Connector —

Box connector — a device for securing a cable, via its sheath or armour, where it enters an enclosure such as an outlet box.

Wire connector — a device that connects two or more conductors together or one or more conductors to a terminal point for the purpose of connecting electrical circuits.

Continuous duty — see **Duty**.

Control circuit — see **Circuit**.

Controller — a device or a group of devices for controlling in some predetermined manner the electric power delivered to the apparatus to which it is connected.

Cord set — an assembly consisting of a suitable length of flexible cord or power supply cable provided with an attachment plug at one end and a cord connector at the other end.

Current-permit — written permission from the inspection department to a supply authority stating that electric energy may be supplied to a particular installation.

Cut-out box — an enclosure of adequate mechanical strength, composed entirely of non-combustible and absorption-resistant material, designed for surface mounting, and having swinging doors or covers secured directly to, and telescoping with, the walls of the box proper.

Damp location — see **Location**.

Dead (as applied to electrical equipment) — the current-carrying parts of electrical equipment are free from any electrical connection to a source of voltage and from electrical charge and do not have a voltage different from that of earth.

Dead front — without live parts exposed to a person on the operating side of the equipment.

Different systems — see **System**.

Disconnecting means — a device, group of devices, or other means whereby the conductors of a circuit can be disconnected from their source of supply.

Dry location — see **Location**.

Duplex receptacle — see **Receptacle**.

Dust-tight — an enclosure constructed so that dust cannot enter it.

Duty — a requirement of service that demands the degree of regularity of the load.

Continuous duty — a requirement of service that demands operation at a substantially constant load for an indefinitely long time.

Intermittent duty — a requirement of service that demands operation for definitely specified alternate intervals of
(a)   load and no-load;
(b)   load and rest; or
(c)   load, no-load, and rest.

Periodic duty — a type of intermittent duty in which the load conditions are regularly recurrent.

Short-time duty — a requirement of service that demands operation at a substantially constant load for a short and definitely specified time.

4

Varying duty — a requirement of service that demands operation at loads and for intervals of time, both of which may be subject to wide variation.

Dwelling unit — one or more rooms for the use of one or more persons as a housekeeping unit with cooking, eating, living, and sleeping facilities.

Electrical contractor — any person, corporation, company, firm, organization, or partnership performing or engaging to perform, either for their or its own use or benefit, or for that of another, and with or without remuneration or gain, any work with respect to an electrical installation or any other work to which this Code applies.

Electrical equipment — any apparatus, appliance, device, instrument, fitting, fixture, machinery, material, or thing used in or for, or capable of being used in or for, the generation, transformation, transmission, distribution, supply, or utilization of electric power or energy, and, without restricting the generality of the foregoing, includes any assemblage or combination of materials or things that is used, or is capable of being used or adapted, to serve or perform any particular purpose or function when connected to an electrical installation, notwithstanding that any of such materials or things may be mechanical, metallic, or non-electric in origin.

Electrical installation — the installation of any wiring in or upon any land, building, or premises from the point(s) where electric power or energy is delivered by the supply authority or from any other source of supply, to the point(s) where such power or energy can be used by any electrical equipment, and the installation includes the connection of any such wiring with any of the electrical equipment and any part of the wiring and also includes the maintenance, alteration, extension, and repair of such wiring.

Electrical metallic tubing — a raceway of metal having circular cross-section into which it is intended that conductors be drawn and that has a wall thinner than that of rigid metal conduit and an outside diameter sufficiently different from that of rigid conduit to render it impracticable for anyone to thread it with standard pipe thread.

Electrical non-metallic tubing — a pliable non-metallic corrugated raceway having a circular cross-section.

Electric elevator — an elevator in which the motion of the car is obtained through an electric motor directly applied to the elevator machinery.

Elevator — a hoisting and lowering mechanism equipped with a car or platform that moves in guides in a substantially vertical direction but not including tiering or piling machines that operate within one storey, or endless belts, conveyors, chains, buckets, or similar devices used for the purpose of elevating materials.

Elevator machinery — the machinery and its equipment used in raising and lowering the elevator car or platform.

Emergency lighting — lighting required by the provisions of the *National Building Code of Canada* for the purpose of facilitating safe exit and access to exit in the event of fire or other emergency.

Explosion-proof — enclosed in a case that is capable of withstanding without damage any explosion that may occur within it of a specified gas or vapour and capable of preventing the ignition of a specified gas or vapour surrounding the enclosure from sparks, flashes, or explosion of the specified gas or vapour within the enclosure.

Exposed (as applied to live parts) — live parts can be inadvertently touched or approached nearer than a safe distance by a person, and the term is applied to parts not suitably guarded, isolated, or insulated.

Exposed (as applied to wiring methods) — not concealed.

Extra-low-voltage — see **Voltage**.

Extra-low-voltage power circuit — see **Circuit**.

Feeder — any portion of an electrical circuit between the service box or other source of supply and the branch circuit overcurrent devices.

Fire-resisting (when applied to a building) — constructed of masonry, reinforced concrete, or equivalent materials.

General-use switch — see **Switch**.

Ground — a connection to earth obtained by a grounding electrode.

Ground fault circuit interrupter — a device whose function is to interrupt, within a predetermined time, the electrical circuit to the load when a current to ground exceeds a predetermined value that is less than that required to operate the overcurrent protective device of the supply circuit (see Appendix B).

Grounded — connected effectively with the general mass of the earth through a grounding path of sufficiently low impedance and having an ampacity sufficient at all times, under the most severe conditions liable to arise in practice, to prevent any current in the grounding conductor from causing a harmful voltage to exist
(a)   between the grounding conductors and neighbouring exposed conducting surfaces that are in good contact with the earth; or
(b)   between the grounding conductors and neighbouring surfaces of the earth itself.

Grounding — a permanent and continuous conductive path to the earth with sufficient ampacity to carry any fault current liable to be imposed on it, and of a sufficiently low impedance to limit the voltage rise above ground and to facilitate the operation of the protective devices in the circuit.

Grounding conductor — the conductor used to connect the service equipment or system to the grounding electrode.

Grounding electrode — a buried metal water-piping system or metal object or device buried in, or driven into, the ground to which a grounding conductor is electrically and mechanically connected.

Grounding system — see **System**.

Guarded — covered, shielded, fenced, enclosed, or otherwise protected by means of suitable covers or casings, barriers, rails or screens, or mats or platforms to remove the liability of dangerous contact or approach by persons or objects.

Hazardous location — see **Location**.

Header — a raceway for electrical conductors, associated with an underfloor raceway or cellular floor system, that provides access to predetermined raceways or cells.

High-voltage — see **Voltage**.

Hoistway — any shaftway, hatchway, well hole, or other vertical opening or space in which an elevator, escalator, or dumbwaiter operates or is intended to operate.

Identified —
(a)   when applied to a conductor, signifies that the conductor has
　　(i)   a white or grey covering; or
　　(ii)   a raised longitudinal ridge(s) on the surface of the extruded covering on certain flexible cords either of which indicates that the conductor is a grounded conductor or a neutral; and
(b)   when applied to other electrical equipment, signifies that the terminals to which grounded or neutral conductors are to be connected have been distinguished for identification by being tinned, nickel-plated, or otherwise suitably marked.

Inaccessible —
(a)   when applied to a room or compartment, signifies that the room or compartment is sufficiently remote from access or placed or guarded so that unauthorized persons cannot inadvertently enter the room or compartment; and
(b)   when applied to electrical equipment, signifies that the electrical equipment is covered by the structure or finish of the building in which it is installed or maintained, or is sufficiently remote from access or placed so that unauthorized persons cannot inadvertently touch or interfere with the equipment.

Indicating switch — see **Switch**.

Industrial establishment — a building or part of a building (other than office or exhibit space) or a part of the premises outside the building where persons are employed in manufacturing processes or in the handling of material, as distinguished from dwellings, offices, and similar occupancies.

Inspection department — an organization legally authorized to enforce this Code and having jurisdiction over specified territory.

Inspector — any person duly appointed by the inspection department for the purpose of enforcing this Code.

Insulated — separated from other conducting surfaces by a dielectric material or air space having a degree of resistance to the passage of current and to disruptive discharge sufficiently high for the condition of use.

Insulating (as applied to non-conducting substances) — capable of bringing about the condition defined as insulated.

Intermittent duty — see **Duty**.

Intrinsically safe — that any spark or thermal effect that may occur in normal use, or under any conditions of fault likely to occur in practice, is incapable of causing an ignition of the prescribed flammable gas, vapour, or dust.

Isolating switch — see **Switch**.

Ladder cable tray — see **Cable tray** (see Appendix B).

Lampholder — a device constructed for the mechanical support of lamps and for connecting them to circuit conductors.

Liquid-tight flexible conduit — see **Conduit**.

Location —

> **Damp location** — an exterior or interior location that is normally or periodically subject to condensation of moisture in, on, or adjacent to electrical equipment and includes partially protected locations under canopies, marquees, roofed open porches, and similar locations.

> **Dry location** — a location not normally subject to dampness, but that may include a location subject to temporary dampness as in the case of a building under construction, provided that ventilation is adequate to prevent an accumulation of moisture.

> **Hazardous location** — premises, buildings, or their parts in which the hazard of fire or explosion exists due to the fact that
> (a)   highly flammable gases, flammable volatile liquid mixtures, or other highly flammable substances are manufactured or used, or are stored in other than original containers;
> (b)   combustible dust or flyings are likely to be present in quantities sufficient to produce an explosive or combustible mixture, or it is impracticable to prevent such dust or flyings from being deposited upon incandescent lamps or from collecting in or upon motors or other electrical equipment in such quantities as to produce overheating through normal radiation being prevented;
> (c)   easily ignitable fibres or materials producing combustible flyings are manufactured, handled, or used in a free open state; or
> (d)   easily ignitable fibres or materials producing combustible flyings are stored in bales or containers but are not manufactured or handled in a free open state.

> **Ordinary location** — a dry location in which, at normal atmospheric pressure and under normal conditions or use, electrical equipment is not unduly exposed to damage from mechanical causes, excessive dust, moisture or extreme temperatures, and in which electrical equipment is entirely free from the possibility of damage through corrosive, flammable, or explosive atmospheres.

> **Outdoor location** — any location exposed to the weather (see Appendix B).

> **Wet location** — a location in which liquids may drip, splash, or flow on or against electrical equipment.

Low-energy power circuit — see **Circuit**.

Low-voltage — see **Voltage**.

Low-voltage protection — a device that operates on the reduction or failure of voltage to cause and maintain the interruption of power to the main circuit.

Low-voltage release — a device that operates on the reduction or failure of voltage to cause interruption of power to the main circuit, but not to prevent its re-establishment on the return of voltage to a safe operating value.

Luminaire — a complete lighting unit designed to accommodate the lamp(s) and to connect the lamp(s) to circuit conductors.

Machine tool, metal cutting — a power-driven machine, not portable by hand, used to remove metal in the form of chips.

Machine tool, metal forming — a power-driven machine, not portable by hand, used to press, forge, emboss, hammer, blank, or shear metals.

Manufactured wiring system — a wiring system containing component parts that are assembled in the process of manufacture and cannot be disassembled at the building site without damage to or destruction of the assembly.

Mineral-insulated cable — a cable having a bare solid conductor(s) supported and insulated by a highly compressed refractory material enclosed in a liquid- and gas-tight metal tube sheathing; the term includes both the regular type (MI) and the lightweight type (LWMI) unless otherwise qualified.

Mobile home — a transportable dwelling unit constructed to be towed on its own chassis (see Appendix B).

Mobile industrial or commercial structure — a transportable structure, other than a mobile home, constructed to be towed on its own chassis (see Appendix B).

Motor-circuit switch — a switch rated in horsepower.

MSG — the Manufacturer's Standard Gauge for uncoated steel.

Multi-outlet assembly — a surface or flush enclosure carrying conductors for extending one 2-wire or multi-wire branch circuit to two or more receptacles of the grounding type that are attached to the enclosure.

Multiple section mobile unit — a single structure composed of separate mobile units, each towable on its own chassis, which, when towed to the site, are coupled together mechanically and electrically to form a single structure.

Multi-winding motor — a motor having multiple and/or tapped windings, intended to be connected or reconnected in two or more configurations, for operation at any one of two or more speeds and/or voltages.

Multi-wire branch circuit — see **Circuit**.

Neutral — that conductor (when one exists) of a polyphase circuit or single-phase, 3-wire circuit which is intended to have a voltage such that the voltage differences between it and each of the other conductors are approximately equal in magnitude and are equally spaced in phase (see Appendix B).

Non-combustible construction — that type of construction in which a degree of fire safety is attained by the use of non-combustible materials for structural members and other building assemblies (see Appendix B).

Non-incendive circuit — see **Circuit**.

Non-relocatable structure — a factory-built unit for use on permanent foundations.

Nonventilated cable tray — see **Cable tray**.

Open (as applied to electrical equipment) — moving parts, windings, or live parts are exposed to accidental contact.

Outdoor location — see **Location**.

Outlet — a point in the wiring installation at which current is taken to supply utilization equipment.

Outline lighting — an arrangement of incandescent lamps or electric discharge tubing to outline or call attention to certain features such as the shape of a building or the decoration of a window.

Overcurrent device — any device capable of automatically opening an electric circuit, under both predetermined overload and short-circuit conditions, either by fusing of metal or by electro-mechanical means.

Overload device — a device affording protection from excess current, but not necessarily short-circuit protection, and capable of automatically opening an electric circuit either by fusing of metal or by electro-mechanical means.

Panelboard — an assembly of buses and connections, overcurrent devices and control apparatus with or without switches, or other equipment constructed for installation as a complete unit in a cabinet.

Panelboard, enclosed — an assembly of buses and connections, overcurrent devices and control apparatus with or without switches, or other equipment installed in a cabinet.

Park model trailer — a recreational vehicle having a gross floor area not exceeding 50 m² when set up (see Appendix B).

Part-winding start motor — a motor arranged for starting by first energizing part of its primary winding and, subsequently, energizing the remainder of this winding in one or more steps, both parts then carrying current.

Periodic duty — see **Duty**.

Permanently connected equipment — equipment that is electrically connected to the supply by means of connectors that can be accessed, loosened, or tightened only with the aid of a tool.

Permit — the official written permission of the inspection department, on a form provided for the purpose, authorizing work to be commenced on any electrical installation.

Plenum — a chamber associated with air-handling apparatus for distributing the processed air from the apparatus (supply plenum) to the supply ducts or for receiving air to be processed by the apparatus (return plenum).

Portable (as applied to electrical equipment) — the equipment is specifically designed not to be used in a fixed position and receives current through the medium of a flexible cord or cable and usually an attachment plug.

Portable ground fault circuit interrupter — a ground fault circuit interrupter that is either of the direct plug-in type or specifically designed to receive current by means of a flexible cord or cable and an attachment plug and that incorporates one or more receptacles for the connection of equipment that is provided with a flexible cord or cable and an attachment plug.

Power supply cord — an assembly consisting of a suitable length of flexible cord or power supply cable provided with an attachment plug at one end.

Protected (as applied mainly to electrical equipment) — such equipment is constructed so that the electrical parts are protected against damage from foreign objects entering the enclosure.

PVC conduit — see **Conduit**.

Qualified person — one familiar with the construction and operation of the apparatus and the hazards involved.

Raceway — any channel designed for holding wires, cables, or busbars, and, unless otherwise qualified in the Rules of this Code, the term includes conduit (rigid and flexible, metal and non-metallic), electrical metallic and non-metallic tubing, underfloor raceways, cellular floors, surface raceways, wireways, cable trays, busways, and auxiliary gutters.

Readily accessible — capable of being reached quickly for operation, renewal, or inspection, without requiring those to whom ready access is a requisite to climb over or remove obstacles or to resort to portable ladders, chairs, etc.

Receptacle — one or more female contact devices, on the same yoke, installed at an outlet for the connection of one or more attachment plugs.

> **Duplex receptacle** — two female contact devices, on the same yoke, installed at an outlet for the connection of two attachment plugs.

> **Single receptacle** — one female contact device, with no other contact device on the same yoke, installed at an outlet for the connection of one attachment plug.

> **Split receptacle** — a duplex receptacle having terminals adapted for connection to a grounded, 3-wire supply, e.g., 120/240 V or 120/208 V.

Recreational vehicle — a portable structure intended as a temporary accommodation for travel, vacation, or recreational use (see Appendix B).

Recreational vehicle park — an area of land designed to accommodate recreational vehicles and park model trailers.

Relocatable structure — a factory-built unit for use without a permanent foundation.

Remote control circuit — see **Circuit**.

Repellent (used as a suffix (e.g., moisture-repellent)) — a material constructed, treated, or surfaced so that liquid will tend to run off and cannot readily penetrate the surface.

Residential occupancy — the occupancy or use of a building or part of a building by persons for whom sleeping accommodation is provided but who are not harboured or detained to receive medical care or treatment or are not involuntarily detained.

Resistant (used as a suffix (e.g., absorption-resistant, moisture-resistant, etc.)) — material constructed, protected, or treated so that it will not be injured readily when subjected to the specific material or condition.

Separate built-in cooking unit — a stationary cooking appliance, including its integral supply leads or terminals and consisting of one or more surface elements or ovens, or a combination of these, constructed so that the unit is permanently built into a counter or wall.

Service, consumer's — all that portion of the consumer's installation from the service box or its equivalent up to and including the point at which the supply authority makes connection.

Service, supply — any one set of conductors run by a supply authority from its mains to a consumer's service.

Δ **Service box** — an approved assembly consisting of an enclosure constructed so that it may be effectually locked or sealed, containing either fuses and a service switch or a circuit breaker, and of such design that either the switch or circuit breaker may be manually operated when the box is closed.

Service room — a room or space provided in a building to accommodate building service equipment and constructed in accordance with the *National Building Code of Canada* or applicable local legislation (see Appendix B, Note to Rule 26-012).

Shockproof (as applied to X-ray and high-frequency equipment) — such equipment is guarded with grounded metal so that no person can come into contact with any live part.

Short-time duty — see **Duty**.

Signal circuit — see **Circuit**.

Single dwelling — a dwelling unit consisting of a detached house, one unit of row housing, or one unit of a semi-detached, duplex, triplex, or quadruplex house.

Single receptacle — see **Receptacle**.

Slow-burning (as applied to conductor insulation) — the insulation has flame-retardant properties.

Soldered — a uniting of metal surfaces by the fusion of a metal alloy, usually of lead and tin.

Special permission — the written authority of the inspection department.

Split receptacle — see **Receptacle**.

Splitter — an enclosure containing terminal plates or busbars having main and branch connectors.

Starter — a controller for accelerating a motor from rest to normal speed and for stopping the motor; the term usually implies inclusion of overload protection.

Supply authority — any person, firm, corporation, company, commission, or other organization supplying electric energy.

Surface raceway — a surface-mounted or pendant enclosure, consisting of one or more channels for the purpose of containing and protecting conductors and intended to accommodate associated fittings, wiring devices, luminaires, and accessories.

Switch — a device for making, breaking, or changing connection in a circuit.

> **General-use switch** — a switch intended for use in general distribution and branch circuits and that is rated in amperes and is capable of interrupting its rated current at rated voltage.

> **Indicating switch** — a switch of such design or marked so that whether it is ON or OFF may be readily determined by inspection.

> **Isolating switch** — a switch intended for isolating either a circuit or some equipment from its source of supply and that is not intended either for establishing or interrupting the flow of current in any circuit.

> **Motor-circuit switch** — a manually operated knife or snap switch rated in horsepower, fused or unfused.

Switchboard — a panel or assembly of panels on which is mounted any combination of switching, measuring, controlling, and protective devices, buses, and connections, designed to successfully carry and rupture the maximum fault current encountered when controlling incoming and outgoing feeders.

System (see Appendix B) —

> **Communication system** — an electrical system whereby voice, sound, or data may be received and/or transmitted and that includes telephone, telegraph, data communications, intercommunications, paging systems, wired music systems, and other systems of similar nature, but excludes alarm systems such as fire,

smoke, or intrusion, radio and television broadcast communication equipment, closed circuit television, or community antenna television systems.

Community antenna distribution system — a distribution system of coaxial cable, together with any necessary amplifiers or other equipment, that is used to transmit television or radio frequency signals typical of a community antenna television (CATV) system.

Different systems — those which derive their energy from different transformers or from different banks of transformers, or from different generators or other sources.

Grounding system — all conductors, clamps, ground clips, ground plates or pipes, and ground electrodes by means of which the electrical installation is grounded.

Theatre — a building, or any portion of a building, that is used for public, dramatic, operatic, motion-picture, or other performances.

Thermal cut-out — a device affording protection from excessive current, but not necessarily short-circuit protection, and containing a heating element in addition to, and affecting, a fusible member that opens the circuit.

Underfloor raceway — a raceway suitable for use in the floor.

Utilization equipment — equipment that utilizes electrical energy for mechanical, chemical, heating, lighting, or similar useful purposes.

Varying duty — see **Duty.**

Vault (transformer vault or **electrical equipment vault)** — an isolated enclosure, either above or below ground, with fire-resisting walls, ceilings, and floors for the purpose of housing transformers and other electrical equipment.

Ventilated cable tray — see **Cable tray** (see Appendix B).

Vessel — any ship or boat or any other description of vessel used or designed to be used in navigation.

Voltage —

Extra-low voltage — any voltage up to and including 30 V.

High voltage — any voltage above 750 V.

Low voltage — any voltage from 31 to 750 V inclusive.

Voltage of a circuit — the greatest root-mean-square (effective) voltage between any two conductors of the circuit concerned.

Voltage-to-ground — the voltage between any given live ungrounded part and any grounded part in the case of grounded circuits, or the greatest voltage existing in the circuit in the case of ungrounded circuits.

Washroom — a room that contains a wash basin(s) and that may contain a water closet(s) but without bathing or showering facilities.

Wet location — see **Location.**

Wire connector — see **Connector.**

Wireway — a raceway consisting of a completely enclosing arrangement of metal troughing and fittings formed and constructed so that insulated conductors may be readily drawn in and withdrawn, or laid in and removed, after the wireway has been completely installed, without injury either to conductors or to their covering.

Section 2 — General Rules
Administrative

2-000 Authority for Rules
By virtue of the authority vested in the inspection department, this Code has been adopted and the inspection department hereby orders and directs its observance.

2-002 Special requirements
Sections devoted to Rules governing particular types of installations are not intended to embody all Rules governing these particular types of installations, but cover only those special Rules or regulations that add to or amend those prescribed in other sections covering installations under ordinary conditions.

2-004 Permit
Electrical contractors or others responsible for carrying out the work shall obtain a permit from the inspection department before commencing work with respect to installation, alteration, repair, or extension of any electrical equipment.

2-006 Application for inspection
An application for inspection shall be filed with the inspection department on a form provided by the latter at the time the permit is obtained.

2-008 Fees
Fees for the permit and inspection in accordance with the schedule prescribed by the inspection department shall be paid at the time the permit is obtained.

2-010 Posting of permit
A copy of the permit shall be posted in a conspicuous place at the work site and shall not be removed until the inspection is completed.

2-012 Notification of inspection
The inspection department shall be notified in writing by the electrical contractor that work is ready for inspection at such time(s) allowing inspection before any work or portion of work is concealed.

2-014 Plans and specifications
Plans and specifications in duplicate or in greater number if required by the inspection department (one copy to be retained by the inspection department) shall be submitted by the owner or an agent to, and acceptance obtained from, the inspection department before work is commenced on
(a)   wiring installations of public buildings, industrial establishments, factories, and other buildings in which public safety is involved;
(b)   large light and power installations and the installation of apparatus such as generators, transformers, switchboards, large storage batteries, etc.; or
(c)   such other installations as may be prescribed by the inspection department.

2-016 Current-permits
Except as provided in Rule 2-018, no reconnection, installation, alteration, or addition shall be connected to any service or other source of electric energy by a supply authority, electrical contractor, or other person, until a current-permit authorizing the supply of electric energy has been obtained from the inspection department.

2-018 Reconnection
A supply authority shall not require a current-permit for reconnection in cases where the service has been cut off for non-payment of bills or a change of occupant provided that there have been no alterations or additions subsequent to the issuance of the last current-permit.

2-020 Reinspection
The inspection department reserves the right to reinspect any installation if and when it considers such action to be necessary.

2-022 Renovation of existing installations
The inspection department may require such changes as may be necessary to be made to existing installations where, through hard usage, wear and tear, or as a result of alterations or extensions, dangerous conditions have developed.

2-024 Use of approved equipment (see Appendix A)

Electrical equipment used in electrical installations within the jurisdiction of the inspection department shall be approved and shall be of a kind or type and rating approved for the specific purpose for which it is to be employed.

2-026 Powers of rejection (see Appendix B)

Even though approval has previously been granted, the inspection department may reject, at any time, any electrical equipment under any of the following conditions:

(a)   the equipment is substandard with respect to the sample on which approval was granted;
(b)   the conditions of use indicate that the equipment is not suitable; or
(c)   the terms of the approval agreement are not being carried out.

2-028 Availability of work for inspection

No electrical work shall be rendered inaccessible by lathing, boarding, or other building construction until it has been accepted by the inspection department.

2-030 Deviation or postponement

In any case where deviation or postponement of these Rules and regulations is necessary, special permission shall be obtained before proceeding with the work, but this special permission shall apply only to the particular installation for which it is given.

2-032 Damage and interference

(1)   No person shall damage any electrical installation or component thereof.
(2)   No person shall interfere with any electrical installation or component thereof except that when, in the course of alterations or repairs to non-electrical equipment or structures, it may be necessary to disconnect or move components of an electrical installation, it shall be the responsibility of the person carrying out the alterations or repairs to ensure that the electrical installation is restored to a safe operating condition as soon as the progress of the alterations or repairs permits.

Technical

General

2-100 Marking of equipment (see Appendix B)

(1)   Each piece of electrical equipment shall bear those of the following markings necessary to identify the equipment and ensure that it is suitable for the particular installation:
    (a)   the maker's name, trademark, or other recognized symbol of identification;
    (b)   catalogue number or type;
    (c)   voltage;
    (d)   rated load amperes;
    (e)   watts, volt amperes, or horsepower;
    (f)   whether for ac, dc, or both;
    (g)   number of phases;
    (h)   frequency in hertz;
    (i)   rated load speed in revolutions per minute;
    (j)   designation of terminals;
    (k)   whether for continuous or intermittent duty;
    (l)   evidence of approval; or
    (m)   other markings necessary to ensure safe and proper operation.

(2)   At the time of installation, each service box shall be marked in a conspicuous, legible, and permanent manner, to indicate clearly the maximum rating of the overcurrent device that may be used for this installation.

(3)   At each distribution point, circuit breakers, fuses, and switches shall be marked, adjacent thereto, in a conspicuous and legible manner to indicate clearly
    (a)   which installation or portion of installation that they protect or control; and
    (b)   the maximum rating of overcurrent device that is permitted.

(4)   The marking on electrical equipment shall not be added to, or changed, to indicate a use under this Code for which the equipment has not been approved.

2-102 Rebuilt equipment (see Appendix B)

(1) Where any electrical machine or apparatus is rebuilt or rewound with any change in its rating or characteristics, it shall be provided with a nameplate giving the name of the person or firm by whom such change was made together with the new marking.

(2) Where the original nameplate is removed, the original manufacturer's name and any original identifying data, such as serial numbers, shall be added to the new nameplate.

Δ (3) Except as provided in Subrule (4), the appropriate requirements of the *Canadian Electrical Code, Part II*, that apply to new electrical equipment shall also apply to rebuilt and rewound equipment unless it is impracticable to comply with such requirements.

(4) Rebuilt or refurbished molded case circuit breakers or molded case switches shall not be considered to be approved for the purpose of Rule 2-024.

2-104 Substitution

Where electrical equipment of the exact size or rating is not procurable for a given purpose, equipment of a larger size or rating that may be consistent with the purpose shall be used, except where use of equipment of a smaller size or rating complies with Rule 2-030.

2-106 Circuit voltage-to-ground — Dwelling units

Branch circuits in dwelling units shall not have a voltage exceeding 150 volts-to-ground except that, where the calculated load on the service conductors of an apartment or similar building exceeds 250 kV•A and where qualified electrical maintenance personnel are available, higher voltages not exceeding the voltage-to-ground of a nominal system voltage of 347/600Y shall be permitted to be used in the dwelling unit to supply the following fixed (not portable) equipment:

(a) space heating, provided that wall-mounted thermostats operate at a voltage not exceeding 300 volts-to-ground;

(b) water heating; and

(c) air conditioning.

2-108 Quality of work

The mechanical arrangement and execution of the work in connection with any electrical installation shall be acceptable.

2-110 Material for anchoring to masonry and concrete

Wood or other similar material shall not be used as an anchor into masonry or concrete for the support of any electrical equipment.

2-112 Corrosion protection for materials used in wiring

(1) Metals used in wiring, such as raceways, cable sheaths and armour, boxes, and fittings, shall be suitably protected against corrosion for the environment in which they are to be used or shall be made of suitable corrosion-resistant material.

(2) Where practicable, dissimilar metals shall not be used where there is a possibility of galvanic action.

2-114 Soldering fluxes

Fluxes used for soldering copper and its alloys shall be of types that are non-corrosive to copper.

2-116 AWG sizes of conductors

Where reference is made in this Code to AWG size, this shall mean the copper AWG size, unless otherwise specified.

2-118 Installation of electrical equipment (see Appendix G)

Electrical equipment shall be installed so as to ensure that after installation there is ready access to nameplates and access to parts requiring maintenance.

2-120 Installation of other than electrical equipment

Equipment or material of other than an electrical nature shall not be installed or placed so close to electrical equipment as to create a condition that is dangerous.

2-122 Use of thermal insulation

(1) Where the hollow spaces between studding, joists, or rafters of buildings are to be filled with thermal insulation, the following restrictions, as applicable, shall apply to the installation of electrical wiring in such spaces:

(a) special care shall be taken to ensure that conductor insulation temperatures are not exceeded due either to mutual heating of adjacent conductors or cables or to reduced heat dissipation through the thermal insulation;

    (b)   if the space is to be filled with a loose or free-flowing material that is non-corrosive, fire-resisting, and non-conductive and that is in compliance with the *National Building Code of Canada*, any type of wiring system recognized by this Code shall be permitted to be used, but special care shall be taken to ensure that there will be no strain on the conductors due to the weight or pressure of the insulating material;

    (c)   if the thermal insulation material, in the form of batts or rigid sheets, is installed prior to the installation of the wiring and secured in place so that there will be no undue pressure on the conductors, no special precaution need be observed;

    (d)   if thermal insulation made of or faced with metal is installed, the wiring shall conform to the following:

        (i)   a 25 mm separation shall be provided between the thermal insulation and the knob-and-tube wiring; and

        (ii)   non-metallic-sheathed cable may be in contact with the insulation; and

    (e)   mineral-insulated cable or aluminum-sheathed cable shall not be used with types of thermal insulation that are liable to have a corrosive action on the sheath.

(2)   Thermal insulation material shall not be sprayed or otherwise introduced into the interior of outlet boxes, junction boxes, or enclosures for other electrical equipment.

2-124 Fire spread (see Appendices B and G)

(1)   Electrical installations shall be made so that the probability of spread of fire through firestopped partitions, floors, hollow spaces, firewalls or fire partitions, vertical shafts, or ventilating or air-conditioning ducts is reduced to a minimum.

(2)   Where a fire separation is pierced by a raceway or cable, any openings around the raceway or cable shall be properly closed or sealed in compliance with the *National Building Code of Canada*.

2-126 Flame spread requirements for electrical wiring and cables (see Appendices B and G)

Electrical wiring and cables installed in buildings shall meet the flame spread requirements of the *National Building Code of Canada* or local building legislation.

2-128 Flame spread requirements for totally enclosed non-metallic raceways (see Appendices B and G)

Totally enclosed non-metallic raceways installed in buildings shall meet the flame spread requirements of the *National Building Code of Canada*.

Δ 2-130 Sunlight resistance requirements (see Appendix B)

Insulated electrical wiring and cables and totally enclosed non-metallic raceways installed and used where exposed to direct rays of the sun shall be specifically approved for the purpose and be so marked.

2-132 Insulation integrity (see Appendix B)

All wiring shall be installed so that, when completed, the system will be free from short-circuits and from grounds except as permitted in Section 10.

2-134 Use of ground fault circuit interrupters

Ground fault circuit interrupters shall be permitted as supplementary protection from shock hazard but shall not be used as a substitute for insulation or grounding except as permitted by Rules 10-408(4) and 26-700(8).

Protection of persons and property

2-200 General

Electrical equipment shall be installed and guarded so that adequate provision is made for the safety of persons and property and for the protection of the electrical equipment from mechanical or other injury to which it is liable to be exposed.

2-202 Guarding of bare live parts

(1)   Bare live parts shall be guarded against accidental contact by means of approved cabinets or other forms of approved enclosures except where the bare live parts are

    (a)   located in a suitable room, vault, or similar enclosed area that is accessible only to qualified persons; or

(b)   as permitted elsewhere by this Code.

(2)   Where electrical equipment has mounted on it, within 900 mm of bare live parts, non-electrical components that require servicing by unqualified persons, suitable barriers or covers shall be provided for the bare live parts.

(3)   Entrances to rooms and other guarded locations containing exposed bare live parts shall be marked with conspicuous warning signs forbidding entry to unqualified persons.

Maintenance and operation

2-300 General requirements for maintenance and operation

(1)   All operating electrical equipment shall be kept in safe and proper working condition.

(2)   Electrical equipment maintained for emergency service shall be periodically inspected and tested as necessary to ensure its fitness for service.

(3)   Infrequently used electrical equipment maintained for future service shall be thoroughly inspected before use in order to determine its fitness for service.

(4)   Defective equipment shall either be put in good order or permanently disconnected.

2-302 Maintenance in hazardous locations

In locations where explosive or flammable materials or gases are present, special precautions shall be observed as follows:

(a)   repairs or alterations shall not be made on any live equipment; and

(b)   fits or seals in enclosures shall be maintained in their original safe condition.

2-304 Disconnection

(1)   No repairs or alterations shall be carried out on any live equipment except where complete disconnection of the equipment is not practicable.

(2)   Three-way or four-way switches shall not be considered as disconnecting means.

(3)   Adequate precautions, such as locks on circuit breakers or switches, warning notices, sentries, or other equally effective means, shall be taken to prevent electrical equipment from being electrically charged when work is being done.

Δ 2-306 Shock and flash protection (see Appendix B)

(1)   Electrical equipment such as switchboards, panelboards, industrial control panels, meter socket enclosures, and motor control centres that are installed in other than dwelling units and are likely to require examination, adjustment, servicing, or maintenance while energized shall be field marked to warn persons of potential electric shock and arc flash hazards.

(2)   The marking referred to in Subrule (1) shall be located so that it is clearly visible to persons before examination, adjustment, servicing, or maintenance of the equipment.

2-308 Working space around electrical equipment

(1)   A minimum working space of 1 m with secure footing shall be provided and maintained about electrical equipment such as switchboards, panelboards, control panels, and motor control centres that are enclosed in metal, except that working space is not required behind such equipment where there are no renewable parts such as fuses or switches on the back and where all connections are accessible from locations other than the back.

(2)   The space referred to in Subrule (1) shall be in addition to the space required for the operation of draw-out-type equipment in either the connected, test, or fully disconnected position and shall be sufficient for the opening of enclosure doors and hinged panels to at least 90°.

(3)   Working space with secure footing not less than that specified in Table 56 shall be provided and maintained around electrical equipment such as switchboards, control panels, and motor control centres having exposed live parts.

(4)   The minimum headroom of working spaces around switchboards or motor control centres where bare live parts are exposed at any time shall be 2.2 m.

2-310 Entrance to, and exit from, working space (see Appendix B)

(1)   Each room containing electrical equipment and each working space around equipment shall have suitable means of egress, which shall be kept clear of all obstructions.

(2)   Where a room or space referred to in Subrule (1) contains equipment that is rated 1200 A or more, or rated over 750 V, and consists of transformers, overcurrent devices, switchgear, or disconnecting means, such equipment shall be arranged so that, in the event of a failure in the equipment, it shall be possible to leave the room or space referred to in Subrule (1) without passing the failure point, except that where this cannot

be done, the working space requirement of Rule 2-308(1) and (2) shall be not less than 1.5 m.

(3)   For the purposes of Subrule (2), the potential failure point is any point within or on the equipment.

(4)   Doors or gates shall be capable of being readily opened from the equipment side without the use of a key or tool.

2-312 Accessibility for maintenance (see Appendix G)

Passageways and working space around electrical equipment shall not be used for storage and shall be kept clear of obstruction and arranged to give authorized persons ready access to all parts requiring attention.

2-314 Illumination of equipment

Adequate illumination shall be provided to allow for proper operation and maintenance of electrical equipment.

2-316 Flammable material near electrical equipment

Flammable material shall not be stored or placed in dangerous proximity to electrical equipment.

2-318 Ventilation (see Appendix B)

Adequate ventilation shall be provided to prevent the development around electrical equipment of ambient air temperatures in excess of those normally permissible for such equipment.

2-320 Drainage

Electrical equipment having provision for draining moisture shall be installed so that the drainage path is not impeded.

Δ ## 2-322 Electrical equipment near combustible gas equipment

When installed outdoors, arc-producing electrical equipment shall not be installed within 1 m of the discharge of a combustible gas relief device or vent.

Enclosures

Δ ## 2-400 Enclosures, type designations, and use (see Appendix B)

(1)   For the purposes of this Code, the following designations of enclosures for electrical equipment other than motors and generators shall be recognized for the intended use as specified in Table 65 and as follows:

   (a)   Type 1 for use indoors in ordinary locations;

   (b)   Type 2 for use indoors where the enclosure may be subject to drops of falling liquid due to condensation or other causes;

   (c)   Type 3R for use outdoors;

   (d)   Type 4 for use where the enclosure may be subject to direct streams of water;

   (e)   Type 5 for use indoors where the atmosphere may contain settling non-hazardous dust, lint, fibres, or flyings; and

   (f)   general-purpose enclosures for use indoors in ordinary locations.

(2)   Other enclosure types tabulated in Table 65 shall be permitted to be substituted for those required in Subrule (1) provided that they

   (a)   offer a degree of protection at least equal to that required by Subrule (1) for the intended use, as indicated in Table 65; and

   (b)   are marked in accordance with Rule 2-402.

(3)   Enclosures for equipment for use in a hazardous location shall be designated in accordance with Rule 18-052.

Δ ## 2-402 Marking of enclosures

(1)   Except for general-purpose enclosures, all enclosures described in Table 65 shall be marked with a type or enclosure designation.

(2)   In addition to the type or enclosure designation specified in Subrule (1), enclosures shall be permitted to be marked with an Ingress Protection (IP) designation.

2-404 Marking of motors

(1)   Drip-proof, weatherproof, and totally enclosed motors for use in non-hazardous locations shall be marked as follows:

   (a)   if a drip-proof motor, with the word "Drip-proof" or the code letters "DP";

   (b)   if a weatherproof motor, with the word "Weatherproof" or the code letters "WP"; and

   (c)   if a totally enclosed motor, with the words "Totally Enclosed" or the code letters "TE".

(2)   Notwithstanding Subrule (1), special-purpose motors that are intended to be used only as components of specific equipment need not be so marked.

Section 4 — Conductors

4-000 Scope

This Section applies to conductors for lighting, appliance, and power supply circuits, and does not apply to other conductors except where specifically referenced in other Sections of this Code.

4-002 Size of conductors

Except for flexible cord, equipment wire, control circuit wire, and cable, conductors shall be not smaller than No. 14 AWG when made of copper and not smaller than No. 12 AWG when made of aluminum.

4-004 Ampacity of wires and cables (see Appendix B)

Δ (1) The maximum current that a copper conductor of a given size and insulation may carry shall be as follows:
   (a) single-conductor and single-conductor metal-sheathed or armoured cable, in a free air run, with a cable spacing not less than 100% of the larger cable diameter, as specified in Table 1;
   (b) 1, 2, or 3 conductors in a run of raceway, or 2- or 3-conductor cable, except as indicated in Subrule (1)(d), as specified in Table 2;
   (c) 4 or more conductors in a run of raceway or cable, as specified in Table 2 with the correction factors applied as specified in Table 5C; and
   (d) single-conductor and 2-, 3-, and 4-conductor cables and single and 2-, 3-, and 4-conductor metal-armoured and metal-sheathed cables, in conductor sizes 1/0 AWG and larger, in an underground run, as calculated by the method of IEEE 835.

Δ (2) The maximum current that an aluminum conductor of a given size and insulation may carry shall be as follows:
   (a) single-conductor and single-conductor metal-sheathed or armoured cable, in a free air run, with a cable spacing not less than 100% of the larger cable diameter, as specified in Table 3;
   (b) 1, 2, or 3 conductors in a run of raceway, or 2- or 3-conductor cable, except as indicated in Subrule (2)(d), as specified in Table 4;
   (c) 4 or more conductors in a run of raceway or cable, as specified in Table 4 with the correction factors applied as specified in Table 5C; and
   (d) single-conductor and 2-, 3-, and 4-conductor cables and single and 2-, 3-, and 4-conductor metal-armoured and metal-sheathed cables, in conductor sizes 1/0 AWG and larger, in an underground run, as calculated by the method of IEEE 835.

(3) A neutral conductor that carries only the unbalanced current from other conductors, as in the case of normally balanced circuits of three or more conductors, shall not be counted in determining ampacities as provided for in Subrules (1) and (2).

(4) When a load is connected between a single-phase conductor and the neutral, or between each of two phase conductors and the neutral, of a three-phase, four-wire system, the common conductor carries a current comparable to that in the phase conductors and shall be counted in determining the ampacities as provided for in Subrules (1) and (2).

(5) The maximum allowable ampacity of neutral supported cable shall be as specified in Tables 36A and 36B.

(6) A bonding conductor shall not be counted in determining the ampacities as provided for in Subrules (1) and (2).

(7) The correction factors specified in this Rule
   (a) shall apply only to, and shall be determined from, the number of power and lighting conductors in a cable or raceway; and
   (b) shall not apply to conductors installed in auxiliary gutters.

(8) The ampacity correction factors of Table 5A shall apply where conductors are installed in an ambient temperature exceeding or anticipated to exceed 30 °C.

(9) Where single conductors having a free air rating are run in contact with each other, the ampacity shall be corrected by applying the factors in Table 5B for up to four conductors in contact, and by utilizing the ampacity of Table 2 or 4 where there are more than four in contact.

(10) Where multi-conductor cables are run in contact with each other for distances exceeding 600 mm, the ampacity of the conductors shall be corrected by applying the factors in Table 5C.

(11) The ampacity of conductors of different temperature ratings installed in the same raceway shall be determined on the basis of the conductor having the lowest temperature rating.

(12) The ampacity of conductors added to a raceway and the ampacity of the conductors already in the raceway shall be determined in accordance with the applicable Subrules.

(13) Where more than one ampacity could apply for a given circuit of single-conductor or multi-conductor cables as a consequence of a transition from an underground portion to a portion above ground, the lower value shall apply except as permitted in Subrule (14).

(14) Where the lower ampacity portion of a cable installation consisting of not more than four conductors in total does not exceed 10% of the circuit length or 3 m, whichever is less, the higher ampacity shall be permitted.

(15) When the load factor of the load is less than 1.00 and is known or can be supported by documentation, the ampacity of conductors derived from Subrules (1)(d) and (2)(d) shall be permitted to be increased by application of that load factor in the calculation of the ampacity.

(16) In consideration of the increased ampacity of any conductor derived in accordance with Subrule (15), no further factors based on load diversity shall be permitted.

Δ (17) The ampacity of nickel or nickel-clad conductors shall be calculated using the method described in IEEE 835.

4-006 Insulated conductors

(1) Insulated conductors shall be of types specified in Table 19 for each specific condition of use, except as may be otherwise required by other Sections of this Code.

(2) Where harmful condensed vapours or liquids of either an acid or alkaline nature, or organic solvents such as hydrocarbons, ketones, esters, alcohols, or their liquid derivatives, can collect on or come in contact with insulation on conductors, such insulation shall be of a type resistant to these materials, or the insulation shall be protected by a sheath of lead or by other material impervious to the corrosive element.

4-008 Sheath currents in single-conductor metal-sheathed cables (see Appendix B)

Δ (1) Where sheath currents in single-conductor cables having continuous sheaths of lead, aluminum, stainless steel, or copper are likely to cause the insulation of the conductors to be subjected to temperatures in excess of the insulation ratings, the cables shall be
  (a) derated to 70% of the current-carrying rating that would otherwise apply;
  (b) derated in accordance with the manufacturer's recommendations and in compliance with Rule 2-030; or
  (c) installed in a manner that prevents the flow of sheath currents.

(2) Circulating currents in single-conductor armoured cable shall be treated in the same manner as sheath currents in Subrule (1).

Δ (3) Where single-conductor mineral-insulated cables are used, all current-carrying conductors shall be grouped together to minimize induced voltage on the sheath.

4-010 Uses of flexible cord

(1) Flexible cord shall be of the types specified in Table 11 for each specific condition of use.

Δ (2) Flexible cord shall be permitted to be used for
  (a) electrical equipment for household or similar use that is intended to be
    (i) moved from place to place; or
    (ii) detachably connected according to a *Canadian Electrical Code, Part II* Standard; and
  (b) electrical equipment for industrial use that must be capable of being moved from place to place for operation;
  (c) pendants;
  (d) wiring of cranes and hoists;
  (e) the connection of stationary equipment to facilitate its interchange, where a deviation is allowed in accordance with Rule 2-030;
  (f) the prevention of transmission of noise and vibration;
  (g) the connection of electrical components between which relative motion is necessary;
  (h) the connection of appliances such as ranges and clothes dryers; and
  (i) both connection, using an attachment plug, and interconnection of data processing systems, provided that the cord is of the extra-hard-usage type.

(3) Flexible cord shall not be used
  (a) as a substitute for the fixed wiring of structures and shall not be
    (i) permanently secured to any structural member;
    (ii) run through holes in walls, ceilings, or floors; or
    (iii) run through doorways, windows, or similar openings;
  (b) at temperatures above the temperature rating of the cord or at temperatures sufficiently low as to be liable to result in damage to the insulation or overall covering; and

(c) for the suspension of any device weighing more than 2.3 kg, unless the cord and device assembly are marked as capable of supporting a weight up to 11 kg.

(4) Flexible cord shall be protected against mechanical damage by an insulating bushing or some other effective means where it enters or passes through the enclosure wall or the partitioning of a device or enters a lampholder.

(5) Where a flexible cord is used as an extension cord or to plug into an appliance or other device, no live parts shall be exposed when one end is connected to a source of supply and the other end is free.

4-012 Sizes of flexible cord

Flexible cord shall be not smaller than a No. 18 AWG copper conductor except for

(a) tinsel cord, which may be No. 27 AWG copper; and

(b) cords for use with specific devices, which may be No. 20 AWG copper.

4-014 Ampacity of flexible cords

(1) The maximum current that two or more copper conductors of given size contained in a flexible cord may carry shall be as follows:

(a) 2 or 3 conductors, as specified in Table 12;

(b) 4, 5, or 6 conductors, 80% of that specified in Table 12;

(c) 7 to 24 conductors inclusive, 70% of that specified in Table 12;

(d) 25 to 42 conductors inclusive, 60% of that specified in Table 12; and

(e) 43 or more conductors, 50% of that specified in Table 12.

(2) Conductors used for bonding equipment to ground and a conductor used as a neutral conductor, which carries only the unbalanced current from other conductors as in the case of a normally balanced circuit of three or more conductors, are not counted in determining ampacities.

4-016 Flexible cord used in show windows or show cases

(1) Flexible cord used in show windows or show cases shall, except for chain fixtures, be of at least hard usage types.

(2) The use of flexible cord to supply current to portable lamps and other devices for exhibition purposes shall be permitted.

4-018 Equipment wire

(1) Equipment wire shall be of a type specified in Table 11 for each specified condition of use.

(2) Equipment wire used as fixture wiring shall be not smaller than a No. 18 AWG copper conductor.

(3) Equipment wire, including its assemblies for applications other than that given in Subrule (2), shall be not smaller than No. 26 AWG copper when rated 300 V and not smaller than No. 24 AWG copper when rated 600 V.

(4) The maximum current that an equipment wire of a given size may carry shall be as specified in Table 12.

4-020 Insulation of neutral conductors

(1) Except as permitted by Rules 6-302, 6-308, 12-302, and 12-318, neutral conductors shall be insulated.

(2) Where insulated neutrals are used, the insulation on the neutral conductors shall have a temperature rating not less than the temperature rating of the insulation on the ungrounded conductors.

4-022 Size of neutral conductor

(1) The neutral conductor shall have sufficient ampacity to carry the unbalanced load.

(2) The maximum unbalanced load shall be the maximum connected load between the neutral and any one ungrounded conductor as determined by Section 8 but subject to the following:

(a) there shall be no reduction in the size of the neutral for that portion of the load that consists of electric discharge lighting; and

(b) except as required otherwise by Item (a), a demand factor of 70% shall be permitted to be applied to that portion of the unbalanced load in excess of 200 A.

(3) The size of a service neutral shall be not smaller than the size of a neutral selected in accordance with Subrule (1) and shall

(a) be not smaller than No. 10 AWG copper or No. 8 AWG aluminum; and

(b) be sized not smaller than a grounded conductor as required by Rule 10-204(3), except in service entrance cable or where the service conductors are No. 10 AWG copper or No. 8 AWG aluminum.

(4) In determining the ampacity of an uninsulated neutral conductor run in a raceway, it shall be considered to be insulated with insulation having a temperature rating not higher than that of the adjacent circuit conductors.

4-024 Common neutral conductor

Provided that when in metal enclosures all conductors of feeder circuits employing a common neutral are contained within the same enclosure, a common neutral shall be permitted to be employed for

(a)   two or three sets of 3-wire, single-phase feeders; or

(b)   two sets of 4-wire, 3-phase feeders.

4-026 Installation of neutral conductor

Where a service, feeder, or branch circuit requires a neutral conductor, it shall be installed

(a)   in all separately enclosed switches and circuit breakers;

(b)   in all centres of distribution associated with the circuit;

(c)   with all connections to the neutral being made in the enclosures and centres; and

(d)   in such a manner that any neutral conductor may be disconnected without disconnecting any other neutral conductor.

4-028 Identification of insulated neutral conductors up to and including No. 2 AWG copper or aluminum

Δ   (1)   Except as permitted in Subrules (2), (3), and (4), all insulated neutral conductors up to and including No. 2 AWG copper or aluminum, and the conductors of flexible cords that are permanently connected to such neutral conductors, shall be identified by a white or grey covering or by three continuous white stripes along the entire length of the conductor.

(2)   Where conductors of different systems are installed in the same raceway, box, or other type of enclosure and the identified circuit conductor of one system is coloured by a white or grey covering, each identified circuit conductor of the other system, if present, shall be provided with a specific identification, and the identification shall be permitted to be an outer covering of white with an identifiable coloured stripe (not green) running along the insulation.

(3)   The covering of the other conductor or conductors shall show a continuous colour contrasting that of an identified conductor; however, in the case of those flexible cords where the identified conductor is identified by a raised longitudinal ridge(s), the other conductors shall have no ridges.

(4)   For multi-conductor cable, the insulated neutral conductor shall be permitted to be permanently marked as the identified conductor by painting or other suitable means at every point where the separate conductors have been rendered accessible and visible by removal of the outer covering of the cable, and the painting or other suitable means of marking the identified conductor shall not be permitted to render illegible the manufacturer's numbering of the conductor.

4-030 Identification of insulated neutral conductors larger than No. 2 AWG copper or aluminum

For insulated neutral conductors other than those mentioned in Rule 4-028(1), identification shall either be continuous, as for No. 2 AWG and smaller, or else each continuous length of conductor shall be suitably labelled or otherwise clearly marked at each end at the time of installation, so that it can be readily identified.

4-032 Identification of type MI neutral conductors

Where mineral-insulated cable is used for neutral conductors and where continuous identification of this type of conductor is, at present, technically impossible in manufacture, each continuous length of conductor shall be permanently and clearly marked at each end at the time of installation, so that it can be readily identified.

4-034 Use of identified conductors

(1)   An identified conductor shall not be used as a conductor for which identification is not required by these Rules; however, in armoured cable, aluminum-sheathed cable, and non-metallic-sheathed cable work the identified conductor shall be permitted to be rendered permanently unidentifiable by painting or other suitable means at every point where the separate conductors have been rendered accessible and visible by removal of the outer covering of the cable.

(2)   Where armoured cable, aluminum-sheathed cable, or non-metallic-sheathed cable containing an identified conductor is used for single-pole, 3-way or 4-way switch loops, it shall not be necessary to render the identified conductor permanently unidentified at the switch if the connections are made so that an unidentified conductor is the return conductor from the switch to the outlet.

(3)   Where armoured cable, aluminum-sheathed cable, or non-metallic-sheathed cable is used so that the identified conductor forms no part of the circuit, the identified conductor shall be cut off short or other suitable means shall be employed to indicate clearly that the identified conductor does not form part of the

circuit; this shall be done at every point where the separate conductors have been rendered accessible and visible by removal of the outer covering of the cable.

(4)　Where conductors of a multi-wire branch circuit are installed, employing an identified conductor, the continuity of the identified conductor shall be independent of device connections, such as lampholders, receptacles, ballasts, etc., so that devices may be disconnected without interrupting the continuity of the identified conductor.

4-036　Colour of conductors

(1)　Insulated grounding or bonding conductors shall
- (a)　have a continuous outer finish that is either green or green with one or more yellow stripes; or
- (b)　if larger than No. 2 AWG, be permitted to be suitably labelled or marked in a permanent manner with a green colour or green with one or more yellow stripes at each end and at each point where the conductor is accessible.

(2)　Conductors coloured or marked in accordance with Subrule (1) shall be used only as grounding or bonding conductors.

(3)　Where colour-coded circuits are required, the following colour coding shall be used, except in the case of service-entrance cable and when Rules 4-030, 4-032, and 6-308 may modify these requirements:
- (a)　1-phase ac or dc (2-wire) — 1 black and 1 red or 1 black and 1 white*† (where identified conductor is required);
- (b)　1-phase ac or dc (3-wire) — 1 black, 1 red, and 1 white*†; and
- (c)　3-phase ac — 1 red (phase A), 1 black (phase B), 1 blue (phase C), and 1 white* (where neutral is required).

*Or grey
†Or white with coloured stripe (see Rule 4-028)

(4)　Where the midpoint of one phase of a 4-wire delta-connected secondary is grounded to supply lighting and similar loads, the conductors shall be colour-coded in accordance with Subrule (3) and the phase A conductor shall be the conductor having the higher voltage-to-ground.

(5)　Where a panelboard is supplied from a 4-wire delta-connected system, the grounded conductor referred to in Subrule (4) shall be located in a compartment provided for single-phase connections and the phase conductor having the higher voltage-to-ground shall be suitably barriered from that compartment.

4-038　Uses of portable power cable

(1)　Portable power cables shall be of types specified in Table 11 for each specific condition of use.

(2)　Portable power cables shall be permitted to be used for
- (a)　electrical equipment that is intended to be
  - (i)　moved from place to place; or
  - (ii)　detachably connected according to a *Canadian Electrical Code, Part II* Standard;
- (b)　wiring of cranes and hoists;
- (c)　the connection of stationary equipment to facilitate its interchange;
- (d)　the connection of electrical components between which relative motion is necessary; and
- (e)　the connection of equipment used in conjunction with travelling amusement rides.

(3)　Portable power cable shall not be used
- (a)　as a substitute for the fixed wiring of structures and shall not be
  - (i)　permanently secured to any structural member;
  - (ii)　run through holes in walls, ceilings, or floors of permanent structures; or
  - (iii)　run through doorways, windows, or similar openings of permanent structures; or
- (b)　at a temperature above the temperature rating of the cable or at a temperature sufficiently low as to be liable to result in damage to the insulation or overall covering.

(4)　Where portable power cable enters or passes through the wall of an enclosure or fitting, it shall be protected in accordance with Rule 12-3022.

4-040　Ampacity of portable power cable

(1)　The maximum current that one or more copper conductors of a given size contained in a portable power cable may carry shall be as specified in Table 12A.

(2)　Conductors used for bonding equipment to ground and a conductor used as a neutral that carries only the unbalanced current from other conductors, as in the case of a normally balanced circuit of three or more conductors, are not counted in determining ampacities.

Section 6 — Services and service equipment

Scope

6-000 Scope
This Section applies to services, service equipment, and metering equipment for
(a)   installations operating at voltages of 750 V or less; and
(b)   installations operating at voltages in excess of 750 V except as modified by the requirements of Section 36.

General

6-100 Special terminology
In this Section the following definition shall apply:

Transformer rated meter mounting device — a meter mounting device with current transformers and with or without test switches mounted in the same enclosure.

6-102 Number of supply services permitted (see Appendix B)
Δ (1)   Two or more supply services of the same voltage shall not be run to any building from the same system of any one supply authority except that additional supply services shall be permitted for supplying
(a)   fire pumps and other emergency systems;
(b)   industrial establishments and other complex structures; or
(c)   completely self-contained occupancies where the occupancies
(i)   are not located one above the other; and
(ii)   have a separate entrance with direct access to ground level.
(2)   When two or more supply services are installed to a building, all service boxes associated with the various consumer's services shall be grouped, where practicable.
(3)   When two or more service boxes installed in accordance with Subrule (2) are not grouped together, a permanent diagram shall be posted on or near each service box indicating the location of all the other service boxes supplying power to the building.

6-104 Number of consumer's services permitted in or on a building
The number of consumer's services of the same voltage and characteristics, terminating at any one supply service, run to, on, or in any building, shall not exceed four, unless there is a deviation allowed in accordance with Rule 2-030.

6-106 Current supply from more than one system
Where an installation, or part of an installation, is to be supplied with current from two or more different systems, the switching equipment controlling the various supplies shall be constructed or arranged so that it will be impossible to accidentally switch on power from one source before power from another has been cut off.

6-108 Supply service from an electric railway system
A supply service shall not be run to a building from an electric railway system using a ground return, unless the building is connected with the operation of an electric railway.

6-110 Three-wire consumer's services
A three-wire consumer's service shall be provided in all cases where more than two 120 V branch circuits are installed, unless such supply is not available from the supply authority.

6-112 Support for the attachment of overhead supply or consumer's service conductors
(see Appendix B)
(1)   A means of attachment shall be provided for all supply or consumer's service conductors.
(2)   The point of attachment of supply or consumer's service conductors shall not exceed 9 m above grade or sidewalk and shall be located such that the clearance of supply conductors at any point above finished grade shall be not less than the following:
(a)   across highways, streets, lanes, and alleys: 5.5 m;
(b)   across driveways to residential garages: 4 m;
(c)   across driveways to commercial and industrial premises: 5 m; and
(d)   across ground normally accessible to pedestrians only: 3.5 m.
(3)   Exposed service conductors that are not higher than windows, doors, and porches shall have a clearance of not less than 1 m from the windows, doors, or porches.

(4) Where service masts are used they shall be of metal and assembled from components suitable for service mast use.

(5) Rigid steel conduit of a minimum nominal size of 63 trade size shall be permitted to be used for the purpose of Subrule (4) provided that all other requirements for a service mast are complied with.

(6) Bolts shall be used for securing the support at the point of attachment, and if attached to wooden structural members, the latter shall be not less than 38 mm in any dimension.

(7) The supply or consumer's service conductor support shall not be attached to the roof of a structure, except as permitted in Subrule (8).

(8) Notwithstanding Subrule (7), it shall be permitted to fasten the upper service mast support and the eye bolt to which a guy wire is attached to a main structural member of the roof such as a roof rafter, a roof truss, or equivalent.

6-114 Methods of terminating conductors at consumer's service

(1) The supply end of a consumer's service shall be equipped with a rain-tight service head except as provided for in Subrules (2) and (3).

(2) Where service cables are employed and are continuous from the supply service to the service equipment, the service head required by Subrule (1) shall be permitted to be omitted.

(3) Where single- or multi-conductor cables are employed, the service head required by Subrule (1) shall be permitted to be omitted provided that
    (a) the cable terminates in a cable termination suitable for exposure to the weather; or
    (b) the cable ends are sealed with self-sealing weather-resistant thermoplastic tape or heat-shrinkable tubing; and
    (c) both single-conductor and multi-conductor cables are bent as necessary so that the conductors emerging from the sealed point of the cable termination point downwards; and
    (d) the cables are held securely in place by a clamp, fitting, or cable termination.

(4) Conductors of different polarity shall be brought out through separately bushed holes of the service head.

(5) Consumer's service conductors shall be installed as specified in Rule 6-302(3).

(6) The overhead supply service conductors and the consumer's service conductors shall be arranged according to the requirements of Rule 6-116 to prevent moisture and water from entering service raceways, cables, or equipment.

6-116 Consumer's service head location

The consumer's service head or equivalent shall be installed
(a) in compliance with the requirements of the supply authority; and
(b) in such a position that the point of emergence of the conductors from the consumer's service head or equivalent is a minimum of 150 mm and a maximum of 300 mm above the support for attachment of the overhead service conductors.

Control and protective equipment

6-200 Service equipment

(1) Except as provided in Subrule (2), each consumer's service shall be provided with a single service box.

(2) More than one service box shall be permitted to be connected to a single consumer's service, provided that
    (a) the subdivision is made in a multiple or dual lug meter mounting device rated at not more than 600 A and 150 volts-to-ground; and
    (b) the meter mounting device is located outdoors.

(3) For the application of Rule 6-104, each subdivision of the meter mounting device shall be considered a consumer's service.

6-202 Subdivision of main consumer's service

In multiple occupancy and in single occupancy multi-rate service, each subdivision of the main consumer's service shall be provided with a separate service box, or equivalent multi-service equipment shall be used, unless there is a deviation allowed in accordance with Rule 2-030 for single occupancy multi-rate services only; where the main consumer's service overcurrent devices adequately protect any subdivision of the main consumer's service, the separate service box for the subdivision so protected shall be permitted to be omitted.

6-204 Fuse enclosure on service boxes

If a service box embodies one or more fuseholders, access to which may be had without opening the door, such fuseholders and their fuses shall be completely enclosed by a separate door, spring-closed, or with a substantial catch.

6-206 Consumer's service equipment location (see Appendices B and G)

(1)   Service boxes or other consumer's service equipment shall be
  (a)   installed in a location that is in compliance with the requirements of the supply authority;
  (b)   readily accessible or have the means of operation readily accessible;
  (c)   not located in coal bins, clothes closets, bathrooms, stairways, rooms in which the temperature normally exceeds 30 °C, dangerous or hazardous locations, locations where the headroom clearance is less than 2 m, nor in any similar undesirable locations;
  (d)   placed within the building being served, unless environmental conditions within the structure are unsuitable, in which case, where there is a deviation allowed in accordance with Rule 2-030, the service box or other consumer's service equipment shall be permitted to be placed on the outside of the building or on a pole and shall be
    (i)   protected from the weather, or be weatherproof; and
    (ii)   protected from mechanical injury if less than 2 m above ground; and
  (e)   as close as practicable to the point where the consumer's service conductors enter the building.
(2)   Notwithstanding Subrule (1)(b), where subject to unauthorized operation, the service disconnecting means shall be permitted to be rendered inaccessible by
  (a)   an integral locking device;
  (b)   an external lockable cover; or
  (c)   location of the service box inside a separate building, room, or enclosure.

6-208 Consumer's service conductors location

(1)   Raceways or cables containing consumer's service conductors shall be located outside buildings unless they are
  (a)   embedded in and encircled by not less than 50 mm of concrete or masonry where permitted by Section 12;
  (b)   directly buried in accordance with Rule 6-300 and located beneath a concrete slab not less than 50 mm thick; or
  (c)   run in a crawl space located underneath a structure, provided that such a crawl space
    (i)   does not exceed 1.8 m in height between the lowest part of the floor assembly and the ground or other surface below it;
    (ii)   is of non-combustible construction; and
    (iii)   is not used for the storage of combustible material.
(2)   Notwithstanding Subrule (1), raceways or cables containing consumer's service conductors shall be permitted to enter the building for connection to a service box.

6-210 Oil switches and oil circuit breakers used as consumer's service switches

(1)   Isolating switches shall be installed on the supply side and interlocked with oil switches and oil circuit breakers except in the case of metal clad equipment, where the primary isolating device shall be considered to be the equivalent of an isolating switch or link.
(2)   Where overcurrent trip coils are used for breakers, one shall be installed on each ungrounded conductor of the circuit; however, if the capacity of the transformers and the extent of the network supplying the service is sufficiently small, and a deviation has been allowed in accordance with Rule 2-030, two trip coils, one in each phase of a 4-wire, 2-phase ungrounded service, shall be permitted to be used.

6-212 Wiring space in enclosures

(1)   Enclosures for circuit breakers and externally operated switches shall not be used as junction boxes, troughs, or raceways for conductors feeding through or tapping off to other apparatus.
(2)   Notwithstanding Subrule (1), service equipment specifically designed for accommodating current monitoring devices shall be permitted.

6-214 Marking of service boxes

If there is more than one service box, each box shall be labelled in a conspicuous, legible, and permanent manner to indicate clearly which installation or portion of an installation it controls.

Wiring methods

6-300 Underground consumer's services

(1) Except where a deviation has been allowed in accordance with Rule 2-030, consumer's service conductors that are run underground to a building from an underground supply system or from a pole line shall be

    (a) installed in rigid conduit, or electrical non-metallic tubing permitted only for the underground portion of the tubing run, and be of a type for use in wet locations in accordance with Table 19; or

    (b) a single- or multiple-conductor cable for service entrance use below ground in accordance with Table 19 provided that

        (i) the installation is in accordance with Rule 12-012; and

        (ii) the cable is without splice or joint from the point of connection at the supply service to the consumer's service equipment except in metering equipment located on the line side of the service box.

(2) Notwithstanding Subrule (1)(b)(ii), joints in the underground portion of an underground consumer's service shall be permitted where such joints are made in accordance with Rule 12-112(4) and joints are required to repair damage to the original installation or to accommodate a pole or service relocation.

(3) Raceways entering a building and forming part of an underground service shall be sealed and shall

    (a) enter the building above ground where practicable;

    (b) be suitably drained; or

    (c) be installed in such a way that moisture and gas will not enter the building.

(4) Consumer's service conduit connected to an underground supply system shall be sealed with a suitable compound to prevent the entrance of moisture or gases.

6-302 Overhead consumer's service conductors

(1) Conductors of a consumer's service that are connected to an overhead supply service at any point above ground on a building or other structure shall be installed in one of the following ways:

    (a) rigid conduit;

    (b) busway;

    (c) steel electrical metallic tubing;

    (d) flexible metal conduit, with lead-sheathed conductors;

    (e) mineral-insulated cable other than the lightweight type;

    (f) aluminum-sheathed cable;

    (g) Type ACWU75 or Type ACWU90 cable;

    (h) Type AC90 cable; or

    (i) Type TECK90 cable.

(2) That portion of the consumer's service conductors on the supply side of the consumer's service head that is run between buildings or structures or on the outside walls of buildings, crossing over or installed on a building roof, shall be permitted to be run as exposed wiring in accordance with Rules 12-302 to 12-318.

(3) The length of consumer's service conductors beyond the service head shall be adequate to enable connection to the supply service conductors or to the conductors referred to in Subrule (2) with a minimum length of 750 mm, and the conductors shall be provided with drip loops.

(4) Consumer's service conductors shall be not less than No. 10 AWG copper wire, nor less than No. 8 AWG aluminum wire.

(5) The insulation on consumer's service conductors shall be suitable for the temperatures that can be experienced in the particular locality.

6-304 Use of mineral-insulated and aluminum-sheathed cable

(1) Mineral-insulated cable and aluminum-sheathed cable may be used for consumer's services, as specified in Rule 6-302,

    (a) in multi-conductor construction; or

    (b) in single-conductor construction in sizes larger than No. 4 AWG copper or aluminum.

(2) Mineral-insulated cable and aluminum-sheathed cable may be exposed and secured directly to the surface over which it is run, but subject to protection as specified in Rule 6-306(b).

6-306 Consumer's service raceways

Consumer's service raceways shall

(a) contain only the consumer's service conductors, and except where a deviation has been allowed in accordance with Rule 2-030, only the conductors of one consumer's service;

26

(b)   be protected against mechanical damage as required by Rule 12-932; and

(c)   if of circular cross-section, have a minimum nominal trade size of 21.

6-308 Use of bare neutral in consumer's service

The neutral conductor of a consumer's service shall be permitted to be bare if this conductor is

(a)   made of copper and is run in a raceway;

(b)   made of aluminum and is run above ground in a non-metallic or an aluminum raceway;

(c)   part of a busway or of a service entrance cable; or

(d)   part of a neutral supported cable used in accordance with Rule 6-302(2).

6-310 Use of joints in consumer's service neutral conductors

The neutral or identified conductor of a consumer's service shall be without joints between the point of connection and the service box or equivalent consumer's service equipment, except a joint shall be permitted where it is made

(a)   by means of a clamp or bolted connection in a meter mounting device or at the service head if exposed wiring is used in accordance with Rule 6-302(2); or

(b)   by a joint underground in accordance with Rule 12-112(4) where such a joint is required to repair damage to the original installation or to accommodate a pole or service relocation.

6-312 Condensation in consumer's service raceway

(1)   The consumer's service raceway entering a building shall be sealed and shall be suitably drained where it enters the building above grade level.

(2)   The consumer's service raceway shall not be terminated on top of the service box except where drained outdoors.

Metering equipment

6-400 Metering equipment

Metering equipment includes any current and potential (voltage) transformers as well as the associated measuring instruments.

6-402 Method of installing meter loops (see Appendix B)

(1)   Meter loops shall be installed so that

(a)   conductors between the service box and the meter are inaccessible to unauthorized persons;

(b)   the wiring method is rigid conduit, flexible metal conduit, electrical metallic tubing, aluminum-sheathed cable, or armoured cable, except where equivalent protection is provided;

(c)   spare conductors not less than 450 mm in length are provided at meter or current transformer connection points; and

(d)   a suitable fitting, or service box with meter backplate, is provided.

(2)   Metering equipment shall be connected on the load side of the service box, except that it may be connected on the supply side where

(a)   no live parts or wiring are exposed;

(b)   the supply is ac and the voltage does not exceed 300 V between conductors; and

(c)   the rating of the consumer's service does not exceed

(i)   200 A for a meter mounting device; or

(ii)   600 A for a transformer rated meter mounting device located outdoors.

6-404 Enclosures for instrument transformers

(1)   Instrument transformers used in conjunction with meters shall be installed in metal enclosures, except where access is to authorized persons only.

(2)   The size of enclosures for instrument transformers shall comply with the requirements of the supply authority.

(3)   Enclosures for current transformers shall be installed on all consumer's services rated in excess of 200 A, except where

(a)   current transformers are an integral part of consumer's service switchgear; or

(b)   the supply authority uses meters that do not require current transformers.

(4)   Enclosures for instrument transformers shall have provision for securing of the transformers to the enclosures.

6-406 Disconnecting provisions for meters

In multiple occupancy and in single occupancy multi-rate service where individual metering is required, the conductors to each meter shall be provided with one of the following:

(a) a separate service box or service equipment; or

(b) a sealable meter fitting.

6-408 Location of meters

(1) Meters and metering equipment shall be
    (a) located as near as practicable to the service box except as provided for in Subrule (2);
    (b) grouped where practicable;
    (c) readily accessible;
    (d) not located in coal bins, clothes closets, bathrooms, stairways, high ambient rooms, dangerous or hazardous locations, nor in any similar undesirable places;
    (e) if mounted outdoors, of weatherproof construction or in weatherproof enclosures; and
    (f) in compliance with the requirements of the supply authority.

(2) Instrument transformers shall be permitted to be outside the consumer's premises and the meter inside the premises, provided that the secondary leads between the instrument transformers and the meter terminal box or test links are continuous and are installed in the same manner as consumer's service conductors, with the exception that a service box with disconnecting switch is not required.

6-410 Space required for meters

The space provided for meters shall comply with the requirements of the supply authority.

6-412 Metering requirements for impedance grounded systems

(1) Equipment and wiring for supply authority metering on impedance grounded systems shall be in compliance with the requirements of the supply authority.

(2) Where a neutral point reference conductor is required for metering on impedance grounded systems, the reference conductor shall be
    (a) insulated for the nominal system voltage;
    (b) isolated from ground throughout its entire length; and
    (c) permitted to be in the same raceway or cable assembly as the consumer's service conductors and to be carried through or extended from the consumer's service box to the metering equipment.

Section 8 — Circuit loading and demand factors

Scope

8-000 Scope
This Section covers:
(a) conductor ampacities and equipment ratings required for consumer's services, feeders, and branch circuits; and
(b) branch circuit positions required for dwelling units.

General

8-100 Current calculations
When calculating currents that will result from loads, expressed in watts or volt amperes, to be supplied by a low-voltage ac system, the voltage divisors to be used shall be 120, 208, 240, 277, 347, 416, 480, or 600 as applicable.

8-102 Voltage drop (see Appendix D)
(1) Voltage drop in an installation shall
   (a) be based upon the calculated demand load of the feeder or branch circuit;
   (b) not exceed 5% from the supply side of the consumer's service (or equivalent) to the point of utilization; and
   (c) not exceed 3% in a feeder or branch circuit.
(2) For the purposes of Subrule (1) the demand load on a branch circuit shall be the connected load, if known; otherwise it shall be 80% of the rating of the overload or overcurrent devices protecting the branch circuit, whichever is smaller.

8-104 Maximum circuit loading (see Appendix B)
(1) The ampere rating of a consumer's service, feeder, or branch circuit shall be the ampere rating of the overcurrent device protecting the circuit or the ampacity of the conductors, whichever is less.
(2) The calculated load in a circuit shall not exceed the ampere rating of the circuit.
(3) The calculated load in a consumer's service, feeder, or branch circuit shall be considered a continuous load unless it can be shown that in normal operation it will not persist for
   (a) a total of more than 1 h in any two-hour period if the load does not exceed 225 A; or
   (b) a total of more than 3 h in any six-hour period if the load exceeds 225 A.
(4) Where a service box, fusible switch, circuit breaker, or panelboard is marked for continuous operation at 100% of the ampere rating of its overcurrent devices, the continuous load as determined from the calculated load shall not exceed
   (a) 100% of the rating of the circuit where the ampacity of the conductors is based on Column 2, 3, or 4 of Table 2 or 4; or
   (b) 85% of the rating of the circuit where the ampacity of the conductors is based on Column 2, 3, or 4 of Table 1 or 3.
(5) Where a service box, fusible switch, circuit breaker, or panelboard is marked for continuous operation at 80% of the ampere rating of its overcurrent devices, the continuous load as determined from the calculated load shall not exceed
   (a) 80% of the rating of the circuit where the ampacity of the conductors is based on Column 2, 3, or 4 of Table 2 or 4; or
   (b) 70% of the rating of the circuit where the ampacity of the conductors is based on Column 2, 3, or 4 of Table 1 or 3.
(6) If other derating factors are applied to reduce the conductor ampacity, the conductor size shall be the greater of that so determined or that determined by Subrule (4) or (5).
(7) Notwithstanding the requirements of Rule 4-004(1)(d) and (2)(d), the ampacity of the underground conductors shall not exceed in any case those determined by Subrules (4)(b) and (5)(b) of this Rule.

8-106 Use of demand factors
(1) The size of conductors and switches computed in accordance with this Section shall be the minimum used except that, if the next smaller standard size in common use has an ampacity not more than 5% less than this minimum, the smaller size conductor shall be permitted.

(2)   In any case other than a service calculated in accordance with Rules 8-200 and 8-202, where the design of an installation is based on requirements in excess of those given in this Section, the service and feeder capacities shall be increased accordingly.

(3)   Where two or more loads are installed so that only one can be used at any one time, the one providing the greatest demand shall be used in determining the calculated demand.

(4)   Where it is known that electric space-heating and air-conditioning loads are installed and will not be used simultaneously, whichever is the greater load shall be used in calculating the demand.

(5)   Where a feeder supplies loads of a cyclic or similar nature such that the maximum connected load will not be supplied at the same time, the ampacity of the feeder conductors shall be permitted to be based on the maximum load that may be connected at any one time.

(6)   The ampacity of conductors of feeders or branch circuits shall be in accordance with the Section(s) dealing with the respective equipment being supplied.

(7)   Notwithstanding the requirements of this Section, the ampacity of the conductors of a feeder or branch circuit need not exceed the ampacity of the conductors of the service or of the feeder from which they are supplied.

(8)   Where additional loads are to be added to an existing service or feeder, the augmented load shall be permitted to be calculated by adding the sum of the additional loads, with demand factors as permitted by this Code to the maximum demand load of the existing installation as measured over the most recent 12-month period, but the new calculated load shall be subject to Rule 8-104(4) and (5).

8-108 Number of branch circuit positions

(1)   For a single dwelling, the panelboard shall provide space for at least the equivalent of the following number of 120 V branch circuit overcurrent devices, including space for two 35 A double-pole overcurrent devices:

(a)   sixteen — of which at least half shall be double-pole, where the required ampacity of the service or feeder conductors does not exceed 60 A;

(b)   twenty-four — of which at least half shall be double-pole
    (i)   where the required ampacity of the service or feeder conductors exceeds 60 A but does not exceed 100 A; or
    (ii)   where the required ampacity of the service or feeder conductors exceeds 100 A but does not exceed 125 A and provision is made for a central electric furnace;

(c)   thirty — of which at least half shall be double-pole
    (i)   where the required ampacity of the service or feeder conductors exceeds 100 A but does not exceed 125 A; or
    (ii)   where the required ampacity of the service or feeder conductors exceeds 125 A but does not exceed 200 A and provision is made for a central electric furnace; and

(d)   forty — of which at least half shall be double-pole, where the required ampacity of the service or feeder conductors exceeds 125 A and the dwelling is not heated by a central electric furnace.

(2)   notwithstanding Subrule (1), sufficient spaces for overcurrent devices shall be provided in the panelboard for the two 35 A double-pole overcurrent devices and for all other overcurrent devices, and at least two additional spaces shall be left for future overcurrent devices.

(3)   for a dwelling unit in an apartment or similar building, the panelboard shall provide space for at least the equivalent of the following number of 120 V branch circuit overcurrent devices, including space for one 35 A double-pole overcurrent device:

(a)   eight — where the required ampacity of the feeder conductors supplying the dwelling unit does not exceed 60 A;

(b)   twelve — where the required ampacity of the feeder conductors supplying the dwelling unit exceeds 60 A.

8-110 Determination of areas

The living area designated in Rules 8-200 and 8-202 shall be determined from inside dimensions and include the sum of

(a)   100% of the area on the ground floor;

(b)   100% of any area used for living purposes on the upper floor; and

(c)   75% of the basement area.

Services and feeders

8-200 Single dwellings (see Appendix B)

(1)   The minimum ampacity of service or feeder conductors supplying a single dwelling shall be based on the greater of Item (a) or (b):

(a)   (i)   a basic load of 5000 W for the first 90 m^2 of living area (see Rule 8-110); plus

(ii)   an additional 1000 W for each 90 m^2 or portion thereof in excess of 90 m^2; plus

(iii) any electric space-heating loads provided for with demand factors as permitted in Section 62 plus any air-conditioning loads with a demand factor of 100%, subject to Rule 8-106(4); plus

(iv)   any electric range load provided for as follows: 6000 W for a single range plus 40% of any amount by which the rating of the range exceeds 12 kW; plus

(v)   any electric tankless water heaters or electric water heaters for steamers, swimming pools, hot tubs, or spas with a demand factor of 100%; plus

(vi)   any loads provided for in addition to those outlined in Items (i) to (v) at 25% of the rating of each load with a rating in excess of 1500 W if an electric range has been provided for, or 100% of the rating of each load with a rating in excess of 1500 W up to a total of 6000 W, plus 25% of the load in excess of 6000 W if an electric range has not been provided for; or

(b)   (i)   100 A where the floor area, exclusive of basement floor area, is 80 m^2 or more; or

(ii)   60 A where the floor area, exclusive of basement floor area, is less than 80 m^2.

(2)   The minimum ampacity of service or feeder conductors from a main service supplying two or more dwelling units of row-housing shall be based on

(a)   Subrule (1), excluding any electric space-heating loads and any air-conditioning loads, with application of demand factors to the loads as required by Rule 8-202(3)(a)(i) to (v); plus

(b)   the requirements of Rule 8-202(3)(b), (c), and (d).

(3)   The total load calculated in accordance with either Subrule (1) or (2) shall not be considered to be a continuous load for application of Rule 8-104.

8-202 Apartment and similar buildings (see Appendix B)

(1)   The minimum ampacity of service or feeder conductors from a main service supplying loads in dwelling units shall be the greater of Item (a) or (b):

(a)   (i)   a basic load of 3500 W for the first 45 m^2 of living area (see Rule 8-110); plus

(ii)   an additional 1500 W for the second 45 m^2 or portion thereof; plus

(iii) an additional 1000 W for each additional 90 m^2 or portion thereof in excess of the initial 90 m^2; plus

(iv)   any electric space-heating loads provided for with demand factors as permitted in Section 62 plus any air-conditioning loads with a demand factor of 100%, subject to Rule 8-106(4); plus

(v)   any electric range load provided for as follows: 6000 W for a single range plus 40% of any amount by which the rating of the range exceeds 12 kW; plus

(vi)   any loads provided for, in addition to those outlined in Items (i) to (v), at

(A)   25% of the rating of each load with a rating in excess of 1500 W, if an electric range has been provided for; or

(B)   25% of the rating of each load with a rating in excess of 1500 W plus 6000 W, if an electric range has not been provided for; or

(b)   60 A.

(2)   The total load calculated in accordance with Subrule (1) and Subrule (3)(a), (b), and (c) shall not be considered to be a continuous load for the application of Rule 8-104.

(3)   The minimum ampacity of service or feeder conductors from a main service supplying two or more dwelling units shall be based on the calculated load obtained from Subrule (1)(a) and the following:

(a)   excluding any electric space-heating loads and any air-conditioning loads, the load shall be considered to be

(i)   100% of the calculated load in the unit having the heaviest load; plus

(ii)   65% of the sum of the calculated loads in the next 2 units having the same or next smaller loads to those specified in Item (i); plus

(iii) 40% of the sum of the calculated loads in the next 2 units having the same or next smaller loads to those specified in Item (ii); plus

(iv)   25% of the sum of the calculated loads in the next 15 units having the same or next smaller loads to those specified in Item (iii); plus

      (v)   10% of the sum of the calculated loads in the remaining units;

   (b)   if electric space heating is used, the sum of all the space-heating loads as determined in accordance with the requirements of Section 62 shall be added to the load determined in accordance with Item (a), subject to Rule 8-106(4);

   (c)   if air conditioning is used, the sum of all the air-conditioning loads shall be added, with a demand factor of 100%, to the load determined in accordance with Items (a) and (b), subject to Rule 8-106(4); and

   (d)   in addition, any lighting, heating, and power loads not located in dwelling units shall be added with a demand factor of 75%.

(4)   The ampacity of feeder conductors from a service supplying loads not located in dwelling units shall be not less than the rating of the equipment installed with demand factors as permitted by this Code.

8-204 Schools

(1)   The minimum ampacity of service or feeder conductors shall be based on the following:

   (a)   a basic load of 50 W/m^2 of classroom area; plus

   (b)   10 W/m^2 of the remaining area of the building based on the outside dimensions; plus

   (c)   electric space-heating, air-conditioning, and power loads based on the rating of the equipment installed.

(2)   Demand factors shall be permitted to be applied as follows:

   (a)   for a building with an area up to and including 900 m^2 based on the outside dimensions:

      (i)   as permitted in Section 62 for any electric space-heating loads provided for; and

      (ii)   75% for the balance of the load; and

   (b)   for a building with an area exceeding 900 m^2 based on the outside dimensions:

      (i)   as permitted in Section 62 for any electric space-heating loads provided for; and

      (ii)   the balance of the load shall be divided by the number of square metres to obtain a load-per-square-metre rating and the demand load may be considered to be the sum of

         (A)   75% of the load per square metre multiplied by 900; and

         (B)   50% of the load per square metre multiplied by the area of the building in excess of 900 m^2.

8-206 Hospitals

(1)   The minimum ampacity of service or feeder conductors shall be based on the following:

   (a)   a basic load of 20 W/m^2 of the area of the building based on the outside dimensions; plus

   (b)   100 W/m^2 for high intensity areas such as operating rooms; plus

   (c)   electric space-heating, air-conditioning, and power loads based on the rating of the equipment installed.

(2)   Demand factors shall be permitted to be applied as follows:

   (a)   for a building with an area up to and including 900 m^2 based on the outside dimensions:

      (i)   as permitted in Section 62 for any electric space-heating loads provided for; and

      (ii)   80% for the balance of the load; and

   (b)   for a building with an area exceeding 900 m^2 based on the outside dimensions:

      (i)   as permitted in Section 62 for any electric space-heating loads provided for; and

      (ii)   the balance of the load shall be divided by the number of square metres to obtain a load-per-square-metre rating and the demand load may be considered to be the sum of

         (A)   80% of the load per square metre multiplied by 900; and

         (B)   65% of the load per square metre multiplied by the area of the building in excess of 900 m^2.

8-208 Hotels, motels, dormitories, and buildings of similar occupancy (see Appendix B)

(1)   The minimum ampacity of service or feeder conductors shall be based on the following:

   (a)   a basic load of 20 W/m^2 of the area of the building based on the outside dimensions; plus

   (b)   lighting loads for special areas such as ballrooms, based on the rating of the equipment installed; plus

   (c)   electric space-heating, air-conditioning, and power loads based on the rating of the equipment installed.

(2)   Demand factors shall be permitted to be applied as follows:

   (a)   for a building with an area up to and including 900 m^2 based on the outside dimensions

      (i)   as permitted in Section 62 for any electric space-heating loads provided for; and

      (ii)   80% for the balance of the load;

    (b)   for a building with an area exceeding 900 m² based on the outside dimensions:
        (i)   as permitted in Section 62 for any electric space-heating loads provided for; and
        (ii)   the balance of the load shall be divided by the number of square metres to obtain a load-per-square-metre rating and the demand load may be considered to be the sum of
            (A)   80% of the load per square metre multiplied by 900; and
            (B)   65% of the load per square metre multiplied by the area of the building in excess of 900 m².

8-210 Other types of occupancy

The minimum ampacity of service or feeder conductors for the types of occupancies specified in Table 14 shall be based on the following:

(a)   a basic load to be calculated on the basis of watts per square metre required by Table 14 for the area served based on the outside dimensions, with application of demand factors as indicated therein; plus

(b)   special loads such as electric space-heating, air-conditioning, power loads, show window lighting, stage lighting, etc., based on the rating of the equipment installed with demand factors permitted by this Code.

8-212 Special lighting circuits

Where a panel is supplying special types of lighting, such as exit lights or emergency lights, which may be located throughout a building making it impossible to calculate the area served, the connected load of the circuits involved shall be used in determining a feeder size.

Branch circuits

8-300 Branch circuits supplying electric ranges

(1)   Conductors of a branch circuit supplying a range in a dwelling unit shall be considered as having a demand of
    (a)   8 kW where the rating of the range does not exceed 12 kW; or
    (b)   8 kW plus 40% of the amount by which the rating of the range exceeds 12 kW.

(2)   For the purpose of Subrule (1), two or more separate built-in cooking units shall be permitted to be considered as one range.

(3)   For ranges or cooking units installed in commercial, industrial, and institutional establishments, the demand shall be considered as not less than the rating.

(4)   The demand loads given in this Rule shall not apply to cord-connected hotplates, rangettes, or other appliances.

8-302 Connected loads

(1)   For show window lighting installations, the demand load shall be determined on the assumption that not less than 650 W/m will be required measured along the base of the window(s), except a lower figure shall be permitted where a deviation has been allowed in accordance with Rule 2-030.

(2)   A load of a cyclic or intermittent nature shall be classified as continuous unless it meets the requirements of Rule 8-104(3).

(3)   The total connected load of a branch circuit supplying one or more units of data processing equipment shall be considered to be a continuous load for the application of Rule 8-104.

Δ 8-304 Maximum number of outlets per circuit

(1)   There shall be not more than 12 outlets on any 2-wire branch circuit except as permitted by other Rules of this Code.

(2)   Such outlets shall be considered to be rated at not less than 1 A per outlet except as permitted by Subrule (3).

(3)   Where the connected load is known, the number of outlets shall be permitted to exceed 12 provided that the load current does not exceed 80% of the rating of the overcurrent device protecting the circuit.

(4)   Where fixed multi-outlet assemblies are used, each 1.5 m or fraction thereof of each separate and continuous length shall be counted as one outlet, but in locations where a number of electrical appliances are likely to be used simultaneously, each 300 mm or fraction thereof shall be counted as one outlet.

Automobile heater receptacles

8-400 Branch circuits and feeders supplying automobile heater receptacles

(1)   In the application of this Rule, the following definitions shall apply:

Controlled — power to the receptacle is cycled by other than a manual operation; and

Restricted — pertaining to the engine block heater only and where the use of an in-car heater shall not be permitted.

(2)　At least one branch circuit protected by an overcurrent device rated or set at not more than 15 A shall be provided for each duplex receptacle or for every two single receptacles.

(3)　Where the loading in each parking space or stall is not restricted or controlled, a separate branch circuit shall be provided for each parking space or stall and the feeder or service conductor shall be considered as having a demand load as follows:

Number of automobile spaces or stalls	Demand load per space or stall (W)
First 30	1200
Next 30	1000
All over 60	800

(4)　Where branch circuits are provided for parking spaces or stalls in which the loading is restricted or controlled, the feeder or service conductors shall be considered as having a demand load as follows:

Number of automobile spaces or stalls	Demand load per space or stall (W)
First 30	650
Next 30	550
All over 60	450

(5)　Parking lots that may be fully occupied under normal usage shall be assigned a greater demand load per space or stall.

Section 10 — Grounding and bonding

Scope and object

10-000 Scope
(1) This Section covers the protection of electrical installations by grounding and bonding.
(2) Insulating, isolating, and guarding may be used as means of affording supplemental protection to grounding or, where permitted in this Code, as a suitable alternative.

10-002 Object
Grounding and bonding as required by this Code shall be done in such a manner as to serve the following purposes:
(a) to protect life from the danger of electric shock, and property from damage by bonding to ground non-current-carrying metal systems;
(b) to limit the voltage on a circuit when it is exposed to higher voltages than that for which it is designed;
(c) in general to limit ac circuit voltages-to-ground to 150 V or less on circuits supplying interior wiring systems;
(d) to facilitate the operation of electrical apparatus and systems; and
(e) to limit the voltage on a circuit that might otherwise occur through exposure to lightning.

System and circuit grounding

10-100 Circuits
Circuits shall be grounded as necessary in accordance with this Section.

10-102 Two-wire direct-current systems
(1) Two-wire direct-current systems supplying interior wiring and operating at not more than 300 V or not less than 50 V between conductors shall be grounded, unless such systems are used for supplying industrial equipment in limited areas and the circuit is equipped with a ground detector.
(2) If such a circuit operates at more than 300 V between conductors and a neutral point can be established so that the maximum difference of voltage between the neutral point and any other point on the system does not exceed 300 V, the neutral shall be permitted to be grounded.

10-104 Three-wire direct-current system
The neutral conductor of all 3-wire direct-current systems supplying interior wiring shall be grounded.

10-106 Alternating-current systems (see Appendix B)
(1) Except as otherwise provided for in this Code, alternating-current systems shall be grounded if
(a) by so doing, their maximum voltage-to-ground does not exceed 150 V; or
(b) the system incorporates a neutral conductor.
(2) Wiring systems supplied by an ungrounded supply shall be equipped with a suitable ground detection device to indicate the presence of a ground fault.

10-108 Electric arc furnace circuits
Circuits supplying electric arc furnaces shall be permitted to, but need not, be grounded.

10-110 Electric crane circuits
Circuits supplying electric cranes operating over combustible fibres in Class III hazardous locations shall not be grounded.

10-112 Isolated circuits
Special circuits shall be permitted to be supplied from the ungrounded secondaries of transformers having the primary and secondary windings separated by a grounded metal shield if
(a) installed under the provisions of other Sections of the Code; or
(b) this is required to recognize a particular accident or fire hazard.

10-114 Circuits of less than 50 V
Circuits of less than 50 V shall be grounded
(a) where run overhead outside of buildings; or
(b) where supplied by transformers energized from
(i) systems of more than 150 volts-to-ground; or
(ii) ungrounded systems unless the circuits are provided in accordance with Rule 10-112.

10-116 Instrument transformer circuits

(1)   Where primary windings of current and voltage instrument transformers are connected to circuits of 300 V or more to ground, the secondary circuits of the transformer shall be grounded.

(2)   Where the transformers are on switchboards, the secondary circuits shall be grounded irrespective of the voltage of the circuits.

Grounding connections for systems and circuits

10-200 Current over grounding and bonding conductors

(1)   Where wiring systems, circuits, electrical equipment, arresters, cable armour, conduit, and other metal raceways are grounded, the grounding shall be arranged so that there is no objectionable passage of current over the grounding conductors.

(2)   The temporary currents that are set up under accidental conditions while the grounding conductors are performing their intended protective functions shall not be considered as objectionable.

(3)   Where, through the use of multiple grounds, an objectionable flow of current occurs over the grounding conductor,

(a)   one or more of the grounds shall be abandoned;

(b)   the location of the grounds shall be changed;

(c)   the continuity of the conductor between the grounding connections shall be suitably interrupted; or

(d)   other effective action shall be taken to limit the current.

10-202 Grounding connections for direct-current systems

Direct-current systems that are to be grounded shall have the grounding connections made at one or more supply stations but not at individual services or elsewhere on interior wiring.

Δ 10-204 Grounding connections for alternating-current systems (see Appendix B)

(1)   Alternating-current circuits that are to be grounded shall have

(a)   a connection to a grounding electrode at each individual service, except as provided for in Rule 10-200;

(b)   the grounding connection made on the supply side of the service disconnecting means either in the service box or in other service equipment, except for areas or buildings housing livestock, the grounding connection shall be permitted to be made within another device specifically intended for the purpose located in the grounding circuit and not more than 3 m from the service equipment;

(c)   at least one additional connection to a grounding electrode at the transformer or elsewhere; and

(d)   no connection between the grounded circuit conductor on the load side of the service disconnecting means and the grounding electrode, except as provided for in Rule 10-208.

(2)   Notwithstanding Subrule (1), for circuits that are supplied from two sources in a common enclosure or grouped together in separate enclosures and employing a tie, a single grounding electrode connection to the tie point of the grounded circuit conductors from each power source shall be permitted.

(3)   Where the system is grounded at any point, the grounded conductor shall

(a)   be run to each individual service;

(b)   have a minimum size as specified for bonding conductors in Table 16;

(c)   also comply with Rule 4-022 where it serves as the neutral; and

(d)   be included in each parallel run where the service conductors are run in parallel.

(4)   Notwithstanding Rule 12-108, the size of the system grounded conductors in each parallel run shall be permitted to be smaller than No. 1/0 AWG.

10-206 Grounding connections for isolated systems (see Appendix B)

(1)   For a wiring system or circuit required to be grounded and not conductively connected to a distribution system, the grounding connection shall be made at the source of supply, or on the supply side of the first switch controlling the system, and

(a)   the grounding conductor shall not be smaller than that specified in Table 17 but in no case does it need to be larger than the largest ungrounded conductor of the system; and

(b)   where two or more systems are employed, a common system grounding conductor shall be installed unless separate grounding electrodes are provided for each system, in which case the grounding electrodes shall be interconnected in accordance with Rule 10-702.

(2)   Notwithstanding Rules 10-802 and 10-806, where a circuit is required to be grounded, and is supplied from a source having a rated output of 1000 V•A or less, the grounding connection shall be permitted to

be made to the grounded metal enclosure of the power supply, or to the bonding conductor within the enclosure.

10-208 Grounding connections for two or more buildings or structures supplied from a single service

Where two or more buildings or structures are supplied from a single service,

(a) the grounded circuit conductor at each of the buildings or structures shall be connected to a grounding electrode and bonded to the non-current-carrying metal parts of the electrical equipment; or

(b) except for buildings housing livestock, the non-current-carrying metal parts of the electrical equipment in or on the building or structure shall be permitted to be bonded to ground by a bonding conductor run with the feeder or branch circuit conductors.

10-210 Conductor to be grounded

(1) For alternating-current wiring systems, the conductor to be grounded shall be as follows:
   (a) single-phase, 2-wire — the identified conductor;
   (b) single-phase, 3-wire — the identified neutral conductor;
   (c) multi-phase systems having one wire common to all phases — the identified neutral conductor;
   (d) multi-phase systems having one phase grounded — the identified conductor; and
   (e) multi-phase systems in which one phase is used as in Item (b) — the identified conductor.

(2) In multi-phase systems in which one phase is used as a single-phase 3-wire system, only one phase shall be grounded.

Conductor enclosure bonding

10-300 Enclosures for service conductors

Service raceways, service cable sheaths, or armouring, if of metal, shall be bonded to ground.

Δ 10-302 Underground service

Notwithstanding Rule 10-300,

(a) where an underground service cable is served from a continuous underground metal-sheathed cable system and the sheath or armour of the service cable is connected to the underground system, the sheath or armour of the service cable shall not be required to be bonded to ground at the building if it is insulated from the interior conduit or piping; and

(b) where metal-sheathed service cable is served from a continuous underground metal-sheathed cable system, is bonded to the underground system, and is contained in an underground service conduit, the conduit shall not be required to be bonded to ground at the building if it is insulated from the interior conduit or piping.

10-304 Other conductor enclosures

(1) Metal enclosures for conductors, other than those referred to in Rule 10-300, shall be bonded to ground except
   (a) in runs of less than 7.5 m that are free from probable contact with ground, grounded metal, metal lath, or conductive thermal insulation, and that, where within reach from grounded surfaces, are guarded against contact by persons; and
   (b) runs used for physical protective sleeving of less than 1.5 m in length, where the installation method is such that it is improbable they will become energized.

(2) Where single-conductor metal-sheathed or armoured cables are installed in raceways of insulating material, in order to prevent the flow of sheath currents in accordance with Rule 4-008(1)(c), the cables shall
   (a) be in separate raceways or supplied with suitable continuous non-conductive jackets;
   (b) have their sheaths or armour bonded together and bonded to ground at the supply end; and
   (c) thereafter have their sheaths or armour isolated from each other and from ground.

Equipment bonding

10-400 Fixed equipment, general

Exposed, non-current-carrying metal parts of fixed equipment shall be bonded to ground if the equipment is

(a) supplied by means of metal-enclosed wiring;
(b) supplied by means of wiring that contains a bonding conductor;
(c) located in a wet location and is not isolated;

(d)   located within reach of a person who can make contact with any grounded surface or object;

(e)   located within reach of a person standing on the ground;

(f)   in a hazardous location;

(g)   in electrical contact with metal, metal foil, or metal lath; or

(h)   to operate with any terminal at more than 150 volts-to-ground, except

    (i)   enclosures for switches or circuit breakers that are accessible to qualified persons only;

    (ii)   metal frames of electrically heated devices that have been exempted in accordance with Rule 2-030 and are permanently and effectively insulated from ground; and

    (iii)   transformers mounted on wooden poles at a height of more than 2.5 m above grade level provided that the installation is in compliance with the requirements of the supply authority.

10-402 Fixed equipment, specific

(1)   Exposed, non-current-carrying metal parts of the following kinds of fixed equipment shall be bonded to ground:

    (a)   frames of motors operating at more than 30 V;

    (b)   cases of controllers for motors;

    (c)   electric equipment of elevators and cranes;

    (d)   electrical equipment in garages, theatres, and motion picture studios, except pendant lampholders on circuits of not more than 150 volts-to-ground;

    (e)   motion-picture projection equipment;

    (f)   electric signs and associated equipment;

    (g)   generator frames in an electrically operated organ, unless the generator is effectively insulated from ground;

    (h)   switchboard frames and structures supporting switching equipment, except that frames of direct-current, single polarity switchboards shall not be required to be bonded to ground if effectively insulated;

    (i)   X-ray equipment used in therapy;

    (j)   equipment supplied by Class 1 and 2 circuits falling within the scope of Section 16 where such circuits require grounding to meet the intent of Rules 10-100 to 10-114; and

    (k)   data processing equipment.

(2)   Electrostatic shields of transformers shall be bonded to ground.

(3)   All non-current-carrying metal parts of lighting fixtures and associated equipment that could become energized shall be bonded to ground if they are

    (a)   exposed; or

    (b)   not exposed, but are in contact with exposed metal parts.

(4)   Electrical equipment, such as livestock waterers, installed in feedlots and open feeding areas shall be bonded to ground by a separate stranded copper bonding conductor not less than No. 6 AWG terminating at a point where the branch circuit receives its supply.

10-404 In non-metallic wiring systems

(1)   Where a non-metallic wiring system is used,

    (a)   a bonding connection shall be provided at all outlets; and

    (b)   metal boxes shall be bonded to ground.

(2)   Where conductors are run in parallel in multiple raceways as permitted in Rule 12-108, a bonding conductor shall be run in each raceway.

10-406 Non-electrical equipment (see Appendix B)

(1)   The following metal parts of non-electrical equipment shall be bonded to ground:

    (a)   frames and tracks of electrically operated cranes;

    (b)   the metal frame of a non-electrically driven elevator car to which electric conductors are attached;

    (c)   hand-operated metal shifting ropes or cables of electric elevators; and

    (d)   metal enclosures such as partitions, grille work, etc., around equipment carrying voltages in excess of 750 V between conductors.

(2)   Where a metal water piping system is installed in a building supplied with electric power and is not used as a grounding electrode required by Rule 10-700,

    (a)   the metal water piping system shall be bonded to the system grounding conductor by means of a copper bonding conductor not smaller than No. 6 AWG; and

    (b)   the bonding conductor shall be attached to the metal water piping system
        (i)   at a location as near to the consumer's electrical service entrance as is practicable; and
        (ii)   at a location where a feeder enters a barn or other building.

(3)   Each continuous metal waste water piping system installed in a building supplied with electric power shall be bonded to the system grounding conductor or to the grounded metal water supply piping by a copper bonding jumper of not less than No. 6 AWG.

(4)   All interior metal gas piping that may become energized shall be made electrically continuous and shall be bonded in accordance with the requirements of Subrule (2).

(5)   In buildings housing livestock, all metal water pipes, stanchions, water bowls, vacuum lines, and other metals that could become energized shall be bonded to ground by a separate stranded copper bonding conductor not smaller than No. 6 AWG except that, where it is necessary to control the effects of stray earth current, a device specifically approved for the purpose, connected in series with the bonding conductor, shall be permitted.

Δ (6)   In rooms that have raised floors of conductive material with electrical wiring under the raised floor, the raised floor assembly shall be bonded with a conductor no smaller than No. 6 AWG copper or the equivalent, in such a manner that the metallic flooring panels or materials form an effective equipotential plane.

10-408 Portable equipment (see Appendix B)

(1)   Exposed non-current-carrying metal parts of portable equipment shall be bonded to ground under the following conditions:
    (a)   when used in hazardous locations unless supplied through an isolating transformer having an ungrounded secondary of not over 50 V;
    (b)   when a Part II Standard requires the appliance or equipment to be provided with a grounding means, or the appliance or the equipment is not acceptable without one;
    (c)   when the equipment is used in damp or wet locations, or by persons standing on the ground, on metal floors, or inside metal tanks or boilers, except where such equipment is supplied through an isolating transformer having an ungrounded secondary of not more than 50 V; or
    (d)   when the equipment operates with any terminal at more than 150 volts-to-ground except
        (i)   motors where guarded; and
        (ii)   where a deviation has been allowed in accordance with Rule 2-030, the metal frames of electrically heated appliances that are impractical to ground but that are permanently and effectively insulated from ground.

(2)   Exposed non-current-carrying metal parts of enclosures of portable X-ray equipment used in therapy shall be bonded to ground except where a deviation has been allowed in accordance with Rule 2-030.

(3)   Notwithstanding Subrules (1) and (2), tools and appliances having double insulation or equivalent protection, and so marked, need not be bonded to ground.

(4)   Notwithstanding Subrule (1), tools and appliances required to have provision for grounding need not be bonded to ground when
    (a)   used only in a location where reliable grounding cannot be obtained; and
    (b)   they are supplied from a double-insulated portable ground fault circuit interrupter of the Class A type.

10-410 Instrument transformer cases

The cases and frames of instrument transformers shall be bonded to ground, but where the primary circuit of a current transformer is not over 150 volts-to-ground and the transformer is used solely to supply current to meters, the case or frame of the current transformer need not be bonded to ground.

10-412 Cases of instruments, meters, and relays — Operating voltage 750 V or less

(1)   Where instruments, meters, and relays
    (a)   are not located on switchboards;
    (b)   operate with windings or working parts at between 300 and 750 volts-to-ground; and
    (c)   are accessible to other than qualified persons,
the cases and other exposed metal parts of the instruments, meters, and relays shall be bonded to ground.

(2)   Where instruments, meters, and relays
    (a)   operate with windings or working parts at 750 V or less to ground;
    (b)   are on switchboards having no live parts on the front of the panels; and
    (c)   operate from current and voltage transformers or are connected directly in the circuit,
the cases of the instruments, meters, and relays shall be bonded to ground.

(3) Where instruments, meters, and relays
    (a)   operate with windings or working parts at 750 V or less to ground;
    (b)   are on switchboards having exposed live parts on the front of the panels; and
    (c)   operate from current and voltage transformers or are connected directly in the circuit,
the cases of the instruments, meters, and relays shall not be bonded to ground and, where the voltage-to-ground exceeds 150 V, mats of insulating rubber or other suitable floor-insulation shall be provided for the operator.

10-414 Cases of instruments, meters, and relays — Operating voltage over 750 V

Where instruments, meters, and relays have current-carrying parts over 750 volts-to-ground, they shall be isolated by elevation or guarded, and their cases shall not be bonded to ground, except that in electrostatic ground detectors the internal ground segments of the instrument shall be connected to the instrument case and bonded to ground, and the detector shall be isolated by elevation.

Methods of grounding

10-500 Effective grounding (see Appendix B)

The path to ground from circuits, equipment, or conductor enclosures shall be permanent and continuous, shall have ample ampacity to conduct safely any currents liable to be imposed on it, and shall have impedance sufficiently low to limit the voltage above ground and to facilitate the operation of the overcurrent devices in the circuit.

10-502 Common grounding conductor

The grounding conductor for circuits shall be permitted to be used as a grounding conductor for equipment, conduit, and other metal raceways, or enclosures for conductors, including service conduit or cable sheath and service equipment.

10-504 Common grounding electrode

Where the alternating-current system is connected to a grounding electrode in or at a building as specified in Rules 10-204 and 10-208, the same electrode shall be permitted to be used to ground conductor enclosures and equipment in or on that building.

Bonding methods

10-600 Clean surfaces

Where a non-conductive protective coating such as paint or enamel is used on the equipment, conduit, couplings, or fittings, such coating shall be removed from threads and other contact surfaces in order to ensure a good electrical connection.

10-602 Dissimilar metals

Where dissimilar metals cannot be avoided at bonding connections as indicated in Rule 2-112(2), connections shall be made using methods or material that will minimize deterioration from galvanic action.

10-604 Bonding at service equipment

The electrical continuity of the grounding circuit at the service equipment shall be assured by one of the means given in Rule 10-606 for the following equipment and enclosures if of metal:
(a)   service raceways or service armour or sheaths;
(b)   all service equipment enclosures containing service entrance conductors including meter fittings, boxes, or the like, interposed in the service raceway or armour; and
(c)   any conduit or armour that forms part of the grounding conductor to the service raceway.

10-606 Means of ensuring continuity at service equipment

(1) Electrical continuity at service equipment shall be assured by
    (a)   the use of bonding conductors, or threaded couplings and threaded bosses on enclosures with joints made up tight where rigid metal conduit is used;
    (b)   the use of bonding conductors or threadless couplings made up tight where electrical metallic tubing is used;
    (c)   the use of bonding conductors or bonding jumpers meeting the requirements of Rules 10-614 and 10-906; or

(d)   other devices (not standard locknuts and bushings) such as grounding bushings equipped with bonding jumpers meeting the requirements of Rule 10-614.

(2)   Notwithstanding Subrule (1)(d), box connectors with standard locknuts shall be permitted for bonding the armour of those types of cable assemblies incorporating an internal bonding conductor where the armour is not permitted to be used for bonding purposes.

10-608 Metal armour or tape of service cable

Where service cable has an uninsulated grounded service conductor in continuous electrical contact with its metal armour or tape, the metal covering shall be considered to be adequately bonded to ground.

10-610 Bonding at other than service equipment

The electrical continuity of metal raceway, metal-sheathed, or armoured cable shall be assured by one of the methods specified in Rules 10-606(1)(a), (b), (c), and (d), or by the use of

(a)   threadless fittings, made up tight with conduit or armoured cable;

(b)   two locknuts, one inside and one outside of boxes and cabinets; or

(c)   one locknut and a metal conduit bushing provided the bushing can be installed so that it is mechanically secure and makes positive contact with the inside surface of the box or cabinet.

10-612 Loosely jointed metal raceways

(1)   The electrical continuity of expansion joints and telescoping sections of raceways shall be assured, and if bonding jumpers are used they shall comply with Rule 10-614.

(2)   Metal trough raceways made up in sections used in connection with sound recording and reproducing equipment shall contain a bonding conductor to which each section shall be bonded.

10-614 Bonding jumpers

(1)   Bonding jumpers shall be

(a)   of copper or other corrosion-resistant material;

(b)   of sufficient size to have an ampacity not less than that required for the corresponding bonding conductor except that for service raceways this ampacity shall be permitted to be determined on the basis of

(i)   Table 41, where the conducting path is supplemented by

(A)   the use of two locknuts and a grounding bushing;

(B)   the use of a conduit or cable connector with a built-in shoulder complete with one locknut and grounding bushing; or

(ii)   the maximum size that the terminal on the grounding bushing will accommodate where single-conductor metal-sheathed cables are employed and the sheaths are attached to a grounded metal plate by connectors, each fitted with a locknut and a grounding bushing;

(c)   attached to cabinets and similar equipment in a manner specified in Rule 10-906; and

(d)   attached in a manner specified in Rule 10-908 where used between grounding electrodes or around water meters and the like.

(2)   Straps, when used for bonding non-current-carrying metal parts, shall be not less than 19 mm in width and not less than 1.4 mm in thickness if of steel or not less than 1.2 mm in thickness if of aluminum or copper.

Δ   ### 10-616 Short section of raceway

Isolated sections of metal raceway or cable armour, if required to be bonded to ground, shall preferably be bonded to ground by connecting to other grounded raceway or armour, but shall be permitted to be bonded to ground in accordance with Rule 10-618.

Δ   ### 10-618 Fixed equipment

(1)   Fixed equipment as specified in Rules 10-400 and 10-402 shall, subject to the provisions of Rule 10-804, be bonded to ground in one of the following ways:

(a)   an effective metallic connection to grounded metal raceways, metal sheath, or cable armour except

(i)   armour as specified in Subrules (2) and (3); and/or

(ii)   sheath of mineral-insulated cable when not of copper or aluminum, as specified in Subrule (4); or

(iii)   where the raceway or cables are run underground, in locations coming within the scope of Section 22, or otherwise subject to corrosion;

(b)   a bonding conductor that is run with circuit conductors as a part of a cable assembly and that may be uninsulated, but, if provided with an individual covering, the covering shall be finished to show a green colour or a green/yellow combination;

(c)   a separate bonding conductor installed in the same way as a bonding conductor for conduit and the like; or

(d)   other means, where a deviation has been allowed in accordance with Rule 2-030.

(2)   The armour of those constructions of armoured cables incorporating a bonding conductor shall not be considered as fulfilling the requirements of a bonding conductor for the purpose of this Rule, and the bonding conductor provided in these cables shall comply with Subrule (1)(b).

(3)   The armour of flexible metal conduit and liquid-tight flexible metal conduit shall not be considered as fulfilling the requirements of a bonding conductor for the purposes of this Rule, and a separate bonding conductor shall be run within the conduit.

(4)   The sheath of mineral-insulated cable, when not of copper or aluminum, shall not be considered as fulfilling the requirements of a bonding conductor for the purposes of this Rule and bonding shall be by one of the methods specified in Subrule (1)(b), (c), or (d).

Δ   **10-620   Portable equipment**

Where the non-current-carrying metal parts of portable equipment are required to be bonded to ground, such bonding shall be obtained by

(a)   the connection of the equipment to a permanent outlet provided with bonding means as required by Rule 10-618 for fixed equipment; and

(b)   the use of one of the following means to obtain bonding continuity for the non-current-carrying metal parts of the equipment:

(i)   the metal enclosure of the conductors feeding the equipment; or

(ii)   a bare conductor, or a green, or green/yellow combination, coloured conductor run with the circuit conductors in flexible cords or power supply cables; and

(c)   the use of a multi-prong plug by which bonding is automatically established.

Δ   **10-622   Pendant equipment**

(1)   Where the non-current-carrying metal parts of pendant equipment are required to be bonded to ground, such bonding shall be obtained by

(a)   connection of the equipment to a permanent outlet provided with a bonding means as required by Rule 10-618 for fixed equipment; and

(b)   the use of one of the following means to provide bonding continuity for the non-current-carrying metal parts of the equipment:

(i)   the metal enclosure of the conductors feeding the equipment; or

(ii)   a bare conductor, or a green, or green/yellow combination, coloured conductor run with the circuit conductors in flexible cords or power supply cables.

(2)   Chains that support electric equipment shall not be used as a means of bonding to ground.

Δ   **10-624   Bonding equipment to grounded circuit conductor**

(1)   The grounded circuit conductor on the load side of the connection to ground shall not be used for bonding equipment, cable armour, or metal raceways to ground, except where a deviation has been allowed in accordance with Rule 2-030.

(2)   The grounded service conductor on the supply side of the service disconnecting means shall be permitted to be used for bonding to ground the metal meter mounting devices and service equipment, and where the grounded service conductor passes through the meter mounting device it shall be bonded to the meter mounting device.

(3)   Notwithstanding Subrule (2), the bonding of the meter mounting device to the grounded service conductor shall not be permitted at a building where a device is installed in the grounding conductor as permitted in Rules 10-204(1)(b) and 10-806(1).

Δ   **10-626   Electrolytic-type water heaters**

Electrolytic-type water heaters connected to a grounded single-phase ac circuit shall be permitted to be used provided that

(a)   a copper bonding conductor of a size given in the second column of Table 16 but in no case less than No. 12 AWG is run connecting the frame of the heater to the grounded conductor of the circuit at the service box; and

(b)   the grounded conductor of the circuit is grounded at the service box to a grounding system.

Grounding electrodes

Δ **10-700 Grounding electrodes** (see Appendix B)
(1)   Grounding electrodes shall consist of
(a)   manufactured grounding electrodes;
(b)   field-assembled grounding electrodes installed in accordance with this Rule; or
(c)   in-situ grounding electrodes forming part of existing infrastructure as defined in this Rule.
(2)   Manufactured grounding electrodes shall
(a)   in the case of a rod grounding electrode, consist of 2 rod electrodes (except for a chemically charged rod electrode where only one need be installed) spaced no less than 3 m apart,
(i)   bonded together with a grounding conductor sized in accordance with Table 17; and
(ii)   driven to the full length of the rod; or
(b)   in the case of a plate electrode, be
(i)   in direct contact with exterior soil at no less than 600 mm below grade level; or
(ii)   encased within the bottom 50 mm of a concrete foundation footing in direct contact with the earth at not less than 600 mm below finished grade.
(3)   A field-assembled grounding electrode shall consist of
(a)   a bare copper conductor not less than 6 m in length, sized in accordance with Table 43 and encased within the bottom 50 mm of a concrete foundation footing in direct contact with the earth at not less than 600 mm below finished grade; or
(b)   a bare copper conductor not less than 6 m in length, sized in accordance with Table 43 and directly buried in earth at least 600 mm below finished grade.
(4)   For the purposes of Rule 2-024, an in-situ grounding electrode shall not be considered electrical equipment and shall provide, at 600 mm or more below finished grade, a surface area exposure to earth equivalent to that of a similar manufactured electrode.
(5)   Where a local condition such as rock or permafrost prevents a rod or a plate grounding electrode from being installed at the required burial depth, a lesser acceptable depth shall be permitted.

Δ **10-702 Spacing and interconnection of grounding electrodes**
Where multiple grounding electrodes exist at a building, including those used for signal circuits, radio, lightning protection, communication, community antenna distribution systems or any other purpose, they shall be
(a)   separated by at least 2 m from each other;
(b)   bonded together with not less than a No. 6 AWG copper conductor protected by location from mechanical injury; and
(c)   in the case of lightning protection systems, bonded together in accordance with Item (b) at or below ground level.

10-704 Railway track as electrodes
Rails or other grounded conductors of electric railway circuits shall not be used as a ground for other than railway lightning arresters and railway equipment, metal conduit, armoured or metal-sheathed cable, metal raceway, and the like; and in no case shall such rails or other grounded conductors of railway circuits be used for grounding interior wiring systems other than those supplied from the railway circuit itself.

Δ **10-706 Use of lightning rod system conductors and grounding electrodes** (see Appendices B and G)
Lightning rod conductors, driven pipes, rods, or other grounding electrodes (excluding metal water-piping systems) used for grounding lightning rod systems shall not be used for grounding wiring systems or other electrical equipment.

Grounding and bonding conductors

10-800 Continuity of grounding and bonding conductors
No automatic cut-out or switch shall be placed in the grounding or bonding conductor of a wiring system unless the opening of the cut-out or switch disconnects all sources of energy.

10-802 Material for system grounding conductors
The grounding conductor of a wiring system, whether also used for grounding electrical equipment or not, shall be permitted to be insulated or bare and shall be of copper.

10-804 Material for bonding conductors

The bonding conductor for equipment and metal raceways and enclosures for conductors shall be one of the following:

(a) a conductor of copper or other corrosion-resistant material, insulated or bare;

(b) a busbar or steel pipe;

(c) rigid metal conduit, except a separate conductor as required by Item (a) shall be installed within the conduit where the conduit is
  (i) made of stainless steel material;
  (ii) directly buried in the earth;
  (iii) located in concrete or masonry slabs in contact with the earth;
  (iv) in any location where material having a deteriorating effect may come in contact with the conduit; or
  (v) installed exposed outdoors where it may be subject to mechanical damage from vehicular traffic;

(d) electrical metallic tubing, except a separate conductor required by Item (a) shall be installed within the tubing where the tubing
  (i) is located in concrete or masonry slabs in contact with the earth; or
  (ii) is in any location where materials having a deteriorating effect may come in contact with the tubing; or
  (iii) is installed outdoors where it may be subject to mechanical damage from vehicular traffic;

(e) the copper or aluminum sheath of mineral-insulated cable or any conductor of a mineral-insulated cable if it is permanently marked at the time of installation so that it can be readily distinguished from conductors that are not used as bonding conductors, except that if the sheath is of aluminum in an underground run or in a location where materials having a deteriorating effect may come in contact with the metal, corrosion-resistant protection suitable for the corrosive condition encountered shall be provided;

(f) the sheath of aluminum-sheathed cable, but if used for underground runs or in locations where materials having a deteriorating effect may come in contact with the metal, corrosion-resistant protection suitable for the corrosive conditions encountered shall be provided; or

(g) other metal raceways or cable armour as provided for in Rule 10-618.

10-806 Installation of system grounding conductors (see Appendix B)

(1) The grounding conductor for a system shall be without joint or splice throughout its length, except in the case of busbars, thermit welded joints, compression connectors applied with a compression tool compatible with the particular connector, or, where it is necessary to control the effects of stray earth current, devices specifically approved for connection in series with the grounding conductor.

(2) A No. 6 AWG or larger copper grounding conductor that is free from exposure to mechanical injury shall be permitted to be run along the surface of the building construction without metal covering or protection, if it is rigidly stapled to the construction; otherwise, it shall be in conduit, electrical metallic tubing, or cable armour.

(3) A No. 8 AWG or smaller grounding conductor shall be in conduit, electrical metallic tubing, or cable armour.

Δ (4) Magnetic materials used to enclose grounding conductors shall be bonded to the grounding conductor at both ends.

(5) Where a grounding conductor is run in the same raceway with other conductors of the system to which it is connected, it shall be insulated, except that where the length of the raceway does not exceed 15 m between pull points and does not contain more than the equivalent of 2 quarter bends between pull points, an uninsulated grounding conductor shall be permitted to be used.

(6) Notwithstanding the requirements of Subrule (2), a grounding conductor No. 6 AWG or larger shall be permitted to be embedded in concrete provided that the points of emergence are located or guarded so as not to constitute exposure to mechanical injury.

10-808 Installation of equipment bonding conductors

(1) The bonding conductor for equipment shall be permitted to be spliced or tapped, but such splices or taps shall be made only within boxes, except in the case of open wiring where they shall be permitted to be made externally from boxes and shall be covered with insulation.

(2) Where more than one bonding conductor enters a box, all such conductors shall be in good electrical contact with each other by securing all bonding conductors under bonding screws, or by connecting them together with a solderless connector and connecting one conductor only to the box by a bonding screw

or a bonding device, and the arrangement shall be such that the disconnection or removal of a receptacle, fixture, or other device fed from the box will not interfere with, or interrupt, the bonding continuity.

(3) Where a bonding conductor is run in the same raceway with other conductors of the system to which it is connected, it shall be insulated, except that where the length of the raceway does not exceed 15 m and does not contain more than the equivalent of 2 quarter bends, an uninsulated bonding conductor shall be permitted to be used.

(4) Where a metal raceway or steel pipe is used as a bonding conductor, the installation shall comply with Section 12.

(5) A copper bonding conductor shall
   (a) if No. 6 AWG or larger and attached securely to the surface on which it is carried, be protected where exposed to mechanical injury; and
   (b) if smaller than No. 6 AWG, or if the installation does not come within the provisions of Item (a) of this Subrule, be installed and protected in the same manner as the circuit conductor for a given installation.

(6) An aluminum bonding conductor shall
   (a) if No. 4 AWG or larger and attached securely to the surface on which it is carried, be protected where exposed to mechanical injury; or
   (b) if smaller than No. 4 AWG, or if the installation does not come within the provisions of Item (a), be installed and protected in the same manner as the circuit conductor for a given installation.

(7) Where a separate bonding conductor is required by this Code to supplement the bonding afforded by a metal raceway, it shall be installed in the same raceway as the circuit conductors.

(8) Where a separate bonding conductor, required by this Code, is run with single-conductor cables, the bonding conductor shall follow the same route as the cables.

10-810 Grounding conductor size for DC circuits

(1) The ampacity of the grounding conductor for a direct-current supply system or generator shall be not less than that of the largest conductor supplied by the system, except that where the grounded circuit conductor is a neutral derived from a balancer winding or a balancer set, the size of the grounding conductor shall be not less than that of the neutral conductor.

(2) The system grounding conductor shall be copper and in no case smaller than No. 8 AWG.

Δ **10-812 Grounding conductor size for AC systems** (see Appendix B)

Table 17 shall be used to determine the size of
(a) a grounding conductor of an AC system;
(b) a common grounding conductor; or
(c) a grounding conductor for service equipment where the AC system is not grounded.

Δ **10-814 Bonding conductor size** (see Appendix B)

(1) The size of a bonding conductor shall be not less than that given in Table 16, but in no case does it need to be larger than the largest ungrounded conductor in the circuit.

(2) Where circuit conductors are paralleled in separate cables or raceways, the bonding conductor shall be permitted to be paralleled and the size of bonding conductor in each parallel run shall not be less than that specified in Table 16 based on the size of the circuit conductors contained in the raceway or cable.

(3) Notwithstanding the requirements of Rule 12-108, the size of the bonding conductor in each parallel run shall be permitted to be smaller than No. 1/0 AWG.

10-816 Bonding conductor size for circuits extended to portable, pendant, or fixed equipment

The bonding conductor size for circuits run to equipment from the outlets, which are bonded in accordance with Rule 10-814, shall be not less than that given in Column 2 or 3 of Table 16 as applicable, except that where flexible cord having copper conductors in sizes No. 16 AWG and smaller is used, the bonding conductor shall be the same size as the circuit conductors.

10-818 Bonding conductor for outline lighting

Isolated non-current-carrying metal parts of outline lighting equipment shall be permitted to be bonded together by a No. 14 AWG conductor of copper or of equal conductance if of other metal, protected from mechanical injury.

10-820 Bonding conductor size for instrument transformers
The bonding conductor for secondary circuits of instrument transformers and for instrument cases shall be not smaller than No. 12 AWG if of copper, or of equal conductance if of other metal.

Grounding and bonding conductor connections

10-900 Bonding conductor connection to raceways
The point of connection of the bonding conductor to interior metal raceways, cable armour, and the like shall be as near as practicable to the source of supply and shall be chosen so that no raceway or cable armour is bonded through a run of smaller size than is called for in Rule 10-814.

10-902 Grounding conductor connection to water pipe electrodes
(1)   Where the grounding electrode is a metal water-piping system to which a common grounding conductor or the grounding conductor of a system is attached, the point of attachment shall be
  (a)   on the street side of the water meter; or
  (b)   on a cold-water pipe of adequate ampacity and as near as practicable to the point of entrance of the water service in the building.
(2)   Where practicable, the point of attachment shall be accessible.
(3)   The metal cold-water system shall be made electrically continuous from the point of attachment of the grounding conductor to the water service entrance by bonding together all parts that contain insulating sections or that may become disconnected at such locations as meters, valves, and unions.

10-904 Grounding conductor connections to other than water pipe electrodes (see Appendix B)
(1)   Where a metal water-piping system is not available, the grounding conductor shall be attached to other electrodes at a point that will assure a permanent ground.
(2)   Where practicable, the point of attachment shall be accessible.

10-906 Bonding conductor connection to circuits and equipment
(1)   The bonding conductor or bonding jumper shall be attached to circuits, conduits, cabinets, equipment, and the like, which are to be bonded by means of lugs, pressure wire connector clamps, or other equally substantial means.
(2)   Connections that depend upon solder shall not be used.
(3)   The bonding conductor shall be secured to every metal box by means of a bonding screw, which shall be used for no other purpose.
(4)   The bonding conductor shall be brought into every non-metallic outlet box in such a manner that it can be connected to any fitting or device that may require bonding to ground.
(5)   Equipment shall be installed so that if the connections between the branch circuit and the internal conductors pass through an access cover, the bonding connection shall remain continuous when the cover is removed.
(6)   A bonding jumper shall be installed to connect the bonding conductor to the grounding terminal of a receptacle and in such a manner that disconnection or removal of the receptacle will not interfere with, or interrupt, grounding continuity.
(7)   In the case of metallically enclosed systems where the grounding path is provided by the metal enclosure, a bonding jumper shall be installed to bond the grounding terminal of the receptacle to the enclosure.
(8)   Notwithstanding Subrules (6) and (7), the bonding jumper, in the case of receptacles having grounding terminals isolated from the mounting strap required for special equipment, shall be permitted to be extended directly back to the distribution panel.
(9)   Notwithstanding Rule 10-808, electronic equipment rated to operate at a supply voltage not exceeding 150 volts-to-ground and that requires a separate bonding conductor shall be permitted to be bonded to ground by an insulated conductor extending directly back to the distribution panel, provided that
  (a)   the separate bonding conductor is enclosed in the same raceway or cable containing the circuit conductors throughout the length of that cable or raceway;
  (b)   the separate bonding conductor is sized not less than as given in Table 16 for each leg of the run, determined by the size of the overcurrent protection for the circuit conductors; and
  (c)   the bonding requirements of Rules 10-304 and 10-400 are met.

10-908 Grounding conductor connection to electrodes
(1) The grounding conductor shall be attached to the grounding electrode by means of
 (a) a bolted clamp;
 (b) a pipe fitting plug or other device, screwed into the pipe or into the fitting;
 (c) copper welding by the thermit process, brazing, or silver solder; or
 (d) other equally substantial means.
(2) Where a bolted clamp is used for a wet location or for direct earth burial, the clamp shall be of copper, bronze, or brass, and the bolts shall be of similar material or of stainless steel.
(3) The grounding conductor shall be attached to the grounding fitting as required by Rule 10-906(1).
(4) Connections that depend on solder shall not be used, except for connections utilizing silver solder.
(5) Not more than one conductor shall be connected to the grounding electrode by a single clamp or fitting, unless the clamp or fitting is specifically designed for multiple conductor connection.

Lightning arresters

10-1000 Lightning arresters on secondary services — 750 V or less
(1) Where a lightning arrester is installed on a secondary service, the connections to the service conductors and to the grounding conductor shall be as short as practicable.
(2) The grounding conductor shall be permitted to be
 (a) the grounded service conductor;
 (b) the common grounding conductor;
 (c) the service equipment grounding conductor; or
 (d) a separate grounding conductor.
(3) The bonding or grounding conductor shall be of copper not smaller than No. 6 AWG.

10-1002 Installation requirements and guarding for lightning arrester grounding conductors
The grounding conductor for lightning arresters shall
(a) when enclosed in metal, be connected to the guard at both ends; and
(b) be installed and protected to meet the requirements of Rule 10-806.

Installation of neutral grounding devices

10-1100 Scope (see Appendix B)
Rules 10-1102 to 10-1108 apply to the installation of neutral grounding devices used for the purpose of controlling the ground fault current or the voltage-to-ground of an alternating-current system.

10-1102 Use (see Appendix B)
(1) Neutral grounding devices shall be permitted to be used only on a system involving a true neutral or an artificial neutral, where line-to-neutral loads are not served.
Δ (2) Where line-to-neutral loads are not served, provision shall be made to automatically de-energize the system on the detection of a ground fault unless the electrical system is operating at 5 kV or less, in which case it shall be permitted to remain energized on the detection of a ground fault provided that
 (a) the ground fault current is controlled at 10 A or less; and
 (b) a visual and/or audible alarm is provided to indicate clearly the presence of the ground fault.
Δ (3) Where line-to-neutral loads are served, provision shall be made to automatically de-energize the system on the occurrence of
 (a) a ground fault;
 (b) a grounded neutral on the load side of the neutral grounding device; or
 (c) a lack of continuity of the conductor connecting the neutral grounding device from the neutral point through the neutral grounding device to the system grounding electrode.

10-1104 Neutral grounding devices
(1) Neutral grounding devices shall be specifically approved for the application.
(2) Only neutral grounding devices with a continuous rating shall be permitted where provision is not made to de-energize the system on the detection of a ground fault.
(3) Neutral grounding devices not having a continuous rating shall be permitted where
 (a) provision is made to automatically de-energize the system on the detection of a ground fault; and

(b) the time rating of the device is coordinated with the time/current rating of the protective devices of the system.

(4) Neutral grounding devices shall have an insulation voltage rating at least equal to the system line-to-neutral voltage.

10-1106 Location of neutral grounding devices and warning signs

(1) All live parts of neutral grounding devices shall be enclosed or guarded in compliance with Rule 2-202.

(2) Neutral grounding devices shall be placed in a location that is accessible only to qualified persons to perform inspection, testing, and maintenance of the neutral grounding device.

(3) Neutral grounding devices shall be placed in a location so that heat dissipation from the device under ground fault conditions will not damage or adversely affect the operation of the device or other equipment.

(4) Where neutral grounding devices are used, warning signs indicating that the system is impedance grounded and the maximum voltage at which the neutral may be operating relative to ground shall be placed at the

(a) transformer or generator, or both;

(b) consumer's service switchgear or equivalent; and

(c) supply authority's metering equipment.

10-1108 Conductors used with neutral grounding devices

(1) The conductor connecting the neutral grounding device to the neutral point of the transformer, generator, or grounding transformer shall be

(a) insulated for the nominal system voltage;

(b) identified white or grey;

(c) sized to conduct the rated current of the neutral grounding device, and in no case less than No. 8 AWG; and

(d) installed in accordance with other appropriate Rules of this Code.

(2) The conductor connecting the neutral grounding device to the neutral point of a transformer, generator, or grounding transformer that shall not be grounded.

(3) The conductor connecting the neutral grounding device to the system grounding electrode shall be

(a) a copper conductor which shall be permitted to be insulated or bare;

(b) identified green if insulated;

(c) sized to conduct the rated current of the neutral grounding device, and in no case less than No. 8 AWG in size; and

(d) installed in accordance with other appropriate Rules of this Code.

Δ (4) Where line-to-neutral loads are served, the size of the conductors connecting to the neutral grounding device shall be sized to conduct the rated current of the neutral grounding device, and they shall in no case be smaller than No. 12 AWG.

Section 12 — Wiring methods

Scope

12-000 Scope (see Appendix B)

Δ (1) The provisions of Section 12 apply to all wiring installations operating at 750 V or less except for
  (a) Class 2 circuits unless otherwise specified in Section 16;
  (b) community antenna distribution and radio and television circuits unless otherwise specified in Section 54; and
  (c) optical fiber cables unless otherwise specified in Section 56;
  (d) communication circuit conductors unless otherwise specified in Section 60; and
  (e) conductors that form an integral part of factory-built equipment.

(2) The provisions of this Section apply also to installations operating at voltages in excess of 750 V except as modified by the requirements of Section 36.

General requirements

12-010 Wiring in ducts and plenum chambers

(1) No electrical equipment of any type, unless specifically approved for the purpose, shall be installed in ducts used to transport dust, loose stock, or flammable vapours.

(2) No electrical equipment, unless specifically approved for the purpose, shall be installed
  (a) in any duct used for vapour removal or for ventilation of commercial-type cooking equipment; or
  (b) in any shaft that is required by regulation to contain only such ducts.

(3) Where conductors are installed in ducts, plenums, or hollow spaces that are used to transport or move air as part of an environmental air system or in a duct or plenum chamber to connect to an integral fan system, the conductors shall be in accordance with the requirements of Rules 2-126 and 12-100.

(4) Notwithstanding Subrule (3), where a plenum or hollow space is created by a suspended ceiling having lay-in panels or tiles, flexible cord not exceeding 3 m in length and terminated with an attachment plug shall be permitted to supply pole type multi-outlet assemblies and recessed fluorescent luminaires, provided that the flexible cord is listed in Table 11 for
  (a) hard usage where connected to either pole type multi-outlet assemblies or recessed fluorescent luminaires, where the voltage does not exceed 300 V;
  (b) extra-hard usage where connected to recessed fluorescent luminaires where the supply voltage does not exceed 750 V; and
  (c) at least 90 °C where supplying recessed fluorescent luminaires.

(5) Where a furnace cold-air return duct is formed by boxing in between joists, wiring methods specified in this Section for use in the particular location shall be permitted to be used.

12-012 Underground installations (see Appendix B)

(1) Direct buried conductors, cables, or raceways shall be installed to meet the minimum cover requirements of Table 53.

(2) The minimum cover requirements shall be permitted to be reduced by 150 mm where mechanical protection is placed in the trench over the underground installation.

(3) Mechanical protection shall consist of one of the following and, when in flat form, shall be wide enough to extend at least 50 mm beyond the conductor, cables, or raceways on each side:
  (a) treated planking at least 38 mm thick; .
  (b) poured concrete at least 50 mm thick;
  (c) concrete slabs at least 50 mm thick;
  (d) concrete encasement at least 50 mm thick; or
  (e) other suitable material.

(4) Direct buried conductors or cables shall be installed so that they run adjacent to each other and do not cross over each other and with a layer of 6 mm (nominal) screened sand or screened earth at least 75 mm deep both above and below the conductors.

(5) Where conductors or cables rise for terminations or splices or where access is otherwise required, they shall be protected from mechanical damage by location or by rigid conduit terminated vertically in the trench and including a bushing or bell end fitting, or other acceptable protection, at the bottom end from 300 mm above the bottom of the trench to at least 2 m above finished grade, and beyond that as may be

required by other Rules of the Code, and with sufficient slack provided in the conductors at the bottom end of the conduit so that the conductors enter the conduit from a vertical position.

(6)   Where a deviation has been allowed in accordance with Rule 2-030, cables buried directly in earth shall be permitted to be spliced or tapped in trenches without the use of splice boxes and such splices and taps shall be made by methods and with material approved for the purpose.

(7)   Raceways or cables, if located in rock, shall be permitted to be installed at a lesser depth entrenched into the rock in a trench not less than 150 mm deep and grouted with concrete to the level of the rock surface.

(8)   Raceways shall be permitted to be installed directly beneath a concrete slab at grade level provided that the concrete slab is not less than a nominal 100 mm in thickness, the location is adequately marked, and the raceway will not be subject to damage during or after installations.

(9)   Any form of mechanical protection that may adversely affect the conductors or cable assemblies shall not be used.

(10)   Backfill containing large rock, paving materials, cinders, large or sharply angular substances, or corrosive material shall not be placed in an excavation where such materials may damage cables, raceways, or other substructures, prevent adequate compaction of fill, or contribute to corrosion of cables, raceways, or other substructures.

(11)   The initial installation shall be provided with a suitable marking tape buried approximately halfway between the installation and grade level, or adequate marking in a conspicuous location to indicate the location and depth of the underground installation.

(12)   For installations not covered by the foregoing requirements of this Rule, the requirements of CSA C22.3 No. 7, or the applicable standard, whichever is more stringent, shall apply.

12-014 Conductors in hoistways

(1)   Where a deviation has been allowed in accordance with Rule 2-030, and where conductors other than those used to furnish energy to the elevator or dumbwaiter are installed in hoistways, they shall be mineral-insulated cable, aluminum-sheathed cable, or armoured cable or be run in rigid metal conduit or flexible metal conduit or electrical metallic tubing.

(2)   The cable, conduit, or tubing referred to in Subrule (1) shall be
(a)   securely fastened to the hoistway construction; and
(b)   arranged so that terminal, outlet, or junction boxes open outside the hoistway except that pull boxes shall be permitted to be installed in long runs for the purpose of supporting or pulling in conductors.

12-016 Lightning rod conductors

Where lightning rod conductors are installed, electrical wiring shall, where practicable, be kept at least 2 m from such conductors except where bonding is provided in accordance with Rule 10-702.

12-018 Entry of raceways and cables into buildings

Holes in outer walls or roofs of buildings through which raceways or cables pass shall be filled to prevent infiltration of moisture.

12-020 Wiring under raised floors for data processing and similar systems

(1)   Flexible cords or cables, liquid-tight flexible conduit, and appliance wiring material with a jacket or overall covering to connect and interconnect data processing and similar systems shall be permitted to be installed under raised floors provided that
(a)   the raised floor is of non-combustible construction, and, if of conductive material, is bonded to ground in accordance with Rule 10-406; and
(b)   the cords or cables terminate in attachment plugs having configurations in accordance with Diagram 2 or that are classified as industrial type, special-use attachment plugs, receptacles, or connectors.

(2)   Branch circuit conductors installed under raised floors to supply receptacles shall be installed in rigid conduit, electrical metallic tubing, flexible metal conduit, armoured cables, or metal-sheathed cable including mineral-insulated cable other than the lightweight type.

Conductors

General

12-100 Types of conductors (see Appendix B)

Conductors installed in any location shall be suitable for the condition of use as indicated in Table 19 for the particular location involved and with particular respect to

(a)   moisture, if any;
(b)   corrosive action, if any;
(c)   temperature;
(d)   degree of enclosure; and
(e)   mechanical protection.

12-102 Thermoplastic-insulated conductors

(1)   Conductors having thermoplastic insulation shall not be installed during any time when the ambient temperature is sufficiently low as to be liable to cause damage to the insulation.
(2)   Such conductors shall not be installed so as to permit flexing or movement of the conductors after installation if the ambient temperature is liable to become low enough to damage the insulation during flexing or movement.

12-104 Flame-tested coverings

Where the insulation on a conductor has a flame-tested covering, the covering shall be removed sufficiently at terminals and splices to prevent creepage of current over it.

12-106 Multi- and single-conductor cables

(1)   Where multi-conductor cable is used, all conductors of a circuit shall be contained in the same multi-conductor cable except that, where it is necessary to run conductors in parallel due to the capacity of an alternating current circuit, additional cables shall be permitted to be used provided that any one such cable includes an equal number of conductors from each phase and the neutral and shall be in accordance with Rule 12-108.
(2)   A multi-conductor cable shall not contain circuits of different systems except as permitted in Rule 12-3030.
(3)   Where single-conductor cables are used, all single-conductor cables of a circuit shall be of the same type and temperature rating and, if run in parallel, shall be in accordance with Rule 12-108.
(4)   Single-conductor armoured cable used as a current-carrying conductor shall be of a type having nonferrous armour.
(5)   A single-conductor cable carrying a current over 200 A shall be run and supported in such a manner that the cable is not encircled by ferrous material.

12-108 Conductors in parallel (see Appendix B)

(1)   Conductors of similar conductivity in sizes No. 1/0 AWG copper or aluminum and larger shall be permitted in parallel provided that they are
    (a)   free of splices throughout the total length;
    (b)   the same circular mil area;
    (c)   the same type of insulation;
    (d)   the same length; and
    (e)   terminated in the same manner.
(2)   The orientation of single-conductor cables in parallel, with respect to each other and to those in other phases, shall be such as to minimize the difference in inductive reactance and the unequal division of current.
(3)   Conductors of similar conductivity in sizes smaller than No. 1/0 AWG copper shall be permitted in parallel to supply control power to indicating instruments and devices, contactors, relays, solenoids, and similar control devices provided that
    (a)   they are contained within one cable;
    (b)   the ampacity of each individual conductor is sufficient to carry the entire load current shared by the parallel conductors; and
    (c)   the overcurrent protection is such that the ampacity of each individual conductor will not be exceeded if one or more of the parallel conductors becomes inadvertently disconnected.
Δ (4)   Where the size of neutral conductors is reduced in conformance with Rule 4-022, neutral conductors smaller than No. 1/0 AWG shall be permitted in circuits run in parallel, provided that they are installed in conformance with the requirements of Subrule (1)(a), (b), (c), (d), and (e).

12-110 Radii of bends in conductors

The radii of bends in conductors shall be sufficiently large to ensure that no injury is done to the conductors or their insulation, covering, or sheathing.

12-112 Conductor joints and splices

(1) Unless made with solderless wire connectors, joints or splices in insulated conductors shall be soldered, but they shall first be made mechanically and electrically secure.

(2) Joints or splices shall be covered with an insulation equivalent to that on the conductors being joined.

(3) Joints or splices in wires and cables shall be accessible.

(4) Splices in underground runs of cable, if required due to damage to the original installation, shall be permitted to be made

    (a) in junction boxes suitably protected from mechanical damage that are located at least 1 m above grade and secured to buildings or to stub poles; or

    (b) notwithstanding the requirements of Subrule (3), by means of splicing devices or materials (kits) for direct earth burial.

12-114 Ends of insulated conductors

When the ends of insulated conductors at switch and fixture outlets and in similar places are not in use, they shall be insulated in the manner prescribed for joints and splices.

12-116 Termination of conductors (see Appendix B)

(1) The portion of stranded conductors to be held by wire-binding terminals or solderless wire connectors shall have the strands confined so that there will be no stray strands to cause either short-circuits or grounds.

(2) Stranded and solid conductors larger than No. 10 AWG shall be terminated in solderless wire connectors or shall be permitted to be soldered into wire connectors specifically approved for the purpose except where prohibited by Section 10.

12-118 Termination and splicing of aluminum conductors

(1) Adequate precaution shall be given to the termination and splicing of aluminum conductors, including the removal of insulation and separators, the cleaning (wire brushing) of stranded conductors, and the compatibility and installation of fittings.

(2) A joint compound, capable of penetrating the oxide film and preventing its reforming, shall be used for terminating or splicing all sizes of stranded aluminum conductors, unless the termination or splice is approved for use without compound and is so marked.

(3) Equipment connected to aluminum conductors shall be specifically approved for the purpose and be so marked except

    (a) where the equipment has only leads for connection to the supply; and

    (b) equipment such as outlet boxes having only grounding terminals.

(4) Aluminum conductors shall not be terminated or spliced in wet locations unless the termination or splice is adequately protected against corrosion.

(5) Field-assembled connections between aluminum lugs and aluminum or copper busbars or lugs, involving bolts or studs 9.5 mm (3/8-in) diameter or larger, shall include as part of the joint any of the following means of allowing for expansion of the parts:

    (a) a conical spring washer;

    (b) a helical spring washer of the heavy series, provided that a flat steel washer of thickness not less than one-sixth of the nominal diameter of the bolt or stud is interposed between the helical washer and any aluminum surface against which it would bear; or

    (c) aluminum bolts or studs, provided that all the elements in the assembled connection are of aluminum.

(6) Connection of aluminum conductors to wiring devices having wire-binding terminal screws, about which conductors can be looped under the head of the screw, shall be made by forming the conductor in a clockwise direction around the screw into three-fourths of a complete loop and only one conductor shall be connected to any one screw.

12-120 Supporting of conductors

(1) Conductors shall be supported so that no damaging strain is imposed on the terminals of any electrical apparatus or devices or on joints or taps.

(2) Conductors in vertical raceways shall be supported independently of the terminal connections and at intervals not exceeding those specified in Table 21 and such supports shall maintain the continuity of the raceway system without damage to the conductors or their covering.

(3) Conductors in raceways shall not hang over the edges of bushings, bends, or fittings of any kind in such a manner that the insulation may be damaged.

Open wiring

12-200 Open wiring Rules
Rules 12-202 to 12-224 apply only to single conductors run as open wiring.

12-202 Types of conductors
Conductors shall be of types specified in Rules 12-100 and 12-102.

12-204 Spacing of conductors
(1) Spacings between conductors and between conductors and adjacent surfaces shall, except as otherwise provided for in this Rule, comply with the following:
  (a) for normally dry locations, the spacings shall be not less than those specified in Table 20;
  (b) where circuits of different voltages are run parallel to each other, the separation between adjacent conductors of the different circuits shall be not less than that specified in Table 20 for conductors of the circuit having the higher voltage; and
  (c) in damp locations, a separation of at least 25 mm shall be maintained between conductors and adjacent surfaces.
(2) In all locations, a separation of at least 25 mm shall be maintained between conductors and adjacent metal piping or conducting materials.
(3) Where conductors are run across the open faces of joists, studs, or timber, the separation between conductors shall be as specified in Rule 12-212.
(4) At connections to fittings and devices or in other cases where it is not practical to maintain the spacing specified in Subrules (1), (2), and (3), the conductors shall be installed in raceways or insulating tubing.

12-206 Conductor supports
(1) Conductors shall be supported rigidly on non-combustible, absorption-resisting insulators.
(2) Split knobs shall not be used to support conductors larger than No. 8 AWG copper or aluminum.
(3) Conductors supported on solid knobs shall be securely tied to the knobs by tie-wires having insulation of the same type as that on the conductors that they secure.
(4) Where used on metal surfaces, thermoplastic-insulated conductors shall not be mounted in split knobs or cleats.

12-208 Conductors on flat surfaces
Where conductors are run on flat surfaces, they shall be supported rigidly at intervals of not more than 1.5 m.

12-210 Material for attachment of conductor supports
Knobs and cleats shall be fastened securely with screws.

12-212 Protection from mechanical injury
(1) Where conductors are supported on or run across the open faces of joists, wall-studs, or other timber or on walls where exposed to mechanical injury, they shall be protected by running-boards, guard-strips, wooden boxing, or sleeves of iron pipe.
(2) Where conductors are not exposed to mechanical injury, they shall be permitted to be run directly from timber to timber, but shall be
  (a) of not less than No. 8 AWG copper or aluminum;
  (b) separated from each other by not less than 150 mm; and
  (c) supported at each timber.
(3) Open wiring shall not be run across the tops of ceiling joists in unfinished attics or similar places.

12-214 Material for running-boards, guard-strips, and boxing
(1) Material for running-boards, guard-strips, and boxing shall be at least 19 mm thick and the edges of running-boards shall project at least 12 mm beyond the insulators on both sides.
(2) Guard-strips shall be at least as high as the insulators and placed as close to the conductors as Table 20 permits.
(3) In wooden boxing, there shall be a clear space of at least 25 mm between conductors and adjacent surfaces, and the ends of boxing not abutting on the structure of the building shall be closed.

12-216 Ends of conductors
(1) Conductors shall not be brought to a dead-end at any fitting distant more than 300 mm from the last supporting insulator.

(2) Where conductors of No. 8 AWG copper or aluminum or larger are run as open wiring, solid knobs or strain insulators shall be used at the ends of the run.

12-218 Conductors passing through walls or floors
Where conductors pass through walls, floors, timbers, or partitions, they shall be installed in raceways or insulating tubing.

12-220 Maintaining clearances
Sub-bases shall be installed under all surface-mounted switches and receptacles unless adequate clearances are otherwise maintained.

12-222 Where open wiring connects to other systems of wiring
Where open wiring is connected to conductors in raceways, armoured cable, or non-metallic-sheathed cable, the junction shall be made in a box, or at, or in, a fitting having a separately bushed hole for each conductor.

12-224 Provision for bonding
Where open wiring is used, provision for bonding to ground shall be made in accordance with the Section 10 requirements.

Exposed wiring on exteriors of buildings and between buildings on the same premises

12-300 Exterior exposed wiring Rules
Rules 12-302 to 12-318 apply only to exposed wiring run on the exterior surfaces of buildings or between buildings on the same premises.

12-302 Types of conductors
Conductors shall be of types suitable for exposure to the weather as indicated in Table 19.

12-304 Location of conductors
(1) Subject to the provisions of Rule 6-112, where the conductors are supported on or in close proximity to the exterior surfaces of buildings, they shall be installed and protected so that they shall not be a hazard to persons or be exposed to mechanical injury, and they shall not be less than 4.5 m from the ground unless a deviation has been allowed in accordance with Rule 2-030.

(2) Where the conductors are exposed to mechanical injury from awnings, swinging signs, shutters, or other movable objects, they shall be run in rigid conduit made watertight.

12-306 Conductor supports
(1) Conductors on the exterior surfaces of buildings shall be supported by brackets, racks, or insulators at intervals of not more than 3 m, and the individual conductors shall be distant at least 150 mm from one another and at least 50 mm from the adjacent surfaces.

(2) Where petticoat insulators are used, they shall be installed at intervals of not more than 4.5 m under normal conditions and at smaller intervals where the conductors are subject to disturbance and shall be located so as to hold the individual conductors at least 300 mm apart and at least 50 mm from adjacent surfaces.

(3) Where the conductors are not exposed to the weather, they shall be permitted to be supported on glass or porcelain knobs placed at intervals of not more than 1.5 m and holding the conductors at least 25 mm from adjacent surfaces.

(4) Where conductors connected to a voltage of 300 V or less are located in proximity to conductors of a higher voltage not exceeding 750 V, the conductors of the higher voltages shall be mounted above and kept at least 300 mm away from the conductors of the lower voltage.

12-308 Minimum size of overhead conductors
Single conductors run aerially between buildings or supports on the same premises in spans exceeding 4.5 m shall be not smaller than
(a) No. 10 AWG copper or No. 6 AWG aluminum for spans of more than 4.5 m but not more than 15 m;
(b) No. 8 AWG copper or No. 4 AWG aluminum for spans of more than 15 m but not more than 30 m; and
(c) No. 6 AWG copper or No. 3 AWG aluminum for spans of more than 30 m but not more than 40 m.

12-310 Clearance of conductors
The conductors shall be located or guarded so that they cannot be reached by a person standing on a fire escape, flat roof, or other portion of a building, and they shall be at least 2.5 m above the highest point of a flat roof or roof that can be readily walked upon and at least 1 m above peaked roofs or the highest point of roofs that cannot

be readily walked upon except that where a deviation has been allowed in accordance with Rule 2-030, they shall be permitted to be less than 2.5 m but not less than 2 m above the highest point of a flat roof or roofs that can be readily walked upon.

12-312 Conductors over buildings
Conductors shall not be installed over buildings unless a deviation has been allowed in accordance with Rule 2-030.

12-314 Conductors on trestles
Where the conductors pass over buildings, they shall, where practicable, be supported on structures not connected to the building but, where not practicable, they shall be supported on and secured to trestles constructed to bear the mechanical force of the conductors.

12-316 Power supply conductors
The conductors of a power supply system attached to the exterior surfaces of buildings shall be at least 300 mm from the conductors of a communication system unless one system is in conduit or is permanently separated from other systems by a continuous fixed non-conductor other than the insulation on the conductors.

12-318 Use of neutral supported cables
When neutral supported cables are used, the following requirements shall apply:
(a) they shall not be mounted directly on any surface;
(b) they shall be secured so that they will be not less than
    (i) 1 m from a building in the case of Types NS-75 and NS-90; or
    (ii) 50 mm from a building in the case of Type NS-75 and NS-90, marked FT1;
(c) they shall be supported in spans of not more than 38 m in length;
(d) the conductors shall be secured to the messenger at all terminations; and
(e) the bare neutral (messenger) when used as a neutral conductor forming part of an electrical circuit shall be
    (i) supplied from a grounded ac system;
    (ii) attached to an insulator at points of support and at terminations; and
    (iii) not connected to or in contact with any grounded surface except as permitted by other Rules of this Code.

Bare busbars and risers

12-400 Where bare busbars may be used
Bare conductors shall not be used as main risers or feeders in buildings unless a deviation has been obtained in accordance with Rule 2-030, and
(a) the building is of non-combustible construction;
(b) the conductors are placed in a chase, channel, or shaft located or guarded so that the conductors are inaccessible;
(c) suitable cut-offs to protect against the vertical spread of fire are provided where floors are pierced;
(d) the mechanical and electrical features of the installation and the conductor supports are suitable and the following specific requirements are used in the case of busbars rated 1200 A or less:
    (i) where flat busbars 6.35 mm or less in thickness are used, the continuous current rating shall not exceed 1000 A per 645 mm^2 of cross-sectional area of copper busbar or 700 A in the case of aluminum busbars; and
    (ii) busbar supports shall be spaced not greater than 750 mm apart, with minimum clearance across insulating surfaces between bars of opposite polarity of not less than 50 mm, and 25 mm between busbars and any grounded surface; and
(e) the resulting installation is acceptable.

Non-metallic-sheathed cable

12-500 Non-metallic-sheathed cable Rules
Rules 12-502 to 12-526 apply only to conductors run as non-metallic-sheathed cable.

12-502 Maximum voltage
Non-metallic-sheathed cable shall not be used where the voltage exceeds 300 V between any two conductors.

12-504 Use of non-metallic-sheathed cable (see Appendix B)

Non-metallic-sheathed cable shall be permitted in or on buildings of combustible construction and in or on other types of construction where acceptable.

12-506 Method of installation (see Appendices B and G)

(1)   The cable shall be run in continuous lengths between outlet boxes, junction boxes, and panel boxes as a loop system and the joints, splices, and taps shall be made in the boxes.

(2)   Where concealed wiring is connected to non-metallic-sheathed cable, the junction shall be made in a box.

(3)   Where open wiring is connected to non-metallic-sheathed cable, the junction shall be made in a box or at, or in, a fitting having a separately bushed hole for each conductor.

(4)   Where non-metallic-sheathed cable is run in proximity to heating sources, transfer of heat to the cable shall be minimized by means of an air space of at least

     (a)   25 mm between the conductor and heating ducts and piping;

     (b)   50 mm between the conductor and masonry or concrete chimneys; or

     (c)   150 mm between the conductor and chimney and flue cleanouts.

(5)   Notwithstanding Subrule (4), a thermal barrier conforming to the requirements of the *National Building Code of Canada* or local building legislation shall be permitted to be installed between the conductor and heating sources to maintain ambient temperature of the conductor at not more than 30 °C.

(6)   Two-conductor cable shall not be stapled on edge.

12-508 Bending and stapling of cable (see Appendix G)

The cable shall not be bent, handled, or stapled so that the insulated conductors or outer covering is damaged.

12-510 Running of cable between boxes and fittings (see Appendix G)

(1)   Where the cable is run between boxes and fittings, it shall be supported by straps or other devices located within 300 mm of every box or fitting and at intervals of not more than 1.5 m throughout the run.

(2)   Cables run through holes in joists or studs shall be considered to be supported.

(3)   Notwithstanding Subrules (1) and (2), where the cable is run as concealed wiring such that it is impracticable to support it, the cable shall be permitted to be fished and need not be supported between boxes and fittings.

12-512 Not to be embedded (see Appendix G)

The cable shall not be buried in plaster, cement, or similar finish.

12-514 Protection on joists and rafters (see Appendix G)

Cables shall not be run on or across

(a)   the upper faces of ceiling joists or the lower faces of rafters in attic or roof spaces, where the vertical distance between the joists and the rafters exceeds 1 m; or

(b)   the lower faces of basement joists, unless suitably protected from mechanical injury.

12-516 Protection for cable in concealed installations (see Appendix G)

(1)   Where the cable is run through studs, joists, or similar members, the outer surfaces of the cable shall be kept distant at least 32 mm from the edges of the members or the cable shall be effectively protected from mechanical injury.

Δ (2)   Where the cable is run through or along metal studs, joists, sheathing, or cladding, it shall be

     (a)   located so as to be effectively protected from mechanical injury both during and after installation;

     (b)   protected where it passes through a member by an insert approved for the purpose and adequately secured in place.

(3)   Where the cable is installed immediately behind a baseboard, it shall be effectively protected from mechanical injury from driven nails.

12-518 Protection for cable in exposed installations (see Appendix G)

Cable used in exposed wiring shall be adequately protected against mechanical damage where it passes through a floor, where it is less than 1.5 m above a floor, or where it is exposed to mechanical damage.

12-520 Fished cable installation

Where the cable is used in concealed wiring and it is impracticable to provide the supports required by Rule 12-510, the cable shall be permitted to be fished.

12-522 Where outlet boxes are not required

(1)   Where the cable is exposed, switch, outlet, and tap devices of insulating material shall be permitted to be used without boxes.

(2)   The openings in the devices shall fit closely around the outer covering of the cable.

(3)   The device shall fully enclose any part of the cable from which any part of the covering has been removed.

(4)   Where the conductors are connected to the devices by binding-screw terminals, there shall be as many screws as there are conductors unless the cables are clamped within the device.

12-524 Types of boxes and fittings

(1)   Boxes and fittings shall be of a type for use with non-metallic-sheathed cable.

(2)   Where grounded metal boxes are not required by these Rules, outlet and switch boxes shall be permitted to be of fire-resisting moulded composition insulating material, furnished with a cover of the same material.

12-526 Provision for bonding

Where non-metallic-sheathed cable is used, provision for bonding to ground shall be made in accordance with the Section 10 requirements.

Armoured cable

12-600 Armoured cable work Rules

Rules 12-602 to 12-618 apply only to armoured cable work.

12-602 Use (see Appendix B)

(1)   Armoured cable shall be permitted to be installed in or on buildings or portions of buildings of either combustible or non-combustible construction.

(2)   Armoured cable shall be of the type listed in Table 19 as suitable for direct burial if used

    (a)   for underground runs;

    (b)   for circuits in masonry or concrete provided that the cable is encased or embedded in at least 50 mm of the masonry or concrete; or

    (c)   in locations where it will be exposed to weather, continuous moisture, excessive humidity, or to oil or other substances having a deteriorating effect on the insulation.

(3)   Notwithstanding Subrule (2), armoured cable that has the armouring made wholly or in part of aluminum shall not be embedded in concrete containing reinforcing steel unless

    (a)   the concrete is known to contain no chloride additives; or

    (b)   the armour has been treated with a bituminous base of paint or other means to prevent galvanic corrosion of the aluminum.

(4)   Where armoured cables are laid in or under cinders or cinder concrete, they shall be protected from corrosive action by a grouting of non-cinder concrete at least 25 mm thick entirely surrounding them unless they are 450 mm or more under the cinders or cinder concrete.

(5)   In buildings of non-combustible construction, armoured cables having conductors not larger than No. 10 AWG copper or aluminum shall be permitted to be laid on the face of the masonry or other material of which the walls and ceiling are constructed and shall be permitted to be buried in the plaster finish for extensions from existing outlets only.

12-604 Protection for armoured cables in lanes

If subject to mechanical injury and unless otherwise protected, steel guards of not less than No. 10 MSG, adequately secured, shall be installed to protect armoured cables less than 2 m above grade in lanes and driveways.

12-606 Use of thermoplastic-covered armoured cable

Armoured cable of the type listed in Table 19 as suitable for direct earth burial and that has a thermoplastic outer covering shall be used only where the outer covering will not be subjected to mechanical injury.

12-608 Continuity of armoured cable

The armour of cables shall be mechanically and electrically continuous throughout and shall be mechanically and electrically secured to all equipment to which it is attached.

12-610 Terminating armoured cable

(1)   Where conductors issue from armour, they shall be protected from abrasion by bushings of insulating material or equivalent devices.

(2) Where conductors are No. 8 AWG or larger, copper or aluminum, such protection shall consist of
   (a) insulated-type bushings, unless the equipment is equipped with a hub having a smoothly rounded throat; or
   (b) insulating material fastened securely in place that will separate the conductors from the armoured cable fittings and afford adequate resistance to mechanical injury.
(3) Where armoured cable is fastened to equipment, the connector or clamp shall be of such design as to leave the insulating bushing or its equivalent visible for inspection.
(4) Where conductors connected to open wiring issue from the ends of armouring, they shall be protected with boxes or with fittings having a separately bushed hole for each conductor.

12-612 Proximity to knob-and-tube and non-metallic-sheathed cable systems
Where armoured cable is used in a building in which concealed knob-and-tube wiring or concealed non-metallic-sheathed cable wiring is installed, the cable shall not be fished if there is a possibility of damage to the existing wiring.

12-614 Radii of bends in armoured cables
(1) Where armoured cables are bent during installation, the radius of the curve of the inner edge of the bends shall be at least 6 times the external diameter of the armoured cable.
(2) Bends shall be made without undue distortion of the armour and without injury to its inner or outer surfaces.

12-616 Concealed armoured cable installation
(1) Where armoured cable is run through studs, joists, or other members, it shall be
   (a) located so that its outer circumference is at least 32 mm from the nearest edge of the members; or
   (b) protected from mechanical injury where it passes through the holes in the members.
(2) Where armoured cable is installed immediately behind baseboards, it shall be protected from mechanical injury from driven nails.

12-618 Running of cable between boxes, etc.
Armoured cable shall be supported between boxes and fittings in accordance with Rule 12-510.

Mineral-insulated and aluminum-sheathed cable

12-700 Mineral-insulated and aluminum-sheathed cable Rules
Rules 12-702 to 12-716 cover the installation of mineral-insulated and aluminum-sheathed cable and amend the other Rules of this Code where they apply.

12-702 Use
(1) Mineral-insulated cable and aluminum-sheathed cable shall be permitted to be installed in or on buildings or portions of buildings of either combustible or non-combustible construction.
(2) Lightweight mineral-insulated cable shall be used only in multi-conductor assemblies.

12-704 Use when embedded
(1) Mineral-insulated cable and round aluminum-sheathed cable, except as noted in Subrule (3), shall be permitted to be used for underplaster extensions from existing outlets only or when encased or embedded in at least 50 mm of masonry or poured concrete.
(2) Except as noted in Subrule (3), flat two-conductor aluminum-sheathed cable shall be permitted to be used for underplaster extensions from existing outlets only or, where a deviation has been allowed in accordance with Rule 2-030, embedded in masonry or concrete.
(3) Cable having an aluminum sheath shall not be embedded in concrete containing reinforcing steel unless
   (a) the concrete is known to contain no chloride additives; or
   (b) the sheath has been treated with a bituminous base paint or other means to prevent galvanic corrosion of the aluminum.

12-706 Method of supporting
(1) Mineral-insulated and aluminum-sheathed cable shall be securely supported by staples, straps, hangers, or similar fittings in such a manner as not to
   (a) injure the sheath of the cable; or
   (b) subject the cable or its termination fittings to undue strain.
(2) Mineral-insulated and aluminum-sheathed cable shall be secured at intervals not exceeding 2 m except where the cable is fished and adequate supports are installed, if needed, adjacent to termination fittings.

(3) When settlement of a structure may occur due to weight of contents, as in certain grain storage occupancies, provision shall be made so that mineral-insulated and aluminum-sheathed cable runs, including their termination fittings, will not be subjected to undue strain.

(4) Mineral-insulated and aluminum-sheathed cable shall be permitted to be run on the surface of walls, partitions, ceilings, or on or across structural members subject to the applicable requirements of Rule 12-710.

12-708 Direct earth burial
Mineral-insulated cable having an aluminum outer sheath and aluminum-sheathed cable in direct contact with the earth shall be provided with a non-metallic jacket or other corrosion-resisting covering.

12-710 Mechanical protection
(1) Where subject to mechanical injury, mineral-insulated and aluminum-sheathed cable shall be suitably protected.

(2) Where mineral-insulated or aluminum-sheathed cable is installed on the face of a wall, partition, ceiling, or structural member within 1.5 m of the floor and in all locations where subject to mechanical injury, as for instance from industrial tractors, other vehicles, equipment, stockpiling, or excessive vibration, a suitable safeguard against such injury shall be provided.

(3) Mineral-insulated or aluminum-sheathed cable shall be protected, located, or arranged so that a 2-1/2-in common nail cannot be driven into it where the cable is
  (a) run through bored or notched holes or grooves in wooden structural members;
  (b) secured directly to the underside of wooden flooring; or
  (c) located behind baseboards or casings.

(4) In order to comply with Subrule (3), the hole, groove, or supporting strap containing the cable shall be permitted to be sufficiently oversized to permit the cable to move a distance equal to at least the radius of the cable.

(5) Where mineral-insulated or aluminum-sheathed cable passes from a point above grade to direct earth burial and is not otherwise protected against mechanical injury, a suitable pipe stub-up shall be arranged to encase the cable to a point, where practicable, at least 300 mm above grade and, in locations where frost heaving may occur, the encasement shall slide freely on the cable so as to avoid damage.

12-712 Radii of bends
(1) The radius of the curve on the inner edge of bends made on mineral-insulated cable shall be not less than 6 times the external diameter of the sheath and shall be made so as not to damage the outer sheath.

(2) The radius of the curve on the inner edge of bends made on smooth aluminum-sheathed cable shall be not less than
  (a) ten times the external diameter of the sheath for cable not more than 19 mm in external diameter;
  (b) twelve times the external diameter of the sheath for cable more than 19 mm but not more than 38 mm in external diameter; and
  (c) fifteen times the external diameter of the sheath for cable more than 38 mm in external diameter.

(3) The radius of the curve on the inner edge of bends made on corrugated aluminum-sheathed cable shall be not less than 9 times the external diameter of the sheath.

12-714 Termination of mineral-insulated cable (see Appendix B)
At all points where mineral-insulated cable terminates,
(a) the end of the cable shall be sealed immediately after stripping to prevent entrance of moisture to the insulation;
(b) each conductor extended beyond the sheath shall be provided with the proper insulation; and
(c) mineral-insulated cable box connectors shall be used.

12-716 Connection to other forms of wiring
Where mineral-insulated or aluminum-sheathed cable is connected to other forms of wiring, the junction shall be made in a box or at, or in, a fitting having a separately bushed hole for each conductor.

Flat conductor cable Type FCC

12-800 Type FCC under-carpet wiring system Rules
Rules 12-802 to 12-824 apply only to the installation of Type FCC under-carpet wiring systems.

12-802 Special terminology (see Appendix B)

In this subsection the following definitions apply:

Bottom shield — a protective layer that is between the floor and the Type FCC cable to protect the cable from physical damage.

Insulating end — an insulator designed to electrically insulate the exposed ends of Type FCC cables.

Metal tape — a metal overlay to prevent physical damage to the Type FCC system.

Top shield — an electrically conductive covering for under-carpet components of a Type FCC system that provides a degree of protection against physical damage and electric shock and may or may not be incorporated as an integral part of a Type FCC cable assembly.

Transition assembly — an assembly specifically approved for the purpose of connecting a Type FCC system to other types of wiring systems.

Type FCC cable — a cable consisting of 3 or more flat separated conductors laid flat and parallel in the same plane and enclosed within an insulating assembly.

Type FCC cable connector — a device used for joining Type FCC cables, with or without the use of a junction box.

Type FCC system — a complete wiring system for installation only under carpet squares and includes cable and associated fittings.

12-804 Use permitted

Type FCC systems shall be permitted to be used only for the extension of general-purpose and appliance branch circuits

(a) in dry or damp locations;

(b) on hard, smooth, continuous floor surfaces made of concrete, ceramic, or composition flooring, wood, or similar materials; and

(c) on floors heated in excess of 30 °C when the FCC system is marked for the purpose.

12-806 Use prohibited

Type FCC systems shall not be used

(a) outdoors or in wet locations;

(b) where subject to corrosive vapours or liquids;

(c) in dwelling units;

(d) in schools, hospitals, or institutional buildings except in office areas;

(e) on walls except where entering transition assemblies;

(f) under permanent-type partitions or walls;

(g) where the voltage exceeds 150 volts-to-ground or 300 V between any two conductors; or

(h) for branch circuits exceeding 3 A.

12-808 Floor covering

Floor-mounted Type FCC cable with associated steel tape, shielding cable connections, and insulating ends shall be covered with carpet squares not exceeding 750 mm and any adhesive used shall be of the release type.

12-810 Connections and terminations

(1) Type FCC cable connections shall be installed so that electrical continuity, insulation, and sealing against dampness and liquid spillage are provided.

(2) Bare ends shall be insulated and sealed by the use of insulating ends.

12-812 Shields

(1) Type FCC systems shall include a bottom shield.

(2) A metal top shield shall be installed over floor-mounted Type FCC cable, connectors, and insulating ends.

12-814 Enclosure and shield continuity (see Appendix B)

Metal shields, tapes, boxes, receptacle housings, and self-contained devices shall be electrically continuous and bonded to ground.

12-816 Connection to other systems

Power feed, bonding, and shield system connections between the Type FCC system and other wiring systems shall be accomplished in a transition assembly intended for surface or recessed mounting.

12-818 Anchoring

Type FCC system components shall be firmly secured to floors and walls by means of

(a)　an adhesive in the case of cables; and

(b)　mechanical fasteners in the case of associated fittings such as outlet boxes and transition assemblies.

12-820 Crossings

A Type FCC cable run shall be permitted to cross over or under another Type FCC cable run or communication flat cable, provided that there is a layer of metal shielding between each of the cables.

12-822 Mechanical protection

(1)　All Type FCC systems installed under carpet squares shall be protected from physical damage by a metal tape completely covering the Type FCC cable and connections.

(2)　Where surface or recessed wall mounting of the Type FCC cable is required to enter transition assemblies, additional mechanical protection shall be provided to prevent damage from items such as nails and screws.

12-824 System height

Except as permitted by Rule 12-820, stacked runs of flat conductor cable shall not be permitted.

Raceways

General

12-900 Raceway Rules

Rules 12-902 to 12-942 apply to raceways and to conductors run in raceways.

12-902 Types of conductors

Conductors shall be of types suitable for use in raceways as indicated in Table 19.

12-904 Conductors in raceways

(1)　Where conductors are placed in metal raceways, all conductors of a circuit shall be contained in the same raceway, or in the same channel of a multiple-channel raceway except that, where it is necessary to run conductors in parallel due to the capacity of an alternating-current circuit, additional enclosures shall be permitted to be used, provided that

(a)　the conductors are installed in accordance with Rule 12-108(1);

(b)　each enclosure includes an equal number of conductors from each phase and the neutral; and

(c)　each enclosure or cable sheath is of the same material and has the same physical characteristics.

(2)　No raceway or compartment of a multiple-channel raceway shall contain conductors that are connected to different power or distribution transformers or other different sources of voltage, except where the conductors

(a)　are separated by the metal armour or metal sheath of cable assemblies of the types listed in Table 19;

(b)　are separated by a barrier of sheet steel not less that 1.34 mm (No. 16 MSG) thick or a flame-retardant non-metallic insulating material not less than 1.5 mm in thickness; or

(c)　are used for the supply and/or control of remote devices, are insulated for at least the same voltage as that of the circuit having the highest voltage, and none of the conductors of the circuits of lower voltages is directly connected to a lighting branch circuit.

12-906 Protection of conductors at ends of raceways

(1)　Bushings or equivalent means shall be used to protect conductors from abrasion where they issue from raceways.

(2)　Where conductors are No. 8 AWG or larger, copper or aluminum, such protection shall consist of

(a)　insulated-type bushings, unless the equipment is equipped with a hub having a smoothly rounded throat; or

(b)　insulating material fastened securely in place that will separate the conductors from the raceway fittings and afford adequate resistance to mechanical injury.

12-908 Inserting conductors in raceways

(1)　Cleaning agents or lubricants of an electrical conducting nature or that might have a deleterious effect on conductor coverings shall not be used when inserting conductors in raceways.

(2)　Lubricants used when inserting conductors in raceways shall be either wire pulling compound, talc, or soapstone.

12-910 Joints or splices within raceways

There shall be no joints or splices in conductors or cables within raceways except in the case of busways, wireways, cable trays, and surface raceways with removable covers.

12-912 Stranding of conductors

Except in the case of conductors used as busbars and mineral-insulated cables, single- or multiple-conductor cables No. 8 AWG or larger, copper or aluminum, when installed in raceways, shall be stranded.

12-914 Electrical continuity of raceways

Metal raceways shall be electrically continuous throughout and electrically secured to all equipment to which they are attached.

12-916 Mechanical continuity of raceways

Raceways shall be mechanically continuous throughout and mechanically secured to all equipment to which they are attached.

12-918 Support of raceways

Raceways shall be supported independently of equipment forming part of the raceway system.

12-920 Removal of fins and burrs of raceways

Fins and burrs shall be removed from the ends of raceways.

12-922 Radii of bends in raceways

Δ (1) Where conductors are drawn into a raceway, the radius of the curve to the centre line of any bend shall be not less than as shown in Table 7.

(2) Bends shall be made without undue distortion of the raceway and without injury to its inner or outer surfaces.

12-924 Junction of open wiring and raceways

Where conductors connected to open wiring issue from the ends of raceways, they shall be protected with boxes or with fittings having a separately bushed hole for each conductor.

12-926 Entry of underground conduits into buildings

Where a conduit enters a building from an underground distribution system, the end of the conduit within the building shall be sealed with a suitable compound to prevent the entrance of moisture and gases.

12-928 Raceways installed underground or where moisture may accumulate

(1) The requirements for Category 1 locations as specified in Section 22 shall be complied with where raceways are installed
  (a) underground;
  (b) in concrete slabs or other masonry in direct contact with moist earth; or
  (c) in other locations where the conductors are subject to moisture.

(2) Where lead-sheathed conductors are used in such locations, a pothead or equivalent device shall be used to protect them from moisture and mechanical injury at their point of issue from the lead sheathing.

12-930 Metal raceways in plaster

In buildings of non-combustible construction where circuits run in metal raceways have conductors not larger than No. 10 AWG copper or aluminum, the circuits shall be permitted to be laid on the face of the masonry or other material of which the walls and ceiling are constructed and shall be permitted to be buried in the plaster finish.

12-932 Protection for raceways in lanes

If subject to mechanical injury and unless otherwise protected, steel guards of not less than No. 10 MSG, adequately secured, shall be installed to protect raceways less than 2 m above grade in lanes and driveways.

12-934 Raceways installed in concrete, cinder concrete, and cinder fill (see Appendix B)

(1) Raceways made wholly or in part of aluminum shall not be embedded in concrete containing reinforcing steel unless
  (a) the concrete is known to contain no chloride additives; or
  (b) the raceway has been treated with a bituminous base paint or other means to prevent galvanic corrosion of the aluminum.

(2)   Where metal raceways are laid in or under cinders or cinder concrete, they shall be protected from corrosive action by a grouting of non-cinder concrete at least 25 mm thick entirely surrounding them unless they are 450 mm or more under the cinders or cinder concrete.

12-936 Raceway completely installed before conductors are installed
(1)   Raceways shall be installed as a complete system before the conductors or cables are installed in them.
(2)   Conductors or cables shall not be drawn into or laid in raceways in a building under construction until the raceway fittings and conductors are reasonably safe from damage due to construction operations.

12-938 Capping of unused raceways
Spare or unused raceways that terminate in enclosures shall be capped.

12-940 Maximum number of bends in raceways
Where it is intended that conductors are to be drawn into a raceway, a run of raceway between outlets or draw-in points shall not have more than the equivalent of four 90° bends including the bends located at an outlet or fitting.

12-942 Metal raceways (see Appendix B)
Electrical metal raceways embedded in parking lot slabs or pavement, road beds, and similar areas subject to vehicular traffic shall comply with the requirements of Rule 2-112(1).

Rigid and flexible metal conduit

12-1000 Rigid and flexible metal conduit Rules
Rules 12-1002 to 12-1014 apply only to the installation of rigid and flexible metal conduit.

12-1002 Use
(1)   Rigid and flexible metal conduit shall be permitted to be installed in or on buildings or portions of buildings of either combustible or non-combustible construction.
(2)   Rigid metal conduit used in damp or wet locations shall be threaded and the joints and fittings shall be made watertight.

12-1004 Minimum size of conduits
No conduits having an internal diameter of less than 16 trade size shall be used except that
(a)   12 trade size flexible metal conduit shall be permitted to be used for runs of not more than 1.5 m for the connection of equipment; and
(b)   12 trade size liquid-tight flexible conduit may be used as permitted by this Code.

12-1006 Conduit threads (see Appendix B)
(1)   Threads of rigid metal conduit shall be tapered.
(2)   External threads performed in the field shall comply with Table 40, using a standard chaser with a taper of 1 to 16.
(3)   Running threads shall not be permitted.
(4)   Notwithstanding Subrule (3), where rigid metal conduit protrudes through the enclosure wall and there are not sufficient threads to accommodate a bushing per Rule 12-906(1), additional threading shall be permitted on the conduit as a continuation of the tapered thread beyond those dimensions specified in Table 40.

12-1008 Thread engagement
The wall thickness of boxes to be drilled and tapped in the field shall be sufficient to ensure thread engagement of at least three complete threads.

12-1010 Maximum spacing of conduit supports
(1)   All rigid metal conduit of one size shall be securely attached to hangers or to a solid surface with the maximum spacings of the points of support not greater than
  (a)   1.5 m for 16 and 21 trade size conduit;
  (b)   2 m for 27 and 35 trade size conduit; and
  (c)   3 m for 41 trade size conduit and larger.
(2)   Where rigid metal conduits of mixed sizes are run in a group, the conduit supports shall be arranged so that the maximum support spacing will be that shown in Subrule (1) for the smallest conduit.

(3) When flexible metal conduit is installed, it shall be secured at intervals not exceeding 1.5 m and within 300 mm on each side of every outlet box or fitting except where flexible metal conduit is fished and except for lengths of not over 900 mm at terminals where flexibility is necessary.

12-1012 Expansion and contraction of conduits (see Appendix B)

(1) In locations subject to extreme temperature changes, provision shall be made for expansion and contraction in long runs of rigid conduit in the form of
  (a) approved expansion joints; or
  (b) in the case of surface-mounted rigid metal conduit only, two 90° bends in the conduit run.
(2) If expansion joints are used with metal raceways, bonding jumpers shall be provided in accordance with Rule 10-614.

12-1014 Conductors in conduit (see Appendix B)

(1) Conduits shall be of sufficient size to permit the conductors to be drawn in and withdrawn without injury to the conductors.
(2) Subrules (3), (4), and (5) refer only to complete systems and not to short sections of conduit used for the protection of portions of open wiring that would otherwise be exposed to mechanical injury.
(3) The maximum number of conductors in one conduit shall not exceed 200.
(4) The maximum number of conductors or multi-conductor cables in one conduit shall be such that the conductors or cables and their coverings will not result in a greater conduit fill than that specified in Table 8, and in this determination
  (a) the interior cross-sectional area for various sizes of conduit shall be those specified in Table 9;
  (b) notwithstanding the requirements of Item (a), the interior resulting cross-section of raceways shall be permitted to be derived from their measured internal dimensions or from manufacturer's listed specifications;
  (c) the diameters and cross-sectional areas of single-conductor bare and insulated wires and multiple-conductor cables shall be obtained by measurement; and
  (d) notwithstanding the requirements of Item (c), the dimensions of single-conductor wires shall be permitted to be obtained from Table 10 for the constructions identified therein.
(5) Notwithstanding the requirements of Subrule (4), the maximum permitted number of conductors of the same size in one conduit shall be permitted to be determined from Table 6 for single conductors of the appropriate construction listed therein.

Rigid PVC conduit and rigid HFT conduit

12-1100 Use

(1) Rigid PVC and HFT conduit shall be permitted for exposed and concealed work above and below ground in accordance with the Rules for threaded rigid metal conduit subject to the provisions of Rules 2-128 and 12-1102 to 12-1122.
(2) Rigid PVC and HFT conduit shall be permitted in cinders or cinder concrete without the grouting referred to in Rule 12-934 being required.

12-1102 Restrictions on use

Rigid PVC conduit shall not be used where enclosed in thermal insulation.

12-1104 Temperature limitations (see Appendix B)

(1) Rigid PVC conduit shall not be used where normal conditions are such that any part of the conduit is subjected to a temperature in excess of 75 °C.
(2) Subrule (1) shall not prevent the use of insulated conductors having temperature ratings in excess of 75 °C, but such conductors shall not have ampacities exceeding those of 90 °C conductors, regardless of their temperature rating.
(3) Rigid HFT conduit shall not be used where normal conditions are such that any part of the conduit is subjected to a temperature in excess of 125 °C.

12-1106 Mechanical protection

Rigid PVC and HFT conduit shall be protected where exposed to mechanical injury either during installation or afterwards.

12-1108 Field bends (see Appendix B)
(1) Rigid PVC conduit shall be permitted to be bent in the field, provided that bending equipment specifically intended for the purpose is used.
(2) The minimum bending radius shall comply with Rule 12-922.
(3) Rigid HFT conduit shall not be bent in the field.

12-1110 Support of luminaires
Rigid PVC boxes shall not be used for the support of luminaires unless they are marked as being suitable for the purpose.

12-1112 Fittings
(1) Rigid PVC and HFT conduit including elbows and bends shall not be threaded but shall be used with adapters and couplings, which shall be applied with solvent cement in the case of PVC and contact cement in the case of HFT.
(2) Female threaded PVC or HFT adapters shall be used together with a metal conduit nipple to terminate at threaded conduit entries in metal enclosures.

12-1114 Maximum spacing of conduit supports
(1) All rigid PVC and HFT conduit of one size shall be securely attached to hangers or to a solid surface with the maximum spacing of the points of supports not greater than
   (a) 750 mm for 16, 21, and 27 trade size conduit;
   (b) 1.2 m for 35 and 41 trade size conduit;
   (c) 1.5 m for 53 trade size conduit;
   (d) 1.8 m for 63 and 78 trade size conduit;
   (e) 2.1 m for 91, 103, and 129 trade size conduit; and
   (f) 2.5 m for 155 trade size conduit.
(2) Where conduits of mixed sizes are run in a group, the conduit supports shall be arranged so that the maximum support spacing will be that shown in Subrule (1) for the smallest conduit.
(3) Except where encased or embedded in at least 50 mm of masonry or poured concrete, conduit shall not be clamped tightly but shall be supported in such a manner as to permit adequate lineal movement to allow for expansion and contraction due to temperature change.

12-1116 Support of equipment
Rigid PVC and HFT conduit shall not be used to support fixtures or other equipment except as permitted by Rule 12-3012(2).

12-1118 Expansion joints (see Appendix B)
Unless the conduit is grouted in concrete, at least one expansion joint shall be installed in any conduit run where the expansion of the conduit due to the maximum probable temperature change during and after installation will exceed 45 mm.

12-1120 Maximum number of conductors
The maximum number of conductors in rigid PVC or HFT conduit shall be determined in accordance with Rule 12-1014.

12-1122 Provision for bonding continuity
A separate bonding conductor shall be installed in rigid PVC and HFT conduit in compliance with Rule 10-404.

Rigid Types EB1 and DB2/ES2 PVC conduit

12-1150 Use permitted (see Appendix B)
Rigid Types EB1 and DB2/ES2 PVC conduit and fittings shall be permitted to be used
(a) for installation underground in accordance with Rule 12-928 except that Type EB1 conduit shall be laid with its entire length encased or embedded in at least a 50 mm envelope of masonry or poured concrete; or
(b) in walls, floors, and ceilings where encased or embedded in at least 50 mm of masonry or poured concrete.

12-1152 Restrictions in use
Rigid Types EB1 and DB2/ES2 conduit and fittings shall not be used above ground except as permitted by Rule 12-1150(b).

12-1154 Temperature limitations (see Appendix B)
Temperature limitations shall comply with Rule 12-1104.

12-1156 Field bends (see Appendix B)
Field bends shall comply with Rule 12-1108.

12-1158 Fittings (see Appendix B)
(1) Rigid Types EB1 and DB2/ES2 PVC conduit including elbows, bends, and other fittings fabricated from rigid Type EB1 and DB2/ES2 PVC conduit shall not be threaded.
(2) Notwithstanding Subrule (1), threaded adapters, acceptable for use in making threaded connections when properly attached to the conduit, shall be permitted to be used.

12-1160 Maximum number of conductors
The maximum number of conductors in rigid Types EB1 and DB2/ES2 PVC conduit shall be in accordance with Rule 12-1014.

12-1162 Method of installation
(1) All cut edges shall be trimmed to remove rough edges.
(2) All joints between conduit lengths and between conduit lengths and bends, adapters, or separate couplings shall be made by a method specified for the purpose.
(3) Rigid Types EB1 and DB2/ES2 PVC conduit shall be secured mechanically to prevent disturbance of their alignment during construction.

12-1164 Split straight conduit
In existing underground or concrete embedded installations only, raceways shall be permitted to be formed using split straight conduit provided that
(a) both halves of each conduit length are properly matched and clamped together to form a close-fitting concrete-tight joint;
(b) each length of conduit is tightly clamped at each end, with additional clamps spaced not more than 900 mm apart; and
(c) clamps made of stainless steel or other corrosion-resistant material are used when not embedded in concrete.

12-1166 Provision for bonding continuity
A separate bonding conductor shall be installed in rigid Types EB1 and DB2/ES2 conduit in compliance with Rule 10-404.

Rigid RTRC conduit

12-1200 Scope
Rules 12-1202 to 12-1220 apply only to the installation of rigid RTRC conduit Type AG and Type BG.

12-1202 Use
(1) Rigid RTRC conduit Type AG and Type BG shall be permitted to be installed
  (a) underground in accordance with Rule 12-012; and
  (b) in walls, floors, and ceilings where encased or embedded in at least 50 mm of masonry or poured concrete.
(2) Rigid RTRC conduit Type AG shall, in addition to the locations permitted in Subrule (1), be permitted for exposed and concealed locations.

12-1204 Restrictions on use (see Appendix B)
Rigid RTRC conduit shall not be used in buildings required to be of non-combustible construction, unless it has a flame spread rating and smoke developed classification as specified in the *National Building Code of Canada*.

12-1206 Mechanical protection
Rigid RTRC conduit shall be provided with mechanical protection where exposed to damage either during installation or afterwards.

12-1208 Field bends
Rigid RTRC conduit shall not be bent in the field.

12-1210 Temperature limitations
Rigid RTRC conduit shall not be used where normal conditions are such that any part of the conduit is subjected to a temperature in excess of 110 °C.

12-1212 Fittings
Rigid RTRC conduit shall not be threaded but shall be used with adapters and couplings specifically designed for the purpose.

12-1214 Expansion joints (see Appendix B)
Except where encased in concrete, at least one expansion joint shall be installed in any conduit run where the expansion of the conduit due to the maximum probable temperature change during and after installation will exceed 45 mm.

12-1216 Conduit supports
Where rigid RTRC conduit Type AG is run in accordance with Rule 12-1202(2), it shall be supported with hangers or clamps
(a)   in such a manner as to permit adequate linear movement to allow for expansion and contraction due to temperature change; and
(b)   with the spacings of the supports not greater than permitted by Rule 12-1010.

12-1218 Maximum number of conductors
The maximum number of conductors in rigid RTRC conduit shall be determined in accordance with Rule 12-1014.

12-1220 Provision for bonding
A separate bonding conductor shall be installed in rigid RTRC conduit in compliance with Rule 10-404.

Liquid-tight flexible conduit

12-1300 Scope
Rules 12-1302 to 12-1306 apply only to liquid-tight flexible conduit.

12-1302 Use of liquid-tight flexible conduit
(1)   Liquid-tight flexible conduit shall be permitted where a flexible connection is required in dry, damp, or wet locations and where permitted by other Sections of this Code.
(2)   Runs of not more than 1.5 m of 12 trade size liquid-tight flexible conduit shall be permitted for the connection of equipment.
(3)   Liquid-tight flexible conduit shall not be used
(a)   where subject to mechanical damage;
(b)   as a general-purpose raceway;
(c)   in lengths greater than that essential for the degree of flexibility required;
(d)   where exposed to gasoline or similar light petroleum solvents, corrosive liquids, or vapours having an injurious effect on the outer jacket;
(e)   under conditions such that the temperature will exceed 60 °C unless marked for a higher temperature; or
(f)   where flexing at low temperatures may cause injury.

12-1304 Maximum number of conductors
(1)   The maximum number of conductors in liquid-tight flexible conduit shall be in accordance with Rule 12-1014.
(2)   For the purposes of Subrule (1), the cross-sectional area of 12 trade size shall be considered as 118 mm^2.

12-1306 Provisions for bonding
A separate bonding conductor shall be installed in liquid-tight flexible conduit in accordance with Section 10.

Electrical metallic tubing

12-1400 Electrical metallic tubing Rules
Rules 12-1402 to 12-1410 apply only to electrical metallic tubing.

12-1402 Use
(1)   Electrical metallic tubing shall be permitted to be used for exposed and concealed work except that it shall not be used
(a)   where it will be subject to mechanical injury either during installation or afterwards;
(b)   where exposed to corrosive vapour except as permitted by Rule 2-112;
(c)   for direct earth burial;

   (d)   in wet locations; and

   (e)   in concrete or masonry slabs in contact with the earth, unless a separate bonding conductor is installed in the tubing.

(2)   Electrical metallic tubing shall be permitted to be installed in or on buildings or portions of buildings of either combustible or non-combustible construction.

12-1404 Supports

Electrical metallic tubing shall be installed as a complete system and shall be securely fastened in place within 1 m of each outlet box, junction box, cabinet, coupling, or fitting, and the spacing between supports shall be in accordance with those specified in Rule 12-1010.

12-1406 Minimum tubing size

The tubing shall have an inside diameter of not less than 16 trade size tubing.

12-1408 Maximum number of conductors

A tube shall not contain more conductors of a given size than are specified in Rule 12-1014.

12-1410 Connections and couplings

Where lengths of electrical metallic tubing are coupled together or connected to boxes, fittings, or cabinets, fittings shall be

   (a)   of the concrete-tight type for installation in poured concrete or in masonry block walls in which cores are filled with concrete or grout;

   (b)   of the rain-tight type for installations exposed to the weather; and

   (c)   of the standard, concrete-tight, or rain-tight type for installation in ordinary locations or buried in plaster or masonry block walls.

Electrical non-metallic tubing

Δ **12-1500 Use**

Subject to the provisions of Rules 2-128 and 12-1502 to 12-1514, the installation of electrical non-metallic tubing shall be permitted

   (a)   underground in accordance with Rule 12-012; and

   (b)   in exposed or concealed locations.

Δ **12-1502 Restriction on use**

Electrical non-metallic tubing shall not be used unless provided with mechanical protection where subject to damage either during or after construction.

12-1504 Supports

Electrical non-metallic tubing shall be securely fastened in place within 1 m of each outlet box, junction box, cabinet, coupling, or fitting, and the spacing between supports shall be not more than 1 m.

12-1506 Maximum number of conductors

A tube shall not contain more conductors of a given size than are specified in Rule 12-1014.

12-1508 Temperature limitations (see Appendix B)

(1)   Electrical non-metallic tubing shall not be used where normal conditions are such that any part of the tubing is subjected to a temperature in excess of 75 °C.

(2)   Subrule (1) shall not prevent the use of insulated conductors having temperature ratings in excess of 75 °C, but such conductors shall not have ampacities exceeding those of 90 °C conductors regardless of their temperature rating.

12-1510 Connections and couplings

(1)   Where lengths of electrical non-metallic tubing are coupled together or connected to boxes, fittings, or cabinets, fittings designed for the purpose shall be used.

(2)   Where lengths of electrical non-metallic tubing are coupled together underground, the couplings shall be applied using a solvent cement suitable for the purpose.

12-1512 Support of equipment

Electrical non-metallic tubing shall not be used to support fixtures or other equipment.

12-1514 Provision for bonding continuity

A separate bonding conductor shall be installed in electrical non-metallic tubing in compliance with Rule 10-404.

Surface raceways

12-1600 Scope
Rules 12-1602 to 12-1614 apply only to surface raceways.

12-1602 Use of surface raceway (see Appendix B)
(1) Surface raceways shall be permitted only for exposed surface installation in dry locations.
(2) Notwithstanding Subrule (1), surface raceways shall be permitted to extend through walls, partitions, and floors provided that
   (a) the raceways are in unbroken lengths where passing through; and
   (b) provisions are made for removing the caps or covers on all exposed portions.
(3) Surface raceways shall not be used where subject to mechanical damage.
(4) Non-metallic surface raceways shall conform with Rule 2-128.

12-1604 Temperature limitations
(1) Surface raceways shall not be used where subject to ambient temperatures in excess of 50 °C unless marked for a higher temperature.
(2) Subrule (1) shall not prevent the use of insulated conductors having temperature ratings in excess of 75 °C, but such conductors shall not have ampacities exceeding those of 75 °C conductors regardless of their temperature ratings.

12-1606 Conductors in surface raceways (see Appendix B)
(1) Conductors shall be of types indicated in Table 19 as being suitable for use in raceways.
(2) The aggregate cross-sectional area of the installed conductors shall not exceed 40% of the minimum available cross-sectional area of the raceway.
(3) The cross-sectional area for conductors in Subrule (2) shall be determined in accordance with Rule 12-1014(4).

12-1608 Maximum voltage
The voltage between conductors contained in surface raceways shall not exceed 300 V unless the raceways are marked for a higher voltage.

12-1610 Joints and splices
Joints and splices shall be permitted in surface raceways having a removable cover that is accessible after installation and shall not fill the raceway to more than 75% of its area at that point.

12-1612 Provisions for bonding
A separate bonding conductor shall be installed in non-metallic surface raceways in compliance with Rule 10-404.

12-1614 Flat cable systems
(1) Flat cables consisting of parallel conductors and side wings formed with integral insulation specifically designed for field installation in metal surface raceways with tap fittings and end cap devices shall be used only
   (a) in branch circuits; and
   (b) in horizontal runs with the conductors uppermost in the raceway.
(2) Metal surface raceways, when used with flat cables, shall be permitted to have covers on the underside omitted when installed out-of-reach.

Underfloor raceways

12-1700 Where underfloor raceways are permitted
(1) Underfloor raceways shall be permitted to be installed under the surface of concrete or other flooring material, but not below the floor.
(2) Underfloor raceways shall not be used
   (a) where they will be exposed to corrosive vapours;
   (b) in commercial garages;
   (c) in storage-battery rooms; or
   (d) on the underside of the floor.

12-1702 Method of installing underfloor raceways

(1)   Underfloor raceways shall be installed in accordance with the manufacturer's instructions in addition to the other requirements of this Rule.

(2)   Underfloor raceways shall be laid so that their centreline coincides with a straight line drawn between the centres of successive junction boxes.

(3)   The raceways shall be mechanically secured to prevent disturbance of the alignment during construction.

(4)   The joints along the edges of the raceways and between the raceways, couplings, and junction boxes, and between the junction box cover plates and cover-rings, shall be filled with waterproof cement.

(5)   The raceways shall be arranged so that there are no low points or traps at the fittings or in the raceway run, and crossings shall be avoided where possible.

12-1704 Fittings for underfloor raceways

(1)   Where underfloor raceways are run at other than right angles, special fittings shall be provided if required.

(2)   The raceways shall be connected to distribution centre and wall outlets by conduit or fittings.

(3)   Dead-ends of the raceways shall terminate in junction boxes or other fittings.

12-1706 Taps and splices in underfloor raceways

Taps and splices in underfloor raceways shall be made only in header access units or in junction boxes.

12-1708 Inserts and junction boxes for underfloor raceways

(1)   Inserts and outlets in underfloor raceways shall be made electrically and mechanically secure.

(2)   Inserts other than the preset type shall be attached to the raceways and, where they are not made mechanically secure by being grouted in separately, they shall not be set until the floor is laid.

(3)   Inserts and junction boxes shall be levelled to the grade of the floor and sealed with watertight plugs.

12-1710 Setting of inserts

When setting inserts or cutting through the walls of underfloor raceways, adequate precautions shall be taken to prevent chips and dirt from falling into the raceway, and special tools designed for the purpose that cannot enter the raceway and injure the conductors shall be used.

12-1712 Discontinued outlets in underfloor raceways

Where an outlet in an underfloor raceway is discontinued, the conductors supplying the outlet shall be removed from the underfloor raceway.

12-1714 Area of conductors in underfloor raceways

(1)   The aggregate cross-sectional area of the conductors and their insulation in an underfloor raceway shall not exceed 40% of the interior cross-sectional area of the raceway.

(2)   Subrule (1) shall not apply where the raceway contains only mineral-insulated cable, aluminum-sheathed cable, armoured cable, or non-metallic-sheathed cable.

(3)   The cross-sectional areas for conductors in Subrule (1) shall be determined in accordance with Rule 12-1014(4).

12-1716 Underfloor raceway junction boxes

Junction boxes shall not be used as outlet boxes in underfloor raceways.

12-1718 Inserts in post- and pre-stressed concrete floors

(1)   Where underfloor distribution raceways are used with post-stressed or pre-stressed poured-in-place floors, they shall be supplied with preset inserts.

(2)   After-set inserts or after-set access units shall not be placed into such a system unless the resulting floor is in compliance with the performance requirements of the *National Building Code of Canada*.

Cellular floors

12-1800 Installation

Cellular floors shall be installed in accordance with the manufacturer's instructions.

12-1802 Conductors in cellular floors

(1)   Conductors shall not be installed in a cellular floor
(a)   where they will be exposed to corrosive vapour;
(b)   in commercial garages; or
(c)   in storage-battery rooms.

(2) Conductors shall not be installed in any cell or header that contains a pipe for steam, water, air, gas, drainage, or other non-electrical service.

(3) Where the cell or header contains such non-electrical services, the cell or header shall be sealed, where practicable.

(4) All conductors of a circuit shall be contained in the same cell of a cellular floor and, except as permitted by Rule 12-3030, the circuits of different systems shall not be contained therein.

12-1804 Maximum conductor size in cellular floors

No conductor larger than No. 0 AWG copper or aluminum shall be installed in a cellular floor unless a deviation has been allowed in accordance with Rule 2-030.

12-1806 Cross-sectional area of cellular floors

(1) Where a cellular floor contains other than mineral-insulated cable, aluminum-sheathed cable, armoured cable, or non-metallic-sheathed cable, the aggregate cross-sectional area of the conductors shall not exceed 40% of the interior area of the header feeding the individual cells.

(2) The cross-sectional areas for conductors in Subrule (1) shall be determined in accordance with Rule 12-1014(4).

12-1808 Taps and splices in cellular floors

Taps and splices in cellular floors shall be made only in header access units or in junction boxes.

12-1810 Cellular floor markers

Where cellular floors are used, a suitable number of markers shall be installed for the future location of cells and for a system identification, and the markers shall extend through the floor.

12-1812 Cellular floor junction boxes

(1) Junction boxes used in cellular floors shall be levelled to floor grade and sealed against the entrance of water.

(2) The junction boxes shall be constructed of metal and shall be electrically continuous with the headers.

(3) Electrical continuity of cellular metal-floor members shall be obtained by spot welding or other equivalent means.

(4) Spot welding shall be done in open spaces between cells and not to the cell walls.

12-1814 Provision for bonding

(1) A separate bonding conductor shall be installed in electrical cells and headers and shall be sized in accordance with Table 16.

(2) Metal headers, cells, and fittings shall be bonded to ground in accordance with Rule 10-500.

12-1816 Cellular floor inserts

(1) Inserts in cellular floors shall be levelled to floor grade and sealed against entrance of water.

(2) Inserts shall be made of metal and shall be electrically continuous with the cellular metal-floor members.

(3) When setting inserts or cutting through cell walls, adequate precautions shall be taken to prevent chips and dirt from falling into the cell and for preventing tools from entering the cells and injuring the conductors within.

12-1818 Cellular floor extensions

Connections from cellular floors to cabinets and extensions from cells to outlets shall be made by means of rigid conduit, flexible metal conduit, or fittings.

12-1820 Cellular floor discontinued outlets

Where an outlet is discontinued, the conductors supplying the outlet shall be removed from the cellular floor.

Auxiliary gutters

12-1900 Where auxiliary gutters are used to supplement wiring spaces

(1) Where auxiliary gutters are used to supplement wiring spaces at meter centres, distribution centres, switchboards, and similar points in interior wiring systems, the gutters shall be permitted to enclose conductors and cables but they shall not be used to enclose busbars, switches, overcurrent devices, or other appliances or apparatus.

(2) The auxiliary gutters shall not extend more than 6 m beyond the equipment that they supplement, and thereafter the conductors shall be permitted to be contained in wireways or busways.

12-1902 Auxiliary gutter supports

Auxiliary gutters shall be securely supported throughout their entire length at intervals of not more than 1.5 m unless the gutter is plainly marked to indicate a greater distance.

12-1904 Auxiliary gutter cross-sectional area

(1)   The aggregate cross-sectional area of the conductors and their insulation at a cross-section of an auxiliary gutter shall not exceed 20% of the cross-sectional area of the gutter at that point.

(2)   A single compartment of an auxiliary gutter shall not contain more than 200 conductors at a cross-section.

(3)   The cross-sectional areas for conductors in Subrule (1) shall be determined in accordance with Rule 12-1014(4).

Busways and splitters

12-2000 Use

(1)   Busways and splitters shall be permitted to be used only for exposed work except as permitted in Subrules (5) and (7).

(2)   Busways and splitters shall not be installed outdoors or in wet or damp locations, unless specifically approved for use in such locations.

(3)   Busways, splitters, and fittings shall not be placed
    (a)   where subject to mechanical injury;
    (b)   where subject to corrosive vapours;
    (c)   in hoistways; or
    (d)   in storage-battery rooms.

(4)   Busways shall be permitted to be used as risers in buildings of non-combustible construction when provided with fire stops in accordance with Rule 2-124.

(5)   Busways shall be permitted in false ceiling spaces, where a deviation has been allowed in accordance with Rule 2-030, provided that
    (a)   ventilation is adequate to prevent development of ambient temperatures in excess of 30 °C; otherwise the rating of the busway shall be reduced to 82%, 71%, and 58% for ambients of 40 °C, 45 °C, or 50 °C respectively, but in no case shall the ambient be higher than 50 °C;
    (b)   any take-off devices located in the false ceiling do not contain overcurrent protection;
    (c)   adequate working space exists between the busway and other services or structural parts;
    (d)   the busway is of the totally enclosed type except that the ventilated type shall be permitted to be used provided that, in addition
        (i)   the busbars are insulated for their full length, including joints between sections, unless provision is made that effectively fully encloses the bare busbars;
        (ii)   the false ceiling is not combustible; and
        (iii)   no combustible material is located within 150 mm of the busway; and
    (e)   if installed in areas used for the building ventilation system, the busway is of the totally enclosed type.

(6)   A splitter with a separate screw or stud for each connection shall be installed, in an accessible location, where two or more conductors are connected to a conductor larger than No. 6 AWG copper or No. 4 AWG aluminum.

(7)   Splitters shall be permitted to be installed flush in a wall provided that they are accessible by removable covers.

12-2002 Extensions from busways and splitters

Rigid conduit, flexible metal conduit, surface raceways, cable trays, electrical metallic tubing, armoured cable, metal-sheathed conductors or cable, or, where necessary, hard-usage cord assemblies shall be used in extensions from busways and splitters and shall be connected to the busway or splitter in a manner appropriate to the material used in accordance with Rule 12-3022.

12-2004 AC circuits in busways and splitters

Where alternating current is used, all conductors of a circuit shall be placed within the same busway, splitter, or section thereof, if the latter is made of magnetic material.

12-2006 Busway and splitter supports

(1)   Busways installed horizontally shall be supported at intervals not greater than 1.5 m unless marked as being suitable for support at greater intervals.

(2)   Busways installed vertically shall be marked as being suitable for vertical installation.

(3) Busways installed vertically shall be supported at each floor and at intervals not greater than 1.5 m unless marked as being suitable for support at greater intervals.

(4) Busways shall be installed so that supports and joints are accessible for maintenance purposes after installation.

(5) Splitters shall be supported at intervals not greater than 1.5 m unless marked as being suitable for support at greater intervals.

12-2008 Method of installation of busways

(1) Where busways extend transversely through dry walls or partitions, they shall pass through the walls or partitions in unbroken lengths and shall be totally enclosed where passing through walls or partitions constructed of combustible material or masonry walls containing voids at the point where the busway passes through.

(2) Busways shall be permitted to extend vertically through floors in dry locations if they are
   (a) totally enclosed where passing through the floor and for the first 300 mm above the floor; and
   (b) provided with fire stops in accordance with Rule 2-124.

(3) Busways shall be provided with adequate protection against mechanical injury and personal contact with live parts for a distance of 2 m above any floor in an area accessible to other than qualified persons.

(4) Dead-ends of busways shall be closed by fittings.

(5) Busways installed outdoors or in parking areas and that are accessible to other than authorized persons shall be of the totally enclosed type.

12-2010 Plug-in devices for busways

When busways supply machine tools, a switch need not be furnished on the machine tool if
(a) a plug-in device having a horsepower rating is used; and
(b) the means of operating the plug-in device is readily within reach of the operator.

12-2012 Reduction in size of busways

Overcurrent protection shall be permitted to be omitted at points where busways are reduced in size, provided that the smaller busway
(a) does not extend more than 15 m;
(b) has a current rating at least equal to one-third the rating or setting of the overcurrent devices next back on the line;
(c) is free from contact with combustible material; and
(d) has an ampacity adequate for the intended load.

12-2014 Length of busways used as branch circuits

(1) Busways that are used as branch circuits, and that are designed so that loads can be connected at any point, shall be limited to lengths such that the circuits will not be overloaded in normal use.

(2) In general, the length of such a run in metres should not exceed the ampere rating of the branch circuit.

12-2016 Manufacturer's identification on busways and splitters

Busways and splitters shall be marked so that the manufacturer's name, trademark, or other recognized symbol of identification shall be readily legible when the installation is completed.

12-2018 Taps in splitters

Taps from busbars or terminal blocks in splitters shall issue from the box on the side thereof nearest to the terminal connections, and the conductors shall not be brought into contact with uninsulated current-carrying parts of opposite polarity.

12-2020 Circuit restrictions in splitters

Splitters shall be used only for the purpose of making connections to the busbars or terminal blocks and shall not be used as a pull box for the conductors of other circuits not connected to the main distribution terminals within the box.

Wireways

12-2100 Where wireways may be used

(1) Wireways shall be permitted to be used only for exposed work and shall not be installed outdoors, or in wet or damp locations, unless specifically approved for such locations.

(2) Wireways and fittings shall not be placed
   (a) where subject to mechanical injury;

(b)   where subject to corrosive vapours;

(c)   in hoistways; or

(d)   in storage-battery rooms.

(3)   Wireways shall be permitted to be used as risers in buildings of non-combustible construction when provided with fire stops in accordance with Rule 2-124.

12-2102 Method of installation of wireways

(1)   Where wireways extend transversely through dry walls or partitions, they shall pass through the walls or partitions in unbroken lengths.

(2)   Wireways shall be securely supported at intervals of not more than 5 ft, unless they are plainly marked to indicate greater distances.

(3)   Dead-ends of wireways shall be closed by fittings.

(4)   Wireways shall be provided with adequate protection against mechanical injury for a distance of 2 m above any floor in an area accessible to other than qualified persons.

12-2104 Conductors in wireways

(1)   Conductors used in wireways shall be the insulated types indicated in Table 19 as being suitable for use in raceways.

(2)   Except as permitted in Subrule (4), wireways shall contain not more than 200 conductors and the aggregate cross-sectional area of the conductors and their insulation shall not exceed 20% of the interior cross-sectional area of the wireway.

(3)   No conductor larger than 500 kcmil copper or 750 kcmil aluminum shall be installed in any wireway.

(4)   Wireways containing only signal and control conductors shall be permitted to contain any number of conductors, but the aggregate cross-sectional area of the conductors and their insulation shall not exceed 40% of the interior cross-sectional area of the wireway.

(5)   The cross-sectional area for conductors in Subrules (2) and (4) shall be determined in accordance with Rule 12-1014(4).

12-2106 Taps and splices in wireways

Where taps and splices are made on feeders or branch circuits within wireways, the connection shall be insulated and shall be accessible.

12-2108 Extensions from wireways

Rigid conduit, flexible metal conduit, surface raceways, cable trays, electrical metallic tubing, armoured cable, metal-sheathed conductors or cable, or, where necessary, hard-usage cord assemblies shall be used in extensions from wireways and shall be connected to the wireway in a manner appropriate to the material used in accordance with Rule 12-3022.

12-2110 AC circuits in wireways

Where alternating current is used, all conductors of a circuit shall be placed within the same wireway, or section thereof, if the latter is made of magnetic material.

12-2112 Manufacturer's identification on wireways

Wireways shall be marked so that the manufacturer's name, trademark, or other recognized symbol of identification shall be readily legible when the installation is completed.

Cable trays

Δ   **12-2200 Method of installation** (see Appendix B)

(1)   Cable trays shall be installed as a complete system using fittings or other means to provide adequate cable support and bending radius before the conductors are installed.

(2)   The maximum design and support spacing shall not exceed the ratings specified by the manufacturer.

(3)   Cable trays shall not pass through walls except where the walls are constructed of non-combustible material.

(4)   Cable trays shall be permitted to extend vertically through floors in dry locations, if provided with fire stops in accordance with Rule 2-124, and if totally enclosed where passing through and for a minimum distance of 2 m above the floor to provide adequate protection from mechanical injury.

(5)   Cable trays shall be adequately supported by non-combustible supports.

(6)   The minimum clearances for cable trays shall be

    (a)   150 mm vertical clearance, excluding depth of cable trays, between cable trays installed in tiers except, where cables of 50 mm diameter or greater may be installed, the clearance shall be 300 mm;

    (b)   300 mm vertical clearance from the top of the cable tray to all ceilings, heating ducts, and heating equipment and 150 mm for short length obstructions; and

    (c)   600 mm horizontal clearance on one side of cable trays mounted adjacent to one another or to walls or other obstructions.

12-2202   Conductors in cable trays (see Appendix B)

(1)   Conductors for use in cable trays shall be as listed in Table 19 and, except as permitted in Subrules (2) and (3), shall have a continuous metal sheath or interlocking armour.

(2)   Type TC tray cable shall be permitted in cable trays in areas of industrial establishments that are inaccessible to the public provided that the cable is

    (a)   installed in conduit, other suitable raceway, or direct buried, when not in cable tray;

    (b)   provided with mechanical protection where subject to damage either during or after installation;

    (c)   no smaller than 1/0 AWG if a single conductor is used; and

    (d)   installed only where qualified persons service the installation.

(3)   Conductors having moisture-resistant insulation and flame-tested non-metal coverings or sheaths of a type listed in Table 19 shall be permitted in ventilated or non-ventilated cable trays where not subject to damage during or after installation in

    (a)   electrical equipment vaults and service rooms; and

    (b)   other locations that are inaccessible to the public and are constructed as a service room where a deviation has been allowed in accordance with Rule 2-030.

(4)   Single conductors shall be fastened to prevent excessive movement due to fault-current magnetic forces.

(5)   Where single conductors are fastened to cable trays, precautions shall be taken to prevent overheating of the fasteners due to induction.

12-2204   Joints and splices within cable trays

Where joints and splices are made on feeders or branch circuits within cable trays, the connectors shall be insulated and shall be accessible.

12-2206   Connection to other wiring methods

Where cable trays are connected to other wiring methods, the arrangement shall be such that the conductors will not be subject to mechanical damage or abrasion, and such that effective bonding will be maintained.

12-2208   Provisions for bonding

(1)   Where metal supports for metal cable trays are bolted to the tray and are in good electrical contact with the grounded structural metal frame of a building, the tray shall be deemed to be bonded to ground.

(2)   Where the conditions of Subrule (1) do not apply, the metal cable tray shall be adequately bonded at intervals not exceeding 15 m and the size of bonding conductors shall be based on the ampacity of the largest ungrounded conductor as specified in Rule 10-814 in the circuits carried by the cable tray.

12-2210   Ampacities of conductors in cable trays

(1)   In ventilated and ladder-type cable trays, where the air space between conductors, cables, or both is maintained at greater than 100% of the largest conductor or cable diameter, the ampacity of the conductors or cables shall be the value specified in Item (a) or (b):

    (a)   single conductors, single-conductor metal-sheathed or armoured cable, and single-conductor mineral-insulated cable, as specified in Tables 1 and 3; and

    (b)   multi-conductor cables as specified in Tables 2 and 4, multiplied by the correction factor in Table 5C for the number of conductors in each cable.

(2)   In ventilated and ladder-type cable trays, where the air space between conductors, cables, or both is maintained at not less than 25% nor more than 100% of the largest conductor or cable diameter, the ampacity of the conductors or cables shall be the value specified in Subrule (1), multiplied by the correction factor specified in Table 5D for the arrangement and number of conductors or cables involved unless a deviation has been allowed in accordance with Rule 2-030 for other correction factors.

(3)   In ventilated and ladder-type cable trays, where the air space between conductors, cables, or both is less than 25% of the largest conductor or cable diameter, and for any spacing in a non-ventilated cable tray, the ampacity of the conductors or cables shall be the value as specified in Table 2 or 4 multiplied by the correction factor specified in Table 5C for the total number of conductors in the cable trays.

(4)   In determining the total number of conductors in the cable tray in Subrule (3), Rule 4-004(7) shall apply.

(5)   Where cable trays are located in room temperatures above 30 °C, the temperature correction factor of Table 5A shall be applied to the ampacities determined from Subrules (1), (2), and (3) as applicable.

Manufactured wiring systems

12-2500 Uses permitted

(1)   A manufactured wiring system shall be permitted to be installed
   (a)   in accessible and dry locations; and
   (b)   in spaces for environmental air when specifically approved for the application, and installed in accordance with Rule 12-010.

(2)   Notwithstanding Subrule (1)(a), a manufactured wiring system shall be permitted to extend into walls for connection to switch and outlet points.

12-2502 Installation

A manufactured wiring system shall be installed in accordance with Rules 12-602 to 12-618.

Δ   # Installations of boxes, cabinets, outlets, and terminal fittings

12-3000 Outlet boxes (see Appendix B)

(1)   A box or an equivalent device shall be installed at every point of outlet, switch, or junction of conduit, raceways, armoured cable, or non-metallic-sheathed cable.

(2)   Non-metallic outlet boxes shall not be used in wiring methods using metal raceways or armoured or metal-sheathed cable, except where the boxes are provided with bonding connections between all conductor entry openings.

(3)   Metal boxes embedded in parking lot slabs or pavement, road beds, and similar areas subject to vehicular traffic shall comply with the requirements of Rule 2-112(1).

(4)   The box shall be provided with a cover or a fixture canopy.

(5)   At least 150 mm of free conductor shall be left at each outlet for making of joints or the connection of fixtures, unless the conductors are intended to loop through lampholders, receptacles, or similar devices without joints.

(6)   Notwithstanding the requirements of Subrule (1), an outlet box shall not be required where equipment has its own integral connection box or has been approved for use as a connection box.

12-3002 Outlet box covers

Cover plates installed on flush-mounted boxes and surface-mounted outlet boxes shall be a type for which each is designed.

12-3004 Terminal fittings

(1)   Where conductors are run from the ends of conduit, armoured cable, surface raceways, or non-metallic-sheathed cable to appliances or open wiring, an outlet fitting or terminal fitting shall be permitted to be used instead of the box required by Rule 12-3000, and the conductors shall be run without splice, tap, or joint within the fitting.

(2)   The fitting shall have a separately bushed hole for each conductor.

(3)   The fittings shall not be used at outlets for fixtures.

12-3006 Terminal fittings behind switchboards

Where conductors issue from conduit behind a switchboard or more than eight conductors issue from a conduit at control apparatus or a similar location, an insulating bushing shall be permitted to be used instead of the box required by Rule 12-3000.

12-3008 Boxes in concrete construction

(1)   Where used in concrete slab construction, ceiling outlet boxes shall have knockouts spaced above the free or lower edge of the boxes a distance of at least twice the diameter of the steel reinforcing bars so that conduit entering the knockouts shall clear the bars without offsetting.

(2)   Sectional boxes shall not be used embedded in concrete or masonry construction.

(3)   Boxes made wholly or in part of aluminum shall not be embedded in concrete containing reinforcing steel unless
   (a)   the concrete is known to contain no chloride additives; or

(b)   the box has been treated with a bituminous base paint or other means to prevent galvanic corrosion of the aluminum.

12-3010 Outlet box supports

(1)   Except as permitted by Subrule (6), boxes and fittings shall be firmly secured to studs, joists, or similar fixed structural units other than wooden, metal, or composition lath, in accordance with this Rule.

(2)   Where ganged sectional boxes are used, they shall be secured to metal supports or to wooden boards at least 19 mm thick that are rigidly secured to the structural units.

(3)   Where boxes having any dimension greater than 100 mm (4 in) are used, they shall be secured on at least two sides or shall be secured to metal supports or to wooden boards at least 19 mm thick that are rigidly secured to the structural units.

(4)   Where boxes are mounted on metal studs, additional support shall be provided to prevent movement of the box after the drywall is installed.

(5)   Mounting nails or screws shall not project into nor pass through the interior of an outlet box unless
(a)   the nails or screws are located so as not to be more than 6.4 mm from the back or ends of the box; and
(b)   the nails or screws are located so that they will not interfere with conductors or connectors.

(6)   This Rule shall not apply to boxes and fittings installed after the studs, joists, or structural units have been concealed.

12-3012 Boxes, cabinets, and fitting supports

(1)   Boxes, cabinets, and fittings shall be fastened securely in place.

(2)   Boxes and fittings having a volume of less than 1640 mL shall be permitted to be attached to a firmly secured exposed raceway by threading or other equally substantial means.

12-3014 Accessibility of junction boxes

(1)   Pull-in, junction, and outlet boxes, cabinets and gutters, and joints in wires and cables shall be accessible.

(2)   A vertical space of 900 mm or more shall be required to provide ready access.

12-3016 Flush boxes, cabinets, and fittings

(1)   The front edges of boxes, cabinets, and fittings installed in walls or ceilings shall not be set in more than 6 mm from the finished surface and, where the walls or ceilings are of wood or other combustible material, shall be flush with the finished surface or shall project from the surface.

(2)   Gaps or open spaces in plaster surfaces of walls or ceilings shall be filled in around the front edges of boxes, cabinets, and fittings.

12-3018 Outlet boxes attached to existing plaster work

Where outlet boxes installed as additions to existing work are mounted directly upon existing plaster surfaces, they shall be fastened securely in place.

12-3020 Outlet boxes, etc., in damp places

Where boxes, cabinets, and fittings are installed in damp places, they shall be placed or constructed so as to prevent moisture from entering and accumulating therein.

12-3022 Entrance of conductors into boxes, cabinets, and fittings (see Appendix B)

(1)   Where conductors pass through the walls of boxes, cabinets, or fittings, provision shall be made to
(a)   protect the insulation on the conductors from injury;
(b)   protect terminal connections from external strain;
(c)   provide electrical continuity between a metal box, cabinet, or fitting and conduit, armour, or metal sheathing of conductors, whether or not the armour or metal sheathing is to be used as a grounding conductor;
(d)   prevent injury to a non-metallic sheath applied over armour or metal sheathing for protection against moisture or corrosion; and
(e)   close the openings through which the conductors pass in such a manner that any remaining opening will not permit entrance of a test rod 6.75 mm in diameter.

(2)   Where conductors run as open wiring enter a box, cabinet, or fitting, they shall pass through insulating bushings or be installed in raceways or insulating tubing.

(3)   Where non-metallic-sheathed cable enters a box, cabinet, or fitting, a box connector, either as a separate device designed for use with such cable or as part of the box, cabinet, or fitting, shall be used to secure the cable in place adequately and without injury to the conductors.

(4)   Where rigid or flexible metal conduit, electrical metallic tubing, or armoured cable enters boxes, cabinets, or fittings, it shall be secured in place in accordance with the requirements of Section 10.

(5)   Where metal-sheathed conductors enter boxes, cabinets, or fittings, the box connector shall be installed in a manner that will meet the requirements of Section 10 without injury to the conductors and shall be of a type for use with the cable.

(6)   Where liquid-tight flexible metal conduit or where flexible metal conduit, armoured cable, or metal-sheathed cable of a type having a non-metallic-sheath over the armour or metal sheath enters a box, cabinet, or fitting, the box connector shall ensure electrical continuity without injury to the non-metallic sheath unless the point of connection is in a dry location free from corrosive atmosphere, where the non-metallic sheath shall be permitted to be stripped back a sufficient distance.

(7)   Where single-conductor cables or conductors enter metal boxes through separate openings, precaution shall be taken to prevent overheating of the metal by induction if the current carried per conductor exceeds 200 A.

(8)   Precautions to be taken to prevent overheating of the metal by induction shall include the use of nonferrous or non-metallic box connectors, locknuts, and bushings and if nonferrous metal plates or insulating plates are field installed, they shall be at least 6 mm thick unless a deviation is allowed in accordance with Rule 2-030.

12-3024 Unused openings in boxes, cabinets, and fittings
Unused openings in boxes, cabinets, and fittings shall be effectively closed by plugs or plates affording protection substantially equivalent to that of the wall of the box, cabinet, or fitting.

12-3026 Extensions from existing outlets
(1)   Where a surface extension is made from an existing outlet of concealed wiring, a box or an extension-ring shall be mounted over the original box and electrically and mechanically secured to it.

(2)   The extension shall then be connected to the box or extension-ring in the manner prescribed by this Section for the method of wiring employed in making the extension.

12-3028 Multi-outlet assemblies
(1)   Multi-outlet assemblies shall be used only in normally dry locations as extensions to wiring systems.

(2)   Multi-outlet assemblies shall not be used in any bathroom, kitchen, or any place where the assembly would be subject to mechanical injury.

(3)   Multi-outlet assemblies shall be permitted to be carried through but not run within dry partitions provided that
    (a)   no outlet falls within the partition;
    (b)   the removal of any cap or cover necessary for proper installation is not prevented; and
    (c)   the assembly is of metal or, if not of metal, is surrounded by metal or the equivalent.

(4)   Multi-outlet assemblies shall not be concealed within the building finish but
    (a)   the back and sides of metal assemblies shall be permitted to be set in plaster applied after the assembly is in place; or
    (b)   the back and sides of non-metallic assemblies shall be permitted to be set in a preformed recess in the building finish; and
    (c)   shall be permitted to be recessed in a baseboard or other wood trim member.

12-3030 Conductors in boxes, cabinets, or fittings
(1)   Conductors that are connected to different power or distribution transformers or other different sources of voltage shall not be installed in the same box, cabinet, or fitting unless
    (a)   a barrier of sheet steel not less than 1.3 mm thick or a flame-retardant non-metallic insulating material not less than 1.6 mm in thickness is used to divide the space into separate compartments for the conductors of each system;
    (b)   the conductors are used for the supply and/or control of remote devices and are insulated for at least the same voltage as that of the circuit having the highest voltage and none of the conductors of the circuits of lower voltages is directly connected to a lighting branch circuit; or
    (c)   the conductors are used for the supply of a double-throw switch in an emergency lighting system.

(2)   Where a barrier is used, it shall be fastened rigidly to the box, cabinet, or fitting, or a device assuring positive separation of the conductors shall be used.

12-3032 Wiring space in enclosures

(1) Enclosures for overcurrent devices, controllers, and externally operated switches shall not be used as junction boxes, troughs, or raceways for conductors feeding through to other apparatus.

(2) Notwithstanding Subrule (1),
    (a) the enclosures identified shall be permitted to be used as junction boxes
        (i) for all installations where a single feeder supplying another enclosure is tapped from it and the connectors used each provide an independent clamping means for each conductor and each clamping means is independently accessible for tightening or inspection; or
        (ii) where wiring is being added to an enclosure forming part of an existing installation and the conductors, splices, and taps do not fill the wiring space at any cross-section to more than 75% of the cross-sectional area of the space; and
    (b) the enclosure identified may be used as a raceway where the conductors are being added to enclosures forming part of an existing installation and all conductors present do not fill the wiring space at any cross-section to more than 40% of the cross-sectional area of the space.

(3) Conductors entering enclosures shall enter such enclosures as near as practicable to their terminal fittings.

12-3034 Maximum number of conductors in a box

(1) Boxes shall be of sufficient size to provide usable space for all insulated conductors contained in the box, subject to the following:
    (a) a conductor running through a box with no connection therein shall be considered as one conductor;
    (b) each conductor entering or leaving a box and connected to a terminal or connector within the box shall be considered as one conductor;
    (c) a conductor of which no part leaves the box shall not be counted; and
    (d) No. 18 and No. 16 AWG fixture wires supplying a lighting fixture mounted on the box containing the fixture wires shall not be counted.

(2) Except as specified in Subrule (3) and subject to the details given in Subrule (1), boxes of the nominal dimensions given in Table 23 shall not contain more insulated conductors of a given size than permitted by the Table, and the number of conductors shall be reduced for each of the following conditions as applicable:
    (a) one conductor if the box contains one or more fixture studs or hickeys;
    (b) one conductor for every pair of wire connectors with insulating caps (no deduction for one wire connector, deduct one conductor for 2- or 3-wire connectors, two conductors for 4- or 5-wire connectors, etc.); or
    (c) two conductors if the box contains one or more flush devices mounted on a single strap.

(3) Where a box contains a device having a dimension greater than 2.54 cm between the mounting strap and back of the device, the total usable space shall be reduced by the space occupied by the device, calculated as 82 cm³ multiplied by the depth of the device in centimetres divided by 2.54 (for example, a device having a depth of 4 cm would occupy a space of 129 cm³, that is, 82 times 4 divided by 2.54).

(4) Subject to the details given in Subrules (1) and (3), boxes having nominal dimensions or volume other than those shown in Table 23 or any box containing insulated conductors of different sizes shall have the amount of usable space per insulated conductor as specified in Table 22, but the number of conductors so calculated shall be reduced for each of the conditions of Subrule (2) as applicable.

(5) The total usable space in a box considered under Table 22 shall be considered to be the internal volume of the box and shall disregard any space occupied by locknuts, bushings, box connectors, or clamps.

(6) Where sectional boxes are ganged, or where plaster rings, extension rings, or raised covers are used in conjunction with boxes, ganged or otherwise, and are marked with their volume measurement, the space in the box shall be the total volume of the assembled sections.

12-3036 Pull box or junction box sizes

(1) For the purposes of Subrule (2), the equivalent cable to trade size of raceway shall be the minimum trade size raceway that would be required for the number and size of conductors in the cable.

(2) Where a pull or junction box is used with raceways containing conductors of No. 4 AWG or larger, or with cables containing conductors No. 4 AWG or larger, the box shall
    (a) for a raceway or cable entering the wall of a box opposite to a removable cover, have a distance from the wall to the cover not less than the trade diameter of the largest raceway or equivalent cable plus 6 times the diameter of the largest conductor;

(b)   for straight pulls or runs of cable, have a length of at least eight times the trade diameter of the largest raceway or equivalent cable; and

(c)   for angle and U pulls or runs of cable,

   (i)   have a distance between each raceway or cable entry inside the box and the opposite wall of the box of at least six times the trade diameter of the largest raceway or equivalent cable, plus the sum of the trade diameters of all other raceways or equivalent cables on the same wall of the box; and

   (ii)   have a distance, as measured in a straight line, between the nearest edges of each raceway or cable entry enclosing the same conductor of at least

      (A)   six times the trade diameter of the raceway or equivalent cable; or

      (B)   six times the trade diameter of the larger raceway or equivalent cable if they are of different sizes.

Section 14 — Protection and control

Scope

14-000 Scope
This Section covers the protection and control of electrical circuits and apparatus installed in accordance with the requirements of this Section and other Sections of this Code.

General requirements

14-010 Protective and control devices required
Electrical apparatus and ungrounded conductors shall, except as otherwise provided for in this Section or in other Sections dealing with specific equipment, be provided with
(a)   devices for the purpose of automatically opening the electrical circuit thereto,
    (i)   if the current reaches a value that will produce a dangerous temperature in the apparatus or conductor; and
    (ii)   in the event of a ground fault, in accordance with Rule 14-102;
(b)   manually operable control devices that will safely disconnect all ungrounded conductors of the circuit at the point of supply simultaneously, except for multi-wire branch circuits that supply only fixed lighting loads or non-split receptacles, and that have each lighting load or receptacle connected to the neutral and one ungrounded conductor; and
(c)   devices that, when necessary, will open the electrical circuit thereto in the event of failure of voltage in such a circuit.

14-012 Ratings of protective and control equipment (see Appendix B)
In circuits of 750 V and less,
(a)   electrical equipment required to interrupt fault currents shall have ratings sufficient for the voltage employed and for the fault current that is available at the terminals; and
(b)   electrical equipment required to interrupt current at other than fault levels shall have ratings sufficient for the voltage employed and for the current it must interrupt.

14-014 Series rated combinations (see Appendix B)
Notwithstanding Rule 14-012(a), a moulded case circuit breaker shall be permitted to be installed in a circuit having an available fault current higher than its rating provided that
(a)   the circuit breaker is a recognized component of an approved series rated combination;
(b)   it is installed on the load side of an overcurrent device that has an interrupting rating at least equal to the available fault current;
(c)   the overcurrent device on the line side of the lower rated circuit breaker is as specified on the equipment in which the lower rated circuit breaker is installed;
(d)   the equipment in which the lower rated circuit breaker is installed is marked with a series combination interrupting rating at least equal to the available fault current; and
(e)   the overcurrent devices installed in a series rated combination are marked at the time of installation in a conspicuous and legible manner to indicate that they must be replaced only with components of the same type and rating.

14-016 Connection of devices
Devices required by this Section shall not be connected in any grounded conductors except where
(a)   the devices simultaneously or previously disconnect all ungrounded conductors;
(b)   an overcurrent device is in a 2-wire circuit having one wire grounded and where there is a possibility that the grounded conductor may assume a voltage difference between itself and ground, due to unreliable grounding conditions of sufficient magnitude to create a dangerous condition; or
(c)   overcurrent devices are located in that part of a circuit that is connected by a 2-pole polarized or unpolarized attachment plug provided that the circuit is rated 15 A, 125 V or less.

Protective devices

General

14-100 Overcurrent protection of conductors (see Appendix B)

Each ungrounded conductor shall be protected by an overcurrent device at the point where it receives its supply of current and at each point where the size of conductor is decreased, except that such protection shall be permitted to be omitted in each of the following cases:

(a)   where the overcurrent device in a larger conductor properly protects the smaller conductor;

(b)   where the smaller conductor

(i)   has an ampacity not less than the combined computed loads of the circuits supplied by the smaller conductor and not less than the ampere rating of the switchboard, panelboard, or control device supplied by the smaller conductor;

(ii)   is not over 3 m long;

(iii)   does not extend beyond the switchboard, panelboard, or control device that it supplies; and

(iv)   is enclosed in non-ventilated raceways, armoured cable, or metal-sheathed cable when not part of the wiring in the switchboard, panelboard, or other control devices;

(c)   where the smaller conductor

(i)   has an ampacity not less than one-third that of the larger conductor from which it is supplied; and

(ii)   is suitably protected from mechanical damage, is not more than 7.5 m long, and terminates in a single overcurrent device rated or set at a value not exceeding the ampacity of the conductor, but beyond the single overcurrent device the conductor shall be permitted to supply any number of overcurrent devices;

(d)   where the conductor

(i)   forms part of the only circuit supplied from a power or distribution transformer rated over 750 V with primary protection in accordance with Rule 26-252(1), (2), and (3) and that supplies only that circuit;

(ii)   terminates at a single overcurrent device with a rating not exceeding the ampacity of the conductor(s) in the circuit; and

(iii)   is protected from mechanical damage;

(e)   where the smaller conductor is No. 14 AWG or larger, is in a control circuit, and is located external to the control equipment enclosure, and

(i)   the rating or setting of the branch circuit overcurrent device is not more than 300% of the ampacity of the control circuit conductor; or

(ii)   the opening of the control circuit would create a hazard;

(f)   where the smaller conductor supplies a transformer, and

(i)   the conductor supplying the primary of the transformer has an ampacity not less than one-third that of the larger conductor;

(ii)   the conductor supplied by the secondary of the transformer has an ampacity not less than the ampacity of the primary conductor multiplied by the ratio of the primary to the secondary voltage;

(iii)   the total length of one primary plus one secondary conductor (the longest, if more than one winding), excluding any portion of the primary conductor that is protected at its own ampacity, does not exceed 7.5 m;

(iv)   the primary and secondary conductors are protected from mechanical damage; and

(v)   the secondary conductor terminates in a single overcurrent device rated or set at a value not exceeding its ampacity; or

(g)   where the smaller conductor

(i)   is supplied by a circuit at not more than 750 V;

(ii)   is supplied from an overhead or underground circuit and is run overhead or underground except where it enters a building;

(iii)   is installed in accordance with the requirements of Section 6; and

(iv)   terminates in service equipment in accordance with Section 6.

14-102 Ground fault protection (see Appendix B)

Δ   (1)   Ground fault protection shall be provided to de-energize all normally ungrounded conductors of a faulted circuit that are downstream from the point or points marked with an asterisk in Diagram 3 in the event of a ground fault in those conductors as follows:

    (a)   in solidly grounded systems rated more than 150 volts-to-ground, less than 750 V phase-to-phase and 1000 A or more; and

    (b)   in solidly grounded systems rated 150 V or less to ground and 2000 A or more.

(2)   Except as permitted by Subrule (8), the maximum setting of the ground fault protection shall be 1200 A and the maximum time delay shall be one second for ground fault currents equal to or greater than 3000 A.

(3)   The ampere rating of the circuits referred to in Subrule (1) shall be considered to be

    (a)   the rating of the largest fuse that can be installed in a fusible disconnecting device;

    (b)   the highest trip setting for which the actual overcurrent device installed in a circuit breaker is rated or can be adjusted; or

    (c)   the ampacity of the main conductor feeding the devices located at points marked with an asterisk in Item 2 of Diagram 3, in the case where no main disconnecting device is provided.

(4)   This protection shall be provided by

    (a)   an overcurrent device that incorporates ground fault protection;

    (b)   a ground fault tripping system consisting of a sensor or sensors, relay, and auxiliary tripping mechanism; or

    (c)   other means.

(5)   The sensor or sensors referred to in Subrule (4) shall be

    (a)   sensors that vectorially totalize the currents in all conductors of the circuit, including the grounded circuit conductor, where one is provided, but excluding any current flowing in the ground fault return current path;

    (b)   sensors that sense ground fault current flowing from the fault to the supply end of the system through the ground return path; or

    (c)   a combination of these two types of sensor.

(6)   Sensors referred to in Subrule (5)(a) shall be permitted to be installed at any point between the supply transformer and the downstream side of the disconnecting means marked with an asterisk in Diagram 3, but if located downstream from this disconnecting means the sensors shall be placed as close as practicable to its load terminals.

(7)   Sensors referred to in Subrule (5)(b) shall be located on each connection between neutral and ground; however, where the neutral is grounded both at the supply transformer and at the switching centre, the sensor at the transformer shall not be required, provided that the maximum pickup setting of the ground fault relay does not exceed 1000 A.

(8)   In ground fault schemes where two or more protective devices in series are used for ground fault coordination, the upstream protective device settings shall be permitted to exceed those specified in Subrule (2) where necessary to obtain the desired coordination, provided that the final downstream ground fault protective device in each circuit required to be protected conforms to the requirements of Subrule (2).

14-104 **Rating of overcurrent devices** (see Appendix B)

The rating or setting of overcurrent devices shall not exceed the allowable ampacity of the conductors that they protect except

(a)   where a fuse or circuit breaker having a rating or setting of the same value as the ampacity of the conductor is not available, the ratings or settings given in Table 13 shall be permitted to be used within the maximum value of 600 A;

(b)   in the case of equipment wire, flexible cord in sizes Nos. 16, 18, and 20 AWG copper, and tinsel cord, which are considered protected by 15 A overcurrent devices; or

(c)   as provided for by other Rules of this Code.

14-106 **Location and grouping**

Overcurrent devices shall be located in readily accessible places, except as provided for elsewhere in this Code, and shall be grouped where practicable.

14-108 **Enclosure of overcurrent devices**

(1)   Overcurrent devices shall be enclosed in cut-out boxes or cabinets unless they form a part of an approved assembly that affords equivalent protection, or unless mounted on switchboards, panelboards, or controllers located in rooms or enclosures free from easily ignitable material and dampness, and accessible only to authorized persons.

(2)   Operating handles of circuit breakers shall be made accessible without opening any door or cover giving access to live parts.

14-110 Grouping of protective devices at distribution centre

(1) Where the number of lighting branch circuits originating from a common enclosure exceeds
   (a) two, in a single-phase three-wire system; or
   (b) three, in a three-phase, four-wire system,
   overcurrent devices protecting such circuits shall be contained in a panelboard.

(2) Where a panelboard is not required, and a fusible switch is used, all overcurrent devices shall have the same rating.

(3) For the purposes of this Rule, each ungrounded conductor of a multi-wire branch circuit shall be counted as a separate circuit.

14-112 Overcurrent devices in parallel

(1) Overcurrent devices shall not be connected in parallel in circuits of 750 V or less.

(2) Notwithstanding Subrule (1), semiconductor fuses having interrupting ratings of 100 000 A and more, 750 V and less, and circuit breakers rated 750 V and less shall be permitted to be connected in parallel provided that they are factory assembled in parallel as a single unit.

Δ ### 14-114 Application of supplementary protectors (see Appendix B)

Supplementary overcurrent protection shall not be used as a substitute for branch circuit overcurrent devices or in place of branch circuit protective devices specified in this Section.

Fuses

14-200 Time-delay and low-melting point fuses

(1) Plug and cartridge fuses of the low-melting point types, including time-delay fuses that also have low-melting points, shall be marked so as to be readily distinguishable.

(2) The marking referred to in Subrule (1) shall be the letter "P" for low-melting point types that do not have time-delay characteristics and the letter "D" for time-delay fuses.

14-202 Use of plug fuses

Plug fuses and fuseholders shall not be used in circuits exceeding 125 V between conductors, except in circuits supplied from a system having a grounded neutral and having no conductor operating at more than 150 volts-to-ground.

14-204 Non-interchangeable fuses (see Appendix B)

(1) Where plug fuses are used in branch circuits, they shall be of such a type and installed so that they are non-interchangeable with a fuse of larger rating.

(2) Where any alterations or additions are made to an existing fusible panelboard, all the plug fuses in the panelboard shall be made to comply with the requirements of Subrule (1), where practicable.

14-206 Fuseholders for plug fuses

Fuseholders for plug fuses shall be of the so-called "covered" type where readily accessible to unauthorized persons.

14-208 Rating of fuses

(1) Plug fuses shall be rated at not more than 30 A.

(2) Standard cartridge fuses shall not be used in capacities larger than 600 A or in circuits at more than 600 V.

(3) The fuses referred to in Rule 14-212(b), (c), and (d) that are used in circuits rated at 750 V or less are not limited in current rating.

(4) Fuses for use in circuits of more than 750 V are not limited in current or voltage rating.

14-210 Fuses and fuseholders

Only fuses and fuseholders of proper rating shall be used, and no bridging or short-circuiting of either component shall be permitted.

14-212 Use of fuses (see Appendix B)

Class C, CA, CB, CC, G, H, J, K, L, R, T, HRCI-MISC, and HRCII-MISC fuses shall be permitted to be used as follows:

(a) Class H fuses, where a standard interrupting rating of 10 000 A symmetrical or less is required;

(b) Class CA, CB, CC, G, J, K, L, R, T, or HRCI-MISC, which have a higher interrupting rating, shall be permitted to be used instead of Class H fuses;

(c) Class C and HRCII-MISC fuses shall be permitted to be used for overcurrent protection only where circuit overload protection is provided by other means; and

(d)   Class C and HRCII-MISC fuses shall be permitted to be used in those applications where this Code permits the installation of fuses greater than the ampere rating of the load, provided that the rating of the Class C or HRCII-MISC fuses does not exceed 85% of the maximum permitted rating.

Circuit breakers

14-300 Circuit breakers, general
(1)   Circuit breakers shall be of the trip-free type.
(2)   Indications shall be provided at the circuit breaker and at the point of operation to show whether the circuit breaker is open or closed.

14-302 Construction of circuit breakers (see Appendix B)
Where circuit breakers are provided for the protection of apparatus or ungrounded conductors, or both, they shall open the circuit in all ungrounded conductors by the manual operation of a single handle and by the action of overcurrent, except
(a)   where single-pole circuit breakers are permitted by Rule 14-010(b); or
(b)   in branch circuits derived from a 3-wire grounded neutral system, two single-pole manually operable circuit breakers shall be permitted to be used instead of a 2-pole circuit breaker, provided that
   (i)   their handles are interlocked with a device as provided by the manufacturer so that all ungrounded conductors will be opened by the manual operation of any handle; and
   (ii)   each circuit breaker has voltage ratings not less than that of the multi-wire branch circuit.

14-304 Non-tamperable circuit breakers
Branch-circuit breakers, unless accessible only to authorized persons, shall be designed so that any alteration by the user of either tripping current or time will be difficult.

14-306 Tripping elements for circuit breakers
Circuit breakers shall be equipped with tripping elements as specified in Table 25.

14-308 Battery control power for circuit breakers
(1)   When power for operating the overcurrent element of a circuit breaker is derived from a battery, the battery voltage shall be continuously monitored.
(2)   If the battery voltage should drop to a value insufficient to operate the circuit breaker overcurrent element
   (a)   the circuit breaker shall automatically trip; or
   (b)   an alarm shall operate continuously until the battery voltage is restored.
(3)   A suitable warning notice shall be placed on or adjacent to the circuit breaker indicating that battery control power must be available before the circuit breaker is closed.

Control devices

General

14-400 Rating of control devices
Control devices shall have ratings suitable for the connected load of the circuits that they control and, with the exception of isolating switches, shall be capable of safely establishing and interrupting such loads.

14-402 Disconnecting means required for fused circuits
Circuits protected by fuses shall be equipped with disconnecting means, integral with or adjacent to the fuseholders, whereby all live parts for mounting fuses can be readily and safely made dead; however, such disconnecting means shall be permitted to be omitted in the case of
(a)   instrument and control circuits on switchboards where the voltage does not exceed 250 V;
(b)   primary circuits of voltage transformers having a primary voltage of 750 V or less, on switchboards; and
(c)   a circuit having only one ungrounded conductor where a plug fuse is used, because a plug fuse can be safely handled while alive in such a circuit.

14-404 Control devices ahead of overcurrent devices
Control devices used in combination with overcurrent devices or overload devices for the control of circuits or apparatus shall be connected so that the overcurrent or overload devices will be dead when the control device is in the open position, except where this is impracticable.

14-406 Location of control devices

(1)   Control devices, with the exception of isolating switches, shall be readily accessible.

(2)   Remotely controlled devices shall be considered to be readily accessible if the means of controlling them are readily accessible.

14-408 Indication of control device positions

Manually operable control devices shall indicate the ON and OFF positions, unless the application of the devices makes this requirement unnecessary.

14-410 Enclosure of control devices

Control devices, unless they are located or guarded in a way that renders them inaccessible to unauthorized persons and prevents fire hazards, shall have all current-carrying parts in enclosures of metal or other fire-resisting material.

14-412 Grouping of control devices

Control devices controlling feeders and branch circuits shall be grouped where practicable.

14-414 Connection to different circuits

(1)   Where electrical equipment is supplied by two or more different transformers or other different sources of voltage, then

(a)   a single disconnecting means that will effectively isolate all ungrounded conductors supplying the equipment shall be provided integral with or adjacent to the equipment; or

(b)   each supply circuit shall be provided with a disconnecting means integral with or adjacent to the equipment, and the disconnecting means shall be grouped together.

(2)   Notwithstanding Subrule (1), disconnecting means integral with or adjacent to equipment need not be provided for control circuits originating beyond the equipment and not exceeding 150 volts-to-ground provided that all associated bare live parts are protected against inadvertent contact by means of barriers.

(3)   Where multiple disconnecting means as in Subrule (1)(b) are provided, suitable warning signs shall be placed on or adjacent to each disconnecting means so that all of the disconnecting means must be opened to ensure complete de-energization of the equipment.

(4)   Where barriers are used as required in Subrule (2), a suitable warning sign shall be placed on or adjacent to the equipment, or on the barriers, indicating that there is more than one source of supply to the equipment.

14-416 Control devices used only for switching

Except as permitted by other Rules in this Code, control devices that perform only switching functions shall disconnect all ungrounded conductors of the controlled circuit when in the OFF position.

Switches

14-500 Operation of switches

Knife switches and other control devices, unless located or guarded in a way that renders them inaccessible to unauthorized persons, shall be constructed so that they may be switched to the OFF position without exposing live parts.

14-502 Mounting of knife switches

(1)   Single-throw knife switches shall be mounted with their bases in a vertical plane.

(2)   Single-throw knife switches shall be mounted so that gravity will not tend to close them.

(3)   Double-throw knife switches may be mounted so that the throw will be either vertical or horizontal but, if the throw is vertical, a positive locking device or stop shall be provided to ensure that the blades remain in the open position when so set, unless the switch is not intended to be left in the open position.

14-504 Maximum rating of switches

Unless of special design, knife switches rated at more than 600 A at 750 V or less shall be used only as isolating switches.

14-506 Connection of switches

Manual single-throw switches, circuit breakers, or magnetic switches shall be connected so that the blades or moving contacts will be dead when the device is in the open position, except that the following need not comply:

(a)   branch-circuit breakers that have all live parts other than terminals sealed and that are constructed so that the line and load connections may be interchanged;

(b)   switchgear that is provided for sectionalizing purposes and has a suitable caution notice attached to the assembly;

(c)   switches that are immersed in a liquid and have a suitable caution notice attached to the outside of the enclosure;

(d)   switches that are designed so that all live parts are inaccessible when the device is in the open position; and

(e)   magnetic switches, when preceded by a circuit breaker or manual switch that is located in the same enclosure or immediately adjacent and is marked to indicate that it controls the circuit to the magnetic switch, unless this is obvious.

14-508 Rating of general-use ac/dc switches (see Appendix B)

AC/DC switches shall be rated as follows:

(a)   for non-inductive loads other than tungsten-filament lamps, switches shall have an ampere rating not less than the ampere rating of the load;

(b)   for tungsten-filament lamp loads, and for combined tungsten-filament and non-inductive loads, switches shall be "T" rated, except where

    (i)   the switches are used in branch-circuit wiring systems in dwelling units; in private hospital or hotel rooms; or in similar locations, but not in public rooms or places of assembly;

    (ii)   the switch controls permanently connected fixtures or lighting outlets in one room only, or in one continuous hallway where the lighting fixtures may be located at different levels or in attics or basements not used for assembly purposes; and

    (iii)   the switch is rated at not less than 10 A, 125 V; 5 A, 250 V; or for the 4-way types, 5 A, 125 V; 2 A, 250 V;

(c)   canopy switches controlling a tungsten-filament lamp load shall be "T" rated or shall have an ampere rating at least three times the ampere rating of the load; and

(d)   for inductive loads, switches shall have an ampere rating of twice the ampere rating of the load.

14-510 Use and rating of manually operated general-use ac switches (see Appendix B)

(1)   Manually operated, general-use switches intended for alternating-current systems and constructed so that they can be installed readily in wiring systems for making and breaking tungsten-filament lighting and power circuits shall be rated as follows:

    (a)   for tungsten-filament lamp loads at 120 V maximum, switches shall have an ampere rating not less than the current rating of the load; and

    (b)   for non-inductive loads and for inductive loads at not less than 75% power factor lag, switches shall have an ampere rating not less than the current rating of the load.

(2)   The current rating of the switches shall be not less than 15 A in conjunction with a voltage rating of 120 or 277 V.

(3)   Switches shall be adapted for mounting in flush-device boxes, surface-type boxes, special boxes, or have complete self-enclosures.

14-512 Manually operated general-use 347 V ac switches (see Appendix B)

(1)   Manually operated general-use 347 V ac switches shall be used only for the control of non-inductive loads other than tungsten-filament lamps and for inductive loads where the power factor is not less than 75% lagging.

(2)   The current rating of the switches shall be not less than 15 A in conjunction with a voltage rating of 347 V.

(3)   The switches designed for mounting in boxes shall not be readily interchangeable with switches referred to in Rules 14-508 and 14-510.

14-514 Manually operated switches in circuits exceeding 300 volts-to-ground

When controlling circuits exceeding 300 volts-to-ground, the switches referred to in Rules 14-508 and 14-512 shall not be ganged or grouped in the same enclosure unless the enclosure provides permanently installed barriers.

Protection and control of miscellaneous apparatus

14-600 Protection of receptacles

Receptacles shall not be connected to a branch circuit having overcurrent protection rated or set at more than the ampere rating of the receptacle except as permitted by other Sections of this Code.

14-602 Additional control devices not necessary

Portable appliances need not be equipped with additional control devices where the appliances are

(a)   rated at not more than 1500 W; and

(b)   provided with cord connectors, attachment plugs, or other means by which they can be disconnected readily from the circuits.

14-604 Outlet control from more than one point

Where switches are used to control an outlet or outlets from more than one point, the switches shall be connected so that all switching is done only in the ungrounded circuit conductor.

14-606 Panelboard overcurrent protection

Δ (1)   Except for panelboards where more than 90% of the overcurrent devices supply feeders or motor branch circuits, every panelboard shall be protected on the supply side by overcurrent devices having a rating not greater than that of the panelboard.

(2)   The overcurrent protection required by Subrule (1) shall be permitted to be in the primary of a transformer supplying the panelboard, provided that the panelboard rating in amperes is not less than the overcurrent rating in amperes multiplied by the ratio of the primary to the secondary voltage.

14-608 Remote-control circuits

Remote-control circuits of remotely controlled apparatus shall be arranged so that they may be conveniently disconnected from their source of supply at the controller, but as an alternative the disconnecting of the apparatus from the supply circuit may be arranged so that it also disconnects the remote-control circuit from the supply circuit.

14-610 Protection of circuits supplying cycling loads

Where fuses protect circuits in which more than 50% of the circuit rating is a cycling load, such as thermostatically controlled electric space heaters, clothes dryers, or water heaters, they shall be time-delay or low-melting point fuses of the type referred to in Rule 14-200 or fuses as referred to in Rule 14-212(b); however, in dwelling units, the fuses referred to in Rule 14-212(b) shall have the same low-melting point characteristics as those referred to in Rule 14-200.

14-612 Transfer equipment for standby power systems

Transfer equipment for standby power systems shall prevent the inadvertent interconnection of normal and standby sources of supply in any operation of the transfer equipment.

Solid-state devices

14-700 Restriction of use

Solid-state devices shall not be used as isolating switches or as disconnecting means.

14-702 Disconnecting means required

(1)   Supplementary disconnecting means shall be provided where failure of or leakage through a solid-state device could result in transfer of energy between two or more power sources.

(2)   The disconnecting means referred to in Subrule (1) shall

(a)   be connected into the circuit in such a way that when opened they will prevent transfer of energy between the different power sources; and

(b)   be provided as an integral part of the solid-state device; or

(c)   be installed as close as practicable and in sight of the solid-state device.

14-704 Warning notices required

Suitable warning notices shall be placed

(a)   on the supplementary disconnecting means required by Rule 14-702 so that

(i)   this disconnecting means shall be opened in the event of a failure of any of the power sources or in the event of servicing of any component in the circuits of the other power sources; and

(ii)   both line and load terminals may be energized when the disconnecting means is open; and

(b)   on all other upstream disconnecting means so that an alternative power source(s) exists in the circuit and that the supplementary disconnecting means must also be opened to prevent the possibility of feedback from the alternative source(s).

Section 16 — Class 1 and Class 2 circuits

General

16-000 Scope
(1) This Section covers
    (a) Class 1 and Class 2 remote-control circuits;
    (b) Class 1 and Class 2 signal circuits;
    (c) Class 1 extra-low-voltage power circuits; and
    (d) Class 2 low-energy power circuits.
(2) This Section does not apply to
    (a) communication circuits as specified in Section 60; and
    (b) circuits forming an integral part of a device.

16-002 Classifications
Circuits covered by this Section are that portion of the wiring system between the load side of the overcurrent device or the power-limited supply and all connected equipment, and shall be classified as follows:
(a) Class 1 — circuits that are supplied from sources having limitations in accordance with Rule 16-100; and
(b) Class 2 — circuits that are supplied from sources having limitations in accordance with Rule 16-200.

16-004 Class 1 extra-low-voltage power circuits
Circuits that are neither remote-control circuits nor signal circuits but that operate at not more than 30 V where the current is not limited in accordance with Rule 16-200 and that are supplied from a transformer or other device restricted in its rated output to 1000 V•A shall be classed as extra-low-voltage power circuits and shall be considered Class 1 circuits.

16-006 Class 2 low-energy power circuits
Circuits that are neither remote-control circuits nor signal circuits but in which the current is limited in accordance with Rule 16-200 shall be classed as low-energy power circuits and shall be considered Class 2 circuits.

16-008 Hazardous locations
Where the circuits or apparatus within the scope of this Section are installed in hazardous locations, they shall also comply with the applicable Rules of Section 18.

16-010 Circuits to safety control devices
Where the failure to operate of a remote-control circuit to a safety control device will introduce a direct fire or life hazard, the remote-control circuit shall be deemed to be a Class 1 circuit.

16-012 Circuits in communication cables
(1) Class 1 circuits shall not be run in the same cable with communication circuits.
(2) Class 2 remote-control and signal circuits or their parts that use conductors in a cable assembly with other conductors forming parts of communication circuits are, for the purpose of this Code, deemed to be communication circuits and shall conform to the applicable Rules of Section 60.

Class 1 circuits

16-100 Limitation of Class 1 circuits
(1) Class 1 extra-low-voltage power circuits shall be supplied from a source having a rated output of not more than 30 V and 1000 V•A.
(2) Class 1 remote-control and signal circuits shall be supplied by a source not exceeding 600 V.

16-102 Methods of installation for Class 1 circuits
The equipment and conductors of Class 1 circuits shall be installed in accordance with the requirements of other appropriate Sections of this Code, except as provided in Rules 16-104 to 16-118.

16-104 Overcurrent protection of Class 1 circuits
(1) Conductors of Class 1 circuits shall be protected against overcurrent in accordance with Section 14 of this Code, except
    (a) where other Rules of this Code specifically permit or require other overcurrent protection; or

(b) where the conductors are of No. 18 or No. 16 AWG copper and extend beyond the equipment enclosure, they shall be protected by overcurrent devices rated at a maximum of 5 A and 10 A respectively.

(2) Where overcurrent protection is installed at the secondary terminals of the transformer and the transformer is suitably enclosed, no overcurrent protection is required on the primary side other than the normal overcurrent protection of the branch circuit supplying the transformer.

16-106 Location of overcurrent devices in Class 1 circuits

(1) In Class 1 circuits, the overcurrent devices shall be located at the point where the conductor to be protected receives its supply.

(2) The overcurrent device shall be permitted to be an integral part of the power supply.

16-108 Class 1 extra-low-voltage power circuit sources including transformers

To comply with the 1000 V•A limitation, Class 1 extra-low-voltage power circuit sources including transformers shall not exceed a maximum power output of 2500 V•A, and the product of the maximum current and maximum voltage shall not exceed 10 000 V•A with the overcurrent protection bypassed.

16-110 Conductor material and sizes

(1) Copper conductors smaller than No. 14 AWG shall be permitted to be used in Class 1 circuits if
    (a) installed in a raceway;
    (b) installed in a cable assembly; or
    (c) within a flexible cord in accordance with Rule 4-010.

(2) Subject to the conditions specified in Subrule (1), conductors shall be not smaller than
    (a) No. 16 AWG for individual conductors pulled in raceways;
    (b) No. 18 AWG for individual conductors laid in raceways; and
    (c) No. 18 AWG for an integral assembly of two or more conductors.

16-112 Insulated conductors for Class 1 wiring

(1) Where conductors larger than No. 16 AWG copper are used in a Class 1 circuit, they shall be of any type shown in Table 19.

(2) Where conductors of No. 18 or No. 16 AWG copper are used in a Class 1 circuit, they shall be equipment wire of the type suitable for such use as shown in Table 11.

16-114 Conductors of different circuits in the same enclosure, cable, or raceway

(1) Different Class 1 circuits shall be permitted to occupy the same enclosure, cable, or raceway without regard to whether the individual circuits are alternating current or direct current, provided that all conductors are insulated for the maximum voltage of any conductor in the enclosure, cable, or raceway.

(2) Power supply conductors and Class 1 circuit conductors shall not be permitted in the same enclosure, cable, or raceway except when connected to the same equipment, and all conductors are insulated for the maximum voltage of any conductor in the enclosure, cable, or raceway.

16-116 Mechanical protection of remote-control circuits

Where mechanical damage to a remote-control circuit would result in a hazardous condition as outlined in Rule 16-010, all conductors of such remote-control circuits shall be installed in conduit, electrical metallic tubing, or be otherwise suitably protected from mechanical injury or other injurious conditions such as moisture, excessive heat, or corrosive action.

16-118 Class 1 circuits extending aerially beyond a building

Class 1 circuits that extend aerially beyond a building shall comply with Rules 12-300 to 12-318.

Class 2 circuits

16-200 Limitations of Class 2 circuits (see Appendix B)

(1) Class 2 circuits, depending upon the voltage, shall have the current limited as follows:
    (a) **0 to 20 V** — circuits in which the open-circuit voltage does not exceed 20 V shall have overcurrent protection rated at not more than 5 A, except that overcurrent protection shall not be required where the current is supplied from
        (i) primary batteries that under short-circuit will not supply a current exceeding 7.5 A after 1 min;
        (ii) a Class 2 circuit transformer;
        (iii) a device having characteristics that will limit the current under normal operating conditions or under fault conditions to a value not exceeding 5 A; or

  (iv) a device having a Class 2 output;

(b) **Over 20 V but not exceeding 30 V** — circuits in which the open-circuit voltage exceeds 20 V but does not exceed 30 V shall have an overcurrent protection rating not exceeding 100/V amperes, where V is the open-circuit voltage, except that the overcurrent protection shall not be required where the current is supplied from

  (i) primary batteries that under short-circuit will not supply a current exceeding 5 A after 1 min;

  (ii) a Class 2 circuit transformer; or

  (iii) a device having characteristics that will limit the current under normal operating conditions or under fault conditions to a value not exceeding 100/V amperes, where V is the open-circuit voltage; or

  (iv) a device having a Class 2 output;

(c) **Over 30 V but not exceeding 60 V** — circuits in which the open-circuit voltage exceeds 30 V but does not exceed 60 V shall have an overcurrent protection rating not exceeding 100/V amperes, where V is the open-circuit voltage, except that the overcurrent protection shall not be required where the current is supplied from

  (i) a Class 2 circuit transformer; or

  (ii) a device having characteristics that will limit the current under normal operating conditions or under fault conditions to a value not exceeding 100/V amperes, where V is the open-circuit voltage;

(d) **Over 60 V but not exceeding 150 V** — circuits in which the open-circuit voltage exceeds 60 V but does not exceed 150 V shall have an overcurrent protection rating not exceeding 100/V amperes, where V is the open-circuit voltage, and in addition shall be equipped with current-limiting means other than overcurrent protection, which will limit the current, either under normal operating conditions or under fault conditions, to a value not exceeding 100/V amperes, where V is the open-circuit voltage.

(2) A device having energy-limiting characteristics may consist of a series resistor of suitable rating or other similar device.

(3) A Class 2 power supply shall not be connected in series or parallel with another Class 2 power source.

16-202 Methods of installation on supply side of overcurrent protection or transformers or other devices for Class 2 circuits

In Class 2 circuits, the conductors and equipment on the supply side of overcurrent protection, transformers, or current-limiting devices shall be installed in accordance with the requirements of other appropriate Sections of this Code.

16-204 Marking

A Class 2 power supply unit shall have permanent markings that shall be readily visible after installation to indicate the class of supply and its electrical rating.

16-206 Overcurrent protection for Class 2 circuits

(1) Overcurrent protection of different ratings shall not be of an interchangeable type.

(2) The overcurrent protection shall be permitted to be an integral part of a transformer or other power-supply device.

16-208 Location of overcurrent devices

Overcurrent devices shall be located at the point where the conductor to be protected receives its supply.

16-210 Conductors for Class 2 circuit wiring (see Appendix B)

(1) Conductors for use in Class 2 circuits shall be of the type suitable for the application as indicated in Table 19 except that where conductors smaller than No. 14 AWG are permitted, use of the types of equipment wire identified in Note (1) to Table 11 shall be permitted, provided that the equipment wires are installed in raceways.

(2) Type ELC conductors shall be limited in use to

  (a) Class 2 circuits operating at 30 V or less;

  (b) dwelling units in buildings of combustible construction;

  (c) dry locations; and

  (d) where concealed or exposed, when not subject to mechanical injury.

(3) Type ELC conductors shall not be permitted for the wiring of heating control circuits or fire safety circuits such as fire alarm or smoke alarm devices.

(4)   Conductors shall be of copper and shall not be smaller than
    (a)   No. 16 AWG for individual conductors pulled into raceways;
    (b)   No. 19 AWG for individual conductors laid in raceways;
    (c)   No. 19 AWG for an integral assembly of two or more conductors;
    (d)   No. 22 AWG for an integral assembly of four or more conductors;
    (e)   No. 24 AWG for an integral assembly of six or more conductors; and
    (f)   No. 26 AWG for an integral assembly of ten or more conductors.
(5)   Notwithstanding Subrule (4)(d), Type ELC wire shall be permitted in an integral assembly of two or more conductors for No. 22 AWG copper wire where the conductors are not pulled into raceways.
(6)   The maximum allowable current shall be as listed in Table 57 for sizes No. 16 AWG and smaller, but in no case shall exceed the current limitations of Rule 16-200.

16-212 Separation of Class 2 circuit conductors from other circuits

(1)   Conductors of Class 2 circuits shall be separated at least 50 mm from insulated conductors of electric lighting, power, or Class 1 circuits operating at 300 V or less, and shall be separated at least 600 mm from any insulated conductors of electric lighting, power, or Class 1 circuits operating at more than 300 V unless for both conditions effective separation is afforded by use of
    (a)   metal raceways for the Class 2 circuits or for the electric lighting, power, and Class 1 circuits subject to the metal raceway being bonded to ground;
    (b)   metal-sheathed or armoured cable for the electric lighting, power, and Class 1 circuit conductors subject to the sheath or armour being bonded to ground;
    (c)   non-metallic-sheathed cable for the electric lighting, power, and Class 1 circuits operating at 300 V or less; or
    (d)   non-metallic conduit, electrical non-metallic tubing, insulated tubing, or equivalent, in addition to the insulation on the Class 2 circuit conductors or the electric lighting, power, and Class 1 circuit conductors.
(2)   Where the electric lighting or power conductors are bare, all Class 2 circuit conductors in the same room or space shall be enclosed in a metal raceway that is bonded to ground and no opening, such as an outlet box, shall be permitted to be located within 2 m of the bare conductors if up to and including 15 kV or within 3 m of bare conductors above 15 kV.
(3)   Unless the conductors of the Class 2 circuits are separated from the conductors of electric lighting, power, and Class 1 circuits by an acceptable barrier, the conductors in Class 2 circuits shall not be placed in any raceway, compartment, outlet box, junction box, or similar fitting with the conductors of electric lighting, power, or Class 1 circuits.
(4)   Subrule (3) shall not apply where the conductors of a power circuit are in the raceway, compartment, outlet box, junction box, or similar fitting for the sole purpose of supplying power to the Class 2 circuits, and all conductors are insulated for the maximum voltage of any conductor in the enclosure, cable, or raceway, except that no Class 2 conductor installed in a raceway, compartment, outlet box, junction box, or similar fitting with such conductors of a power circuit shall show a green-coloured insulation, unless such Class 2 conductor is completely contained within a sheathed or jacketed cable assembly throughout the length that is present in such a raceway or enclosure.

16-214 Conductors of different Class 2 circuits in the same cable, enclosure, or raceway

Conductors of two or more Class 2 circuits shall be permitted within the same cable, enclosure, or raceway provided that all conductors in the cable, enclosure, or raceway are insulated for the maximum voltage of any conductor.

16-216 Penetration of a fire separation

Conductors of a Class 2 circuit extending through a fire separation shall be installed so as to limit fire spread in accordance with Rule 2-124.

16-218 Conductors in vertical shafts and hoistways

Class 2 conductors and cables installed in a vertical shaft or hoistway shall meet the requirements of Rules 2-124 and 2-126.

16-220 Class 2 conductors and equipment in ducts and plenum chambers

Class 2 conductors and equipment shall not be placed in ducts or plenum chambers except as permitted by Rules 2-126 and 12-010.

16-222 Equipment located on the load side of overcurrent protection, transformers, or current-limiting devices for Class 2 circuits (see Appendix B)

(1)   Equipment located on the load side of overcurrent protection, transformers, or current-limiting devices for Class 2 circuits shall

    (a)   for Class 2 circuits operating at not more than 42.4 V peak or dc be acceptable for the particular application; and

    (b)   for Class 2 circuits operating at more than 42.4 V peak or dc be arranged so that no live parts are accessible to unauthorized persons.

(2)   Notwithstanding Subrule (1), lighting fixtures, electromedical equipment, equipment for hazardous locations, and thermostats incorporating heat anticipators shall be approved.

16-224 Class 2 circuits extending beyond a building

Where Class 2 circuits extend beyond a building and are run in such a manner as to be subject to accidental contact with lighting or power conductors operating at a voltage exceeding 300 V between conductors, the conductors of the Class 2 circuits shall also meet the requirements of Section 60.

16-226 Underground installations

(1)   Underground installations of Class 2 circuits shall be installed in accordance with Rule 12-012.

(2)   Direct buried Class 2 circuits shall maintain a minimum horizontal separation of 300 mm from other underground systems except when installed in accordance with Subrule (3).

(3)   Direct buried Class 2 circuits shall be permitted to be placed at random separation in a common trench with power circuits that are for the sole purpose of supplying power to the Class 2 circuits, provided that

    (a)   the Class 2 circuit is in a metal-sheathed cable, with sheath bonded to ground;

    (b)   the power circuit operates at 750 V or less; and

    (c)   all conductors are insulated for the maximum voltage of any conductor in the trench.

Section 18 — Hazardous locations

Scope and introduction

Δ **18-000 Scope** (see Appendices B, F, and J)
(1) This Section applies to locations in which electrical equipment and wiring are subject to the conditions indicated by the following classifications.
(2) This Section supplements or amends the general requirements of this Code.
(3) For additions, modifications, renovations to, or operation and maintenance of existing facilities employing the Division system of classification for Class I locations, the continued use of the Division system of classification shall be permitted.
(4) Where the Division system of classification is used for Class I locations, as permitted by Subrule (3), the Rules for Class I locations found in Annex J18 of Appendix J shall apply.

Δ **18-002 Special terminology** (see Appendix B)
In this Section, the following definitions apply:

Cable gland — a device or combination of devices intended to provide a means of entry of a cable or flexible cord into an enclosure situated in a hazardous location and that also provides strain relief and shall be permitted to provide sealing characteristics where required, either by an integral means or when combined with a separate sealing fitting.

Cable seal — a seal that is installed at a cable termination to prevent the release of an explosion from an explosion-proof enclosure and that minimizes the passage of gases or vapours at atmospheric pressure.

Conduit seal — a seal that is installed in a conduit to prevent the passage of an explosion from one portion of the conduit system to another and that minimizes the passages of gases or vapours at atmospheric pressure.

Degree of protection — the measures applied to the enclosures of electrical apparatus to ensure
(a) the protection of persons against contact with live or moving parts inside the enclosure and protection of apparatus against the ingress of solid foreign bodies; and
(b) the protection of apparatus against ingress of liquids.

Explosive gas atmosphere — a mixture with air, under atmospheric conditions, of flammable substances in the form of gas, vapour, or mist in which, after ignition, combustion spreads throughout the unconsumed mixture.

Explosive limits — the lower and upper percentage by volume of concentration of gas in a gas-air mixture that will form an ignitable mixture.

    LEL — lower explosive limit.

    UEL — upper explosive limit.

Methods of protection — defined methods to reduce the risk of ignition of explosive gas atmospheres.

Non-incendive circuit — a circuit in which any spark or thermal effect that may occur under normal operating conditions or due to opening, shorting, or grounding of field wiring is incapable of causing an ignition of the prescribed flammable gas or vapour.

Normal operation — the situation when the plant or equipment is operating within its design parameters.

Primary seal — a seal that isolates process fluids from an electrical system and that has one side of the seal in contact with the process fluid.

Secondary seal — a seal that is designed to prevent the passage of process fluids at the pressure it will be subjected to upon failure of the primary seal.

Protective gas — the gas used to maintain pressurization or to dilute a flammable gas or vapour.

18-004 Classification
Hazardous locations shall be classified according to the nature of the hazard, as follows:
(a) Class I locations are those in which flammable gases or vapours are or may be present in the air in quantities sufficient to produce explosive gas atmospheres;
(b) Class II locations are those that are hazardous because of the presence of combustible or electrically conductive combustible dusts; and

94

(c)   Class III locations are those that are hazardous because of the presence of easily ignitable fibres or flyings, but in which such fibres or flyings are not likely to be in suspension in air in quantities sufficient to produce ignitable mixtures.

Δ   **18-006 Division of Class I locations** (see Appendices B and J)
Class I locations shall be further divided into three Zones based upon frequency of occurrence and duration of an explosive gas atmosphere as follows:
(a)   Zone 0, consisting of Class I locations in which explosive gas atmospheres are present continuously or are present for long periods;
(b)   Zone 1, consisting of Class I locations in which
    (i)   explosive gas atmospheres are likely to occur in normal operation; or
    (ii)   the location is adjacent to a Class I, Zone 0 location, from which explosive gas atmospheres could be communicated.
(c)   Zone 2, consisting of Class I locations in which
    (i)   explosive gas atmospheres are not likely to occur in normal operation and, if they do occur, they will exist for a short time only; or
    (ii)   the location is adjacent to a Class I, Zone 1 location, from which explosive gas atmospheres could be communicated, unless such communication is prevented by adequate positive-pressure ventilation from a source of clean air, and effective safeguards against ventilation failure are provided.

Δ   **18-008 Division of Class II locations** (see Appendix B)
Class II locations shall be further divided into two Divisions as follows:
(a)   Division 1, consisting of Class II locations in which
    (i)   combustible dust is or may be in suspension in air continuously, intermittently, or periodically under normal operating conditions in quantities sufficient to produce explosive or ignitable mixtures;
    (ii)   the abnormal operation or failure of equipment might
        (A)   cause explosive or ignitable mixtures to be produced; and
        (B)   provide a source of ignition through simultaneous failure of electrical equipment, operation of protection devices, or from other causes; or
    (iii)   combustible dusts having the property of conducting electricity may be present.
(b)   Division 2, consisting of Class II locations in which
    (i)   combustible dust may be in suspension in the air as a result of infrequent malfunctioning of handling or processing equipment, but such dust would be present in quantities insufficient to
        (A)   interfere with the normal operation of electrical or other equipment; and
        (B)   produce explosive or ignitable mixtures, except for short periods of time; or
    (ii)   where combustible dust accumulations on, in, or in the vicinity of the electrical equipment, may be sufficient to interfere with the safe dissipation of heat from electrical equipment or may be ignitable by abnormal operation or failure of electrical equipment.

18-010 Division of Class III locations (see Appendix B)
Class III locations shall be further divided into two Divisions as follows:
(a)   Division 1, consisting of Class III locations in which readily ignitable fibres or materials producing combustible flyings are handled, manufactured, or used; and
(b)   Division 2, consisting of Class III locations in which readily ignitable fibres other than those in process of manufacture are stored or handled.

General

18-050 Electrical equipment (see Appendix B)
(1)   Where electrical equipment is required by this Section to be approved for use in hazardous locations, it shall also be approved for the specific gas, vapour, mist, or dust that will be present.
(2)   For equipment approved with a method of protection permitted in a Class I location, such approval shall be permitted to be indicated by one or more of the following atmospheric group designations:
    (a)   **Group IIC**, consisting of atmospheres containing acetylene, carbon disulphide, or hydrogen, or other gases or vapours of equivalent hazard;
    (b)   **Group IIB**, consisting of atmospheres containing acrylonitrile, butadiene, diethyl ether, ethylene, ethylene oxide, hydrogen sulphide, propylene oxide, or unsymmetrical dimethyl hydrazine (UDMH), or other gases or vapours of equivalent hazard;

(c) **Group IIA**, consisting of atmospheres containing acetaldehyde, acetone, cyclopropane, alcohol, ammonia, benzine, benzol, butane, ethylene dichloride, gasoline, hexane, isoprene, lacquer solvent vapours, naphtha, natural gas, propane, propylene, styrene, vinyl acetate, vinyl chloride, xylenes, or other gases or vapours of equivalent hazard;

(d) **Group II**, consisting of all Group II gases;

(e) **Group IIXXXXX**, where XXXXX is a chemical formula or chemical name suitable for that specific gas only.

(3) For equipment approved for Class I, Division 1 or 2, the specific gas shall be permitted to be indicated by one or more of the following atmospheric group designations:

(a) **Group A**, consisting of atmospheres containing acetylene;

(b) **Group B**, consisting of atmospheres containing butadiene, ethylene oxide, hydrogen (or gases or vapours equivalent in hazard to hydrogen, such as manufactured gas), or propylene oxide;

(c) **Group C**, consisting of atmospheres containing acetaldehyde, cyclopropane, diethyl ether, ethylene, hydrogen sulphide, or unsymmetrical dimethyl hydrazine (UDMH), or other gases or vapours of equivalent hazard;

(d) **Group D**, consisting of atmospheres containing acetone, acrylonitrile, alcohol, ammonia, benzine, benzol, butane, ethylene dichloride, gasoline, hexane, isoprene, lacquer solvent vapours, naphtha, natural gas, propane, propylene, styrene, vinyl acetate, vinyl chloride, xylenes, or other gases or vapours of equivalent hazard.

(4) Notwithstanding Subrule (3)(b), where the atmosphere contains

(a) butadiene, Group D equipment shall be permitted to be used if such equipment is isolated in accordance with Rule 18-108(1) by sealing all conduit 16 trade size or larger; or

(b) ethylene oxide or propylene oxide, Group C equipment shall be permitted to be used if such equipment is isolated in accordance with Rule 18-108(1) by sealing all conduit 16 trade size or larger.

(5) For equipment approved for Class II locations, approval for the specific dust shall be permitted to be indicated by one or more of the following atmospheric group designations:

(a) **Group E**, consisting of atmospheres containing combustible metal dust, including aluminum, magnesium, and their commercial alloys, and other metals of similarly hazardous characteristics;

(b) **Group F**, consisting of atmospheres containing carbon black, coal, or coke dust;

(c) **Group G**, consisting of atmospheres containing flour, starch, or grain dust, and other dusts of similarly hazardous characteristics.

18-052 Marking (see Appendix B)

(1) Electrical equipment intended for use in Class I hazardous locations shall be permitted to be marked with the following:

(a) the letters "Ex" or "EEx";

(b) the symbol(s) to indicate a method(s) of protection used;

(c) the gas group as specified in Rule 18-050(2); and

(d) the temperature rating in accordance with Subrule (4) for equipment of the heat-producing type.

(2) Notwithstanding Subrule (1), electrical equipment approved for

(a) Class I or Class I, Division 1 or 2 locations shall be permitted to be marked with the class and the group described in Rule 18-050(3), or the specific gas or vapour for which it has been approved;

(b) Class I, Division 2 only, shall be permitted to be so marked.

(3) Electrical equipment approved for Class II and III hazardous locations shall be permitted to be so marked and, for Class II locations, with the group or specific dust for which it has been approved.

(4) Electrical equipment permitted for use in Class I locations shall be permitted to be marked with

(a) the maximum surface temperature; or

(b) one of the following temperature codes to indicate the maximum surface temperature:

Temperature code	Maximum surface temperature
T1	450 °C
T2	300 °C
T2A	280 °C
T2B	260 °C
T2C	230 °C
T2D	215 °C
T3	200 °C
T3A	180 °C
T3B	165 °C
T3C	160 °C
T4	135 °C
T4A	120 °C
T5	100 °C
T6	85 °C

(5) If no maximum surface temperature marking is shown on Class I equipment approved for the class and group, the equipment, if of the heat-producing type, shall be considered as having a maximum surface temperature of 100 °C or less for the purpose of compliance with Rule 18-054.

(6) Electrical equipment approved for operation at ambient temperatures exceeding 40 °C shall, in addition to the marking specified in Rule 18-052(3), be marked with the maximum ambient temperature for which the equipment is approved and with the maximum surface temperature of the equipment at that ambient temperature.

18-054 Temperature (see Appendix B)

(1) In Class I hazardous locations, equipment shall not be installed in an area where vapours or gases are present that have an ignition temperature less than the maximum external temperature of the equipment as referred to in Rules 18-052(4) and (5).

(2) Where the equipment is not required to be approved for hazardous locations, the maximum external temperature referred to in Subrule (1) shall be the surface temperature at any point, internal or external, of the equipment.

18-056 Non-essential electrical equipment

(1) No electrical equipment shall be used in a hazardous location, unless the equipment is essential to the processes being carried on therein.

(2) Service equipment, panelboards, switchboards, and similar electrical equipment shall, where practicable, be located in rooms or sections of the building in which hazardous conditions do not exist.

18-058 Rooms, sections, or areas

Each room, section, or area, including motor- and generator-rooms and rooms for the enclosure of control equipment, shall be considered as a separate location for the purpose of determining the classification of the hazard.

18-060 Equipment rooms

(1) Where walls, partitions, floors, or ceilings are used to form hazard-free rooms or sections, they shall be
   (a) of substantial construction;
   (b) built of or lined with non-combustible material; and
   (c) such as to ensure that the rooms or sections will remain free from hazards.

(2) Where a non-hazardous location within a building communicates with a Class I, Zone 2 location, a Class II location, or a Class III location, the locations shall be separated by close-fitting, self-closing, approved fire doors.

(3) For communication from a Class I, Zone 1 location, the provisions of Rule 18-006(c)(ii) shall apply.

18-062 Metal-covered cable (see Appendix B)

(1) Where exposed overhead conductors supply mineral-insulated cable in a hazardous location, surge arresters shall be installed to limit the surge voltage level to 5 kV on the cable.

(2) Where single-conductor metal-covered cable is used in hazardous locations, it shall be installed in such a manner as to prevent sparking between cable sheaths or between cable sheaths and metal bonded to ground, and

    (a) cables in the circuit shall be clipped or strapped together, in a manner that will ensure good electrical contact between metal coverings, at intervals of not more than 1.8 m, and the metal coverings shall be bonded to ground; or

    (b) cables in the circuit shall have the metal coverings continuously covered with insulating material and the metal coverings shall be bonded to ground at the point of termination in the hazardous location only.

18-064 Pressurized equipment or rooms (see Appendix B)

Electrical equipment and associated wiring in Class I locations shall be permitted to be located in enclosures or rooms constructed and arranged so that a protective gas pressure is effectively maintained, in which case the provisions of Rules 18-100 to 18-180 of this Section need not apply.

18-066 Intrinsically safe and non-incendive electrical equipment and wiring (see Appendices B and F)

(1) Where electrical equipment is approved as intrinsically safe and associated circuits are designed and installed as intrinsically safe for the intended hazardous location, they shall be permitted and the provisions of Rules 18-100 to 18-376 of this Section need not apply.

(2) Where electrical equipment is approved as non-incendive and associated circuits are designed and installed as non-incendive, they shall be permitted in Class I, Zone 2 locations and the provisions of Rules 18-152 to 18-158 need not apply.

(3) Raceways or cable systems for intrinsically safe and non-incendive wiring and equipment in Class I locations shall be installed to prevent migration of gas or vapour to other locations.

(4) The conductors in intrinsically safe and non-incendive circuits shall not be placed in any raceway, compartment, outlet, junction box, or similar fitting with the conductors of any other system unless the conductors of the two systems are separated by a suitable mechanical barrier.

18-068 Cable trays

Cable trays shall not be used to support cables in hazardous locations except where

    (a) the type of cable is approved in the Rules of this Section for use in the particular hazardous location;

    (b) the type of cable is approved for use in cable trays in accordance with Rule 12-2202; and

    (c) there can be no hazardous accumulation of combustible process dust or fibre in or upon the cable, the cable tray, or the supports.

Δ ### 18-070 Combustible gas detection (see Appendices B and H)

Electrical equipment suitable for non-hazardous locations shall be permitted to be installed in a Class I, Zone 2 hazardous location and electrical equipment suitable for Class I, Zone 2 hazardous locations shall be permitted to be installed in a Class I, Zone 1 hazardous location provided that

    (a) no specific equipment suitable for the purpose is available;

    (b) the equipment, during its normal operation, does not produce arcs, sparks, or hot surfaces, capable of igniting an explosive gas atmosphere; and

    (c) the location is continuously monitored by a combustible gas detection system that will

        (i) activate an alarm when the gas concentration reaches 20% of the lower explosive limit;

        (ii) activate ventilating equipment or other means designed to prevent the concentration of gas from reaching the lower explosive limit when the gas concentration reaches 20% of the lower explosive limit, where such ventilating equipment or other means is provided;

        (iii) automatically de-energize the electrical equipment being protected when the gas concentration reaches 40% of the lower explosive limit, where the ventilating equipment or other means referred to in Item (ii) is provided;

        (iv) automatically de-energize the electrical equipment being protected when the gas concentration reaches 20% of the lower explosive limit, where the ventilating equipment or other means referred to in Item (ii) cannot be provided; and

(v)   automatically de-energize the electrical equipment being protected upon failure of the gas detection instrument.

Δ   **18-072   Flammable gas or liquid seals**
Electrical equipment containing a seal intended to prevent flammable gases or liquids from reaching the electrical housing or conduit system shall not be used at pressures in excess of the marked maximum working pressure (MWP).

Δ   **18-074   Bonding in hazardous locations**
(1)   Exposed non-current-carrying metal parts of electrical equipment, including the frames or metal exteriors of motors, fixed or portable lamps or other utilization equipment, lighting fixtures, cabinets, cases, and conduit shall be bonded to ground using
(a)   bonding conductors sized in accordance with Rule 10-814; or
(b)   rigid metal conduit with threaded couplings and threaded bosses on enclosures with joints made up tight.
(2)   Notwithstanding Subrule (1), where raceways or cable assemblies incorporate an internal bonding conductor, box connectors with standard locknuts shall be permitted to bond the metallic armour or raceway.

Class I locations

Installations in Class I, Zone 0 locations

Δ   **18-090   Equipment and wiring** (see Appendices B and F)
(1)   Except as provided for in Subrules (2) and (3), electrical equipment and wiring shall not be installed in a Class I, Zone 0 hazardous location.
(2)   Electrical equipment that is approved as intrinsically safe, type i or ia, shall be permitted in Class I, Zone 0 locations.
(3)   Intrinsically safe circuits and wiring shall be designed for the application and shall be installed in accordance with the design.

Δ   **18-092   Sealing Class I, Zone 0** (see Appendix B)
(1)   Secondary seals shall be provided, between devices containing a primary seal and conduit or cable seals, where failure of a single component in the device containing the primary seal could allow passage of process fluids.
(2)   Where secondary seals are installed, drains, vents or other devices intended to make primary seal leakage obvious shall be installed.
(3)   Conduit seals shall be provided where the conduit leaves the Class I, Zone 0 location with no box, coupling, or fitting in the conduit run between the seal and the point at which the conduit leaves the location, except that a rigid unbroken conduit that passes completely through a Class I, Zone 0 area, with no fittings less than 300 mm beyond each boundary, need not be sealed provided that the termination points of the unbroken conduit are in non-hazardous areas.
(4)   Cable seals shall be provided on cables at the first point of termination after entry into the Zone 0 location.

Installations in Class I, Zone 1 locations

18-100   Equipment in Class I, Zone 1 locations (see Appendices B and F)
Where required by other Rules of this Code, electrical equipment installed in a Class I, Zone 1 location shall be approved
(a)   for Class I or Class I, Division 1 locations; or
(b)   as providing one or more of the following methods of protection:
(i)   intrinsically safe i, ia, or ib;
(ii)   flame-proof d;
(iii)   increased safety e;
(iv)   oil immersed o;
(v)   pressurized p;
(vi)   powder-filled q;
(vii)   encapsulation m.

18-102 Transformers and capacitors, Class I, Zone 1

Transformers and electrical capacitors shall comply with the requirements of Rule 18-100 or shall be installed in electrical equipment vaults in accordance with Rules 26-350 to 26-356, and

(a) there shall be no door or other connecting opening between the vault and the hazardous area;

(b) the vault shall be provided with adequate ventilation;

(c) vent openings or vent ducts shall lead to a safe location outside the building containing the vault;

(d) vent openings and vent ducts shall be of sufficient area to relieve pressure caused by explosions within the vault; and

(e) every portion of a vent duct within the building shall be constructed of reinforced concrete.

18-104 Meters, instruments, and relays, Class I, Zone 1

(1) Where practicable, meters, instruments, and relays, including kilowatt-hour meters, instrument transformers and resistors, rectifiers, and thermionic tubes, shall be located outside the hazardous location.

(2) Where it is not practicable to install meters, instruments, and relays outside Class I, Zone 1 locations, they shall comply with the requirements of Rule 18-100.

18-106 Wiring methods, Class I, Zone 1 (see Appendix B)

(1) The wiring method shall be threaded rigid metal conduit or cables approved for hazardous locations with associated cable glands that comply with the requirements of Rule 18-100.

(2) Explosion-proof or flame-proof boxes, fittings, and joints shall be threaded for connection to conduit and cable glands.

(3) Threaded joints that are required to be explosion-proof or flame-proof shall be permitted to be either tapered or straight and shall comply with the following:

(a) tapered threads shall have at least five fully engaged threads, and running threads shall not be used;

(b) where straight threads are used in Groups IIA and IIB atmospheres, they shall have at least five fully engaged threads; and

(c) where straight threads are used in Groups IIC atmospheres, they shall have at least eight fully engaged threads.

(4) Where threadforms differ between the equipment and the wiring system, approved adapters shall be used.

(5) Conduit and cable entries into increased safety "e" enclosures shall be made in such a manner as to maintain the degree of protection provided by the enclosure.

(6) Cables shall be installed and supported in a manner to avoid tensile stress at the cable glands.

(7) Where flexible fittings are used for connection at motor terminals and similar places, they shall be of a type approved for the location.

Δ 18-108 Sealing, Class I, Zone 1 (see Appendix B)

(1) Secondary seals shall be provided between devices containing a primary seal and conduit or cable seals, where failure of a single component in the device containing the primary seal could allow passage of process fluids.

(2) Where secondary seals are installed, drains, vents or other devices intended to make primary seal leakage obvious shall be installed.

(3) Conduit seals shall be provided in conduit systems where

(a) the conduit enters an explosion-proof or flame-proof enclosure containing devices that may produce arcs, sparks or high temperatures and shall be located as close as practicable to the enclosure, or as marked on the enclosure, but not further than 450 mm from the enclosure; or

(b) the conduit is 53 trade size or larger and enters an explosion-proof or flame-proof enclosure housing terminals, splices, or taps, and shall be located no further than 450 mm from the enclosure; or

(c) the conduit leaves the Class I, Zone 1 location with no box, coupling, or fitting in the conduit run between the seal and the point at which the conduit leaves the location, except that a rigid unbroken conduit that passes completely through a Class I, Zone 1 area with no fittings less than 300 mm beyond each boundary, need not be sealed provided that the termination points of the unbroken conduit are in non-hazardous areas; or

(d) the conduit enters an enclosure that is not required to be explosion-proof or flame-proof, except that a seal is not required where an unbroken and continuous run of conduit connects two enclosures that are not required to be explosion-proof or flame-proof.

(4) Only explosion-proof or flame-proof unions, couplings, reducers, and elbows that are not larger than the trade size of the conduit shall be permitted between the sealing fitting and an explosion-proof or flame-proof enclosure.

(5) Cable seals shall be provided in a cable system where
    (a) the cable enters an enclosure required to be explosion-proof or flame-proof; or
    (b) the cable enters an enclosure not required to be explosion-proof or flame-proof; and
        (i) the cable leaves the Zone 1 area and is less than 10 m in length; or
        (ii) the other end of the cable terminates in a Zone 2 or non-hazardous location in which a negative atmospheric pressure greater than 0.2 kPa exists.

(6) Where secondary seals, cable seals, or conduit seals are required, they shall conform to the following:
    (a) the seal shall be made
        (i) in a field-installed sealing fitting or cable gland that shall be accessible and shall comply with the requirements of Rule 18-100; or
        (ii) in a sealing fitting provided as part of an enclosure approved for the area and where the seal is factory made, the enclosure shall be marked to indicate that such a seal is provided;
    (b) splices and taps shall not be made in fittings intended only for sealing with compound, nor shall other fittings in which splices or taps are made be filled with compound;
    (c) where there is a probability that liquid or other condensed vapour may be trapped within enclosures for control equipment or at any point in the raceway system, approved means shall be provided to prevent accumulation or to permit periodic draining of such liquid or condensed vapour; and
    (d) where there is a probability that liquid or condensed vapour may accumulate within motors or generators, joints and conduit systems shall be arranged to minimize entrance of liquid, but if means to prevent accumulation or to permit periodic draining are judged necessary, such means shall be provided at the time of manufacture and shall be deemed an integral part of the machine.

(7) Runs of cables, each having a continuous sheath, either metal or non-metal, shall be permitted to pass through a Class I, Zone 1 location without seals.

(8) Cables that do not have a continuous sheath, either metal or non-metal, shall be sealed at the boundary of the Zone 1 location.

18-110 Switches, motor controllers, circuit breakers, and fuses, Class I, Zone 1

Switches, motor controllers, circuit breakers, and fuses, including push buttons, relays, and similar devices, shall be provided with enclosures, and the enclosure in each case together with the enclosed apparatus shall be approved as a complete assembly and shall comply with the requirements of Rule 18-100.

18-112 Control transformers and resistors, Class I, Zone 1

Transformers, impedance coils, and resistors used as or in conjunction with control equipment for motors, generators, and appliances and the switching mechanism, if any, associated with them, shall comply with the requirements of Rule 18-100.

18-114 Motors and generators, Class I, Zone 1 (see Appendix B)

Motors, generators, and other rotating electrical machines shall comply with the requirements of Rule 18-100.

18-116 Ignition systems for gas turbines, Class I, Zone 1 (see Appendix B)

Ignition systems for gas turbines shall comply with the requirements of Rule 18-100.

18-118 Lighting fixtures, Class I, Zone 1

(1) Fixtures for fixed and portable lighting shall be approved as complete assemblies in accordance with the requirements of Rule 18-100 and shall be clearly marked to indicate the maximum wattage of lamps for which they are approved.

(2) Fixtures intended for portable use shall be specifically approved as complete assemblies for that use.

(3) Each fixture shall be protected against physical damage by a suitable guard or by location.

(4) Pendant fixtures shall be
    (a) suspended by and supplied through threaded rigid conduit stems, and threaded joints shall be provided with set screws or other effective means to prevent loosening; and
    (b) for stems longer than 300 mm, provided with permanent and effective bracing against lateral displacement at a level not more than 300 mm above the lower end of the stem, or provided with flexibility in the form of a fitting or flexible connector approved for the purpose and for the location not more than 300 mm from the point of attachment to the supporting box or fitting.

(5) Boxes, box assemblies, or fittings used for the support of lighting fixtures shall be approved for the purpose and shall comply with the requirements of Rule 18-100.

18-120 Utilization equipment, fixed and portable, Class I, Zone 1

(1)   Utilization equipment, fixed and portable, including electrically heated and motor-driven equipment, shall comply with the requirements of Rule 18-100.

(2)   Ground fault protection shall be provided to de-energize all normally ungrounded conductors of an electric heat tracing cable set with the ground fault trip setting adjusted to allow normal operation of the heater.

18-122 Flexible cords, Class I, Zone 1

Flexible cords shall be permitted to be used only for connection between a portable lamp or other portable utilization equipment and the fixed portion of its supply circuit and, where used, shall

(a)   be of a type approved for extra-hard usage;

(b)   contain, in addition to the conductors of the circuit, a bonding conductor; and

(c)   be provided with glands that comply with the requirements of Rule 18-100 where the flexible cord enters a box, fitting, or enclosure.

18-124 Receptacles and attachment plugs, Class I, Zone 1

Receptacles and attachment plugs shall be of the type providing for connection to the bonding conductor of the flexible cord and shall comply with the requirements of Rule 18-100.

18-126 Conductor insulation, Class I, Zone 1

Where condensed vapours or liquids may collect on or come in contact with the insulation on conductors, such insulation shall be of a type approved for use under such conditions or the insulation shall be protected by a sheath of lead or by other approved means.

18-128 Signal, alarm, remote-control, and communication systems, Class I, Zone 1

Signal, alarm, remote-control, and communication systems shall conform to the following:

(a)   all apparatus and equipment shall comply with the requirements of Rule 18-100; and

(b)   all wiring shall comply with Rules 18-106 and 18-108.

18-130 Live parts, Class I, Zone 1

No live parts of electrical equipment or of an electrical installation shall be exposed.

Installations in Class I, Zone 2 locations

18-150 Equipment in Class I, Zone 2 locations (see Appendices B and F)

Where required by other Rules of this Code, electrical equipment installed in a Class I, Zone 2 location shall be

(a)   approved for Class I, Division 2 locations;

(b)   approved as non-incendive;

(c)   approved as providing a method of protection "n";

(d)   equipment specifically allowed in Rules 18-066 and 18-152 through 18-178; or

(e)   equipment permitted in Zone 1.

18-152 Process instrumentation, communication, and remote-control equipment, Class I, Zone 2

Process instrumentation, communication, and remote-control equipment shall comply with the requirements of Rule 18-150 except that transformers, solenoids, and other windings that do not incorporate sliding or make-and-break contacts, or heat-producing resistance devices, are not required to comply with the requirements of Rule 18-150.

18-154 Transformers and capacitors, Class I, Zone 2

Installation of transformers and capacitors shall be permitted provided that they do not contain arcing or spark-producing components.

18-156 Wiring methods, Class I, Zone 2 (see Appendix B)

(1)   The wiring method shall be

    (a)   threaded metal conduit;

    (b)   cables approved for hazardous locations;

    (c)   Type TC cable, installed in cable tray in accordance with Rule 12-2202;

    (d)   Type ACWU cable;

    (e)   control and instrument cables with an interlocking metallic armour and a continuous jacket in control circuits (Type ACIC); or

    (f)   Type CIC cable (non-armoured control and instrumentation cable) installed in cable tray in accordance with the installation requirements of Rule 12-2202(2), where
        (i)   the voltage rating of the cable is not less than 300 V;
        (ii)   the circuit voltage is 150 V or less; and
        (iii) the circuit current is 5 A or less.

(2)   Explosion-proof or flame-proof boxes, fittings, and joints shall be threaded for connection to conduit and cable glands.

(3)   Threaded joints that are required to be explosion-proof or flame-proof shall be permitted to be either tapered or straight and shall comply with the following:
    (a)   tapered threads shall have at least five fully engaged threads, and running threads shall not be used;
    (b)   where straight threads are used in Groups IIA and IIB atmospheres, they shall have at least five fully engaged threads; and
    (c)   where straight threads are used in Group IIC atmospheres, they shall have at least eight fully engaged threads.

(4)   Where thread forms differ between the equipment and the wiring system, approved adapters shall be used.

(5)   Cables shall be installed and supported in a manner to avoid tensile stress at the cable glands.

(6)   Where it is necessary to use flexible connections at motor terminals and similar places, flexible metal conduit shall be permitted.

(7)   Boxes, fittings, and joints need not be explosion-proof or flame-proof except as required by the Rules in this Section.

(8)   Cable glands shall be compatible with the degree of protection and explosion protection provided by the enclosure that the cable enters, where the area classification and environmental conditions require these degrees of protection.

Δ **18-158 Sealing, Class I, Zone 2** (see Appendix B)

(1)   Secondary seals shall be provided, between devices containing a primary seal and conduit or cable seals, where failure of a single component in the device containing the primary seal could allow passage of process fluids.

(2)   Where secondary seals are installed, drains, vents or other devices intended to make primary seal leakage obvious shall be installed.

(3)   Conduit seals shall be provided in a conduit system where
    (a)   the conduit enters an enclosure that is required to be explosion-proof or flame-proof and shall be located as close as practicable to the enclosure, or as marked on the enclosure, but not further than 450 mm from the enclosure;
    (b)   the conduit leaves the Class I, Zone 2 location with no box, coupling, or fitting in the conduit run between the seal and the point at which the conduit leaves the location, except that a rigid unbroken conduit that passes completely through a Class I, Zone 2 area with no fittings less than 300 mm beyond each boundary, need not be sealed provided that the termination points of the unbroken conduit are in non-hazardous areas; or
    (c)   the conduit leaves a Class I, Zone 2 location outdoors; the seal may be located more than 300 mm beyond the Class I, Zone 2 boundary provided that it is located on the conduit prior to entering an enclosure or building.

(4)   Only explosion-proof or flame-proof unions, couplings, reducers, and elbows that are not larger than the trade size of the conduit shall be permitted between the sealing fitting and an explosion-proof or flame-proof enclosure.

(5)   Cable seals shall be provided in a cable system where
    (a)   the cable enters an enclosure required to be explosion-proof or flame-proof; or
    (b)   the cable enters an enclosure not required to be explosion-proof or flame-proof; and
        (i)   the cable leaves the Zone 2 area and is less than 10 m in length; or
        (ii)   the other end of the cable terminates in a non-hazardous location in which a negative atmospheric pressure greater than 0.2 kPa exists.

(6)   Where a run of conduit enters an enclosure that is required to be explosion-proof or flame-proof, every part of the conduit from the seal to that enclosure shall comply with Rule 18-106.

(7)   Runs of cables, each having a continuous sheath, either metal or non-metal, shall be permitted to pass through a Class I, Zone 2 location without seals.

(8)   Cables that do not have a continuous sheath, either metal or non-metal, shall be sealed at the boundary of the Zone 2 location.

(9)   Where seals are required, Rule 18-108(6) shall apply.

18-160 Switches, controllers, and circuit breakers, Class I, Zone 2 (see Appendix B)

(1)   Switches, controllers, and circuit breakers shall be provided with enclosures and shall comply with the requirements of Rule 18-150.

(2)   Notwithstanding Subrule (1), switches, controllers, and circuit breakers that are approved for the location shall be permitted to be provided with general-purpose enclosures.

18-162 Isolating switches, Class I, Zone 2

Isolating switches shall conform to the following:

(a)   they shall be interlocked with their associated current-interrupting devices such that they cannot be opened under load; and

(b)   they shall be permitted to have enclosures of the general-purpose type, provided that they are unfused.

Δ ### 18-164 Fuses for motors, appliances, and portable lamps, Class I, Zone 2

Where fuses are used in Class I, Zone 2 locations for the protection of motors, appliances, and portable lamps,

(a)   a standard plug or cartridge fuse shall be permitted to be used if placed within an explosion-proof or flame-proof enclosure; or

(b)   a fuse installed within a general-purpose enclosure shall be permitted provided that the operating element of the fuse
     (i)   is immersed in oil or other suitable liquid;
     (ii)   is enclosed within a hermetically sealed chamber; or
     (iii) is a non-indicating, filled, current-limiting type.

18-166 Sets of fuses or circuit breakers for fixed lighting, Class I, Zone 2 (see Appendix B)

(1)   In this Rule, "sets of fuses" means a group containing as many fuses as are required to perform a single protective function in a circuit, but excluding fuses conforming to Rule 18-164.

(2)   Where not more than
     (a)   ten sets of approved enclosed fuses; or
     (b)   ten circuit breakers that are not used as switches for the normal operation of the lamps
     are installed in Class I, Zone 2 locations for the protection of a branch circuit or a feeder circuit that supplies only lamps in a fixed position, the enclosures for the fuses or circuit breakers shall be permitted to be of the general-purpose type.

18-168 Motors and generators, Class I, Zone 2 (see Appendix B)

(1)   Motors, generators, and other rotating electrical machines that incorporate arcing, sparking, or heat-producing resistance components shall be explosion-proof or flame-proof unless these arcing, sparking, heat-producing components are provided with enclosures that comply with the requirements of Rule 18-100.

(2)   Motors, generators, and other rotating electrical machines that do not incorporate arcing, sparking, or heat-producing components shall be permitted to be of the open or non-explosion-proof type.

18-170 Ignition systems for stationary internal combustion engines, Class I, Zone 2 (see Appendix B)

Ignition systems for stationary internal combustion engines shall comply with the requirements of Rule 18-150.

18-172 Lighting fixtures, Class I, Zone 2

(1)   Lighting fixtures shall conform to the following:
     (a)   portable lamps shall conform to Rule 18-118(1) and (2); and
     (b)   fixed lighting shall be
          (i)   protected from physical damage by suitable guards or by location; and
          (ii)   in compliance with the requirements of Rule 18-150.

(2)   Pendant fixtures shall be
     (a)   suspended by threaded rigid conduit stems or by other approved means; and
     (b)   for stems longer than 300 mm, provided with permanent and effective bracing against lateral displacement at a level not more than 300 mm above the lower end of the stem, or flexibility in the form of a fitting or flexible connector approved for the purpose shall be provided not more than 300 mm from the point of attachment to the supporting box or fitting.

(3)   Boxes, box assemblies, or fittings used for the support of lighting fixtures shall be approved for the purpose.

(4)   Switches that are part of an assembled fixture or of an individual lampholder shall conform to Rule 18-160.

(5)   Starting and control equipment for electric-discharge lighting equipment incorporating arcing, sparking, or heat-producing devices shall be provided with enclosures that comply with the requirements of Rule 18-100.

18-174 Utilization equipment, fixed and portable, Class I, Zone 2

(1)   Electrically heated utilization equipment, whether fixed or portable, shall comply with the requirements of Rule 18-100.

(2)   Motors of motor-driven utilization equipment shall conform to Rule 18-168.

(3)   Switches, circuit breakers, and fuses forming part of or used in connection with utilization equipment shall conform to Rules 18-160 to 18-164.

18-176 Flexible cords, Class I, Zone 2

Flexible cords shall be permitted to be used only for connection between permanently mounted lighting fixtures, portable lamps, or other portable utilization equipment and the fixed portion of supply circuits and, where used, shall

(a)   be of a type approved for extra-hard usage;

(b)   contain, in addition to the circuit conductors, a bonding conductor; and

(c)   be provided with a sealing gland where the flexible cord enters a box, fitting, or enclosure that is required to be explosion-proof or flame-proof.

Δ **18-178 Receptacles and attachment plugs, Class I, Zone 2**

Receptacles and attachment plugs shall comply with the requirements of Rule 18-150.

18-180 Live parts, Class I, Zone 2

No live parts of electrical equipment or of an electrical installation shall be exposed.

Class II locations

Installations in Class II, Division 1 locations (see Appendix E)

18-200 Transformers and capacitors, Class II, Division 1

(1)   Transformers and electrical capacitors that contain a liquid that will burn shall be installed in electrical equipment vaults in accordance with Rules 26-350 to 26-356, and

(a)   doors or other openings communicating with the hazardous area shall have self-closing fire doors on both sides of the wall, and the doors shall be carefully fitted and provided with suitable seals (such as weatherstripping) to minimize the entrance of dust into the vault;

(b)   vent openings and ducts shall communicate only with the air outside the building; and

(c)   suitable pressure-relief openings communicating only with the air outside the building shall be provided.

(2)   Transformers and electrical capacitors that do not contain a liquid that will burn shall be

(a)   installed in electrical equipment vaults conforming to Subrule (1); or

(b)   approved as a complete assembly including terminal connections for Class II locations.

(3)   No transformer or capacitor shall be installed in a location where dust from magnesium, aluminum, aluminum bronze powders, or other metals of similarly hazardous characteristics may be present.

18-202 Wiring methods, Class II, Division 1 (see Appendix B)

(1)   The wiring method shall be threaded rigid metal conduit or cables approved for hazardous locations with associated cable glands approved for the particular hazardous location.

(2)   Boxes, fittings, and joints shall be threaded for connection to conduit or cable glands and boxes and fittings shall be approved for Class II locations.

(3)   Cables shall be installed and supported in a manner to avoid tensile stress at the cable glands.

(4)   Where flexible connections are necessary, they shall be provided by

(a)   flexible connection fittings approved for the location;

(b)   liquid-tight flexible conduit with fittings approved for the location; or

(c)   extra-hard-usage flexible cord and cable glands approved for the location.

(5)   Where flexible connections are subject to oil or other corrosive conditions, the insulation of the conductors shall be of a type approved for the condition or shall be protected by means of a suitable sheath.

18-204 Sealing, Class II, Division 1

Where a raceway provides communication between an enclosure that is required to be dust-tight and one that is not, the entrance of dust into the dust-tight enclosure through the raceway shall be prevented by
(a)   a permanent and effective seal;
(b)   a horizontal section not less than 3 m long in the raceway; or
(c)   a vertical section of raceway not less than 1.5 m long and extending downward from the dust-tight enclosure.

18-206 Switches, controllers, circuit breakers, and fuses, Class II, Division 1

Switches, motor controllers, circuit breakers, and fuses, including push buttons, relays, and similar devices, shall be provided with a dust-tight enclosure approved for Class II locations.

18-208 Control transformers and resistors, Class II, Division 1

Transformers, impedance coils, and resistors used as or in conjunction with control equipment for motors, generators, or electric appliances and the overcurrent devices or switching mechanisms, if any, associated with them shall be provided with a dust-tight enclosure approved for Class II locations.

18-210 Motors and generators, Class II, Division 1 (see Appendix B)

Motors, generators, and other rotating electrical machines shall be approved for Class II locations.

18-212 Ventilating pipes, Class II, Division 1 (see Appendix B)

(1)   Every vent pipe for a motor, generator, or other rotating electrical machine or for enclosures for electrical apparatus or equipment shall
    (a)   be of metal not less than 0.52 mm (No. 24 MSG) thick or of an equally substantial non-combustible material;
    (b)   lead directly to a source of clean air outside a building;
    (c)   be screened at the outer end to prevent the entrance of small animals or birds; and
    (d)   be protected against mechanical damage and corrosion.
(2)   Every vent pipe and its connection to a motor or to a dust-tight enclosure for other equipment or apparatus shall be dust-tight throughout its entire length.
(3)   The seams and joints of every metal vent pipe shall be
    (a)   riveted and soldered;
    (b)   bolted and soldered;
    (c)   welded; or
    (d)   rendered dust-tight by some other equally effective means.
(4)   No exhaust pipe shall discharge inside a building.

18-214 Utilization equipment, fixed and portable, Class II, Division 1

Utilization equipment, fixed and portable, including electrically heated and motor-driven equipment shall be approved for Class II locations.

18-216 Lighting fixtures, Class II, Division 1

(1)   Fixtures for fixed and portable lighting shall be approved as complete assemblies for Class II locations and shall be clearly marked to indicate the maximum wattage of lamps for which they are approved.
(2)   Fixtures intended for portable use shall be specifically approved as complete assemblies for that use.
(3)   Each fixture shall be protected against physical damage by a suitable guard or by location.
(4)   Pendant fixtures shall be
    (a)   suspended by threaded rigid conduit stems or chains with approved fittings or by other approved means, which shall not include a flexible cord as the supporting medium, and threaded joints shall be provided with set screws or other effective means to prevent loosening;
    (b)   for rigid stems longer than 300 mm, provided with permanent and effective bracing against lateral displacement at a level not more than 300 mm above the lower end of the stem, or provided with flexibility in the form of a fitting or flexible connector approved for the purpose and for the location not more than 300 mm from the point of attachment to the supporting box or fitting;
    (c) .   where wiring between an outlet box or fitting and the fixture is not enclosed in conduit, provided with a flexible cord approved for extra-hard usage and suitable seals where the cord enters the fixture and the outlet box or fitting.
(5)   Boxes, box assemblies, or fittings used for the support of lighting fixtures shall be approved for the purpose and Class II locations.

18-218 Flexible cords, Class II, Division 1

Flexible cords used shall

(a) be of a type approved for extra-hard usage; and

(b) contain a bonding conductor in addition to the conductors of the circuit; and

(c) be provided with glands approved for the class and group to prevent the entrance of dust at the point where the cord enters a box or fitting that is required by this Section to be dust-tight.

18-220 Receptacles and attachment plugs, Class II, Division 1

Receptacles and attachment plugs shall be approved for Class II locations.

18-222 Signal, alarm, remote-control, and communication systems, meters, instruments, and relays, Class II, Division 1

Signal, alarm, remote-control, and communication systems, and meters, instruments, and relays shall conform to the following:

(a) all apparatus and equipment shall be provided with enclosures approved for Class II locations, except that

    (i) devices that carry or interrupt only a voice current shall not be required to be provided with such enclosures; and

    (ii) current-breaking contacts that are immersed in oil or enclosed in a chamber sealed against the entrance of dust shall be permitted to be provided with a general-purpose enclosure if the prevailing dust is electrically non-conductive; and

(b) all wiring shall comply with Rules 18-202 and 18-204.

18-224 Live parts, Class II, Division 1

No live parts of electrical equipment or of an electrical installation shall be exposed.

Installations in Class II, Division 2 locations (see Appendix E)

18-250 Transformers and capacitors, Class II, Division 2

(1) Transformers and electrical capacitors that contain a liquid that will burn shall be installed in electrical equipment vaults in accordance with Rules 26-350 to 26-356.

(2) Transformers and electrical capacitors that contain a liquid that will not burn shall be

    (a) installed in electrical equipment vaults in accordance with Rules 26-350 to 26-356; or

    (b) approved for Class II locations.

(3) Dry core transformers installed in Class II, Division 2 locations shall

    (a) be installed in electrical equipment vaults in accordance with Rules 26-350 to 26-356; or

    (b) have their windings and terminal connections enclosed in tight housings without ventilating or other openings and operate at not more than 750 V.

18-252 Wiring methods, Class II, Division 2 (see Appendix B)

(1) The wiring method shall be

    (a) threaded metal conduit;

    (b) cables approved for hazardous locations with associated cable glands approved for the particular hazardous location;

    (c) Type TC cable installed in cable tray in accordance with Rule 12-2202, enclosed in rigid conduit or another acceptable wiring method wherever it leaves the cable tray;

    (d) Type ACWU cable, with associated cable glands approved for the particular location; or

    (e) control and instrument cables with an interlocking metallic armour and a continuous jacket in control circuits (Type ACIC), with associated cable glands approved for the requirements of the enclosure that it enters.

(2) Boxes and fittings in which taps, joints, or terminal connections are made shall be either an Enclosure Type 4 or 5, or

    (a) be provided with telescoping or close-fitting covers, or other effective means to prevent the escape of sparks or burning material; and

    (b) have no openings, such as holes for attachment screws, through which, after installation, sparks or burning material might escape, or through which exterior accumulations of dust or adjacent combustible material might be ignited.

(3) Cables shall be installed and supported in a manner to avoid tensile stress at the cable glands.

(4) Where it is necessary to use flexible connections, the provisions of Rule 18-202(4) and (5) shall apply.

18-254 Sealing, Class II, Division 2
Sealing of raceways shall conform to Rule 18-204.

18-256 Switches, controllers, circuit breakers, and fuses, Class II, Division 2
Enclosures for switches, motor controllers, circuit breakers, and fuses, including push buttons, relays, and similar devices, shall be either an Enclosure Type 4 or 5, or

(a) be equipped with telescoping or close-fitting covers, or with other effective means to prevent the escape of sparks or burning material; and

(b) have no openings, such as holes for attachment screws, through which, after installation, sparks or burning material might escape, or through which exterior accumulations of dust or adjacent combustible material might be ignited.

18-258 Control transformers and resistors, Class II, Division 2
(1) Switching mechanisms, including overcurrent devices, used in conjunction with control transformers, impedance coils, and resistors shall be provided with enclosures conforming to Rule 18-256.

(2) Where not located in the same enclosure with switching mechanisms, control transformers and impedance coils shall be provided with tight housings without ventilating openings.

(3) Resistors and resistance devices shall have dust-tight enclosures approved for Class II locations, except that where the maximum normal operating temperature of the resistor will not exceed 120 °C, non-adjustable resistors and resistors that are part of an automatically timed starting sequence may have enclosures conforming to Subrule (2).

18-260 Motors and generators, Class II, Division 2 (see Appendix B)
(1) Except as provided in Subrule (2), motors, generators, and other rotating electrical machinery shall be
    (a) approved for Class II, or Class II, Division 2 locations; or
    (b) ordinary totally enclosed pipe-ventilated or totally enclosed fan-cooled, subject to the following:
        (i) equipped with integral overheating protection in accordance with Rule 28-314; and
        (ii) if drain holes or other openings are provided, they shall be closed with threaded plugs.

(2) Where accumulations of nonconductive, nonabrasive combustible dust are or will be moderate and if machines can be easily reached for routine cleaning and maintenance, the following shall be permitted to be installed:
    (a) standard open-type machines without sliding contacts, centrifugal or other types of switching mechanisms (including motor overcurrent, overload, and overtemperature devices), or integral resistance devices;
    (b) standard open-type machines with such contacts, switching mechanisms, or resistance devices enclosed within dust-tight housings without ventilating or other openings; and
    (c) self-cleaning textile motors of the squirrel-cage type.

18-262 Ventilation pipes, Class II, Division 2 (see Appendix B)
(1) Vent pipes for motors, generators, or other rotating electrical machinery, or for enclosures for electrical apparatus or equipment, shall conform to Rule 18-212(1).

(2) Vent pipes and their connections shall be sufficiently tight to prevent the entrance of appreciable quantities of dust into the ventilated equipment or enclosure, and to prevent the escape of sparks, flame, or burning material that might ignite accumulations of dust or combustible material in the vicinity.

(3) Where metal vent pipes are used, lock seams and riveted or welded joints shall be permitted to be used and, where some flexibility is necessary, for example at connections to motors, tight-fitting slip joints shall be permitted to be used.

18-264 Utilization equipment, fixed and portable, Class II, Division 2
(1) Electrically heated utilization equipment, whether fixed or portable, shall be approved for Class II locations.

(2) Motors of motor-driven utilization equipment shall conform to Rule 18-260.

(3) The enclosure for switches, circuit breakers, and fuses shall conform to Rule 18-256.

(4) Transformers, impedance coils, and resistors forming part of or used in connection with utilization equipment shall conform to Rule 18-258(2) and (3).

(5) Where portable utilization equipment is permitted to be used in Class II, Division 1 locations and in Class II, Division 2 locations, it shall conform to Rule 18-214.

18-266 Lighting fixtures, Class II, Division 2
(1) Lighting fixtures shall conform to the following:

(a) portable lamps shall be approved as complete assemblies for Class II locations and shall be clearly marked to indicate the maximum wattage of lamps for which they are approved; and

(b) fixed lighting shall
  (i) be protected from physical damage by suitable guards or by location;
  (ii) provide enclosures for lamps and lampholders that shall be designed to minimize the deposit of dust on lamps and to prevent the escape of sparks, burning material, or hot metal; and
  (iii) be clearly marked to indicate the maximum wattage of lamps for which they are permitted to be used without exceeding a maximum exposed surface temperature of 165 °C under normal conditions of use.

(2) Pendant fixtures shall be
  (a) suspended by threaded rigid conduit stems or chains with approved fittings, or by other approved means, which shall not include flexible cord as the supporting medium;
  (b) for rigid stems longer than 300 mm, provided with permanent and effective bracing against lateral displacement at a level not more than 300 mm above the lower end of the stem, or provided with flexibility in the form of a fitting or flexible connector approved for the purpose not more than 300 mm from the point of attachment to the supporting box or fitting; and
  (c) where wiring between an outlet box or fitting and the fixture is not enclosed in conduit, provided with a flexible cord approved for extra-hard usage.

(3) Boxes, box assemblies, or fittings used for the support of lighting fixtures shall be approved for that purpose.

(4) Starting and control equipment for mercury vapour and fluorescent lamps shall conform to Rule 18-258.

18-268 Flexible cords, Class II, Division 2

Flexible cords shall conform to Rule 18-218.

18-270 Receptacles and attachment plugs, Class II, Division 2

Receptacles and attachment plugs shall be

(a) of a polarized type that affords automatic connection to the bonding conductor of the flexible supply cord; and

(b) designed so that the connection to the supply circuit cannot be made or broken while live parts are exposed.

18-272 Signal, alarm, remote-control, and communication systems, meters, instruments, and relays, Class II, Division 2

Signal, alarm, remote-control, and communications systems, and meters, instruments, and relays shall conform to the following:

(a) contacts that interrupt other than voice currents shall be enclosed in conformity with Rule 18-256;

(b) the windings and terminal connections of transformers and choke coils that may carry other than voice currents shall be provided with tight enclosures without ventilating openings; and

(c) resistors, resistance devices, thermionic tubes, and rectifiers that may carry other than voice currents shall be provided with dust-tight enclosures approved for Class II locations, except that where the maximum normal operating temperature of thermionic tubes, non-adjustable resistors, or rectifiers will not exceed 120 °C, such devices shall be permitted to have tight enclosures without ventilating openings.

18-274 Live parts, Class II, Division 2

No live parts of electrical equipment or of an electrical installation shall be exposed.

Class III locations

Installations in Class III, Division 1 locations (see Appendix E)

18-300 Transformers and capacitors, Class III, Division 1

Transformers and electrical capacitors shall conform to Rule 18-250.

18-302 Wiring methods, Class III, Division 1 (see Appendix B)

(1) The wiring method shall be threaded rigid metal conduit, electrical metallic tubing, or cables approved for hazardous locations with associated cable glands approved for the particular hazardous location.

(2) Boxes and fittings in which taps, joints, or terminal connections are made shall be either an Enclosure Type 5 or shall

(a)   be provided with telescoping or close-fitting covers, or other effective means to prevent the escape of sparks or burning material; and

(b)   have no openings, such as holes for attachment screws, through which, after installation, sparks or burning material might escape, or through which adjacent combustible material might be ignited.

(3)   Cables shall be installed and supported in a manner to avoid tensile stress at the cable glands.

(4)   Where it is necessary to use flexible connections, the provisions of Rule 18-202(4) and (5) shall apply.

18-304 Switches, controllers, circuit breakers, and fuses, Class III, Division 1

Enclosures for switches, motor controllers, circuit breakers, and fuses, including push buttons, relays, and similar devices, shall be either an Enclosure Type 5 or shall be provided with tight enclosures designed to minimize entrance of fibres and flyings, and shall

(a)   be equipped with telescoping or close-fitting covers, or with other effective means to prevent the escape of sparks or burning material; and

(b)   have no openings, such as holes for attachment screws, through which, after installation, sparks or burning material might escape, or through which exterior accumulations of fibres or flyings or adjacent combustible material might be ignited.

18-306 Control transformers and resistors, Class III, Division 1

Transformers, impedance coils, and resistors used as or in conjunction with control equipment for motors, generators, and appliances shall conform to Rule 18-258, with the exception that, when these devices are in the same enclosure with switching devices of such control equipment, and are used only for starting or short-time duty, the enclosure shall conform to the requirements of Rule 18-304.

18-308 Motors and generators, Class III, Division 1 (see Appendix B)

(1)   Except as provided in Subrule (2), motors, generators, and other rotating electrical machinery shall be

    (a)   totally enclosed non-ventilated;

    (b)   totally enclosed pipe-ventilated; or

    (c)   totally enclosed fan-cooled.

(2)   Where only moderate accumulations of lint and flyings are likely to collect on, or in the vicinity of, a rotating electrical machine and the machine is readily accessible for routine cleaning and maintenance, it shall be permissible to install in the location:

    (a)   standard open-type machines without sliding contacts, centrifugal, or other types of switching mechanisms, including motor overload devices;

    (b)   standard open-type machines that have contacts, switching mechanisms, or resistance devices enclosed within tight housings without ventilating or other openings; or

    (c)   self-cleaning textile motors of the squirrel-cage type.

(3)   Motors, generators, or other rotating electrical machinery of the partially enclosed or splash-proof type shall not be installed in Class III locations.

18-310 Ventilating pipes, Class III, Division 1 (see Appendix B)

(1)   Vent pipes for motors, generators, or other rotating electrical machinery or for enclosures for electrical apparatus or equipment shall conform to Rule 18-212(1).

(2)   Vent pipes and their connections shall be sufficiently tight to prevent the entrance of appreciable quantities of fibres or flyings into the ventilated equipment or enclosure, and to prevent the escape of sparks, flame, or burning material that might ignite accumulations of fibres or flyings or combustible material in the vicinity.

(3)   Where metal vent pipes are used, lock seams and riveted or welded joints shall be permitted to be used and, where some flexibility is necessary, tight-fitting slip joints shall be permitted to be used.

18-312 Utilization equipment, fixed and portable, Class III, Division 1

(1)   Electrically heated utilization equipment, whether fixed or portable, shall be approved for Class III locations.

(2)   Motors of motor-driven utilization equipment shall conform to Rule 18-308.

(3)   The enclosures for switches, motor controllers, circuit breakers, and fuses shall conform to Rule 18-304.

18-314 Lighting fixtures, Class III, Division 1

(1)   Lighting fixtures shall conform to the following:

    (a)   portable lamps shall

        (i)   be equipped with handles;

        (ii)   be protected with substantial guards;

  (iii)   have lampholders of the unswitched type with no exposed metal parts and without provision for receiving attachment plugs; and

  (iv)   in all other aspects comply with Item (b);

(b)   fixed lighting shall

  (i)   provide enclosures for lamps and lampholders that shall be designed to minimize entrance of fibres and flyings and to prevent the escape of sparks, burning material, or hot metal;

  (ii)   be clearly marked to indicate the maximum wattage lamp that is permitted to be used without exceeding a maximum exposed surface temperature of 165 °C under normal conditions of use.

(2)   Lighting fixtures that may be exposed to physical damage shall be protected by a suitable guard.

(3)   Pendant fixtures shall comply with Rule 18-266(2).

(4)   Boxes, box assemblies, or fittings used for the support of lighting fixtures shall be approved for that purpose.

(5)   Starting and control equipment for mercury vapour and fluorescent lamps shall comply with Rule 18-306.

18-316 Flexible cords, Class III, Division 1

Flexible cords shall comply with Rule 18-218.

18-318 Receptacles and attachment plugs, Class III, Division 1

Receptacles and attachment plugs shall comply with Rule 18-270.

18-320 Signal, alarm, remote-control, and communication systems, Class III, Division 1

Signal, alarm, remote-control, and communication systems shall comply with Rule 18-272.

18-322 Electric cranes, hoists, and similar equipment, Class III, Division 1

Where installed for operation over combustible fibres or accumulations of flyings, travelling cranes and hoists for material handling, travelling cleaners for textile machinery, and similar equipment shall conform to the following:

(a)   the power supply to contact conductors shall be isolated from all other systems, ungrounded, and equipped with a recording ground detector that will give an alarm and will automatically de-energize the contact conductors in case of a fault to ground, or with a ground fault indicator that will give a visual and audible alarm and maintain the alarm as long as power is supplied to the system and the ground fault remains;

(b)   contact conductors shall be located or guarded so as to be inaccessible to other than authorized persons and shall be protected against accidental contact with foreign objects;

(c)   current collectors shall conform to the following:

  (i)   they shall be arranged or guarded to confine normal sparking and to prevent escape of sparks or hot particles;

  (ii)   to reduce sparking, two or more separate surfaces of contact shall be provided for each contact conductor; and

  (iii)   reliable means shall be provided to keep contact conductors and current collectors free of accumulations of lint or flyings; and

(d)   control equipment shall comply with Rules 18-304 and 18-306.

18-324 Storage-battery charging equipment, Class III, Division 1

Storage-battery charging equipment shall be located in separate rooms built or lined with substantial non-combustible materials constructed so as to adequately exclude flyings or lint and shall be well ventilated.

18-326 Live parts, Class III, Division 1

No live parts of electrical equipment or of an electrical installation shall be exposed, except as provided in Rule 18-322.

Installations in Class III, Division 2 locations (see Appendix E)

18-350 Transformers and capacitors, Class III, Division 2

Transformers and capacitors shall conform to Rule 18-250.

18-352 Wiring method, Class III, Division 2

The wiring method in Class III, Division 2 locations shall conform to Rule 18-302 except that in sections, compartments, or areas used solely for storage and containing no machinery, open wiring on insulators in accordance with Rules 12-202 to 12-224 shall be permitted to be used, provided that, where conductors are installed elsewhere than in roof spaces and remote from physical damage, they shall be protected as required by Rules 12-212 and 12-214.

18-354 Switches, controllers, circuit breakers, and fuses, Class III, Division 2
Enclosures for switches, motor controllers, circuit breakers, and fuses shall conform to Rule 18-304.

18-356 Control transformers and resistors, Class III, Division 2
Transformers, impedance coils, and resistors used as or in conjunction with control equipment for motors, generators, and appliances shall conform to Rule 18-306.

18-358 Motors and generators, Class III, Division 2 (see Appendix B)
Motors, generators, and other rotating machinery shall conform to Rule 18-308.

18-360 Ventilating pipes, Class III, Division 2 (see Appendix B)
Ventilating pipes shall conform to Rule 18-212(1).

18-362 Utilization equipment, fixed and portable, Class III, Division 2
Fixed or portable utilization equipment shall conform to Rule 18-312.

18-364 Lighting fixtures, Class III, Division 2
Lighting fixtures shall conform to Rule 18-314.

18-366 Flexible cords, Class III, Division 2
Flexible cords shall conform to Rule 18-218.

18-368 Receptacles and attachment plugs, Class III, Division 2
Receptacles and attachment plugs shall conform to Rule 18-270.

18-370 Signal, alarm, remote-control, and communication systems, Class III, Division 2
Signal, alarm, remote-control, and communication systems shall conform to Rule 18-272.

18-372 Electric cranes, hoists, and similar equipment, Class III, Division 2
Electric cranes, hoists, and similar equipment shall be installed as prescribed by Rule 18-322.

18-374 Storage-battery charging equipment, Class III, Division 2
Storage-battery charging equipment shall be located in rooms conforming to Rule 18-324.

18-376 Live parts, Class III, Division 2
No live parts of electrical equipment or of an electrical installation shall be exposed, except as provided in Rule 18-322.

Section 20

Flammable liquid and gas dispensing and service stations,
garages, bulk storage plants, finishing processes, and aircraft hangars

Section 20 — Flammable liquid and gas dispensing and service stations, garages, bulk storage plants, finishing processes, and aircraft hangars

20-000 Scope (see Appendices G and J)

(1) This Section supplements or amends the general requirements of this Code and applies to installations as follows:
- (a) gasoline dispensing and service stations — Rules 20-002 to 20-014;
- (b) propane dispensing, container filling, and storage — Rules 20-030 to 20-042;
- (c) compressed natural gas refuelling stations and compressor and storage facilities — Rules 20-060 to 20-072;
- (d) commercial garages — repairs and storage — Rules 20-100 to 20-114;
- (e) residential storage garages — Rules 20-200 to 20-206;
- (f) bulk storage plants — Rules 20-300 to 20-312;
- (g) finishing processes — Rules 20-400 to 20-414; and
- (h) aircraft hangars — Rules 20-500 to 20-522.

(2) For additions, modifications, or renovations to, or operation and maintenance of existing facilities employing the Division system of classification for Class I locations, the continued use of the Division system of classification shall be permitted.

(3) Where the Division system of classification is used for Class I locations, as permitted by Subrule (2), the Rules for Class I locations found in Annex J20 of Appendix J shall apply.

(4) Notwithstanding Subrule (3), equipment permitted in the Rules for installation in Class I, Zone 2 locations shall also be permitted for installations in Class I, Division 2 locations.

(5) The definitions stated in Rule 18-002 shall also apply to Section 20.

Gasoline dispensing and service stations

20-002 General

(1) Rules 20-004 to 20-014 apply to electrical apparatus and wiring installed in gasoline dispensing and service stations and other locations where gasoline or other similar volatile flammable liquids are dispensed or transferred to the fuel tanks of self-propelled vehicles.

(2) Other areas used as lubritoriums, service rooms and repair rooms, and offices, salesrooms, compressor rooms, and similar locations shall conform to Rules 20-100 to 20-114 with respect to electrical wiring and equipment.

20-004 Hazardous areas (see Appendix B)

(1) Except as provided for in Subrule (3), the space within a dispenser enclosure up to 1.2 m vertically above its base, including the space below the dispenser that may contain electrical wiring and equipment, shall be considered a Class I, Zone 1 location.

(2) The space within a nozzle boot of a dispenser shall be considered a Class I, Zone 0 location.

(3) The space within a dispenser enclosure above the Class I, Zone 1 location as specified in Subrule (1) or spaces within a dispenser enclosure isolated from the Zone 1 location by a solid vapour-tight partition or by a solid nozzle boot but not completely surrounded by a Zone 1 location shall be considered a Class I, Zone 2 location.

(4) The space within 450 mm horizontally from the Zone 1 location within the dispenser enclosure as specified in Subrule (1) shall be considered a Class I, Zone 1 location.

(5) The space outside the dispenser within 450 mm horizontally from the opening of a solid nozzle boot located above the vapour-tight partition shall be considered a Class I, Zone 2 location, except that the classified area need not extend beyond the plane in which the boot is located.

(6) In an outside location, any area beyond the Class I, Zone 1 area (and in buildings not suitably cut off) within 6 m horizontally from the exterior enclosure of any dispenser shall be considered a Class I, Zone 2 location that extends to a level 450 mm above driveway or ground level.

(7) In an outside location, any area beyond the Class I, Zone 1 location (and in buildings not suitably cut off) within 3 m horizontally from any tank fill-pipe shall be considered a Class I, Zone 2 location that extends upward to a level 450 mm above driveway or ground level.

(8)   Electrical wiring and equipment, any portion of which is below the surface of areas defined as Class I, Zone 1 or Zone 2 in Subrule (1), (4), (6), or (7), shall be considered within a Class I, Zone 1 location that extends at least to the point of emergence above grade.

(9)   Areas within the vicinity of tank vent-pipes shall be classified as follows:

(a)   the spherical volume within a 900 mm radius from the point of discharge of any tank vent-pipe shall be considered a Class I, Zone 1 location and the volume between the 900 mm to 1.5 m radius from the point of discharge of a vent shall be considered a Class I, Zone 2 location;

(b)   for any vent that does not discharge upward, the cylindrical volume below both the Zone 1 and Zone 2 locations extending to the ground shall be considered a Class I, Zone 2 location; and

(c)   the hazardous area shall not be considered to extend beyond an unpierced wall.

(10)   Areas within lubrication rooms shall be classified as follows:

(a)   the area within any pit or space below grade or floor level in a lubrication room shall be considered a Class I, Zone 1 location, unless the pit or space below grade is beyond the hazardous areas specified in Subrules (6), (7), and (9), in which case the pit or space below grade shall be considered a Class I, Zone 2 location;

(b)   notwithstanding Item (a), for each floor below grade that is located beyond the hazardous area specified in Subrules (6), (7), and (9) and where adequate ventilation is provided, a Class I, Zone 2 location shall extend up to a level of only 50 mm above each such floor;

(c)   the area within the entire lubrication room up to 50 mm above the floor or grade, whichever is higher, and the area within 900 mm measured in any direction from the dispensing point of a hand-operated unit dispensing volatile flammable liquids shall be considered a Class I, Zone 2 location.

20-006 Wiring and equipment within hazardous areas

(1)   Electrical wiring and equipment within the hazardous areas defined in Rule 20-004 shall conform to Section 18 requirements.

(2)   Where dispensers are supplied by rigid metal conduit, a union and a flexible fitting shall be installed between the conduit and the dispenser junction box in addition to any sealing fittings required by Section 18.

(3)   The flexible metal fitting required by Subrule (2) shall be installed in a manner that allows relative movement of the conduit and the dispenser.

(4)   Where dispensers are supplied by a cable approved for hazardous locations, provisions shall be made to separate the cable from the dispenser junction box without rendering ineffective the explosion-proof cable seal.

20-008 Wiring and equipment above hazardous areas

Wiring and equipment above hazardous areas shall conform to Rules 20-106 and 20-110.

20-010 Circuit disconnects

Each circuit leading to or through a dispensing pump shall be provided with a switching means that will disconnect simultaneously from the source of supply all ungrounded conductors of the circuit.

20-012 Sealing

(1)   Seals as required by Section 18 shall be provided in each conduit run entering or leaving a dispenser or any cavities or enclosures in direct communication with a dispenser.

(2)   Additional seals shall be provided in conformance with Rules 18-108 and 18-158, and the requirements of Rules 18-108(1)(a)(iii) and 18-158(1)(a)(ii) shall include horizontal and vertical boundaries.

20-014 Bonding

All non-current-carrying metal parts of dispensing pumps, metal raceways, and other electrical equipment shall be bonded to ground in accordance with Section 10.

Propane dispensing, container filling, and storage

20-030 Scope (see Appendix B)

Rules 20-032 to 20-042 apply to locations where propane is dispensed or transferred to the fuel tanks of self-propelled vehicles or to portable containers and to locations where propane is stored or transferred from rail cars or tanker vehicles to storage containers.

20-032 Special terminology

In this Subsection, the following definitions apply:

Section 20

Flammable liquid and gas dispensing and service stations,
garages, bulk storage plants, finishing processes, and aircraft hangars

Container refill centre — a facility such as a propane service station that is open to the public and where propane is dispensed into containers or the fuel tanks of motor vehicles and that consists of propane storage containers, piping, and pertinent equipment including pumps and dispensing devices.

Filling plant — a facility such as a bulk propane plant, the primary purpose of which is the distribution of propane, that receives propane in tank car or truck transport for storage and/or distribution in portable containers or tank trucks, that has bulk storage, and that usually has container filling and truck loading facilities on the premises.

Propane — any material that is composed predominantly of the following hydrocarbons either by themselves or as mixtures: propane, propylene, butane (normal butane or iso-butane), and butylene.

20-034 Hazardous areas
In container refill centres and in filling plants, the hazardous areas shall be classified as listed in Table 63.

20-036 Wiring and equipment in hazardous areas
(1) All electrical wiring and equipment in the hazardous areas referred to in Rule 20-034 shall conform to the requirements of Section 18.
(2) Where dispensing devices are supplied by rigid metal conduit, the requirements of Rule 20-006(2) and (3) shall be met.

20-038 Sealing
(1) Seals shall be installed as required by Section 18 and the requirements shall be applied to horizontal as well as vertical boundaries of the defined hazardous locations.
(2) Seals for dispensing devices shall be provided as required by Rule 20-012.

20-040 Circuit disconnects
Each circuit leading to or through a propane dispensing device or pump shall be provided with a switching means that will disconnect simultaneously all ungrounded conductors of the circuit from the source of supply.

20-042 Bonding
All non-current-carrying metal parts of equipment and raceways shall be bonded to ground in accordance with Section 10.

Compressed natural gas refuelling stations and compressor and storage facilities

20-060 Scope (see Appendix B)
(1) Rules 20-062 to 20-072 apply to locations where compressed natural gas is dispensed to the fuel tanks of self-propelled vehicles and to associated compressor and storage facilities.
(2) The Rules in this Section do not apply to vehicle refuelling appliances installed in accordance with CAN/CSA-B149.1 that do not have storage facilities.

20-062 Hazardous areas
(1) The areas surrounding compressors shall be classified as follows:
   (a) in an outdoor location the space within 4.5 m in all directions from the compressor shall be considered to be a Class I, Zone 2 location;
   (b) if the compressor is enclosed, the space within the compressor enclosure shall be considered a Class I, Zone 1 location;
   (c) if the compressor is enclosed, the space within 3 m in all directions from non-gas-tight, non-welded seams and openings in the enclosure shall be considered a Class I, Zone 2 location;
   (d) a compressor shall be regarded as enclosed when it is sheltered by a building or enclosure having four sides, a roof, and limited ventilation; and
   (e) when a gas-tight wall is located within the distances specified in Items (a) and (c), the distances shall be measured around the end of the wall, over the wall, or through any doors, windows, or openings in the wall.
(2) The areas surrounding a natural gas dispensing point located outdoors shall be classified as follows:
   (a) for fast fill dispensing, the space within 3 m in all directions from the dispensing point shall be considered a Class I, Zone 2 location;
   (b) for slow fill dispensing, the space within 1.5 m in all directions from the dispensing point shall be considered a Class I, Zone 2 location; and

(c) the distances specified in Subrule (2)(a) and (b) shall be measured from the breakaway coupling at the transition point between rigid piping and the refuelling hose.

(3) For dispensing devices, the entire space within the dispenser enclosure and the space below the dispenser shall be considered a Class I, Zone 1 location.

20-064 Hazardous areas surrounding gas storage facilities

The electrical classification of areas surrounding gas storage facilities shall be as indicated in Table 64.

20-066 Wiring and equipment in hazardous areas

(1) All electrical wiring and equipment in the hazardous areas defined in Rules 20-062 and 20-064 shall comply with the requirements of Section 18.

(2) Where dispensing devices are supplied with rigid metal conduit, the requirements of Rule 20-006(2) and (3) shall be met.

20-068 Sealing

(1) Seals shall be installed as required by Section 18, and the requirements shall be applied to horizontal as well as vertical boundaries of the defined hazardous locations.

(2) Seals for dispensing devices shall be provided as required by Rule 20-012.

20-070 Circuit disconnects

Each circuit leading to a compressor or a dispensing device shall be provided with a switching means that will disconnect simultaneously from the source of supply all ungrounded conductors of the circuit.

20-072 Bonding

All non-current-carrying metal parts of equipment and raceways shall be bonded to ground in accordance with Section 10.

Commercial garages — Repairs and storage

20-100 Scope

Rules 20-102 to 20-114 apply to locations used for service and repair operations in connection with self-propelled vehicles in which volatile flammable liquids or flammable gases are used for fuel or power, and locations in which more than three such vehicles are, or may be, stored at one time.

20-102 Hazardous areas

(1) For each floor at or above grade, the entire area up to a level 50 mm above the floor shall be considered a Class I, Zone 2 location.

(2) For each floor below grade, the entire area up to a level 50 mm above the bottom of outside doors or other openings that are at, or above, grade level shall be considered a Class I, Zone 2 location except that where adequate ventilation is provided, the hazardous location shall extend up to a level of only 50 mm above each such floor.

(3) Notwithstanding Subrule (2), in storage garages only the area up to a level of 50 mm above each floor that is below grade shall be considered a Class I, Zone 2 location.

(4) Any pit or depression below floor level shall be considered a Class I, Zone 2 location that extends up to the floor level.

(5) Adjacent areas in which hazardous vapours are not likely to be released, such as stockrooms, switchboard rooms, and other similar locations having floors elevated at least 50 mm above the adjacent garage floor, or separated from the garage floor by tight-fitting barriers such as curbs, ramps, or partitions at least 50 mm high, shall not be classed as hazardous.

20-104 Wiring and equipment in hazardous areas

Within hazardous areas as defined in Rule 20-102, wiring and equipment shall conform to the applicable requirements of Section 18.

20-106 Wiring above hazardous areas

(1) All fixed wiring above hazardous areas shall be in accordance with Section 12 and suitable for the type of building and occupancy.

(2) For pendants, flexible cord of the hard-usage type shall be used.

(3) For connection of portable luminaires, portable motors, or other portable utilization equipment, flexible cord of the hard-usage type shall be used.

Section 20
Flammable liquid and gas dispensing and service stations,
garages, bulk storage plants, finishing processes, and aircraft hangars

20-108 Sealing

(1) Seals shall be installed as required by Section 18, and the requirements of Rule 18-158(1)(a)(ii) shall include horizontal and vertical boundaries.

(2) Raceways embedded in a masonry floor or buried beneath a floor shall be considered within the hazardous area above the floor if any connections or extensions lead into or through such an area.

20-110 Equipment above hazardous areas

(1) Fixed equipment that is less than 3.6 m above the floor level and that may produce arcs, sparks, or particles of hot metal, such as cut-outs, switches, charging panels, generators, motors, or other equipment (excluding receptacles and luminaires) having make-and-break or sliding contacts, shall be the totally enclosed type or constructed to prevent escape of sparks or hot metal particles.

(2) Permanently installed luminaires that are located over lanes through which vehicles are commonly driven or that may otherwise be exposed to physical damage shall be located not less than 3.6 m above floor level unless they are of the totally enclosed type or constructed to prevent escape of sparks or hot metal particles.

(3) Portable luminaires shall

    (a) be of the totally enclosed gasketted type, equipped with handle, lampholder, hook, and substantial guard attached to the lampholder or handle, and all exterior surfaces that may come in contact with battery terminals, wiring terminals, or other objects shall be of non-conducting materials or shall be effectively protected with an insulating jacket;

    (b) be the unswitched type; and

    (c) not be provided with receptacles for attachment plugs.

20-112 Battery-charging equipment

Battery chargers and their control equipment, and batteries being charged, shall not be located within the hazardous areas classified in Rule 20-102.

20-114 Electric vehicle charging

(1) Flexible cords used for charging shall be the extra-hard-usage type.

(2) Connectors shall have a rating not less than the ampacity of the cord and in no case less than 50 A.

(3) Connectors shall be designed and installed so that they will break apart readily at any position of the charging cable, and live parts shall be guarded from accidental contact.

(4) No connector shall be located within the hazardous area defined in Rule 20-102.

(5) Where plugs are provided for direct connection to vehicles, the point of connection shall not be within a hazardous area as defined in Rule 20-102.

(6) Where a cord is suspended from overhead, it shall be arranged so that the lowest point of sag is at least 150 mm above the floor.

(7) Where the vehicle is equipped with a plug that will readily pull apart, and where an automatic arrangement is provided to pull both cord and plug beyond the range of mechanical damage, no additional connector shall be required in the cable or outlet.

Residential storage garages

20-200 Scope

Rules 20-202 to 20-206 apply to a building or part of a building in which not more than three vehicles of the type described in Rule 20-100 are, or may be, stored, but that will not normally be used for service or repair operations on stored vehicles.

20-202 Non-hazardous location

Where the lowest floor is at or above adjacent grade or driveway level, and where there is at least one outside door at or below floor level, the garage area shall not be classed as a hazardous location.

20-204 Hazardous location

Where the lowest floor is below adjacent grade or driveway level, the following shall apply:

(a) the entire area of the garage or of any enclosed space that includes the garage shall be classified as a Class I, Zone 2 location up to a level 50 mm above the garage floor; and

(b) adjacent areas in which hazardous vapours or gases are not likely to be released, and where floors are elevated at least 50 mm above the garage floor or separated from the garage floor by tight curbs or partitions at least 50 mm high, shall not be classed as hazardous.

20-206 Wiring
(1)   Wiring above the hazardous locations shall conform to Section 12.
(2)   Wiring in the hazardous locations shall conform to Section 18.

Bulk storage plants

20-300 Scope
Rules 20-302 to 20-312 apply to locations where gasoline or other similar volatile flammable liquids are stored in tanks having an aggregate capacity of one carload or more, and from which such products are distributed (usually by tank truck).

20-302 Hazardous areas
(1)   Areas containing pumps, bleeders, withdrawal fittings, meters, and similar devices that are located in pipelines handling flammable liquids under pressure shall be classified as follows and meet the following requirements:
    (a)   indoor areas having adequate ventilation shall be considered Class I, Zone 2 locations within a 1.5 m distance extending in all directions from the exterior surface of such devices as well as 7.5 m horizontally from any surface of these devices and extending upwards to 900 mm above floor or grade level, provided that the following conditions are met:
        (i)   design of the ventilation systems takes into account the relatively high relative density of the vapours;
        (ii)   where openings are used in outside walls, they are of adequate size and located at floor level unobstructed except by louvres or coarse screens; and
        (iii)   where natural ventilation is inadequate, mechanical ventilation is provided;
    (b)   indoor areas not having adequate ventilation in accordance with Subrule (1)(a) shall be considered Class I, Zone 1 locations within a 1.5 m distance extending in all directions from the exterior surface of such devices as well as 7.5 m horizontally from any surface of the device and extending upward 900 mm above floor or grade level; and
    (c)   outdoor areas shall be considered Class I, Zone 2 locations within a 900 mm distance extending in all directions from the exterior surface of such devices as well as up to 450 mm above grade level within 3 m horizontally from any surface of the devices.
(2)   Areas where flammable liquids are transferred shall be classified as follows:
    (a)   in outdoor areas or where adequate ventilation is provided in indoor areas in which flammable liquids are transferred to individual containers, such areas shall be considered a Class I, Zone 1 location within 900 mm of the vent or fill opening extending in all directions and a Class I, Zone 2 location within the area extending between a 900 mm and 1.5 m radius from the vent or fill opening extending in all directions, and including the area within a horizontal radius of 3 m from the vent or fill opening and extending to a height of 450 mm above floor or grade levels; or
    (b)   where adequate ventilation is not provided in indoor areas in which flammable liquids are transferred to individual containers, such areas shall be considered a Class I, Zone 1 location.
(3)   Areas in outside locations where loading and unloading of tank vehicles and tank cars takes place shall be classified as follows:
    (a)   the area extending 900 mm in all directions from the dome when loading through an open dome or from the vent when loading through a closed dome with atmospheric venting shall be considered a Class I, Zone 1 location;
    (b)   the area extending between a 900 mm and a 1.5 m radius from the dome when loading through an open dome or from the vent when loading through a closed dome with atmospheric venting shall be considered a Class I, Zone 2 location;
    (c)   the area extending within 900 mm in all directions from a fixed connection used in bottom loading or unloading, loading through a closed dome with atmospheric venting, or loading through a closed dome with a vapour recovery system shall be considered a Class I, Zone 2 location, except that in the case of bottom loading or unloading this classification shall also be applied to the area within a 3 m radius from the point of connection and extending 450 mm above grade; and
    (d)   the internal space of tank vehicles and tank cars shall be a Zone 0 location.
(4)   Areas within the vicinity of above-ground tanks shall be classified as follows:
    (a)   the area above the roof and within the shell of a floating roof type tank shall be considered a Class I, Zone 1 location;

Section 20
Flammable liquid and gas dispensing and service stations,
garages, bulk storage plants, finishing processes, and aircraft hangars

(b)   for all types of above-ground tanks:
    (i)   the area within 3 m from the shell, ends, and roof of other than a floating roof shall be considered a Class I, Zone 2 location; and
    (ii)   where dikes are provided, the area inside the dike and extending upwards to the top of the dike shall be considered a Class I, Zone 2 location;

(c)   the area within 1.5 m of a vent opening and extending in all directions shall be considered a Class I, Zone 1 location;

(d)   the area between 1.5 m and 3 m of a vent opening and extending in all directions shall be considered a Class I, Zone 2 location; and

(e)   the vapour space above a liquid in a storage tank shall be considered a Zone 0 location.

(5)   Pits and depressions shall be classified as follows:

(a)   any pit or depression, any part of which lies within a Zone 1 or Zone 2 location unless provided with adequate ventilation, shall be considered a Class I, Zone 1 location;

(b)   any such areas, when provided with adequate ventilation, shall be considered a Class I, Zone 2 location; and

(c)   any pit or depression not within a Zone 1 or Zone 2 location as defined in this Section but that contains piping, valves, or fittings shall be considered a Class I, Zone 2 location.

(6)   Garages in which tank vehicles are stored or repaired shall be considered a Class I, Zone 2 location up to 450 mm above floor or grade level, unless conditions warrant more severe classification or a greater extent of the hazardous area.

(7)   Buildings such as office buildings, boiler rooms, etc., that are outside the limits of hazardous areas as defined in this Section and that are not used for handling or storage of volatile flammable liquids or containers for such liquids shall not be considered hazardous locations.

20-304 Wiring and equipment in hazardous areas
All electrical wiring and equipment in hazardous areas defined in Rule 20-302 shall conform to the requirements of Section 18.

20-306 Wiring and equipment above hazardous areas
(1)   Wiring installed above a hazardous location shall conform to the requirements of Section 12 and be suitable for the type of building and the occupancy.

(2)   Fixed equipment that may produce arcs, sparks, or particles of hot metal, such as lamps and lampholders, cut-outs, switches, receptacles, motors, or other equipment having make-and-break or sliding contacts, shall be of the totally enclosed type or constructed to prevent the escape of sparks or hot metal particles.

(3)   Portable lamps or utilization equipment and the flexible cords supplying them shall conform to the requirements of Section 18 for the class of location above which they are connected or used.

20-308 Sealing
(1)   Seals shall be installed in accordance with Section 18 and shall be applied to horizontal as well as vertical boundaries of the defined hazardous locations.

(2)   Buried raceways under defined hazardous areas shall be considered within such areas.

20-310 Gasoline dispensing
Where gasoline dispensing is carried on in conjunction with bulk station operations, the applicable provisions of Rules 20-002 to 20-014 inclusive shall apply.

20-312 Bonding
All non-current-carrying metal parts of equipment and raceways shall be bonded to ground in accordance with Section 10.

Finishing processes

20-400 Scope
Rules 20-402 to 20-414 apply where paints, lacquers, or other flammable finishes are regularly or frequently applied by spraying, dipping, brushing, or by other means, and where volatile flammable solvents or thinners are used or where readily ignitable deposits or residues from such paints, lacquers, or finishes may occur.

20-402 Hazardous locations
(1)   The following areas shall be considered Class I, Zone 1 locations:
(a)   where adequate ventilation is provided, the interiors of spray booths and their exhaust ducts;

(b) all space within 6 m horizontally in any direction, extending to a height of 1 m above the goods to be painted, from spraying operations more extensive than touch-up spraying and not conducted within the spray booth, and as otherwise shown in Diagram 5;

(c) all space within 6 m horizontally in any direction from dip tanks and their drain boards, such space extending to a height of 1 m above the dip tank and drain board; and

(d) all other spaces where hazardous concentrations of flammable vapours are likely to occur.

(2) For spraying operations within an open-faced spray booth, the extent of the Class I, Zone 2 location shall extend not less than 1.5 m from the open face of the spray booth, and as otherwise shown in Diagram 4.

(3) For spraying operations confined within a closed spray booth or room or for rooms where hazardous concentrations of flammable vapours are likely to occur, such as paint mixing rooms, and as otherwise shown in Diagram 10, the space within 1 m in all directions from any openings in the booth or room shall be considered a Class I, Zone 2 location.

(4) All space within the room but beyond the limits for Class I, Zone 1 as classified in Subrule (1) for extensive open spraying, and as otherwise shown in Diagram 5, for dip tanks and drain boards, and for other hazardous operations, shall be considered to be Class I, Zone 2 locations.

(5) Adjacent areas that are cut off from the defined hazardous area by tight partitions without communicating openings, and within which hazardous vapours are not likely to be released, shall be permitted to be classed as non-hazardous.

(6) Drying and baking areas provided with adequate ventilation and effective interlocks to de-energize all electrical equipment not approved for Class I locations in case the ventilating equipment is inoperative shall be permitted to be classed as non-hazardous.

(7) Notwithstanding the requirements of Subrule (1)(b), where adequate ventilation with effective interlocks is provided at floor level, and as otherwise shown in Diagram 6:

(a) the space within 1 m horizontally in any direction from the goods to be painted and such space extending to a height of 1 m above the goods to be painted shall be considered a Class I, Zone 1 location; and

(b) all space between a 1 m and a 1.5 m distance above the goods to be painted and all space within 6 m horizontally in any direction beyond the limits for the Class I, Zone 1 location shall be considered a Class I, Zone 2 location.

(8) Notwithstanding the requirements of Subrule (2), where a baffle of sheet metal of not less than No. 18 MSG is installed vertically above the front face of an open-face spray booth to a height of 1 m or to the ceiling, whichever is less, and extending back on the side edges for a distance of 1.5 m, the space behind this baffle shall be considered a non-hazardous location.

(9) Notwithstanding the requirements of Subrule (3), where a baffle of sheet metal of not less than No. 18 MSG is installed vertically above an opening in a closed spray booth or room to a height of 1 m or to the ceiling, whichever is less, and extends horizontally a distance of 1 m beyond each side of the opening, the space behind the baffle shall be considered a non-hazardous location.

20-404 Ventilation and spraying equipment interlock

The spraying equipment for a spray booth shall be interlocked with the spray booth ventilation system so that the spraying equipment is made inoperable when the ventilation system is not in operation.

20-406 Wiring and equipment in hazardous areas

(1) All electrical wiring and equipment within the hazardous areas as defined in Rule 20-402 shall conform to the requirements of Section 18.

(2) Unless specifically approved for both readily ignitable deposits and the flammable vapour location, no electrical equipment shall be installed or used where it may be subject to a hazardous accumulation of readily ignitable deposits or residue.

(3) Illumination of readily ignitable areas through panels of glass or other transparent or translucent materials is permissible only where

(a) fixed lighting units are used as the source of illumination;

(b) the panel is non-combustible and effectively isolates the hazardous area from the area in which the lighting unit is located;

(c) the panel is of a material or is protected so that breakage is unlikely; and

(d) the arrangement is such that normal accumulations of hazardous residue on the surface of the panel will not be raised to a dangerous temperature by radiation or conduction from the source of illumination.

Section 20

*Flammable liquid and gas dispensing and service stations,
garages, bulk storage plants, finishing processes, and aircraft hangars*

(4) Portable electric lamps or other utilization equipment shall

    (a)   not be used within a hazardous area during operation of the finishing process;

    (b)   be of a type specifically approved for Class I locations when used during cleaning or repairing operations.

(5) Notwithstanding Subrule (2),

    (a)   totally enclosed and gasketted lighting shall be permitted to be used on the ceiling of a spray room where adequate ventilation is provided; and

    (b)   infrared paint drying units shall be permitted to be utilized in a spray room if the controls are interlocked with those of the spraying equipment so that both operations cannot be performed simultaneously, and if portable, the paint drying unit shall not be brought into the spray room until spraying operations have ceased.

20-408 Fixed electrostatic equipment

Electrostatic spraying and detearing equipment shall conform to the following:

(a) no transformers, power packs, control apparatus, or other electrical portions of the equipment except high-voltage grids and their connections shall be installed in any of the hazardous areas defined in Rule 20-402, unless of a type specifically approved for the location;

(b) high-voltage grids or electrodes shall be

    (i)   located in suitable non-combustible booths or enclosures provided with adequate ventilation;

    (ii)   rigidly supported and of substantial construction; and

    (iii)   effectively insulated from ground by means of non-porous, non-combustible insulators;

(c) high-voltage leads shall be

    (i)   effectively and permanently supported on suitable insulators;

    (ii)   effectively guarded against accidental contact or grounding; and

    (iii)   provided with automatic means for discharging any residual charge to ground when the supply voltage is interrupted;

(d) where goods are being processed,

    (i)   they shall be supported on conveyors in such a manner that minimum clearance between goods and high-voltage grids or conductors cannot be less than twice the sparking distance; and

    (ii)   a conspicuous sign indicating the sparking distance shall be permanently posted near the equipment;

(e) automatic controls shall be provided that will operate without time delay to disconnect the power supply and to signal the operator in the event of

    (i)   stoppage of ventilating fans;

    (ii)   failure of ventilating equipment;

    (iii)   stoppage of the conveyor carrying goods through the high-voltage field;

    (iv)   occurrence of a ground or of an imminent ground at any point on the high-voltage system; or

    (v)   reduction of clearance below that specified in Item (d); and

(f) adequate fencing, railings, or guards that are electrically conducting and effectively bonded to ground shall be provided for safe isolation of the process, and signs shall be permanently posted designating the process area as dangerous because of high voltage.

20-410 Electrostatic hand spraying equipment

Electrostatic hand spray apparatus and devices used with such apparatus shall conform to the following:

(a) the high-voltage circuits shall be intrinsically safe and not produce a spark of sufficient intensity to ignite any vapour-air mixtures, nor result in an appreciable shock hazard to anyone coming in contact with a grounded object;

(b) the electrostatically charged exposed elements of the hand gun shall be capable of being energized only by a switch that also controls the paint supply;

(c) transformers, power packs, control apparatus, and all other electrical portions of the equipment, with the exception of the hand gun itself and its connections to the power supply, shall be located outside the hazardous area;

(d) the handle of the spray gun shall be bonded to ground by a metallic connection and be constructed such that the operator in normal operating position is in intimate electrical contact with the handle in order to prevent buildup of a static charge on the operator's body;

(e) all electrically conductive objects in the spraying area shall be bonded to ground and the equipment shall carry a prominent permanently installed warning regarding the necessity for this bonding feature;

(f) precautions shall be taken to ensure that objects being painted are maintained in metallic contact with the conveyor or other grounded support, and they shall include the following:
   (i) hooks shall be regularly cleaned;
   (ii) areas of contact shall be sharp points or knife edges; and
   (iii) points of support of the object shall be concealed from random spray where feasible and, where the objects being sprayed are supported from a conveyor, the point of attachment to the conveyor shall be located so as to not collect spray material during normal operation; and

(g) the spraying operation shall take place within a spray area that is adequately ventilated to remove solvent vapours released from the operation and the electrical equipment shall be interlocked with the ventilation of the spraying area so that the equipment cannot be operated unless the ventilation system is in operation.

20-412 Wiring and equipment above hazardous areas

(1) All fixed wiring above hazardous areas shall conform to Section 12.
(2) Equipment that may produce arcs, sparks, or particles of hot metal, such as lamps and lampholders for fixed lighting, cut-outs, switches, receptacles, motors, or other equipment having make-and-break or sliding contacts, where installed above a hazardous area or above an area where freshly finished goods are handled, shall be of the totally enclosed type or constructed to prevent the escape of sparks or hot metal particles.

20-414 Bonding

All metal raceways and all non-current-carrying metal portions of fixed or portable equipment, regardless of voltage, shall be bonded to ground in accordance with Section 10.

Aircraft hangars

20-500 Scope

Rules 20-502 to 20-522 apply to locations used for storage or servicing of aircraft in which gasoline, jet fuels, or other volatile flammable liquids, or flammable gases are used but shall not include those locations used exclusively for aircraft that have never contained such liquids or gases, or that have been drained and properly purged.

20-502 Hazardous areas

(1) Any pit or depression below the level of the hangar floor shall be considered a Class I, Zone 1 location, which shall extend up to the floor level.
(2) The entire area of the hangar including any adjacent and communicating areas not suitably cut off from the hangar shall be considered a Class I, Zone 2 location up to a level 450 mm above the floor.
(3) The area within 1.5 m horizontally from aircraft power plants, aircraft fuel tanks, or aircraft structures containing fuel shall be considered a Class I, Zone 2 location that extends upward from the floor to a level 1.5 m above the upper surface of wings and of engine enclosures.
(4) Adjacent areas in which hazardous vapours are not likely to be released, such as stock rooms, electrical control rooms, and other similar locations, shall be permitted to be classed as non-hazardous when adequately ventilated and when effectively cut off from the hangar itself in accordance with Rule 18-060.

20-504 Wiring and equipment in hazardous areas

(1) All fixed and portable wiring and equipment that is or may be installed or operated within any of the hazardous locations defined in Rule 20-502 shall conform to the requirements of Section 18.
(2) All wiring installed in or under the hangar floor shall conform to the requirements for Class I, Zone 1 locations.
(3) Wiring systems installed in pits, or other spaces in or under the hangar floor shall be provided with adequate drainage and shall not be placed in the same compartment with any other service except piped compressed air.
(4) Attachment plugs and receptacles in hazardous locations shall be explosion-proof, or shall be designed so that they cannot be energized while the connections are being made or broken.

20-506 Wiring not within hazardous areas

(1) All fixed wiring in a hangar not within a hazardous area as defined in Rule 20-502 shall be installed in metal raceways or shall be armoured cable, Type MI cable, or aluminum-sheathed cable, except that wiring in a non-hazardous location as set out in Rule 20-502(4) shall be permitted to be of any type recognized in Section 12 as suitable for the type of building and the occupancy.

Section 20
Flammable liquid and gas dispensing and service stations,
garages, bulk storage plants, finishing processes, and aircraft hangars

(2) For pendants, flexible cord of the hard-usage type and containing a separate bonding conductor shall be used.

(3) For portable utilization equipment and lamps, flexible cord approved for hard usage and containing a separate bonding conductor shall be used.

(4) Suitable means shall be provided for maintaining continuity and adequacy of the bonding between the fixed wiring system and the non-current-carrying metal portions of pendant fixtures, portable lamps, and other portable utilization equipment.

20-508 Equipment not within hazardous areas

(1) In locations other than those described in Rule 20-502, equipment that is less than 3 m above wings and engine enclosures of aircraft and that may produce arcs, sparks, or particles of hot metal, such as lamps and lampholders for fixed lighting, cut-outs, switches, receptacles, charging panels, generators, motors, or other equipment having make-and-break or sliding contacts, shall be of totally enclosed type or constructed so as to prevent the escape of sparks or hot metal particles, except that equipment in areas described in Rule 20-502(4) shall be permitted to be the general-purpose type.

(2) Lampholders of metal shell, fibre-lined types shall not be used for fixed lighting.

(3) Portable lamps that are used within a hangar shall comply with Rule 18-118.

(4) Portable utilization equipment that is, or may be, used within a hangar shall be a type suitable for use in Class I, Zone 2 locations.

20-510 Stanchions, rostrums, and docks

(1) Electric wiring, outlets, and equipment including lamps on, or attached to, stanchions, rostrums, or docks that are located, or likely to be located, in a hazardous area as defined in Rule 20-502(3) shall conform to the requirements for Class I, Zone 2 locations.

(2) Where stanchions, rostrums, and docks are not located, or are not likely to be located, in a hazardous area as defined in Rule 20-502(3), wiring and equipment shall conform to Rules 20-506 and 20-508, except for the following:
   (a) receptacles and attachment plugs shall be the locking type that will not break apart readily; and
   (b) wiring and equipment, not more than 450 mm above the floor in any position, shall conform to Subrule (1).

(3) Mobile stanchions with electrical equipment conforming to Subrule (2) shall carry at least one permanently affixed warning sign, to the effect that the stanchions be kept 1.5 m clear of aircraft engines and fuel tank areas.

20-512 Sealing

(1) Seals shall be installed in accordance with Section 18 and shall apply to horizontal as well as to vertical boundaries of the defined hazardous areas.

(2) Raceways embedded in a masonry floor or buried beneath a floor shall be considered within the hazardous area above the floor when any connections or extensions lead into or through the hazardous area.

20-514 Aircraft electrical systems

Aircraft electrical systems shall be de-energized when the aircraft is stored in a hangar and, whenever possible, while the aircraft is undergoing maintenance.

20-516 Aircraft battery-charging and equipment

(1) Aircraft batteries shall not be charged when installed in an aircraft located inside or partially inside a hangar.

(2) Battery chargers and their control equipment shall not be located or operated within any of the hazardous areas defined in Rule 20-502 but shall be permitted to be located or operated in a separate building or in an area complying with Rule 20-502(4).

(3) Mobile chargers shall carry at least one permanently affixed warning sign stating that the chargers be kept 1.5 m clear of aircraft engines and fuel tank areas.

(4) Tables, racks, trays, and wiring shall not be located within a hazardous area, and shall conform to the provisions of Section 26 pertaining to storage batteries.

20-518 External power sources for energizing aircraft

(1) Aircraft energizers shall be designed and mounted so that all electrical equipment and fixed wiring are at least 450 mm above floor level, and they shall not be operated in a hazardous area as defined in Rule 20-502(3).

(2) Mobile energizers shall carry at least one permanently affixed sign to the effect that the energizer be kept 1.5 m clear of aircraft engines and fuel tank areas.

(3) Aircraft energizers shall be equipped with polarized external power plugs and with automatic controls to isolate the ground power unit electrically from the aircraft in case excessive voltage is generated by the ground power unit.

(4) Flexible cords for aircraft energizers and ground support equipment shall be of the extra-hard-usage type and shall include a bonding conductor.

20-520 Mobile servicing equipment with electrical components

(1) Mobile servicing equipment such as vacuum cleaners, air compressors, air movers, etc., having electrical wiring and equipment not suitable for Class I, Zone 2 locations shall
   (a) be designed and mounted so that all such wiring and equipment is at least 450 mm above the floor;
   (b) not be operated within the hazardous areas defined in Rule 20-502(3); and
   (c) carry at least one permanently affixed warning sign stating that the equipment be kept 1.5 m clear of aircraft engines and fuel tank areas.

(2) Flexible cords used for mobile equipment shall be of the extra-hard-usage type and shall include a bonding conductor.

(3) Attachment plugs and receptacles shall provide for the connection of the bonding conductor to the raceway system.

(4) Equipment shall not be operated in areas where maintenance operations likely to release hazardous vapours are in progress, unless the equipment is at least suitable for use in a Class I, Zone 2 location.

20-522 Bonding

All metal raceways, and all non-current-carrying metal portions of fixed or portable equipment, regardless of voltage, shall be bonded to ground in accordance with Section 10.

Section 22
Locations in which corrosive liquids, vapours,
or excessive moisture are likely to be present

Section 22 — Locations in which corrosive liquids, vapours, or excessive moisture are likely to be present

General

22-000 Scope
This Section applies to electrical equipment and installations in locations in which corrosive liquids, vapours, or excessive moisture are likely to be present, and supplements or amends the general requirements of this Code.

22-002 Category definitions (see Appendix B)
Locations covered in this Section shall be classified as follows:
(a) **Category 1** — the location is one in which moisture in the form of vapour or liquid is present in quantities that are liable to interfere with the normal operation of electrical equipment, whether the moisture is caused by condensation, the dripping or splashing of liquid, or otherwise; and
(b) **Category 2** — the location is one in which corrosive liquids or vapours are likely to be present in quantities that are likely to interfere with the normal operation of electrical equipment.

22-004 Application of category definitions
Where the expressions "Category 1" or "Category 2" do not appear in any Rule in this Section, the Rule shall apply to both categories.

Equipment

22-100 Essential equipment only (see Appendix B)
(1) Only electrical equipment that is essential for the processes being carried on in a room or section of a building shall be installed in Category 1 and Category 2 locations.
(2) Service equipment, motors, panelboards, switchboards, and other electrical equipment shall, where practicable, be installed in rooms or sections of the building that are not Category 1 or Category 2 locations.
(3) Enclosures containing moulded case circuit breakers shall not be located in a Category 2 location unless marked as suitable for the application.

22-102 Type of construction
(1) Where the electrical equipment is, or is likely to be, partially or wholly submerged, it shall be a submersible type of construction.
(2) Where the electrical equipment is, or is likely to be, subjected to direct streams of liquid under pressure, it shall be a watertight type of construction.
(3) Where the electrical equipment is, or is likely to be, exposed to corrosive vapours, it shall be a corrosion-resistant type of construction.
(4) Where the electrical equipment is, or is likely to be, exposed to splashing of water, it shall be of a weatherproof or watertight type of construction.
(5) Where the electrical equipment is, or is likely to be, exposed only to the falling or condensing of moisture, it shall be a drip-proof, weatherproof, or watertight type of construction.
(6) Where a protective coating on electrical equipment is, or may be, exposed to corrosive liquids or vapour, the coating shall be suitable for the corrosive condition.

22-104 Pendant lampholders
(1) Pendant lampholders shall be of the weatherproof type and hung from insulated stranded copper conductors of not less than No. 14 AWG.
(2) Where the pendant conductors exceed 900 mm in length, they shall be twisted together.

22-106 Fixtures
(1) Every lighting fixture in a Category 1 location shall be constructed so that water cannot enter or accumulate within the fixture.
(2) Every lighting fixture in a Category 2 location shall be totally enclosed, gasketted, and of a corrosion-resistant type of construction.

22-108 Receptacles, plugs, and cords for portable equipment
(1)   Every receptacle and attachment plug for portable equipment shall be
   (a)   of the weatherproof type; and
   (b)   provided with grounding terminals and conductors properly bonded to ground.
(2)   Flexible cords or power supply cables for portable equipment shall contain a bonding conductor and be the outdoor type suitable for hard usage as indicated in Table 11.

Wiring

22-200 Wiring method in Category 1 locations
(1)   Where conductors are exposed to moisture in a Category 1 location, they shall
   (a)   if used in exposed wiring, be the types specified in Table 19
      (i)   for exposed wiring in wet locations; or
      (ii)   for exposed wiring where exposed to the weather, provided that they are located more than 1.5 m horizontally or 2.5 m vertically from floors, decks, balconies, or stairs; and
   (b)   if used in conduit, be of the types specified in Table 19 for use in raceways in wet locations.
(2)   Non-metallic-sheathed cable of the NMW or NMWU type shall be permitted to be used in a Category 1 location.
(3)   Armoured cable and aluminum-sheathed cable installed in a Category 1 location shall be the type listed in Table 19 for direct earth burial.
(4)   Split knobs or cleats shall not be used in a Category 1 location.
(5)   Mineral-insulated cable may be used in a Category 1 location, but if the cable is secured to walls it shall be spaced at least 6 mm from the wall at each point of support.
(6)   Aluminum conductors shall not be used in Category 1 locations unless the termination or joint is adequately sealed against ingress of moisture.

22-202 Wiring method in Category 2 locations
(1)   Where conductors are exposed to corrosive liquids or vapours in a Category 2 location, they shall
   (a)   if used in exposed wiring, be a type with corrosion-resistant protection and be located more than 1.5 m horizontally or 2.5 m vertically from floors, decks, balconies, or stairs; and
   (b)   if used in conduit, be of a type with corrosion-resistant protection.
(2)   Non-metallic-sheathed cable of the NMW or NMWU type shall be permitted to be used in a Category 2 location.
(3)   Surface metal raceways, underfloor raceways, bare conductors, armoured cable, except where permitted in Table 19 for exposure to corrosive action, wireways, busways, and split knobs, shall not be used in Category 2 locations.
(4)   Mineral-insulated cable shall be permitted to be used in a Category 2 location if the corrosive action is not of such a nature as to cause deterioration of the outer sheath.
(5)   Aluminum-sheathed cable shall be permitted to be used in a Category 2 location provided that it has suitable corrosion-resistant protection where necessary.
(6)   Aluminum conductors shall not be used in Category 2 locations unless the termination or joint is adequately sealed against ingress of corrosive liquids or vapours.

22-204 Wiring methods in buildings housing livestock or poultry (see Appendix B)
(1)   Wiring in buildings housing livestock or poultry shall be the type listed in Table 19 for wet locations.
(2)   Where non-metallic-sheathed cable is used in buildings housing livestock and poultry it shall be the NMW or NMWU type.
(3)   Notwithstanding Subrules (1) and (2), wiring listed in Table 19 for damp locations shall be permitted in buildings housing livestock or poultry when provided with adequate ventilation.
(4)   Aluminum conductors shall not be used in buildings housing livestock or poultry.
(5)   Non-metallic-sheathed cables shall be mechanically protected against damage by rodents where it is less than 300 mm above a surface where rodents may be present.

22-206 Rinks
(1)   Conductors run as open wiring in accordance with Rules 12-200 to 12-224 shall be permitted to be used for the lighting of curling or skating rink areas that are subject to condensation, provided that the conductors are suitable for wet locations as indicated by Table 19.

126

Section 22
Locations in which corrosive liquids, vapours,
or excessive moisture are likely to be present

(2) The wiring method used in waiting rooms and other portions of rinks shall be in accordance with Section 12 based on the area and moisture conditions involved.

(3) Rink areas that are provided with positive mechanical ventilation capable of changing the air at least three times per hour shall be permitted to be regarded as dry locations.

Drainage, sealing, and exclusion of moisture and corrosive vapour

22-300 Drip loops

Where exposed conductors or non-metallic-sheathed cables enter into or issue from a Category 1 or Category 2 location, the conductors shall pass through the wall of the location in an upward direction from the Category 1 or Category 2 location and in the case of exposed conductors, shall be in non-combustible, non-absorptive insulating tubes.

22-302 Drainage, sealing, and exclusion of moisture

(1) Where conduit is used, it shall be
   (a) arranged to drain at frequent intervals to suitable locations;
   (b) equipped with approved fittings that permit the moisture to drain out of the system;
   (c) installed to give 12 mm clearance from the supporting surface when either conduit or supporting surface is metallic; and
   (d) sealed to prevent the migration of corrosive vapour where due to the location of equipment, such migration is considered possible.

(2) Where a conduit or aluminum-sheathed cable leaves a warm room and enters a cooler atmosphere, it shall be sealed off to prevent breathing and subsequent condensation and shall be done in such a manner that condensate will not be trapped at the seal.

(3) Every joint in a conduit in a Category 1 location shall be watertight.

(4) Every cabinet and fitting in a Category 1 location shall be
   (a) of splash-proof or drip-proof construction;
   (b) placed so as to prevent moisture or water from entering and accumulating within; and
   (c) mounted to give at least 12 mm clearance from the supporting surface when either enclosure or supporting surface is metallic.

Circuit control

22-400 Circuit control

Every circuit in a Category 1 or Category 2 location shall, where practicable, be arranged so that the current-carrying conductors may be entirely cut off from the supply of electrical power or energy at a convenient point outside the location.

Materials

22-500 Corrosion-resistant material

All conduits, metal enclosures, and fittings, including every bolt and screw used to secure electrical equipment, shall be protected by or be of material resistant to the specific corrosive environment.

Bonding

22-600 Exposed metal parts

Exposed, non-current-carrying metal parts of fixed or portable equipment shall be bonded to ground in accordance with Section 10.

Sewage lift and treatment plants

Δ **22-700 Scope**

(1) Rules 22-702 to 22-710 apply to the installation of electrical facilities in sewage lift and pumping stations, and in primary and secondary sewage treatment plants where the environment could contain multiple hazards such as moisture, corrosion, explosions, fire, and atmospheric poisoning.

(2) Rules 22-702 to 22-710 do not apply to methane generation facilities associated with some treatment facilities.

Δ **22-702 Special terminology**
In this Subsection, the following definitions apply:

Continuous positive pressure ventilation — a ventilation system capable of maintaining a positive pressure in a room or area and of changing the air in the room or area at least six times an hour with means for detecting ventilation failure.

Dry well — the location below ground designed to accommodate equipment associated with wastewater pumping and isolated from the wet well location to prevent the migration of gases and vapours into the dry well.

Suitably cut off — an area rendered impermeable and cut off from an adjoining area with no means of liquid, gas or vapour communication between the areas at atmospheric pressure.

Wet well — the location below ground where the raw sewage is collected and temporarily stored before passing through the lift pumps or being processed in a treatment plant.

Δ **22-704 Classification of areas** (see Appendix B)
(1)   Sewage lift and treatment plants shall be classified for
        (a)   hazardous areas in accordance with Section 18; and
        (b)   corrosive liquids, vapours or moisture in accordance with this Section.
(2)   Wet wells provided with adequate continuous positive pressure ventilation shall be considered to be Class I, Zone 2.
(3)   Except as permitted by Subrule 5(c), all locations below ground suitably cut off from locations in which sewage gases may be present shall be considered to be Category 1.
(4)   All locations in which sewage gases may be present in explosive concentrations shall be considered to be hazardous areas and Category 2.
(5)   The following areas shall be permitted to be classified as ordinary locations:
        (a)   all locations suitably cut off from a Category 2 location and not classified as a Category 1 location;
        (b)   all locations not suitably cut off from a Category 2 location but with adequate continuous positive pressure ventilation; and
        (c)   dry well locations below ground where adequate heating and adequate continuous positive pressure ventilation is installed.

Δ **22-706 Wiring methods**
(1)   Wiring methods within hazardous areas shall be in accordance with Section 18.
(2)   Wiring methods in a Category 1 or a dry Category 2 location shall be in accordance with Rules 22-200 and 22-202, respectively.
(3)   Wiring methods in a wet or damp Category 2 location shall be in accordance with Rule 22-202, with the following exceptions:
        (a)   rigid steel conduit and electrical metallic tubing shall not be used;
        (b)   armoured cable, mineral-insulated cable, and aluminum-sheathed cable shall be permitted to be used provided that the cable is spaced from walls by at least 12 mm, has a corrosion resistant jacket, and the cable connectors are adequately sealed against ingress of corrosive liquids or vapours; and
        (c)   grounding and bonding conductors shall be insulated or otherwise protected from corrosion and the point of connection to ground, if exposed to a corrosive atmosphere, shall be protected from corrosion or be of a material resistant to the specific corrosive environment.
(4)   Conduits installed from the wet well to an electrical enclosure shall be sealed with a suitable compound to prevent the entrance of moisture, vapour or gases into the enclosure.

Δ **22-708 Electrical equipment**
(1)   Electrical equipment installed in hazardous areas shall be in accordance with Section 18.
(2)   Electrical equipment installed in a Category 1 or a dry Category 2 location shall be in accordance with the applicable requirements of this Code.
(3)   Electrical equipment installed in a wet or damp Category 2 location shall be in accordance with the applicable requirements of the Code, with the following exceptions:
        (a)   receptacles shall be fitted with self-closing covers, and if of the duplex type, have individual covers over each half of the receptacle;
        (b)   lighting switches shall have weatherproof covers;
        (c)   unit emergency lighting equipment and emergency lighting control units, other than remote lamps, shall not be located in such locations;

Section 22
*Locations in which corrosive liquids, vapours,
or excessive moisture are likely to be present*

(d)　heating equipment shall be approved for such locations or installed outside the corrosive location;

(e)　motors shall be totally enclosed and fan cooled and shall not incorporate dissimilar metals relative to the motor frame and connection box; and

(f)　electrical equipment in wet well areas shall not contain devices that will cause an open arc or spark during normal operation.

(4)　Ventilation fans shall not be located within the wet well, and fan blades shall be of spark-resistant material.

(5)　Areas provided with continuous positive pressure ventilation shall be interlocked to de-energize all electrical equipment not approved for a Class I location in case the ventilating equipment is inoperative.

Δ 22-710 Grounding of structural steel
Structural steel below ground in contact with the surrounding earth shall be bonded to the system ground.

Section 24 — Patient care areas

24-000 Scope (see Appendix B)

(1) This Section applies to the installation of
  (a) electrical wiring and equipment within patient care areas of health care facilities; and
  (b) those portions of the electrical systems of health care facilities designated as essential electrical systems.
(2) Except as noted in Rules 24-104(7) and 24-108, this Section does not apply to installations of electrical communication systems as covered in Section 60, nor to radio and television installations as covered in Section 54.
(3) This Section supplements or amends the general requirements of this Code.

24-002 Special terminology (see Appendix B)

In this Section, the following definitions apply:

Anaesthetizing location — any area of a health care facility where the induction and maintenance of general anaesthesia are routinely carried out in the course of the examination or treatment of patients.

Applied part — the part or parts of medical electrical equipment, including the patient leads, that come intentionally into contact with the patient to be examined or treated.

Basic care area — a patient care area where body contact between a patient and medical electrical equipment is neither frequent nor usual.

Body contact — an intentional contact at the skin surface or internally, but no direct contact to the heart.

Cardiac contact — an intentional contact directly to the heart by means of an invasive procedure.

Casual contact — contact by voluntary action with a device that has no applied part and is not intended to be connected to a patient.

Conditional branch — that portion of an essential electrical system in which circuits require power restoration by emergency service within 24 h, depending on special circumstances such as environmental or climatic conditions.

Critical care area — a patient care area that is an anaesthetizing location, or in which cardiac contact between a patient and medical electrical equipment is frequent or normal.

Delayed vital branch — that portion of an essential electrical system in which the circuits require power restoration within 2 min.

Emergency power system — a power system that is supplied from an emergency supply and connected to feed essential systems.

Emergency supply — one or more in-house generators of electricity intended to be available in the event of a failure of all other supplies and capable of supplying all the essential loads.

Essential electrical system — an electrical system that has the capability of restoring and sustaining a supply of electrical energy to specified loads in the event of a loss of the normal supply of energy.

Hazard index — for a given set of conditions in an isolated power system, the current, expressed in milliamperes and consisting of resistive and capacitive leakage and fault currents, that would flow through a low impedance if the low impedance were to be connected between either isolated conductor and ground.

Health care facility — a set of physical infrastructure elements that are intended to support the delivery of specific health-related services.

Intermediate care area — a patient care area in which body contact between a patient and medical electrical equipment is frequent or normal.

Isolated system — an electrical distribution system in which no circuit conductor is connected directly to ground.

Line isolation monitor — a device that measures and displays the total hazard index of an isolated electrical system and provides warning when the index reaches a preset limit.

Normal supply — the main electrical supply into a building or a building complex; it may consist of one or more consumer services capable of supplying all loads in the building or building complex.

Patient — a person undergoing medical investigation or treatment.

Patient care area — an area intended primarily for the provision of diagnosis, therapy, or care.

Patient care environment — a zone in a patient care area that has been pre-selected for the accommodation of a patient bed, table, or other supporting mechanism, and for the accommodation of equipment involved in patient treatment, and that includes the space within the room 1.5 m beyond the perimeter of the bed, table, or other supporting mechanism in its normal location and to within 2.3 m of the floor.

Patient care environment bonding point — a common bus, in a patient care environment, that is bonded to ground, and that serves as a common point to which equipment and other bonding connections can be made by means of a group of jacks.

Total hazard index — the hazard index of a given isolated system with all appliances, including the line isolation monitor, connected.

Vital branch — that portion of an essential electrical system in which the circuits require power restoration within 10 s.

Patient care areas

24-100 Rules for patient care areas (see Appendix B)
Rules 24-102 through 24-116 shall apply to those patient care areas that have been designated as
(a)   basic care areas;
(b)   intermediate care areas; or
(c)   critical care areas.

24-102 Circuits in basic care areas (see Appendix B)
Δ (1)   The branch circuits supplying receptacles or permanently connected equipment in basic care areas shall be supplied from a grounded distribution system.
(2)   Branch circuit conductors shall be copper and shall be sized not smaller than No. 12 AWG.
(3)   A branch circuit that supplies receptacles or permanently connected medical electrical equipment, including parts of the essential electrical system within a patient care environment, shall supply loads only within such environments.
(4)   All branch circuits for a patient care environment shall be supplied from
(a)   a single panelboard; or
(b)   two panelboards, provided that one of the panelboards is a part of an essential electrical system.
(5)   Branch circuits shall be supplied at not more than 150 volts-to-ground, unless designated for special-purpose use (e.g., to supply mobile X-ray, laser, and similar equipment) or for permanently connected equipment.

24-104 Bonding to ground in basic care areas (see Appendix B)
(1)   Bonding conductors shall be insulated unless they are
(a)   installed in non-metallic conduit; or
(b)   incorporated into a cable assembly that is constructed in such a manner that contact between any metal shield or armour, if it is present, and a bare bonding conductor is not possible.
(2)   All receptacles and other permanently connected equipment shall be bonded to ground by copper equipment bonding conductors, sized not smaller than the minimum size required for circuit conductors, and run in accordance with Rule 10-808 or run with the circuit conductors in accordance with the following:
(a)   each multi-wire branch circuit shall be provided with its own equipment bonding conductor;
(b)   except as permitted by Items (c) and (d), each 2-wire branch circuit supplying a receptacle in a patient care environment shall be provided with its own equipment bonding conductor;
(c)   when the receptacles in a patient care environment are supplied from two 2-wire branch circuits in the same raceway, a single equipment bonding conductor shall be permitted to be shared by the two circuits; or
(d)   when receptacles intended for a pair of adjacent patient care environments are supplied by three 2-wire branch circuits and one of the circuits is intended to be shared by both environments, the three circuits shall be permitted to share two equipment bonding conductors.

(3)   Utilization equipment bonding conductors required by Subrules (2), (6), and (7) shall terminate either at the panelboard supplying the branch circuits to the patient care environment from which they arise or on a separately installed busbar that is bonded to that panelboard.

(4)   Where branch circuits for a patient care environment are supplied from two panelboards as permitted by Rule 24-102(4), the panelboards shall be bonded together with a single copper equipment bonding conductor sized in accordance with Table 16, but in no case smaller than a No. 6 AWG.

(5)   Each panelboard described in Rule 24-102(4) shall be bonded to ground by a copper utilization equipment bonding conductor that is
   (a)   installed in the same raceway as the circuit conductors supplying that panelboard or installed in accordance with Rule 10-808, and sized in accordance with Table 16; or
   (b)   incorporated into the assembly of the cable supplying that panelboard.

(6)   Each item of 3-phase equipment shall be bonded to ground with a copper equipment bonding conductor that is
   (a)   sized in accordance with Table 16, but in no case smaller than No. 12 AWG; and
   (b)   connected to its own terminal at the equipment and the panelboard.

(7)   If they could become energized, exposed non-current-carrying metal parts of communication, radio, or television equipment, other than telephone sets, in a patient care environment, shall be bonded to ground using a copper equipment bonding conductor sized in accordance with Subrule (6), by
   (a)   connection to the bonding screw in the communication section of a barriered and ganged metal outlet box that serves a patient care environment; or
   (b)   connection to an equipment bonding conductor or bonding busbar for that patient care environment as identified in Subrule (3).

Δ   **24-106 Receptacles in basic care areas** (see Appendix B)
(1)   Receptacles intended for a given patient care environment shall be located to minimize the likelihood of their inadvertent use for a patient care environment for which they are not intended.

(2)   Receptacles located in areas that are routinely cleaned using liquids that normally splash against the walls shall be installed not less than 300 mm above the floor.

(3)   Receptacles located in bathrooms or washrooms shall be
   (a)   located within 1.5 m of the wash basin;
   (b)   located outside of any bathtub enclosure or shower stall.

(4)   Receptacles intended for housekeeping equipment and other non-medical loads shall be so identified.

(5)   Except for receptacles described in Subrule (3), all 15 A and 20 A non-locking receptacles shall be hospital grade.

(6)   All receptacles that are part of an essential electrical system shall be coloured red, and no other receptacles shall be so coloured.

24-108 Other equipment in basic care areas (see Appendix B)
Notwithstanding the requirements of Rule 60-400, emergency signalling and similar equipment manufactured in conformance with the additional watertightness requirements of CSA C22.2 No. 125 or the CAN/CSA-C22.2 No. 601 series of standards, and intended for use in shower stalls and bathtub enclosures, shall be permitted to be installed at normal heights within such stalls and enclosures.

Δ   **24-110 Circuits in intermediate and critical care areas** (see Appendix B)
The branch circuits supplying receptacles or other permanently connected equipment in intermediate or critical care areas shall be supplied from either a grounded system meeting the requirements of Rule 24-102, or an isolated system meeting the requirements of Rule 24-200, except that all branch circuits supplying loads within patient care environments, other than those supplying multiphase equipment, shall be 2-wire circuits.

24-112 Bonding to ground in intermediate and critical care areas (see Appendix B)
(1)   Bonding to ground in intermediate and critical care areas shall conform to Rule 24-104 whether the supply is derived from a grounded or an isolated system.

(2)   If a patient care environment bonding point is provided, it shall be bonded to the panelboard serving the patient care environment with which it is associated by either
   (a)   a bonding jumper connecting it to the bonding terminal in an enclosure that accommodates the bonding point along with receptacles for a patient care environment; or
   (b)   a copper conductor that is installed for that specific purpose and is run in the same raceway as the equipment bonding conductors serving that patient care environment.

24-114 Receptacles in intermediate and critical care areas (see Appendix B)

Receptacles in intermediate and critical care areas shall

(a) meet the requirements of Rule 24-106; and

(b) where supplied from an isolated system, be identified as such.

Δ **24-116 Receptacles subject to standing fluids on the floor or drenching of the work area** (see Appendix B)

All receptacles in areas subject to standing fluids on the floor or drenching of the work area shall be

(a) protected by a ground fault circuit interrupter of the Class A type; or

(b) supplied by an isolated system conforming to Rule 24-200.

Isolated systems

24-200 Rules for isolated systems (see Appendix B)

Δ (1) Rules 24-202 through 24-208 shall apply to isolated systems installed under the provisions of Rules 24-110 and 24-116.

(2) In a patient care environment supplied by an isolated system, branch circuits supplying only fixed lighting fixtures and permanently connected medical electrical equipment shall be permitted to be supplied by a conventional grounded system, provided that wiring for grounded and isolated circuits does not occupy the same raceway.

24-202 Sources of supply (see Appendix B)

(1) The means of supply to an isolated system shall be

(a) the secondary of one or more isolating transformers having no direct electrical connection between primary and secondary windings;

(b) a motor-generator set; or

(c) a suitably isolated, battery-powered inverter supply.

(2) Where more than one single-phase isolated power system serves a single patient care environment, the grounding buses of all of these systems shall be bonded together with a copper bonding conductor

(a) having a total impedance not greater than 0.2 Ω; and

(b) sized not smaller than that permitted by Table 16.

24-204 Single-phase isolated circuits (see Appendix B)

(1) Except where Rule 24-206 applies, isolated circuits shall meet the requirements of Subrules (2) through (7).

(2) Isolated circuits shall

(a) not be deliberately grounded, except through the impedance of an isolation-sensing device (e.g., isolation monitor);

(b) have circuit conductors of one of the following types:

(i) RW75 EP;

(ii) RW75 XLPE;

(iii) RW90 EP; or

(iv) RW90 XLPE;

(c) have the insulation of one circuit conductor coloured orange and the other coloured brown;

(d) have the orange-insulated conductor connected to the nickel screw of receptacles;

(e) have overcurrent devices that will open all ungrounded conductors simultaneously;

(f) be installed in non-metallic raceways conforming to flame spread requirements in accordance with Rule 2-128.

(3) Any disconnecting means controlling an isolated circuit shall safely and simultaneously disconnect all ungrounded conductors.

(4) Single-phase isolated circuits shall be 2-wire circuits with copper equipment bonding conductors, operating at voltages (rms) between conductors not exceeding

(a) 300 V for special-use receptacles and for permanently connected equipment; and

(b) 150 V for other receptacles.

(5) A single-phase isolated system shall include automatic means (a line isolation monitor), with an indicator located where visible to persons using the system, to monitor the impedance-to-ground of the system together with any loads connected to it.

(6) At the time of installation the total impedance (capacitive and resistive) between ground and each energized conductor of a single-phase isolated system shall exceed 500 000 Ω without utilization equipment or the line isolation monitor connected.

(7) Where a single-phase isolated system is employed, it shall supply

(a) general-purpose receptacles in

(i) a single anaesthetizing location;

(ii) one or more patient care environments in a single room; or

(iii) a maximum of two patient care environments in separate but adjacent rooms, provided that the alarm indicator clearly identifies the patient care environments affected by the fault; or

(b) special-purpose receptacles at different locations or in different patient care environments, provided that the system is used only for one purpose and is arranged so that only one receptacle can be energized at a time.

24-206 Individually isolated branch circuits (see Appendix B)

A single-phase isolated system that supplies only a single load via a single branch circuit shall meet the requirements of Rule 24-204(2) through (6), except that

(a) overcurrent devices need not be installed in the isolated circuit; and

(b) the use of a line isolation monitor shall be optional.

24-208 Three-phase isolated systems (see Appendix B)

A 3-phase isolated system shall

(a) supply

(i) permanently connected medical equipment; or

(ii) special-purpose receptacles in one or more anaesthetizing locations or patient care environments, provided that the system is used only for the one purpose and is arranged so that only one receptacle can be energized at a time;

(b) meet the requirements of Rule 24-204(2)(a), (b), (e), and (f);

(c) have its circuit conductors identified as follows:

(i) isolated conductor No. A — orange;

(ii) isolated conductor No. B — brown; and

(iii) isolated conductor No. C — yellow; and

(d) meet the requirements of Rule 24-204(3).

Essential electrical systems

24-300 Rules for essential electrical systems (see Appendix B)

Rules 24-302 through 24-306 shall apply to those portions of a health care facility electrical system in which the interruption of a normal supply of power would jeopardize the effective and safe care of patients, with the object of reducing hazards that might arise from such an interruption.

24-302 Circuits in essential electrical systems (see Appendix B)

(1) An essential electrical system shall comprise circuits that supply loads designated by the health care facility administration as being essential for the life, safety, and care of the patient and the effective operation of the health care facility.

(2) An essential electrical system shall comprise at the minimum a vital branch and may also include a delayed vital branch or a conditional branch, or both.

(3) The wiring of the essential electrical system shall be kept entirely independent of all other wiring and equipment and shall not enter a fixture, raceway, box, or cabinet occupied by other wiring except where necessary

(a) in transfer switches; and

(b) in emergency lighting fixtures supplied from two sources.

24-304 Transfer switches (see Appendix B)

(1) All transfer switches shall comply with the requirements of the supply authority.

(2) Automatic transfer switches used in essential electrical systems shall conform to the requirements of CSA C22.2 No. 178 and, in addition, shall

(a) be electrically operated and mechanically held; and

(b) include means for safe manual operation.

(3)   Manual transfer switches shall conform to the following:

  (a)   the switching means shall be mechanically held and the operation shall be by direct manual control or by electrical remote manual control utilizing control power from the supply to which the load is being transferred;

  (b)   a manual transfer switch that is operated by electrical remote manual control shall include a means for safe manual mechanical operation;

  (c)   reliable mechanical interlocking (and, in the case of a switch operated by electrical remote manual control, electrical interlocking) to prevent interconnection of the normal and the emergency supplies of power shall be inherent in the design of a manual transfer switch; and

  (d)   a manual transfer switch shall include a readily visible mechanical indicator showing the switch position.

(4)   The vital and delayed vital branches shall be connected to the emergency power supply by means of one or more automatic transfer switches.

(5)   The conditional branch shall be connected to the emergency power supply by either a manual or an automatic transfer switch.

24-306   Emergency supply

(1)   An emergency supply shall be one or more generator sets driven by a prime mover that is located on the health care facility premises in a fire-resistant enclosure or room, in accordance with CAN/CSA-Z32, and located such that the possibility of flooding and damage is minimized.

(2)   The prime mover of the generating set, as specified in Subrule (1), shall be capable of operating independently of supplies of water and fuel from public utilities.

Section 26 — Installation of electrical equipment

General

26-000 — *Reserved for future use.*

26-002 Connection to identified terminals or leads
Wherever a device having an identified terminal or lead is connected in a circuit having an identified conductor, the identified conductor shall be connected to the identified terminal or lead.

26-004 Equipment over combustible surfaces (see Appendix B)
Where there is a combustible surface directly under stationary or fixed electrical equipment, that surface shall be covered with a steel plate at least 1.6 mm thick, which shall extend not less than 150 mm beyond the equipment on all sides, if
(a)   the equipment is marked to require such protection; or
(b)   the equipment is open on the bottom.

26-006 Installation of ventilated enclosures
Ventilated enclosures shall be installed in a manner that does not restrict ventilation.

26-008 Sprinklered equipment (see Appendix B)
Where electrical equipment vaults or electrical equipment rooms are sprinklered, the electrical equipment contained in such vaults or rooms shall be protected where needed by non-combustible hoods or shields arranged to minimize interference with the sprinkler protection.

26-010 Outdoor installations
(1)   Outdoor installations of apparatus, unless housed in suitable enclosures, shall be surrounded by suitable fencing in accordance with Rules 26-300 to 26-324.
(2)   Outdoor equipment shall be bonded to ground.

26-012 Dielectric liquid-filled equipment — Indoors (see Appendices B and G)
(1)   Dielectric liquid-filled electrical equipment containing more than 23 L of liquid in one tank, or more than 69 L in a group of tanks, shall be located in an electrical equipment vault.
(2)   Except as permitted in Subrule (4), dielectric liquid-filled electrical equipment containing 23 L of liquid or less in one tank, or 69 L or less in a group of tanks, shall be
(a)   installed in a service room conforming to the requirements of the *National Building Code of Canada*;
(b)   provided with a metal pan or concrete curbing capable of collecting and retaining all the liquid of the tank or tanks;
(c)   isolated from other apparatus by fire-resisting barriers, with metal-enclosed equipment considered as providing segregation and isolation; and
(d)   separated from other dielectric liquid-filled electrical equipment by such a distance that, if the liquid in such equipment were spread at a density of 12 L/m², the areas so covered would not overlap; these areas being deemed to be circular if the tank (or group of tanks) is in an open area, semi-circular if the tank is against a wall, and quarter-sector if the tank is in a corner.
(3)   Notwithstanding Subrules (1) and (2), motor starters shall be permitted to have these quantities of liquids doubled.
(4)   Notwithstanding Subrule (2), capacitors filled with flammable liquids of 14 L or less in each tank shall not be required to be installed in an electrical equipment vault nor in a service room, provided that
(a)   a metal pan or concrete curbing that is capable of collecting and retaining all the liquid of the tank or tanks is installed;
(b)   no other dielectric liquid-filled electrical equipment nor any combustible surface or material is within 4.5 m unless segregated by fire-resisting barriers, with metal-enclosed equipment considered as providing segregation; and
(c)   each capacitor tank is provided with overcurrent protection to minimize rupture of the case.

26-014 Dielectric liquid-filled equipment — Outdoors (see Appendix B)
(1)   Dielectric liquid-filled electrical equipment containing more than 46 L in one tank, or 137 L in a group of tanks, and installed outdoors shall not, except as permitted by Subrule (3), be located within 6 m of
(a)   any combustible surfaces or material on a building;
(b)   any door or window; or

(c)   any ventilation inlet or outlet.

(2)   The dimension referred to in Subrule (1) shall be the shortest line-of-sight distance from the face of the container containing the liquid to the building or part of the building in question.

(3)   Notwithstanding the requirements of Subrule (1), the equipment shall be permitted to be installed within 6 m of any item listed in Subrule (1)(a), (b), and (c) provided that a wall or barrier with non-combustible surfaces or material is constructed between the equipment and that item.

(4)   Where dielectric liquid-filled electrical equipment containing more than 46 L in one tank, or 137 L in a group of tanks, is installed outdoors it shall

(a)   be inaccessible to unauthorized persons;

(b)   not obstruct firefighting operations;

(c)   if installed at ground level, be located on a concrete pad draining away from structures or be in a curbed area filled with coarse crushed stone; and

(d)   not have open drains for the disposal of the liquid in the proximity of combustible construction or materials.

Isolating switches

26-100 Location of isolating switches

(1)   Isolating switches shall be permitted to be located so that a hook stick is required to operate them.

(2)   Isolating switches shall be plainly marked to minimize the chance that they will be opened under load, unless

(a)   they are located or guarded so that they are inaccessible to unauthorized persons; or

(b)   they are interlocked so that they cannot normally be opened under load.

Circuit breakers

26-120 Indoor installation of circuit breakers

(1)   Dielectric liquid-filled circuit breakers installed indoors shall be installed in accordance with Rule 26-012.

(2)   Circuit breakers installed in electrical equipment vaults shall be operable without opening the door of the vault.

Fuses and fusible equipment

26-140 Installation of fuses

Fuses shall be located so that

(a)   their operation will not result in injury to persons or damage to property or other equipment; and

(b)   they can be readily inserted or removed.

Δ **26-142 Fusible equipment**

Fusible equipment shall employ low-melting point fuses of the type referred to in Rule 14-200 or fuses as referred to in Rule 14-212(b) when connected to conductors whose ampacity is based on Table 1 or 3 or on Column 4 of Table 2 or 4, unless equipment using other types of fuses is marked as being suitable for such use.

Capacitors

26-200 Capacitors exempted

The requirements of Rules 26-202 to 26-222 shall not apply to capacitors that form component parts of factory-assembled electrical equipment nor to surge protective capacitors.

26-202 Capacitors installed indoors

Dielectric liquid-filled capacitors located indoors shall be installed in accordance with Rule 26-012.

26-204 Guarding of capacitors

All live parts of capacitors shall be inaccessible to unauthorized persons.

26-206 Grounding of capacitors

Non-current-carrying metal parts of capacitors shall be bonded to ground.

26-208 Conductor size for capacitors

(1) The ampacity of capacitor feeder circuits and branch circuits shall be not less than 135% of the rated current of the capacitor.

(2) Where a branch circuit supplies two or more capacitors, the overcurrent device protecting the conductors of the branch circuit shall be considered as protecting the taps made thereto to supply single capacitors, provided that

    (a) the tap is not more than 7.5 m long; and

    (b) its conductors comply with Subrule (1) and also have an ampacity not less than one-third that of the branch circuit conductors from which they are supplied.

26-210 Overcurrent protection (see Appendix B)

An overcurrent device, rated or set as low as practicable without causing unnecessary opening of the circuit, but not exceeding 250% of the rated current of the capacitor, shall be provided in each ungrounded conductor of a capacitor feeder or branch circuit, unless a deviation has been allowed in accordance with Rule 2-030.

26-212 Disconnecting means for capacitor feeders or branch circuits

(1) A disconnecting means shall be provided in each ungrounded conductor connected to each capacitor bank in order that the capacitors can be made dead without having to disconnect other loads.

(2) The disconnecting means shall be within sight of and not more than 9 m from the capacitor unless the disconnecting means can be locked in the open position.

(3) A warning notice shall be affixed to the disconnecting means used on circuits having capacitors only, stating that

    (a) the circuit has capacitors; and

    (b) a waiting period of 5 min is necessary when the circuit is opened, after which the capacitors shall be discharged before handling.

26-214 Rating of the disconnecting means for capacitor feeders or branch circuits

The disconnecting means for a capacitor feeder or branch circuit shall be rated not less than 135% of the rated current of the capacitor.

26-216 Rating of contactors for capacitor feeders or branch circuits

Contactors used for the switching of capacitors shall have a current rating not less than the following percentage of the rated capacitor current:

(a) open-type contactor: 135%; and

(b) enclosed-type contactor: 150%.

26-218 Special provisions for motor circuit capacitors

(1) Where a capacitor is connected on the load side of a motor circuit disconnecting means

    (a) individual disconnecting means for the capacitor need not be provided;

    (b) the rating of the disconnecting means, the overcurrent device, and the size of the motor-circuit conductors need not be greater than would be required without the capacitor; and

    (c) the ampacity of the conductors connecting the capacitor to the motor circuit shall be in accordance with Rule 26-208 and shall be not less than one-third that of the motor circuit conductors.

(2) Where a capacitor is connected on the load side of a motor controller

    (a) the rating of the capacitor shall not exceed the value required to raise the no-load power factor of the motor to unity;

    (b) the rating or setting of the overload device shall be reduced to a value corresponding with the current obtained with the improved power factor;

    (c) individual overcurrent protection for the capacitor need not be provided;

    (d) the motor shall not be subject to star-delta starting, auto-transformer starting, or switching service such as plugging, rapid reversals, reclosings, jogging, or other similar operations that generate overvoltages and overtorques; and

    (e) time-delay devices shall be installed in the motor control circuit of motors driving high inertia loads, so that the motor cannot be restarted until the residual voltage is reduced to 10% of the nominal value.

26-220 Transformers supplying capacitors

The volt-ampere rating of a transformer supplying a capacitor shall not be less than 135% of the capacitor volt-ampere rating.

26-222 Drainage of stored charge of capacitors

(1)   Capacitors shall be provided with a means of draining the stored charge.

(2)   The draining means shall be such that the residual voltage will be reduced to 50 V or less after the capacitor is disconnected from the source of supply
(a)   within 1 min in the case of capacitors rated at 750 V or less; and
(b)   within 5 min in the case of capacitors rated at more than 750 V.

(3)   The discharge circuit shall be
(a)   permanently connected to the terminals of the capacitor bank; or
(b)   provided with automatic means of connecting it on removal of voltage from the line.

(4)   The discharge circuit shall not be switched or connected by manual means.

(5)   Motors, transformers, or other electrical equipment capable of constituting a suitable discharge path, connected directly to capacitors without the interposition of a switch or overcurrent device, constitute a suitable discharge path.

Transformers

26-240 Transformers — General

(1)   In this Subsection, "transformer" means a single-phase transformer, a polyphase transformer, or a bank of two or three single-phase transformers connected to operate as a polyphase transformer.

(2)   Transformers shall be constructed so that all live parts are enclosed unless they are installed to be inaccessible to unauthorized persons.

(3)   Conductors used for connection to air-cooled (dry-type) transformers shall be permitted to enter a transformer enclosure through the top only where the transformer is marked to permit such entry.

(4)   Transformers shall be protected from mechanical damage.

(5)   Dielectric liquid-filled transformers shall be mounted so that there is an air space of 150 mm between transformers, and between transformers and adjacent surfaces of combustible material except the plane on which the transformer is mounted.

Δ 26-242 Outdoor transformer installations

(1)   Except as permitted by Subrule (2), where transformers, including their conductors and control and protective equipment, are installed outdoors, they shall
(a)   be installed in accordance with Rule 26-014 if they are dielectric liquid-filled;
(b)   have the bottom of their platform not less than 3.6 m above ground if they are isolated by elevation;
(c)   have the entire installation surrounded by a suitable fence in accordance with Rules 26-300 to 26-324 if they are not isolated by elevation or not housed in suitable enclosures;
(d)   have conspicuously posted, suitable warning signs indicating the highest voltage employed except where there is no exposed live part.

(2)   Dielectric liquid-filled pad-mounted distribution transformers shall be installed at least 3 m from any combustible surface or material on a building and at least 6 m from any window, door or ventilation inlet or outlet on a building, except where
(a)   a wall or barrier with non-combustible surfaces or material is constructed between the transformer and any door, window, ventilation opening, or combustible surface; or
(b)   the transformer is protected by an internal current-limiting fuse and equipped with a pressure relief device, with working spaces around the transformer of at least 3 m on the access side and on all other sides:
(i)   1 m for three-phase transformers; and
(ii)   0.6 m for single-phase transformers.

26-244 Transformers mounted on roofs (see Appendix B)

(1)   Except as permitted by Subrule (2), dielectric liquid-filled transformers installed on the roof of a building shall be located in an electrical equipment vault in accordance with Rules 26-350 to 26-356, and adequately supported by means of non-combustible construction.

(2)   Transformers containing a non-propagating liquid, suitable for the purpose, having a flash point not less than 275 °C, installed on the roof of a building need not be located in an electrical equipment vault but shall not be placed adjacent to doors or windows, nor within 4.5 m of discharge vents for flammable fumes or combustible or electrically conductive dusts.

26-246 Dielectric liquid-filled transformers indoors (see Appendix B)

(1)   Except as permitted by Subrule (2), dielectric liquid-filled transformers shall be installed in accordance with Rule 26-012.

(2)   Transformers containing a non-propagating liquid, suitable for the purpose, having a flash point not less than 275 °C, that are located indoors shall be installed in an electrical equipment vault, unless the following conditions are met:

(a)   the transformer is protected from mechanical damage either by location or guarding;

(b)   a pressure relief vent is provided where the rating exceeds 25 kV•A at 25 Hz or 37.5 kV•A at 60 Hz;

(c)   a means of absorbing gases generated by arcing inside the case, or a pressure relief vent connected to the outdoors, is provided where the transformer is installed in a poorly ventilated location;

(d)   where the voltage rating exceeds 15 000 V, the transformer is installed in a service room accessible only to authorized persons; and

(e)   the transformer is provided with a metal pan or concrete curbing capable of collecting and retaining all the liquid of the tank or tanks.

26-248 Dry-core, open-ventilated type transformers

(1)   Transformers of the dry-core open-ventilated type shall be mounted so that there is an air space of not less than 150 mm between transformer enclosures and between a transformer enclosure and any adjacent surface except floors.

(2)   Notwithstanding Subrule (1), where the adjacent surface is a combustible material, the minimum permissible separation between the transformer enclosure and the adjacent surface shall be 300 mm.

(3)   Notwithstanding Subrule (1), where the adjacent surface is the wall on which the transformer is mounted, the minimum permissible separation between the enclosure and the mounting wall shall be 6 mm if the adjacent surface is made of

(a)   non-combustible material;

(b)   combustible material adequately protected by non-combustible heat insulating material other than sheet metal; or

(c)   combustible material shielded by grounded sheet metal with an air space of not less than 50 mm between the sheet metal and the combustible material.

(4)   Dry-type transformers not of the sealed type shall not be installed below grade level unless adequate provision is made to prevent flooding.

(5)   Dry-type transformers not of the sealed type shall be installed in such a manner that water or other liquids cannot fall onto the windings.

26-250 Disconnecting means for transformers

A disconnecting means shall be installed in the primary circuit of each power and distribution transformer.

Δ   26-252 Overcurrent protection for power and distribution transformer circuits rated over 750 V

(1)   Except as permitted in Subrules (2), (3), and (4), each ungrounded conductor of the transformer feeder or branch circuit supplying the transformer shall be provided with overcurrent protection

(a)   rated at not more than 150% of the rated primary current of the transformer in the case of fuses; and

(b)   rated or set at not more than 300% of the rated primary current of the transformer in the case of breakers.

(2)   Where 150% of the rated primary current of the transformer does not correspond to a standard rating of a fuse, the next higher standard rating shall be permitted.

(3)   An individual overcurrent device shall not be required where the feeder or branch circuit overcurrent device provides the protection specified in this Rule.

(4)   A transformer having an overcurrent device on the secondary side rated or set at not more than the values in Table 50 or a transformer equipped with coordinated thermal overload protection by the manufacturer shall not be required to have an individual overcurrent device on the primary side, provided that the primary feeder overcurrent device is rated or set at not more than the values in Table 50.

Δ **26-254 Overcurrent protection for power and distribution transformer circuits rated 750 V or less, other than dry-type transformers**

(1) Except as permitted in Subrules (2), (3), (4), (5), and (6), each ungrounded conductor of the transformer feeder or branch circuit supplying the transformer shall be provided with overcurrent protection rated or set at not more than 150% of the rated primary current of the transformer.

(2) Where the rated primary current of a transformer is

    (a) 9 A or more, and 150% of this current does not correspond to a standard rating of a fuse or non-adjustable circuit breaker, the next higher standard rating shall be permitted; or

    (b) less than 9 A, an overcurrent device rated or set at not more than 167% of the rated primary current shall be permitted, except that where the rated primary current is less than 2 A, an overcurrent device rated or set at not more than 300% of the rated primary current shall be permitted.

(3) An individual overcurrent device shall not be required where the feeder or branch circuit overcurrent device provides the protection specified in this Rule.

(4) A transformer having an overcurrent device on the secondary side rated or set at not more than 125% of the rated secondary current of the transformer shall not be required to have an individual overcurrent device on the primary side, provided that the primary feeder overcurrent device is rated or set at not more than 300% of the rated primary current of the transformer.

(5) Notwithstanding Subrule (4), where the rated secondary current of a transformer is

    (a) 9 A or more, and 125% of this current does not correspond to a standard rating of a fuse or non-adjustable circuit breaker, the next higher standard rating shall be permitted; or

    (b) less than 9 A, an overcurrent device rated or set at not more than 167% of the rated secondary current shall be permitted.

(6) A transformer equipped with coordinated thermal overload protection by the manufacturer and arranged to interrupt the primary current shall not be required to have an individual overcurrent device on the primary side if the primary feeder overcurrent device is rated or set at a value

    (a) not more than 6 times the rated current of the transformer for a transformer having not more than 7.5% impedance; or

    (b) not more than 4 times the rated current of the transformer for a transformer having more than 7.5% but not more than 10% impedance.

Δ **26-256 Overcurrent protection for dry-type transformer circuits rated 750 V or less** (see Appendix B)

(1) Except as permitted in Subrule (2), each ungrounded conductor of the transformer feeder or branch circuit supplying the transformer shall be provided with overcurrent protection rated or set at not more than 125% of the rated primary current of the transformer, and this primary overcurrent device shall be considered as protecting secondary conductors rated at 125% or more of the rated secondary current.

(2) Notwithstanding Subrule (1), a transformer having an overcurrent device on the secondary side set at not more than 125% of the rated secondary current of the transformer shall not be required to have an individual overcurrent device on the primary side, provided that the primary feeder overcurrent device is set at not more than 300% of the rated primary current of the transformer.

(3) Where a value not exceeding 125% of the rated primary current of the transformer as specified in Subrule (1) does not correspond to the standard rating of the overcurrent device, the next higher standard rating shall be permitted.

26-258 Conductor size for transformers

(1) The conductors supplying transformers shall have an ampacity rating

    (a) not less than 125% of the rated primary current of the transformer for a single transformer; or

    (b) not less than the sum of the rated primary currents of all transformers plus 25% of the rated primary current of the largest transformer for a group of transformers operated in parallel or on a common feeder.

(2) The secondary conductors connected to transformers shall have an ampacity rating

    (a) not less than 125% of the rated secondary current of the transformer for a single transformer; or

    (b) not less than 125% of the sum of the rated secondary currents of all the transformers operated in parallel.

(3) Notwithstanding Subrules (1) and (2), primary and secondary conductors shall be permitted to have an ampacity rating not less than that required by the demand load, provided that they are protected in accordance with Rules 14-100 and 14-104.

(4)   Where multi-rating transformers are used, the primary and secondary conductors shall have an ampacity rating not less than 125% of the rated primary and secondary current of the transformer at the utilization voltage.

26-260 Overcurrent protection of instrument voltage transformers (see Appendix B)

(1)   Except under the conditions of Subrules (2), (3), and (4), instrument voltage transformers shall have primary fuses rated not more than
   (a)   10 A for low-voltage circuits; and
   (b)   3 A for high-voltage circuits.
(2)   Primary fuses shall not be installed where they would be connected in the grounded primary neutral connection of "Y" or "Open Y" connected voltage transformers.
(3)   Primary fuses shall be permitted to be omitted
   (a)   where the transformers are protected by adequate power fuses or other adequate protective devices for clearing equipment failure, and convenient means are provided for disconnecting the transformers on the primary side;
   (b)   where voltage transformers and meters, operating at low voltage and installed in suitable enclosures, are used in place of self-contained meters; or
   (c)   where both voltage and current transformers are supplied by the manufacturer in a single enclosure filled with an insulating medium, which may be air for use on low-voltage circuits if the enclosure is non-combustible, and where
      (i)   the primary terminals outside the enclosure are common to both voltage and current transformers; and
      (ii)   the enclosures are installed outdoors, if filled with an insulating medium that will burn in air.

26-262 Marking of transformers

Each transformer shall be provided with a nameplate bearing the following marking:
   (a)   manufacturer's name;
   (b)   rating in kilovolt amperes;
   (c)   rated full-load temperature rise;
   (d)   primary and secondary voltage ratings;
   (e)   frequency in hertz;
   (f)   liquid capacity, if of the liquid-filled type;
   (g)   type of liquid to be used;
   (h)   rated impedance, if of the power or distribution type; and
   (i)   basic impulse insulation level (BIL) for transformers rated 2.5 kV voltage class and higher.

26-264 Auto-transformers

(1)   In this Rule, "auto-transformers" means transformers in which part of the turns are common to both primary and secondary alternating-current circuits.
(2)   Auto-transformers shall not be connected to interior wiring systems, other than a wiring system or circuit used wholly for motor purposes, unless
   (a)   the system supplied contains an identified grounded conductor solidly connected to a similar identified grounded conductor of the system supplying the auto-transformer;
   (b)   the auto-transformer is used for starting or controlling an induction motor; or
   (c)   the auto-transformer supplies a circuit wholly within the apparatus that contains the auto-transformer.
(3)   Where an auto-transformer is used for starting or controlling an induction motor, it may be included in a starter case or it may be installed as a separate unit.
(4)   Notwithstanding Subrule (2), auto-transformers shall be permitted for fixed voltage transformation in circuits not incorporating a grounded circuit conductor.

26-266 Zero sequence filters (see Appendix B)

(1)   In this Rule, a "zero sequence filter" means a zig-zag or otherwise wound transformer installed to reduce unbalanced current in a 3-phase, 4-wire circuit.
(2)   Ampacities of conductors supplying zero sequence filters in conformance with Rule 4-004 shall be based on the neutral conductor being a current-carrying conductor.
(3)   Phase conductors shall have an ampacity of at least 125% of the rated primary current.
(4)   The neutral conductor shall have an ampacity equal to at least 125% of the neutral current rating.

(5)   Overcurrent protection for the filter shall not exceed 125% of the rated primary current.

(6)   The overcurrent protection required by Subrule (5) shall be equipped with an integral device arranged to activate a warning signal or alarm when operation of the overcurrent protection occurs.

Fences

26-300 General

Rules 26-302 to 26-324 apply to fences for guarding electrical equipment, especially transformers, when located outdoors.

26-302 Clearance of equipment

(1)   The minimum clearance between the fence and unguarded live parts shall be in accordance with Table 33.

(2)   The minimum clearance between the fence and enclosures containing live parts shall be 1.1 m.

(3)   The clearance shall provide adequate working space around the equipment, taking into consideration the space required for draw-out types of equipment and the opening of enclosure doors.

26-304 Height of fence

The fence, excluding barbed wire, shall be not less than 1.8 m high.

26-306 Barbed wire

The fence shall be topped with not less than three strands of barbed wire.

26-308 Setting of posts

(1)   Posts shall be set at a depth of 1.1 m for end, gate, and corner posts and 1 m for intermediate posts wherever ground conditions permit.

(2)   Where ground conditions do not permit this depth, extra bracing or concrete footings shall be provided.

(3)   Concrete footings may be required for metal posts in any case.

(4)   The spacing between posts shall be 3 m maximum.

(5)   End, gate, and corner posts shall be adequately braced against strain.

26-310 Gates

(1)   Gates shall preferably open outwardly but, if it is necessary that they open inwardly, they shall not come into contact with the frame or enclosure of any electrical equipment when open.

(2)   Gates shall be adequately braced as necessary and double gates shall be used where the width of opening exceeds 1.5 m.

(3)   Centre stops shall be provided for double gates.

(4)   Gates shall have provision for securing with padlocks.

26-312 Chain link fabric

(1)   Chain link fabric shall be securely attached to all posts and gate frames.

(2)   Chain link fabric shall be reinforced as necessary at top and bottom to prevent distortion.

(3)   Chain link fabric shall extend to within 50 mm of the ground.

(4)   Chain link fabric shall be
    (a)   made of galvanized steel wire not less than 3.6 mm in diameter;
    (b)   have a mesh not greater than 50 mm; and
    (c)   be not less than 1.8 m in width.

26-314 Use of wood

Where wood slats are acceptable, they shall
(a)   extend to within 50 mm of the ground;
(b)   be placed on the outside of the stringers; and
(c)   be spaced not more than 40 mm apart, except that where the frame or enclosure of any electrical equipment is less than 2 m from the fence there shall be no spacing permitted.

26-316 Posts

(1)   Metal posts shall be
    (a)   made of galvanized steel;
    (b)   88.9 mm specified outside diameter nominal pipe size (11.31 kg/m) for corner, end, and gate posts; and
    (c)   60.3 mm specified outside diameter nominal pipe size (5.44 kg/m) for intermediate posts.

(2)   Wood posts shall be not less than 140 × 140 mm and shall be suitably protected against decay.

26-318 Top rails
Top rails shall
(a)   be made of galvanized steel;
(b)   have a 42.2 mm specified outside diameter nominal pipe size (3.35 kg/m); and
(c)   be provided with suitable expansion joints where necessary.

26-320 Wood stringers
Wood stringers shall be not less than 38 × 140 mm nominal size if two are used and not less than 38 × 89 mm nominal size if three are used.

26-322 Wood slats
Wood slats shall be not less than 19 × 89 mm nominal size.

26-324 Preservative treatment
(1)   Steel or iron parts shall be either hot dip galvanized or electroplated with non-ferrous metal.
(2)   Wood shall be impregnated, treated, or well painted before assembly and, where in contact with the earth or concrete, shall be impregnated or otherwise suitably treated against decay.

Electrical equipment vaults

26-350 General
(1)   For the purposes of Rules pertaining to the construction of electrical equipment vaults, the single word "vault(s)" shall be understood to have the same meaning as "electrical equipment vault(s)."
(2)   Vaults shall not be used for storage purposes.

26-352 Vault size
Vaults shall be of such dimensions as to accommodate the installed equipment with at least the minimum clearances specified in the pertinent Sections of this Code.

26-354 Electrical equipment vault construction (see Appendices B and G)
Every electrical equipment vault, including the doors, ventilation, and drainage, shall be constructed in accordance with the applicable requirements of the *National Building Code of Canada*.

26-356 Illumination
(1)   Each vault shall be provided with adequate lighting, controlled by one or more switches located near the entrance.
(2)   Lighting fixtures shall be located so that they may be relamped without danger to personnel.
(3)   Each vault shall have a grounding-type receptacle installed in accordance with Rule 26-700 and located in a convenient location inside the vault and near the entrance.

Panelboards

26-400 Panelboards in dwelling units (see Appendix B)
(1)   A panelboard shall be installed in every dwelling unit except for dwelling units in hotels and motels, and dwelling units that
(a)   are not individually metered for electrical power consumption; and
(b)   have been created by subdivision of a single dwelling.
(2)   Every panelboard installed in accordance with Subrule (1) shall have a single supply protected by overcurrent devices and this supply shall be capable of being disconnected without disconnecting the supply to any other dwelling unit.

26-402 Location of panelboards (see Appendix G)
(1)   Panelboards shall not be located in coal bins, clothes closets, bathrooms, stairways, high ambient rooms, dangerous or hazardous locations, nor in any similar undesirable places.
(2)   Panelboards in dwelling units shall be installed as high as possible, with no overcurrent device operating handle positioned more than 1.7 m above the finished floor level.

Lightning arresters

26-500 Use and location of lightning arresters
(1) Lightning arresters shall be installed in every distributing substation in locations where lightning disturbances are of frequent occurrence and no other adequate protection is provided.
(2) Lightning arresters installed for the protection of utilization equipment
   (a) shall be permitted to be installed either inside or outside the building or enclosure containing the equipment to be protected; and
   (b) shall be isolated by elevation, enclosed, or made otherwise inaccessible to unauthorized persons.

26-502 Indoor installations of lightning arresters
(1) Where lightning arresters are installed in a building, they shall be located well away from all equipment other than that which they protect and from passageways and combustible parts of buildings.
(2) Where lightning arresters containing oil are installed in a building, they shall be separated from other equipment by walls conforming to electrical equipment vault construction requirements in accordance with Rules 26-350 to 26-356.

26-504 Outdoor installations of lightning arresters
Where arresters containing oil are located outdoors, means of draining or absorbing oil shall be provided by
(a) ditches or drains; or
(b) paving the yard in which the arrester is contained with cinders or other absorbent material to an adequate depth.

26-506 Choke coils for lightning arresters
Where choke coils are used in connection with a lightning arrester, the coils shall be installed between the lightning arrester tap and the apparatus to be protected.

26-508 Connection of lightning arresters
The connection between arrester and line conductor shall be
(a) made of copper wire or cable not smaller than No. 6 AWG;
(b) as short and as straight as practicable with a minimum of bends; and
(c) free of sharp bends and turns.

26-510 Insulation of lightning arrester accessories
The insulation from ground and from other conductors for accessories such as gap electrodes and choke coils shall be at least equal to the insulation required at other points of the circuit.

26-512 Grounding of lightning arresters
Lightning arresters shall be grounded in accordance with Section 10.

Low-voltage surge protective devices

26-520 Low-voltage surge protectors (see Appendix B)
(1) Except as provided for in Subrule (2), where low-voltage surge protective devices are to be connected to a consumer's service, they shall be installed outdoors at least 12.5 mm from combustible material
   (a) at the service head supplying the consumer's service;
   (b) at any supply point on the overhead distribution;
   (c) on the load side of a self-contained utility revenue meter socket, provided that the socket is fitted with approved lugs for the termination; or
   (d) on any outdoor distribution enclosure supplied from underground distribution.
(2) Low-voltage surge protective devices shall be permitted to be connected to an overcurrent device or to a branch circuit supplying utilization equipment in the building.

Storage batteries

26-540 Scope
(1) Rules 26-542 to 26-552 apply to the installation of storage batteries.
(2) Rule 26-554 applies to the installation of electrical equipment, other than storage batteries, in a battery room.

26-542 Special terminology

In this Subsection, the following definitions apply:

Sealed cell or **battery** — a storage battery that has no provision for the addition of water or electrolyte or for the external measurement of electrolyte specific gravity.

Storage battery — a battery consisting of more than one rechargeable cell of the lead-acid, alkaline, or other electrochemical types.

26-544 Location of storage batteries

Batteries with exposed live parts shall be kept in a room or enclosure accessible only to authorized personnel.

26-546 Ventilation of battery rooms or areas (see Appendix B)

(1)   Storage battery rooms or areas shall be adequately ventilated.

(2)   Storage batteries shall not be subjected to ambient temperatures greater than 45 °C or less than the freezing point of the electrolyte.

26-548 Battery vents

(1)   Vented cells shall be equipped with flame arresters.

(2)   Sealed cells shall be equipped with pressure release vents.

26-550 Battery installation

(1)   Battery trays, racks, and other surfaces on which batteries are mounted shall be
    (a)   level;
    (b)   protected against corrosion from the battery electrolyte;
    (c)   except as permitted in Subrule (5), covered with an insulating material having a dielectric strength of at least 1500 V;
    (d)   of sufficient strength to carry the weight of the battery; and
    (e)   designed to withstand vibration and sway where appropriate.

(2)   Battery cells shall be spaced a minimum of 10 mm apart.

(3)   Battery cells having conductive containers shall be installed on non-conductive surfaces.

(4)   Sealed cells and multi-compartment sealed batteries having conductive containers shall have an insulating support if a voltage is present between the container and ground.

(5)   Cells and multi-compartment vented storage batteries, with covers sealed to containers of non-conductive, heat-resistant material, shall not require additional insulating support.

(6)   Batteries having a nominal voltage greater than 150 V and with cells in rubber or composition containers shall be sectionalized into groups of 150 V or less.

26-552 Wiring to batteries

(1)   The wiring between cells and batteries and between the batteries and other electrical equipment shall be
    (a)   bare conductors, which shall not be taped;
    (b)   open wiring;
    (c)   a jacketed flexible cord;
    (d)   mineral-insulated cable, provided that it is adequately protected against corrosion where it may be in direct contact with acid or acid spray; or
    (e)   aluminum-sheathed cable, provided that it has suitable corrosion-resistant protection where necessary.

(2)   Where wiring is installed in rigid conduit or electrical metallic tubing
    (a)   the conduit or tubing shall be of corrosion-resistant material or other materials suitably protected from corrosion;
    (b)   the end of the raceway shall be tightly sealed with sealing compound, rubber tape, or other material to resist the entrance of electrolyte by spray or creeping;
    (c)   the conductor shall issue from the raceway through a substantial glazed insulating bushing;
    (d)   at least 300 mm of the conductor shall be free from the raceway where connected to a cell terminal; and
    (e)   the raceway exit shall be located at least 300 mm above the highest cell terminal to reduce electrolyte creepage or spillage entering the raceway.

26-554 Wiring methods and installation of equipment in battery rooms

The installation of wiring and equipment in a battery room shall be in accordance with the requirements for a dry location.

Arc lamps

26-600 Location of arc lamps
(1) Outdoor arc lamps, attached to a building and supplied from the interior installation, shall be suspended at least 2.5 m above the ground level.
(2) Indoor arc lamps shall be hung out of reach or shall be protected from mechanical damage.

26-602 Conductors to arc lamps
(1) Leads to arc lamps shall have an ampacity of approximately 150% of the normal current of the lamp.
(2) The leads shall be stranded where
    (a)   the size exceeds No. 14 AWG; and
    (b)   the lamp suspension provides for raising and lowering.

26-604 Overcurrent protection for arc lamps
An overcurrent device shall be provided for each arc lamp or series of lamps.

26-606 Resistors or regulators
(1) Resistors or regulators shall be enclosed in non-combustible cases and located away from readily combustible material.
(2) Incandescent lamps shall not be used as resistors or regulators.

26-608 Globes and spark arresters
(1) Arc lamps other than those of the enclosed arc type shall be equipped with globes and spark arresters.
(2) Globes shall be guarded by wire netting having a mesh of not more than 32 mm.

Resistance devices

26-640 Location of resistance devices
Resistance devices, including wiring to the resistance elements, shall be installed so that danger of igniting adjacent combustible material is reduced to a minimum.

26-642 Conductors for resistance devices
Insulated conductors used for connection between resistance elements and controllers, unless used for infrequent motor starting, shall conform to the following:
(a)   they shall be as indicated in Table 19 as being suitable for the temperature involved and in no case less than 90 °C;
(b)   conductors having an approved flame-retardant outer covering may be grouped where the voltage between any two conductors in the group does not exceed a maximum of 75 V.

26-644 Use of incandescent lamps as resistance devices
(1) Incandescent lamps shall be permitted to be used
    (a)   as protective resistors for automatic controllers; or
    (b)   where a deviation has been allowed in accordance with Rule 2-030, as resistors in series with other devices provided that the resulting installation is acceptable.
(2) Where incandescent lamps are used as resistors, they shall
    (a)   be mounted in porcelain lampholders on non-combustible supports;
    (b)   be arranged so that they cannot be subjected to a voltage greater than that for which they are rated;
    (c)   be provided with a permanently attached nameplate showing the wattage and voltage of the lamp to be used in each lampholder;
    (d)   not carry or control the main current; and
    (e)   not constitute the regulating resistance of the device.

Receptacles

26-700 General (see Appendix B)
(1) Receptacle configurations shall be in accordance with Diagrams 1 and 2, except
    (a)   for receptacles used on equipment solely for interconnection purposes;
    (b)   for receptacles for specific applications as required by other Rules of this Code; or
    (c)   where other configurations are suitable.

(2)   Unless otherwise acceptable, receptacles having configurations in accordance with Diagrams 1 and 2 shall be connected only to circuits having a nominal system voltage corresponding to the rating of the configurations.

(3)   Receptacles connected to circuits having different voltages, frequencies, or types of current (ac or dc) on the same premises shall be designed so that attachment plugs used on such circuits are not interchangeable.

(4)   Receptacles with exposed terminals shall be used only in fittings, metal troughs, and similar devices.

(5)   Receptacles located in floors shall be enclosed in floor boxes.

(6)   Receptacles rated 30 A or more and installed facing downward shall have provision for locking or latching to prevent unintentional detachment.

(7)   Where grounding-type receptacles are used in existing installations to replace the ungrounded type, the grounding terminal shall be effectively bonded to ground and one of the following methods shall be permitted to be used:
    (a)   connection to a metal raceway or cable sheath that is bonded to ground;
    (b)   connection to the system ground by means of a separate bonding conductor; or
    (c)   bonding to an adjacent grounded metal cold-water pipe.

Δ (8)   Notwithstanding Subrule (7), at existing outlets where a grounding means does not exist in the receptacle enclosure, grounding-type receptacles without a bonding conductor shall be permitted to be installed, provided that each receptacle is
    (a)   protected by a ground fault circuit interrupter of the Class A type that is an integral part of this receptacle;
    (b)   supplied from a receptacle containing a ground fault circuit interrupter of the Class A type; or
    (c)   supplied from a circuit protected by a ground fault circuit interrupter of the Class A type.

(9)   A bonding conductor shall not be extended from any receptacle protected by a ground fault circuit interrupter in accordance with Subrule (8) to any other outlet.

(10)   After installation,
    (a)   receptacle faces shall project a minimum of 0.4 mm from metal or conductive faceplates;
    (b)   any openings around the receptacle or cover shall be such that a rod 6.75 mm in diameter will not enter; and
    (c)   receptacles, faceplates, and covers shall not prevent the use of an attachment plug in the manner or use for which the attachment plug is approved.

Δ (11)   Receptacles having CSA configuration 5-15R or 5-20RA installed within 1.5 m of sinks (wash basins complete with drainpipe), bathtubs or shower stalls shall be protected by a ground fault circuit interrupter of the Class A type except where the receptacle is
    (a)   intended for a stationary appliance designated for the location; and
    (b)   located behind the stationary appliance such that it is inaccessible for use with general-purpose portable appliances.

26-702 Receptacles exposed to the weather

(1)   Receptacles exposed to the weather shall be provided with weatherproof cover plates, except that, when these receptacles are installed facing downward, at an angle of 45° or less from the horizontal, standard metallic cover plates may be used.

(2)   Where receptacles exposed to the weather are installed in surface-mounted outlet boxes, the cover plates shall be held in place by four screws or by some other equivalent means.

(3)   Where receptacles exposed to the weather are installed in flush-mounted outlet boxes, the boxes shall be installed in accordance with Rule 12-3016 and the cover plates shall be fitted to make a proper weatherproof seal.

Receptacles for residential occupancies

Δ **26-710 General** (see Appendices B and G)

This Rule applies to receptacles for all residential occupancies (including dwelling units and single dwellings) as follows:
    (a)   for the purposes of this Rule, "finished wall" means any wall finished to within 450 mm of the floor with drywall, wood panelling, or like material;
    (b)   for the purposes of this Rule, all receptacles shall be CSA Configuration 5-15R or 5-20RA (T-slot) (see Diagram 1);

(c)   receptacles shall not be mounted facing up in the work surfaces or counters in the kitchen or dining area;

(d)   where receptacles (15 A split or 20 A T-slot) are installed on a side of a counter work surface in a kitchen designed for use by persons with disabilities, such receptacles shall not be considered as substituting for the receptacles required by Rule 26-712(d);

(e)   at least one duplex receptacle shall be provided

    (i)   in each space intended to accommodate a washing machine;

    (ii)   in each laundry room or area in addition to any receptacle specified in Item (i) above;

    (iii)   in each utility room; and

    (iv)   in any unfinished basement area;

(f)   at least one receptacle shall be installed in each bathroom and washroom with a wash basin(s) and shall be located within 1 m of any one wash basin;

(g)   receptacles installed in bathrooms shall, where practicable, be located at least 1 m but in no case less than 500 mm from the bathtub or shower stall, this distance being measured horizontally between the receptacle and the bathtub or shower stall, without piercing a wall, partition, or similar obstacle;

(h)   where a cord-connected hydromassage bathtub conforming to Rule 68-302 is intended to be installed, one receptacle located not less than 300 mm from the floor shall be provided for use with the hydromassage bathtub and shall be inaccessible to the hydromassage bathtub occupant;

(i)   a receptacle shall not be placed in a cupboard, cabinet, or similar enclosure except where

    (i)   the receptacle is an integral part of a factory-built enclosure; or

    (ii)   the receptacle is provided for use with a specific type of appliance that is intended for use within the enclosure; or

    (iii)   the receptacle is intended for use with a microwave oven;

(j)   except for cord-connected dishwashers, in-line water heaters, garbage disposal units, and other similar appliances, receptacles installed in cupboards, cabinets, or similar enclosures in accordance with Item (i)(ii) shall be de-energized unless the enclosure door is in the fully opened position;

(k)   any receptacle that is part of a lighting fixture or appliance, or that is located within cabinets or cupboards as permitted by Item (i) or that is located more than 1.7 m above the floor shall not be considered as any of the receptacles required by this Rule;

(l)   where a switched duplex receptacle is used instead of a light outlet and fixture, the receptacle shall be considered as one of the wall-mounted receptacles meeting the requirements of Rule 26-712(a) provided that only half of the receptacle is switched;

(m)   at least one receptacle shall be provided for each cord-connected central vacuum system, where the complete duct for such a central vacuum system is installed;

(n)   public corridors and public stairways in buildings of residential occupancies shall have at least one duplex receptacle in each 10 m of length or fraction thereof; and

(o)   except for automobile heater receptacles provided in conformance with Rule 8-400, all receptacles installed outdoors and within 2.5 m of finished grade shall be protected with a ground fault circuit interrupter of the Class A type.

Δ   **26-712 Receptacles for dwelling units** (see Appendices B and G)

This Rule applies to receptacles for dwelling units (including single dwellings) as follows:

(a)   except as otherwise provided for in this Code, in dwelling units, duplex receptacles shall be installed in the finished walls of every room or area, other than kitchens, bathrooms, hallways, laundry rooms, water closet rooms, utility rooms, or closets, so that no point along the floor line of any usable wall space is more than 1.8 m horizontally from a receptacle in that or an adjoining space, such distance being measured along the floor line of the wall spaces involved;

(b)   at least one duplex receptacle shall be provided in each area, such as a balcony or porch, that is not classified as a finished room or area in accordance with Item (a);

(c)   the usable wall space referred to in Item (a) shall include a wall space 900 mm or more in width but shall not include doorways, areas occupied by a door when fully opened, windows that extend to the floor, fireplaces, or other permanent installations that would limit the use of the wall space;

(d)   in dwelling units there shall be installed in each kitchen

    (i)   one receptacle for each refrigerator;

    (ii)   where a gas supply piping or a gas connection outlet has been provided for a free-standing gas range, one receptacle behind the intended gas range location not more than 130 mm from the floor and as near midpoint as is practicable, measured along the floor line of the wall space intended for the gas range;

(iii)   a sufficient number of receptacles (15 A split or 20 A T-slot) along the wall behind counter work surfaces (excluding sinks, built-in equipment, and isolated work surfaces less than 300 mm long at the wall line) so that no point along the wall line is more than 900 mm from a receptacle measured horizontally along the wall line;

(iv)   at least one receptacle (15 A split or 20 A T-slot) installed at each permanently fixed island counter space with a long dimension of 600 mm or greater and a short dimension of 300 mm or greater;

(v)   at least one receptacle (15 A split or 20 A T-slot) installed at each peninsular counter space with a long dimension of 600 mm or greater and a short dimension of 300 mm or greater;

(vi)   at least one duplex receptacle in a dining area forming part of a kitchen;

(e)   the receptacles specified in Item (d) shall not be located

   (i)   on the area of the wall directly behind the kitchen sink; or

   (ii)   on the area of the counter directly in front of the kitchen sink; and

(f)   no point in a hallway within a dwelling unit shall be more than 4.5 m from a duplex receptacle as measured by the shortest path that the supply cord of an appliance connected to the receptacle would follow without passing through an opening fitted with a door.

26-714 Receptacles for single dwellings (see Appendices B and G)

This Rule applies to receptacles for single dwellings only as follows:

(a)   for each single dwelling, at least one duplex receptacle shall be installed outdoors so as to be readily accessible from ground or grade level for the use of appliances that need to be used outdoors; and

(b)   at least one duplex receptacle shall be provided for each car space in a garage or carport of a single dwelling.

Branch circuits for residential occupancies

Δ 26-720 General

This Rule applies to branch circuits for all residential occupancies (including dwelling units and single dwellings) as follows:

(a)   each receptacle installed for a refrigerator shall be supplied by a branch circuit that does not supply any other outlets, except a recessed clock receptacle intended for use with an electric clock;

(b)   at least one branch circuit shall be provided solely for receptacles installed in the laundry room or area; and

(c)   at least one branch circuit shall be provided solely for receptacles installed in the utility room;

(d)   each receptacle installed in a cupboard, wall cabinet, or enclosure for the use of a microwave oven in accordance with Rule 26-710(i) shall be supplied by a branch circuit that does not supply any other outlets, and this circuit shall not be considered as forming part of the circuits required under Rule 26-722(b);

(e)   a separate branch circuit shall be provided solely to supply power to each central vacuum system; and

(f)   the ampere rating of the branch circuit wiring supplying receptacles with CSA Configuration 5-20RA shall be not less than 20 A.

Δ 26-722 Branch circuits for dwelling units

This Rule applies to branch circuits for dwelling units (including single dwellings) as follows:

(a)   branch circuits from a panelboard installed in accordance with Rule 26-400 shall not be connected to outlets or electrical equipment in any other dwelling unit;

(b)   except as may be permitted by Items (c) and (d), at least two branch circuits shall be provided for receptacles (15 A split or 20 A T-slot) installed for kitchen counters of dwelling units in accordance with Rule 26-712(d)(iii), (iv), and (v); and

   (i)   no more than two receptacles shall be connected to a branch circuit; and

   (ii)   no other outlets shall be connected to these circuits;

(c)   notwithstanding Item (b), where the provisions of Rule 26-712(d)(iii) require only one receptacle, only one branch circuit need be provided;

(d)   notwithstanding Item (b)(i), receptacles identified in Rule 26-710(d) shall be permitted to be connected to those receptacles required by Rule 26 712(d)(iii), even though the circuit already supplies two receptacles;

(e)   receptacles installed in a dining area forming part of a kitchen of a dwelling unit shall be supplied by a branch circuit that does not supply any other outlets, except that a receptacle required by Rule 26-712(d)(ii) shall also be permitted to be supplied by this branch circuit;

(f)   branch circuits that supply receptacles installed in sleeping facilities of a dwelling unit shall be protected by an arc-fault circuit interrupter; and

(g)   for the purpose of Item (f), "arc-fault circuit interrupter" means a device intended to provide protection from the effects of arc-faults by recognizing characteristics unique to arcing and functioning to de-energize the circuit when an arc-fault is detected.

Δ   **26-724 Branch circuits for single dwellings**
This Rule applies to branch circuits for single dwellings only as follows:

(a)   outdoor receptacles readily accessible from ground level and installed in accordance with Rule 26-714(a) shall be supplied from at least one branch circuit dedicated for those outdoor receptacles; and

(b)   at least one branch circuit shall be provided solely for the receptacles in a carport or garage of a single dwelling except that the lighting fixtures and garage door operator for these areas may be connected to this circuit.

Electric heating and cooking appliances

26-740 Location of non-portable appliances
Non-portable electric heating and cooking appliances shall be installed so that the danger of igniting adjacent combustible material is reduced to a minimum.

26-742 Separate built-in cooking units
Tap conductors feeding individual built-in cooking units from a single branch circuit shall be permitted to be smaller than the branch circuit conductors, provided that the tap conductors

(a)   are not more than 7.5 m in length;

(b)   have an ampacity not less than the ampere rating of the built-in cooking unit they supply; and

(c)   have an ampacity not less than one-third the ampere rating of the branch circuit overcurrent device.

26-744 Supply connections for appliances
(1)   Electric heating and cooking appliances shall have only one point of connection for supply.

(2)   Where an electric clothes dryer having an input in excess of 1500 W at 115 V but not exceeding 30 A is intended to be installed in a dwelling unit, a receptacle of CSA Configuration 14-30R, as shown in Diagram 1, shall be installed for the supply of energy to the appliance.

(3)   An electric clothes dryer having an input in excess of 1500 W at 115 V but not exceeding 30 A, and used in a dwelling unit, shall be cord-connected by means of a cord and attachment plug of CSA Configuration 14-30P to the receptacle referred to in Subrule (2).

(4)   Where a free-standing electric range, having a calculated demand of 50 A or less, is intended to be installed in a dwelling unit, a receptacle of CSA Configuration 14-50R, as shown in Diagram 1, shall be installed for the supply of electric energy to the appliance.

(5)   The receptacle required by Subrule (4) shall be installed

(a)   above the finished floor at a height not exceeding 130 mm to the centre of the receptacle;

(b)   as near midpoint as is practicable, measured along the floor line of the wall space intended for the electric range; and

(c)   with the U-ground slot orientated to either side.

(6)   In a dwelling unit, a free-standing electric range having a calculated demand of 50 A or less shall be cord-connected by means of a cord and attachment plug of CSA Configuration 14-50P.

(7)   Appliances that are intended for connection by a wiring method as specified in Section 12 shall be permitted to be cord-connected using an attachment plug and receptacle.

(8)   The receptacles required by Subrules (2) and (4) shall be flush-mounted wherever practicable.

26-746 Appliances exceeding 1500 W
(1)   Every electric heating and cooking appliance rated at more than 1500 W shall be supplied from a branch circuit used solely for one appliance, except that more than one appliance may be connected to a single-branch circuit provided that the following is used:

(a)   a multiple-throw manually operated device that will permit only one such appliance to be energized at one time; or

(b)   an automatic device that will limit the total load to a value that will not cause operation of the overcurrent devices protecting the branch circuit.

(2)   Every electric heating and cooking appliance rated at more than 1500 W shall be controlled by an indicating switch that shall be permitted to be in the circuit or on the appliance, except that

(a)   if the rating of the appliance does not exceed 30 A, an attachment plug and receptacle shall be permitted to be used instead of a switch; and

(b)   if the appliance has more than one individual heating element, each controlled by a switch, no main switch need be provided.

(3)   For the purpose of this Rule, two or more separate built-in cooking units shall be considered as one appliance.

26-748 Signals for heated appliances

Where glue pots, soldering irons, or appliances intended to be applied to combustible materials are used in other than dwelling units,

(a)   each appliance or group of appliances shall be provided with an indicating switch and a red pilot light; or

(b)   each appliance shall be equipped with an integral temperature-limiting device, in which case the pilot light shall be permitted to be omitted where a deviation has been allowed in accordance with Rule 2-030.

26-750 Installation of storage-tank water heaters

(1)   Electric storage-tank water heaters, other than those having a tank open to the atmosphere, shall be controlled by means of a temperature-regulating device and shall also be provided with secondary protection that will open if the water attains a temperature of 96 °C.

(2)   The temperature-regulating device referred to in Subrule (1) shall regulate the temperature of the water so that it does not exceed 90 °C.

(3)   Electric storage-tank water heaters shall be located so that the electric supply connections, service covers, and nameplate markings will be accessible after completion of the building structure.

(4)   Every electric storage-tank water heater shall be supplied from a branch circuit used solely for the heater.

26-752 Infrared drying lamps

The following requirements shall apply to the installation of infrared drying lamps:

(a)   branch circuits shall be protected in accordance with Section 14;

(b)   lampholders of the medium-base, unswitched, porcelain type or other types approved for the service shall be permitted to be used with lamps rated at 300 W or less;

(c)   screwshell lampholders shall not be used with lamps rated at more than 300 W unless approved for the purpose; and

(d)   in industrial occupancies, lampholders shall be permitted to be operated in series on circuits of more than 150 volts-to-ground where adequate spacings for the higher circuit voltage are provided.

26-754 Control of ventilation of commercial cooking equipment

Where a fan is used to ventilate commercial cooking equipment, the control for the fan motor shall be readily accessible, within reach of the cooking equipment, and external to the ventilation duct or hood.

26-756 Induction and dielectric heating equipment

(1)   Overcurrent devices shall meet the requirements of Section 14, except in circuits supplying non-motor-generator equipment the overcurrent device shall be permitted to be rated or set at not more than 200% of the ampacity of the circuit conductors.

(2)   A readily accessible disconnecting means having a rating in accordance with Section 28 shall be provided for each generator or group of generators at a single location.

(3)   The supply circuit switch shall be permitted to be used as the disconnecting means if the circuit supplies only one generator.

(4)   Exposed non-current-carrying metal parts of each piece of equipment shall be bonded to a common bonding point that shall be bonded to ground.

26-758 Bare element water heaters

(1)   A water heater with a bare heater element immersed in water shall be

(a)   supplied from a grounded system;

(b)   permanently connected to a branch circuit that supplies no other equipment; and

(c)   protected by a ground fault circuit interrupter of the Class A type.

(2)   A water heater with a bare heater element immersed in water shall not be located within 1.5 m of the point of utilization of the heated water.

(3)   A water heater with a bare heater element immersed in water shall be bonded to ground in accordance with Section 10.

Heating equipment

26-800 Scope
Rules 26-802 to 26-808 apply to circuits supplying power for the operation and control of non-portable heating equipment that uses solid, liquid, or gaseous fuel.

26-802 Mechanical protection of conductors
All branch circuit or tap conductors within 1.5 m from the floor shall be adequately protected from mechanical injury.

26-804 Fuel burner safety controls (see Appendix B)
Fuel burner safety controls shall be installed in accordance with the requirements of CSA C22.2 No. 3.

26-806 Heating equipment rated 117 kW and less (see Appendix B)
(1) Except as permitted by Subrule (3), all electric power for the heating unit and associated equipment operating in connection with it shall be obtained from a single-branch circuit that shall be used for no other purpose.
(2) For the purpose of this Rule, circulating pumps and similar equipment need not be considered as associated equipment, provided that such equipment is not essential for the safe operation of the heating unit.
(3) Subrule (1) does not apply to a water heater using a gaseous fuel.
(4) The branch circuit shall be permitted to be tapped as necessary to supply the various pieces of associated equipment, but there shall be no overcurrent protection supplied in the tap to any piece of associated equipment the operation of which is essential to the proper operation of the heating unit, unless the control equipment is of such a nature that the heating unit will be shut down if the associated equipment fails to function due to the operation of the overcurrent device.
(5) Suitable disconnecting means shall be provided for the branch circuit.
(6) The disconnecting means shall be permitted to be a branch circuit breaker at the distribution panelboard, provided that the panelboard is located between the furnace and the point of entry to the area where the furnace is located.
(7) Where a separate switch is required, due to the unsuitable location of the branch circuit breaker, it shall
    (a) not be located on the furnace nor in a location that can be reached only by passing close to the furnace; and
    (b) be marked to indicate the equipment it controls.

26-808 Heating equipment rated at more than 117 kW
(1) All electric power for the heating unit and associated equipment operating in connection with it shall be obtained from a single feeder or branch circuit that shall not be used for other purposes.
(2) A suitable disconnecting means shall be provided for the feeder or branch circuit.

Pipe organs

26-900 Installation of electrically operated pipe organs
(1) Organ blower motors, when located remote from the organ console, shall be provided with a pilot lamp located at the organ console.
(2) A receptacle shall be provided in the organ loft to facilitate the use of a portable lamp.

Submersible pumps

26-950 Special terminology
In this Subsection, the following definitions apply:

Deep well submersible pump — a submersible pump intended for use in a well casing or similar protective enclosure that does not have provision for electrical connection by conduit.

Submersible pump — a pump-motor combination where the enclosed electrical equipment is intended to operate submerged in water.

26-952 General
Submersible pumps shall be installed in accordance with the manufacturer's instructions and Rule 26-954 or 26-956 as applicable.

26-954 Deep well submersible pumps installed in wells

Deep well submersible pumps installed in wells shall comply with the following:

(a) the power supply conductors or cable run from the well head to the pump shall be

   (i)   Types RWU75, RWU90, TWU, and TWU75 single conductors or twisted assemblies of these types, suitable for handling at −40 °C; or

   (ii) Type SOW, G, G-GC, W, or the equivalent portable power cable;

(b) the supply conductors or cable shall be suitably supported at intervals not exceeding 3 m to the discharge pipe;

(c) supply conductors or cable shall be run from the well head to the main distribution panelboard in accordance with the requirements of Section 12; and

(d) pumps shall be bonded to ground in accordance with Section 10 except that when the discharge pipe is of metal and is continuous from the pump to the well head, the equipment bonding conductor shall be permitted to be terminated by connection to a discharge pipe at the well head location.

26-956 Submersible pumps installed in lakes, rivers, and streams

Δ (1) Except as provided in Subrule (2), submersible pumps installed in lakes, rivers, and streams and at similar locations shall comply with the following:

(a) the voltage supplying the submersible pump shall not exceed 150 volts-to-ground;

(b) the pump motor shall be bonded to ground by a conductor that is

   (i)   sized in accordance with Rule 10-814;

   (ii)   integral with the supply cable, or within the same protective enclosure as the power supply conductors if single conductors are used;

   (iii) of the same type of insulation as the supply conductors; and

   (iv)   terminated adjacent to the location where the branch circuit conductors receive their supply;

(c) the wiring method to the pump shall be

   (i)   Type RWU75, RWU90, TWU, or TWU75 or equivalent single conductor or twisted assemblies of these types, suitable for handling at −40 °C, enclosed in a plastic water pipe or in rigid PVC conduit; or

   (ii) Type SOW, G, G-GC, W, or the equivalent portable power cable;

(d) ground fault protection shall be provided to de-energize all normally ungrounded conductors supplying the submersible pump with a ground fault current trip setting adjusted to function as low as practicable to permit normal operations of the pump, but in no case shall the ground fault current setting be greater than 10 mA for an operating time period not exceeding 2.7 s; and

(e) the supply conductors or cables shall be run from an outdoor connection facility, above or below ground, to the main distribution panelboard in accordance with the requirements of Section 12.

(2) Submersible pumps operating at voltages exceeding 150 volts-to-ground, but not exceeding 5.5 kV, shall be installed only in lakes, rivers, and streams where a deviation has been allowed in accordance with Rule 2-030, and

(a) the electrical installation shall be maintained by a qualified electrical maintenance staff; and

(b) the area around the submersible pump shall be protected from access by the public by fencing, cribbing, or isolation and so marked.

Data processing

26-1000 Permanently connected data processing units

Branch circuits supplying permanently connected data processing units shall not supply any other types of loads.

Section 28 — Motors and generators

Scope

28-000 Scope
This Section applies to the installation, wiring methods, conductors, and protection and control of electric motors and generators and supplements or amends the general requirements of this Code.

General

28-010 Special terminology
In this Section, the following definitions apply:

Hermetic refrigerant motor-compressor — a compressor unit in which the compressor and motor are housed within a single container structure with no external shaft or shaft seals, or the motor is housed within a container structure integral with the compressor structure, so that the motor windings operate within a refrigerant atmosphere.

Locked rotor current rating — a current rating marked on electric equipment or, where not marked, is deemed to be equal to six times the full load current rating from the nameplate of the equipment or from Table 44 or 45 as applicable.

Non-continuous duty motor — a motor having characteristics or ratings described in Section 0, Definitions, as **Short-time duty**, **Intermittent duty**, **Periodic duty**, and **Varying duty**.

Rated load current (for a hermetic refrigerant motor-compressor) — a value marked on a hermetic motor-compressor intended for use where applicable to ascertain wiring, protection, and control for the unit.

Service —

Continuous duty service — any application of a motor where the motor can operate continuously with load under any normal or abnormal condition of use.

Non-continuous duty service — an application of a motor where the apparatus driven by the motor has the characteristics described in Section 0, Definitions, as **Short-time duty**, **Intermittent duty**, **Periodic duty**, and **Varying duty**.

Service factor — a multiplier that, when applied to the rated horsepower of an ac motor, to the rated armature current of a dc motor, or to the rated output of a generator, indicates a permissible loading that may be carried continuously at rated voltage and frequency.

28-012 Guarding
Exposed live parts of motors and controllers operating at 50 V or more between terminals shall be guarded against accidental contact by means of enclosures or by location, except that stationary motors having commutators, collectors, and brush rigging located inside of motor end brackets and not conductively connected to supply circuits operating at more than 150 volts-to-ground shall be permitted to have live parts exposed.

28-014 Methods of guarding
Methods of guarding of motors having exposed live parts are by
(a)   installation in a room or enclosure that is accessible only to authorized persons;
(b)   installation on a suitable balcony, gallery, or platform elevated and arranged to exclude other than qualified persons;
(c)   elevation by 2.5 m or more above the floor; or
(d)   guard rails if the motor operates at 750 V or less.

28-016 Ventilation
(1)   Adequate ventilation shall be provided to prevent the development around motors of ambient air temperatures exceeding 40 °C for integral horsepower motors and 30 °C for fractional horsepower motors.
(2)   Notwithstanding Subrule (1), motors suitable for use in higher ambient temperatures shall be specifically marked for the temperatures in which they will operate.
(3)   In locations where dust or flying material will collect in or on motors in quantities that interfere with the ventilating or cooling of motors, thereby causing dangerous temperatures, suitable types of enclosed motors that will not overheat under prevailing conditions shall be used.

Wiring methods and conductors

28-100 Stationary motors (see Appendix B)
The wiring method for stationary motors shall be in accordance with the applicable requirements of Sections 12 and 36.

28-102 Portable motors
Connections to portable motors shall be permitted with flexible cord, which shall have a serviceability not less than that of Type S cord, unless the motor forms part of a motor-operated device.

28-104 Motor supply conductor insulation temperature rating and ampacity
(1) Supply conductors to a motor connection box shall have an insulation temperature rating equal to or greater than that required by Table 37, unless the motor is marked otherwise and their ampacity is based on a 75 °C conductor insulation rating except for Class A rated motors only, where their ampacity shall be permitted to be based on a 90 °C insulation rating, when 90 °C wire is used as circuit conductors to the motor.

(2) Where Table 37 requires insulation temperature ratings in excess of 75 °C, the motor supply conductors shall be not less than 1.2 m long and shall terminate in a location not less than 600 mm from any part of the motor except that for motors rated 100 hp or larger, their terminations shall be not less than 1.2 m from any part of the motor.

(3) For ambients higher than 30 °C, the supply conductor insulation rating shall be increased at least by the difference between the ambient and 30 °C.

28-106 Conductors — Individual motors
(1) The conductors of a branch circuit supplying a motor for use on continuous duty service shall have an ampacity not less than 125% of the full load current rating of the motor.

(2) The conductors of a branch circuit supplying a motor for use on non-continuous duty service shall have an ampacity not less than the current value obtained by multiplying the full load current rating of the motor by the applicable percentage given in Table 27 for the duty involved, or for varying duty service where a deviation has been allowed in accordance with Rule 2-030 by a percentage less than that specified in Table 27.

(3) Tap conductors supplying individual motors from a single set of branch circuit overcurrent devices supplying two or more motors shall have an ampacity at least equal to that of the branch circuit conductors, except that where the tap conductors do not exceed 7.5 m in length, they shall be permitted to be sized in accordance with Subrule (1) or (2) provided that the ampacity so determined is not less than 1/3 of the ampacity of the branch circuit conductors.

28-108 Conductors — Two or more motors
(1) Conductors supplying a group of two or more motors shall have an ampacity not less than
  (a) 125% of the full load current rating of the motor having the largest full load current rating plus the full load current ratings of all the other motors in the group, where all motors in the group are for use on continuous duty service;
  (b) The total of the calculated currents determined in accordance with Rule 28-106(2) for each motor, where all motors in the group are for use on non-continuous duty service; or
  (c) The total of the following, where the group consists of two or more motors for use on both continuous and non-continuous duty service:
    (i) 125% of the current of the motor having the largest full load current rating for use on continuous duty service;
    (ii) the full load current ratings of all other motors for use on continuous duty service; and
    (iii) the calculated current determined in accordance with Rule 28-106(2) for motors for use on non-continuous duty service.

(2) Where the circuitry is interlocked in order to prevent all motors of the group from running at the same time, the size of the conductors feeding the group shall be permitted to be determined for the motor, or group of motors operating at the same time, having the largest rating selected as determined in Subrule (1).

(3) Demand factors shall be permitted to be applied where the character of the motor loading justifies reduction of the ampacity of the conductor to less than the ampacity specified in Subrule (1), provided that
  (a) the conductors have sufficient ampacity for the maximum demand load; and
  (b) the rating or setting of the overcurrent devices protecting them are in accordance with Rule 28-204(4).

156

28-110 Feeder conductors

(1) Where a feeder supplies both motor loads and other loads, the ampacity of the conductors shall be calculated in accordance with Rules 28-106 and 28-108 plus the requirements of the other loads.

(2) The ampacity of a tap from a feeder to a single set of overcurrent devices protecting a motor branch circuit shall be not less than that of the feeder, except that the ampacity of the tap shall be permitted to be calculated in accordance with Rules 28-106 and 28-108 if the tap

(a) does not exceed 3 m in length and is enclosed in metal; or

(b) does not exceed 7.5 m in length, has an ampacity not less than 1/3 that of the feeder, and is suitably protected from mechanical damage.

28-112 Secondary conductors

(1) Conductors connecting the secondaries of wound rotor motors to their controllers shall have an ampacity not less than

(a) 125% of the rated full load secondary current for motors used on continuous duty service; or

(b) the percentage of rated full load specified in Table 27 for motors used on non-continuous duty service.

(2) Ampacities of conductors connecting secondary resistors to their controllers shall be not less than that determined by applying the appropriate percentage in Table 28 to the maximum current that the devices are required to carry.

Overcurrent protection

28-200 Branch circuit overcurrent protection

Each ungrounded conductor of a motor branch circuit shall be protected by an overcurrent device complying with the following:

(a) a branch circuit supplying a single motor shall be protected, except as permitted by Item (c), by using an overcurrent device of rating not to exceed the values in Table 29 using the rated full load current of the motor;

(b) notwithstanding Item (a), an overcurrent device having a minimum rating or setting of 15 A shall be permitted even though it exceeds the values specified in Table 29;

(c) instantaneous trip (magnetic only) circuit interrupters shall be permitted where applied in accordance with Rule 28-210;

(d) where the overcurrent devices as determined in Item (a) will not permit the motor to start, the rating or setting of the overcurrent device shall be permitted to be increased as follows:

  (i) a non-time-delay fuse:

    (A) not in excess of 400% of the motor full load current for fuses rated up to 600 A; or

    (B) not in excess of 300% of the motor full load current for fuses rated 601 to 6000 A;

  (ii) a time-delay fuse to a maximum of 225% of the motor full load current;

  (iii) an inverse time circuit breaker:

    (A) not in excess of 400% of the motor full load current for breakers rated up to 100 A; or

    (B) not in excess of 300% of the motor full load current for breakers rated greater than 100 A;

(e) where the overcurrent device is a thermal magnetic breaker that has separate instantaneous trip settings, the instantaneous trip setting shall not be greater than that specified in Rule 28-210; and

(f) for a branch circuit supplying two or more motors, the rating or setting of the overcurrent device shall not exceed the maximum value permitted by Rule 28-206.

28-202 Overcurrent protection marked on equipment

Where branch circuit protective device characteristics and ratings or settings are specified in the marking of motor control equipment, they shall not be exceeded, notwithstanding any greater rating or setting permitted by Rule 28-200.

28-204 Feeder overcurrent protection

(1) For a feeder supplying motor branch circuits only, the ratings or settings of the feeder overcurrent device shall not exceed a maximum value calculated by determining the maximum rating or setting of the overcurrent device permitted by Rule 28-200 for the motor that is permitted the highest rated overcurrent devices of any motor supplied by the feeder, and adding to that value the sum of the full load current ratings of all other motors that will be in operation at the same time.

(2) Where a feeder supplies a group of motors, two or more of which are required to start simultaneously, and the feeder overcurrent devices as calculated in accordance with Subrule (1) are not sufficient to allow the motors to start, the rating or setting of the feeder overcurrent devices shall be permitted to be increased as necessary, to a maximum that does not exceed the rating permitted for a single motor having a full load current rating not less than the sum of the full load current ratings of the greatest number of motors that start simultaneously, provided that this value does not exceed 300% of the ampacity of the feeder conductors.

(3) Where a feeder supplies one or more motor branch circuits together with other loads, the overcurrent protection ʀequired shall be determined by calculating the overcurrent protection required for the motor circuits and adding thereto the requirements of the other loads supplied by the feeder.

(4) Where a demand factor has been applied as permitted in Rule 28-108(3), the rating or setting of the overcurrent device(s) protecting a feeder shall not exceed the ampacity of the feeder, except as permitted by Rule 14-104 and Table 13.

28-206 Grouping of motors on a single branch circuit

Two or more motors shall be permitted to be grouped under the protection of a single set of branch circuit overcurrent devices having a rating or setting calculated in accordance with Rule 28-204(1) provided that the protection conforms to one of the following:

(a) the rating or setting of the overcurrent devices does not exceed 15 A;

(b) protection is provided for the control equipment of the motors by having the branch circuit overcurrent devices rated or set at

    (i) values not in excess of those marked on the control equipment for the lowest rated motor of the group, as suitable for the protection of that control equipment; or

    (ii) in the absence of such markings, values not in excess of 400% of the full load current of the lowest rated motor;

(c) the motors are used on a machine tool or woodworking machine and the following conditions exist:

    (i) the control equipment is arranged so that all contacts that open motor primary circuits are in enclosures, either forming part of the machine base or for separate mounting, that have a wall thickness not less than 1.69 mm for steel, 2.4 mm for malleable cast iron, or 6.3 mm for other cast metal, that have hinged doors with substantial catches, and that have no openings to the floor or the foundation on which the machine rests; and

    (ii) the rating or setting of the branch circuit overcurrent protection is not greater than that permitted by Table 29 for the full load current rating of the largest motor in the group, plus the sum of the full load current ratings of all other motors in the group that may be in operation at one time, but in no case more than 200 A at 250 V or less or 100 A at voltages from 251 to 750 V; or

(d) all the motors are operated by a single controller, as provided for in Rule 28-500(3)(d);

(e) where a deviation is allowed in accordance with Rule 2-030 for the group of motors that form part of the coordinated drive of a single machine or process, wherein the failure of one motor to operate creates a hazard unless all the other motors in the group are stopped; or

(f) the motors are contained within and form part of refrigerant equipment on a 120 V branch circuit protected at not more than 20 A where each motor is rated not more than 1 hp and has a full load current rating of not more than 6 A.

28-208 Size of fuseholders

Where fuses are used for motor branch circuit or feeder protection, the fuseholders shall be not of a size smaller than those required to accommodate fuses of the maximum rating permitted by Table 29 except that fuseholders of a smaller size shall be permitted to be used

(a) where Rule 28-202 is applicable;

(b) where fuses having time delay appropriate for the starting characteristics of the motor are used, in which case the fuseholders shall not be smaller than those required to accommodate fuses rated at 125% of the full load current of the motor; or

(c) in the case of a circuit supplying a group of motors, where the fuseholders accommodate fuses of a size calculated by taking 150% of the largest motor current and adding to this value the applicable full load currents of all other motors in the group that may be in operation at the same time.

28-210 Instantaneous-trip circuit breakers (see Appendix B)

When used for branch circuit protection, instantaneous-trip circuit breakers shall be

(a) part of a combination motor starter or controller that also provides overload protection; and either

158

(b) rated or adjusted, for an ac motor, to trip at not more than 1300% of the motor full load current or at not more than 215% of the motor-locked rotor current, where given, except that ratings or settings for trip currents need not be less than 15 A; or

(c) rated or adjusted, for a dc motor rated at 50 hp or less, to trip at not more than 250% of the motor full load current, or for a dc motor rated at more than 50 hp, to trip at not more than 200% of the motor full load current.

Overload and overheating protection

28-300 Overload protection required
The branch circuit conductors and control equipment of each motor shall have overload protection, except as permitted by Rule 28-308.

28-302 Types of overload protection
(1) Overload devices shall be either
    (a) a separate overload device that is responsive to motor current and that shall be permitted to combine the function of overload and overcurrent protection if it is capable of protecting the circuit and motor under both overload and short-circuit conditions; or
    (b) a protective device, integral with the motor and responsive to motor current or to motor current and temperature, provided that such a device will protect the circuit conductors and control equipment as well as the motor.
(2) Fuses used as separate overload protection of motors shall be time-delay fuses of the type referred to in Rule 14-200.

28-304 Number and location of overload devices (see Appendix B)
(1) The number and location of current responsive devices shall, unless otherwise required, be as follows:
    (a) if fuses are used, one in each ungrounded conductor; or
    (b) if devices other than fuses are used, as specified in Table 25.
(2) Unless a deviation has been allowed in accordance with Rule 2-030, where current responsive devices are used for the overload protection of 3-phase motors, such devices shall consist of three current responsive elements that shall be permitted to be
    (a) connected directly in the motor circuit conductors as required by Subrule (1); or
    (b) fed by two or three current transformers connected so that all three phases will be protected.

28-306 Rating or trip selection of overload devices (see Appendix B)
(1) Overload devices responsive to motor current, if of the fixed type, shall be selected or rated or, if of the adjustable type, shall be set to trip at not more than the following:
    (a) 125% of the full load current rating of a motor having a marked service factor of 1.15 or greater; or
    (b) 115% of the full load current rating of a motor that does not have a marked service factor or where the marked service factor is less than 1.15.
(2) Where a motor overload device is connected so that it does not carry the total current designated on the motor nameplate, such as for wye-delta starting, the percentage of motor nameplate current applying to the selection or setting of the overload device shall be clearly marked on the motor starter shown in the motor starter manufacturer's overload selection table.

28-308 Overload protection not required
Overload protection shall not be required for motors complying with any of the following:
(a) a manually started motor rated at 1 hp or less that is continuously attended while in operation, and that is on a branch circuit having overcurrent protection rated or set at not more than 15 A, or on an individual branch circuit having overcurrent protection as required by Table 29 if it may be readily determined from the starting location that the motor is running;
(b) a motor constructed so that it cannot be overloaded;
(c) a motor whose operating requirements are such that it is impracticable to obtain proper overload protection; or
(d) an automatically started motor having a rating of 1 hp or less forming part of an assembly equipped with other safety controls that protect the motor from damage due to stalled-rotor current and on which a nameplate, located so that it is visible after installation, indicates that such protection features are provided.

28-310 Shunting of overload protection during starting

Overload protection shall be permitted to be shunted or cut out of the circuit during the starting period, provided that the device by which the protection is shunted or cut out cannot be left in the starting position and that the overcurrent device is in the motor circuit during the starting period.

28-312 Automatic restarting after overload

Where automatic restarting of a motor after a shutdown on overload could cause injury to persons, the overload or overheating devices protecting the motor shall be arranged so that automatic restarting cannot occur.

28-314 Overheating protection required (see Appendix B)

Each motor shall be provided with overheating protection, except as permitted by Rule 28-318.

28-316 Types of overheating protection (see Appendix B)

Where required by Rule 28-314, overheating protection shall be provided by devices integral with the motor and responsive to both motor current and temperature or to motor temperature only, and shall be arranged to cut off power to the motor, or where a deviation has been allowed in accordance with Rule 2-030, to activate a warning signal when the temperature exceeds the safe limit for the motor.

28-318 Overheating protection not required

Overheating protection shall not be required

(a)   where the motor circuit requires no overload protection under Rule 28-308; or

(b)   where overload protective devices required by Rule 28-302 adequately protect the motor against overheating due to excess current, and the motor is in a location where

    (i)   ambient temperatures are not more than 10 °C higher than those at the location of the overload devices; and

    (ii)   dust or other conditions will not interfere with the normal dissipation of heat from the motor.

Undervoltage protection

28-400 Undervoltage protection required for motors (see Appendix B)

Motors shall be disconnected from the source of supply in case of undervoltage by one of the following means, unless it is evident that no hazard will be incurred through lack of such disconnection:

(a)   when automatic restarting is liable to create a hazard, the motor control device shall provide low-voltage protection; or

(b)   when it is necessary or desirable that a motor stop on failure or reduction of voltage and automatically restart on return of voltage, the motor control device shall provide low-voltage release.

Control

28-500 Control required

(1)   Except as permitted by Subrule (3), each motor shall be provided with a motor starter or controller for starting and stopping it that has a rating in horsepower not less than the rating of the motor it serves.

(2)   A motor controller need not open the circuit in all ungrounded conductors to a motor unless it also serves as a disconnecting means.

(3)   The motor starter or controller specified in Subrule (1) shall not be required for motors applied as follows:

    (a)   a single-phase portable motor rated at 1/3 hp or less connected by means of a receptacle and attachment plug rated not in excess of 15 A 125 V;

    (b)   a motor controlled by a manually operated general-use switch complying with Rule 14-510 having an ampere rating not less than 125% of the full load current rating of the motor;

    (c)   a 2-wire portable ac or dc motor having a rating not in excess of 1/3 hp, 125 V controlled by a horsepower rated single pole motor switch;

    (d)   two or more motors that are required to operate together shall be permitted to be operated from a single controller specifically approved for such purpose; or

    (e)   for a motor where the controller is specifically approved for use with that motor, it need not be rated in horsepower.

28-502 Control location

A motor controlled manually, either directly or by remote control of a motor starter, shall have the means to operate of the controller located as follows:

(a)   the controller shall be located such that safe operation of the motor and the machinery driven by it is assured, or the motor and the machinery shall be guarded or enclosed to prevent accidents due to contact of persons with live or moving parts; or

(b)   where compliance with Item (a) is not practicable because of the type, size, or location of the motor or machinery and its parts, devices shall be provided at each point where the danger of accidents exists by which means the machine or parts of the machine may be stopped in an emergency.

28-504 Starters having different starting and running positions

(1)   Manual motor starters having different starting and running positions shall be constructed so that they cannot remain in the starting position.

(2)   Magnetic motor starters having different starting and running positions shall be constructed so that they cannot remain in the starting position under normal operating conditions.

28-506 Grounded control circuit

When power for a control circuit for a motor controller is obtained conductively from a grounded system, the control circuit shall be so arranged that an accidental ground in the wiring from the controller to any remote or signal device will not

(a)   start the motor; or

(b)   prevent the stopping of the motor by the normal operation of any control or safety device in the control circuit.

Disconnecting means

28-600 Disconnecting means required

(1)   Except as permitted by Subrules (2) and (3), a separate disconnecting means shall be provided for
    (a)   each motor branch circuit;
    (b)   each motor starter or controller; and
    (c)   each motor.

(2)   A single disconnecting means shall be permitted to serve more than one of the functions described in Subrule (1).

(3)   A single disconnecting means shall be permitted to serve two or more motors and their associated starting and control equipment grouped on a single branch circuit.

28-602 Types and ratings of disconnecting means (see Appendix B)

(1)   A disconnecting means for a motor branch circuit shall be
    (a)   a manually operable fused or unfused motor circuit switch that complies with Rule 14-010(b) and has a horsepower rating not less than that of the motor it serves;
    (b)   a moulded case switch or circuit breaker that complies with Rule 14-010(b) and has a current rating not less than 115% of the full load current rating of the motor it serves;
    (c)   an instantaneous-trip circuit breaker that complies with Rules 14-010(b) and 28-210;
    (d)   an equivalent device that opens all ungrounded conductors of the branch circuit simultaneously and is capable of safely making and interrupting the locked rotor current of the connected load;
    (e)   a single plug fuse for a branch circuit having one grounded conductor feeding a two-wire single-phase or dc motor rated at not more than 1/3 hp, provided that it is used only as an isolating means and is not used to interrupt current; or
    (f)   the draw-out feature of a high-voltage motor starter or controller of the draw-out type that complies with Rule 14-010(b), provided that it is used only as an isolating means and is not used to interrupt current.

(2)   A disconnecting means serving a group of motors on a single branch circuit shall have
    (a)   a current rating not less than 115% of the full load current rating of the largest motor in the group plus the sum of the full load current ratings of all the other motors in the group that may be in operation at the same time; and
    (b)   a horsepower rating not less than the largest motor in the group if a motor circuit switch is used.

(3)   A disconnecting means for a motor, motor starter, or controller shall comply with Subrule (1), except that
    (a)   an isolating switch or a general-use switch used as an isolating switch, if lockable in the open position, marked as required by Rule 26-100(2) and having a current rating not less than 115% of the full load

current rating of the motor it serves shall be permitted to serve as the disconnecting means for a motor or motor starter

    (i)   rated at more than 100 hp if for 3-phase operation; or

    (ii)   rated at more than 50 hp if for other than 3-phase operation;

(b)   a manually operated across-the-line type of motor starter marked "Suitable for Motor Disconnect" shall be permitted to serve as both starter and disconnecting means for

    (i)   a single motor provided that it has a horsepower rating not less than the single motor it serves; or

    (ii)   a group of motors provided that it has a horsepower rating not less than the largest motor in the group and a current rating not less than 115% of the full load current of the largest motor in the group plus the sum of the full load currents of all the other motors in the group that may be in operation at the same time;

(c)   an attachment plug shall be permitted to serve as a disconnecting means for a portable motor and its starting and control equipment, provided that

    (i)   the attachment plug and receptacle has a current rating not less than the ampacity of the minimum size conductors permitted for the motor branch circuit or tap in which they are connected and are used only as an isolating means and not to interrupt current; or

    (ii)   the attachment plug and receptacle is used as permitted by Rule 28-500(3);

(d)   the draw-out feature of a high-voltage starter or controller of the draw-out type shall be permitted to serve as the disconnecting means for the motor or controller providing it is used only as an isolating means and is not used to interrupt current;

(e)   a manually operated general-use ac switch complying with the requirements of Rule 14-510 having a current rating not less than 125% of the full load current of the motor and that need not be horsepower rated shall be permitted to be used as a disconnecting means for a single phase motor; and

(f)   a fused or unfused motor circuit switch shall be permitted to be used as a disconnecting means for a group of motors served from a single circuit and need not have a rating greater than that necessary to accommodate the proper rating of fuse required for the fused switch, provided that it has

    (i)   a horsepower rating not less than that of the largest motor in the group; and

    (ii)   a current rating not less than 115% of the full load current of the largest motor in the group plus the sum of the full load currents of all the other motors in the group that may be in operation at the same time.

(4)   Disconnecting means shall not be of a type that is electrically operated either automatically or by remote control.

28-604 Location of disconnecting means

(1)   Motor branch circuit disconnecting means described in Rule 28-602(1)(a), (b), (c), and (d) shall

    (a)   be located at the distribution centre from where the motor branch circuit originates; and

    (b)   where intended to serve as a single disconnecting means for a motor branch circuit, motor, and controller or starter shall also be

        (i)   located in accordance with Subrule (3); or

        (ii)   capable of being locked in the open position by a lock-off device approved for the purpose and be clearly labelled to describe the load or loads connected.

(2)   Motor branch circuit disconnecting means described in Rule 28-602(1)(f) shall be located in accordance with Subrule (3).

(3)   Except as required in Subrule (5), motor and motor starter or controller disconnecting means shall be located

    (a)   within sight of and within 9 m of the motor and the machinery driven by it; and

    (b)   within sight of and within 9 m of the motor starter or controller.

(4)   Notwithstanding Subrule (3), where a motor or group of motors is fed from a single branch circuit in which the branch circuit disconnecting means is not capable of being acceptably locked in the open position and where the motor disconnecting means is a manually operable across-the-line type of motor starter, the motor disconnecting means shall be permitted to be located beyond the limits defined in Subrule (3) provided that it is capable of safely making and interrupting the locked rotor current of the connected load, is capable of being locked in the open position, and it can be demonstrated that the location specified in Subrule (3) is clearly impracticable.

(5)   Motor disconnecting means for air-conditioning and refrigeration equipment shall be located within sight of and within 3 m of the equipment.

(6)   Disconnecting means shall be readily accessible or have the means for operating them readily accessible.

(7)   Motor-driven machinery of a movable or portable type for industrial use shall have a motor circuit switch or circuit breaker mounted on the machine and accessible to the operator.

Hermetic refrigerant motor-compressors

28-700 Rules for hermetic refrigerant motor-compressors
Rules 28-702 to 28-714 apply to hermetic motor-compressors, hereafter referred to as motor-compressors, and supplement or amend the general Rules of this Section.

28-702 Marking
Motor-compressors, or equipment that incorporates them, shall be marked as required by Rule 2-100; specifically, the marking shall show the rated-load current and the locked rotor current rating.

28-704 Horsepower rated equipment
(1)   Horsepower rated equipment used for the control of motor-compressors and not having a locked rotor current rating shall be given an equivalent locked rotor current rating equal to 6 times the full load current rating.

(2)   Where the full load current rating is not marked, an equivalent full load current rating shall be determined from the horsepower rating by referring to Table 44 or 45, as applicable.

28-706 Conductor ampacity
The ampacity of conductors of a branch circuit supplying a motor-compressor, or equipment consisting of one or more motor-compressors and other loads, shall be based upon the marked rated load current of the motor-compressor or equipment and shall comply with the general requirements of this Section.

28-708 Overcurrent protection
(1)   Except as permitted in Subrule (2), each ungrounded conductor of a branch circuit feeding a motor-compressor shall be protected by an overcurrent device rated or set at not more than 50% of the locked rotor current of the motor-compressor, unless such a device will not permit the motor-compressor to start, in which case the rating or setting shall be permitted to be increased to a value not exceeding 65% of the locked rotor current of the motor-compressor.

(2)   Subrule (1) shall not be deemed to require use of overcurrent devices rated or set at less than 15 A.

28-710 Overload protection
The branch circuit conductors and control equipment for each motor-compressor shall be provided with overload protection complying with Rules 28-302 to 28-306, except that

(a)   the rating or setting of overload relays shall not exceed 140% of the marked rated load current of the motor-compressor;

(b)   the rating or setting of other overload devices, such as fuses, shall not exceed 125% of the marked rated load current of the motor-compressor; and

(c)   approved assemblies consisting of one or more motor-compressors with or without other loads in combination shall be permitted to include the overload protection as part of the approved assembly.

28-712 Control equipment
(1)   Control equipment used for the control of motor-compressors shall have
   (a)   either a marked or an equivalent locked rotor current rating not less than that of the motor-compressor that it controls; and
   (b)   either a marked or an equivalent full load current rating not less than that of the rated load current of the motor-compressor that it controls.

(2)   In all other respects, control equipment for motor-compressors shall be in accordance with Rules 28-500, 28-502, and 28-506.

28-714 Disconnecting means
(1)   The disconnecting means serving a motor-compressor shall have
   (a)   a continuous duty current rating not less than 115% of the rated load current of the motor-compressor; and
   (b)   an interrupting capacity, or an equivalent locked rotor current rating, as determined in accordance with Rule 28-704, not less than the locked rotor current rating of the motor-compressor.

(2) Where one disconnecting means serves one or more motor-compressors together with other loads, the disconnecting means shall have

    (a) a continuous duty current rating not less than 115% of the rated load current of the motor or motor-compressor having the largest rated load current plus the sum of the rated load currents and full load currents of all other loads that may be in operation at the same time; and

    (b) an interrupting capacity or equivalent locked rotor current rating as determined in accordance with Rule 28-704 not less than the locked rotor current rating of the motor or motor-compressor having the largest marked or equivalent locked rotor current rating plus the sum of the full load current rating of all other loads that may be in operation at the same time.

Multi-winding and part-winding-start motors

28-800 Rules for multi-winding and part-winding-start motors
Rules 28-802 to 28-812 apply to the installation of multi-winding and part-winding-start motors.

28-802 Permanent connection
Where a multi-winding motor is used with windings connected in a permanent configuration, it shall be treated as a single winding motor with ratings corresponding to the winding configuration used.

28-804 Conductor sizes
(1) The circuit conductors on the supply side of the controller for a multi-winding or part-winding-start motor shall be of a size specified by Rule 28-106 for the largest full load current of any winding configuration provided by the controller as connected.

(2) Each conductor run from the controller to the motor shall be of the size specified by Rule 28-106 for the largest full load current of any winding or winding configuration that it must supply.

28-806 Overcurrent protection
(1) Each ungrounded conductor on the supply side of the controller shall be protected by an overcurrent device rated or set in accordance with Rule 28-200 for the largest full load current rating of any winding configuration provided by the controller as connected.

(2) Each ungrounded conductor run from the controller to the motor shall be protected by an overcurrent device rated or set in accordance with Rule 28-200 for the largest full load current of any winding configuration served by the conductor so protected, unless the overcurrent device required by Subrule (1) adequately protects it.

(3) Notwithstanding Subrules (1) and (2), if the motor is a part-winding-start motor, a single set of overcurrent devices on the supply side of the controller shall be permitted to protect both windings and if a time-delay fuse is used it shall be permitted to have a maximum rating of 150% of full load current.

28-808 Overload protection
(1) Each winding or configuration shall be provided with overload protection in accordance with Rules 28-300 to 28-310 inclusive, rated or set at not more than 125% of the full load current rating of the winding or configuration so protected.

(2) For a part-winding-start motor, separate overload devices need not be supplied for each winding, provided that overload devices are

    (a) located in the circuit, feeding that winding that is used for starting;

    (b) arranged to de-energize both windings when an overload occurs; and

    (c) selected in accordance with the motor or equipment manufacturer's recommendation.

28-810 Controls
Each multi-winding or part-winding-start motor shall be provided with starting and control equipment in accordance with Rules 28-500, 28-502, and 28-506, except that

    (a) the controller shall be specifically approved for use with the motor that it controls;

    (b) where separate control equipment is provided for each winding or configuration, the individual controllers shall be rated in horsepower (or locked rotor current) not less than the rating of the winding or configuration controlled by each, and interlocks shall be provided where necessary to prevent simultaneous operation of controllers not intended to be so operated; or

    (c) the starting and control equipment for each primary winding of a part-winding-start motor shall have a horsepower (or locked rotor current) rating not less than that of the motor, unless specifically approved for use with that motor.

28-812 Disconnecting means

Each multi-winding motor and its control equipment shall be provided with disconnecting means in accordance with Rules 28-600 to 28-604 except that, for the purpose of Rule 28-602, the horsepower (or locked rotor current) rating of the motor shall be that for the winding or configuration having the largest horsepower (or locked rotor current) rating, and the full load current rating of the motor shall be that for the winding or configuration having the largest full load current rating.

Protection and control of generators

28-900 Disconnecting means required for generators

Generators shall be equipped with an indicating switch or circuit breaker by means of which the generator and all protective devices and control apparatus are able to be disconnected entirely from the circuits supplied by the generator, except where
(a)   the driving means for the generator can be readily shut down; and
(b)   the generator is not arranged to operate in parallel with another generator or other source of voltage.

28-902 Protection of constant-voltage generators

(1)   Constant-voltage generators, whether direct-current or alternating-current, shall be protected from excess current by overcurrent devices, except that
   (a)   where the type of apparatus used and the nature of the system operated make protective devices inadvisable or unnecessary, the protective devices need not be provided; or
   (b)   where an alternating-current generator and a transformer are located in the same building and are intended to operate as a unit for stepping up or stepping down voltage, the protective devices shall be permitted to be connected to the primary or to the secondary of the transformer.
(2)   Subrule (1) shall not apply to exciters for alternating-current machines.

28-904 Generator not driven by electricity

Where a generator not driven by electricity supplies a 2-wire grounded system, the protective device shall be capable of disconnecting the generator from both conductors of the circuit.

28-906 Balancer sets

Where a 3-wire direct-current system is supplied by 2-wire generators operated in conjunction with a balancer set to obtain a neutral, the system shall be equipped with protective devices that disconnect the system in the event of an excessive unbalancing of voltages.

28-908 Three-wire direct-current generators

(1)   Three-wire direct-current generators, whether shunt or compound wound, shall be equipped with
   (a)   a 2-pole circuit breaker with two tripping elements; or
   (b)   a 4-pole circuit breaker connected in the main-and-equalizer leads and tripped by two tripping elements.
(2)   The circuit breaker shall be connected so that it is actuated by the entire armature current.
(3)   One tripping element shall be connected in each armature lead.

Section 30 — Installation of lighting equipment

30-000 Scope
(1) This Section applies to
- (a) the installation of lighting equipment in general — Rules 30-100 to 30-110;
- (b) indoor lighting equipment and the portion of an installation that is inside of buildings — Rules 30-200 to 30-910;
- (c) outdoor lighting equipment and the portion of an installation that is outside of buildings, where protection of the installation and safety from shock hazard are the main concern and the fire hazard is secondary — Rules 30-1000 to 30-1120; and
- (d) extra-low-voltage lighting systems — Rules 30-1200 to 30-1208.

(2) This Section supplements or amends the general requirements of this Code.

30-002 Special terminology (see Appendix B)
In this Section, the following definitions apply:

Cabinet lighting system — a complete, extra-low-voltage lighting assembly consisting of a plug-in power supply having a Class 2 output, luminaires, wiring harness, and connectors, intended for surface or recessed mounting under a shelf or similar structure or in an open or closed cabinet.

Cable lighting system — a permanently connected extra-low-voltage lighting system that comprises an isolating-type transformer with bare secondary conductors for connection to one or more luminaire heads.

Landscape lighting system — an extra-low-voltage outdoor lighting system.

Recessed luminaire — a luminaire that is designed to be either wholly or partially recessed in a mounting surface.

Recessed luminaire, Type IC (intended for insulation contact) — a recessed luminaire designed for installation in a cavity filled with thermal insulation and permitted to be in direct contact with combustible materials and insulation.

Recessed luminaire, Type IC, inherently protected (intended for insulation contact) — a recessed luminaire that does not require a thermal protective device and cannot exceed the maximum allowable temperatures under all conceivable operating conditions.

Recessed luminaire, Type Non-IC (not intended for insulation contact) — a recessed luminaire designed for installation in a cavity with minimum dimensions and spacings to thermal insulation and combustible material.

Recessed luminaire, Type Non-IC, marked spacings (not intended for insulation contact) — a recessed luminaire designed for installation in a cavity where the clearances to combustible building members and thermal insulation are specified by the manufacturer.

Undercabinet lighting system — a complete, extra-low-voltage lighting assembly consisting of a plug-in power supply having a Class 2 output, luminaires, wiring harness, and connectors, intended for surface mounting only under a shelf or similar structure or in an open or closed cabinet.

General

30-100 General
Rules 30-100 to 30-110 cover general requirements that apply to
- (a) the installation of luminaires, lampholders, incandescent filament lamps, and electric discharge lamps; and
- (b) the wiring and electric equipment used in conjunction with this installation.

30-102 Voltage
(1) Branch circuit voltage shall not exceed 150 volts-to-ground in dwelling units.
(2) Branch circuit voltage shall not exceed a nominal system voltage of 347/600Y in other than dwelling units.

30-104 Protection (see Appendix B)
Luminaires, lampholders, and lighting track shall not be connected to a branch circuit protected by overcurrent devices rated or set at more than
- (a) 15 A in dwelling units;
- (b) 15 A in other than dwelling units, where the input voltage exceeds 347 V nominal;
- (c) 20 A in other than dwelling units, where the input voltage does not exceed 347 V nominal; or

(d)   40 A in other than dwelling units where the load is from
    (i)   luminaires with lampholders of the incandescent mogul base type;
    (ii)   high-intensity discharge (HID) luminaires, with or without auxiliary lighting systems, where the input voltage does not exceed 120 V nominal;
    (iii)   tungsten halogen luminaires with double-ended lampholders, where the input voltage does not exceed 240 V nominal; or
    (iv)   luminaires provided with an integral overcurrent device rated at not more than 15 A, where the input voltage does not exceed 120 V nominal.

30-106 Overcurrent protection of high-intensity discharge lighting equipment

Overcurrent protection shall not be provided in a high-intensity discharge luminaire or separate ballast box unless the combination is approved for the purpose and so marked.

30-108 Polarization of luminaires

The identified conductor shall be attached to the luminaire terminal or wire that has been distinguished for identification or otherwise suitably marked unless the luminaire is approved as suitable for connection to line-to-line voltages.

30-110 Bonding of lighting equipment

Non-current-carrying metal parts of luminaires and associated equipment shall be bonded to ground in accordance with Section 10.

Location of lighting equipment

30-200 Near or over combustible material

(1)   Luminaires installed where combustible material is liable to be stored shall be equipped with shades or guards to limit the temperature to which the combustible material may be subjected to a maximum of 90 °C.
(2)   Luminaires installed under the conditions of Subrule (1) shall be of the unswitched type.
(3)   Where luminaires are installed over readily combustible material, every luminaire shall be controlled by an individual wall switch, but a wall switch may control more than one luminaire if every luminaire is located at least 2.5 m above floor level, or located or guarded so that the lamps cannot be readily removed or damaged.
(4)   Switches and luminaires installed under the conditions of Subrule (1) shall have no exposed wiring.

30-202 In show windows

(1)   Except for luminaires installed in accordance with Rule 30-1206, or luminaires of the chain suspension type, no luminaire having exposed wiring shall be used in a show window.
(2)   No lampholder having a paper or fibre lining shall be used in a show window.
(3)   Exposed flexible cord or equipment wire shall not be used to supply permanently installed luminaires in show cases or wall cases.

30-204 In clothes closets

(1)   Every luminaire installed in a clothes closet shall be located on the ceiling or on the front wall above the door of the closet, unless mounted on the trim or sidewall of the doorway and approved for the application.
(2)   Lampholders and luminaires of the pendant or suspended type, and lampholders and luminaires of the bare lamp type shall not be installed in clothes closets.

Installation of lighting equipment

30-300 Live parts

Luminaires, lampholders, and associated equipment shall be installed so that no live part is exposed to contact while they are in use.

30-302 Supports

(1)   Every luminaire shall be securely supported.
(2)   Where a luminaire weighs more than 2.7 kg or exceeds 400 mm in any dimension, it shall not be supported by the screwshell of the lampholder.

(3) Where the weight of a luminaire does not exceed 13 kg, the luminaire shall be permitted to be supported by a wall outlet box attached directly to the building structure or by a wall outlet box attached to a bar hanger.

(4) Where the weight of a luminaire does not exceed 23 kg, the luminaire shall be permitted to be supported by a ceiling outlet box attached directly to the building structure or by a ceiling outlet box attached to a bar hanger.

(5) Where the weight of a luminaire prohibits the installation methods specified in Subrule (3) or (4), the luminaire shall be supported
    (a) independently of the outlet box; or
    (b) by a fixture hanger provided with an integral outlet box suitable for the purpose.

(6) Rigid PVC boxes shall not be used for the support of luminaires unless they are marked as being suitable for the purpose.

30-304 Outlet boxes to be covered

Every outlet box used with lighting equipment shall be provided with a cover or covered by a luminaire-canopy, outlet-box-type luminaire, or other device.

30-306 Wiring space

(1) Every luminaire-canopy and outlet box shall be installed so as to provide adequate space for conductors and connections.

(2) Every luminaire shall be constructed and installed so that conductors in the luminaire and outlet box are not subjected to temperatures greater than those for which the conductors are rated.

30-308 Circuit connections

(1) Every luminaire shall be installed so that the connections between the luminaire conductors and the branch circuit conductors may be inspected without disconnecting any part of the wiring unless the connection employs a plug and receptacle.

(2) Luminaires weighing more than 4.5 kg shall be installed so that the branch circuit wiring connections and the bonding connections will be accessible for inspection without removing the luminaire supports.

(3) Branch circuit conductors within 75 mm of a ballast within the ballast compartment shall have a maximum allowable conductor temperature of not less than 90 °C.

Δ (4) Each fluorescent luminaire installed on branch circuits with voltages exceeding 150 volts-to-ground shall be
    (a) provided with a disconnecting means integral with the luminaire that simultaneously opens all circuit conductors between the branch circuit conductors and the conductors supplying the ballast(s); and
    (b) marked in a conspicuous, legible, and permanent manner adjacent to the disconnecting means, identifying the specific purpose.

30-310 Luminaire as a raceway

(1) Branch circuit conductors run through a luminaire shall be contained in a raceway that is an integral part of the luminaire and that meets the requirements for a surface raceway, except that the conductors of a 2-wire, 3-wire, or 4-wire branch circuit supplying the luminaires may be carried through luminaires marked as suitable for continuous row mounting.

(2) Ballasts located within luminaires referred to in Subrule (1) shall be deemed to be sources of heat and the conductors supplying the luminaires shall
    (a) have a voltage rating not less than 600 V;
    (b) have a temperature rating not less than 90 °C;
    (c) be of a type listed in
        (i) Table 19, as being suitable for use in raceways; or
        (ii) Table 11, as being suitable for use in accordance with this Rule, provided that the conductors are not smaller than No. 14 AWG and do not extend beyond the luminaires through raceways more than 2 m long.

(3) Notwithstanding Subrule (2), non-metallic-sheathed cable shall be permitted to be used for supplying the luminaires, provided that it has a temperature rating of 90 °C.

30-312 Combustible shades and enclosures (see Appendix G)

Every luminaire having a combustible shade or enclosure shall be installed to provide an adequate air space between the lamps and the combustible shade or enclosure.

30-314 Minimum height of low luminaires (see Appendix G)

(1) Where a rigid luminaire is located at a height of less than 2.1 m above the floor and is readily accessible, the luminaire shall be protected from mechanical injury by a guard or by location.

(2) A short flexible drop light or luminaire shall be permitted to be used in place of the rigid luminaire in Subrule (1).

30-316 Luminaires exposed to flying objects

Where luminaires are installed in gymnasiums or similar locations where the lamps are normally exposed to damage from flying objects, the lamp shall be guarded by one of the following means:

(a) metal reflectors that effectively protect the lamps;

(b) metal screens; or

(c) enclosures of armoured glass or suitable plastic material.

30-318 Luminaires in damp or wet locations

(1) Luminaires installed in damp or wet locations shall be approved for such locations and be so marked.

(2) Luminaires suitable for use in wet locations shall be permitted to be used in damp locations as well.

30-320 Lighting equipment in damp locations or near grounded metal

(1) Where luminaires are installed in damp locations or within 2.5 m vertically or 1.5 m horizontally of laundry tubs, plumbing fixtures, steam pipes, or other grounded metal work or grounded surfaces, the luminaires shall be controlled by a wall switch, except as permitted in Subrule (2).

(2) Outlet-box-type luminaires marked for use in damp locations and luminaires marked for use in wet locations, with an integral switch, shall be permitted to be installed under the conditions of Subrule (1).

Δ (3) Switches (including wall switches) controlling luminaires covered by Subrules (1) and (2) shall

(a) be located not less than 1 m from a bathtub or shower stall (this distance measured horizontally between the switch and the bathtub or shower stall) without piercing a wall, partition, or similar obstacle; or

(b) if the condition in Item (a) is not practicable, be located not less than 500 mm from a bathtub or shower stall and be protected by a ground fault circuit interrupter of the Class A type.

30-322 Totally enclosed gasketted luminaires

Incandescent totally enclosed gasketted luminaires, unless marked as suitable for the purpose, shall not be mounted on a combustible ceiling.

Wiring of lighting equipment

30-400 Wiring of luminaires

All electrical wiring on or within a luminaire shall be

(a) neatly arranged without excess wiring;

(b) not exposed to mechanical injury; and

(c) arranged so that it is not subjected to temperatures above those for which it is rated.

30-402 Colour coding

Notwithstanding the requirements of Sections 0, 4, and 10 with regard to the colours used for distinguishing and identifying conductors, a continuous-coloured tracer in the braid of an individual braided conductor shall be permitted for the supply conductors of a luminaire, the colour of the tracer being black, white, and green for the ungrounded, identified, and bonding conductors respectively.

30-404 Conductor insulation

Luminaires shall be wired with conductors at least No. 18 AWG, having insulation suitable for the voltage and temperatures to which the conductors may be subjected.

30-406 Conductors on movable parts

(1) Stranded conductors shall be used on chain-type luminaires and other movable parts of lighting equipment.

(2) Conductors shall be arranged so that the weight of the luminaire or the movable parts does not place undue tension on the connections.

(3) All conductors that supply movable parts of lighting equipment shall be protected against mechanical injury.

30-408 Wiring of ceiling outlet boxes

(1) Branch circuit conductors having insulation suitable for 90 °C shall be used for wiring of ceiling outlet boxes on which a luminaire is mounted, except for boxes in wet locations where Type NMW or NMWU cables are used.

(2) For purposes of compliance with this Rule, the ampacity of 90 °C wire shall be limited to the ampacity of 60 °C wire.

30-410 Wiring of show window luminaires

(1) Where show window luminaires are closely spaced, they shall be permitted to be connected to a conductor suitable for the purpose that is listed in Table 11, with a temperature rating of not less than 125 °C.

(2) The connection of show window luminaires to the circuit conductors shall be in a junction box.

(3) The junction box shall be maintained at a sufficient distance from the luminaire to ensure that the circuit conductors are not subjected to temperatures in excess of their rating.

30-412 Tap connection conductors

No. 14 AWG copper tap connection conductors shall be permitted for a single luminaire and for luminaires mounted in a continuous row as specified in Rule 30-310(1), on a branch circuit protected by an overcurrent device rated or set at 20 A, provided that the tap connection conductors

(a) have an ampacity not less than the rating of the single luminaire or the luminaires mounted in a continuous row as specified in Rule 30-310(1); and

(b) do not exceed 7.5 m in length.

Luminaires in buildings of residential occupancy

30-500 Lighting equipment at entrances (see Appendix G)

An exterior luminaire controlled by a wall switch located within the building shall be provided at every entrance to buildings of residential occupancy.

30-502 Luminaires in dwelling units (see Appendix G)

(1) Except as provided in Subrule (2), a luminaire controlled by a wall switch shall be provided in kitchens, bedrooms, living rooms, utility rooms, dining rooms, bathrooms, water-closet rooms, vestibules, and hallways in dwelling units.

(2) Where a receptacle controlled by a wall switch is provided in bedrooms or living rooms, such rooms need not conform to the requirements in Subrule (1).

30-504 Stairways (see Appendix G)

(1) Every stairway shall be lighted.

(2) Except as provided in Subrule (3), 3-way wall switches located at the head and foot of every stairway shall be provided to control at least one luminaire for stairways with four or more risers in dwelling units.

(3) The stairway lighting for basements that do not contain finished space nor lead to an outside entrance or built-in garage, and that serve not more than one dwelling unit, is permitted to be controlled by a single switch located at the head of the stairs.

30-506 Basements (see Appendix G)

(1) A luminaire shall be provided for each 30 m^2 or fraction thereof of floor area in unfinished basements.

(2) The luminaire required in Subrule (1) that is located nearest the stairs shall be controlled by a wall switch located at the head of the stairs.

30-508 Storage rooms (see Appendix G)

A luminaire shall be provided in storage rooms.

30-510 Garages and carports (see Appendix G)

(1) A luminaire shall be provided for an attached, built-in, or detached garage or carport.

(2) Except as provided in Subrule (3), luminaires required in Subrule (1) shall be controlled by a wall switch near the doorway.

Δ (3) Where the luminaire required in Subrule (1) is ceiling-mounted above an area not normally occupied by a parked car, or is wall-mounted, a luminaire with a built-in switch accessible to an adult of average height shall be permitted to be used.

(4) Where a carport is lighted by a luminaire at the entrance to a dwelling unit, additional carport lighting is not required.

Lampholders

30-600 Connections to lampholders
The identified conductor, if present, shall be connected to the lampholder screwshell.

30-602 Switched lampholders used on unidentified circuits
Where lampholders of the switched type are used on unidentified 2-wire circuits tapped from the ungrounded conductors of multi-wire circuits, the switching devices of the lampholders shall disconnect both conductors of the circuit simultaneously.

30-604 Luminaires with pull-type switch mechanisms
On luminaires employing pull-type switch mechanisms, the operating means shall be
(a)   cords made of insulating materials;
(b)   cords made of insulating materials or chains with links made of insulating material, connected to metal chains as close as possible to where the chains emerge from the enclosure; or
(c)   metal chains without insulating links, provided that the lampholder is approved as not requiring insulating links.

30-606 Lampholders in wet or damp locations
Where lampholders are installed in wet or damp locations, they shall be the weatherproof type.

30-608 Pendant lampholders
(1)   Where pendant lampholders having permanently attached leads are used with other than festoon wiring, they shall be hung from separate stranded rubber- or thermoplastic-insulated pendant conductors that are connected directly to the circuit conductors but supported independently thereof.
(2)   Where thermoplastic-insulated pendant conductors are used in locations where they may be subjected to temperatures lower than –10 °C, they shall be of a type approved for the purpose.
(3)   Where the pendant conductors supply mogul or medium-base screwshell lampholders, they shall be not smaller than No. 14 AWG.
(4)   Where the pendant conductors supply intermediate or candelabra-base lampholders other than approved Christmas-tree and decorative lighting outfits, the conductors shall be not smaller than No. 18 AWG.
(5)   Where the pendant conductors are longer than 900 mm, they shall be twisted together.

Electric-discharge lighting systems operating at 1000 V or less

30-700 Rules for discharge lighting systems 1000 V or less
Rules 30-702 to 30-712 apply to electrical equipment used with electric-discharge lighting systems operating at 1000 V or less.

30-702 Oil-Filled transformers
Transformers of the oil-filled type shall not be used.

30-704 Direct-Current equipment
Luminaires shall not be installed on a direct-current circuit unless they are equipped with auxiliary equipment and resistors designed for direct-current operation, and the luminaires are so marked.

30-706 Voltages — Dwelling units
Where equipment has an open-circuit voltage of more than 300 V, it shall not be installed in dwelling units unless the equipment is designed so that no live parts are exposed during the insertion or removal of lamps.

30-708 Auxiliary equipment
(1)   Reactors, capacitors, resistors, and other auxiliary equipment shall be
(a)   enclosed within the luminaire;
(b)   enclosed within an accessible, permanently installed, metal cabinet where remote from the luminaire; or
(c)   of a type suitable for use without an additional enclosure.
(2)   Adequate provision shall be made for the dissipation of heat from enclosed auxiliary equipment and the conductors supplying the auxiliary equipment.
(3)   The metal cabinet, if not part of the luminaire, shall be installed as close as possible to the luminaire.
(4)   Where display cases are not permanently installed, no part of a secondary circuit shall be included in more than one case.

Δ **30-710 Control** (see Appendix B)
(1)   The luminaires and lamp installations shall be controlled by a switch, circuit breaker, or contactor.
(2)   Where a switch is used, it shall
    (a)   be approved for the purpose and marked for the control of electric lighting systems operating at 1000 V or less;
    (b)   have a current rating not less than twice the current rating of the lamps or transformers;
    (c)   be of a type approved with the assembly;
    (d)   be a manually operated general-use ac switch complying with Rule 14-510; or
    (e)   be a manually operated general-use 347 V ac switch complying with Rule 14-512.
(3)   Where a circuit breaker is used, it shall
    (a)   comply with the requirements of Rule 14-104; and
    (b)   in the case of 15 A and 20 A branch circuits at 347 V and less supplying fluorescent luminaires, the circuit breaker shall be suitable for such switching duty and shall be marked "SWD".
(4)   Where a contactor is used, it shall
    (a)   be approved for the purpose and marked for the control of electric lighting systems operating at 1000 V or less; or
    (b)   have a current rating of not less than twice the current rating of the lamps or transformers.

30-712 Branch circuit capacity
(1)   Where lighting branch circuits supply luminaires employing ballasts, transformers, or auto-transformers, the load on the branch circuits shall be computed on the basis of the total amperes of the units and not on the watts of the lamps.
(2)   The aggregate capacity of luminaires connected to a lighting branch circuit shall not exceed 80% of the branch circuit overcurrent protection.

Electric-discharge lighting systems operating at more than 1000 V

30-800 Rules for discharge lighting systems — More than 1000 V
Rules 30-802 to 30-822 apply to electrical equipment used with electric-discharge lighting systems operating at more than 1000 V.

30-802 Voltages — Dwelling units
Where equipment has an open-circuit voltage of more than 1000 V, it shall not be installed in dwelling units.

Δ **30-804 Control** (see Appendix B)
(1)   The luminaires and lamp installations shall be controlled singly or in groups by an externally operated switch or circuit breaker that opens all ungrounded primary conductors.
(2)   The switch or circuit breaker shall
    (a)   be installed within sight of the luminaires or lamps; or
    (b)   be provided with a means for locking it in the open position.
(3)   The switch shall
    (a)   be approved for the purpose and marked for the control of electric lighting systems operating at more than 1000 V;
    (b)   have a current rating of not less than twice the current rating of the transformer or transformers controlled by it;
    (c)   be of a type approved with the assembly;
    (d)   be a manually operated general-use ac switch complying with Rule 14-510; or
    (e)   be a manually operated general-use 347 V ac switch complying with Rule 14-512.
(4)   The circuit breaker shall comply with the requirements of Rule 14-104.

30-806 Transformer rating
(1)   Every transformer and ballast shall have a secondary open-circuit voltage of not more than 15 000 V, except that every transformer and ballast of the open core-and-coil type shall have a secondary open-circuit voltage of not more than 7500 V.
(2)   The secondary current rating shall be not more than 240 mA, except that where the secondary open-circuit voltage exceeds 7500 V, the secondary current rating shall be not more than 120 mA.

30-808 Liquid-filled transformers
Transformers of the liquid-filled type shall not be used unless they are filled with a nonflammable liquid.

30-810 Transformers — Secondary connection

(1)   The high-voltage windings of transformers operating at more than 1000 V shall not be connected in series or in parallel, but where each of two transformers has one end of its high-voltage winding grounded and connected to the enclosure, the high-voltage windings shall be permitted to be connected in series to form the equivalent of a midpoint-grounded transformer.

(2)   The grounded end of each high-voltage winding shall be connected by an insulated stranded copper conductor not smaller than No. 14 AWG.

30-812 Location of transformers

(1)   Transformers operating at more than 1000 V shall be accessible for servicing or replacement.

(2)   The transformers shall be installed as near to the lamps as practicable.

(3)   The transformers shall be located so that adjacent combustible materials are not subjected to temperatures in excess of 90 °C.

30-814 Wiring method

(1)   The secondary conductors shall be luminous-tube-sign cable approved for the purpose and for the voltage of the circuit.

(2)   Not more than a total of 6 m of cable shall be run in a metal raceway from a transformer.

(3)   Not more than a total of 16 m of cable shall be run in a non-metallic raceway from a transformer.

(4)   The conductors shall be installed in conformance with Section 34.

30-816 Transformer loading

Where the lamps are connected to a transformer, they shall be of such lengths and characteristics as not to cause a condition of continuous over-voltage on the transformer.

30-818 Lamp supports

(1)   Lamps operating at more than 1000 V shall be supported in the manner required by Section 34.

(2)   The lamps shall not be installed where they are exposed to mechanical injury.

30-820 Lamp terminals and lampholders

(1)   Parts that must be removed for lamp replacement shall be hinged or fastened in a secure manner.

(2)   Lamp terminals and lampholders shall be designed so that the tubing can be replaced with the minimum exposure of bare live parts during re-lamping.

(3)   The designs referred to in Subrule (2) need not afford protection against "space discharge" shocks as tubes are replaced by trained maintenance staff.

30-822 Marking

Every luminaire and every secondary circuit of tubing having an open-circuit voltage of more than 1000 V shall be clearly and legibly marked in letters and figures not less than 25 mm high with the words "CAUTION...V", the rated open-circuit voltage being inserted in figures in the space between the words.

Recessed luminaires (see Appendix G)

30-900 General

Rules 30-900 to 30-910 apply to the installation of luminaires recessed in cavities in ceilings or walls.

30-902 Spacings for Non-IC type luminaires

Except as provided in Rules 30-904 and 30-908, the recessed portion of every recessed luminaire marked "Type Non-IC" shall be at least 13 mm from combustible material at every point other than the point of support, and thermal insulation shall not be installed closer than 76 mm to the luminaire.

30-904 Spacings for Non-IC — Marked spacings type luminaires

The recessed portion of every recessed luminaire marked "Type Non-IC, marked spacings" shall be installed to maintain a minimum spacing from thermal insulation and combustible material at every point other than the point of support in accordance with the manufacturer's spacings marked on the luminaire.

30-906 Luminaires designed for insulation contact

The recessed portion of every recessed luminaire marked "Type IC" or "Type IC, inherently protected" shall be permitted to be in contact with combustible material or blanketed with thermal insulation.

30-908 Luminaires designed for non-combustible surfaces contact only
A recessed luminaire marked as suitable for installation on a non-combustible surface shall be installed only on a non-combustible material.

30-910 Wiring of recessed luminaires (see Appendix B)
(1)   The temperature rating of insulation of conductors other than branch circuit conductors used to wire recessed luminaires shall comply with the conductor temperature rating marked on the luminaire.
(2)   The temperature rating of insulation of the branch circuit conductors run directly to the luminaire shall be in compliance with the conductor temperature rating marked on the luminaire.
(3)   Tap connection conductors shall be installed in a raceway extending at least 450 mm but not more than 2 m from the luminaire and terminated in an outlet box conforming to Subrule (4).
(4)   The outlet box referred to in Subrule (3) shall be
  (a)   accessible as required by Rule 12-3014;
  (b)   located not less than 30 cm from the luminaire; and
  (c)   located within 35 cm from an opening intended for access.
(5)   Where access to the outlet box referred to in Subrule (4) is through the opening for mounting the luminaire, this opening shall not be less than a circle of 180 cm^2, with no dimension less than 15 cm.
(6)   Where the luminaire opening referred to in Subrule (5) is smaller than 15 cm in any direction, access to the outlet box referred to in Subrule (4) shall be through some other opening not less than a square or rectangle of 400 cm^2, with no dimension less than 20 cm.
(7)   A supply connection box that is an integral part of the luminaire shall
  (a)   be accessible in accordance with Rule 12-3014; and
  (b)   if access is through the opening for mounting the luminaire, the following requirements shall be met:
    (i)   the electrical components of the luminaire shall be capable of extraction through the opening for service, and the components shall include ballasts, transformers, thermal protectors, and wire connections in the supply connection box; and
    (ii)   the cover of the supply connection box shall be capable of removal by hand tool, held below the ceiling.
(8)   A supply connection box that is an integral part of the luminaire shall not have branch circuit conductors pass directly through the junction box unless the luminaire is marked as suitable for the purpose.

Permanent outdoor floodlighting installations

30-1000 General (see Appendix B)
(1)   Rules 30-1002 to 30-1036 apply to permanent outdoor installations of floodlights that are mounted on poles or towers.
(2)   These Rules are based on the understanding that authorized persons may replace lamps but all other maintenance will be done by qualified persons.
(3)   Rules 30-1000 to 30-1120 cover only that portion of the installation that is outside the buildings.

30-1002 Service equipment
(1)   Service equipment shall comply with Section 6 for low-voltage installations, and with Section 36 for high-voltage installations.
(2)   Where indoor equipment is installed outdoors, it shall be installed in a weatherproof enclosure.

30-1004 Wiring methods — Underground
(1)   Wiring underground shall be run
  (a)   in rigid steel or rigid aluminum conduit;
  (b)   in non-metallic underground conduit;
  (c)   as mineral-insulated cable or aluminum-sheathed cable; or
  (d)   as conductors or cable assemblies suitable for direct earth burial as indicated in Table 19 or, where a deviation has been allowed in accordance with Rule 2-030, for service entrance below ground as indicated in Table 19.
(2)   Conductors in conduit shall be of types indicated in Table 19 as being suitable for use in wet locations.
(3)   Conductors buried directly in the earth shall be installed in accordance with Rule 12-012.
(4)   Suitable corrosion-resistant protection shall be provided for aluminum-sheathed cable and aluminum conduit, and also for mineral-insulated cable, if used where materials coming in contact with the cable may have a deteriorating effect on the sheath.

30-1006 Wiring methods on poles

(1) All electrical equipment on the pole shall be controlled by a switch that can be locked in the OFF position, and each pole shall be provided with a prominent sign warning against climbing the pole until the switch is off, unless all conductors and live parts other than those used for pole-top wiring are guarded against accidental contact in one of the following ways:

(a) the conductors are run in rigid or flexible metal conduit, as mineral-insulated cable, or up the centre of steel, aluminum, or hollow concrete poles;

(b) the conductors and live parts are kept at least 1 m from the climbing ladder or climbing steps;

(c) barriers are provided between conductors or live parts, or both, and the climbing ladder to reduce the likelihood of contact by the climber.

(2) Conductors run up the centre of poles shall be supported to prevent injury to the conductors inside the pole and to prevent undue strain on the conductors where they leave the pole.

(3) Where vertical conductors, cables, and grounding conductors are within 2.5 m of locations accessible to unauthorized persons, they shall be provided with a covering that gives mechanical protection.

(4) On wood poles, for grounding conductors from lightning arresters, the protective covering specified in Subrule (3) shall be of wood moulding or other insulating material giving equivalent protection.

30-1008 Disconnecting means at individual poles

Notwithstanding Rule 14-402, a disconnecting means is not required adjacent to an inline fuseholder used at individual poles, provided that

(a) the fuseholder is the weatherproof type having load-breaking capability;

(b) the maximum number of fuseholders at any one pole shall be two on a single-phase system and shall be three on a three-phase system;

(c) the fuseholder is of a design and is wired such that any exposed fuse parts are retained by the load side portion of the fuseholder when it is open; and

(d) the load is connected between the live conductor and the identified conductor.

30-1010 Overcurrent protection of pole-top branch circuits

Notwithstanding Rule 30-104, pole-top branch circuits shall be permitted to have overcurrent protection rated or set at no more than 100 A.

30-1012 Pole-top branch circuit wiring

Pole-top branch circuit wiring, exclusive of leads provided with the floodlights to which they are connected, shall be run

(a) as lead-sheathed cable or rubber- or thermoplastic-insulated moisture-resistant types of conductors installed in rigid conduit;

(b) as mineral-insulated cable or aluminum-sheathed cable; or

(c) where a deviation has been allowed in accordance with Rule 2-030, as insulated or uninsulated exposed wiring, provided that

(i) the wiring is supported on suitable insulators;

(ii) the wiring is controlled by a switch that can be locked in the OFF position; and

(iii) the pole is provided with a prominent sign warning against climbing it until the switch is off.

30-1014 Joints

(1) Open taps and joints shall be permitted to be made in pole-top exposed wiring, provided that the joint or tap is given insulation equivalent to that on the conductors joined.

(2) There shall be no joints or splices concealed within conduit.

30-1016 Location of transformers

Transformers shall comply with the following:

(a) if mounted on floodlight poles, all live parts shall be guarded as required by Rule 30-1006;

(b) if mounted on poles, the bottom of the transformer shall be at least 5 m above locations accessible to unauthorized persons; and

(c) if located on platforms on the ground, they shall be completely enclosed to prevent access by unauthorized persons or they shall be surrounded by a protecting fence that shall comply with the requirements of Rules 26-300 to 26-324.

30-1018 Overcurrent protection of transformers

Overcurrent protection of transformers shall be in accordance with Section 26.

30-1020 Switching of floodlights
Switches controlling floodlights shall comply with the following:
(a)   a switch on the primary side of a transformer shall be capable of making and interrupting the full load on the transformer;
(b)   switches controlling floodlights from the secondary side of a transformer shall have a current rating not less than 125% of the current requirements of the floodlights controlled;
(c)   switches shall be capable of being operated without exposing the operator to danger of contact with live parts, either by remote operation or by proper guarding; and
(d)   switches shall be capable of being locked in the OFF position.

30-1022 Grounding of circuits at 300 V or less
Circuits operating at voltages of 300 V or less between conductors shall be grounded.

30-1024 Grounding of circuits above 300 V
Circuits operating at voltages above 300 V shall be permitted to be grounded and in compliance with the requirements of the supply authority.

30-1026 Material for grounding and bonding conductors
Grounding and bonding conductors shall be of the material specified in Rules 10-802 and 10-804.

30-1028 Grounding methods
(1)   A grounded secondary circuit shall be grounded in accordance with Section 10.
(2)   The secondary grounded circuit conductor shall be permitted to be grounded by an interconnection to the primary grounded circuit conductor, provided that
    (a)   the primary is grounded at the transformers; and
    (b)   interconnection is made only at the transformer.

30-1030 Grounding and bonding of non-current-carrying metal parts
(1)   All non-current-carrying metal parts within 2.5 m of ground or at locations where unauthorized persons may stand shall be bonded to ground by a separate bonding conductor sized in accordance with Table 16.
(2)   Except for isolated metal parts, such as crossarm braces, bolts, insulator pins, and the like, non-current-carrying metal parts of electrical equipment at the pole top shall be bonded together and, if within reach of any grounded metal, shall be grounded.
(3)   The size of grounding or bonding conductor shall be as specified in Rule 10-814.

30-1032 Installation of lightning arresters
Where lightning arresters are installed, they shall be in accordance with Rules 10-1000 and 10-1002 with the addition that a common grounding conductor and common electrode shall be permitted to be used for grounding primary and secondary neutrals and lightning arresters.

30-1034 Types of equipment permitted
Floodlights, secondary wiring, conduit, conduit fittings, and distribution panelboards shall be approved for the purpose, and other electrical pole-top equipment shall be of a type suitable for the purpose.

30-1036 Climbing steps
Where it is necessary to climb the pole to replace lamps, permanent climbing steps shall be provided and the lowest permanent step shall be not less than 3.7 m above locations accessible to unauthorized persons.

Exposed wiring for permanent outdoor lighting

30-1100 General
Rules 30-1102 to 30-1120 apply to exposed wiring for permanent outdoor lighting other than floodlighting, where the circuits are run between buildings, between poles, or between buildings and poles.

30-1102 Conductors
Conductors shall be stranded copper not less than No. 12 AWG, and shall be
(a)   of a type suitable for exposed wiring where exposed to the weather as specified in Table 19;
(b)   of the rubber-insulated type suitable for exposed wiring where exposed to the weather as specified in Table 19, when lampholders of a type that punctures the insulation and makes contact with the conductors are used; or
(c)   of the moisture-resistant rubber-insulated type suitable for exposed wiring where exposed to the weather as specified in Table 19, if cabled together and used with messenger cables.

176

30-1104 Use of insulators

(1) Conductors shall be securely attached to insulators at each end of the run if a messenger is not used and at intermediate points of support if there are any.

(2) Insulators at the ends of runs shall be of the strain type unless the conductors are supported by messenger cables.

(3) Split knobs shall not be used.

30-1106 Height of conductors

Conductors supplying lamps in parking lots, used car lots, drive-in establishments, and similar commercial areas shall be maintained such that the conductors or the bottom of a lamp fed from the conductors, whichever is lower, shall have a clearance of not less than 4 m above grade at any point in a run except that, where a driveway or thoroughfare exists, this clearance shall be not less than 5 m.

30-1108 Spacing from combustible material

Conductors and lampholders shall be maintained at a distance not less than 1 m from any combustible material, except for branch circuit conductors at the point of connection to buildings or poles.

30-1110 Spacing of conductors

Conductors shall be separated at least 300 mm from each other by means of insulating spacers at intervals of not more than 4.5 m unless the conductors are secured to and supported by messenger cables.

30-1112 Lampholders

(1) Lampholders shall be of weatherproof types.

(2) Lampholders shall be of types having either
   (a) permanently attached leads; or
   (b) terminals of a type that puncture the insulation and make contact with the conductors.

(3) Lampholders having permanently attached leads shall have the connections to the circuit wires staggered where a cabled assembly is used.

30-1114 Protection of lampholders

Notwithstanding Rule 30-104, lampholders shall be permitted to be connected to branch circuits protected by overcurrent devices rated or set at not more than 30 A, provided that the lampholders are

(a) for incandescent lamps;

(b) of the unswitched type; and

(c) rated not less than 660 W.

30-1116 Use of messenger cables

(1) Messenger cables shall be used to support the conductors
   (a) if lampholders having permanently attached leads are used, and the span exceeds 12 m; and
   (b) in all cases where lampholders having terminals that puncture the insulation are used.

(2) Messenger cable shall be securely attached at each end of the run and shall be grounded in accordance with Section 10.

(3) Conductors shall be permanently attached to the messenger.

30-1118 Construction of messenger cables

(1) Messenger cables shall be of galvanized steel, copper-coated steel, or stainless steel, and shall be of stranded construction with not less than seven strands.

(2) Galvanized steel shall have a coating of not less than 45 g/m².

(3) The effective ultimate strength of a messenger cable shall be not less than 3 times the calculated maximum working load, including loading due to ice loads and wind loads, and in no case shall the individual strands be less than
   (a) 1.17 mm in diameter in the case of galvanized or copper-coated wire; or
   (b) 1.11 mm in diameter in the case of stainless steel wire.

30-1120 Branch circuit loading and protection

(1) Branch circuits shall be protected by overcurrent devices rated at not more than 30 A.

(2) The total load on a branch circuit shall not exceed 80% of the rating or setting of the overcurrent devices.

Extra-low-voltage lighting systems

30-1200 Rules for extra-low-voltage lighting systems
Rules 30-1202 to 30-1208 apply to extra-low-voltage lighting systems.

30-1202 Sources of supply
(1) Extra-low-voltage lighting systems shall be supplied from branch circuits operating at not more than 150 volts-to-ground.

(2) The extra-low-voltage portion of the system shall be supplied from the secondary of an isolating transformer approved for the purpose having no direct electrical connection between the primary and secondary windings.

(3) The extra-low-voltage portion of the system shall not be grounded.

30-1204 Installation of landscape lighting systems
(1) Flexible cord shall be permitted to be used on the secondary side of the transformer and be permitted to be secured to structural members and run through holes.

(2) Electrical connections shall be permitted to be made without an enclosure where not exposed to mechanical damage.

30-1206 Installation of cable lighting systems
(1) Cable lighting systems shall only be permitted in dry locations.

(2) Cable lighting systems shall not be installed in bathrooms.

(3) Conductors of extra-low-voltage circuits shall be rigidly supported.

(4) Conductors shall not be installed in contact with combustible materials and not run through walls, ceilings, floors, or partitions.

(5) Uninsulated conductors shall not be installed less than 2.2 m from the floor.

30-1208 Installation of cabinet and undercabinet lighting systems
(1) Notwithstanding Rule 4-010(3), flexible cords on the secondary side of the power supply shall be permitted to be secured to structural members of cabinets and run through cabinet holes.

(2) Electrical connections shall be permitted to be made without an enclosure where not exposed to mechanical damage.

Section 32 — Fire alarm systems and fire pumps
(See Appendix G)

32-000 Scope (see Appendix B)
(1) This Section applies to the installation of electrical local fire alarm systems and fire pumps required by the *National Building Code of Canada*.
(2) The requirements of this Section supplement or amend the general requirements of this Code.

Fire alarm systems

32-100 Conductors
(1) Conductors shall be of copper and shall have an ampacity adequate to carry the maximum current that can be provided by the circuit.
(2) Stranded conductors with more than 7 strands shall be bunch-tinned or terminated in compression connectors.
(3) Conductors shall have an insulation rating not less than 300 V and shall be not smaller than
    (a) No. 16 AWG for individual conductors pulled into raceways;
    (b) No. 19 AWG for individual conductors laid in raceways;
    (c) No. 19 AWG for an integral assembly of two or more conductors; and
    (d) No. 22 AWG for an integral assembly of four or more conductors.
(4) Conductors shall be suitable for the purpose of the type listed in Table 19, except that individual conductors smaller than No. 14 AWG copper installed in a raceway shall be equipment wire of the type listed in Table 11.

32-102 Wiring method
(1) All conductors of a fire alarm system shall be
    (a) installed in metal raceway of the totally enclosed type;
    (b) incorporated in a cable having a metal armour or sheath;
    (c) installed in rigid non-metallic conduit, where embedded in at least 50 mm of masonry or poured concrete or installed underground; or
    (d) installed in electrical non-metallic tubing, where embedded in at least 50 mm of masonry or poured concrete.
(2) Notwithstanding Subrule (1), conductors installed in buildings of combustible construction in accordance with the Rules of Section 12 shall be permitted to be
    (a) non-metallic-sheathed cable;
    (b) fire alarm and signal cable; or
    (c) installed in a totally enclosed non-metallic raceway.
(3) The conductors shall be installed to be entirely independent of all other wiring and shall not enter a raceway, box, or enclosure occupied by other wiring, except as may be necessary for connection to
    (a) the point of supply;
    (b) a signal;
    (c) an ancillary device; or
    (d) a communication circuit.
(4) All wiring of a communication system connected to a fire alarm system to extend the fire alarm system beyond the building shall conform to the applicable Rules of Section 60.
(5) All conductors contained in the same raceway or cable shall be insulated for the highest voltage in the raceway or cable.

32-104 Equipment bonding
(1) Exposed non-current-carrying metal parts of electrical equipment including outlet boxes, conductor enclosures, raceway, and cabinets shall be bonded to ground in accordance with Section 10.
(2) Where a non-metallic wiring system is used, a bonding conductor shall be incorporated in each cable and shall be sized in accordance with Rule 10-814(1).

32-106 Electrical supervision
Wiring to dual terminals and dual splice leads shall be independently terminated to each terminal or splice lead.

32-108 Current supply

(1)   A fire alarm system shall be supplied by a separate circuit connected as close as practicable, without violating other Rules of this Code, to
    (a)   the load terminals of the service box;
    (b)   the secondary terminals of the transformer, where transformation is necessary in order to supply a utilization voltage required by the fire alarm system; or
    (c)   the terminals of a transfer switch, where the fire alarm system receives emergency power from an emergency power source that also supplies other electrical equipment.

(2)   Overcurrent devices and disconnecting means for the separate circuit supplying a fire alarm system shall be clearly identified as the fire alarm power supply in a permanent, conspicuous, and legible manner and the disconnecting means shall be coloured red and lockable in the ON position.

32-110 Installation of smoke alarm devices in dwelling units (see Appendices B and G)

The following requirements apply to the installation of smoke alarms in dwelling units:

(a)   A smoke alarm shall be supplied from a lighting circuit, or from a circuit that supplies a mix of lighting and receptacles, and in any case shall not be installed
    (i)   where prohibited by Rules 26-720 to 26-724; and
    (ii)   where the circuit is protected by a GFCI or AFCI;

(b)   there shall be no disconnecting means between the smoke alarm device and the overcurrent device for the branch circuit;

(c)   the wiring method for the smoke alarm device, including any interconnection of units and their associated equipment, shall be in accordance with Rules 32-100 and 32-102; and

(d)   notwithstanding Item (c), where a smoke alarm circuit utilizes a Class 2 power supply for the interconnection of the smoke alarms and their associated equipment, Class 2 wiring methods shall be permitted in buildings of combustible construction, provided that the conductors are installed in accordance with Rules 12-506 to 12-524 inclusive.

Δ

Fire pumps

32-200 Conductors (see Appendices B and G)

Conductors from the emergency power source to a fire pump shall

(a)   have an ampacity not less than
    (i)   125% of the full load current rating of the motor, where an individual motor is provided with the fire pump; and
    (ii)   125% of the sum of the full load currents of the fire pump, jockey pump, and the fire pump auxiliary loads, where two or more motors are provided with the fire pump; and

(b)   be protected against fire exposure to provide continued operation in compliance with the *National Building Code of Canada*.

32-202 Wiring method (see Appendices B and G)

All conductors to fire pump equipment shall be

(a)   installed in metal raceways of the totally enclosed type;
(b)   incorporated in a cable, having a metal armour or sheath, of a type listed in Table 19;
(c)   installed in rigid non-metallic conduit where embedded in at least 50 mm of masonry or poured concrete or installed underground; or
(d)   installed in electrical non-metallic tubing where embedded in at least 50 mm of masonry or poured concrete.

32-204 Service box for fire pumps (see Appendix G)

(1)   A separate service box conforming to Rule 32-206 shall be permitted for fire pump equipment.
(2)   Notwithstanding Rule 6-102(2), a service box for fire pump equipment shall be permitted to be located remote from other service boxes.

32-206 Disconnecting means and overcurrent protection (see Appendices B and G)

(1)   No device capable of interrupting the fire pump circuit, other than a circuit breaker specifically approved for fire pump service, shall be placed between the service box and a fire pump transfer switch or a fire pump controller.
(2)   The circuit breaker referred to in Subrule (1) shall be labelled in a conspicuous, legible, and permanent manner identifying it as the fire pump power supply.

(3)   The circuit breaker referred to in Subrule (1) shall be permitted to be used in the separate service box described in Rule 32-204.

(4)   Where the circuit breaker conforming to this Rule is installed in an emergency supply circuit between the emergency power source and the fire pump transfer switch, the rating or setting of the circuit breaker shall comply with Rule 28-200.

(5)   Where the circuit breaker conforming to this Rule is installed in a normal supply circuit upstream of the fire pump controller, the rating or setting of the circuit breaker shall be not less than the overcurrent protection that is provided integral with the fire pump controller.

32-208 Transfer switch (see Appendix G)

(1)   Where an on-site electrical transfer switch is used to provide emergency power supply to fire pump equipment, such a transfer switch shall be
   (a)   located in a barriered compartment of the fire pump controller, or in a separate enclosure adjacent to the controller;
   (b)   labelled in a conspicuous, legible, and permanent manner identifying it as the fire pump automatic transfer switch; and
   (c)   approved for fire pump service.

(2)   Where more than one fire pump is provided with emergency power as described in Subrule (1), a separate transfer switch shall be provided for each fire pump.

32-210 Overload and overheating protection (see Appendix G)

The branch circuit conductors and control conductors or equipment of a fire pump shall not require overload or overheating protection and shall be permitted to be protected by the motor branch circuit overcurrent device(s).

32-212 Ground fault protection (see Appendices B and G)

Ground fault protection shall not be installed in a fire pump circuit.

Section 34 — Signs and outline lighting

34-000 Scope
(1) This Section applies to signs and outline lighting in which the sources of light are
  - (a) incandescent lamps;
  - (b) fluorescent lamps;
  - (c) high-voltage luminous discharge tubes, commonly known as cold-cathode or neon tubes;
  - (d) high-intensity discharge lamps; and
  - (e) other light emitting sources, such as the LED.

(2) The requirements of this Section supplement or amend the general requirements of this Code.

34-002 Special terminology
In this Section, the following definitions apply:

GTO sleeving — a flexible polymeric sleeve intended to enclose luminous tube sign GTO cable operating at not more than 7500 volts-to-ground and intended to be installed within an approved raceway.

Neon supply — a transformer or electronic power supply intended to operate high-voltage luminous discharge tubing.

Sign — an assembly consisting of electrical parts designed to attract attention by illumination, animation, or other electrical means, singly or in combination.

General requirements

34-100 Disconnecting means
Each sign and outline lighting system installation, other than the portable type, shall be provided with a disconnecting means that shall
(a) open all ungrounded conductors;
(b) be suitable for conditions of installation such as exposure to weather;
(c) be integral with the sign or outline lighting, or be located within sight and within 9 m of the sign or outline lighting installation; and
(d) be capable of being locked in the open position where it is located out of the line of sight or more than 9 m from the sign.

34-102 Rating of disconnecting means and control devices
Switches, flashers, and similar devices controlling neon supplies, transformers, and ballasts shall be either of a type approved for the purpose or have a current rating not less than twice the current rating of the neon supply, transformer, or ballasts.

34-104 Thermal protection
Ballasts of the thermally protected type shall be required for all signs and outline lighting that employ fluorescent lamps, except where the ballasts are of the simple reactance type.

34-106 Location
(1) Signs and outline lighting systems shall be located so that
  - (a) any person working on them is not likely to come into contact with overhead conductors;
  - (b) no part of the sign or outline lighting or its support will interfere with normal work operations performed on electrical and communication utility lines as defined by the utility;
  - (c) no part of the sign or outline lighting or its support is in such proximity to overhead conductors as to constitute a hazard; and
  - (d) no part of the sign or outline lighting, other than its support, is less than 2.2 m above grade.

(2) Notwithstanding Subrule (1)(d), freestanding signs, indoor signs, and outline lighting including installations in show windows and similar locations shall be permitted to be mounted with electrical components less than 2.2 m above grade where approved for the location or where mechanical protection is provided to prevent persons or vehicles from coming into contact with the electrical components of the sign.

34-108 Supporting means
(1) Poles, masts, standards, or devices designed as supports that are for use as electrical raceways shall be approved for the purpose.
(2) The devices referred to in Subrule (1), when used for mechanical support only, shall be suitable.

34-110 Bonding

(1) Except as permitted in Subrules (2) and (3), all conductive non-current-carrying parts of a sign or outline lighting installation, as well as non-electrical equipment to which the sign is mounted, shall be bonded to ground in accordance with the requirements of Section 10.

(2) Small metal parts not exceeding 50 mm in any direction, not likely to be energized, and spaced at least 19 mm from neon tubing shall not require bonding.

(3) Metal wire ties used to secure neon tubing supports shall not require bonding.

34-112 Protection of sign leads

Sign leads that pass through the surfaces or partitions of the sign structure shall be protected by non-combustible moisture-absorption-resisting bushings.

34-114 Fuseholders and flashers

Fuseholders, flashers, etc., shall be enclosed in suitable electrical enclosures, unless they form part of an approved assembly, and shall be accessible without the necessity of removing obstructions or otherwise dismantling the sign.

Enclosures

34-200 Enclosures (see Appendix B)

Neon supplies, switches, timers, relays, sequencing units, and other similar devices shall be enclosed in suitable electrical enclosures unless they form part of an approved assembly.

34-202 Protection of uninsulated parts

Doors or covers accessible to unauthorized persons that give access to uninsulated parts of signs or outline lighting shall be either provided with interlock switches that, on the opening of the doors or covers, disconnect the primary circuit, or shall be fastened so that the use of other than ordinary tools will be necessary to open them.

Neon supplies

Δ 34-300 Maximum secondary voltage for neon supplies

The rated secondary open circuit voltage of a neon supply shall not exceed 15 000 V and shall not exceed 7500 volts-to-ground.

Δ 34-302 Secondary-circuit ground-fault protection

Neon supplies other than the following types shall have secondary-circuit ground-fault protection:

(a) transformers with isolated secondaries and with a maximum open circuit voltage of 6000 V or less between any combination of leads or terminals; and

(b) transformers with integral porcelain or glass secondary housing for neon tubing and requiring no field wiring of the secondary circuit.

34-304 Open-type neon supplies

Open-type neon supplies, such as a core-and-coil type transformer, shall be used only in dry locations.

34-306 Neon supplies for damp or wet locations

(1) Neon supplies used in damp locations shall be the damp or wet type.

(2) Neon supplies used in wet locations shall be the wet type.

(3) Neon supplies installed in a sign body, sign enclosure, or separate enclosure shall be the damp or wet type.

(4) Neon supplies installed in a location where protected from direct exposure to water and the weather by a building structure shall be the damp or wet type.

34-308 Neon supply installation

(1) Neon supplies shall be installed in locations such that they are accessible and capable of being removed and replaced.

(2) Where that location is in an attic, bulkhead, or similar locations, there shall be an access door not less than 900 mm × 600 mm and a passageway not less than 900 mm high by 600 mm wide, with a suitable permanent walkway not less than 300 mm wide extending from the point of entry to each component.

(3) Neon supplies shall be rigidly secured to the enclosure in which they are housed in a manner to prevent rotation, and the enclosure shall be rigidly secured to structural members.

34-310 Neon supply overcurrent protection

(1)   Each neon supply shall be protected by an overcurrent device, rated at a maximum of 30 A, except that two or more neon supplies may be protected by one overcurrent device, provided that the load does not exceed that prescribed by Rule 8-104.

(2)   Where additional overcurrent devices for the individual protection of neon supplies in signs are used, they may be placed either inside or outside the sign structure.

(3)   Where exposed to the weather, overcurrent devices protecting neon supplies shall be the weatherproof type.

34-312 High-voltage output connection

The high-voltage outputs of neon supplies shall not be connected in parallel nor in series with the output of any other neon supply.

Wiring methods

34-400 High-voltage wiring methods

(1)   High-voltage cables shall be installed in
   (a)   neon supply enclosures;
   (b)   sign enclosures;
   (c)   flexible metal conduit 16 trade size or larger;
   (d)   flexible non-metallic conduit;
   (e)   rigid conduit; or
   (f)   except for surface raceways and ENT, all other types of raceways that are approved for the purpose.

(2)   Notwithstanding Rule 12-1302(3)(c), high-voltage cables shall be permitted to be installed in liquid-tight flexible conduit 16 or larger trade size with compatible connectors in lengths required but not exceeding that permitted by Rule 34-404.

(3)   In a midpoint-return connected sign, the cables from the ends of gas-tubes to the neon supply's midpoint-return shall be high-voltage cables rated for the maximum voltage in the output circuit.

(4)   There shall be no sharp bends in high-voltage cables and the bends shall have radii no less than specified in Table 15.

(5)   Where high-voltage cables are installed in non-metallic conduit, the separation of the conduit from conducting or combustible material shall be
   (a)   at least 38 mm for installations operating at 100 Hz or less; or
   (b)   at least 44 mm for installations operating at more than 100 Hz.

Δ (6)   Notwithstanding Subrule (1), where cable used for high-voltage wiring of signs and outline lighting is exposed, the cable shall
   (a)   be run inside GTO sleeving to a point at least 50 mm inside the raceway (where a raceway is provided); and
   (b)   be spaced at least 38 mm from conducting or combustible material for installations operating at 100 Hz or less or at least 44 mm for installations operating at more than 100 Hz; and
   (c)   not exceed 300 mm in length.

(7)   Secondary wiring for field wired signs and outline lighting shall be a minimum of No. 18 AWG.

(8)   Only one high-voltage conductor shall be installed in a conduit.

(9)   Where high-voltage cable enters or leaves conduit in a damp or wet location, the penetration shall be made watertight.

34-402 High-voltage cables in show windows and similar locations

Where high-voltage cables used with signs hang freely in the air and are not enclosed in raceways, as in show windows and similar locations, they shall

(a)   be enclosed in approved GTO sleeving;

(b)   have a separation of at least 38 mm from combustible and conducting material;

(c)   be located so that they are not susceptible to mechanical damage; and

(d)   not be used to support any part of the sign.

34-404 Length of high-voltage cable from neon supplies

(1)   The length of high-voltage cable from the high-voltage terminal of a neon supply to the first neon tube shall be
   (a)   not more than 6 m when the cable is installed in metal raceway; and

     (b)   not more than 16 m when the cable is installed in non-metallic raceway.

(2)   All other sections of high-voltage cable in a neon tubing circuit shall be as short as practicable.

34-406 Connections of high-voltage cables

Connections of high-voltage cables to neon tubing shall be inaccessible to unauthorized persons and made by means of

(a)   an electrode receptacle approved for the location; or

(b)   a connection to the neon tube in an enclosure approved for the location, provided that

     (i)   the insulation of all conductors extends not less than 100 mm beyond the raceway for damp or wet locations; or

     (ii)   the insulation of all conductors extends not less than 65 mm beyond the raceway for dry locations.

34-408 Bonding of metal electrode assembly housing and metal parts

(1)   Flexible metal conduit and liquid-tight flexible metal conduit used to enclose the high-voltage cable between an electrode receptacle assembly and a neon supply or between one electrode receptacle assembly and another shall be permitted to serve as the bonding means for the metal electrode receptacle assembly, provided that the conduit terminates in a connector suitable for ensuring a secure bonding connection.

Δ (2)   Where non-metallic conduit is used to enclose high-voltage cables, the bonding conductor required to bond metal electrode receptacle assemblies, metal parts of a sign, or other metal to which the sign is mounted shall be installed exterior to the non-metallic conduit

     (a)   at least 38 mm from the conduit for installations operating at 100 Hz or less; or

     (b)   at least 44 mm from the conduit for installations operating at more than 100 Hz; and

     (c)   not smaller than No. 12 AWG.

Section 36 — High-voltage installations

General

36-000 Scope (see Appendix B)

(1) This Section applies to installations operating at voltages in excess of 750 V.

(2) The supply authority and the inspection department must be consulted before proceeding with any such installation.

(3) This Section supplements or amends the applicable general requirements of this Code for installations operating at voltages of 750 V or less.

(4) This Section does not affect construction details of factory-fabricated assemblies approved under the *Canadian Electrical Code, Part II*.

36-002 Special terminology

In this Section, the following definitions apply:

Boundary fence — a fence forming the boundary of a property or area, but not part of a station fence enclosure.

Ground grid conductor — the horizontally buried conductor used for interconnecting ground rods or similar equipment that form the station ground electrode.

Maximum ground fault current — the magnitude of the greatest fault current that may flow between the grounding grid and the surrounding earth during the life of the installation.

Potential rise of ground grid — the product of the ground grid resistance and the maximum ground fault current that flows between the station ground grid and the remote earth.

Station — an assemblage of equipment at one place, including any necessary housing, for the conversion or transformation of electrical energy and for connection between two or more circuits.

Step voltage — the potential difference between two points on the earth's surface separated by a distance of one pace, assumed to be 1 m in the direction of maximum voltage gradient.

Touch voltage — the potential difference between a grounded metal structure and a point on the earth's surface separated by a distance equal to normal maximum horizontal reach.

36-004 Guarding

Live parts of electrical equipment shall be accessible to authorized persons only.

36-006 Warning notices (see Appendix B)

Δ (1) A permanent, legible warning notice carrying the wording "DANGER — HIGH VOLTAGE" or "DANGER...V" shall be placed in a conspicuous position

    (a) at electrical equipment vaults, electrical equipment rooms, areas, or enclosures;

    (b) on all high-voltage conduits and cables at points of access to conductors;

    (c) on all cable trays containing high-voltage conductors with the maximum spacing of warning notices not to exceed 10 m;

    (d) on exposed portions of all high-voltage cables at a spacing not to exceed 10 m; and

    (e) on a station fence required by Rule 26-010

        (i) located immediately adjacent to the locks on all access gates;

        (ii) installed at all outside corners formed by the fence perimeter; and

        (iii) installed at intervals not exceeding 15 m of horizontal distance.

(2) Permanent legible signs shall be installed at isolating equipment and shall warn against operating that equipment while carrying current, unless the equipment is interlocked so that it cannot be operated under load.

(3) Suitable warning signs shall be erected in a conspicuous place adjacent to fuses and shall warn operators not to replace fuses while the supply circuit is energized.

(4) Where the possibility of feedback exists,

    (a) each group-operated isolating switch or disconnecting means shall bear a warning notice to the effect that contacts on either side of the device may be energized; and

    (b)   a permanent, legible, single-line diagram of the station switching arrangement, clearly identifying each point of connection to the high-voltage section, shall be provided in a conspicuous location within sight of each point of connection.

(5)   Where metal enclosed switchgear is installed,

    (a)   a permanent, legible, single-line diagram of the switchgear shall be provided in a conspicuous location within sight of the switchgear and this diagram shall clearly identify interlocks, isolation means, and all possible sources of voltage to the installation under normal or emergency conditions, including all equipment contained in each cubicle, and the marking on the switchgear shall cross-reference the diagram;

    (b)   permanent, legible signs shall be installed on panels or doors that give access to live parts, warning of the danger of opening while energized;

    (c)   where the panel gives access only to parts that can be de-energized and visibly isolated by the supply authority, the warning shall add that access is limited to the supply authority or following an authorization of the supply authority;

    (d)   notwithstanding Item (a), where the equipment consists solely of a single cubicle or metal-enclosed unit substation containing only one set of high-voltage switching devices, diagrams are not required.

Wiring methods

36-100 Conductors (see Appendix B)

(1)   Bare conductors or insulated conductors not enclosed in grounded metal shall be used only

    (a)   outdoors;

    (b)   in electrical equipment vaults constructed in accordance with Rules 26-350 to 26-356;

    (c)   in cable trays in accordance with Subrule (2)(d); or

    (d)   in electrical equipment rooms accessible only to authorized persons.

(2)   Except as permitted in Subrule (1)(b), (c), and (d), conductors used indoors or attached to buildings outdoors shall be as follows:

    (a)   installed in metal conduit;

    (b)   electrical metallic tubing;

    (c)   metal enclosed busways;

    (d)   cables having a continuous metal sheath, steel wire armour, or of the interlocking armour type; or

    (e)   Type TC tray cable installed in cable tray in accordance with Rule 12-2202.

(3)   High-voltage type TC cables shall not be installed in the same cable tray with low-voltage conductors, except where the high-voltage TC cables are separated from the low-voltage conductors by a barrier of sheet metal not less than 1.34 mm thick (No. 16 MSG).

(4)   The location of conductors encased or embedded in concrete or masonry shall be indicated by permanent markers set in the walls, floors, or ceilings at intervals of not more than 3 m.

(5)   Where the coverings are of a conductive nature they shall be stripped back from the terminals sufficiently to prevent leakage of current.

(6)   Service conductors shall have a mechanical strength not less than that of No. 6 AWG hard drawn copper.

36-102 Radii of bends

The minimum bending radii measured at the innermost surface of the bend for permanent training of cables during installation shall be as shown in Table 15.

36-104 Shielding of thermoset insulated conductors (see Appendix B)

(1)   Except as permitted in Subrules (2), (3), and (4), shielding shall be provided over the thermoset insulation of each permanently installed conductor with or without fibrous covering or non-metallic jacket, operating at circuit voltages above 2000 V phase-to-phase.

(2)   Shielding need not be provided for conductors having thermoset insulation where they are run underground in raceways or directly buried in the soil and operating at circuit voltages not exceeding 3000 V phase-to-phase, provided that the insulation or the non-metallic jacket, if provided, is the ozone- and discharge-resistant type.

(3)   Shielding need not be provided for conductors having thermoset insulation where the circuit voltage does not exceed 5000 V phase-to-phase, where the conductors are installed on insulators and bound together, in electrical equipment rooms, electrical equipment vaults, metal-enclosed switchgear assemblies, and similar permanently dry locations where the conductor run does not exceed 15 m.

(4)   Shielding need not be provided for conductors having thermoset insulations that are
    (a)   intended for operation at not more than 5000 V phase-to-phase;
    (b)   intended and installed for permanent duty; and
    (c)   provided in either single- or multi-conductor cable construction with
        (i)   a metal sheath;
        (ii)   metal armour of the interlocking type, the wire type, or the flat-tape type; or
        (iii)   totally enclosed metal raceways where installed above ground in dry locations.
(5)   Subject to Rule 10-304, metal sheaths, shielding, armour, conduit, and fittings shall be bonded together and connected to ground.

36-106 Supporting of exposed conductors

Bare conductors and insulated conductors, unless enclosed in or in contact with grounded metal, shall be mounted on suitable insulating supports capable of withstanding the short-circuit stresses liable to be imposed by the supply system.

36-108 Spacing of exposed conductors

(1)   Bare conductors, insulated conductors, and other bare live parts, unless enclosed in or in contact with grounded metal, other than those within or at the point of connection to apparatus or devices, shall be spaced to provide a clearance under all operating conditions, in accordance with Tables 30 and 31, between
    (a)   live parts of opposite polarity; and
    (b)   live parts and all other structural parts other than the conductor supports.
(2)   Where the conductors mentioned in Subrule (1) are connected to apparatus or devices having terminal spaces less than those shown in Tables 30 and 31, the conductors shall be spread out to attain the required spacings at the first point of support beyond such terminals.

36-110 Guarding of live parts and exposed conductors

(1)   Bare conductors, insulated conductors unless enclosed in or in contact with grounded metal, and other bare live parts shall be
    (a)   accessible only to authorized persons; and
    (b)   isolated by elevation or by barriers.
(2)   Where the conductors or live parts mentioned in Subrule (1) are isolated by elevation, the elevations and clearances maintained shall be as those specified in Tables 32, 33, and 34, except that
    (a)   for voltages in excess of those specified in Tables 32, 33, and 34, the elevations and clearances maintained shall be in accordance with the requirements of CAN/CSA-C22.3 No. 1; and
    (b)   for conductors crossing highways, railways, communication lines, and other locations not covered in this Code, the elevations and clearances maintained shall be in accordance with the requirements of CAN/CSA-C22.3 No. 1, or the applicable standard, whichever is greater.
(3)   For a given span, clearances specified in Table 34 shall be increased by 1% of the amount by which the span exceeds 50 m.

36-112 Terminating facilities

Suitable terminating facilities shall be provided to protect cables from harm due to moisture or mechanical damage.

36-114 Joints in sheathed conductors or cables

(1)   Splices or taps in sheathed conductors or cables shall have the conductor or cable covered with insulation and shall have shielding, when used, electrically and mechanically equivalent to that on the conductors or cables joined.
(2)   For conductors or cables having a metal or conducting sheath, provision shall be made for continuity of the sheath over the splice or tap, unless the joint is made in a suitable splicing box that maintains the continuity of the bonding path.

36-116 Elevator shafts

(1)   High-voltage conductors shall not be installed in elevator shafts.
(2)   The conductors shall be permitted to be installed in conduit embedded in the masonry walls of the hoistway but the conduit shall be surrounded throughout the entire length of its run by not less than 50 mm of masonry or concrete.

Control and protective equipment

36-200 Service equipment location
Service equipment shall be installed in a location that is in compliance with the requirements of the supply authority and, in the case of a building, shall be at the point of service entrance.

36-202 Rating and capacity
The type and ratings of circuit breakers, fuses, and switches, including the trip settings of circuit breakers and the interrupting capacity of overcurrent devices, shall be
(a)   in compliance with Rule 14-012(a) and (b);
(b)   in compliance with the requirements of the supply authority for consumer's service equipment; and
(c)   sized in accordance with the appropriate Rules of this Code for transformers, capacitors, motors, and other electrical equipment.

36-204 Overcurrent protection
(1)   Each consumer's service, operating unit of apparatus, feeder, and branch circuit shall be provided with overcurrent protection having adequate rating and interrupting capacity in all ungrounded conductors by one of the following:
    (a)   a circuit breaker;
    (b)   fuses preceded by a group-operated visible break load-interrupting device capable of making and interrupting its full load rating and that may be closed with safety to the operator with a fault on the system; or
    (c)   fuses, preceded by a group-operated visible break air-break switch that is capable of interrupting the magnetizing current of the transformer installation, that may be closed with safety to the operator with a fault on the system, and that, to prevent its operation under load, is interlocked with the transformer's secondary load interrupting device.
(2)   Fuses shall be accessible to authorized persons only.

36-206 Indoor installation of circuit breakers, switches, and fuses
(1)   Circuit breakers, switches, and fuses installed indoors shall be an enclosed type unless installed in a room of non-combustible construction.
(2)   In addition to the requirements of Subrule (1), dielectric liquid-filled equipment located indoors shall be installed in accordance with Rules 26-012 and 26-246.

36-208 Interlocking of fuse compartments
Compartments containing fuses shall have the cover (or door) interlocked with the isolating or disconnecting means so that
(a)   access cannot be had to the fuses unless the isolating or disconnecting means immediately ahead of the fuses is in the de-energized position; and
(b)   the switch cannot be placed in the closed position until the fuse compartment has been closed.

36-210 Protection and control of instrument transformers
(1)   Instrument voltage transformers shall have overcurrent protection as required by Rule 26-260.
(2)   A suitable disconnecting means shall be provided on the supply side of fuses used for the protection of instrument voltage transformers.

36-212 Outdoor installations
(1)   High-voltage switches not of the metal-enclosed type that are assembled in the field shall be spaced according to Table 35.
(2)   Horn-gap switches shall be mounted in a horizontal position and be capable of being locked in the open position.
(3)   High-voltage fuses shall be spaced according to Table 35.

36-214 Disconnecting means
(1)   Where conductors fed from a station enter a building, either
    (a)   a load-breaking device shall be installed indoors at the entry of the conductors to the building; or
    (b)   a load-breaking device at the supply station shall be capable of being tripped or operated from within the building.

(2) Unless of the draw-out type, each circuit breaker and each load-break switch having contacts that are not visible for inspection in both the open and closed positions shall be provided with a group-operated isolating switch on the supply side that shall be

    (a) provided with the means for adequate visible inspection of all contacts in both the open and closed position;

    (b) interlocked so that it cannot be operated under load; and

    (c) provided with positive position indicators.

(3) Where more than one source of voltage exists in a station consisting of two or more interconnected sections operating at high-voltage or where there is another possibility of feedback, a visible point of connection, meeting the requirements of Subrule (2), shall be provided in all circuits where the possibility of feedback between sections exists.

Grounding and bonding

36-300 Material and minimum size of grounding conductor and ground grid conductor and connections (see Appendix B)

(1) Except as provided for in Subrule (2), bare copper conductors shall be used for grounding purposes and shall be not smaller than those specified in Rules 36-302 to 36-310 and Table 51.

Δ (2) Notwithstanding the requirement of Subrule (1), a galvanized steel, copper-weld, or other conductor shall be permitted for grounding purposes, provided that

    (a) its current-carrying rating is equal to or greater than that of the copper conductor specified in Rules 36-302 to 36-310;

    (b) consideration is given to galvanic action if such conductors are buried in the ground or come in contact with dissimilar metals; and

    (c) the method of bolting or connecting such conductors to each other and to other surfaces is such as to maintain the required current-carrying capacity for the life of the electrode design.

36-302 Station ground electrode (see Appendix B)

(1) Every outdoor station shall be grounded by means of a station ground electrode that shall meet the requirements of Rule 36-304 and shall

    (a) consist of a minimum of four driven ground rods not less than 3 m long and 19.0 mm in diameter spaced at least the rod length apart and, where practicable, located adjacent to the equipment to be grounded;

    (b) have the ground rods interconnected by ground grid conductors not less than No. 2/0 AWG bare copper buried to a maximum depth of 600 mm below the rough station grade and a minimum depth of 150 mm below the finished station grade; and

    (c) have the station ground grid conductors in Item (b) connected to all non-current-carrying metal parts of equipment and structures and shall form a loop around the equipment to be grounded, except that

        (i) a portion of the loop shall be permitted to be omitted where an obstacle such as a wall prevents a person from standing on the corresponding side or sides of the equipment; and

        (ii) loops formed by the rebar in a reinforced concrete slab are adequate when the rebar members are interconnected and reliably connected to all other parts of the station ground electrode.

(2) Where a deviation has been allowed in accordance with Rule 2-030, a buried station ground electrode other than that described in Subrule (1) shall be permitted to be used.

(3) Where it is not practicable to locate the station ground electrode adjacent to the station as described in Subrule (1), the station ground electrode shall be permitted to be remote from the station, and

    (a) two grounding conductors of a minimum of No. 2/0 AWG copper shall connect the ground electrode to the station equipment in such a way that should one grounding conductor or ground electrode be damaged, no single metal structure or equipment frame may become isolated; and

    (b) in locations with system short-circuit currents exceeding 30 000 A, the grounding conductor wire size shall be increased and shall be such that it will not suffer thermal damage or be a fire hazard under the severest fault conditions occurring on the system.

(4) Every indoor station shall be grounded by means of a station ground electrode

    (a) in accordance with Subrule (1), (2), or (3); or

(b)   if it is not practicable to ground an indoor station in accordance with Subrule (1), (2), or (3) and the indoor station receives its supply from a main station on the same property, the station equipment shall be connected to the main station ground electrode in accordance with Subrule (3).

(5)   All parts of the indoor station that are required to be grounded shall be connected together by copper conductors of not less than No. 2/0 AWG.

(6)   The reinforcing steel members to be found in building foundations and concrete platforms shall be permitted to be included as part of the station ground electrode design, provided that

(a)   no insulating film separates the concrete from the surrounding soil;

(b)   the maximum expected fault current magnitude and duration will not result in thermal damage to the steel members or the concrete structure; and

(c)   the steel members are connected to the rest of the station ground electrode with not less than 2 copper conductors of not less than No. 2/0 AWG in such a way that, should one grounding conductor or ground electrode be damaged, no single metal structure or equipment frame may become isolated; and

(d)   the ground electrode design is made assuming that the concrete resistivity is greater than or equal to that of the surrounding soil.

36-304 Station ground resistance (see Appendix B)

(1)   The maximum permissible resistance of the station ground electrode shall be determined by the maximum available ground fault current injected into the ground by the station ground electrode or by the maximum fault current in the station, and the ground resistance shall be such that under all soil conditions that exist in practice (e.g., wet, dry, and frozen conditions), the maximum ground fault current conditions shall limit the potential rise of all parts of the station ground grid to 5000 V; whereas in special circumstances where this level cannot be reasonably achieved, a higher voltage up to the maximum insulation level of the communication equipment shall be permitted where a deviation has been allowed in accordance with Rule 2-030.

(2)   In addition to the requirements of Subrule (1), the touch and step voltage at the edge, within, and around the station grounding electrode, including all areas in which metallic structures electrically connected to the station are to be found, shall not exceed the tolerable values specified in Table 52.

(3)   When a station ground electrode design is selected according to the procedure delineated in Appendix B and when it is proven that the station parameters used in the procedure are valid, this electrode design shall be deemed to meet the requirements of Subrules (1) and (2).

(4)   After completion of construction, the resistance of the station ground electrode at each station shall be measured and changes shall be made if necessary to verify and ensure that the maximum permissible resistance of Subrule (1) is not exceeded.

(5)   Where the safety of persons is dependent upon the integral presence of a ground surface covering layer, such as crushed rock or asphalt, the ground surface covering layer shall exist throughout the station grounding electrode area, including all areas in which metallic structures electrically connected to the station are to be found and shall extend at least 1 m beyond the station grounding electrode area on all sides.

36-306 Station exemption

Where the phase-to-phase voltage is less than or equal to 7500 V and a ground surface covering layer with a minimum thickness of 150 mm is installed and maintained as specified in Rule 36-304(5) and it can be demonstrated that the potential rise (GPR) of a station shall not exceed the tolerable touch and step voltages specified in Table 52 during the lifetime of the station, the following exemptions shall apply:

(a)   no soil resistivity measurements need be made at the station site;

(b)   notwithstanding Rule 36-304(2), no analysis is required to prove that touch and step voltages within the station grounding electrode area shall not exceed tolerable values; and

(c)   notwithstanding Rule 36-304(4), neither the resistance of the station ground electrode nor the touch voltage near the centre or corner of the ground electrode need be measured after completion of construction.

36-308 Connections to the station ground electrode (see Appendix B)

(1)   All non-current-carrying metal equipment and structures forming part of the station shall be grounded to the station ground electrode to prevent the buildup of dangerous potential differences between the equipment or structures and the nearby earth.

(2) All metal items forming part of the station shall be connected to the station ground electrode as follows:

    (a) metal structures:

        (i) single columns or pedestal-type (pipe, etc.) structures shall be grounded by a grounding conductor not less than No. 2/0 AWG copper; and

        (ii) single and multi-bay structures shall be bonded to ground at each column by a bonding conductor not less than No. 2/0 AWG copper;

    (b) apparatus mounted on metallic or non-metallic structures:

        (i) tanks or frames of transformers, generators, motors, circuit breakers, reclosers, instrument transformers, switchgear, and other equipment shall be grounded by grounding conductors of not less than No. 2/0 AWG copper;

        (ii) metal bases of all gang-operated switches shall be grounded by a grounding conductor of not less than No. 2/0 AWG copper (for switch handles see Rule 36-310); and

        (iii) the grounding of metal bases of single-pole fuse cut-outs and isolating switches on wood structures is optional;

    (c) lightning arresters:

        (i) the lightning arresters shall be connected to the station ground electrode by a conductor of not less than No. 2/0 AWG copper;

        (ii) lightning arrester grounding conductors shall be as short, straight, and direct as practicable; and

        (iii) where lightning arresters are for the protection of high-voltage cable and cable sheath, the lightning arrester grounding conductor shall be connected to metal potheads and/or metal sheath or armour or shielding of all cables;

    (d) a metal water main inside or adjacent to the station ground electrode area shall be grounded by at least one copper conductor of not less than No. 2/0 AWG copper, at intervals not exceeding 12 m;

    (e) the non-current-carrying parts of metal equipment, such as

        (i) cable sheaths, cable armour, shield, ground wires, potheads, raceways, pipe work, screen guards, and switchboards, shall be grounded by copper conductor of not less than No. 4 AWG;

        (ii) meter, instrument, and relay cases, when mounted on insulated panels, shall be grounded by a copper conductor of not less than No. 10 AWG;

        (iii) the metal frame and all exposed metal work on buildings within or forming part of the station, shall be grounded to the station ground electrode by a minimum of No. 2/0 AWG copper in at least two places and at intervals not exceeding 12 m along the building perimeter;

    (f) steel rails of railway spur tracks entering an outdoor station ground electrode area by a copper conductor of not less than No. 2/0 AWG with the part of the spur track located outside the station ground electrode area properly isolated from the station ground electrode or grounded or both in order that touch voltages along the track not exceed the tolerable values specified in Table 52.

(3) Where it is proven that touch and step potentials around a building shall not exceed the tolerable values specified in Table 52, no loop need be installed around the building.

(4) A transmission line overhead ground wire shall be connected to the station ground electrode with a grounding conductor of not less than No. 2/0 AWG copper that, notwithstanding Rule 36-300(1), shall be permitted to be insulated.

(5) A line neutral conductor on grounded neutral systems shall be connected to the station ground electrode by a grounding conductor having an ampacity not less than the neutral conductor.

(6) A transformer neutral on grounded systems shall be connected to the station ground electrode by a copper grounding conductor of not less than No. 2/0 AWG that shall also have sufficient ampacity to carry the maximum ground fault current of the transformer in accordance with Table 51, and this grounding conductor shall be in addition to the requirement of Subrule (2)(b)(i).

(7) Connections to the items in Subrules (2)(d), (4), and (5) shall be made through removable connectors that will permit isolation from the station ground electrode for the purpose of station ground grid resistance measurement.

36-310 Gang-operated switch handle grounds

(1) The operating handle of all gang-operated switches not enclosed in metal housings shall be grounded by one of the following methods:

    (a) an approved, multi-revolution grounding device shall be connected to the station ground electrode by a conductor having a current-carrying capacity of not less than No. 2/0 AWG copper; or

(b) the operating shaft shall be grounded to the station ground electrode by a combination of extra-flexible conductor, braid, and/or stranded conductor of not less than No. 2/0 AWG copper.

(2) In addition to the requirements of Subrule (1), the touch voltage shall be maintained at a tolerable level as specified in Table 52 at the location where the operator is normally standing and shall be done as follows:

(a) by the use of a metallic gradient control mat connected to the operating handle grounding conductor as required in Subrule (1) by two separate conductors, each not less than No. 2/0 AWG copper; and

(b) the gradient control mat shall

(i) be positioned so that the operator will not be required to step from the mat during the operation of the switch;

(ii) be placed on a minimum of 150 mm of crushed stone on the ground;

(iii) have dimensions approximately 1.2 m × 1.8 m; and

(iv) be permitted to be covered by a layer of crushed stone, asphalt, or concrete not exceeding 150 mm in depth.

36-312 Grounding of metallic fence enclosures of outdoor stations

(1) The fence shall be located at least 1 m inside the perimeter of the station ground electrode area.

(2) The station ground electrode shall be connected to the fence by a tap conductor at each end post, corner post, and gate post, and at intermediate posts at intervals not exceeding 12 m by a conductor of not less than No. 2/0 AWG copper.

(3) The tap conductor at each hinge gate post shall be clamped or bonded to the gate frame by a copper braid or a flexible copper conductor of at least No. 2/0 AWG.

(4) The tap conductor shall be connected to the fence post, the bottom tension wire, the fence fabric (for which the conductor may be woven in at least two places), the top rail, and each strand of barbed wire, with the connection to the bottom tension wire, the fence fabric, and barbed wire strands made with bolted or equivalent connectors, and with the top rail connections bonded at every joint by a jumper equivalent to No. 2/0 AWG copper.

(5) When there is a metal boundary fence in proximity to the station fence, the touch voltages within 1 m of all parts of the boundary fence shall not exceed the tolerable values specified in Table 52.

Section 38 — Elevators, dumbwaiters, material lifts, escalators, moving walks, lifts for persons with physical disabilities, and similar equipment

38-001 Scope (see Appendix B)

This Section applies to the installation of electrical equipment and wiring for elevators, dumbwaiters, material lifts, escalators, moving walks, lifts for persons with physical disabilities, and similar equipment, and supplements or amends the general requirements of this Code.

38-002 Special terminology (see Appendix B)

In this Section, the following definitions apply:

Motor controller — the operative units of the control system comprising the starter device(s) and power conversion equipment used to drive an electric motor, or the pumping unit used to power hydraulic control equipment.

Operating device — the car switch, push buttons, key or toggle switch(es), or other devices used to activate the controller.

Signal equipment — audible and visual equipment such as chimes, gongs, lights, and displays that convey information to the user.

38-003 Voltage limitations (see Appendix B)

The circuit voltage shall not exceed 300 V unless otherwise permitted in Items (a) to (c):

(a)   branch circuits to door operator controllers and door motors, and branch circuits and feeders to motor controllers, driving machine motors, machine brakes, and motor-generator sets shall have a circuit voltage not in excess of 750 V;

(b)   branch lighting circuits shall comply with the requirements of Section 30;

(c)   branch circuits for heating and air-conditioning equipment located on the car shall not have a circuit voltage in excess of 750 V.

38-004 Live parts enclosed

All live parts of electrical apparatus in hoistways, at the landings or in or on the cars of elevators, dumbwaiters, material lifts, and lifts for persons with physical disabilities, or in the wellways or at the landings of escalators or moving walks, shall be enclosed to protect against accidental contact.

38-005 Working clearances

(1)   The minimum headroom in working spaces around controllers, disconnecting means, and other electrical equipment shall be 2000 mm.

Δ (2)   The working space requirements of Rules 2-308 and 38-005(1) need not apply where conditions of maintenance and supervision ensure that only authorized persons have access to such areas, and where

(a)   equipment in Items (i) to (iv) is equipped with flexible cables to all external connections to allow its repositioning for compliance with the working space requirements of Rule 2-308:

(i)   controllers and disconnecting means for dumbwaiters, escalators, moving walks, material lifts, and lifts for persons with physical disabilities installed in the same space with the driving machine;

(ii)   controllers and disconnecting means for elevators, installed in the hoistway or on the car;

(iii)   controllers for door operators; and

(iv)   other electrical equipment installed in the hoistway or on the car;

(b)   live parts of the equipment are suitably guarded, isolated, or insulated, and the equipment can be examined, adjusted, serviced, or maintained while energized without removal of this protection;

(c)   electrical equipment is not required to be examined, adjusted, serviced, or maintained while energized; or

(d)   uninsulated parts are extra-low voltage or do not exceed 60 V dc.

38-011 Insulation and types of conductors

(1)   Conductors in hoistways, in or on cars or platforms, in wellways, and in machine rooms shall be the types listed in Table 11 or 19 and shall meet the requirements of Rule 2-126.

(2)   The conductors to the hoistway door interlocks from the hoistway riser shall meet the requirements of Rule 2-126.

Section 38
Elevators, dumbwaiters, material lifts, escalators, moving walks,
lifts for persons with physical disabilities, and similar equipment

(3)   The voltage rating of insulation of all conductors shall be suitable for the voltage to which the conductors are subjected and shall have an insulation voltage at least equal to the maximum nominal circuit voltage applied to any conductor within the enclosure, cable, or raceway.

(4)   Travelling cables used as flexible connections between the car or counterweight and the raceway shall be the types of elevator cable listed in Table 11 or other types approved for the purpose.

38-012 Minimum size of conductors

(1)   In travelling cables, the minimum size conductors shall be

(a)   for lighting circuits, No. 14 AWG copper except that smaller conductors shall be permitted to be used in parallel provided that the ampacity is equivalent to at least that of No. 14 AWG copper; and

(b)   for all operating, control, signal, and extra-low-voltage lighting circuits, No. 20 AWG copper.

(2)   Except as specified in Subrule (1), the minimum size of conductors for operating, control, signal, and communications circuits shall be No. 26 AWG copper.

38-013 Ampacity of feeder and branch-circuit conductors (see Appendices B and G)

(1)   With generator field control, the conductor ampacity shall be based on the nameplate current rating of the driving motor of the motor-generator set that supplies power to the driving machine motor.

(2)   Conductors shall have an ampacity in accordance with Items (a) to (d):

(a)   **Conductors supplying a single motor:** Conductors supplying a single motor shall have an ampacity not less than the percentage of motor nameplate current required by Rule 28-106 and Table 27.

(b)   **Conductors supplying a single motor controller:** Conductors supplying a single motor controller shall have an ampacity not less than the motor controller nameplate current rating, plus all other connected loads.

(c)   **Conductors supplying a single power transformer:** Conductors supplying a single power transformer shall have an ampacity not less than the nameplate current rating of the power transformer, plus all other connected loads.

(d)   **Conductors supplying more than one motor, motor controller, or power transformer:** Conductors supplying more than one motor, motor controller, or power transformer shall have an ampacity not less than the sum of the nameplate current ratings of the equipment plus all other connected loads. The ampere ratings of motors to be used in the summation shall be determined as required by Rule 28-108 and Table 62.

38-014 Feeder demand factor

Feeder conductors of less ampacity than that required by Rule 38-013 shall be permitted subject to the requirements of Table 62.

38-015 Motor controller rating

The motor controller rating shall comply with Rule 28-500(1), except that the rating shall be permitted to be less than the nominal rating of the driving machine motor when the controller inherently limits the available power to the motor and is marked "power limited".

38-021 Wiring methods (see Appendix B)

Elevators

(1)   Unless otherwise permitted in Items (a) through (d), conductors and optical fibers located in hoistways, machinery spaces, control spaces, in or on cars, and in machine rooms and control rooms, not including travelling cables connecting the car or counterweight and hoistway wiring, shall be installed in rigid metal conduit, electrical metallic tubing, rigid PVC conduit, or wireways, except that mineral-insulated cable, aluminum-sheathed cable, or armoured cable shall be permitted if not subject to mechanical damage.

Hoistways

(a)   In hoistways the following wiring methods shall also be permitted if not subjected to mechanical damage:

(i)   flexible metal conduit or liquid-tight flexible conduit shall be permitted in hoistways between risers and limit switches, interlocks, operating devices, or similar devices;

(ii)   cables used in Class 1 extra-low-voltage and Class 2 low-energy circuits, including but not limited to hoistway cable, extra-low-voltage cable, extra-low-voltage control cable, communication cable, fire alarm and signal cable, multi-conductor jacketed thermoplastic-insulated cable, and hard-usage and extra-hard-usage cables shall be permitted to be installed between risers and signal equipment and operating devices, provided that the cables are supported and protected from physical damage and are of a jacketed and flame-tested type;

    (iii)   flexible cords and cables that are components of approved equipment and used in extra-low-voltage circuits (30 V or less) shall be permitted in lengths not exceeding 2 m, provided that the cords and cables are supported and protected from physical damage and are of a jacketed and flame-tested type;

    (iv)   flexible metal conduit, liquid-tight flexible metal conduit, liquid-tight flexible non-metallic conduit, or flexible cords and cables, or conductors grouped together and taped or corded that are part of listed equipment, a driving machine, or a driving machine brake, shall be permitted in the hoistway in lengths not exceeding 2 m without being installed in a raceway and where located to be protected from physical damage and if of a flame-tested type.

Cars

(b)   On cars, the following wiring methods shall also be permitted:

    (i)   flexible metal conduit or liquid-tight flexible conduit not exceeding 2 m in length shall be permitted on cars where located to be free from oil and if securely fastened in place;

    (ii)   extra-hard-usage cords and hard-usage cords (see Table 11) shall be permitted as flexible connections between the fixed wiring on the car and devices on the car doors or gates, and extra-hard-usage cords only shall be permitted as flexible connections for the top-of-car operating device or the car-top work light;

    (iii)   cables with smaller conductors and other types and thickness of insulation and jackets than extra-hard usage or hard usage, used as flexible connections between the fixed wiring on the car and devices on the car doors or gates, shall be permitted as flexible connections between the fixed wiring on the car and devices on the car doors or gates, if approved for extra-hard usage or hard usage;

    (iv)   flexible cords and cables that are components of approved equipment and used in extra-low-voltage circuits (30 V or less) shall be permitted in lengths not exceeding 2 m, provided that the cords and cables are supported and protected from physical damage and are of a jacketed and flame-tested type;

    (v)   flexible metal conduit, liquid-tight flexible metal conduit, liquid-tight flexible non-metallic conduit, or flexible cords and cables, or conductors grouped together and taped or corded that are part of listed equipment, a driving machine, or a driving machine brake, shall be permitted on the car assembly in lengths not to exceed 2 m without being installed in a raceway and where located to be protected from physical damage and if of a flame-tested type.

Within machine rooms, control rooms and machinery spaces, and control spaces

(c)   Within machine rooms, control rooms and machinery spaces, and control spaces, the following wiring methods shall also be permitted:

    (i)   flexible metal conduit or liquid-tight flexible conduit shall be permitted between control panels and machine motors, machine brakes, motor-generator sets, disconnecting means, or pumping unit motors and valves;

    (ii)   where motor-generators, machine motors, or pumping unit motors and valves are located adjacent to or underneath control equipment and are provided with extra-length terminal leads, such leads shall be permitted to be extended to connect directly to controller terminal studs without regard to the current-carrying capacity requirements of Section 28, provided that the conductors are

        (A)   not over 2 m long;

        (B)   bound together and supported at intervals not more than 1 m; and

        (C)   not located where they would be subject to physical damage;

    (iii)   auxiliary gutters shall be permitted in machine and control rooms between controllers, starters, and similar apparatus;

    (iv)   flexible cords and cables that are components of approved equipment and used in extra-low-voltage circuits (30 V or less) shall be permitted in lengths not to exceed 2 m, provided that the cords and cables are supported and protected from physical damage and are of a jacketed and flame-tested type.

Counterweights

(d)   Flexible metal conduit, liquid-tight flexible conduit, or flexible cords and cables, or conductors grouped together and taped or corded that are part of approved equipment, a driving machine, or a driving machine brake, shall be permitted on the counterweight assembly in lengths that do not exceed 2 m

Section 38

Elevators, dumbwaiters, material lifts, escalators, moving walks,
lifts for persons with physical disabilities, and similar equipment

without being installed in a raceway if they are located to be protected from physical damage and are a flame-tested type.

Escalators

(2) Conductors and optical fibers in escalator and moving walk wellways shall be installed in rigid metal conduit, flexible metal conduit, liquid-tight flexible conduit, electrical metallic tubing, rigid PVC conduit, or wireways or shall be mineral-insulated cable, aluminum-sheathed cable, or armoured cable, if not subject to physical damage, unless otherwise permitted in Items (a) and (b):

    (a) cables used in Class 1 extra-low-voltage and Class 2 low-energy circuits, including extra-low-voltage cable, extra-low-voltage control cable, communication cable, fire alarm and signal cable, multi-conductor jacketed thermoplastic-insulated cable, and hard-usage and extra-hard-usage cables shall be permitted to be installed between risers and signal equipment and operating devices, provided that the cables are supported and protected from physical damage and are a jacketed and flame-tested type;

    (b) flexible cords and cables that are components of approved equipment and used in extra-low-voltage circuits (30 V or less) shall be permitted in lengths not exceeding 2 m, provided that the cords and cables are supported and protected from physical damage and are a jacketed and flame-tested type.

Lifts for persons with physical disabilities

(3) Conductors and optical fibers located in hoistways, runways, and machinery spaces and in machine and control rooms of dumbwaiters, material lifts, and lifts for persons with physical disabilities shall be installed in rigid metal conduit, electrical metallic tubing, rigid PVC conduit, or wireways; or, if not subject to physical damage, shall be mineral-insulated cable, aluminum-sheathed cable, armoured cable, flexible metal conduit, or liquid-tight flexible conduit, unless otherwise permitted in Items (a) and (b):

    (a) cables used in Class 1 extra-low-voltage and Class 2 low-energy circuits, including but not limited to hoistway cable, extra-low-voltage cable, extra-low-voltage control cable, communication cable, fire alarm and signal cable, multi-conductor jacketed thermoplastic-insulated cable, and hard-usage and extra-hard-usage cables shall be permitted to be installed between risers and signal equipment and operating devices, provided that the cables are supported and protected from physical damage and are a jacketed and flame-tested type;

    (b) flexible cords and cables that are components of approved equipment and used in extra-low-voltage circuits (30 V or less) shall be permitted in lengths not exceeding 2 m, provided that the cords and cables are supported and protected from physical damage and are a jacketed and flame-tested type.

38-022 Branch circuits for car lighting, receptacles, ventilation, accessories, heating, and air conditioning

(1) At least one branch circuit shall be provided solely for the car lights, receptacles, auxiliary lighting power source, accessories, and ventilation on each car.

(2) Where air-conditioning and heating units are installed on the car, they shall be supplied by separate branch circuits.

(3) The overcurrent device protecting each branch circuit shall be located in the machine room or control room/machinery space or control space.

38-023 Branch circuits for machine room or control room/machinery space or control space lighting and receptacle(s) (see Appendix B)

(1) A separate branch circuit shall supply the machine room or control room/machinery space or control space lighting and receptacle(s).

(2) Required lighting shall not be connected to the load side terminals of a ground fault circuit interrupter.

(3) A machine room or control room/machinery space or control space lighting switch shall be provided and shall be within easy reach of the point of entry.

(4) At least one 125 V, single-phase, duplex receptacle, connected to a 15 A branch circuit, having a configuration in accordance with Diagram 1, shall be provided in each machine room or control room and machinery space or control space.

38-024 Branch circuit for hoistway pit lighting and receptacles (see Appendix B)

(1) A separate branch circuit shall supply the hoistway pit lighting and receptacles.

(2)   Required lighting shall not be connected to the load side terminals of a ground fault circuit interrupter receptacle(s).

(3)   A lighting switch shall be provided and shall be located so as to be readily accessible from the pit access door.

(4)   At least one 125 V, single-phase, duplex receptacle connected to a 15 A branch circuit shall be provided in the hoistway pit.

38-025 Branch circuits for other utilization equipment

(a)   Separate branch circuits shall supply other utilization equipment not identified in Rules 38-022, 38-023, and 38-024, but used in conjunction with equipment identified in Rule 38-001.

(b)   The overcurrent devices protecting the branch circuits shall be located in the machinery room or control room/machinery space or control space.

38-032 Metal wireways and non-metallic wireways

See Rule 12-1014 and Table 8.

38-033 Number of conductors in raceways

See Rule 12-1014.

38-034 Supports

Supports for cables or raceways in a hoistway or in an escalator or moving walk wellway or a hoistway or runway for a material lift or lift for persons with physical disabilities shall be securely fastened to the guide rail, escalator or moving walk truss, or to the hoistway, wellway, or runway construction.

38-035 Auxiliary gutters

See Rules 12-1900, 12-1902, and 12-1904.

38-036 Grouping of conductors

Optical fiber cables, shielded cables, and conductors for operating devices, power, motor, heating, air-conditioning, operating, control, signal, telephone, fire alarm, and lighting circuits shall be permitted to be run in the same raceway system or travelling cable, provided that all conductors are insulated for the maximum voltage found in the cable or raceway system.

38-037 Wiring in hoistways, machine rooms, control rooms, and machinery spaces and control spaces

Unless a deviation has been permitted in accordance with Rule 2-030, only conductors used in connection with operation of the elevator, dumbwaiter, escalator, moving walk, material lift, or lift for persons with physical disabilities, including supply or feeder conductors, wiring for signals, hoistway fire detection, communication with the car, and for lighting and ventilating the car, shall be permitted to be installed inside hoistways, runways, machine rooms, control rooms, machinery spaces and control spaces, or escalator wellways. (See also Rule 12-014.)

38-041 Suspension of travelling cables (see Appendix B)

(1)   Travelling cables shall be suspended at the car and hoistway ends, or counterweight end where applicable, to reduce to a minimum the strain on the individual copper conductors.

(2)   Travelling cables shall be supported by one of the following means:

   (a)   by its steel supporting member(s);

   (b)   by looping the cables around supports for unsupported lengths less than 30 m; or

   (c)   by suspending from the supports by a means that automatically tightens around the cable when tension is increased for unsupported lengths up to 60 m.

38-042 Hazardous locations

All electrical equipment installed in hazardous locations shall comply with Section 18.

38-043 Location of and protection for cables

(1)   Travelling cable supports shall be located to reduce to a minimum the possibility of damage due to the cables coming in contact with the hoistway construction or equipment in the hoistway.

(2)   Where necessary, suitable guards shall be provided to protect the cables against damage.

38-044 Installation of travelling cables

Travelling cable to the car or counterweight shall be permitted to be installed in the hoistway and on the car and counterweight as fixed wiring without the use of conduit or other raceway, provided that it is suitably supported and protected from damage.

Section 38

*Elevators, dumbwaiters, material lifts, escalators, moving walks,
lifts for persons with physical disabilities, and similar equipment*

38-051 Disconnecting means (see Appendix B)

(1)   A single disconnecting means shall be provided for the opening of all ungrounded conductors of each of the following:

(a)   the drive motor and its ventilation and control circuits in each elevator, escalator, dumbwaiter, or lift for persons with physical disabilities operating individually or as one of a group; and

(b)   the branch circuit(s) supplying the lighting and ventilation, heating, and air conditioning in each car, and such circuit(s) shall not be controlled by the disconnecting means described in Item (a).

(2)   Each disconnecting means shall be an externally operated fusible switch or a circuit breaker and shall be equipped with means for locking it in the open position.

(3)   Where circuit breakers are used as a disconnect means, they shall not be opened automatically by a fire alarm system.

(4)   Means shall be provided on the switch or circuit breaker to indicate the disconnected position.

(5)   The disconnecting means shall be located where it is visible on entry to the machinery area and readily accessible to authorized persons.

(6)   When the disconnecting means required by Subrule (1)(a) is not visible from, or is located more than 9 m from the motor controller(s), an additional manually operable motor controller disconnecting switch, whose opening is not solely dependent on springs, shall

(a)   be installed so that it is visible from, or adjacent to, the remote equipment;

(b)   open all ungrounded conductors; and

(c)   be capable of being locked in the open position.

(7)   (a)   Driving machines or controllers other than motor controllers not within sight of the disconnecting means shall be provided with a manually operated switch installed in the control circuit to prevent starting.

(b)   The manually operated switch(es) shall be installed adjacent to this equipment.

(8)   Where there is more than one driving machine in a machine room, the disconnecting means shall be numbered to correspond to the identifying number of the driving machine that it controls.

(9)   The disconnecting means shall be provided with a sign to identify the location of the supply side overcurrent protective device.

(10)   (a)   No provision shall be made to automatically close this disconnecting means.

(b)   Power shall be restored only by manual means.

(11)   The disconnecting means serving an escalator or moving walk controller shall be installed in the same location as the controller.

(12)   Where multiple driving machines are connected to a single elevator, escalator, moving walk, or pumping unit, there shall be one disconnecting means to disconnect the motor(s) and control valve operating magnets.

(13)   Where the driving machine of an electric elevator, dumbwaiter, material lift, or lift for persons with disabilities or the hydraulic machine of a hydraulic elevator, dumbwaiter, material lift, or lift for persons with disabilities is located in a remote machine room or remote machinery space, or the motor-generator set is located in a remote machine room or remote machinery space, a single means for disconnecting all ungrounded main power supply conductors shall be provided that is visible from the machine and capable of being locked in the open position.

38-052 Power from more than one source

(1)   **Single car and multi-car installations:** On single car and multi-car installations, equipment receiving electrical power from more than one source shall be provided with a disconnecting means, within sight of the equipment served, for each source of electrical power.

(2)   **Warning sign for multiple disconnecting means:** Where multiple disconnecting means are used and parts of the controllers remain energized from a source other than the one disconnected, a clearly legible warning sign reading "Warning — Parts of the controller are not de-energized by this switch" shall be mounted on or next to the disconnecting means.

(3)   **Interconnection of multi-car controllers:** Where interconnections between controllers are necessary for the operation of the system on multi-car installations that remain energized from a source other than the one disconnected, a warning sign in accordance with Subrule (2) shall be mounted on or next to the disconnecting means.

38-053 Car light, receptacle(s), and ventilation disconnecting means

(1)   Elevators, dumbwaiters, material lifts, and lifts for persons with physical disabilities shall have a single means for disconnecting all ungrounded car light, receptacle, and ventilation power-supply conductors for that car.

(2)   The disconnecting means shall be an enclosed externally operable fused motor circuit switch or circuit breaker capable of being locked in the open position and shall be located in the machine room or control room for that car, unless there is no machine room or control room, in which case the disconnecting means shall be located in the same space as the disconnecting means required by Rule 38-051.

(3)   The disconnecting means shall be numbered to correspond to the identifying number of the car whose light source it controls.

(4)   The disconnecting means shall be provided with a sign to identify the location of the supply side overcurrent protective device.

38-054 Heating and air-conditioning disconnecting means

(1)   Elevators, dumbwaiters, material lifts, and lifts for persons with physical disabilities shall have a single means for disconnecting all ungrounded car heating and air-conditioning power-supply conductors for that car.

(2)   The disconnecting means shall be an enclosed externally operable fused motor circuit switch or circuit breaker capable of being locked in the open position and shall be located in the machine room or control room for that car, unless there is no machine room, in which case the disconnecting means shall be located in the same space as the disconnecting means required by Rule 38-051.

(3)   The disconnecting means shall be numbered to correspond to the identifying number of the car whose heating and air-conditioning source it controls.

(4)   The disconnecting means shall be provided with a sign to identify the location of the supply side overcurrent protective device.

38-055 Utilization equipment disconnecting means

(1)   Each branch circuit for other utilization equipment (see Rule 38-025) shall have a single means for disconnecting all ungrounded conductors.

(2)   The disconnecting means shall be capable of being locked in the open position and shall be located in the machine room or control room/machine space or control space.

(3)   Where there is more than one branch circuit for other utilization equipment, the disconnecting means shall be numbered to correspond to the identifying number of the equipment served.

(4)   The disconnecting means shall be provided with a sign to identify the location of the supply side overcurrent protective device.

38-061 Overcurrent protection

(1)   Overcurrent protection for operating and control circuits shall be provided in accordance with Section 14.

(2)   Overcurrent protection for signal circuits shall be provided in accordance with Section 16.

(3)   Class 2 extra-low-voltage, low-energy circuits shall comply with Section 16.

(4)   Each ac drive motor for an elevator, dumbwaiter, escalator, moving walk, material lift, or lift for persons with physical disabilities, and each ac drive motor of a motor-generator set supplying current to the machine-drive motor, shall be provided with overload protection in accordance with Rule 28-302.

(5)   Overload devices shall be provided for each dc machine-drive motor where

    (a)   the motor-generator set provides power to two or more drive motors;

    (b)   the capacity of the motor-generator set is such that the protection provided in accordance with Subrule (1) is inadequate; or

    (c)   the drive motor of a variable-voltage machine is subject to overcurrent at reduced voltage during levelling.

(6)   The overload devices required by Subrule (5)(c) shall be permitted to be omitted where a time-delay relay is provided in the levelling circuit to disconnect the power supply at the motor-generator set within an interval that will prevent damage to motor windings.

38-062 Selective coordination

The overcurrent protection shall be coordinated with any upstream overcurrent protective device.

38-071 Guarding equipment

Elevator, dumbwaiter, escalator, moving walk, material lift, and lift for persons with physical disabilities driving machines, motor-generator sets, motor controllers, and disconnecting means shall be installed in a room or space

Section 38

Elevators, dumbwaiters, material lifts, escalators, moving walks,
lifts for persons with physical disabilities, and similar equipment

set aside for that purpose that is secured against unauthorized access, unless otherwise permitted in Items (a) to (d):

(a)   motor controllers shall be permitted outside the spaces specified in this Rule, provided that they are in enclosures with doors or removable panels capable of being locked in the closed position, and the disconnecting means is located adjacent to or is an integral part of the motor controller;

(b)   motor controller enclosures for escalators or moving walks shall be permitted in the balustrade on the side away from the moving steps or treadway;

(c)   where provided as an integral part of the motor controller, the disconnecting means shall be operable without opening the enclosure;

(d)   elevators with driving machines located on the car, counterweight, or in the hoistway and driving machines for dumbwaiters, material lifts, and lifts for persons with physical disabilities shall be permitted outside the spaces specified in this Rule.

38-081 Bonding of raceways to cars
Metal raceways, armoured cable, metallic-sheathed cable, or mineral-insulated cable attached to cars shall be bonded to the metal parts of the car that they contact.

38-082 Bonding of equipment
The frames of all motors, generators, machines, and controllers, and the metal enclosures for all electrical equipment in or on the car or in the hoistway, shall be bonded to ground in accordance with Section 10.

38-083 Bonding of non-electric elevators
For elevators other than electric having any electric conductors attached to the car, the metal frame of the car, where normally accessible to persons, shall be bonded to ground in accordance with Section 10.

38-084 Bonding of escalators, moving walks, and lifts for persons with physical disabilities
Metal parts of escalators, moving walks, and lifts for persons with physical disabilities shall be bonded to ground in accordance with Section 10.

38-085 Ground fault circuit interrupter protection for personnel
(1)   Each 125 V, single-phase receptacle installed in pits, hoistways, elevator and enclosed vertical platform lift car tops, and escalator or moving walk wellways shall be of the Class A ground fault circuit interrupter type.

(2)   All 125 V, single-phase receptacles installed in machine rooms and machinery spaces shall have Class A ground fault circuit interrupter type protection.

(3)   A single receptacle supplying a permanently installed sump pump shall not require ground fault circuit interrupter protection.

38-091 Emergency power (see Appendix B)
(1)   An elevator shall be permitted to operate from an emergency power supply in the event of normal power supply failure.

(2)   For elevator systems that regenerate power back into a power source that is unable to absorb the regenerative power under overhauling elevator load conditions, a means shall be provided to absorb this power.

(3)   Other building loads, such as power and lighting, shall be permitted as the energy absorption means required in Subrule (2), provided that such loads are automatically connected to the emergency power system operating the elevators and are large enough to absorb the elevator regenerative power.

(4)   The disconnecting means required by Rule 38-051 shall disconnect the emergency power source and the normal power source.

(5)   Where an additional power source is connected to the load side of the disconnecting means, the disconnecting means required in Rule 38-051 shall be provided with an auxiliary contact that is positively opened mechanically, the opening not being solely dependent on springs, and connected in the control circuit to prevent movement of the car when the disconnecting means is open.

Section 40 — Electric cranes and hoists

40-000 Scope

(1) This Section covers features of the installations of electrical equipment providing circuits for electric cranes, hoists, and monorails, and supplements or amends the general requirements of this Code.

(2) This Section does not cover equipment and wiring of cranes, hoists, and monorails that are assembled and erected in the field, which shall comply with CSA C22.2 No. 33.

40-002 Supply conductor sizes

The size of conductors in raceways or cables supplying main contact conductors or supplying equipment directly shall be determined from Table 58.

40-004 Conductor protection

(1) Conductors supplying main contact conductors shall be in rigid conduit, electrical metallic tubing, armoured cable, mineral-insulated cable, or aluminum-sheathed cable, except as otherwise provided for in Rule 40-018.

(2) Conductors supplying the equipment directly shall comply with Subrule (1) unless a flexible connection is required, in which case an armoured or unarmoured cable, festoon cable or flexible cord, with take-up devices where necessary to prevent damage to the cable or cord and to keep it clear of the operating floor, shall be permitted.

40-006 Overcurrent protection

Conductors supplying main contact conductors or supplying the equipment directly where there are no main contact conductors shall be provided with overcurrent protection in accordance with the requirements of Rule 28-200 for the motor load plus an allowance in accordance with Rule 14-104 for any other loads if the size of conductors has been increased to provide capacity for the other loads.

40-008 Disconnecting means

Suitable means for disconnecting all ungrounded conductors of the circuit simultaneously shall be

(a) provided within sight of the main contact conductors or within sight of the equipment if there are no main contact conductors; and

(b) accessible and operable from the ground or from the floor over which the equipment operates.

40-010 Main contact conductors

(1) Bare main contact conductors shall have an ampacity not less than that of the conductors supplying them and, if wire is used, in no case shall these conductors be smaller than

    (a) No. 4 AWG copper or No. 2 AWG aluminum if the length of contact conductor is 18 m or less; or

    (b) No. 2 AWG copper or No. 1/0 AWG aluminum if the length of contact conductor is greater than 18 m, unless the intermediate insulating supports are a clamp type capable of providing some strain relief.

(2) Bare main contact conductors shall be permitted to be of hard drawn copper or aluminum wire or shall be permitted to be made of steel or other suitable metal in the form of tees, angles, T-rails, or other rigid shapes.

40-012 Spacing of main contact conductors

(1) Bare main contact conductor wires shall be supported so that

    (a) they are separated, centre-to-centre, as follows:

        (i) not less than 150 mm for other than monorail hoists, if installed in a horizontal plane;

        (ii) not less than 75 mm for monorail hoists, if installed in a horizontal plane; or

        (iii) not less than 200 mm, if installed in other than a horizontal plane; and

    (b) the extreme limit of displacement does not bring them within less than 38 mm of the surface wired over.

(2) Rigid main contact conductors shall be supported so that there is an air space of not less than 25 mm between conductors, between conductors and adjacent collectors, and between conductors and the surface wired over.

40-014 Supporting of main contact conductors

(1) Bare main contact conductor wires shall be secured at each end to strain insulators and shall be supported on insulating supports placed at intervals not exceeding 6 m except that, where building conditions make

such intervals impossible, the interval between insulating supports shall be permitted to be increased to a maximum of 12 m if the separation between contact conductors is increased proportionately.

(2)   Rigid main contact conductors shall be secured to insulating supports spaced at intervals of not more than 80 times the vertical dimension of the conductor, but in no case greater than 4.5 m.

40-016 Joints in rigid contact conductors

Joints in rigid main contact conductors shall be made so as to ensure proper ampacity without overheating.

40-018 Use of track as a conductor

Monorail, tramrail, or crane runway tracks shall be permitted to be used as a main contact conductor or as a supply circuit conductor for one phase of a 3-phase alternating-current circuit if

(a)   the power for all phases is obtained from an isolating transformer;

(b)   the voltage does not exceed 300 V;

(c)   the rail serving as a conductor is effectively bonded to ground, preferably at the transformer, with permissive additional grounding by the fittings used for the suspension or attachment of the rail to the building structure; and

(d)   any joints in the rail meet the requirements of Rule 40-016.

40-020 Guarding of contact conductors

(1)   Contact conductors shall be guarded so that inadvertent contact cannot be made with bare current-carrying parts or they shall be incorporated in an enclosed contact assembly.

(2)   Guarding of bare contact conductors shall not be required where a clearance of at least 6 m between such conductors and grade, floor, or any working surface is provided and maintained.

40-022 Contact conductors not to supply other equipment

Contact conductors shall not be used as feeders for any equipment other than that essential for the operation of the cranes, hoists, or monorails that they supply.

40-024 Bonding (see Appendix B)

(1)   All exposed non-current-carrying metal parts shall be bonded to ground.

(2)   Tracks shall be bonded to ground as required by Rule 10-406 or 40-018.

(3)   The flexible supply connection permitted in Rule 40-004(2) shall incorporate a bonding conductor.

Section 42 — Electric welders

General

42-000 Scope

This Section applies to the installation of electric welders and supplements or amends the general requirements of this Code.

42-002 Special terminology

In this Section, the following definitions apply:

Actual primary current — the current drawn from the supply circuit during each welder operation at the particular heat tap and control setting used.

Duty cycle — the ratio of the time during which the welder is loaded to the total time required for one complete operation.

Rated primary current — the kilovolt-ampere rating of the welder as shown on its nameplate, multiplied by 1000 and divided by the rated primary voltage shown on the nameplate.

42-004 Receptacles and attachment plugs

Where a welder is cord-connected, the rating of the receptacle and attachment plug shall be permitted to be less than the rating of the overcurrent devices protecting them, but not less than the ampacity of the supply conductors required for the welder.

Transformer arc welders

42-006 Supply conductors

(1)   The supply conductors for an individual transformer arc welder shall have an ampacity of not less than the value obtained by multiplying the rated primary current of that welder in amperes by a factor of
    (a)   1.00, 0.95, 0.89, 0.84, 0.78, 0.71, 0.63, 0.55, or 0.45 for welders having a duty cycle of 100, 90, 80, 70, 60, 50, 40, 30, and 20% or less respectively; or
    (b)   0.75 for a welder having a time rating of 1 h.

(2)   The supply conductors for a group of transformer arc welders shall have an ampacity not less than the sum of the currents determined for each welder in the group in accordance with Subrule (1) multiplied by a demand factor of
    (a)   100% of the two largest calculated currents of the welders in the group; plus
    (b)   85% of the third largest calculated current of the welders in the group; plus
    (c)   70% of the fourth largest calculated current of the welders in the group; plus
    (d)   60% of the calculated currents of all remaining welders in the group.

(3)   Lower values than those given in Subrule (2) shall be permitted in cases where the work is such that a high operating duty cycle for individual welders is impossible.

42-008 Overcurrent protection for transformer arc welders

(1)   Each transformer arc welder shall have overcurrent protection rated or set at not more than 200% of the rated primary current of the welder, unless the overcurrent device protecting the supply conductors meets this requirement.

(2)   Each ungrounded conductor supplying a transformer arc welder shall have overcurrent protection rated or set at not more than 200% of the allowable ampacity of the conductor as specified in Table 1, 2, 3, or 4, except that the next higher rating or setting may be used where
    (a)   the nearest standard rating of the overcurrent device is less than the rating or setting otherwise required by this Rule; or
    (b)   the rating or setting otherwise required by this Rule results in too frequent opening of the overcurrent device.

(3)   The maximum rating or setting of an overcurrent device protecting a feeder supplying a group of transformer arc welders shall not exceed a value calculated by determining the maximum rating or setting of overcurrent device permitted by Subrules (1) and (2) for the welder allowed the highest overcurrent protection and adding to this value the sum of ampacities as calculated by Rule 42-006 for all other welders in the group.

42-010 Disconnecting means

(1) A disconnecting means shall be provided in the supply connections of each welder that is not equipped with a disconnecting means mounted as an integral part of the welder.

(2) The disconnecting means shall be a switch or circuit breaker and its rating shall be not less than necessary to accommodate overcurrent protection as specified under Rule 42-008.

Motor-generator arc welders

42-012 Conductors, protection, and control of motor-generator arc welders

(1) The Rules of Sections 4 and 28 shall apply to motor-generator arc welders, except that

    (a) the motors shall be permitted to be marked in amperes only; and

    (b) where the controller is built in as an integral part of the motor-generator set, the controller need not be separately marked provided that the necessary data is on the motor nameplate.

(2) The supply conductors for an individual motor-generator arc welder shall have an ampacity of not less than the value obtained by multiplying the rated primary current of that welder by a factor of

    (a) 1.00, 0.96, 0.91, 0.86, 0.81, 0.75, 0.69, 0.62, or 0.55 for welders having a duty cycle of 100, 90, 80, 70, 60, 50, 40, 30, and 20% or less respectively; or

    (b) 0.80 for a welder having a time rating of 1 h.

(3) The supply conductors for a group of motor-generator arc welders shall have an ampacity not less than the sum of the currents determined for each welder in the group in accordance with Subrule (2) multiplied by a demand factor of

    (a) 100% of the two highest calculated currents of the welders in the group; plus

    (b) 85% of the third largest calculated current of the welders in the group; plus

    (c) 70% of the fourth largest calculated current of the welders in the group; plus

    (d) 60% of the calculated currents for all remaining welders in the group.

(4) Lower values than those given in Subrule (3) shall be permitted in cases where the work is such that a high operating duty cycle for individual welders is impossible.

Resistance welders

42-014 Supply conductors for resistance welders

The ampacity of supply conductors shall be as follows:

(a) where an individual seam resistance welder or an individual automatically fed resistance welder is operated at different times at different values of primary current or duty cycle, the supply conductors shall have an ampacity of not less than 70% of the rated primary current of the welder;

(b) where an individual manually operated non-automatic resistance welder is operated at different times at different values of primary current or duty cycle, the ampacity of the supply conductors shall be not less than 50% of the rated primary current of the welder;

(c) where an individual resistance welder operates at known and constant values of actual primary current and duty cycle, the supply conductors shall have an ampacity of not less than the value obtained by multiplying the actual primary current by a factor of 0.71, 0.63, 0.55, 0.50, 0.45, 0.39, 0.32, 0.27, or 0.22 for duty cycles of 50, 40, 30, 25, 20, 15, 10, 7.5, and 5% or less respectively;

(d) where there is a group of resistance welders, the supply conductors shall have an ampacity of not less than

    (i) the sum of the values obtained from Item (a), (b), or (c) for the largest welder in the group; and

    (ii) 60% of the values obtained for all of the other welders in the group.

42-016 Overcurrent protection

(1) Every resistance welder shall have overcurrent protection rated or set at not more than 300% of the rated primary current of the welder unless the overcurrent device protecting the supply conductors gives equivalent protection.

(2) Every ungrounded conductor of a resistance welder shall have overcurrent protection rated or set at not more than 300% of the allowable ampacity of the conductor as specified in Table 1, 2, 3, or 4, except that the next higher rating or setting may be used where

    (a) the nearest standard rating of the overcurrent devices is less than the rating or setting required by this Rule; or

    (b) the rating or setting required by this Rule results in too frequent opening of the overcurrent device.

(3)   The maximum rating or setting of an overcurrent device protecting a feeder supplying a group of resistance welders shall not exceed a value calculated by determining the maximum rating or setting of overcurrent device permitted by Subrules (1) and (2) for the welder allowed the highest overcurrent protection and adding to this value the sum of ampacities as calculated by Rule 42-014 for all other welders in the group.

42-018   Control of resistance welders

Every resistance welder shall have installed in its supply circuit a switch or circuit breaker, rated at not less than the rating of the conductors as determined by Rule 42-014, whereby the welder and its control equipment can be isolated from the supply circuit.

42-020   Nameplate data for resistance welders

Every resistance welder shall be provided with a nameplate giving the manufacturer's name, primary voltage, frequency, rated kilovolt amperes at 50% duty cycle, maximum and minimum open-circuit secondary voltage, short-circuit secondary current at maximum secondary voltage, and the specified throat and gap setting.

Section 44 — Theatre installations

Scope

44-000 Scope
This Section applies to electrical equipment and installations in buildings or parts of a building designed, intended, or used for dramatic, operatic, motion picture, or other shows, and it supplements or amends the general requirements of this Code.

44-002 Motion picture studios and projectors
Motion picture studios and projectors shall comply with the requirements of Section 48.

General

44-100 Travelling shows
Electrical equipment used by a travelling theatrical company, circus, or other travelling show, whether or not the performance is held within a theatre, shall not be used for the initial performance of any stand until a permit has been obtained from the inspection department.

44-102 Wiring method
(1)   Wiring in stage and stage wing areas, orchestra pits, and projection booths shall be in rigid metal conduit, electrical metallic tubing, mineral-insulated cable, flexible metal conduit, armoured cable, lead-sheathed armoured cable, or aluminum-sheathed cable, except that
    (a)   other wiring methods shall be permitted for temporary work; and
    (b)   flexible cord or cable shall be permitted in accordance with other Rules in this Section.
(2)   Surface raceways shall not be used on the stage side of the proscenium wall.
(3)   Wiring in areas other than those listed in Subrule (1) shall be in accordance with the requirements of the appropriate Sections of this Code.

44-104 Number of conductors in raceways
For border or stage pocket circuits or for remote-control circuits,
(a)   the number of conductors run in rigid metal conduit or electrical metallic tubing shall not exceed that shown in Rule 12-1014; and
(b)   conductors run in auxiliary gutters or metal wireways shall have a total cross-sectional area not exceeding 20% of the cross-sectional area of the gutter or wireway.

44-106 Aisle lights in moving picture theatres
Circuits for aisle lights located under seats shall be permitted to supply 30 outlets provided that the size of lamp that can be used with each outlet is limited by barriers or the equivalent to 25 W or less.

Fixed stage switchboards

44-200 Stage switchboards to be dead front
Stage switchboards shall be
(a)   the dead-front type; and
(b)   protected above with a suitable metal guard or hood extending the full length of the board and completely covering the space between the wall and the board to protect the latter from falling objects.

44-202 Guarding stage switchboards
(1)   Where a stage switchboard has exposed live parts on the back of the board, it shall be enclosed by the walls of the building, by wire mesh grills, or by other acceptable methods.
(2)   The entrance to the enclosures shall have a self-closing door.

44-204 Switches
Switches shall be the enclosed type and externally operated.

44-206 Pilot lamp on switchboards
(1)   A pilot lamp shall be installed within every switchboard enclosure.
(2)   The pilot lamp shall be connected to the circuit supplying the switchboard so that the opening of the master switch does not cut off the supply to the lamp.

(3) The lamp shall be on an independent circuit protected by an overcurrent device rated or set at not more than 15 A.

44-208 Fuses

Fuses on switchboards shall be
(a) either the plug or cartridge type; and
(b) provided with enclosures in addition to the switchboard enclosure.

44-210 Overcurrent protection

(1) All circuits leaving the switchboard shall have an overcurrent device connected in each ungrounded conductor.
(2) Notwithstanding Rule 30-104, a luminaire having an input voltage at not more than 120 V nominal shall be permitted to be protected by an overcurrent device rated or set at not more than 100 A.

44-212 Dimmers

(1) Dimmers shall be connected so as to be dead when their respective circuit switches are open.
(2) Dimmers that do not open the circuit may be connected in a grounded neutral conductor.
(3) The terminals of dimmers shall be enclosed.
(4) Dimmer faceplates shall be arranged so that accidental contact cannot readily be made with the faceplate contacts.

44-214 Control of stage and gallery pockets

Stage and gallery pockets shall be controlled from the switchboard.

44-216 Conductors

(1) Stage switchboards equipped with resistive or transformer-type dimmer switches shall be wired with conductors having insulation suitable for the temperature generated in those switchboards and in no case less than 125 °C.
(2) The conductors shall have an ampacity not less than that of the switch or overcurrent device to which they are connected.
(3) Holes in the metal enclosure through which conductors pass shall be bushed.
(4) The strands of the conductor shall be soldered together before they are fastened under a clamp or binding screw.
(5) Where a conductor of No. 8 AWG or of a larger size is connected to a terminal,
    (a) it shall be soldered into a lug; or
    (b) a solderless connector shall be used.

Portable switchboards on stage

44-250 Construction of portable switchboards

(1) Portable switchboards shall be placed within enclosures of substantial construction but may be arranged so that the enclosure is open during operation.
(2) There shall be no live parts exposed within the enclosure, except those on dimmer faceplates.

44-252 Supply for portable switchboards

(1) Portable switchboards shall be supplied by means of flexible cord or cable, Type S, SO, or ST, terminating within the switchboard enclosure in an externally operated, enclosed, fused master switch.
(2) The master switch shall be arranged to cut off current from all apparatus within the enclosure except the pilot light.
(3) The flexible cord or cable shall have sufficient ampacity to carry the total load current of the switchboard.
(4) The ampere-rating of the fuses of the master switch shall be not greater than the total load current of the switchboard.

Fixed stage equipment

44-300 Footlights

(1) Where footlights are wired in rigid metal conduit or electrical metallic tubing, every lampholder shall be installed in an individual outlet box.
(2) Where footlights are not wired in rigid metal conduit or electrical metallic tubing, the wiring shall be installed in a metal trough.

44-302 Metal work

(1) The metal work for footlights, borders, proscenium sidelights, and strips shall be not less than 0.78 mm thick.

(2) The metal work for bunches and portable strips shall be not less than 0.53 mm thick.

44-304 Clearances at terminals

The terminals of lampholders shall be separated from the metal of the trough by at least 13 mm.

44-306 Mechanical protection of lamps in borders, etc.

Borders, proscenium sidelights, and strips shall be constructed so that the flanges of the reflectors or other suitable guards protect the lamps from mechanical injury and from accidental contact with scenery or other combustible material.

44-308 Suspended fixtures

Borders and strips shall be suspended so as to be electrically and mechanically safe.

44-310 Connections at lampholders

Conductors shall be soldered to the terminals of lampholders unless other suitable means are provided to obtain positive and reliable connection under severe vibration.

44-312 Ventilation for mogul lampholders

Where the lighting devices are equipped with mogul lampholders, the lighting devices shall be constructed with double walls and with adequate ventilation between the walls.

44-314 Conductor insulation for field assembled fixtures

Foot, border, proscenium, and portable strip light fixtures assembled in the field shall be wired with conductors having insulation suitable for the temperature at which the conductors will be operated and in no case less than 125 °C.

44-316 Branch circuit overcurrent protection

Branch circuits for footlights, border lights, and proscenium sidelights shall have overcurrent protection in accordance with Rule 30-104.

44-318 Pendant lights rated more than 100 W

Where a pendant lighting device contains a lamp or group of lamps of more than 100 W capacity, it shall be provided with a guard of not more than 13 mm mesh arranged to prevent damage from falling glass.

44-320 Cable for border lights

(1) Flexible cord or cable for border lights shall be Type S, SO, or ST.

(2) The flexible cord or cable shall be fed from points on the grid iron or from other acceptable overhead points but shall not be fed from side walls.

(3) The flexible cord or cable shall be arranged so that strain is taken from clamps and binding screws.

(4) Where the flexible cord or cable passes through a metal or wooden enclosure, a metal bushing shall be provided to protect the cord.

(5) Terminals or binding posts to which flexible cords or cables are connected inside the switchboard enclosure shall be located to permit convenient access to the cords or cables.

44-322 Wiring to arc pockets

Where the wiring to arc pockets is in rigid metal conduit or electrical metallic tubing, the end of the conduit or tubing shall be exposed at a point approximately 300 mm away from the pocket, and the wiring shall be continued in flexible metal conduit in the form of a loop at least 600 mm long, with sufficient slack to permit the raising or lowering of the box.

44-324 Receptacles in gallery pockets

At least one receptacle having a rated capacity of not less than 30 A shall be installed in the gallery of theatres in which dramatic or operatic performances are staged.

44-326 Receptacles and plugs

(1) Receptacles intended for the connection of arc lamps shall
   (a) have an ampere rating not less than 35 A; and
   (b) be supplied by copper conductors not smaller than No. 6 AWG.

(2) Receptacles intended for the connection of incandescent lamps shall
   (a) have an ampere rating not less than 15 A; and

(b)   be supplied by conductors not smaller than No. 12 AWG copper or No. 10 AWG aluminum.

(3)   Plugs for arc and incandescent receptacles shall not be interchangeable.

44-328 Curtain motors
Curtain motors shall be the enclosed type.

44-330 Flue damper control
(1)   Where stage flue dampers are released by an electrical device, the circuit operating the device shall, in normal operation, be closed.

(2)   The circuit shall be controlled by at least two single-pole switches enclosed in metal boxes with self-closing doors without locks or latches.

(3)   One switch shall be placed at the electrician's station and the other at a location that is acceptable.

(4)   The device shall be
  (a)   designed for the full voltage of the circuit to which it is connected, no resistance being inserted;
  (b)   located in the loft above the scenery; and
  (c)   enclosed in a suitable metal box with a tight self-closing door.

Portable stage equipment

44-350 Fixtures on scenery
(1)   Fixtures attached to stage scenery shall be
  (a)   the internally wired type; or
  (b)   wired with flexible cord or cable suitable for hard usage as listed in Table 11.

(2)   The fixtures shall be secured firmly in place.

(3)   The stems of the fixtures shall be carried through to the back of the scenery and shall have a suitable bushing on the end.

44-352 String or festooned lights
(1)   Joints in the wiring of string or festooned lights shall be staggered where practicable.

(2)   Where the lamps of string or festooned lights are enclosed in paper lanterns, shades, or other devices of combustible material, they shall be equipped with lamp guards.

44-354 Flexible conductors for portable equipment
Conductors for arc lamps, bunches, or other portable equipment shall be flexible cord types suitable for extra-hard usage, as shown in Table 11, but for separate miscellaneous portable devices operated under conditions where the conductors are not exposed to severe mechanical injury, flexible cord types suitable for other than hard usage, as shown in Table 11, may be used.

44-356 Portable equipment for stage effects
Portable equipment for stage effects shall be a type acceptable for the purpose and shall be located so that flames, sparks, or hot particles cannot come in contact with combustible material.

Section 46 — Emergency systems, unit equipment, and exit signs

46-000 Scope

(1) This Section applies to the installation, operation, and maintenance of emergency systems and unit equipment intended to supply illumination and to emergency systems intended to supply power, in the event of failure of the normal supply, where required by the *National Building Code of Canada*.

(2) This Section applies to the wiring of exit signs.

(3) The requirements of this Section supplement or amend the general requirements of this Code.

General

46-100 Capacity

Emergency systems and unit equipment shall have adequate capacity and rating to ensure the satisfactory operation of all connected equipment when the principal source of power fails.

46-102 Instructions

(1) Complete instructions for the operation and care of an emergency system or unit equipment that shall specify testing at least once every month to ensure security of operation shall be posted on the premises in a frame under glass.

(2) The form of instructions and their locations shall be in compliance with the *National Building Code of Canada*.

46-104 Maintenance

Where batteries are used as a source of supply, the batteries shall be kept

(a) in proper condition; and

(b) fully charged at all times.

46-106 Arrangement of lamps

(1) Emergency lights shall be arranged so that the failure of any one lamp will not leave in total darkness the area normally illuminated by it.

(2) No appliance or lamp, other than those required for emergency purposes, shall be supplied by the emergency circuits.

46-108 Method of wiring (see Appendices B and G)

(1) Except as permitted by Subrule (2) and Rule 46-304(3), all conductors of systems, equipment, and devices installed in accordance with this Section shall be

(a) installed in metal raceway of the totally enclosed type;

(b) incorporated in a cable having a metal armour or sheath;

(c) installed in rigid non-metallic conduit where embedded in at least 50 mm of masonry or poured concrete or installed underground; or

(d) installed in electrical non-metallic tubing where embedded in at least 50 mm of masonry or poured concrete.

(2) Conductors installed in buildings of combustible construction in accordance with Rules 12-506 to 12-520 shall be permitted to be incorporated in a non-metallic-sheathed cable.

(3) Conductors of emergency systems and conductors between unit equipment and remote lamps shall be kept entirely independent of all other conductors and equipment and shall not enter a fixture, raceway, box, or cabinet occupied by other conductors except where necessary

(a) in transfer switches; and

(b) in exit signs and emergency lighting fixtures supplied from two sources.

Emergency systems

46-200 Emergency systems (see Appendix B)

Rules 46-202 to 46-210 apply only to emergency systems from central standby supplies.

46-202 Supply (see Appendix G)

(1) The emergency supply shall be a standby supply consisting of

    (a) a storage battery of the rechargeable type having sufficient capacity to supply and maintain at not less than 91% of full voltage the total load of the emergency circuits for the time period required by the *National Building Code of Canada,* but in no case less than 30 min, and equipped with a charging means to maintain the battery in a charged condition automatically; or

    (b) a generator driven by a dependable prime mover.

(2) Automobile batteries and lead batteries not of the enclosed glass-jar type are not considered suitable under Subrule (1) and shall be used only where a deviation has been allowed in accordance with Rule 2-030.

Δ (3) Where a generator is used, it shall be

    (a) of sufficient capacity to carry the load;

    (b) arranged to start automatically without failure and without undue delay upon the failure of the normal power supply of the equipment connected to this generator; and

    (c) in conformance with CAN/CSA-C282, except for a generator installed in health care facilities as described in Rule 24-306.

46-204 Control

(1) The current supply for emergency systems shall be controlled by automatic transfer equipment that energizes the emergency system upon failure of the normal current supply and that is accessible only to authorized persons.

(2) An automatic light-actuated device, approved for the purpose, shall be permitted to be used to control separately the lights located in an area that is adequately illuminated during daylight hours without the need for artificial lighting.

46-206 Overcurrent protection

(1) No device capable of interrupting the circuit, other than the overcurrent device for the current supply of the emergency system, shall be placed ahead of the branch circuit overcurrent devices.

(2) The branch circuit overcurrent devices shall be accessible only to authorized persons.

46-208 Audible and visual trouble-signal devices

(1) Every emergency system shall be equipped with audible and visual trouble-signal devices that warn of derangement of the current source or sources and that indicate when the emergency load is supplied from batteries or generators.

(2) Audible trouble signals shall be permitted to be wired so that

    (a) they can be silenced, but a red warning or trouble light shall continue to provide the protective function; and

    (b) when the system is restored to normal, the audible signal will

        (i) sound, indicating the need to restore the silencing switch to its normal position; or

        (ii) reset automatically so as to provide sound for any subsequent operation of the emergency system.

46-210 Remote lamps

Lamps shall be permitted to be mounted at some distance from the current supply that feeds them, but the voltage drop in the wiring feeding such lamps shall not exceed 5% of the applied voltage.

Unit equipment

46-300 Unit equipment (see Appendix B)

Rules 46-302 to 46-306 apply to individual unit equipment for emergency lighting only.

46-302 Mounting of equipment

Each unit equipment shall be mounted with the bottom of the enclosure not less than 2 m above the floor, wherever practicable.

46-304 Supply connections

(1) Receptacles to which unit equipment is to be connected shall be not less than 2.5 m above the floor, where practicable, and shall be not more than 1.5 m from the location of the unit equipment.

(2) Unit equipment shall be permanently connected to the supply if

    (a) the voltage rating exceeds 250 V; or

    (b) the marked input rating exceeds 24 A.

(3)   Where the ratings in Subrule (2) are not exceeded, the unit equipment shall be permitted to be connected using the flexible cord and attachment plug supplied with the equipment.

(4)   Unit equipment shall be installed in such a manner that it will be automatically actuated upon failure of the power supply to the normal lighting in the area covered by that unit equipment.

46-306 Remote lamps (see Appendix B)

(1)   The circuit conductors to remote lamps shall be of such size that the voltage drop does not exceed 5% of the marked output voltage of the unit equipment, or such other voltage drop for which the performance of unit equipment is certified when connected to the specific remote lamp being installed.

(2)   Remote lamps shall be suitable for remote connection and shall be included in the list of lamps provided with the unit equipment.

(3)   The number of lamps connected to a single unit equipment shall not result in a load in excess of the watts output rating marked on the equipment for the emergency period required by the *National Building Code of Canada*, and the load shall be computed from the information in the list of lamps referred to in Subrule (2).

Exit signs

46-400 Exit signs (see Appendices B and G)

(1)   Where exit signs are connected to an electrical circuit, that circuit shall be used for no other purpose.

(2)   Notwithstanding Subrule (1), exit signs shall be permitted to be connected to a circuit supplying emergency lighting in the area where these exit signs are installed.

(3)   Exit signs in Subrules (1) and (2) shall be illuminated by an emergency power supply where emergency lighting is required by the *National Building Code of Canada*.

Section 48 — Motion picture studios, projection rooms, film exchanges including film-vaults, and storehouses for pyroxylin plastic and nitrocellulose X-ray and photographic film

48-000 Scope
(1) This Section supplements or amends the general requirements of this Code and applies to
    (a) motion picture studios, projection rooms, exchanges, factories, and laboratories; and
    (b) any building or portion of a building in which motion picture films, pyroxylin plastic and nitrocellulose X-ray and photographic films are manufactured, projected, developed, printed, rewound, repaired, or stored.
(2) This Section does not apply where only slow-burning (cellulose-acetate or equivalent) film is used.

48-002 Wiring method
The wiring method, unless otherwise specified in this Section, shall be rigid conduit, steel electrical metallic tubing, or mineral-insulated cable, except that portable cables or flexible cord may be used on studio stages and other locations where fixed wiring methods are impracticable.

48-004 Lamp outlets
Lamp outlets on walls shall consist of lampholders mounted in outlet boxes and equipped with open-end guards securely fastened to the cover of the box.

48-006 Pendant lamps
Pendant lamps shall be suspended by means of reinforced cord, armoured cord, or armoured cable, and shall be protected by guards or metal shades.

48-008 Portable lamps
For portable lamps other than those used as properties in a motion picture set on a studio stage or similar location, the lampholders shall be
(a) unswitched;
(b) of composition or metal-sheathed porcelain; and
(c) provided with a guard hook and handle.

48-010 Flexible cords
Type S, SO, or ST cord shall be used on portable lamps and equipment.

48-012 Patching table fixtures
At film patching tables all lighting fixtures, except lamps forming an integral part of patching table equipment, shall be of the totally enclosed gasketted type.

48-014 Motors and generators
Motors and generators having brushes or sliding contacts, other than those used on studio stages or installed in accordance with Rule 48-032, shall be of approved dust-tight or enclosed types.

48-016 Storage batteries
Storage batteries shall comply with the requirements of Rules 26-540 to 26-554.

48-018 Pyroxylin plastic storage rooms
In rooms used for the storage of pyroxylin plastic no receptacle or attachment plugs shall be installed.

Film-vaults

48-020 Equipment in film-vaults
No electrical equipment other than that necessary for fixed lighting shall be installed in film-vaults.

48-022 Wiring methods in film-vaults (see Appendix B)
(1) The wiring method in film-vaults shall be rigid conduit or mineral-insulated cable only, with threaded joints at couplings, boxes, and fittings.
(2) Conduit or cable shall not run directly from vault to vault, but only from the switch to the lighting fixture within the vault.

214

Section 48

Motion picture studios, projection rooms, film exchanges including film-vaults, and
storehouses for pyroxylin plastic and nitrocellulose X-ray and photographic film

(3)   Conduit shall be sealed off near the switch enclosure with a fitting and compound approved for the purpose.

48-024 Lighting fixtures in film-vaults
(1)   Lighting fixtures in film-vaults shall be of the explosion-proof type approved for use in Class I, Group C hazardous locations and shall have metal cages or guards protecting the globes.
(2)   The fixtures shall be located as close as practicable to the ceiling to prevent damage through the handling of film containers.

48-026 Circuits in film-vaults
(1)   Fixtures shall be controlled by a double-pole switch located outside the film-vault.
(2)   A red pilot light shall be provided to indicate when the switch is closed and shall be located outside the film-vault.
(3)   Wiring shall be arranged so that when the switch is off, all conductors within the film-vault will be dead.

Motion picture projection rooms

48-028 Flexible cords in projection rooms
Type S, SJ, SO, or ST flexible cords shall be used on portable equipment in motion picture projection rooms.

48-030 Lamps in projection rooms
Incandescent lamps in projection rooms or booths shall be provided with a lamp guard unless otherwise protected by non-combustible shades or other enclosures.

48-032 Arc lamp current supply
Motor-generator sets, frequency changers, transformers, rectifiers, rheostats, and similar equipment for the supply or control of current to arc lamps or projectors shall be located in a room separate from the projection room.

48-034 Ventilation
Exhaust ventilation fans for the projection room shall be controlled from inside the projection room.

Section 50 — Solar photovoltaic systems

50-000 Scope (see Appendix B)

(1) This Section applies to the installation of solar photovoltaic systems except where the voltage and current are limited in accordance with Rule 16-200(1)(a) and (b).

(2) The requirements of this Section supplement or amend the general requirements of this Code.

50-002 Special terminology (see Appendix B)

In this Section, the following definitions apply:

AC module — a complete, environmentally protected assembly of interconnected solar cells, inverter, and other components designed to generate ac power.

Array — a mechanical integrated assembly of modules or panels with a support structure and foundation, tracking, and other components as required, to form a power producing unit.

Interconnected system — a solar photovoltaic system that operates in parallel with, and that may deliver power to, another system such as a supply authority system.

Module — the smallest complete, environmentally protected assembly of interconnected solar cells.

Panel — an assembly of modules, mechanically fastened together and prewired to form a self-contained unit.

Photovoltaic output circuit — circuit conductors between the photovoltaic source circuit(s) and the power conditioning unit or direct-current utilization equipment.

Photovoltaic power source — an array or aggregate of arrays that generates direct-current power at system voltage and current.

Photovoltaic source circuit — conductors between modules and from modules to the common connection point(s) of the direct-current system.

Power conditioning unit — equipment that is used to change voltage level or waveform, or otherwise alter or regulate the output of a photovoltaic power source.

Power conditioning unit output circuit — conductors between the power conditioning unit and the connection to the service, distribution, or utilization equipment.

Solar cell — the basic photovoltaic device that generates electricity when exposed to light.

Solar photovoltaic systems — the total components and subsystems that in combination convert solar energy into electrical energy suitable for connection to a utilization load.

Stand-alone system — a solar photovoltaic system that supplies power independently of another system.

Δ **50-004 Marking** (see Appendix B)

A permanent marking shall be provided at an accessible location at the disconnecting means for the photovoltaic output circuit specifying the following:

(a) rated operating current and voltage;

(b) rated open-circuit voltage; and

(c) rated short-circuit current.

50-006 Voltage rating of photovoltaic source circuits

The voltage rating of photovoltaic source circuits shall be the rated open-circuit voltage of the photovoltaic power source multiplied by 125%.

50-008 Current rating of photovoltaic source circuits

Where overcurrent protection is not provided, the current rating of a photovoltaic power source circuit shall be the rated short-circuit current of all available photovoltaic power sources multiplied by 125%.

50-010 Overcurrent protection for apparatus and conductors

(1) Overcurrent protection shall be provided for all photovoltaic conductors and apparatus in compliance with the requirements of Section 14, except that individual overcurrent protection devices shall not be required where the available short-circuit current is not greater than the rated ampacity of the apparatus or conductor.

(2) Overcurrent devices for photovoltaic source circuits shall be accessible and, where practicable, shall be grouped.

50-012 Disconnecting means (see Appendix B)

(1) Means shall be provided to disconnect all equipment, including the power conditioning unit, from all ungrounded conductors of all sources.

(2) Where the equipment in Subrule (1) is energized from more than one source, the disconnecting means shall comply with Rules 14-414 and 14-700.

Δ (3) Any photovoltaic output circuit rated 50 V and greater shall have means to disable and isolate it.

50-014 Wiring method

Notwithstanding Section 12, flexible cords of a type specified in Table 11 for extra-hard usage shall be permitted to interconnect modules within an array.

50-016 Attachment plugs and similar wiring devices (see Appendix B)

Attachment plugs and similar wiring devices shall be permitted to connect flexible cord between modules and panels where

(a) there are no exposed energized parts whether the devices are connected or disconnected;

(b) the devices are polarized;

(c) the devices have a configuration that is not interchangeable with receptacles or attachment plugs of other systems on the premises;

(d) the devices are of the locking type;

(e) the devices are rated for the voltage and current of the circuit in which they are installed; and

(f) the devices provide strain relief.

50-018 Module connection arrangement

The connections to a module or panel shall be arranged so that removal of a module or panel from a photovoltaic source circuit does not interrupt a bonding conductor to other photovoltaic source equipment.

Δ 50-020 Interconnected system connection

(1) Connection from the power conditioning unit to the supply authority of an interconnected system shall comply with Section 84.

(2) A power conditioning unit used for interconnection shall be approved for this specific purpose.

(3) Each system connection shall be permitted to be made at a dedicated circuit breaker or fusible disconnecting means at the load side of the service box.

Section 52 — Diagnostic imaging installations

52-000 Scope

(1) This Section applies to the installation of X-ray and other diagnostic imaging equipment operating at any frequency, and supplements or amends the general requirements of this Code.

(2) Nothing in this Section shall be construed as specifying safeguards against direct, stray, or secondary radiation emitted by the equipment.

52-002 Special terminology

In this Section, the following definitions apply:

Long-time rating when applied to X-ray or **computerized tomography equipment** — a rating that is applicable for an operating period of 5 min or more.

Momentary rating when applied to X-ray or **computerized tomography equipment** — a rating that is applicable for an operating period of not more than 20 s.

52-004 High-voltage guarding

(1) High-voltage parts shall be mounted within metal enclosures that are bonded to ground, except when installed in separate rooms or enclosures, where a suitable switch shall be
   (a) provided to control the circuit supplying diagnostic imaging equipment; and
   (b) arranged so that it will be in an open position except when the door of the room or enclosure is locked.

(2) High-voltage parts of diagnostic imaging equipment shall be insulated from the enclosure.

(3) Conductors in the high-voltage circuits shall be of the shockproof type.

(4) A milliammeter, if provided, shall be
   (a) connected, if practicable, in the lead that is bonded to ground; or
   (b) guarded if connected in the high-voltage lead.

52-006 Connections to supply circuit

(1) Permanently connected diagnostic imaging equipment shall be connected to the power supply by means of a wiring method meeting the general requirements of this Code, except that apparatus properly supplied by branch circuits not larger than a 30 A branch circuit may be supplied through a suitable plug and hard-usage cable or cord.

(2) Mobile diagnostic imaging equipment of any capacity shall be permitted to be connected to its power supply by suitable temporary connections and hard-usage cable or cord.

52-008 Disconnecting means

(1) A disconnecting means of adequate capacity for at least 50% of the input required for the momentary rating or 100% of the input required for the long-time rating of X-ray or computerized tomography equipment, whichever is greater, shall be provided in the supply circuit.

(2) A disconnecting means of adequate capacity shall be provided in a location readily accessible from the radiation control.

(3) For apparatus requiring a 120 V branch circuit fused at 30 A or less, a plug and receptacle of proper size shall be permitted to serve as a disconnecting means.

52-010 Transformers and capacitors

(1) Transformers and capacitors forming a part of diagnostic imaging equipment shall not be required to conform to the requirements of Section 26 of this Code.

(2) Capacitors shall be provided with an automatic means for discharging and grounding the plates whenever the transformer primary is disconnected from the source of supply, unless all current-carrying parts of the capacitors and of the conductors connected with them are
   (a) at least 2.5 m from the floor, and are inaccessible to unauthorized persons; or
   (b) within metal enclosures that are bonded to ground or within enclosures of insulating material if within 2.5 m of the floor.

52-012 Control

(1) For stationary equipment, the low-voltage circuit of the step-up transformer shall contain an overcurrent device that
   (a) has no exposed live parts;
   (b) protects the radiographic circuit against fault conditions under all operating conditions; and

(c)   is installed as a part of, or adjacent, to the equipment.

(2)   Where in Subrule (1) the design of the step-up transformer is such that branch fuses having a current rating lower than the current rating of the overcurrent device are required for adequate protection for fluoroscopic and therapeutic circuits, they shall be added for protection of these circuits.

(3)   For portable equipment, the requirements of Subrules (1) and (2) shall apply, but the overcurrent device shall be located in or on the equipment; however, except that no current-limiting device is required when the high-voltage parts are within a single metal enclosure that is provided with a means for bonding to ground.

(4)   Where more than one piece of equipment is operated from the same high-voltage circuit, each piece or group of equipment, as a unit, shall be provided with a high-voltage switch or equivalent disconnecting means.

52-014 Bonding

Non-current-carrying parts of tube stands, tables, and other apparatus shall be bonded to ground in conformity with the requirements of Section 10.

52-016 Ampacity of supply conductors and rating of overcurrent protection

(1)   The ampacity of supply conductors and the rating of overcurrent protection devices shall not be less than
   (a)   the long-time current rating of X-ray or computerized tomography equipment; or
   (b)   50% of the maximum momentary current rating required by X-ray or computerized tomography equipment on a radiographic setting.

(2)   The ampacity of conductors and the rating of overcurrent protection devices for two or more branch circuits supplying X-ray or computerized tomography units shall not be less than
   (a)   the sum of the long-time current rating of all X-ray or computerized tomography units that are intended to be operated at any one time; or
   (b)   the sum of 50% of the maximum momentary current rating for X-ray or computerized tomography equipment on a radiographic setting for the two largest units, plus 20% of the maximum current rating of the other units.

Section 54 — Community antenna distribution and radio and television installations

54-000 Scope
(1) This Section supplements or amends the general requirements of this Code and applies to
   (a) community antenna distribution;
   (b) equipment for the reception of radio and television broadcast transmission; and
   (c) equipment employed in the normal operation of a radio station licensed by the Government of Canada as an experimental amateur radio station.
(2) This Section does not apply to equipment and antennas used for broadcast transmission and for coupling carrier current to power line conductors.
(3) In Subrule (2), "broadcast" means one-way communication other than by community antenna distribution.

54-002 Special terminology
In this Section, the following definitions apply:

Cable distribution plant — a coaxial cable system with passive devices, amplifiers, or power sources, as covered by CAN/CSA-C22.3 No. 1 and C22.3 No. 7, that is used to deliver radio and television frequency signals and power associated with the community antenna distribution equipment.

Customer distribution circuit — a coaxial cable with passive devices, amplifiers, or power sources that is used to deliver radio and television frequency signals and power from the cable distribution plant that has current-limiting devices.

Customer service enclosure — a cabinet not accessible by the customer that is placed on the outside or inside wall of the building to house community antenna television (CATV) equipment.

Multitap — a passive device that extends radio and television frequency signals and that may extend current limited power, from the cable distribution plant to the customer distribution circuit associated with the community antenna distribution circuit.

Power blocking device — an approved unit that is used to prevent power other than radio and television frequency signals from extending to the outgoing coaxial cable.

54-004 Community antenna distribution (see Appendix B)
(1) Community antenna distribution applies to coaxial cable circuits employed to distribute radio and television frequency signals typical of a community antenna television (CATV) system.
(2) Rules 54-100 to 54-704 apply to community antenna distribution installations.

54-006 Equipment
Equipment referred to in this Section shall not require approval in accordance with Rule 2-024, except where specifically noted in this Section as requiring approval.

54-008 Receiving equipment and amateur transmitting equipment Rules
Rules 54-800 to 54-1006 apply to
(a) radio and television receiving equipment; and
(b) amateur radio transmitting equipment.

54-010 Circuits in communication cables
Community antenna distribution circuits, or their parts, that use conductors in a cable assembly with other conductors forming parts of communication circuits are, for the purpose of this Code, deemed to be communication circuits and shall conform to the applicable Rules of Section 60, except that the requirements for protectors and grounding for the coaxial cables shall meet the requirements of this Section.

Community antenna distribution

54-100 Conductors
(1) The conductors used in community antenna distribution shall consist of coaxial cable having a central inner conductor and an outer conductive shield of circular cross-section.
(2) Conductors placed within buildings shall be the approved types as specified in Table 19.

54-102 Voltage and current limitations (see Appendix B)

(1)   Coaxial cable shall be permitted to be used for connection between the cable distribution plant and the customer service enclosure, or between two customer service enclosures, and for providing power to associated community antenna distribution circuits, provided that the following requirements are met:

(a)   for a single dwelling, the open circuit voltage does not exceed 90 V, and the maximum current is limited to 100/V amperes, up to and including the customer service enclosure;

(b)   for a building with multiple occupancies, the open circuit voltage does not exceed 90 V, and the maximum current is limited to 10 A, up to and including the customer service enclosure;

(c)   the current supply is from an approved amplifier, transformer, or other approved device having energy-limiting characteristics;

(d)   the cable distribution plant complies with the applicable requirements of CAN/CSA-C22.3 No. 1 and C22.3 No. 7;

(e)   the power does not extend beyond the customer service enclosure unless extending to another customer service enclosure, and not to the coaxial cable extending to the customer electrical equipment;

(f)   the customer service enclosure is grounded and all coaxial cables entering are bonded to ground;

(g)   the customer service enclosure is provided with a lock or similar closing device and contains power blocking devices to prevent the coaxial cable to the customer electrical equipment from being energized; and

(h)   all customer distribution circuits fed from a common multitap that is capable of delivering power to the customer distribution circuit are provided with power blocking devices to prevent the coaxial cable to the customer electrical equipment from being energized.

(2)   Coaxial cable for the connection between the customer service enclosure and a point located at least 1 m from the customer electrical equipment shall be permitted to be energized by an approved 0 to 30 V Class 2 transformer or power supply within the premises, provided that power blocking devices are installed to prevent the connection at the customer electrical equipment from being energized.

54-104 Hazardous locations

Where the circuits or equipment within the scope of this Section are installed in hazardous locations, they shall also comply with the applicable Rules of Sections 18, 20, and 24.

54-106 Inspection by an inspector

(1)   Community antenna distribution circuits employed by an electrical utility, or a communication utility operating within the scope of Section 60, shall not in the exercise of its function as a utility be subject to inspection by an inspector.

(2)   Where the community antenna distribution circuit derives power for operation from an electric supply circuit, the transformer, amplifier, or other current-limiting device used at the junction of the community antenna distribution and electric supply circuit shall be subject to inspection by an inspector.

54-108 Supports

Where conductors are attached to, or supported on, buildings, the attachment or supporting equipment shall be acceptable for the purpose.

Protection

54-200 Grounding of outer conductive shield of a coaxial cable (see Appendix B)

(1)   Where coaxial cable is exposed to lightning or to accidental contact with lightning arrester conductors or power conductors operating at a voltage exceeding 300 volts-to-ground, the outer conductive shield of the coaxial cable shall be grounded at the building as close to the point of cable entry as possible.

(2)   Where the outer conductive shield of a coaxial cable is grounded, no other protective device shall be required.

(3)   Grounding of a coaxial cable shield by means of a protective device shall be permitted, provided that the device does not interrupt the grounding system within the building.

54-202 Provision of protector (see Appendix B)

Where a protective device is provided, it shall be

(a)   approved for the purpose;

(b)   located in or on the building as near as practicable to the point of cable entry;

(c)   located external to any hazardous location as defined in Sections 18, 20, and 24, and away from the immediate vicinity of flammable or explosive materials;

(d)   mounted on a flame-retardant, absorption-resisting insulating base; and

(e)   covered if located outdoors.

Grounding

54-300 Grounding conductor

(1)   The grounding conductor for the outer conductive shield of a coaxial cable or the protector shall be insulated.

(2)   The grounding conductor shall be made of copper.

(3)   The grounding conductor shall be not smaller than No. 14 AWG.

(4)   The grounding conductor shall have an ampacity at least equal to, or greater than, that of the outer conductive sheath of the exposed coaxial cable.

(5)   Where two or more coaxial cables that have outer conductive shields differing in size and ampacity join at a common connection to the grounding conductor, the ampacity of the grounding conductor shall at least equal or exceed the ampacity of the largest coaxial outer conductive shield.

(6)   The grounding conductor shall be run from the protector or the coaxial cable shield to the grounding electrode in as straight a line as possible.

(7)   The grounding conductor shall be protected when exposed to mechanical damage.

54-302 Grounding electrode

(1)   Grounding electrodes shall conform to Rule 10-700 except that the minimum driven length of a rod electrode shall be 2 m.

(2)   Artificial grounding electrodes for community antenna distribution shall be spaced and bonded with other electrodes in accordance with Rule 10-702.

54-304 Grounding electrode connection

The grounding conductor shall be attached to a grounding electrode, as required in Rule 10-908

(a)   directly; or

(b)   to a wire lead permanently connected to the ground rod electrode in a manner specified in CSA C83.

Conductors within buildings

54-400 Separation from other conductors

(1)   Conductors of community antenna distribution circuits shall be separated at least 50 mm from insulated conductors of electric lighting, power, or Class 1 circuits operating at 300 V or less, and shall be separated at least 600 mm from any insulated conductor of an electric lighting, power, or Class 1 circuit operating at more than 300 V unless effective separation is afforded by use of the following:

(a)   grounded metal raceways for the community antenna distribution circuits, or for the electric lighting, power, and Class 1 circuits;

(b)   grounded metal-sheathed or armoured cable for the electric lighting, power, and Class 1 conductors; or

(c)   raceways of a non-metal type as permitted in Section 12, in addition to the insulation on the community antenna distribution circuit conductors, or the electric lighting, power, and Class 1 circuit conductors.

(2)   Where the electric lighting or power conductors are bare, all community antenna distribution conductors in the same room or space shall be enclosed in a grounded metal raceway and no opening, such as an outlet box, shall be located within 2 m of bare conductors if up to and including 15 kV or within 3 m of bare conductors above 15 kV.

(3)   The conductors of a community antenna distribution circuit shall not be placed in any raceway, compartment, outlet box, junction box, or similar fitting that contains conductors of electric light, power, or Class 1 circuits, unless

(a)   the conductors of the community antenna distribution circuit are separated from the electric light, power, or Class 1 circuit conductors by a barrier that conforms to Rule 12-904(2) for conductors in raceways or Rule 12-3030(1) for conductors in boxes, cabinets, and fittings; or

(b)   the power or Class 1 conductors are placed solely for the purpose of supplying power to the community antenna distribution circuit.

54-402 Conductors in a vertical shaft
Conductors of a community antenna distribution circuit in a vertical shaft shall be in a totally enclosed non-combustible raceway.

54-404 Penetration of a fire separation
Conductors of a community antenna distribution circuit extending through a fire separation shall be installed so as to limit fire spread in accordance with Rule 2-124.

54-406 Community antenna distribution conductors in ducts and plenum chambers
Community antenna distribution conductors shall not be placed in ducts or plenum chambers except as permitted by Rules 2-126 and 12-010.

54-408 Raceways
Raceways shall be installed in accordance with the requirements of Section 12.

Equipment

54-500 Community antenna distribution amplifiers and other power sources
(1) Amplifiers and other devices that supply current to a community antenna distribution circuit from an electric supply circuit shall be approved for the purpose.
(2) Where amplifiers and other power devices are connected to an electric supply circuit and enclosed in a cabinet, the cabinet shall be positioned to be readily accessible and shall be adequately ventilated.
(3) The chassis and cabinets of the community antenna distribution amplifier or other power sources, the outer conductive shield of the coaxial cables, and the metal conduit or the metal cable sheath enclosing the electric supply conductors shall all be connected to the system ground with a minimum of No. 6 AWG copper conductor.
(4) Where a cabinet containing an amplifier or other power device is mounted accessible to the public, it shall be provided with a lock or similar closing device.

54-502 Exposed equipment and terminations
Exposed community antenna distribution equipment and/or associated terminations shall be located in a suitable room or similar area as required by Rule 2-202, separate from electrical light or power installations, except where necessary to place them in a joint-use room, in which case a minimum separation of 900 mm from electrical equipment requiring adjustment and maintenance shall be provided and maintained.

54-504 Equipment grounding
Non-powered equipment and enclosures, or equipment powered exclusively by the coaxial cable, shall be considered as grounded where they are effectively connected to the grounded outer conductive coaxial cable shield.

Conductors outside of buildings

54-600 Overhead conductors on poles
The installation of overhead community antenna distribution conductors in proximity with power conductors on poles and in aerial spans between buildings, poles, and other structures shall be established in conformance with the provisions of the *Canadian Electrical Code, Part III*.

54-602 Overhead conductors on roofs
(1) Community antenna distribution conductors passing over buildings shall be kept at least 2.5 m above any roof that may be readily walked upon.
(2) Community antenna distribution conductors shall not be attached to the upper surfaces of roofs or be run within 2.5 m, measured vertically, of a roof, unless a deviation has been allowed in accordance with Rule 2-030.
(3) A deviation in accordance with Rule 2-030 shall not be necessary where the building is a garage or other auxiliary building of one storey.

54-604 Conductors on buildings
(1) Community antenna distribution conductors on buildings shall be separated from insulated light or power conductors not in cable or conduit by at least 300 mm, unless the conductors are permanently separated by a continuous and firmly fixed non-metal type raceway as permitted in Section 12 in addition to the insulation on the conductors.

(2)   Community antenna distribution conductors subject to accidental contact with light or power conductors operating at voltages exceeding 300 V, and attached exposed to buildings, shall be separated from combustible material by being supported on glass, porcelain, or other insulating material acceptable for the purpose, except that such separation is not required where the outer conductive sheath of the coaxial cable is grounded.

(3)   Community antenna distribution conductors attached to buildings shall not conflict with other communication conductors attached to the same building, and sufficient clearances shall be provided so that there will be no unnecessary interference to maintenance operations, and in no case should the conductors, strand, or equipment of one system cause abrasion to the conductors, strand, or equipment of the other system.

54-606 Conductors entering buildings

The community antenna distribution conductors shall enter the building either through a non-combustible, non-absorptive insulating bushing, or through a metal raceway, except that the insulating bushing or raceway may be omitted where the entering conductors pass through masonry or are acceptable for the purpose.

54-608 Lightning conductors

A separation of at least 2 m shall be maintained between conductors of a community antenna distribution circuit on buildings and lightning conductors.

54-610 Swimming pools

Where conductors are installed over or adjacent to swimming pools, they shall be placed in accordance with Rules 68-054 and 68-056.

Underground circuits

54-700 Direct buried systems

Where community antenna distribution conductors are direct buried, the sheath shall be suitable for direct burial and the conductor shall be

(a)   installed outside of the same vertical plane that contains differing underground conductors other than communication conductors, except when installed in accordance with Item (f);

(b)   maintained at a minimum horizontal separation of 300 mm from differing underground conductors other than communication conductors, except when installed in accordance with Item (f);

(c)   placed at a minimum depth of 600 mm, unless rock bottom is encountered at a lesser depth, in which case a minimum depth of 450 mm shall be permitted, except that for service wires under parkways and lawns the depth may be reduced to 450 mm;

(d)   placed with a layer of sand 75 mm deep, both above and below the cable, if in rocky or stony ground;

(e)   placed at a minimum depth of 900 mm under an area that is subject to vehicular traffic, except that the depth may be reduced to 600 mm provided that there is mechanical protection that consists of

(i)   treated plank at least 38 mm thick or other suitable material that shall be placed over the conductor or cable after first backfilling with 75 mm of sand or earth containing no rocks or stones; or

(ii)   a conduit suitable for earth burial placed to facilitate cable replacement and to minimize traffic vibration damage; and

(f)   equipped with a metal shield when placed in a common trench involving random separation with power supply cables or wiring operations at 750 V or less, in which case the community antenna distribution conductors shall not cross under the supply cables.

Δ 54-702 Underground raceway

Where community antenna distribution conductors are placed in underground raceway systems

(a)   the raceway, including laterals, shall be separated from those used for the electric power system by not less than 50 mm of concrete or 300 mm of well-tamped earth;

(b)   the raceway shall be located to maintain a minimum depth of 600 mm in areas subject to vehicular traffic and 450 mm in all other areas, except that where rock bottom is encountered at lesser depth the raceway shall be encased in concrete;

(c)   the raceway shall not terminate in the same maintenance hole, and the conductors or cable assembly shall not be placed in the same maintenance hole, used for an electric power system, unless all requirements of Clause 5 of CSA C22.3 No. 7 are adhered to;

(d)   the conductors shall not be placed in the same raceway containing electric lighting, power, or Class 1 circuit conductors;

(e)   the cable sheath shall be suitable for wet locations; and

(f)   raceways entering a building and forming part of an underground installation shall be sealed with a suitable compound in such a way that moisture and gas will not enter the building and shall

    (i)   enter the building above ground where practicable; or

    (ii)   be suitably drained.

54-704 Underground block distribution

Where the entire street circuit is run underground and the circuit is placed to prevent contact with electric lighting, power, or Class 1 circuits of more than 300 V, insulating bushings or raceways as specified in Rule 54-606 shall not be required where the circuit conductors enter a building.

Receiving equipment and amateur transmitting equipment

54-800 Lightning arresters for receiving stations

(1)   A lightning arrester shall be provided for each lead-in conductor from an outdoor antenna to a receiving station except where such a lead-in conductor is protected by a continuous grounded metal shield between the antenna and the point of entrance to the building.

(2)   Lightning arresters for receiving stations shall be located outside the building or inside the building between the point of entrance of the lead-in and the radio set or transformer, and as near as practicable to the entrance of the conductors to the building.

(3)   Lightning arresters for receiving stations shall not be located near combustible material nor in a hazardous location.

54-802 Lightning arresters for transmitting stations

Each conductor of a lead-in to a transmitting station from an outdoor antenna shall be provided with a lightning arrester or other suitable means that will drain static charges from the antenna system except:

(a)   where protected by a continuous metal shield that is grounded; or

(b)   where the antenna is grounded.

Grounding for receiving equipment and amateur transmitting equipment

54-900 Material for grounding conductor

The grounding conductor shall be of copper, aluminum alloy, copper-clad steel, bronze, or other corrosion-resistant material unless otherwise specified.

54-902 Insulation of grounding conductor

The grounding conductor shall be permitted to be uninsulated.

54-904 Support for grounding conductor

The grounding conductor shall be securely fastened in place and may be directly attached to the supporting surface without the use of insulating supports.

54-906 Mechanical protection of grounding conductor

The grounding conductor shall be protected where exposed to mechanical injury.

54-908 Grounding conductor to be run in a straight line

The grounding conductor shall be run in as straight a line as is practicable from the lightning arresters or antenna mast, or both, to the grounding electrode.

54-910 Grounding electrode

The grounding conductor shall be connected to a grounding electrode as specified in Section 10.

54-912 Grounding conductors

The grounding conductor shall be permitted to be run either inside or outside the building.

54-914 Size of protective ground

The size of the protective grounding conductor for receiving and transmitting stations providing ground connection for mast and lightning arresters shall be in accordance with Section 10.

54-916 Common ground

A single grounding conductor shall be permitted to be used for both protective and operating purposes but must be installed so that disconnection of the operating ground will not affect the protective ground circuit.

54-918 Equipment in hospitals

Exposed non-current-carrying metal parts, if they could become energized, of radio and television equipment installed in basic, intermediate, and critical care areas of hospitals as defined in Section 24 shall also be grounded to conform with Rule 24-104(7).

54-920 Radio noise suppressors

Radio interference eliminators, interference capacitors, or radio noise suppressors connected to power supply leads shall be of a type approved for the purpose and shall not be exposed to mechanical damage.

54-922 Grounding of antennas

Masts, metal support structures, and antenna frames for receiving stations shall be grounded in accordance with Section 10.

Transmitting stations

54-1000 Enclosure of transmitters

Transmitters shall be enclosed in a metal frame or grille, or thoroughly shielded or separated from the operating space by a barrier or other equivalent means.

54-1002 Grounding of transmitters

All exposed metal parts of transmitters, including external metal handles and controls accessible to the operating personnel and accessories such as microphone stands, shall be grounded.

54-1004 Interlocks on doors of transmitters

All access doors of transmitters shall be provided with interlocks that will disconnect all voltages in excess of 250 V when any access door is opened.

54-1006 Amplifiers

Audio-amplifiers that are located outside the transmitter housing shall be suitably housed and shall be located to be readily accessible and adequately ventilated.

Section 56 — Optical fiber cables

Scope

56-000 Scope
This Section applies to the installation of optical fiber cables in conjunction with electrical systems and supplements or amends the general requirements of this Code.

General

56-100 Special terminology
In this Section, the following definition applies:

Optical fiber cable — a cable consisting of one or more optical fibers that transmits modulated light for the purpose of control, signalling, or communications.

56-102 Types
Optical fiber cables shall be grouped into the following three types:
(a) nonconductive cables that contain no metal members and no other electrically conductive materials;
(b) conductive cables that contain non-current-carrying conductive members such as metal strength members, metal vapour barriers, or metal sheaths or shields; and
(c) hybrid cables that contain both optical fiber cables and current-carrying electrical conductors.

56-104 Approvals
(1) Optical fiber cables placed within buildings shall be of the types specified in Table 19.
(2) Optical fiber cables outside of buildings shall be suitable for outdoor installation.

56-106 Acceptance of inspector
Installations of optical fiber cables by an electrical utility or a communication utility in the exercise of its function as a utility shall not be subject to the acceptance of an inspector.

Installation methods

56-200 Nonconductive optical fiber cables (see Appendix B)
(1) Nonconductive optical fiber cables shall not be permitted to occupy the same raceway with conductors of electric light, power, or Class 1 circuits, unless
  (a) the nonconductive optical fiber cables are functionally associated with the electric light, power, or Class 1 circuit not exceeding 750 V; and
  (b) the number and size of nonconductive optical fiber cables and other types of electric conductors in the raceway meet with the applicable requirements for the electrical wiring method.
(2) Nonconductive optical fiber cables shall not be permitted to occupy the same cabinet, panel, outlet box, or similar enclosure housing the electric terminals of an electric light, power, or Class 1 circuit, unless
  (a) the nonconductive optical fiber cables are functionally associated with the electric light, power, or Class 1 circuit not exceeding 750 V, and the number and size of optical fiber cables and other types of electric conductors in the enclosure meet with the applicable requirements for the electrical wiring method; or
  (b) the nonconductive optical fiber cables are factory-assembled in the enclosure.
(3) Notwithstanding Subrules (1) and (2), nonconductive optical fiber cables shall be permitted to occupy the same raceway, cabinet, panel, outlet box, or similar enclosure with functionally associated electric circuits exceeding 750 V for industrial sites if installed and maintained by qualified persons.

56-202 Conductive optical fiber cables (see Appendix B)
(1) Conductive optical fiber cables shall be permitted to occupy the same raceway with any of the following systems:
  (a) Class 2 circuits in accordance with Section 16;
  (b) communication circuits in accordance with Section 60; or
  (c) community antenna distribution and radio and television circuits in accordance with Section 54.
(2) Conductive optical fiber cables shall not be permitted to occupy the same raceway, panel, cabinet, or similar enclosure housing electric light, power, or Class 1 circuits.

(3) Conductive optical fiber cables shall not be permitted to occupy the same cabinet, panel, outlet box, or similar enclosure housing the electrical terminals of a Class 2, communications, community antenna distribution, or radio and television circuit, unless

    (a) the conductive optical fiber cables are functionally associated with the Class 2, communication, community antenna distribution, or radio and television circuit; or

    (b) the conductive optical fiber cables are factory assembled in the enclosure.

(4) The conductive non-current-carrying members of conductive optical fiber cables shall be grounded in accordance with Section 10.

56-204 Hybrid cables

(1) Optical fibers shall be permitted within the same hybrid cable for electric light, power, or Class 1 circuit conductors not exceeding 750 V, or within the same hybrid cable for Class 2, communications, community antenna, or radio and television circuit conductors, provided that the functions of the optical fibers and the electrical conductors are associated.

(2) Hybrid cables shall be classed as electrical cables in accordance with the type of electrical circuit in the conductors and shall be installed in accordance with the Code Rules applicable to the electrical circuit conductors.

56-206 Penetration of a fire separation

Optical fiber cables extending through a fire separation shall be installed to limit fire spread in accordance with Rule 2-124.

56-208 Optical fiber cables in a vertical shaft (see Appendix B)

(1) Optical fiber cables in a vertical shaft shall be in a totally enclosed non-combustible raceway.

(2) Notwithstanding Subrule (1), conductive and non-conductive optical fiber cables shall be permitted to be installed in a vertical shaft without a totally enclosed non-combustible raceway provided that these cables meet the flame spread requirements of the *National Building Code of Canada* or local building legislation for buildings of non-combustible construction.

56-210 Optical fiber cables in ducts and plenum chambers

Optical fiber cables shall not be placed in ducts or plenum chambers except as permitted by Rules 2-126 and 12-010.

56-212 Raceways

Raceways shall be installed in accordance with the requirements of Section 12.

56-214 Grounding of entrance cables (see Appendix B)

Where conductive optical fiber cables are exposed to lightning or accidental contact with electrical light or power conductors, the metal members of the conductive optical fiber cable shall be grounded in the building as close as possible to the point of cable entry.

Section 58

Note — This Section has been deleted.

Section 60 — Electrical communication systems

Scope

60-000 Scope (see Appendix B)
(1)   This Section applies to the installation of communication systems.
(2)   The requirements of this Section supplement or amend the general requirements of this Code.

General

60-100 Special terminology
In this Section, the following definition applies:

Exposed plant — the circuit or any portion of it subject to lightning strikes, to voltage exceeding 300 V rms due to accidental contact with electric lighting or power conductors, to induction from power line unbalance operation or faults, and to ground potential rise.

60-102 Use of approved equipment
Electrical equipment used in the installation of a communication system shall be approved
(a)   if connected to exposed plant;
(b)   if the equipment is connected to a telecommunication network, unless the equipment is specifically allowed by other Rules in this Section; or
(c)   if required by other Rules in this Section.

60-104 Circuits in communication cables (see Appendix B)
Radio and television circuits, remote control circuits, and fire alarm circuits or parts of this equipment shall be
(a)   permitted to use conductors in a communication building entrance cable assembly having other conductors used as communication circuits;
(b)   deemed to be communication circuits for those portions of circuits that use conductors within the communication building entrance cable assembly; and
(c)   suitably protected at the point of interface connection with the communication cable conductors.

60-106 Hazardous locations
Where the wiring or electrical equipment within the Scope of this Section is installed in hazardous locations as defined in Section 18, 20, or 24, it shall also comply with the applicable Rules of these Sections.

60-108 Inspection by an inspector
(1)   Communication circuits employed by an electrical or communication utility in the exercise of its function as a utility shall not be subject to inspection by an inspector.
(2)   Where the communication circuit derives power for operation from a supply circuit, the transformer or other current-limiting device used at the junction of the communication and the supply circuit shall be subject to inspection by an inspector.

60-110 Approved transformers
Where transformers or other devices supply current to a communication circuit from an electric supply circuit, the transformers or other devices shall be of a type approved for the service.

Protection

60-200 Provision of primary protectors (see Appendix B)
(1)   A primary protector shall be provided on each electrical communication circuit, except as permitted in Subrule (4).
(2)   The primary protector shall be located in, on, or immediately adjacent to the structure or building served and as close as practicable to the point at which the conductors enter or attach.
(3)   The primary protector shall not be located in any hazardous location as defined in Sections 18, 20, and 24, nor in the immediate vicinity of flammable or explosive materials.
(4)   A primary protector need not be provided if no portion of the circuit is considered exposed plant.

60-202 Primary protector requirements
(1)   The primary protector shall connect between each line conductor and ground and, when required by Subrule (3), have a fuse connected in series in each line conductor.

(2) Fuseless primary protectors shall be permitted

  (a) on circuits that enter a building in a cable with a grounded metal sheath or shield, provided that the conductors in the cable safely fuse at currents less than the ampacity of the primary protector and the primary protector grounding conductor;

  (b) on circuits served by insulated conductors extending to a building from a grounded metal-sheathed or shielded cable, provided that the conductors in the cable safely fuse at currents less than the ampacity of the primary protector, the associated insulated conductors, and the primary protector grounding conductor;

  (c) on circuits served by insulated conductors, extending to a building from other than grounded metal-sheathed or shielded cable, provided that

    (i) the primary protector grounding conductor is grounded in conformance with Rule 60-704; and

    (ii) the connections of the insulated conductors extending from the building to the exposed plant, or the conductors of the exposed plant, safely fuse at currents less than the ampacity of the primary protector, the associated insulated conductors, and the primary protector grounding conductor; or

  (d) on circuits in a cable with a grounded metal sheath or shield that are subject to lightning strikes but are otherwise not exposed plant.

(3) Where the requirements of Subrule (2) are not met, fused primary protectors shall be placed.

(4) Primary protectors having exposed live parts shall be located in a suitable room or similar area as required by Rule 2-202, separate from electrical light or power installations, except where necessary to place them in a joint-use room, in which case a minimum separation of 900 mm from electrical equipment requiring adjustment and maintenance shall be provided and maintained.

60-204 Protection for communication circuits in high-voltage stations (see Appendix B)

Equipment for the protection of communication circuits used for control and signalling in high-voltage stations shall be suitable for the application.

Inside conductors

60-300 Conductor installations

Rules 60-302 to 60-334 apply to the installation of inside communication conductors.

60-302 Raceways

Raceways for communication conductors shall be installed in accordance with the requirements of Section 12 and, if metal, shall be grounded in accordance with Section 10.

60-304 Insulation

Wire and cable used for communication systems shall be of the approved types as specified in Table 19.

60-306 Grounding of conductors with an outer metal covering

Where a conductor or cable is equipped with an outer metal covering, the covering shall be grounded.

60-308 Separation from other conductors

(1) The conductors of an electrical communication system in a building shall be separated at least 50 mm from any insulated conductor of a Class 1 circuit or an electric light or power system operating at 300 V or less, and shall be separated at least 600 mm from any insulated conductor or an electric light or power system operating at more than 300 V, unless

  (a) one system is in grounded metal raceways, metal-sheathed cable, or grounded armoured cable;

  (b) the Class 1 circuit or electric light or power system operating at 300 V or less utilizes a hard-usage or extra-hard-usage flexible cord as specified in Table 11; or

  (c) both systems are permanently separated by a continuous, firmly fixed non-metal raceway in addition to the insulation of the conductors.

(2) Where the light or power conductors are bare, all communication conductors in the same room or space shall be enclosed in a grounded metal raceway and no opening, such as an outlet box, may be located within 2 m of bare conductors if up to and including 15 kV or within 3 m of bare conductors above 15 kV.

(3) The conductors of an electrical communication system shall not be placed in any outlet box, junction box, raceway, or similar fitting or compartment that contains conductors of electric light or power systems or of Class 1 circuits (as defined in Rule 16-002), unless

(a) the communication conductors are separated from the other conductors by a suitable partition; or

(b) the power or Class 1 conductors are placed solely for the purpose of supplying power to the communication system, or for connection to remote-control equipment, except that no communication conductors installed in an outlet box, junction box, raceway, or similar fitting or compartment that contains such conductors of power or Class 1 circuits shall show a green-coloured insulation, unless such a communication conductor is completely contained within a sheathed- or jacketed-cable assembly throughout the length that is present in such raceways or enclosures.

(4) The conductors of an electrical communication system in a building shall not be placed in a shaft with the conductors of an electric light or power system, unless

(a) the conductors of all systems are insulated and are separated by at least 50 mm; or

(b) the conductors of either system are encased in non-combustible tubing.

60-310 Penetration of a fire separation

Conductors of communication circuits extending through a fire separation shall be installed to limit fire spread in accordance with Rule 2-124.

60-312 Communication cables in hoistways

(1) Special permission shall be required to install communication conductors in hoistways.

(2) All conductors, except travelling cables, shall be totally enclosed in continuous metal raceway.

(3) Pullboxes required for communication interconnection shall be located outside the hoistway.

60-314 Communication conductors in ducts and plenum chambers

Communication conductors shall not be placed in ducts or plenum chambers except as permitted by Rules 2-126 and 12-010.

60-316 Data processing systems

The interconnecting cables used in data processing systems shall be permitted to contain power and communication conductors where such cables are specifically approved for the purpose.

60-318 Conductors under raised floors

Conductors of communication circuits shall be permitted to be installed without additional mechanical protection under a raised floor, provided that

(a) the raised floor is of suitable non-combustible construction;

(b) a minimum separation of 50 mm is provided and maintained where the conductors are used to serve data processing systems and are placed parallel to any other power supply wiring; and

(c) the conductors serve the equipment located only on the floor above the raised floor, where the space under the raised floor is used as an air plenum.

60-320 Conductors in concealed installations

Where the ends of cables or conductors are not terminated on a device, they shall be capped or taped.

60-322 Type CFC under-carpet wiring system Rules

Rules 60-324 to 60-334 apply to the installation of Communication Flat Cable type (CFC) systems.

60-324 Use permitted

Type CFC system wiring shall be permitted to be used

(a) only under carpet squares not exceeding 750 mm, and any adhesive used shall be of the release type;

(b) as an extension of conventional wiring to serve areas or zones, and each run of wiring from the transition point shall not exceed 15 m;

(c) on hard, smooth continuous floor surfaces made of concrete if sealed, ceramic, or composition flooring, wood, or similar material;

(d) in dry or interior damp locations; and

(e) on floors heated in excess of 30 °C only if approved and identified for that purpose.

60-326 Use prohibited

Type CFC system wiring shall not be used

(a) outdoors or in wet locations;

(b) where subject to corrosive vapours or liquids;

(c) in hazardous locations;

(d) in dwelling units;

(e) in hospitals or institutional buildings except in office areas;

(f) on walls except when entering the transition point; or

(g)   under permanent-type partitions or walls.

60-328 Floor protective coverings

Type CFC system wiring shall be covered with abrasion-resistant tape, secured to the floor, to completely cover all cables, corners, and bare conductor ends.

60-330 Coverings

Type CFC system wiring shall be permitted to cross over or under each other, and over or under power supply type FCC system wiring provided that there is a layer of grounded metal shielding between the FCC and CFC system cables as required in Rule 12-820.

60-332 System height

Type CFC system wiring shall not be stacked on top of each other except as required to enter the transition point.

60-334 Grounding of shields

Type CFC system wiring equipped with a metal shield shall be grounded.

Equipment

Δ 60-400 Communication equipment in bathrooms and in areas adjacent to pools

(1)   Communication equipment located in a bathroom shall be permanently fixed on the wall and shall be located so that no part may be reached or used from the bath or from the shower enclosure; however, it shall be permitted to be actuated by means of a cord with an insulating link.

(2)   Communication jacks shall not be located in a bathroom.

(3)   Communication equipment located in areas adjacent to pools shall be installed in accordance with Rule 68-070.

60-402 Equipment in air ducts, plenums, or suspended ceilings (see Appendix B)

Communication equipment and terminals shall not be placed in ducts, plenums, or hollow spaces that are used to transport environmental air nor in suspended ceiling areas, except that where a duct, plenum, or hollow space is created by a suspended ceiling having lay-in panels or tiles, connecting blocks that are a non-protective type shall be permitted to be installed if they are placed in an accessible enclosure.

60-404 Exposed equipment and terminations

Exposed communication equipment and/or associated terminations shall be located, as required by Rule 2-202, in a suitable room or similar area that is separate from electrical light or power installations, except where it is necessary to place them in a joint-use room, in which case a minimum separation of 900 mm from electrical equipment requiring adjustment and maintenance shall be provided and maintained.

60-406 Ground start circuits

Communication circuits connected to a telecommunication network and having return paths via local ground or other circuitry that similarly could present a fire hazard shall be provided with a current-limiting device installed in or adjacent to the equipment of a type recommended by the equipment manufacturer as suitable for the application, which will limit the current under normal operating conditions and under fault conditions, to prevent fire hazards.

60-408 Communication systems in hospitals

If they could become energized, exposed non-current-carrying metal parts, of communications equipment, other than telephone sets, installed in general, intermediate, and critical care areas of hospitals as defined in Section 24, shall also be grounded to conform with Rule 24-104(7).

Outside conductors

60-500 Overhead conductors on poles

The installation of overhead communication conductors on poles in proximity with power conductors shall be established in accordance with the *Canadian Electrical Code, Part III*.

60-502 Overhead conductors on roofs

(1)   Communication conductors passing over buildings shall be kept at least 2.5 m above any roof that may be readily walked upon.

(2)   Communication conductors shall not be attached to the upper surfaces of roofs or be run within 2 m, measured vertically, of a roof unless a deviation has been allowed in accordance with Rule 2-030.

(3)   A deviation in accordance with Rule 2-030 shall not be necessary where the building is a garage or other auxiliary building of one storey.

60-504 Circuits requiring primary protectors

Communication circuits that require primary protectors in accordance with Rule 60-200 of this Code shall comply with Rules 60-506 to 60-514.

60-506 Wire insulation

In a communication circuit requiring a primary protector, each wire shall have rubber or thermoplastic insulation and shall

(a)   have a protective jacket placed over individual or groups of wires, which may be integral with the insulation; and

(b)   be suitable for the application and in accordance with the manufacturer's recommendations.

60-508 Cable insulation

(1)   Wires within a cable used for communication circuits requiring primary protectors shall be permitted to have paper, thermoplastic, or other suitable insulation.

(2)   The cable shall be of a type suitable for the application and in accordance with the manufacturer's recommendations, with

(a)   a metal sheath; or

(b)   a composite sheath having a metal shield and overall outer protective rubber or thermoplastic jacket; or

(c)   a protective rubber or thermoplastic jacket without a metal shield.

60-510 Communication conductors on buildings

(1)   Communication conductors on buildings shall be separated from insulated light or power conductors not in conduit by at least 300 mm unless permanently separated by a continuously and firmly fixed non-metal raceway, in addition to the insulation on the conductors.

(2)   Where the light or power conductors are bare, the communication conductors shall be in the lower position and, in order to provide adequate working space, the clearance given in Subrule (1) shall be increased to a minimum of 600 mm from a conductor operating at 750 V or less.

(3)   Communication conductors subject to accidental contact with light or power conductors operating at voltages exceeding 300 V, and attached exposed to buildings, shall be separated from combustible material by being supported on glass, porcelain, or other suitable insulating material, except that such separation is not required where fuses are omitted as provided for in Rule 60-202(2), or where conductors are used to extend circuits to a building from a cable having a grounded metal sheath or shield.

60-512 Communication conductors entering buildings

Where a primary protector is installed inside the building, the communication conductors shall enter the building either through a non-combustible, non-absorptive insulating bushing, or through a metal raceway, except that the insulating bushing shall be permitted to be omitted where the entering conductors

(a)   are in metal-sheathed or shielded cable;

(b)   pass through masonry;

(c)   are without fuses in the primary protectors as provided for in Rule 60-202(2); or

(d)   are used to extend circuits to a building from a cable having a grounded metal sheath or shield.

60-514 Communication conductors entering mobile homes

Communication conductors shall enter the mobile home using the facility provided by Rule 70-106.

60-516 Lightning conductors

A separation of at least 2 m shall, where practicable, be maintained between conductors of communication circuits on buildings and lightning conductors.

60-518 Swimming pools

Where wires or cables are installed over or adjacent to swimming pools, they shall be placed in accordance with Rules 68-054 and 68-056.

Underground circuits

Δ **60-600 Direct buried systems**

Where communication conductors or cable assemblies are direct buried, they shall be suitable for direct burial and the conductor or cable assembly shall

(a)   not be installed in the same vertical plane with other underground systems, except when installed in accordance with Item (g);

(b)   maintain a minimum horizontal separation of 300 mm from other underground systems, except when installed in accordance with Item (g);

(c)   be not less than 600 mm deep, unless rock bottom is encountered at a lesser depth, in which case a minimum depth of 450 mm shall be permitted, except that for service wires under parkways and lawns the depth may be reduced to 450 mm;

(d)   be placed with a layer of sand 75 mm deep, both above and below the cable, if in rocky or stony ground;

(e)   be not less than 900 mm deep under an area that is subject to vehicular traffic, except that the depth shall be permitted to be reduced to 600 mm if mechanical protection is provided that consists of

    (i)   treated plank at least 38 mm thick or other suitable material, which shall be placed over the conductor or cable after first backfilling with 75 mm of sand or earth containing no rocks or stones; or

    (ii)   a conduit suitable for earth burial placed to facilitate cable replacement and to minimize traffic vibration damage;

(f)   not be placed in a common trench involving random separation with power supply cables or conductors operating at over 750 V, except when the applicable requirements of CSA C22.3 No. 7 are adhered to and the power supply cables' or conductors' operating voltage does not exceed 22 kV line to ground; and

(g)   have a metal sheath when placed in a common trench involving random separation with power supply cables or conductors, in which case the communication conductor or cable assembly shall not cross under the supply cables.

Δ **60-602 Underground raceway**

Where communication conductors or cable assemblies are placed in underground raceway systems

(a)   the raceway, including laterals, shall be separated from those used for the electric power system by not less than 50 mm of concrete or 300 mm of well-tamped earth;

(b)   the raceway shall be located to maintain a minimum depth of 600 mm in areas subject to vehicular traffic and 450 mm in all other areas, except that where rock bottom is encountered at a lesser depth the raceway shall be encased in concrete;

(c)   the raceway shall not terminate in the same maintenance hole, and the conductors or cable assembly shall not be placed in the same maintenance hole, used for the electric power system, unless all requirements of Clause 5 of CSA C22.3 No. 7 are adhered to;

(d)   the cables shall not be placed in the same raceway containing electric lighting or power supply cables;

(e)   the cables shall be suitable for wet locations; and

(f)   raceways entering a building and forming part of an underground installation shall be sealed with a suitable compound in such a way that moisture and gas will not enter the building and shall

    (i)   enter the building above ground where practicable; or

    (ii)   be suitably drained.

60-604 Underground block distribution

Where the entire street circuit is run underground and the part of the circuit within the block is placed so that it is not liable to contact with electric lighting or power circuits of more than 300 V

(a)   no primary protector is required as specified in Rule 60-200;

(b)   the insulation requirements of Rules 60-506 and 60-508 shall not apply;

(c)   conductors need not be placed on insulating supports as specified in Rule 60-510(3); and

(d)   where the conductors enter the building, no bushings are required as specified in Rule 60-512.

Grounding

60-700 Bonding of cable sheath (see Appendix B)

Where cables, either aerial or underground, enter buildings, the metal sheath or shield of the cable shall be bonded to ground as close as practicable to the point of entrance or shall be interrupted as close as practicable to the point of entrance by an insulating joint or equivalent device.

60-702 Cable sheath bonding conductor (see Appendix B)

The cable sheath bonding conductor required by Rule 60-700 shall have an ampacity at least equal to, or greater than, that of the outer conductive sheath of the exposed cable, except that the bonding conductor shall not be required to be larger than No. 6 AWG copper.

60-704 Primary protector grounding conductor (see Appendix B)

The grounding conductor used to ground primary protectors specified in Rule 60-202 shall be copper and shall

(a)   have rubber or thermoplastic insulation;

(b)   be not smaller than the required grounding conductor specified in Table 59;

(c)   be run from the primary protector to the point of connection described in Rule 60-706 in as straight a line as possible; and

(d)   be guarded from mechanical injury, where necessary.

60-706 Grounding electrode (see Appendix B)

(1)   The grounding conductor shall preferably be connected to a water pipe grounding electrode, as close to the point of entrance as possible.

(2)   Where the water pipe is not readily available and the grounding conductor of the power consumer's service is connected to the water pipe at the building, the primary protector grounding conductor shall be permitted to be connected to the metal conduit, service equipment enclosures, or to the grounding conductor of the power consumer's service.

(3)   In the absence of a water pipe, the communication primary protector grounding conductor shall be permitted to be connected to an effectively grounded metal structure, or to a ground rod or pipe driven into permanently damp earth, but

(a)   steam, gas, or hot water pipes, or lightning rod conductors shall not be used as grounding electrodes; and

(b)   a driven rod or pipe used for grounding power circuits shall not be used as a communication primary protector grounding electrode unless it is connected to the grounded conductor of a multi-grounded power neutral.

(4)   Where a driven ground rod or pipe is used as a grounding electrode for an electrical communication system, it shall be separated by at least 2 m from any other electrode, including those used for power circuits, radio, lightning rods, or any other purpose, and shall be connected only to that of the power circuits in accordance with Rule 10-702.

(5)   The normal length of a driven ground rod used as the grounding electrode for a communication station primary protector is 1.5 m, but where the normal rod would not reach moist soil when installed, a rod of suitable additional length shall be used.

60-708 Grounding electrode connection

The grounding conductor shall be attached to the grounding electrode as required in Rule 10-908

(a)   directly; or

(b)   to a wire lead permanently connected to the ground rod electrode in a manner specified in CSA C83.

60-710 Bonding of electrodes

A copper conductor not smaller than No. 6 AWG shall be connected between communication and power grounding electrodes when separate artificial grounding electrodes are required as described in Rule 60-706.

Section 62 — Fixed electric space and surface heating systems

Scope

62-000 Scope

(1)   This Section applies to

    (a)   fixed electric space-heating systems for heating rooms and similar areas; and

    (b)   fixed surface heating systems for pipe heating, melting of snow or ice on roofs or concrete or asphalt surfaces, soil heating, and similar applications other than space heating.

(2)   The requirements of this Section supplement or amend the general requirements of this Code.

General

62-100 General Rules

Rules 62-102 to 62-130 apply to both fixed space and surface heating installations.

62-102 Special terminology

In this Section, the following definitions apply:

Central unit — any heating unit (or group of units assembled to form a complete unit) permanently installed in such a way that it can convey heat to rooms or areas using air, liquid, or vapour flowing through pipes or ducts; and including duct heaters.

Fixture — any heating unit (or group of units assembled so as to form a complete unit) permanently installed in such a manner that it can be removed or replaced without removing or damaging any part of the building structure.

Heating device — any form of electrical heating device, including cable, fixture, panel, and strip system.

Heating device set — a heating device assembled with the associated parts necessary to connect it to a source of electrical supply.

Heating panel — a rigid or non-rigid laminated plane section in which the heating element consisting of a continuously parallel resistive material, a series resistive material, or a parallel-series resistive material is embedded between or in sheets of electrical insulating material.

Heating panel set — a heating panel together with cold leads or a non-heating portion.

Parallel heating cable — a cable incorporating heating elements connected in parallel either continuously or intermittently such that the watt density along the length of the cable is not altered by changes in the cable length.

Parallel heating cable set — the combination of a parallel heating cable and associated parts necessary to connect it to a source of electrical supply.

Sauna heater — a device that is designed for heating air and that is installed permanently in a special room to produce a hot atmosphere with generally a relatively low humidity, although brief excursions to relatively high humidity may take place.

Series heating cable — a cable using a series resistance conductor(s).

Series heating cable set — the combination of a series heating cable and a means of connecting it to a source of electrical supply where the combination is assembled by the manufacturer.

62-104 Special locations

Heating equipment that is installed in hazardous locations or subject to wet or corrosive conditions shall be marked as suitable for the particular location.

62-106 Terminal connections

(1)   Connections to heating equipment shall be made in terminal fittings or boxes, and equipment shall be installed so that connections between circuit conductors and equipment conductors are accessible without disturbing any part of the wiring.

(2)   Where the connections of Subrule (1) are made in terminal fittings, they shall be contained in an enclosure of non-combustible material.

(3) Where the temperature at the point of connection between branch circuit conductors and heating unit exceeds 60 °C, the branch circuit conductors shall be installed in accordance with Rule 30-910.

62-108 Branch circuits

(1) Branch circuit conductors used for the supply of energy to heating equipment shall
   (a) be used solely for such equipment;
   (b) have an ampacity not less than that of the connected load supplied; and
   (c) conductors having insulation suitable for the temperatures encountered shall be used for branch circuits supplying a heating unit.

(2) For the purpose of this Rule, an approved unit that combines heating with ventilating or lighting equipment, or both, shall be considered to be heating equipment.

Δ (3) Notwithstanding Subrule (1), where a heat lamp is not the sole source of heat, it shall be permitted to be used in a luminaire approved for the purpose or in a box-mount-type luminaire, where the luminaire is supplied from a general-use branch circuit.

62-110 Installation of fixtures

(1) Fixtures shall be installed so that
   (a) the proper radiation of heat shall not be obstructed by any portion of the building structure;
   (b) adjacent combustible material shall not be subjected to temperatures in excess of 90 °C.

(2) Where a fixture is recessed in non-combustible material in a building of concrete, masonry, or equally non-combustible construction, the non-combustible material shall be permitted to be subjected to temperatures not exceeding 150 °C, but the fixture shall be plainly marked as suitable for the service.

(3) Fixtures weighing more than 4.54 kg shall be installed so that the wiring connections in the outlet box or its equivalent will be accessible for inspection without removing the fixture supports.

(4) Where the weight of a fixture does not exceed 13 kg, the fixture shall be permitted to be supported by a wall outlet box attached directly to the building structure or by a wall outlet box attached to a bar hanger.

(5) Where the weight of a fixture does not exceed 23 kg, the fixture shall be permitted to be supported by a ceiling outlet box attached directly to the building structure or by a ceiling outlet attached to a bar hanger.

(6) Where the weight of a fixture prohibits the installation methods specified in Subrules (4) and (5), the fixture shall be supported
   (a) independently of the outlet box; or
   (b) by a fixture hanger provided with an integral outlet box suitable for the purpose.

(7) When fixtures are installed less than 5.5 m above the floor in an arena, gymnasium, or similar location where they may be exposed to damage from flying objects, the heating elements shall be the metal-sheathed type, or the fixture shall be suitable for the application.

62-112 Fixtures as raceways

(1) No fixture shall be used as a raceway for circuit conductors unless the fixture is marked for this use.

(2) Notwithstanding Subrule (1), use of the wiring channel of a baseboard heating unit to contain the wiring for the interconnection of adjacent baseboard units on the same branch circuit shall be permitted if the units are marked for this use.

62-114 Overcurrent protection and grouping (see Appendix B)

(1) Every fixture, cable set, heating panel set, or parallel heating set having an input of more than 30 A shall be supplied by a branch circuit that supplies no other equipment.

(2) In buildings for residential occupancy, two or more fixtures, cable sets, or heating panel sets shall be permitted to be connected to a branch circuit used for space heating, provided that the branch circuit overcurrent devices are rated or set at not more than 30 A.

(3) In other than buildings for residential occupancy,
   (a) two or more fixtures, cable sets, heating panel sets, or parallel heating sets shall be permitted to be grouped on a branch circuit, and the branch circuit overcurrent devices shall not be set or rated in excess of 60 A unless a deviation has been allowed in accordance with Rule 2-030 to use overcurrent devices having a higher setting or rating;
   (b) where three fixtures, cable sets, heating panel sets, or parallel heating sets are grouped on a branch circuit in a balanced 3-phase arrangement, the branch circuit overcurrent devices shall be permitted to be set or rated in excess of 60 A.

(4) Where two or more fixtures, cable sets, heating panel sets, or parallel heating sets are grouped on a single branch circuit, the non-heating leads of cable sets and taps to cable sets, fixtures, and strip systems shall

(a) have an ampacity not less than one-third the rating of the branch circuit overcurrent device; and

(b) be not more than 7.5 m in length.

(5) Where the heating portion of a cable set is not totally embedded in non-combustible material, the rating or setting of the branch circuit overcurrent devices shall not exceed 15 A.

(6) Where a service, or feeder, or branch circuit is used solely for the supply of energy to heating equipment, the load, as determined using Rule 62-116, shall not exceed

(a) 100% of the rating or setting of the overcurrent devices protecting the service conductors, feeder conductors, or branch circuit conductors when the service box, fusible switch, circuit breaker, or panelboard is marked for continuous operation at 100% of the ampere rating of its overcurrent devices; or

(b) 80% of the rating or setting of the overcurrent devices protecting the service conductors, feeder conductors, or branch circuit conductors when the service box, fusible switch, circuit breaker, or panelboard is marked for continuous operation at 80% of the ampere rating of its overcurrent devices.

(7) Service, feeder, or branch conductors supplying only fixed resistance heating loads shall be permitted to have an ampacity less than the rating or setting of the circuit overcurrent protection, provided that their ampacity is

(a) not less than the load; and

(b) at least 80% of the rating or setting of the circuit overcurrent protection.

(8) Notwithstanding Subrule (7)(b), where 125% of the allowable ampacity of a conductor does not correspond to a standard rating of the overcurrent device, the next higher standard rating shall be permitted.

62-116 Demand factors for service conductors and feeders

(1) Where service conductors or feeders are used solely for the supply of energy to heating equipment, they shall have an ampacity not less than the sum of the current ratings of all the equipment they supply.

(2) Notwithstanding Subrule (1), where a heating installation in a building for residential occupancy is provided with automatic thermostatic control devices in each room or heated area, the ampacity of service conductors or feeders supplying heating equipment only shall be based on the following:

(a) the first 10 kW of connected heating load at 100% demand factor; plus

(b) the balance of the connected heating load at 75% demand factor.

(3) Where service conductors or feeders supply a combined load of heating and other equipment, they shall have an ampacity consisting of the following:

(a) in the case of buildings for residential occupancy, the sum of the heating load as computed by Subrule (2) plus the combined loads of other equipment with demand factors as applicable in Section 8; or

(b) in the case of other occupancies, 75% of the total connected heating load plus the combined loads of the other equipment with demand factors as applicable in Section 8 for the type of occupancy.

(4) Notwithstanding Subrule (3)(b), where the combined load with applicable demand factors of other than heating equipment is less than 25% of the connected heating load on a service or feeder, no demand factor shall be applicable to the heating portion of the load.

62-118 Temperature control devices

(1) Temperature control devices rated to operate at line voltage shall have a current rating at least equal to the sum of the current ratings of the equipment they control.

(2) Temperature control devices that can be turned automatically or manually to a marked OFF position and that either interrupt line current directly or control a contactor or similar device that interrupts line current shall open all ungrounded conductors of the controlled heating circuit when in the OFF position.

(3) Where the liquid to be heated is a fuel or other flammable product, temperature controls shall be installed to ensure that the liquid temperature does not exceed the minimum flash point of the liquid.

62-120 Construction of series heating cable sets

Series heating cable sets shall be complete assemblies, including both the heating portion and the non-heating end leads, and shall have permanent markings as required located on one or both of the non-heating leads not more than 75 mm from the supply end of a non-heating lead.

62-122 Installation of series heating cable sets

(1)   The heating portion of a series heating cable set shall not be shortened, and any cable set that does not bear its original markings shall be considered to have been shortened and will be rejected unless the installer can prove, by instrument measurements, that the characteristics of the series heating cable set have not been altered.

(2)   The entire length of the heating portion, including connections to non-heating leads, shall be installed within the heating area.

(3)   Series heating cable sets shall be installed so that the temperature on any part will not exceed 90 °C except as permitted in Rule 62-304(1).

(4)   The heating portions of series heating cable sets shall not be run closer than 200 mm to any outlet to which a lighting fixture or other heat-producing equipment is liable to be connected.

(5)   Where series heating cable sets without metal shields or sheaths are installed, metal structures or materials used for the support of such series heating cable sets shall be bonded to ground.

(6)   Where a series heating cable set is liable to accidental contact with conductive material that is not effectively bonded to ground, the heating portion of the series heating cable set shall have a metal shield or sheath.

(7)   Metal shields and sheaths of series heating cable sets shall be bonded to ground.

62-124 Field repair, modification or assembly of series heating sets (see Appendix B)

Notwithstanding Rule 62-122, at industrial establishments where conditions of maintenance and supervision ensure use by qualified persons, series heating cable sets shall be permitted to be

(a)   field-repaired or spliced with splice kits supplied by the cable manufacturer, provided that the total length of the heating portion of the sets is not changed by more than 3%; and

(b)   Field-modified or field-assembled with splice and termination kits supplied by the cable manufacturer, provided that

   (i)   a permanent tag with the new design information is installed;

   (ii)   a permanent record of the new design information is retained;

   (iii)   tests for insulation resistance and verification of the finished cable set resistance are made; and

   (iv)   the design, or modifications to a design, are done by a qualified person and reviewed by the cable manufacturer.

62-126 Field-assembled series heating cable sets for embedding in concrete indoors (see Appendix B)

(1)   Heating devices for embedding in concrete indoors shall indicate a wet rating where the application is likely to be subjected to wet conditions.

(2)   Notwithstanding Rules 62-120 and 62-122(1), series heating cable sets forming part of a heating cable system for embedding in concrete indoors and approved for assembly at the time of installation shall be permitted to be installed.

(3)   The electrical rating of the series heating cable sets referred to in Subrule (1) shall be marked in the junction box that is provided as part of the system and that encloses the connection between the branch circuit conductors and the non-heating end leads.

(4)   Notwithstanding Rules 62-128 and 62-218(2), the series heating cable sets referred to in Subrule (1), subject to the conditions of approval, shall be permitted to be installed with the joint between the heating portion and the non-heating end leads in the supply junction box forming part of the system, provided that the heating portion is contained within a raceway between the point where it leaves the concrete and enters the box.

62-128 Non-heating end leads of series heating cable sets and heating panel sets

(1)   The non-heating end leads of series heating cable sets and heating panel sets shall be installed in accordance with the requirements of Section 12 for the type of conductors employed.

(2)   Where the heating element of a series heating cable set is embedded in a concrete or similar floor, the non-heating end leads, if not the metal-sheathed type, shall be run from within the concrete to the junction box in rigid conduit, electrical metallic tubing, or other approved raceway, which shall terminate in a horizontal run within the concrete and have a bushing or equivalent fitting to prevent abrasion of the conductors where they emerge.

62-130 Heating panel and heating panel sets

(1)   Heating panels shall be complete assemblies including terminal fittings.

(2) Heating panel sets shall be complete assemblies including the terminal fittings and the non-heating leads.

(3) Single-conductor non-heating leads of heating panels and panel sets installed as concealed wiring shall be the types permitted in Table 19.

Electric space-heating systems

62-200 Electric space heating
Rules 62-202 to 62-226 apply to fixed electric space-heating systems for heating rooms and similar areas.

62-202 Temperature control
Each enclosed area within which a heater is located shall have a temperature control device.

62-204 Connections to branch circuit conductors
(1) A cable set or heating panel used for interior space heating shall have non-heating end leads for connection to branch circuit conductors.

(2) For the heating panel referred to in Subrule (1), the non-heating end leads shall be permitted to be attached at the time of installation in accordance with the manufacturer's instructions.

62-206 Proximity of other wiring (see Appendix B)
Wiring of other circuits located

(a) above heated ceilings shall be spaced not less than 50 mm above the ceiling and shall be considered as operating at an ambient temperature of 50 °C, unless thermal insulation having a minimum thickness of 50 mm is interposed between the wiring and the ceiling;

(b) in heated concrete slabs shall be spaced not less than 50 mm from the heating cables and shall be considered as operating at an ambient temperature of 40 °C.

62-208 Installation of central units
(1) Central units shall be installed so that there is reasonable accessibility for repair and maintenance.

(2) Central units shall be installed

(a) in an area that is large compared with the physical size of the unit unless specifically approved for installation in an alcove or closet; and

(b) to comply with the clearances from combustible materials as specified on the nameplate.

62-210 Wattage of heating panels and panel sets
In accordance with the manufacturer's instructions and Rule 62-214, the heating portion of the heating panels and panel sets, when in contact with gypsum board or plaster lath or when embedded in plaster, shall not have a watt density that will produce an exposed ceiling surface temperature in excess of the limiting temperature of the ceiling finish materials used.

62-212 Location of heating panels or heating panel sets
(1) The heating portion of heating panels or heating panel sets shall not be

(a) installed in or behind any wall surface, nor in any location where it may be subject to mechanical injury either during or after construction; or

(b) run through walls, partitions, floors, or similar portions of structures.

(2) The heating panels or heating panel sets shall be permitted to be in contact with thermal insulation but shall not be run in or through thermal insulation.

Δ 62-214 Installation of heating panels and heating panel sets
(1) Field-made connections necessary to assemble heating panel sets shall be permitted to be inaccessible provided that they are accessible before ceiling finishing materials are applied and the connectors and enclosures are part of the heating panel sets.

(2) Nailing or stapling of the heating panels and heating panel sets to the ceiling or to the floor joist, headers, or nailing strips shall be done only through the unheated strips provided for this purpose.

(3) Heating panels and heating panel sets shall not be cut through or nailed through any point closer than 6 mm to the element.

(4) The heating portion of the heating panels and heating panel sets shall not be installed closer than 200 mm to any outlet to which a lighting fixture or other heat-producing equipment is liable to be connected.

(5) Heating panels and heating panel sets shall not be installed above or below cupboards, walls, or other obstructions.

(6) Heating panels and heating panel sets shall be permitted in ceilings and below floors of clothes closets if they are provided with an independent temperature control.

(7)   Heating panels and heating panel sets in ceilings shall be parallel to and secured to the lower face of joists, headers, or nailing strips.

(8)   Heating panels and heating panel sets below floors shall be installed parallel to the joist so that the minimum air gap between the bottom of the subfloor and heating panels and heating panel sets is 50 mm.

(9)   The ceiling finish material shall be secured so that nails or other fastening devices do not pierce the heating panels.

(10) Heating panels and heating panel sets in ceilings shall be installed at least 300 mm from all walls and partitions.

(11) Branch circuits supplying heating panels and heating panel sets shall be marked by a warning label supplied by the heating panel and heating panel set manufacturer and affixed to the panelboard by the installer, stating that the ceiling or floor (as applicable) supplied by the branch circuit contains live wiring and should not be penetrated by nails, screws, or similar devices.

62-216 Heating cable sets in ceilings

Heating cable sets installed in ceilings shall be permitted to be marked as being specifically intended for this application.

62-218 Series heating cable sets in cement or plaster ceilings

(1)   Series heating cable sets installed in cement or plaster shall be secured in place on the undercoat, gypsum board, or plaster lath at intervals not over 600 mm by fastening devices that are suitable for the temperature involved and not likely to damage the cable.

(2)   The entire length of the heating portion, including the connections to the non-heating leads, shall be completely embedded in non-combustible material.

(3)   Where series heating cable sets are installed in plastered ceilings, the plaster shall be a thermally non-insulating sand plaster, or equivalent, having a nominal thickness of not less than 13 mm.

62-220 Series heating cable sets in "dry-board" installations

(1)   For "dry-board" installations, the cable shall be installed parallel to the joist or nailing strips, leaving a clear space of not less than 25 mm wider than the width of the lower face of the joist, header, or nailing strip, between centres of adjacent cable runs.

(2)   Crossing of joists by cable shall be done only at the ends of the joists except where a deviation has been allowed in accordance with Rule 2-030.

(3)   After the heating cable is installed

(a)   the entire ceiling below the cable shall be covered with gypsum board not exceeding 13 mm in thickness;

(b)   the voids between the upper layer of gypsum board and the surface layer of gypsum board shall be filled with thermally conducting plaster or other suitable material; and

(c)   the surface layer of gypsum board shall be mounted so that the nails or other fastenings do not pierce the heating cable.

62-222 Wattage rating and spacing of series heating cable sets (see Appendix B)

(1)   Series heating cable sets, when in contact with gypsum board or plaster lath, or when embedded in plaster or sand that is in contact with gypsum board or plaster lath, shall not

(a)   have a rating in excess of 9 W/m of the heating portion; and

(b)   be spaced closer than 50 mm on centres.

(2)   Series heating cable sets, when embedded in concrete or poured masonry, shall not

(a)   have a rating in excess of 65 W/m of heating portion, unless no adjacent heating cable is closer than 450 mm when up to 100 W/m may then be used;

(b)   be spaced closer than 25 mm on centres; and

(c)   have watts per square metre in excess of 430 W.

62-224 Location of series heating cable sets

The heating portions of series heating cable sets shall not be

(a)   installed in or behind any wall surface, nor in any other location where they may be subject to mechanical injury either during or after construction;

(b)   installed in, nor concealed behind, any surface having wood lath, wood panelling, or similar combustible material;

(c)   run through walls, partitions, floors, or similar structures; or

(d)   run in or through any thermal insulation.

62-226 Installation of under-floor covering heating panel sets

(1)   The heating panel sets shall be
  (a)   installed on floor surfaces that are smooth and flat;
  (b)   installed indoors in dry locations;
  (c)   completely covered by tiles, carpet squares, carpet, or similar floor covering; and
  (d)   provided with ground fault protection to de-energize the panel sets with a ground fault setting sufficient to allow their normal operations.

(2)   The heating panel sets shall not be
  (a)   cut through or nailed; or
  (b)   installed closer than 200 mm to any heating supply duct, other heating panels, or any other sources of heat.

(3)   Notwithstanding Rule 12-806(c), Type FCC non-heating leads shall be permitted to be used in dwelling units for connecting under-floor covering heating panel sets to the branch circuit.

(4)   Branch circuits supplying the heating panel sets shall be marked by a warning label supplied by the heating panel set manufacturer and affixed to the panelboard by the installer stating that the floor supplied by the branch circuit contains live wiring and should not be penetrated by nails, screws, or similar devices.

Electric surface heating systems

62-300 Electric surface heating (see Appendix B)

(1)   Rules 62-302 to 62-316 apply to fixed surface heating systems for pipe heating, melting of snow or ice on roofs or concrete or asphalt surfaces, soil heating, and similar applications other than space heating.

(2)   Only heating cable sets having a metal braid or sheath shall be permitted to be installed in accordance with Rules 62-308 and 62-312.

(3)   Only heating panel sets having a metal covering over the heat source on the side opposite to the one in contact with the surface to be heated shall be permitted to be installed in accordance with Rule 62-314.

Δ (4)   Ground fault protection shall be provided to de-energize all normally ungrounded conductors of electric heating cable sets and heating panel sets, with a ground fault setting sufficient to allow normal operation of the heater.

(5)   In establishments where conditions of maintenance and supervision ensure that only qualified persons will service the installed systems and that continued circuit operation is necessary for safe operation of equipment or processes, alarm indications of ground fault shall be permitted in place of the requirements of Subrule (4).

62-302 Installation of fixtures

(1)   If located where they will be exposed to rainfall, fixtures shall be provided with a weatherproof enclosure.
(2)   All exposed metal surfaces of fixtures shall be bonded to ground.

62-304 Installation of heating devices — General

(1)   Heating devices shall be installed so that adjacent materials will not be subjected to temperatures in excess of 90 °C unless a deviation has been allowed in accordance with Rule 2-030 for the use of higher temperatures and the heating devices are approved for such higher temperature.

(2)   No heating device shall be installed closer than 13 mm to any exposed combustible surface unless the cable has a metal shield or sheath and is provided with a positive temperature control that will limit the surface temperature of the heating device to a value not exceeding 72 °C.

62-306 Heating cables and heating panels installed below the heated surface

(1)   Heating cables and heating panels installed outdoors under the surface of driveways, sidewalks, and similar locations shall
  (a)   have a metal shield or sheath over the heating portion;
  (b)   be surrounded by non-combustible material throughout their length, including the point of connection to the non-heating leads;
  (c)   when embedded in concrete, be embedded to a depth of at least 50 mm, the concrete being reinforced except in sidewalks, and have a minimum depth of 150 mm where subject to vehicular traffic, or 100 mm where not subject to vehicular traffic;
  (d)   when embedded in asphalt
    (i)   be embedded

(A) at least 25 mm after first being covered with iron or steel mesh not less than No. 10 gauge or greater than 100 mm mesh; or

(B) at least 25 mm after first being fastened securely to an asphaltic or equivalent base slab not less than 25 mm thick at intervals not exceeding 750 mm; and

(ii) be installed so that adjacent runs of cable are 150 mm or less apart and be rated at not more than 82 W/m;

(iii) be located not less than 300 mm from the edge of the driveway where no curbs are provided; and

(iv) be supported on a substantial base of concrete or well-compacted crushed stone at least 150 mm deep.

(2) Non-metallic heating cables or heating panels installed indoors shall be not less than 25 mm from any uninsulated metal bodies located below the surface to be heated.

(3) Where heating cables or heating panels do not have a metal sheath or shield, all uninsulated metal located at or below the surface to be heated shall be bonded to ground.

(4) The metal braid or sheath of each heating cable set shall be bonded to ground.

62-308 Heating cable sets installed on or wrapped around surfaces

(1) Heating cable sets installed on or wrapped around surfaces shall be secured in place by suitable fastening devices that will not damage the heating unit and that are suitable for the temperature involved.

(2) Heating cable sets wrapped over valves or expansion joints in pipes shall be installed in such a manner as to avoid damage when movement occurs at these areas.

(3) The metal braid or sheath of each heating cable set shall be bonded to ground.

62-310 Parallel heating cable sets

(1) Parallel heating cable sets shall be assembled and installed in accordance with the manufacturer's instructions.

(2) Branch circuits used to supply energy to parallel heating cable sets shall have a nominal voltage rating of 750 V or less.

(3) Metal structures or materials used for their support or on which parallel heating cable sets are installed shall be bonded to ground in accordance with Section 10.

62-312 Heating cable sets installed in or on non-metallic pipes, ducts, or vessels

(1) Heating cable sets intended for use in or on non-metallic pipes, ducts, or vessels shall be installed in accordance with the manufacturer's instructions.

(2) The temperature of the pipe, duct, or vessel shall be controlled by a thermostat or other equivalent means in such a manner that the temperature shall be low enough to eliminate the danger of damage to the pipe, duct, or vessel.

(3) Internal heating of pipes, ducts, or vessels shall be limited to those not containing sewage solids or flammable liquids.

(4) Where the pipes, ducts, or vessels are heated by an internal heating cable set, the heating cable set shall be provided with a non-heating section that shall pass through a suitable gland.

(5) The metal braid or sheath of each heating cable set shall be bonded to ground.

62-314 Heating panel sets installed on tanks, vessels, or pipes for industrial application

(1) Heating panel sets shall be secured in place by suitable fastening devices.

(2) The metal covering of each heating panel set shall be bonded to ground.

62-316 Marking

Pipes, ducts, or vessels with electric heating shall be suitably marked to indicate that they are electrically traced if the systems are not readily visible throughout the length.

Other heating systems

62-400 Heating cable sets installed in pipes, ducts, or vessels (see Appendix B)

(1) Heating cable sets installed in pipes, ducts, or vessels shall be a type rated for immersion.

(2) Where practicable, heating cable sets installed in pipes, ducts, or vessels shall be secured in place by suitable fastening devices that will not damage the heating cable set.

(3) Where the heating cable set passes through the pipe, duct, or vessel, it shall pass through a suitable gland.

(4) Where a metal raceway is required for the non-heating leads of a heating cable set installed in a pipe, duct, or vessel, it shall be installed so that it will not become flooded in the event of the failure of the gland required by Subrule (3).

(5) Ground fault protection shall be provided to de-energize all normally ungrounded conductors of electric heating cable sets with a ground fault setting sufficient to allow normal operation of the heater.

62-402 Pipeline resistance heating

Pipeline resistance heating equipment shall conform to the following:

(a) voltage applied to the piping shall not exceed 30 V, and the supply shall be from an isolating-type transformer;

(b) no part of the extra-low-voltage circuit, including the conductors and the piping in the loop used for heating, shall be bonded to ground;

(c) pipe hangers shall have insulating bushings, or be made of insulating material;

(d) pipes shall have a minimum clearance of 100 mm from adjacent material, and from each other, except from hangers or supports;

(e) where pipes pass through walls, floors, or ceilings, they shall be bushed with insulating bushings or have 100 mm of clearance as required in Item (d);

(f) vertical runs shall be supported every 6 m or at each floor, whichever distance is less, with insulating hangers, and shall be firestopped at each floor;

(g) horizontal runs shall be supported at least every 3 m;

(h) pipes used as heating elements shall be electrically insulated and guarded or shielded;

(i) pipes shall be protected from mechanical damage or installed in such a manner that the building beams or framing provide mechanical protection;

(j) all pipes used for conductors in the electrical circuit shall be of the same diameter and made of the same material; and

(k) joints shall be at least as electrically conductive as the adjacent piping such as provided by welding or bonding.

62-500 Heaters for sauna rooms (see Appendix B)

(1) Heaters for sauna rooms shall be marked as being suitable for the purpose.

(2) Sauna heaters shall be installed in rooms that are built in accordance with the nameplate size specifications and shall be fastened securely in place to ensure that the minimum safe clearances indicated on the nameplate are not reduced.

(3) Each sauna heater shall be controlled by a thermostat or other temperature regulating device installed in accordance with the manufacturer's instructions.

(4) Sauna heaters shall not be installed below shower heads or water spray devices.

(5) Each sauna heater shall be controlled by a timed cut-off switch having a maximum time setting of one hour, with no override feature, which, if not forming part of the sauna heater or cabinet, shall be mounted on the outside wall of the room containing the sauna heater and shall disconnect all ungrounded conductors in the circuit supplying the heater.

Section 64

Note — This Section has been deleted.

Section 66
Amusement parks, midways, carnivals, film and TV sets,
TV remote broadcasting locations, and travelling shows

Section 66 — Amusement parks, midways, carnivals, film and TV sets, TV remote broadcasting locations, and travelling shows

Scope and application

66-000 Scope
(1) This Section applies to the installation of electrical equipment utilizing any source of power, including generator sets, used for amusement parks, midways, carnivals, and other events of a temporary nature held indoors, outdoors, or in tents, such as film and TV sets, TV remote broadcasting locations, home shows, live theatre, and travelling shows.

(2) The installation of electrical equipment forming part of an amusement ride shall comply with CAN/CSA-Z267.

(3) The requirements of this Section supplement or amend the general requirements of this Code.

66-002 Special terminology
In this Section, the following definitions apply:

Amusement park — a tract of land used as a temporary or permanent location for amusement rides and structures.

Amusement ride — a device or combination of devices designated or intended to entertain or amuse people by physically moving them.

Concession — a structure or a combination of structures erected for the purpose of entertaining or amusing people with games or shows and for the dispensing of food, souvenirs, and tickets, by sale or for any other purpose.

General

66-100 Supporting of conductors
(1) Only decorative lighting, signal, communication, and control circuits shall be supported on structures that support amusement rides.

(2) The decorative lighting and control circuits of one amusement ride shall not be installed on a supporting structure of another ride.

(3) Overhead conductors shall have a vertical clearance to finished grade of not less than the following:
    (a) across highways, streets, lanes, and alleys: 5.5 m;
    (b) across areas accessible to vehicles: 5 m; and
    (c) across areas accessible to pedestrians: 3.5 m.

66-102 Protection of electrical equipment
Electrical equipment shall be protected in accordance with Rule 2-200.

Grounding

66-200 Grounding
(1) The service and electrical distribution shall be grounded in accordance with Section 10 of this Code.

(2) Notwithstanding Rule 10-908(1)(a), grounding electrodes for mobile generators shall be permitted to be connected using single-conductor plug-in locking-type connectors.

66-202 Equipment bonding
(1) Exposed non-current-carrying metal parts of fixed electrical equipment such as motor frames, starters, and switch boxes; parts of rides, concessions, and ticket booths; and moving electrically operated equipment shall be bonded to ground by
    (a) means of the bonding conductor in the supply cord; or
    (b) connection to a separate insulated flexible copper bonding conductor, not less than No. 6 AWG, that is connected to the grounded circuit conductor at the service disconnect.

(2) Cord-connected operator-controlled remote stations shall be bonded to ground.

Services and distribution

66-300 Service equipment
(1)   Service equipment shall be of a size suitable for the connected load.
(2)   Where accessible to unauthorized persons, enclosures for service equipment shall be lockable.
(3)   Generators shall not be accessible to unauthorized persons.

66-302 Mounting of service equipment
Service equipment shall be mounted on a solid backing and be
(a)   located so as to be protected from the weather;
(b)   installed in a weatherproof enclosure; or
(c)   of weatherproof construction.

66-304 Distribution equipment
(1)   Each concession and ride shall be provided with a fused disconnect switch or circuit breaker.
(2)   Where accessible to unauthorized persons, enclosures for switches, panelboards, and splitters shall be lockable.

Wiring methods and equipment

66-400 Wiring methods
(1)   Except as permitted in Rules 66-450 to 66-458, wiring methods shall be in accordance with Section 12 and suitable for the condition of use.
(2)   Cords, cables, conduits, and other electrical equipment shall be protected from physical damage.
(3)   Cords shall be of the hard-usage type, in good repair, and
    (a)   provided with strain relief where they enter into enclosures and plug-in connectors; and
    (b)   if exposed to the weather, be of a type suitable for outdoor use; and
    (c)   where plug-in connections are used,
        (i)   have connectors and receptacles that are rated in amperes and designed so that differently rated devices cannot be connected together;
        (ii)   have the female connector attached to the load end of the cord; and
        (iii)   be polarized if an ac multi-conductor connector is used.
(4)   Notwithstanding Subrule (3)(c)(ii), for single-conductor cables the grounded conductor and the bonding conductor shall be permitted to have the female half connected to the supply end of the cord.
(5)   With the exception of amusement parks, midways, carnivals, home shows, and tent meetings, receptacles rated 15 A, conforming to CSA configuration 5-15R rated 120 V, hospital grade, and protected by a fuse or circuit breaker rated not greater than 20 A, shall be permitted for temporary lighting installations within the Scope of this Section of the Code, where the loads are of an intermittent nature.
(6)   Temporary wiring for portable stage equipment shall be in accordance with Rules 44-350, 44-352, and 44-356.

66-402 Equipment
(1)   Lighting streamers shall be made up of extra-hard-usage outdoor flexible cord with weatherproof lampholders having either
    (a)   terminals of a type that puncture the insulation and make contact with the conductors; or
    (b)   permanently attached leads connected to the cord.
(2)   Fluorescent luminaires shall not be mounted end-to-end unless they are marked for that purpose.
(3)   Incandescent lampholders shall be of the screwshell type.
(4)   Notwithstanding Subrule (3), for film and TV sets and TV remotes, bayonet-type lampholders shall be permitted.
(5)   Utilization equipment intended for use outdoors shall be suitable for the location, unless precautions are taken to protect it from inclement weather.

Single-conductor cables

66-450 Single-conductor cables
Single-conductor cables shall be permitted in sizes No. 4 AWG and larger, provided that they are
(a)   rated for the circuit voltage and suitable for the intended application;

(b)   of a matched set with the same length for all conductors of the circuit including the bonding conductor; and

(c)   covered or guarded so as not to present a tripping hazard in pedestrian walkways or roadways.

66-452 Fault current limiting

Where the available fault current exceeds 10 000 A, systems employing single-conductor cables, except where installed as fixed wiring, shall be supplied by means of current-limiting overcurrent devices to prevent inadvertent movement of the cables.

66-454 Free air ampacity

(1)   Single-conductor cables shall be rated in accordance with Section 4.

(2)   Notwithstanding Subrule (1), for temporary installations, bundled single-conductor cables of any one circuit shall be permitted to be free air rated without correction factors if different circuits are separated by at least one cable bundle diameter.

66-456 Single-conductor cable connections

(1)   Connections to single-conductor cables shall not be accessible to unqualified persons.

(2)   Plug-in connectors for single-conductor cables shall

    (a)   be a locking type; and

    (b)   incorporate a mechanical interlock to prevent wrong connections or be colour-coded.

(3)   Single-conductor cables shall not be connected in parallel except as a means of reducing voltage drop, and cables so connected shall have overcurrent protection sized to protect the cable having the smallest ampacity as though it were used alone.

66-458 Bonding

Each circuit incorporating single-conductor cables shall include a bonding conductor that shall be run with the circuit conductors.

Motors

66-500 Motors

Motors, including protection and control, shall be installed in accordance with Section 28.

66-502 Location

Motors shall be installed only in dry locations unless they are a type specifically marked for the location or are suitably protected.

66-504 Portable motors

Connections to portable motors shall be permitted to be made with flexible cord that shall have a serviceability not less than Type SOW for outdoor use.

Section 68 — Pools, tubs, and spas

Scope

68-000 Scope
(1)   This Section applies to
    (a)   electrical installations and electrical equipment in or adjacent to pools; and
    (b)   non-electrical metal accessories in a pool or within 3 m of the inside wall of a pool.
(2)   A pool shall be deemed to include
    (a)   permanently installed and storable swimming pools;
    (b)   hydromassage bathtubs;
    (c)   spas and hot tubs;
    (d)   wading pools;
    (e)   baptismal pools; and
    (f)   decorative pools.
(3)   The requirements of this Section supplement or amend the general requirements of this Code.

General

68-050 Special terminology
In this Section, the following definitions apply:

Decorative pool — a pool that could be used as a wading pool, that is larger than 1.5 m in any dimension, and that is readily accessible to the public.

Dry-niche luminaire — a luminaire intended for installation in the wall of the pool in a niche that is sealed against the entry of pool water by a fixed lens.

Forming shell — a structure intended for mounting in a pool structure to support a wet-niche luminaire assembly.

Hydromassage bathtub — a permanently installed bathtub having an integral or remote water pump or air blower, and having a fill and drain water system, and includes therapeutic pools.

Leakage current collector — a device designed to provide a path to ground for leakage current originating from devices in contact with pool water.

Permanently installed swimming pool — a pool constructed in such a manner that it cannot be disassembled for storage.

Spa or **hot tub** — a pool or tub designed for the immersion of persons in heated water circulated in a closed system incorporating a filter, heater, pump, and with or without a motor-driven blower but not intended to be filled and drained with each use.

Storable swimming pool — a pool constructed in such a manner that it may be readily disassembled for storage and reassembled to its original integrity.

Wet-niche luminaire — a luminaire intended for installation in a forming shell mounted in a pool structure where the luminaire will be completely surrounded by pool water.

68-052 Electrical wiring or equipment in pool walls or water
Electrical wiring or equipment shall not be installed in the walls nor in the water of pools except as permitted by this Section.

68-054 Overhead wiring (see Appendix B)
(1)   No pool shall be placed under or near overhead wiring and no overhead wiring shall be placed over or near a pool unless the installation complies with the requirements of this Rule.
(2)   There shall not be any overhead wiring above the pool, diving structure, observation stand, tower, or platform, or above the area extending 3 m horizontally from the pool edge except as permitted by Subrules (3) and (4).
(3)   Insulated communication conductors, communication antenna distribution conductors, and neutral supported cables not exceeding 750 V shall be permitted to be located over a pool, diving structure,

observation stand, tower, or platform, or above the area extending 3 m horizontally from the pool edge, provided that there is a clearance (measured radially) of at least 4.5 m.

(4)   Conductors other than those covered by Subrule (3) and operating at not more than 50 kV phase-to-phase shall be permitted to be located above a pool, diving structure, observation stand, tower, or platform, or above the area extending 3 m horizontally from the pool edge, provided that there is a clearance (measured radially) of at least 7.5 m.

68-056 Underground wiring
The horizontal separation between the inside walls of a pool and underground conductors, except for bonding conductors or conductors supplying electrical equipment associated with the pool and protected by a ground fault circuit interrupter, shall be not less than that shown in Table 61.

68-058 Bonding to ground (see Appendix B)
(1)   The metal parts of the pool and of other non-electrical equipment associated with the pool such as piping, pool reinforcing steel, ladders, diving board supports, and fences within 1.5 m of the pool shall be bonded together and to non-current-carrying metal parts of electrical equipment such as decorative-type pool luminaires and lighting equipment not located in a forming shell, forming shells, metal screens of shields for underwater speakers, conduit, junction boxes, and the like by a copper bonding conductor.

(2)   Pool reinforcing steel shall be bonded with a minimum of four connections equally spaced around the perimeter.

(3)   Notwithstanding Subrule (2), where reinforcing steel is encapsulated with a non-conductive compound, provisions shall be made for an alternative means to eliminate voltage gradients that would otherwise be provided by unencapsulated, bonded reinforcing steel.

(4)   Bonding conductors for pools shall be
  (a)   not smaller than No. 6 AWG for permanently installed pools and for all in-ground pools; or
  (b)   as required by Table 16 for all other pools.

(5)   Metal sheaths and raceways shall not be relied upon as the bonding medium and a separate copper bonding conductor shall be used, except that a metal conduit between a forming shell and its associated junction box shall be permitted to be used as the bonding medium, provided that the forming shell and junction box are installed in the same structural section.

Δ (6)   The bonding conductor from the junction box referred to in Rule 68-060 shall be run to the panelboard supplying pool electrical equipment, and if smaller than No. 6 AWG shall be installed and mechanically protected in the same manner as the circuit conductors.

(7)   The bonding conductor in Subrule (4) shall be of copper and not smaller than that required by Table 16, except that the bonding conductor for an in-ground pool shall be not smaller than No. 6 AWG.

(8)   Notwithstanding Subrule (1), the metal parts of a pool need not be bonded to ground or to each other where the electrical equipment associated with the pool is
  (a)   not located within 3 m of the pool;
  (b)   suitably separated from the pool by a fence, wall, or other barrier; or
  (c)   approved without a bonding conductor.

68-060 Junction and deck boxes (see Appendix B)
(1)   Junction boxes shall be permitted to be submerged in decorative pools provided the boxes are marked for such usage.

(2)   Junction boxes installed on the supply side of conduits extending to forming shells, referred to hereinafter as deck boxes, shall be specifically approved for the purpose.

(3)   Deck boxes shall be provided with a means for independently terminating at least three bonding conductors inside the box and one No. 6 AWG bonding conductor outside the box.

(4)   Deck boxes shall not contain the conductors of any circuits other than those used exclusively to supply the underwater equipment.

(5)   Deck boxes shall be provided with electrical continuity between every connected metal conduit and the bonding terminals by means of copper, brass, or other corrosion-resistant metal that is integral with the box.

(6)   Deck boxes shall be installed
  (a)   above the normal water level of the pool;
  (b)   so that the top of the box is located at or above the finished level of the pool deck;
  (c)   in such a manner or location that the box will not be an obstacle; and
  (d)   in such a manner that any water on the deck will drain away from the box.

(7) Junction boxes and conduit shall be watertight and provided with a packing seal that will seal around the cord and effectively prevent water from entering the box through the conduit from the forming shell.

68-062 Transformers and transformer enclosures (see Appendix B)

(1) Transformers shall not be located within 3 m of the inside wall of the pool unless suitably separated from the pool area by a fence, wall, or other permanent barrier that will make the transformer not accessible to persons using the pool area.

(2) A metal shield, if provided between the primary and secondary windings of a transformer, shall be bonded to ground.

(3) Audio isolation transformers shall
   (a) be connected between the audio output terminals of each amplifier and any loudspeaker that is located within 3 m of the pool wall;
   (b) be located in or adjacent to the amplifier with which they are used; and
   (c) have an audio output voltage of not more than 75 V rms.

68-064 Receptacles

(1) Receptacles shall not be located within 1.5 m of the inside walls of the pools.

(2) Receptacles located between 1.5 m and 3 m of the inside walls of a pool shall be protected by a ground fault circuit interrupter.

(3) In maintaining the dimensions referred to in this Rule, the distance to be measured is the shortest path that the power supply cord of an appliance connected to the receptacle would follow without piercing a building floor, wall, or ceiling.

68-066 Luminaires and lighting equipment

(1) Wet-niche or submersible luminaires shall
   (a) be mounted in forming shells that shall have provision for a suitable connection to the wiring method used;
   (b) unless specifically approved and marked for submersion at a greater depth, not be submersed in the pool water at a depth of more than 600 mm, such distance being measured from the centre of the lens face of the luminaire to the normal water level; and
   (c) operate with neither the supply voltage to the luminaire nor its associated ballast or transformer, if applicable, nor the secondary open-circuit voltage of the ballast or transformer exceeding 150 V during either starting or operating conditions.

(2) Notwithstanding Subrule (1)(a), wet-niche or submersible luminaires installed in a decorative pool need not be mounted in a forming shell but shall have provision for a suitable connection to the wiring method used.

(3) Where dry-niche luminaires are installed to be accessible from a walkway or a service tunnel outside the walls of the pool or from a closed, drained recess in the walls of the pool, neither the supply voltage to the fixture nor its associated ballast or transformer shall exceed 300 V during either starting or operating conditions.

(4) Dry-niche luminaires shall be accessible for maintenance
   (a) from a service tunnel or walkway outside the walls of the pool; or
   (b) through a handhole in the deck of the pool to a closed, drained recess in the wall of the pool.

(5) Metal parts of luminaires in contact with the pool water shall be of brass or other suitable corrosion-resistant material.

(6) Luminaires installed below, or within 3 m of, the pool surface or walls, and not suitably separated from the pool area by a fence, wall, or other permanent barrier, shall be electrically protected by a ground fault circuit interrupter.

(7) Standards or supports for luminaires shall not be installed within 3 m of the inside walls of a swimming pool unless such luminaires are protected by ground fault circuit interrupters.

(8) Forming shells for lamps supplied from a grounded circuit or a circuit operating at a voltage exceeding 30 V shall be metal and have provision for a threaded connection to a rigid metal conduit.

68-068 Ground fault circuit interrupters (see Appendix B)

(1) Except as permitted in Subrule (2), ground fault circuit interrupters required by the Rules of this Section shall be of the Class A type.

(2) Where ground fault circuit interrupters of the Class A type are not available due to rating, the equipment shall be permitted to be protected by a ground fault circuit interrupter that will clear a ground fault within the time specified for a Class A type interrupter.

(3) Ground fault circuit interrupters shall be permanently connected.

(4) Ground fault circuit interrupters shall be permitted to be applied to a feeder, a branch circuit, or an individual device.

(5) A warning sign shall be located beside the switches controlling circuits electrically protected by ground fault circuit interrupters advising that the circuits have this protection and that the equipment shall be tested regularly.

(6) Ground fault circuit interrupters shall be installed
  (a) in a location that will facilitate the testing required in Subrule (5);
  (b) not closer than 3 m to the pool water except as permitted by Item (c); and
  (c) not closer than 3 m to the pool water in a spa or hot tub and not closer than 1.5 m to a hydromassage bathtub, unless the ground fault circuit interrupter is an integral part of an approved factory-built spa, hot tub, or hydromassage bathtub or is located behind a barrier that will prevent the occupant of the pool from contacting the device.

(7) Except as permitted by Rule 68-070, the following equipment shall be protected by a ground fault circuit interrupter:
  (a) electrical equipment placed in the water in the pool;
  (b) audio amplifiers connected to loudspeakers in the pool water;
  (c) electrical equipment located within the confines of the pool walls or within 3 m of the inside walls of the pool and not suitably separated from the pool area by a fence, wall, or other permanent barrier; and
  (d) receptacles located in wet areas of a building, and associated with the pool, such as locker and change rooms.

Δ **68-070 Other electrical equipment**

(1) Loudspeakers installed beneath the pool surface shall be
  (a) mounted in a recess in the wall or floor of the pool and enclosed by a separate, rigid, corrosion-resistant metal screen; and
  (b) connected to their audio isolating transformers by ungrounded wiring.

(2) Communication equipment installed within 3 m of the inside walls of the pool shall be
  (a) permanently fixed on the wall and located so that no part is within 1.5 m of the inside walls of the pool or can be used from the pool, unless actuated by means of a cord with an insulating link; or
  (b) separated from the pool area by a fence, wall or other permanent barrier.

(3) Notwithstanding Subrule (2), communication jacks shall not be installed within 3 m from the inside walls of the pool.

Permanently installed swimming pools

68-100 Wiring method

(1) Rigid conduit of copper or other corrosion-resistant metal or rigid PVC conduit shall be provided between the forming shell of luminaires installed below the pool surface and the junction box referred to in Rule 68-060.

(2) The wiring method between the wet-niche luminaires and the junction boxes referred to in Rule 68-060 shall be flexible cord suitable for use in wet locations and supplied as a part of the luminaire.

(3) Where Subrules (1) and (2) do not apply, any suitable wiring method specified in Section 12 shall be permitted to be used.

(4) Conductors on the load side of each ground fault circuit interrupter shall be kept entirely independent of all other wiring that is not protected in that way and shall not enter a luminaire, raceway, box, or cabinet occupied by other wiring except for panelboards that house the interrupters.

(5) Conduits in the walls and deck of a swimming pool shall be installed so that suitable drainage is provided.

Storable swimming pools

68-200 Electrical equipment

No electrical equipment shall be located in the pool water or on the pool wall unless specifically approved for the purpose.

68-202 Pumps

(1) Swimming pool pumps shall be
   (a) supplied from a permanently installed receptacle located not less than 1.5 m nor more than 7.5 m from the pool wall; and
   (b) protected by a ground fault circuit interrupter if located within 3 m of the inside walls of the pool and not suitably separated from the pool area by a fence, wall, or other permanent barrier.
(2) Swimming pool pumps located within 3 m of the pool walls shall be specifically approved for the purpose.

Hydromassage bathtubs

Δ 68-300 General

Rules 68-302 to 68-308 apply to the installation of permanently and cord-connected hydromassage bathtubs.

68-302 Protection

Electrical equipment forming an integral part of a hydromassage bathtub shall be protected by a ground fault circuit interrupter of the Class A type.

68-304 Control

(1) A hydromassage bathtub shall be controlled by an on-off device located in accordance with Subrule (2).
(2) Electric controls associated with a hydromassage bathtub shall be located behind a barrier or shall be located not less than 1 m horizontally from the wall of the hydromassage bathtub, unless they are an integral part of an approved factory-built hydromassage bathtub.

Δ 68-306 Receptacle for a cord-connected hydromassage bathtub (see Appendix B)

(1) Where a cord-connected hydromassage bathtub conforming to Rule 68-302 is intended to be installed:
   (a) one receptacle located at not less than 300 mm from the floor shall be provided for the use with the hydromassage bathtub; and
   (b) the receptacle shall be inaccessible to the hydromassage bathtub occupant.
(2) The appropriate warning label supplied by the cord-connected hydromassage bathtub manufacturer shall be affixed to the receptacle specified in Subrule (1).
(3) Notwithstanding Subrule (2), a warning label is not required where a single receptacle is used.

68-308 Other electric equipment

Luminaires, switches, receptacles, and other electrical equipment not directly associated with a hydromassage bathtub shall be installed in accordance with the Rules of this Code covering the installation of that equipment in bathrooms.

Spas and hot tubs

68-400 General

Rules 68-402 to 68-408 apply to the installation of spas and hot tubs.

68-402 Bonding to ground

(1) Metal parts of spas and hot tubs shall be bonded together and to ground in accordance with Rule 68-058.
(2) Notwithstanding Subrule (1), metal rings or bands used to secure staves of wooden hot tubs need not be bonded.

68-404 Controls and other electrical equipment (see Appendix B)

(1) Controls for a spa or hot tub shall be located behind a barrier or not less than 1 m horizontally from the spa or hot tub, unless they are an integral part of an approved factory-built spa or hot tub.
(2) Receptacles shall be installed in accordance with Rule 68-064.
(3) Luminaires shall be installed in accordance with Rule 68-066.
Δ (4) Except for a spa or hot tub installed at a dwelling unit, an emergency shutoff switch shall be installed for each spa or hot tub that
   (a) disconnects the motors supplying power to the closed water circulating systems;

(b) is independent of the controls for a spa or hot tub;

(c) is located at a point readily accessible to the users and within sight of and within 15 m of the spa or hot tub;

(d) is labelled in a conspicuous, legible, and permanent manner identifying it as the "emergency" shutoff switch; and

(e) is equipped with audible and visual trouble-signal devices that give immediate warning upon actuation of the emergency shutoff switch.

68-406 Leakage current collectors

(1) Leakage current collectors shall be installed in all water inlets and in all water outlets of a field-assembled spa or hot tub so that all water flows through the leakage current collectors.

(2) A leakage current collector shall be

(a) a section of corrosion-resistant metal tubing at least five times as long as its diameter, provided with a corrosion-resistant lug, in a run of non-metallic pipe; or

(b) a device providing equal protection as in Item (a) when it is an integral part of a spa or hot tub that is factory built for field installation or assembly.

(3) Leakage current collectors shall be electrically insulated from the spa or hot tub and shall be bonded to the control panel or the main service ground with a copper bonding conductor.

(4) Notwithstanding Subrule (1), leakage current collectors shall not be required in a system in which the only electrical component is a pump marked as an insulated wet end pump.

(5) The bonding conductor for leakage current collectors shall be not smaller than required by Table 16 where the bonding conductors are mechanically protected in the same manner as the circuit conductors, or a minimum No. 6 AWG copper conductor.

68-408 Field-assembled units (see Appendix B)

(1) Spas and hot tubs field assembled with individual components shall be installed in accordance with Rules 68-400 to 68-406 and Subrules (2) and (3).

(2) Individual components such as pumps, heaters, and blowers shall be specifically approved for use with spas or hot tubs.

(3) Air blowers shall be installed above the tub rim, or other means shall be used to prevent water from contacting blower live parts.

Section 70 — Electrical requirements for factory-built relocatable structures and non-relocatable structures

Scope

70-000 Scope
(1) Rules 70-100 to 70-130 apply to relocatable structures (factory-built) towable on their own chassis, for use without permanent foundations, having provision for connection to utilities and include
    (a) mobile homes; and
    (b) mobile commercial and industrial structures.
(2) Rules 70-200 to 70-204 apply to non-relocatable structures (factory-built) for use on permanent foundations and include
    (a) housing (residential); and
    (b) commercial and industrial structures.
(3) These Rules do not apply to recreational vehicles covered by CAN/CSA-Z240 RV Series.
(4) This Section supplements or amends the general requirements of this Code.

Relocatable structures

70-100 Equipment
Electrical components, including those connected in Class 1 extra-low-voltage power circuits (e.g., lighting fixtures) and in Class 2 extra-low-voltage circuits, shall conform to the requirements of the *Canadian Electrical Code, Part II* and be suitable for the application.

70-102 Method of connection
(1) Subject to the conditions of Subrule (2), the method of connection to the supply circuit shall be
    (a) connection to an overhead or underground supply;
    (b) a power supply cord or cord set; or
    (c) a length of flexible cord, or cord or cable without an attachment plug.
(2) For mobile homes the method of connection to the power supply shall be directly to an overhead or underground supply, except where a deviation has been allowed in accordance with Rule 2-030.

70-104 Connection to an overhead or underground supply
(1) Where the supply connection is directly to an overhead or underground supply, a conduit nipple or a length of rigid conduit shall be provided and shall
    (a) project from the structure through the exterior wall, roof, or floor to permit attachment of a conduit fitting;
    (b) have a suitable cap on the exposed end;
    (c) terminate at the disconnecting means, at an intermediate box, or, for other than mobile homes, at the distribution equipment if a disconnecting means is not provided; and
    (d) be of sufficient size to accommodate copper conductors of a calculated ampacity for the load involved, except
        (i) where the structure is specifically designed for connection by conductors other than copper; or
        (ii) as specified in Subrule (3).
(2) For mobile homes the conduit shall project so that it is readily accessible for power supply connection.
(3) For mobile homes the size of conduit shall be not less than that specified in Table 48.
(4) Where it is intended or likely that the system grounding conductor be run separately, a non-metallic raceway shall be installed at the time of manufacture for this purpose.

70-106 Service for communication systems (see Appendix B)
All mobile homes shall be provided with a length of raceway, 16 trade size or larger, for use as a communication service that shall
(a) project from the structure a minimum of 75 mm through the floor;
(b) terminate at least 300 mm above the finished floor in a wall or partition in a standard switch or outlet box complete with cover;
(c) where of metal, be bonded to the frame of the mobile home; and
(d) have a suitable cap on the exposed end of the conduit stub.

Section 70
Electrical requirements for factory-built
relocatable structures and non-relocatable structures

70-108 Power supply cord or cord set

(1) Where a power supply cord or cord set is used except as provided for in Subrule (4), the cord shall
  (a) be provided as part of the mobile structure;
  (b) have an ampacity not less than the ampere rating of the attachment plug;
  (c) be the extra-hard-usage type suitable for outdoor use as specified in Table 11;
  (d) have separate identified and bonding conductors;
  (e) be not less than 7.5 m in length, as measured from the attachment plug to the point of entrance to the unit;
  (f) if a permanently connected power supply cord, terminate at the main disconnecting means in the unit or at a box in or on the unit, suitable space being provided in the unit for storage of the cord when not in use to protect it from damage; and
  (g) have a suitable grounding-type attachment plug with an ampere rating as follows:
    (i) for applications covered by Section 8, not less than that of the service conductor ampacity required in Section 8; or
    (ii) for other applications, not less than that for which it is approved.
(2) Bushings of rubber, unless of an oil-resistant compound, shall not be used in locations where they are exposed to mechanical injury.
(3) Where a cord set is used, a male receptacle shall be provided on the unit and shall
  (a) be of weatherproof construction unless adequately protected or enclosed;
  (b) have a contact arrangement that will mate with the cord connector on the cord; and
  (c) have a current rating not less than that of the main overcurrent protection.
(4) Where provided for in Rule 70-102(2), a cord or cord set shall be permitted to be used for mobile homes, provided that it
  (a) is not smaller than No. 6 AWG;
  (b) has an attachment plug moulded to the cord with configuration designated as CSA 14-50P (3-pole, 4-wire, 125/250 V, 50 A); and
  (c) enters where it will not be subject to mechanical damage.

70-110 Disconnecting means and main overcurrent protection

(1) Except as provided for in Subrule (2), each structure shall be provided with
  (a) a service box or a combined service and distribution box located within the structure with provision for grounding the neutral;
  (b) main overcurrent protection having a current rating at least equal to the minimum ampacity of the consumer's service as determined in accordance with Section 8 but in no case less than 50 A for mobile homes and not exceeding the ampacity of the supply conductors actually used except as permitted by Rule 14-104; and
  (c) an identified conductor that shall be
    (i) connected to ground within the mobile structure if a power supply cord or cord set is not provided; or
    (ii) isolated from ground if a power supply cord or cord set is used.
(2) For other than mobile homes, the structure shall be permitted to be provided with distribution equipment instead of the type of service equipment listed in Subrule (1) where such service equipment is provided in the supply to the unit.

70-112 Location of service or distribution equipment

Service or distribution equipment shall be
(a) readily accessible;
(b) not located in clothes closets unless in its own compartment, in bathrooms, in stairways, or in any similar or undesirable location;
(c) located within the structure with consideration being given to the possibility of the formation of condensation;
(d) located as close as practicable to the point where the supply conductors enter the structure; and
(e) of the circuit-breaker type if in other than extra-low-voltage circuits and if mounted less than 1.5 m above the floor, in which case it shall be protected from mechanical injury.

70-114 Wiring methods — General

(1) The wiring method shall be as specified in Section 12 except where flexible cords are permitted in Rule 70-116 or for Class 2 circuits.

(2) Surfaces against which conductors are in contact shall be smooth and entirely free from sharp edges and burrs that may cause abrasion of the insulation on the conductors.

(3) Where cable is required to be protected from mechanical injury by Rules 12-516, 12-616, and 12-710, plates or tubes of sheet steel of at least No. 16 MSG or the equivalent, secured in place, shall be used to protect the cable from driven nails, screws, or staples.

(4) Cable run through holes in joists or studs shall be considered to be secured for the purposes of Rules 12-510 and 12-618.

(5) Unless provided with insulation suitable for the highest voltage involved, insulated conductors of low-voltage and extra-low-voltage circuits shall be separated by barriers, or shall be segregated by clamping, routing, or equivalent means that will ensure permanent separation and shall in any case be separated or segregated from bare live parts of the other circuit.

(6) For the purposes of Subrule (5), the outer covering of non-metallic-sheathed cable shall be considered to be a suitable barrier.

(7) Bare live parts, including terminals of electrical equipment in extra-low-voltage circuits other than Class 2 circuits, shall be enclosed in accordance with Rule 2-202(1).

(8) Conductors for extra-low-voltage Class 2 circuits shall be Type LVT, low-energy safety control cable, or the equivalent, and if protected by fuses, in accordance with Rule 16-200, the fuses shall not be interchangeable with those of higher ratings.

70-116 Wiring methods — Swing-out and expandable room sections

(1) The means used to make electrical connections between a swing-out or expandable room section and the wiring in the main section of the structure shall be located or protected so that there is no likelihood of damage to the interconnecting means when the section is extended or retracted, or when the structure is in transit.

(2) A flexible cord or power supply cable shall be used as an interconnecting means where flexibility is involved and shall
   (a) be of the extra-hard-usage type;
   (b) have an ampacity suitable for the connected load but in no case be smaller than No. 14 AWG;
   (c) be of the outdoor type if it has thermoplastic insulation or is exposed to the weather; and
   (d) incorporate a bonding conductor.

(3) A plug, connector, or fitting used in conjunction with a flexible cord for electrical interconnections shall have an electrical rating suitable for the maximum connected load and if located outside of the mobile home shall be protected from the weather or other adverse conditions (including when the structure is in transit).

70-118 Wiring methods — Multiple section mobile units

(1) Provision shall be made for interconnection of circuits in each section of multiple section units.

(2) The means of interconnection shall be such that no bare live parts of a low-voltage circuit are exposed to accidental contact should any section be temporarily energized before the other sections are in place.

70-120 Branch circuits — Mobile homes

(1) Circuits other than those referred to in Rules 26-746, 26-750, 26-806, 26-808, and 62-108 supplying permanently connected appliances shall be permitted to have additional outlets, but not receptacles, provided that these outlets are for fans, stationary lighting fixtures, or similar permanently connected appliances.

(2) The outlets referred to in Subrule (1) shall be considered to have a demand of 1 A each, except where the load is known to be greater, and in no case shall the total load exceed 80% of the rating of the overcurrent device protecting the circuit.

(3) Notwithstanding Rule 8-104, a circuit supplying an electric water heater having an input not more than 1500 W at 115 V or 3000 W at 230 V shall be permitted to have overcurrent protection rated or set at 15 A.

(4) In determining compliance with Rule 62-108(2), fans on oil or gas heaters that are not required for the operation of the heaters, and are rated not more than 3 A, shall not be required to be on individual branch circuits.

Δ 70-122 Receptacles, switches, and lighting fixtures (see Appendix B)

(1) In applying Rule 26-712(a), a hallway need not be considered a room.

(2) Switches of the pull-type including those for fans and lights shall conform to Rule 30-604.

Section 70
Electrical requirements for factory-built
relocatable structures and non-relocatable structures

(3) Where a ceiling-mounted, rigid luminaire is located at a height of less than 2 m above the floor and is readily accessible, the luminaire shall be protected from mechanical injury by a guard or by location.

(4) A receptacle installed on the underside of a mobile home, intended to supply a heating cable set(s) for freeze protection of plumbing pipes, shall

    (a) be protected by a ground fault circuit interrupter of the Class A type; and

    (a) be labelled in a conspicuous, legible, and permanent manner identifying it for the supply of heating cable set(s) for plumbing pipes.

70-124 Ventilating fans used in kitchen areas

(1) The motor of any fan installed in the kitchen area above or in the vicinity of cooking equipment and located in the air stream shall be the totally enclosed type unless specifically approved for this application.

(2) For the purposes of Subrule (1), the "area above or in the vicinity of cooking equipment" shall be

    (a) that portion of any wall located within 1.2 m of the cooking surface, as measured from any point on the cooking surface, regardless of the height of the walls; and

    (b) that portion of the ceiling defined by a rectangle having sides parallel to the edges of the cooking surface and located within 1.2 m of a vertical projection of the cooking surface, as measured from any point on this projection, regardless of the height of the ceiling.

(3) For the purposes of Subrule (2), the "cooking surface" of a built-in oven is the area of a bottom-hinged door of a size required to close the oven opening, when such a door is in the fully opened (horizontal) position; and for a free-standing stove or range (with or without an oven) or a built-in counter top surface element unit, the "cooking surface" is the entire top surface of the unit, including the backsplash (if any).

(4) For the purposes of Subrules (1), (2), and (3), if any full-height wall or partition is located within the space defined above, the space beyond this full height shall not be included in this restriction.

70-126 Grounding and bonding

(1) All major exposed metal parts that may become energized, including the water, gas, and waste plumbing, the roof and outer metal covering, the chassis, and metal circulating air ducts, shall be in good electrical contact with one another.

(2) The metal roof and exterior covering shall be considered bonded as required by Subrule (1)

    (a) if the metal panels overlap one another and are securely attached to the wood or metal frame parts by metal fasteners; and

    (b) if bonded to the chassis by metal fasteners or by a metal strap.

(3) All exposed non-current-carrying metal parts of a swing-out or expandable room section shall be reliably bonded to the exposed non-current-carrying metal parts of the main section of the mobile unit.

(4) The grounding or bonding conductors of the low-voltage wiring system other than the chassis shall not be used to carry current of any extra-low-voltage circuit.

(5) Grounding and bonding connections and terminals shall be

    (a) made of non-ferrous metal or plated steel;

    (b) used for no other purpose than grounding or bonding except for bonding between the chassis and skin where assembly screws may be used;

    (c) protected from mechanical injury; and

    (d) readily accessible for inspection and maintenance.

(6) Bare grounding and bonding conductors shall be located so that there is no danger of contact with live parts, but if their location or flexibility is such that separation from live parts is not assured, they shall be insulated by taping or sleeving.

(7) The major exposed metal parts described in Subrule (1) shall be bonded to ground with a bonding conductor from the metal chassis directly to

    (a) the neutral terminal of the service box for structures built in conformance with Rule 70-110(1); or

    (b) the bonding terminal in the distribution equipment for structures built in conformance with Rule 70-110(2).

(8) The bonding conductor in Subrule (7) may be insulated or bare and shall be

    (a) made of copper;

    (b) protected from salt spray;

    (c) not smaller than that specified in Table 41 where the values in the first column in Table 41 shall correspond to the rated input current of the structure;

    (d) located so that it is not subject to mechanical injury; and

    (e) suitably secured within 300 mm of the attachment to the chassis.

(9)   Bonding conductors other than those referred to in Subrule (7) shall have adequate ampacity but in no case less than that of a No. 14 AWG copper conductor.

70-128 Marking

(1)   Units connecting to the main power supply shall be marked in a permanent manner in a place where the details will be readily visible with the following information as required by Rule 2-100:
  (a)   manufacturer's name, trademark, trade name, or other recognized symbol of identification;
  (b)   model, style, or type designation;
  (c)   nominal voltage of the system to which the unit is to be connected (e.g., 120, 120/240, etc.);
  (d)   rated frequency; and
  (e)   rated input current in amperes.

(2)   For purposes of Subrule (1)(e), the rated input current in amperes shall be
  (a)   the ampere rating of the main overcurrent protection, if provided;
  (b)   the ampere rating of the distribution equipment, if no main overcurrent protection and no power supply cord is provided; or
  (c)   the ampere rating of the attachment plug, if provided.

(3)   Markings adjacent to the main and branch circuit overcurrent devices shall be provided in accordance with Rule 2-100(3).

(4)   For multiple section mobile homes, or structures, each section shall be suitably and permanently marked to identify the other sections to be used with it to form a single structure.

(5)   Unless it is otherwise clearly evident, instructions shall be provided on the main section of multiple section mobile homes or structures to indicate the interconnections necessary to complete the installation.

70-130 Tests (see Appendix B)

(1)   The following tests shall be performed on the complete assembly at the factory:
  (a)   **Continuity** — All circuits, including grounding or bonding circuits, shall be tested for continuity.
  (b)   **Insulation resistance** — The insulation resistance between live parts and ground at the completion of a 1 min application of a 500 V dc test voltage shall be not less than that specified in Table 24.

(2)   As an alternative to the insulation resistance test specified in Subrule (1)(b), an ac dielectric strength test shall be permitted to be performed, in which case an ac voltage of 900 V shall be applied for 1 min (or 1080 V for 1 s) between all live parts and non-current-carrying metal parts without breakdown occurring.

(3)   In performing either the insulation resistance or the dielectric strength test, the neutral shall be disconnected from ground for the test and be reconnected afterwards.

Non-relocatable structures (factory-built)

70-200 General

Rules 70-100, 70-112, 70-114, 70-118, 70-122, 70-124, 70-126, 70-128, and 70-130 shall also apply to non-relocatable structures.

70-202 Connection to overhead and underground supply

Provision shall be made at the factory for the electrics in the structure to be connected either to an overhead or underground power supply through conduit nipples or the equivalent and supports that shall be
(a)   of sufficient size to accommodate conductors having the minimum ampacity determined by Section 8 of this Code; and
(b)   limited in number to meet the limitations set out in Rules 6-102 and 6-200.

70-204 Service and distribution equipment

(1)   Provision shall be made at the factory for the installation either at the factory or on the job site of a service box or other service equipment in the structure that shall be
  (a)   in a readily accessible location within the building;
  (b)   as close as practicable to the point where the service conductors enter the building; and
  (c)   within the individual units where multiple occupancy residential condominium or row-house structures are involved; or
  (d)   in a central location accessible to all tenants in all other cases.

(2)   Each complete structure shall be provided with distribution equipment.

258

Section 72 — Mobile home and recreational vehicle parks

Scope and application

72-000 Scope

(1)   Rules 72-100 to 72-112 apply to services and distribution facilities for mobile home and recreational vehicle parks.

(2)   This Section supplements or amends the general requirements of this Code.

General

72-100 Service

Each mobile home and recreational vehicle park and/or consumer's service shall be provided with service equipment in accordance with the applicable requirements of Section 6 of this Code.

72-102 Demand factors for service and feeder conductors

(1)   The minimum ampacity of the consumer's service and feeder conductors for mobile home parks shall be based on the requirements of
   (a)   Rule 8-200 with respect to service or feeder conductors supplying an individual mobile home; and
   (b)   Rule 8-202 with respect to service or feeder conductors supplying more than one mobile home.

(2)   The minimum ampacity of the consumer's service and feeder conductors in the case of recreational vehicle parks shall be calculated on the basis of the ampere rating of the receptacles and by applying the following demand factors:
   (a)   100% of the sum of the first five receptacles having the highest ampere ratings; plus
   (b)   75% of the sum of the ampere ratings of the next ten receptacles having the same or next smaller ratings to those specified in Item (a); plus
   (c)   50% of the sum of the ampere ratings of the next ten receptacles having the same or next smaller ratings to those specified in Item (b); plus
   (d)   25% of the sum of the ampere ratings of the remainder of the receptacles.

(3)   For the purpose of Subrule (2), each duplex receptacle supplied from a multi-wire branch circuit shall be counted as two receptacles.

(4)   For the purpose of Subrule (2), where receptacles of different ratings are installed on one lot, the receptacle having the highest ampere rating shall serve as a basis for calculation.

72-104 Feeders

Feeders between the park consumer's service equipment and the park distribution centres shall be permitted to be installed in accordance with the applicable requirements for service conductors.

72-106 Overcurrent devices and disconnecting means for recreational vehicles

(1)   The branch circuit for each receptacle for a recreational vehicle lot shall be preceded by an individual overcurrent device not exceeding the rating of the receptacle involved and by a suitable disconnecting means.

(2)   The disconnecting means shall be accessible.

72-108 Overcurrent devices and disconnecting means for mobile homes

(1)   The circuit for each mobile home lot shall be preceded by an individual overcurrent device not exceeding the rating of the equipment involved and by a suitable disconnecting means.

(2)   All supply facilities for overcurrent devices and disconnecting means for mobile homes shall be within enclosures of weatherproof construction if installed outdoors.

(3)   The disconnecting means shall be accessible.

72-110 Connection facilities for recreational vehicles and mobile homes (see Appendix B)

(1)   Where receptacles are installed on recreational vehicle lots, they shall be of the following types:
   (a)   15 A, 125 V, 2-pole, 3-wire Type 5-15R receptacle;
   (b)   20 A, 125 V, 2-pole, 3-wire Type 5-20RA receptacle;
   (c)   30 A, 125 V, 2-pole, 3-wire NEMA Standard WD6, Figure TT receptacle (see Appendix B); or
   (d)   50 A, 125/250 V, 3-pole, 4-wire Type 14-50R receptacle.

(2) Each mobile home lot shall have provision for a permanent connection to the mobile unit, except that for mobile homes having main overcurrent protection of 50 A, a 50 A, 125/250 V, 3-pole, 4-wire Type 14-50R receptacle shall be permitted where a deviation has been allowed in accordance with Rule 2-030.

(3) Receptacles, when mounted in other than a horizontal plane, shall be oriented so that the U-ground slot is uppermost.

Δ (4) The receptacle described in Subrule (1)(a) or (b) shall be protected by a ground fault circuit interrupter of the Class A type.

72-112 Power supply cords

(1) Power supply cords shall be permitted only for the connection of recreational vehicles where the cords are not subject to severe physical abuse or extended periods of use.

(2) Power supply cords or cord sets shall be permitted only for the connection of a mobile home when the lot is equipped with a 50 A, 3-pole, 4-wire Type 14-50R receptacle and a deviation has been allowed in accordance with Rule 2-030.

Section 74 — Airport installations

Δ

74-000 Scope

(1)   This Section applies to the installation of series-type constant-current circuitry supplying airport visual aid systems.

(2)   The requirements of this Section supplement or amend the general requirements of this Code.

74-002 Special terminology

In this Section, the following definitions apply:

Ground anchor — a post set into the ground and supporting the lighting fixture.

Ground counterpoise — a conductor installed over lighting cables for the purpose of interconnecting the system ground electrodes and providing lightning protection for the cables.

Pullpit — a below-grade junction box used as a cable pulling point, to house transformers or series lighting cable splices.

Series isolating transformer — a transformer used in airport series lighting circuits to maintain continuity of the primary circuit when the continuity of the secondary circuit is interrupted.

74-004 Wiring methods (see Appendix B)

(1)   Series cables for 6.6 A systems shall be type ASLC and shall be installed in accordance with the requirements of Rule 12-012.

(2)   For aircraft and vehicle visual aid systems on public areas of airports, or that extend beyond the airport property, the installation of buried cables shall be in accordance with the requirements of Rule 12-012.

(3)   For installations covered by this Section of the Code, in areas not accessible to the public, single conductors and cable assemblies shall be of the type indicated in Table 19 as suitable for direct earth burial and shall be installed as follows:

   (a)   when installed in a raceway, be no less than 450 mm deep;

   (b)   when direct buried, be no less than 450 mm deep with a layer of sand or screened earth extending at least 75 mm above and below the conductors, if in rocky or stony ground; and

   (c)   when installed under runways, taxiways, aprons, and roads, mechanical protection shall be provided in the form of rigid conduit or a system of concrete-encased underground raceways installed a minimum of 600 mm deep.

(4)   When installed within a concrete or asphalt surface, Type ASLC shall be installed in a raceway.

(5)   Series cables for 6.6 A systems directly buried in a trench shall have at least

   (a)   75 mm lateral separation between cables of different series circuits;

   (b)   300 mm lateral separation from low-voltage and control cables;

   (c)   75 mm vertical separation in crossovers on the same system; and

   (d)   300 mm vertical separation from low-voltage cables crossing over, with the low-voltage cables in the upper position.

(6)   Each cable of a series circuit shall be identified with a cable marker indicating the circuit origin at each point where the cables are accessible including maintenance holes, pull pits, and similar locations.

74-006 Direct burial transformers

(1)   Series isolating transformers when direct buried in a trench, shall be installed such that the transformer body and primary leads are at a minimum depth of 450 mm below grade.

(2)   The secondary conductors shall be colour-coded, with one conductor being identified.

(3)   The secondary connectors shall be polarized with the identified conductor connected to the larger pin or receptacle.

(4)   The identified conductor shall be grounded.

74-008 Series lighting systems

Series lighting systems shall be installed with a ground counterpoise.

74-010 Ground counterpoise

(1)   Ground counterpoise conductors shall be soft copper wire not smaller than No. 8 AWG and shall

   (a)   be solid bare wire where installed in earth; or

   (b)   be insulated and have a green finish if installed underground in raceways.

(2) The ground counterpoise when installed in earth shall be
  (a) placed 75 mm above all cable in a trench;
  (b) run in zig-zag pattern when outer cables are more than 150 mm apart, crossing cables at 300 mm intervals measured along the trench;
  (c) placed 75 mm over non-metallic conduit containing groups of cables; and
  (d) placed under any protective covering used.

(3) The counterpoise shall be connected to
  (a) the ground anchor of each anchor-mounted light unit;
  (b) the grounded secondary conductor of each series isolating transformer;
  (c) the sheath of metal-sheathed cables and the armour of armoured cables where used to supply light units;
  (d) the ground electrodes at all regulators, towers, and lighting equipment that the counterpoise system serves;
  (e) the ground electrode in each maintenance hole through which the counterpoise conductor passes;
  (f) metallic pull pits, lids or covers; and
  (g) non-current-carrying metallic parts of inset lights.

(4) Where counterpoise conductors of different systems come together or cross each other, they shall be bonded together at those points.

Section 76 — Temporary wiring
(See Appendix G)

76-000 Scope
(1)   This Section of the Code covers temporary wiring installations for buildings or projects under construction or demolition and experimental or testing facilities of a temporary nature.
(2)   The requirements of this Section supplement or amend the general requirements of this Code.

76-002 Conductors
(1)   Conductors shall be
    (a)   of a type in accordance with Section 12; or
    (b)   power supply cable or flexible cord of the outdoor type suitable for extra-hard usage as indicated in Table 11.
(2)   Conductors shall be insulated except as permitted by Rules 6-308, 10-802, and 10-806.
(3)   Service conductors shall be installed in accordance with Sections 6, 10, and 36.
Δ (4)   Overhead conductors shall be aerially supported on poles or other equally substantial means with the spacing of supports not to exceed the maximum span length allowable for the type of conductors used.

Δ 76-004 Grounding and bonding
All grounding and bonding shall be in accordance with Section 10.

Δ 76-006 Service entrance equipment
Where the service equipment is installed in an outdoor location, the equipment shall
(a)   be accessible to authorized persons only;
(b)   be capable of being locked;
(c)   be protected against weather and mechanical damage; and
(d)   not exceed 200 A where mounted on a single pole.

76-008 Distribution centres
(1)   Distribution centres shall have a sufficient number of branch circuits and be of adequate capacity to serve the connected load without overloading any branch circuits and without violating the requirements of Section 14.
(2)   Distribution centres shall be installed in a weatherproof building or be of weatherproof construction.
Δ (3)   Distribution centres including portable ones shall be mounted in an upright position.

76-010 Feeders
(1)   Feeders supplying distribution centres shall be installed in armoured cable or the equivalent.
Δ (2)   Notwithstanding Subrule (1), feeders to portable distribution centres may be permitted to be flexible cord or power supply cable of the outdoor type suitable for extra-hard usage as indicated in Table 11 and containing a bonding conductor.
(3)   Feeders shall be protected at all times from mechanical damage and protected by suitable overcurrent protective devices and controlled by suitable disconnecting means.

76-012 Branch circuits
Δ (1)   Non-metallic-sheathed cable shall be permitted to be used for branch circuits, provided that it is installed in accordance with Rules 12-500 to 12-526.
(2)   Lighting branch circuits shall be kept entirely separate from power branch circuits.
(3)   The installation and type of luminaires or lampholders shall comply with Section 30.
(4)   Each lighting branch circuit shall be protected by a circuit breaker set in accordance with Rule 30-104 and the connected load shall not exceed 80% of the circuit breaker rating.
(5)   Power branch circuits shall be provided as follows:
    (a)   separate branch circuits sized and protected by circuit breakers in accordance with Section 28 shall be provided for motor loads exceeding that encountered from general use hand-held tools;
    (b)   separate branch circuits for known loads such as electric heating shall be protected by circuit breakers set at a value so that the load connected does not exceed 80% of the rating of the breaker; and
    (c)   general use receptacle power branch circuits shall be protected by a circuit breaker set at a value not exceeding the lowest rating of any receptacle connected on the branch circuit.

76-014 Interconnections

Temporary installations shall be constructed as separate installations and shall not be interconnected with any of the circuits of the permanent installations except by special permission.

Δ 76-016 Receptacles

15 A and 20 A receptacles installed to provide power for buildings or projects under construction or demolition shall be protected by ground fault circuit interrupters of the Class A type.

Section 78 — Marinas, yacht clubs, marine wharves, structures, and fishing harbours

78-000 Scope
This Section supplements and amends the general Sections of this Code and applies to the following installations:
- (a) marinas, yacht clubs, and similar establishments, including fixed or floating piers, that are used for the construction, repair, storage, launching, berthing, and fuelling of small craft; and
- (b) facilities for marine wharves, structures, and fishing harbours.

Marinas and yacht clubs

78-050 General
Rules 78-052 to 78-066 inclusive apply to electrical installations in marinas and yacht clubs.

78-052 Receptacles
(1) Receptacles installed on fixed or floating piers, docks, or wharves shall conform to either Diagram 1 or 2.

(2) Receptacles shall be made of corrosion-resistant materials.

(3) Receptacles shall be located above the permanent or maximum normal water level so they cannot become immersed in water and shall be protected from splashing.

(4) 15 A and 20 A, single-phase, 125 V receptacles other than those supplying shore power to boats shall be protected by ground fault circuit interrupters of the Class A type.

(5) Receptacles of configuration 5-15R, intended to supply shore power to boats and installed outdoors or on fixed or floating piers, docks, or wharves, shall be protected by ground fault circuit interrupters of the Class A type.

78-054 Branch circuits
Each receptacle that supplies shore power to boats shall be supplied by an individual branch circuit that supplies no other equipment.

78-056 Feeders and services
(1) The load for each feeder and service supplying receptacles installed on fixed or floating piers, docks, or wharves, and intended to supply shore power to boats shall be calculated on the basis of the ampere rating of the receptacles and by applying the following demand factors:
- (a) 100% of the sum of the first four receptacles having the highest ampere ratings; plus
- (b) 65% of the sum of the ampere ratings of the next four receptacles having the same or next smaller ratings to those specified in Item (a); plus
- (c) 50% of the sum of the ampere ratings of the next five receptacles having the same or next smaller ratings to those specified in Item (b); plus
- (d) 25% of the sum of the ampere ratings of the next sixteen receptacles having the same or next smaller ratings to those specified in Item (c); plus
- (e) 20% of the sum of the ampere ratings of the next twenty receptacles having the same or next smaller ratings to those specified in Item (d); plus
- (f) 15% of the sum of the ampere ratings of the next twenty receptacles having the same or next smaller ratings to those specified in Item (e); plus
- (g) 10% of the sum of the ampere ratings of the remainder of the receptacles.

(2) Where a service or a feeder supplies receptacles as in Subrule (1) plus other loads, the capacity of the conductors shall be calculated in accordance with Subrule (1) plus the other loads in accordance with the Rules of the Code.

78-058 Wiring methods
(1) The wiring method, where exposed to the weather or splashing of water, shall be
- (a) corrosion-resistant rigid metal conduit or rigid PVC conduit;
- (b) mineral-insulated cable having a copper sheath;
- (c) non-metallic-sheathed cable of the NMWU type;
- (d) armoured cable having moisture-resistant insulation and overall corrosion protection; or
- (e) metal-sheathed cable having overall corrosion protection.

(2) Where flexibility is required, outdoor flexible cord suitable for at least hard usage as specified in Table 11 shall be used.

78-060 Grounding and bonding
Grounding and bonding requirements shall be in accordance with Section 10, except that an equipment bonding conductor of copper shall be used.

78-062 Wiring over and under navigable water
Wiring over and under navigable water shall not contravene the Government of Canada's *Navigable Waters Protection Act*, R.S. 1985, c. N-22.

78-064 Gasoline dispensing stations
Requirements shall be in accordance with Section 20 of this Code except that when considering hazardous areas, the grade or ground level shall be the highest water surface and the specific hazardous area shall include the total tidal movement space.

78-066 Communication systems
Where communication systems and circuits are installed, they shall conform to Section 60 of this Code.

Marine wharves, structures, and fishing harbours

78-100 General
Rules 78-054, 78-056, 78-062, 78-064, 78-066, and 78-102 to 78-112 inclusive apply to electrical installations on marine wharves, structures, and fishing harbours.

78-102 Receptacles
(1)   Where receptacles are installed on fixed or floating piers, docks, or wharves in fishing harbours or on marine structures, they shall be
    (a)   15 A, single or duplex, locking or non-locking type conforming to Diagram 1 or 2;
    (b)   20 A up to and including 60 A, single-locking type, conforming to Diagram 2 or special-purpose pin and sleeve type; or
    (c)   over 60 A, single, special-purpose pin and sleeve type.

(2)   Receptacles shall be fabricated from materials resistant to a salt spray and shall be provided with weatherproof enclosures.

(3)   15 A and 20 A, single-phase, 125 V receptacles other than those supplying shore power to boats shall be protected by ground fault circuit interrupters of the Class A type.

78-104 Wiring methods
(1)   The wiring method, where exposed to the weather or splashing of water or salt spray, shall be
    (a)   corrosion-resistant rigid metal conduit, rigid RTRC conduit, or rigid PVC conduit;
    (b)   mineral-insulated cable having a copper sheath;
    (c)   non-metallic-sheathed cable of the NMWU type; or
    (d)   armoured or metal-sheathed cables of the types listed in Table 19 as suitable for exposed wiring in wet locations.

(2)   To allow for tidal movement, an outdoor flexible cord suitable for wet locations and at least hard usage as listed in Table 11 or equivalent, and supported at both ends of gangways to floats by means capable of gripping the cable in reaction to tension due to the weight of the cable or a pull on the cable shall be used.

(3)   Conduit, cable, and overhead wiring shall be installed to avoid mechanical damage and shall be routed to avoid conflict with other potential users of the wharf or structure.

(4)   Conduit, cable, and wiring systems shall be installed to prevent damage from wave action, ice, storm damage, and mooring hooks and lines.

(5)   Fastening hardware shall be galvanized steel, stainless steel, PVC-coated steel, brass, or other materials with similar corrosion-resistant properties.

78-106 Grounding and bonding
(1)   Grounding and bonding requirements shall be in accordance with Section 10, except that bonding conductors of copper not smaller than No. 12 AWG shall be used.

(2)   For electrical systems on wharves located in areas where it is impractical to install a shore-based grounding electrode because of poor earth conductivity, an underwater grounding grid conforming to one of the following methods shall be permitted:
    (a)   on structures with steel piling where the piles are founded in the harbour bottom and continually immersed in salt water, it shall be permissible to ground to the piling, provided that the connections

266

are readily accessible and the grounding conductor is mechanically protected throughout its length; or

(b) on structures that do not conform to Subrule (2)(a), it shall be permissible to connect the grounding conductor to a steel plate electrode, minimum 10 mm thick and 0.36 m² in area; and

(i) the grounding conductor shall be connected to the plate electrode using a thermit-weld connection and shall be mechanically protected to a point 2 m below the normal low tide elevation; and

(ii) the plate electrode shall be founded on the harbour bottom on the lee side of the wharf where the lee side is determined from the prevailing winds.

78-108 Corrosion-resistant materials (see Appendix B)

Corrosion-resistant materials or materials treated to render them resistant to corrosion shall be used for outdoor locations.

78-110 Wharf facilities

All electrical wiring and equipment shall be located to avoid interference with docking of vessels, unloading and loading of vessels, and operation of wharf equipment and trucks.

78-112 Equipment location

(1) Electrical equipment shall be

(a) located above the wharf deck and protected from wave action, ice, storm damage, and mooring lines;

(b) located in such a manner as to minimize risk of damage from wave action and splashing; and

(c) located to avoid impact from docking vessels and vehicular traffic on the wharf.

(2) Receptacles, communication systems, equipment, and other electrical apparatus that may be subject to mechanical damage from boats, vehicles, or other apparatus shall be protected by mounting the equipment in robust shrouds or kiosks constructed of metal, concrete bollards, plywoods, fibreglass, or shall be protected by other equivalent methods.

Section 80 — Cathodic protection

80-000 Scope (see Appendix B)
(1) This Section applies to the installation of impressed current cathodic protection systems.
(2) The requirements of this Section supplement or amend the general requirements of this Code.

80-002 Wiring methods for direct current conductors
(1) DC wiring in non-hazardous areas shall conform to the requirements of Section 12 of this Code except that wiring below ground shall be permitted to be
   (a) buried at a depth of not less than 450 mm; or
   (b) buried at a depth of not less than 200 mm where installed in a raceway or where mechanical protection is provided in accordance with Rule 12-012(3).
(2) DC wiring in hazardous areas shall conform to the requirements of Sections 18 and 20.
(3) Notwithstanding Rule 20-004(8), underground dc wiring below a Class I area shall be permitted to be installed in accordance with Subrule (1), provided that
   (a) the wiring is in threaded rigid metal conduit where it emerges from the ground; and
   (b) the conduit is sealed where it emerges from the ground and at other locations as required by Rule 18-108 or 18-158.

80-004 Conductors
(1) Conductors for dc cathodic protection wiring shall be not smaller than No. 12 AWG and shall be suitable for the conditions of use as indicated in Table 19 for the particular location where they are installed.
(2) Notwithstanding Subrule (1), conductors smaller than No. 12 AWG shall be permitted for instrumentation and reference electrode leads.

80-006 Splices, taps, and connections (see Appendix B)
(1) Splices and taps shall be permitted to be made in dc wiring below ground, provided that
   (a) the splice or tap is made by welding, by a positive compression tool, by crimping and soldering, or by means of a copper, bronze, or brass cable connector; and
   (b) the splice or tap is effectively sealed against moisture by taping or by some other method that is at least as effective as the original insulation of the conductor.
(2) Where exposed to the weather, splices and taps in dc wiring shall be in accordance with Subrule (1).
(3) Connections to piping shall be made by means of
   (a) thermit welding;
   (b) a clamp constructed of the same material as the piping; or
   (c) a clamp constructed of material that is anodic to the piping.
(4) Connections to tanks or other structures shall be made by means of a welded stud, thermit welding, or other permanent means.
(5) Underground connections and connections exposed to the weather shall be sealed against moisture by the application of a material resistant to the specific corrosive environment.

80-008 Branch circuit
The branch circuit supplying the rectifier shall be
(a) in accordance with the requirements of Section 12 of this Code;
(b) provided solely for the cathodic protection system rectifier; and
(c) supplied from a switch or circuit breaker that is capable of being locked in the ON position.

Δ 80-010 Operating voltage
When a cathodic protection system has a maximum available voltage of more than 50 V, the voltage difference between any exposed point of the protected system or any point in the vicinity of the anodes and any point 1 m away on the earth's surface shall not exceed 10 V.

80-012 Warning signs and drawings
(1) Tanks, pipes, or structures protected by a cathodic protection system shall bear a marking, either on the structure, or on a tag attached to the conductor close to the connection to the structure, warning that the connection is not to be disconnected unless the power source is turned off.
(2) A sign shall be placed in a conspicuous location adjacent to the disconnecting means for any electrical apparatus connected to the cathodically protected structures, advising that before equipment or piping is replaced or modified
   (a) cathodic protection shall be turned off; and

(b) a temporary cathodic protection bypass conductor, sized for the maximum available current, shall be installed.

(3) Notwithstanding Subrule (2), in a non-hazardous location the required sign shall be permitted to advise the use of a temporary conductor, sized for the maximum available current, to bypass the location where equipment or piping is to be replaced or modified, as an alternative to turning off the cathodic protection.

(4) A drawing showing the location of underground wiring, polarity, and anodes shall be provided inside the rectifier cabinet or in a location near the cabinet.

(5) When the immersed surfaces of a storage or process container are cathodically protected, a notice shall be placed in a conspicuous location adjacent to the entrance way advising that the cathodic protection system must be turned off before entering the container.

Section 82 — Closed-loop and pre-closed-loop power distribution

82-000 Scope

(1) This Section applies to the installation of closed-loop and pre-closed-loop power distribution systems.

(2) The requirements of this Section supplement or amend the general requirements of this Code.

82-002 Special terminology

In this Section, the following definitions apply:

Closed-loop power distribution system — a power distribution system jointly controlled by signalling between the energy-controlling equipment and the utilization equipment.

Pre-closed-loop power distribution system — a power distribution system that can be readily converted to a closed-loop power distribution system.

82-004 Approval

All components of a closed-loop power distribution system, including conductors, shall be specifically approved for the purpose.

82-006 Control

(1) Outlets forming part of a closed-loop power distribution system shall not be energized unless any utilization equipment plugged into them first exhibits a nominal-operation acknowledgement.

(2) Outlets forming part of a closed-loop power distribution system shall be disconnected when any of the following conditions occur:

(a) a nominal-operation acknowledgement signal is not being received from the utilization equipment connected to that outlet;

(b) a ground-fault condition exists; or

(c) an overcurrent condition exists.

(3) In the event of a controller malfunction, all associated outlets shall be de-energized.

82-008 Branch circuits for both closed-loop and pre-closed-loop systems in dwelling units

(1) As an alternative to the multi-wire branch circuits required by Rule 26-722(b), it shall be permitted to provide 20 A two-wire circuits for the receptacles installed for the kitchen counter work spaces, provided that

(a) at least two such circuits are provided;

(b) the rating of the overcurrent device protecting each circuit is 20 A; and

(c) the ampacity of the branch circuit conductor is 20 A.

(2) The 20 A circuits permitted by Subrule (1) shall not supply any other outlets.

(3) The receptacles shall meet the requirements of Rule 82-010.

82-010 Receptacles in both closed-loop and pre-closed-loop systems in dwelling units

Where the alternative in Rule 82-008(1) has been chosen, it shall be permitted to substitute a duplex receptacle having the CSA Configuration 5-20 RA and rated at 20 A for the split receptacles required by Rule 26-712(d)(iii) provided that all other requirements of Rule 26-712(d)(iii) are met, except that no point along the wall line is more than 610 mm from a receptacle measured horizontally along that wall line.

82-012 Protection of ungrounded conductors

Approved devices providing equivalent overcurrent protection in closed-loop power distribution systems shall be permitted to substitute for fuses or circuit breakers.

82-014 Not interchangeable

(1) Receptacles, cord connector bodies, and attachment plugs used in a closed-loop power distribution system shall be constructed so that they are not interchangeable with other receptacles, cord connector bodies, and attachment plugs.

(2) Notwithstanding Subrule (1), receptacles intended for use on closed-loop power distribution systems shall be permitted to accept attachment plugs corresponding to Diagrams 1 and 2, provided that the receptacles incorporate a means for detecting the inserting of the attachment plug, so that the safety features of the closed-loop power system are retained.

82-016 Power limitation in a control circuit

Control circuits forming part of a closed-loop power distribution system shall be current-limited in accordance with Rule 16-200.

82-018 Cables and conductors

(1)   Power and control conductors shall be permitted within common jackets provided that the conductor insulation voltage rating is not less than the maximum nominal circuit voltage rating of any conductors in the jacket and the cable is of the type listed in Table 19.

(2)   The individual conductors of a hybrid cable shall conform to the requirements of this Code covering their current, voltage, and insulation ratings.

(3)   Hybrid cables incorporating optical fibers shall be installed in accordance with Section 56.

(4)   Power and control conductors forming part of a closed-loop power distribution system shall be permitted to occupy the same cabinet, panelboard, outlet box, or similar enclosure, provided that only connectors specifically approved for hybrid cabling are used.

82-020 Outlet box

Notwithstanding the requirements of Rule 12-3000, an outlet box shall not be required where a component of a closed-loop power distribution system has been specifically approved for use as a connection box.

Δ

Section 84 — Interconnection of electric power production sources

84-000 Scope (see Appendix B)

This Section supplements or amends the general sections of this Code and applies to the installation of electric power production sources interconnected with a supply authority system.

84-002 General requirement (see Appendix B)

The interconnection arrangements shall be in accordance with the requirements of the supply authority.

84-004 Interconnection

The outputs of interconnected electric power production sources shall provide protection against back-feed into a supply authority system fault.

84-006 Synchronization

Electric power production sources shall be equipped with the necessary means to establish and maintain a synchronous condition without adverse effect on the interconnected system.

84-008 Loss of supply authority voltage (see Appendix B)

(1) Unless an alternative procedure is followed in accordance with the requirements of the supply authority, electric power production sources shall, upon loss of voltage in one or more phases of the supply authority system,

(a) be automatically disconnected from all ungrounded conductors of the supply authority system that the electric power production source feeds; and

(b) not be reconnected until the normal voltage of the supply authority system is restored.

(2) An inverter suitable for interconnection with electric power production sources and designed to serve as a disconnection device shall be permitted to be used to meet the requirement of Subrule (1) if approved by the supply authority.

84-010 Overcurrent protection

Equipment and conductors that are energized from both directions shall be provided with overcurrent protection from each source of supply.

84-012 Transformer overcurrent protection

Overcurrent protection for a transformer that is energized from both directions shall be provided in accordance with Section 26 by considering first one side of the transformer, then the other side of the transformer, as the primary.

84-014 System protection devices

Each interconnected electric power production source installation shall be provided with such additional devices as are necessary for system stability and equipment protection.

84-016 Ground fault protection

Ground fault protection shall be provided in accordance with Rule 14-102.

84-018 Loss of electric power production source voltage

An electric power production source shall, upon loss of voltage in one or more of its phases, automatically disconnect all phases from the interconnected system.

84-020 Disconnecting means — Electric power production source

Disconnecting means shall be provided to disconnect simultaneously all ungrounded conductors of any electric power production source of an interconnected system from all circuits supplied by the electric power production source equipment.

84-022 Disconnecting means — Supply authority system (see Appendix B)

Disconnecting means shall be provided to disconnect simultaneously all the electric power production sources from the supply authority system.

84-024 Disconnecting means — General (see Appendix B)

(1) Disconnecting means shall

(a) be capable of being energized from both sides;

(b) plainly indicate whether in the open or closed position;

(c) have contact operation verifiable by direct visible means if required by the supply authority;

(d) have provision for being locked in the open position;

(e) conform to Sections 14, 28, and 36 of this Code if it includes an overcurrent device;

(f) be capable of being opened at rated load;

(g) be capable of being closed with safety to the operator with a fault on the system;

(h) disconnect all ungrounded conductors of the circuit simultaneously;

(i) bear a warning to the effect that inside parts can be energized when the disconnecting means is open; and

(j) be readily accessible.

(2) Where a main fusible disconnecting means is used, an isolating switch shall be provided to allow the fuses to be dead during handling.

84-026 Isolating means

Means shall be provided to isolate equipment that is energized from both directions from all ungrounded conductors of each source of supply.

84-028 Grounding (see Appendix B)

(1) The grounding means at the service entrance shall be permitted to serve as the grounding means for the electric power production source, and the grounding shall be in accordance with Sections 10 and 36.

(2) Notwithstanding Subrule (1), a direct-current power source connected through a solid-state inverter shall not be grounded unless the inverter AC power is separated from the supply authority system by means of an isolating transformer.

84-030 Warning notice and diagram (see Appendix B)

(1) A warning notice of an interconnected system shall be installed in a conspicuous place at the supply authority disconnecting means of Rule 84-022 and the supply authority meter location.

(2) A single-line, permanent, legible diagram of the interconnected system shall be installed in a conspicuous place at the supply authority disconnecting means.

Section 86 — Electric vehicle charging systems

Scope

86-000 Scope

(1) This Section applies to the installation of the electrical conductors and equipment external to an electric vehicle that connect an electric vehicle to a source of electric current by conductive or inductive means, and to the installation of equipment and devices related to electric vehicle charging.

(2) The requirements of this Section supplement or amend the general requirements of this Code.

General

86-100 Special terminology

In this Section, the following definitions apply:

Electric vehicle — an automotive-type vehicle for highway use that

(a) includes passenger automobiles, buses, trucks, vans, etc., primarily powered by an electric motor(s) that draws current from a rechargeable storage battery, fuel cell, photovoltaic array, or other source of electric current; and

(b) excludes electric motorcycles, scooters for persons with disabilities and similar type vehicles, and off-road self-propelled electric vehicles, such as industrial trucks, hoists, lifts, transports, golf carts, airline ground support equipment, tractors, boats, etc.

Electric vehicle charging equipment — the apparatus and conductors, including the electric vehicle connectors, attachment plugs, and all other fittings and devices, specifically used for the purpose of delivering current supply from the premises wiring to the electric vehicle.

Electric vehicle connector — a conductive or inductive device that, by insertion into an inlet on the electric vehicle, establishes connection to an electric vehicle.

Electric vehicle inlet — a conductive or inductive device that is permanently affixed to the electric vehicle and that by coupling into the connector, establishes a connection to the current supply.

86-102 Hazardous locations (see Appendix B)

Where electric vehicle charging equipment or wiring within the Scope of this Section is installed in hazardous locations as specified in Section 18 or 20, it shall comply with the applicable Rules of those Sections.

86-104 Voltages

The nominal ac system voltages used to supply equipment covered in this Section shall not exceed 750 V.

Equipment

86-200 Warning sign

Permanent, legible signs shall be installed at the point of connection of the electric vehicle charging equipment to the branch circuit wiring, warning against operation of the equipment without sufficient ventilation as recommended by the manufacturer's installation instructions.

86-202 Marking

Couplings and inlets shall be specifically approved for the purpose and marked accordingly.

Control and protection

86-300 Branch circuits

Electric vehicle charging equipment rated at 20 A or more shall be supplied by a separate branch circuit that supplies no other loads except ventilation equipment intended for use with the electric vehicle supply equipment.

86-302 Connected load

The total connected load of a branch circuit supplying electric vehicle charging equipment and the ventilation equipment permitted by Rule 86-300 shall be considered continuous for the purposes of Rule 8-104.

86-304 Disconnecting means

(1) A separate disconnecting means shall be provided for each installation of electric vehicle charging equipment rated at 60 A or more, or more than 150 volts-to-ground.

274

(2) The disconnecting means required in Subrule (1) shall be
  (a) on the supply side of the point of connection of the electric vehicle charging equipment;
  (b) located within sight of and accessible to the electric vehicle charging equipment; and
  (c) capable of being locked in the open position.

Electric vehicle charging equipment locations

86-400 Indoor charging sites (see Appendix B)

(1) Indoor sites shall be permitted to include, but not be limited to, integral, attached, and detached residential garages, enclosed or underground parking structures, repair and non-repair commercial garages, agricultural buildings, and similar rooms or locations where the electric vehicle connector can couple to the electric vehicle.

(2) Where the electric vehicle charging equipment requires ventilation
  (a) adequate ventilation shall be provided in each indoor charging site;
  (b) the electric vehicle charging equipment shall be electrically interlocked with the ventilation equipment so that the ventilation equipment operates with the electric vehicle charging equipment; and
  (c) if the supply to the ventilation equipment is interrupted, the electric vehicle charging equipment shall be made inoperable.

86-402 Outdoor charging sites

Outdoor charging sites shall be permitted to include, but not be limited to, residential carports and driveways, curbsides, open parking structures, parking lots, commercial charging facilities, and similar locations.

Tables

Table 1
Allowable ampacities for single copper conductors in free air (based on ambient temperature of 30 °C*)

(See Rules 4-004, 8-104, 12-2210, 26-142, 42-008, and 42-016 and Tables 5A, 5B, and 19.)

Size, AWG or kcmil	Allowable ampacity†					
	60 °C‡	75 °C‡	85–90 °C‡	110 °C‡	125 °C‡	200 °C‡
			Types R90, RW90, T90 NYLON Single-conductor mineral-insulated cables§			
	Type TW	Types RW75, TW75		See Note (3)	See Note (3)	Bare wire
14	20	20	20	40	40	45
12	25	25	25	50	50	55
10	40	40	40	65	70	75
8	55	65	70	85	90	100
6	80	95	100	120	125	135
4	105	125	135	160	170	180
3	120	145	155	180	195	210
2	140	170	180	210	225	240
1	165	195	210	245	265	280
0	195	230	245	285	305	325
00	225	265	285	330	355	370
000	260	310	330	385	410	430
0000	300	360	385	445	475	510
250	340	405	425	495	530	—
300	375	445	480	555	590	—
350	420	505	530	610	655	—
400	455	545	575	665	710	—
500	515	620	660	765	815	—
600	575	690	740	855	910	—
700	630	755	815	940	1005	—
750	655	785	845	980	1045	—
800	680	815	880	1020	1085	—
900	730	870	940	—	—	—
1000	780	935	1000	1165	1240	—
1250	890	1065	1130	—	—	—
1500	980	1175	1260	1450	—	—
1750	1070	1280	1370	—	—	—
2000	1155	1385	1470	1715	—	—
Col. 1	Col. 2	Col. 3	Col. 4	Col. 5	Col. 6	Col. 7

*See Table 5A for the correction factors to be applied to the values in Columns 2 to 7 for ambient temperatures over 30 °C.

†The ampacity of single-conductor aluminum-sheathed cable is based on the type of insulation used on the copper conductor.

(Continued)

Table 1 (Concluded)

‡*These are maximum allowable conductor temperatures for single conductors run in free air and may be used in determining the ampacity of other conductor types in Table 19, which are so run, as follows: From Table 19 determine the maximum allowable conductor temperature for that particular type; then from Table 1 determine the ampacity under the column of corresponding temperature rating.*

§*These ratings are based on the use of 90 °C insulation on the emerging conductors and for sealing. Where a deviation has been allowed in accordance with Rule 2-030, mineral-insulated cable may be used at higher temperatures without decrease in allowable ampacity, provided that insulation and sealing material approved for such higher temperature is used.*

Notes:

(1) *The ratings of Table 1 may be applied to a conductor mounted on a plane surface of masonry, plaster, wood, or any material having a conductivity not less than 0.4 W/(m°C).*

(2) *See Table 5B for correction factors where from 2 to 4 conductors are present and in contact.*

(3) *These ampacities are applicable only under special circumstances where the use of insulated conductors having this temperature rating is acceptable.*

Table 2
Allowable ampacities for not more than 3 copper conductors in raceway or cable (based on ambient temperature of 30 °C*)

(See Rules 4-004, 8-104, 12-2210, 26-142, 42-008, and 42-016 and Tables 5A, 5C, 19, and D3.)

Size, AWG or kcmil	Allowable ampacity†‡‡					
	60 °C‡	75 °C‡	85–90 °C‡	110 °C‡	125 °C‡	200 °C‡
			Types R90, RW90, T90 NYLON			
	Type TW	Types RW75, TW75	Mineral-insulated cable**	See Note	See Note	See Note
14	15	15	15	30	30	30
12	20	20	20	35	40	40
10	30	30	30	45	50	55
8	40	45	45	60	65	70
6	55††	65	65	80	85	95
4	70	85	85	105	115	120
3	80	100	105	120	130	145
2	100	115	120	135	145	165
1	110	130	140	160	170	190
0	125	150	155	190	200	225
00	145	175	185††	215	230	250
000	165	200	210	245	265	285
0000	195	230	235	275	310	340
250	215	255	265	315	335	—
300	240	285	295	345	380	—
350	260	310	325	390	420	—
400	280	335	345	420	450	—
500	320	380	395	470	500	—
Col. 1	Col. 2	Col. 3	Col. 4	Col. 5	Col. 6	Col. 7

(Continued)

Table 2 (Concluded)

| Size AWG, kcmil | Allowable ampacity†‡‡ | | | | | |
| | 60 °C‡ | 75 °C‡ | 85–90 °C‡ | 110 °C‡ | 125 °C‡ | 200 °C‡ |
	Type TW	Types RW75, TW75	Types R90, RW90, T90 NYLON Mineral-insulated cable**	See Note	See Note	See Note
600	355	420	455	525	545	—
700	385	460	490	560	600	—
750	400	475	500	580	620	—
800	410	490	515	600	640	—
900	435	520	555	—	—	—
1000	455	545	585	680	730	—
1250	495	590	645	—	—	—
1500	520	625	700	785	—	—
1750	545	650	735	—	—	—
2000	560	665	775	840	—	—
Col. 1	Col. 2	Col. 3	Col. 4	Col. 5	Col. 6	Col. 7

*See Table 5A for the correction factors to be applied to the values in Columns 2 to 7 for ambient temperatures over 30 °C.

†The ampacity of aluminum-sheathed cable is based on the type of insulation used on the copper conductors.

‡These are maximum allowable conductor temperatures for 1, 2, or 3 conductors run in a raceway, or 2 or 3 conductors run in a cable and may be used in determining the ampacity of other conductor types in Table 19, which are so run, as follows: From Table 19 determine the maximum allowable conductor temperature for that particular type; then from Table 2 determine the ampacity under the column of corresponding temperature rating.

**These ratings are based on the use of 90 °C insulation on the emerging conductors and for sealing. Where a deviation has been allowed in accordance with Rule 2-030, mineral-insulated cable may be used at higher temperatures without decrease in allowable ampacity, provided that insulation and sealing material approved for such higher temperature is used.

Δ ††For 3-wire 120/240 V and 120/208 V service conductors for single dwellings, or for feeder conductors supplying single dwelling units of row housing of apartment and similar buildings, and sized in accordance with Rules 8-200(1), 8-200(2), and 8-202(1), the allowable ampacity for sizes No. 6 and No. 2/0 AWG shall be 60 A and 200 A respectively. In this case, the 5% adjustment of Rule 8-106(1) cannot be applied.

‡‡See Table 5C for the correction factors to be applied to the values in Columns 2 to 7 where there are more than 3 conductors in a run of raceway or cable.

Note: These ampacities are applicable only under special circumstances where the use of insulated conductors having this temperature rating is acceptable.

Table 3
Allowable ampacities for single aluminum conductors in free air (based on ambient temperature of 30 °C*)

(See Rules 4-004, 8-104, 12-2210, 26-142, 42-008, and 42-016 and Tables 5A and 5B.)

Size AWG, kcmil	Allowable ampacity†					
	60 °C‡	75 °C‡	85–90 °C‡	110 °C‡	125 °C‡	200 °C‡
	Type TW	Types RW75, TW75	Types R90, RW90, T90 NYLON	See Note (3)	See Note (3)	Bare wire
12	20	20	20	40	40	45
10	30	30	30	50	55	60
8	45	45	45	65	70	80
6	60	75	80	95	100	105
4	80	100	105	125	135	140
3	95	115	120	140	150	165
2	110	135	140	165	175	185
1	130	155	165	190	205	220
0	150	180	190	220	240	255
00	175	210	220	255	275	290
000	200	240	255	300	320	335
0000	230	280	300	345	370	400
250	265	315	330	385	415	—
300	290	350	375	435	460	—
350	330	395	415	475	510	—
400	355	425	450	520	555	—
500	405	485	515	595	635	—
600	455	545	585	675	720	—
700	500	595	645	745	795	—
750	515	620	670	775	825	—
800	535	645	695	805	855	—
900	580	700	750	—	—	—
1000	625	750	800	930	990	—
1250	710	855	905	—	—	—
1500	795	950	1020	1175	—	—
1750	875	1050	1125	—	—	—
2000	960	1150	1220	1425	—	—
	Col. 2	Col. 3	Col. 4	Col. 5	Col. 6	Col. 7

Col. 1

*See Table 5A for the correction factors to be applied to the values in Columns 2 to 7 for ambient temperatures over 30 °C.

†The ampacity of single-conductor aluminum-sheathed cable is based on the type of insulation used on the aluminum conductor.

‡These are maximum allowable conductor temperatures for single conductors run in free air and may be used in determining the ampacity of other conductor types in Table 19, which are so run, as follows: From Table 19 determine the maximum allowable conductor temperature for that particular type; then from Table 3 determine the ampacity under the column of corresponding temperature rating.

Notes:

(1) The ratings of Table 3 may be applied to a conductor mounted on a plane surface of masonry, plaster, wood, or any material having a conductivity not less than 0.4 W/(m°C).

(2) See Table 5B for correction factors where from 2 to 4 conductors are present and in contact.

(3) These ampacities are applicable only under special circumstances where the use of insulated conductors having this temperature rating is acceptable.

Table 4
Allowable ampacities for not more than 3 aluminum conductors in raceway or cable (based on ambient temperature of 30°C*)

(See Rules 4-004, 8-104, 12-2210, 26-142, 42-008, and 42-016 and Tables 5A, 5C, and D3.)

Size AWG, kcmil	Allowable ampacity†§					
	60 °C‡	75 °C‡	85–90 °C‡	110 °C‡	125 °C‡	200 °C‡
	Type TW	Types RW75, TW75	Types R90, RW90, T90 NYLON	See Note	See Note	See Note
12	15	15	15	25	30	30
10	25	25	25	35	40	45
8	30	30	30	45	50	55
6	40	50	55**	60	65	75
4	55	65	65	80	90	95
3	65	75	75	95	100	115
2	75	90	95**	105	115	130
1	85	100	105	125	135	150
0	100	120	120	150	160	180
00	115	135	145	170	180	200
000	130	155	165	195	210	225
0000	155	180	185**	215	245	270
250	170	205	215	250	270	—
300	190	230	240	275	305	—
350	210	250	260	310	335	—
400	225	270	290	335	360	—
500	260	310	330	380	405	—
600	285	340	370	425	440	—
700	310	375	395	455	485	—
750	320	385	405	470	500	—
800	330	395	415	485	520	—
900	355	425	455	—	—	—
1000	375	445	480	560	600	—
1250	405	485	530	—	—	—
1500	435	520	580	650	—	—
1750	455	545	615	—	—	—
2000	470	560	650	705	—	—
Col. 1	Col. 2	Col. 3	Col. 4	Col. 5	Col. 6	Col. 7

*See Table 5A for the correction factors to be applied to the values in Columns 2 to 7 for ambient temperatures over 30 °C.

†The ampacity of aluminum-sheathed cable is based on the type of insulation used on the aluminum conductors.

(Continued)

Table 4 (Concluded)

‡*These are maximum allowable conductor temperatures for 1, 2, or 3 conductors run in a raceway, or 2 or 3 conductors run in a cable and may be used in determining the ampacity of other conductor types in Table 19, which are so run, as follows: From Table 19 determine the maximum allowable conductor temperature for the particular type; then from Table 4 determine the ampacity under the column of corresponding temperature rating.*

§*See Table 5C for the correction factors to be applied to the values in Columns 2 to 7 where there are more than 3 conductors in a run of raceway or cable.*

Δ ***For 3-wire 120/240 V and 120/208 V service conductors for single dwellings, or for feeder conductors supplying single dwelling units of row housing of apartment and similar buildings, and sized in accordance with Rules 8-200(1), 8-200(2) and 8-202(1), the allowable ampacity for sizes No. 6, No. 2, and No. 4/0 AWG shall be 60 A, 100 A, and 200 A respectively. In this case, the 5% adjustment of Rule 8-106(1) cannot be applied.*

Note: *These ampacities are applicable only under special circumstances where the use of insulated conductors having this temperature rating is acceptable.*

Table 5A
Correction factors applying to Tables 1, 2, 3, and 4 (ampacity correction factors for ambient temperatures above 30 °C)

(See Rules 4-004(8) and 12-2210 and Tables 1, 2, 3, 4, 57, and 58.)

| Ambient temperature, °C | Correction factor | | | | | |
	60 °C Type TW	75 °C Types RW75, TW75	85–90 °C Types R90, RW90, T90 NYLON	110 °C See Note (2)	125 °C See Note (2)	200 °C See Note (2)
40	0.82	0.88	0.90	0.94	0.95	1.00
45	0.71	0.82	0.85	0.90	0.92	1.00
50	0.58	0.75	0.80	0.87	0.89	1.00
55	0.41	0.65	0.74	0.83	0.86	1.00
60	—	0.58	0.67	0.79	0.83	0.91
70	—	0.35	0.52	0.71	0.76	0.87
75	—	—	0.43	0.66	0.72	0.86
80	—	—	0.30	0.61	0.69	0.84
90	—	—	—	0.50	0.61	0.80
100	—	—	—	—	0.51	0.77
120	—	—	—	—	—	0.69
140	—	—	—	—	—	0.59
Col. 1	Col. 2	Col. 3	Col. 4	Col. 5	Col. 6	Col. 7

Notes:

(1) *These correction factors apply, column for column, to Tables 1, 2, 3, and 4. The correction factors in Column 2 also apply to Table 57.*

(2) *The ampacity of a given conductor type at these higher ambient temperatures is obtained by multiplying the appropriate value from Table 1, 2, 3, or 4 by the correction factor for that higher temperature.*

(3) *These ampacities are applicable only under special circumstances where the use of insulated conductors having this temperature rating is acceptable.*

Table 5B
Correction factors for Tables 1 and 3 (where from 2 to 4 single conductors are present and in contact)

(See Rule 4-004(9) and Tables 1 and 3.)

Number of conductors	Correction factors
2	0.90
3	0.85
4	0.80

Notes:

(1) *Where four conductors form a 3-phase-with-neutral system, the values for three conductors may be used. Where three conductors form a single-phase, 3-wire system, the values for two conductors may be used.*

(2) *Where more than four conductors are in contact, the ratings for conductors in raceways shall be used.*

Table 5C
Ampacity correction factors for Tables 2 and 4

(See Rules 4-004 and 12-2210 and Tables 2 and 4.)

Number of conductors	Ampacity correction factor
1–3	1.00
4–6	0.80
7–24	0.70
25–42	0.60
43 and up	0.50

Table 5D
Current rating correction factors where spacings are maintained (ventilated and ladder-type cable trays)

(See Rule 12-2210.)

Number of conductors or cables horizontally	1	2	3	4	5	6
Vertically (layers)						
1	1.00	0.93	0.87	0.84	0.83	0.82
2	0.89	0.83	0.79	0.76	0.75	0.74

Table 6

Maximum number of conductors of one size in trade sizes of conduit or tubing

(See Rule 12-1014(5).)

Conductor Type	Size, AWG or kcmil	16	21	27	35	41	53	63	78	91	103	116	129	155
600 V Without jacket	14	8	15	25	43	59	97	139	200	200	200	200	200	200
	12	6	11	19	33	45	74	106	164	200	200	200	200	200
	10	5	8	14	24	33	55	78	121	162	200	200	200	200
R90XLPE	8	2	4	7	13	18	30	43	67	90	116	146	183	200
RW75XLPE	6	1	3	5	10	13	22	32	50	67	86	108	136	196
RW90XLPE	4	1	2	4	7	10	16	23	36	48	62	78	98	142
	3	1	1	3	6	8	14	19	30	41	53	66	83	120
	2	1	1	3	5	7	11	16	25	34	44	55	70	101
	1	1	1	1	3	5	8	12	19	25	33	41	52	75
	1/0	0	1	1	3	4	7	10	16	21	27	34	44	63
	2/0	0	1	1	2	3	6	8	13	17	23	29	36	53
	3/0	0	1	1	1	3	5	7	11	14	19	24	30	44
	4/0	0	0	1	1	1	4	6	9	12	15	20	25	36
	250	0	0	1	1	1	3	4	7	10	13	16	21	30
	300	0	0	1	1	1	2	4	6	8	11	14	18	25
	350	0	0	0	1	1	2	3	5	7	9	12	16	23
	400	0	0	0	1	1	1	3	5	6	8	11	14	20
	450	0	0	0	1	1	1	3	4	6	8	10	13	18
	500	0	0	0	1	1	1	2	4	5	7	9	11	17
	600	0	0	0	0	0	1	1	3	4	5	7	9	13
	700	0	0	0	0	1	1	1	3	4	5	6	8	12
	750	0	0	0	0	0	1	1	2	3	4	6	8	11
	800	0	0	0	0	0	1	1	2	3	4	5	7	10
	900	0	0	0	0	1	1	1	1	3	4	5	6	9
	1000	0	0	0	0	0	1	1	1	2	3	4	6	9
	1250	0	0	0	0	1	1	1	1	1	3	3	5	7
	1500	0	0	0	0	0	0	1	1	1	1	3	4	6
	1750	0	0	0	0	0	0	1	1	1	1	2	3	5
	2000	0	0	0	0	0	0	1	1	1	1	1	3	4

(Continued)

Table 6 (Continued)

Size of conduit or tubing		16	21	27	35	41	53	63	78	91	103	116	129	155
Conductor	Size, AWG or kcmil													
Type														
1000 V Without jacket	14	5	10	16	28	39	64	92	142	190	200	200	200	200
	12	4	8	13	23	31	52	74	114	153	197	200	200	200
	10	3	6	10	18	24	40	57	88	118	153	191	200	200
R90XLPE	8	2	4	7	13	18	30	43	67	90	116	146	183	200
RW75XLPE	6	1	2	4	8	11	18	26	40	54	70	88	110	159
RW90XLPE	4	1	1	3	6	8	13	19	30	40	52	65	82	118
	3	1	1	3	5	6	11	16	26	34	44	56	70	102
	2	1	1	2	4	6	10	14	22	29	38	47	60	86
	1	0	1	1	3	4	7	10	15	20	26	33	42	60
	1/0	0	1	1	2	3	6	8	13	17	22	28	35	51
	2/0	0	1	1	1	3	5	7	11	15	19	24	30	43
	3/0	0	0	1	1	2	4	6	9	12	16	20	25	37
	4/0	0	0	1	1	1	3	5	8	10	13	17	21	31
	250	0	0	1	1	1	3	4	6	8	11	14	17	25
	300	0	0	0	1	1	2	3	5	7	9	12	15	22
	350	0	0	0	1	1	1	3	5	6	8	11	13	20
	400	0	0	0	1	1	1	2	4	6	7	9	12	17
	450	0	0	0	1	1	1	2	4	5	7	8	11	16
	500	0	0	0	1	1	1	1	3	5	6	8	10	15
	600	0	0	0	1	1	1	1	3	4	5	7	8	12
	700	0	0	0	0	1	1	1	2	3	4	6	7	11
	750	0	0	0	0	1	1	1	2	3	4	5	7	10
	800	0	0	0	0	1	1	1	2	3	4	5	7	10
	900	0	0	0	0	1	1	1	1	3	4	5	6	9
	1000	0	0	0	0	1	1	1	1	2	3	4	5	8
	1250	0	0	0	0	0	1	1	1	1	3	3	4	6
	1500	0	0	0	0	0	1	1	1	1	2	3	3	5
	1750	0	0	0	0	0	0	1	1	1	1	2	3	4
	2000	0	0	0	0	0	0	0	1	1	1	1	3	4

(Continued)

Table 6 (Continued)

Size of conduit or tubing →		16	21	27	35	41	53	63	78	91	103	116	129	155
Conductor	**Size, AWG or kcmil**													
Type														
600 V	14	5	10	16	28	39	64	92	142	190	200	200	200	200
With jacket	12	4	8	13	23	31	52	74	114	153	197	200	200	200
	10	3	6	10	18	24	40	57	88	118	153	191	200	200
R90XLPE	8	1	3	6	10	14	24	34	53	71	91	115	144	200
RW75XLPE	6	1	1	3	6	9	15	21	33	45	58	72	91	132
RW90XLPE	4	1	1	3	5	7	11	16	25	34	44	55	69	101
R90EP	3	1	1	2	4	6	10	14	22	30	38	48	60	87
RW75EP	2	1	1	1	3	5	8	12	19	25	33	41	52	75
RW90EP	1	0	1	1	2	4	6	8	13	17	22	28	36	52
	1/0	0	1	1	1	3	5	7	11	15	19	24	31	44
	2/0	0	0	1	1	2	4	6	9	13	16	21	26	38
	3/0	0	0	1	1	1	3	5	8	11	14	18	22	32
	4/0	0	0	1	1	1	3	4	7	9	12	15	19	27
	250	0	0	0	1	1	1	3	5	7	9	11	14	21
	300	0	0	0	0	1	1	3	4	6	8	10	12	18
	350	0	0	0	0	1	1	2	4	5	7	9	11	16
	400	0	0	0	1	1	1	1	3	5	6	8	10	15
	450	0	0	0	0	1	1	1	3	4	6	7	9	13
	500	0	0	0	0	0	1	1	3	4	5	7	8	12
	600	0	0	0	0	0	1	1	2	3	4	5	7	10
	700	0	0	0	0	0	1	1	1	3	4	5	6	9
	750	0	0	0	0	0	1	1	1	3	3	4	6	8
	800	0	0	0	0	0	1	1	1	2	3	4	5	8
	900	0	0	0	0	0	1	1	1	2	3	4	5	7
	1000	0	0	0	0	0	1	1	1	1	3	3	4	7
	1250	0	0	0	0	0	0	1	1	1	1	2	3	5
	1500	0	0	0	0	0	0	1	1	1	1	1	3	4
	1750	0	0	0	0	0	0	0	1	1	1	1	2	4
	2000	0	0	0	0	0	0	0	1	1	1	1	1	3

(Continued)

Table 6 (Continued)

Conductor Type	Size, AWG or kcmil	16	21	27	35	41	53	63	78	91	103	116	129	155
RWU90XLPE	14	4	7	11	20	28	46	66	102	136	175	200	200	200
TWU	12	3	6	9	17	23	38	54	84	113	145	182	200	200
TWU75	10	2	4	8	13	18	30	44	68	91	117	147	184	200
	8	1	2	4	8	11	18	26	40	53	69	87	109	157
	6	1	1	3	6	8	14	20	31	42	55	68	86	124
	4	1	1	2	5	6	11	15	24	32	42	52	66	95
	3	1	1	1	4	5	9	13	21	28	36	46	57	83
	2	1	1	1	3	5	8	11	18	24	31	39	49	72
	1	0	1	1	2	3	6	9	13	18	23	29	37	54
	1/0	0	1	1	1	3	5	7	11	16	20	25	32	46
	2/0	0	1	1	1	2	4	6	10	13	17	21	27	39
	3/0	0	0	1	1	1	3	5	8	11	14	18	23	33
	4/0	0	0	1	1	1	3	4	7	9	12	15	19	28
	250	0	0	0	1	1	2	3	6	8	10	12	16	23
	300	0	0	0	1	1	1	3	5	7	9	11	14	20
	350	0	0	0	1	1	1	3	4	6	8	10	12	18
	400	0	0	0	1	1	1	2	3	5	7	9	11	16
	450	0	0	0	1	1	1	1	3	5	6	8	10	15
	500	0	0	0	0	1	1	1	3	4	6	7	9	13
	600	0	0	0	0	0	1	1	2	3	6	6	7	11
	700	0	0	0	0	0	1	1	1	3	5	5	6	10
	750	0	0	0	0	0	1	1	1	3	4	5	6	9
	800	0	0	0	0	0	1	1	1	3	4	5	6	9
	900	0	0	0	0	0	1	1	1	2	3	4	5	8
	1000	0	0	0	0	0	1	1	1	1	3	4	5	7
	1250	0	0	0	0	0	0	1	1	1	2	3	4	6
	1500	0	0	0	0	0	0	1	1	1	1	2	3	5
	1750	0	0	0	0	0	0	1	1	1	1	1	3	4
	2000	0	0	0	0	0	0	0	0	1	1	1	2	4

(Continued)

Table 6 (Continued)

Size of conduit or tubing		16	21	27	35	41	53	63	78	91	103	116	129	155
Conductor														
Type	Size, AWG or kcmil													
TW	14	8	15	25	43	59	97	139	200	200	200	200	200	200
TW75	12	6	11	19	33	45	74	106	164	200	200	200	200	200
	10	5	8	14	24	33	55	78	121	162	200	200	200	200
	8	2	4	7	13	18	30	43	67	90	116	146	183	200
	6	1	2	4	8	11	18	26	40	54	70	88	110	159
	4	1	1	3	6	8	13	19	30	40	52	65	82	118
	3	1	1	3	5	7	11	16	26	34	44	56	70	102
	2	1	1	2	4	6	10	14	22	29	38	47	60	86
	1	0	1	1	3	4	7	10	15	20	26	33	42	60
	1/0	0	1	1	2	3	6	8	13	17	22	28	35	51
	2/0	0	1	1	1	3	5	7	11	15	19	24	30	43
	3/0	0	0	1	1	2	4	6	9	12	16	20	25	37
	4/0	0	0	1	1	1	3	5	8	10	13	17	21	31
	250	0	0	1	1	1	2	4	6	8	11	13	17	25
	300	0	0	0	1	1	1	3	5	7	9	12	15	21
	350	0	0	0	0	1	1	3	5	6	8	10	13	19
	400	0	0	0	0	1	1	2	4	6	7	9	12	17
	450	0	0	0	1	1	1	2	4	5	7	8	11	15
	500	0	0	0	1	1	1	1	3	5	6	8	10	14
	600	0	0	0	0	1	1	1	3	4	5	6	8	11
	700	0	0	0	1	0	1	1	2	3	4	5	7	10
	750	0	0	0	0	0	1	1	1	3	4	5	6	9
	800	0	0	0	0	0	1	1	1	3	4	5	6	9
	900	0	0	0	0	0	1	1	1	2	3	4	5	8
	1000	0	0	0	0	0	1	1	1	2	3	4	5	7
	1250	0	0	0	0	0	1	1	1	1	2	3	4	5
	1500	0	0	0	0	0	0	1	1	1	1	2	3	5
	1750	0	0	0	0	0	0	1	1	1	1	2	3	4
	2000	0	0	0	0	0	0	0	1	1	1	1	2	4

(Continued)

Table 6 (Concluded)

Size of conduit or tubing		16	21	27	35	41	53	63	78	91	103	116	129	155
Conductor Type	Size, AWG or kcmil													
TWN75	14	12	22	36	62	85	140	200	200	200	200	200	200	200
	12	9	16	26	45	62	102	145	200	200	200	200	200	200
T90 Nylon	10	5	10	16	28	38	63	90	139	187	200	200	200	200
	8	3	5	9	16	22	36	52	80	100	138	173	200	200
	6	1	4	6	11	16	26	37	58	80	100	125	157	200
	4	1	2	4	7	9	16	23	35	47	61	77	96	140
	3	1	1	3	6	8	13	19	30	40	52	65	82	118
	2	1	1	2	5	7	11	16	25	34	43	55	69	99
	1	1	1	2	3	5	8	12	18	25	32	40	51	73
	1/0	0	1	1	3	4	7	10	15	21	27	34	42	62
	2/0	0	1	1	2	3	6	8	13	17	22	28	35	51
	3/0	0	1	1	1	3	5	7	11	14	19	23	29	43
	4/0	0	0	1	1	1	4	5	9	12	15	19	24	35
	250	0	0	1	1	1	3	4	7	10	12	16	20	29
	300	0	0	0	1	1	2	4	6	8	11	13	17	25
	350	0	0	0	1	1	2	3	5	7	9	12	15	22
	400	0	0	0	1	1	1	3	5	6	8	10	13	19
	450	0	0	0	1	1	1	2	4	6	7	9	12	17
	500	0	0	0	1	1	1	2	4	5	7	9	11	16

Notes:

(1) The calculated values in this Table are based on conventional concentric Class B stranded conductors.

(2) The calculated values in this Table are based on metallic conduit. Other types of raceway of the same nominal size may have different dimensions.

(3) Some raceways are required to contain a separate bonding or grounding conductor. No allowance is made for extra conductors in this Table.

Δ

Table 7
Radius of conduit or tubing bends

(See Rule 12-922.)

Size of conduit or tubing	Minimum radius to centre of conduit or tubing, mm
16	102
21	114
27	146
35	184
41	210
53	241
63	267
78	330
91	381
103	406
129	610
155	762

Table 8
Maximum allowable per cent conduit and tubing fill

(See Rules 12-1014 and 38-032.)

	Maximum conduit and tubing fill, %				
	Number of conductors or multi-conductor cables				
	1	2	3	4	Over 4
Conductors or multi-conductor cables (not lead-sheathed)	53	31	40	40	40
Lead-sheathed conductor or multi-conductor cables	55	30	40	38	35

Table 9
Cross-sectional areas of conduit and tubing
(See Rule 12-1014.)

Nominal conduit size	Internal diameter, mm	Cross-sectional area of conduit and tubing, mm²							
		100%	55%	53%	40%	38%	35%	31%	30%
16	15.8	196	107.8	103.9	78.41	74.49	68.61	60.77	58.81
21	20.9	344	189.2	182.3	137.6	130.7	120.4	106.7	103.2
27	26.6	557.6	306.7	295.5	223	211.9	195.2	172.9	167.3
35	35.1	965	530.7	511.4	386	366.7	337.7	299.1	289.5
41	40.9	1313	722.4	696.1	525.4	499.1	459.7	407.2	394
53	52.5	2165	1191	1147	866	822.7	757.7	671.1	649.5
63	62.7	3089	1699	1637	1236	1174	1081	957.5	926.7
78	77.9	4769	2623	2528	1908	1812	1669	1479	1431
91	90.1	6379	3508	3381	2551	2424	2233	1977	1914
103	102.3	8213	4517	4353	3285	3121	2875	2546	2464
116	114.5	10 288	5659	5453	4115	3910	3601	3189	3086
129	128.2	12 907	7099	6841	5163	4905	4517	4001	3872
155	154.1	18 639	10 251	9879	7456	7083	6524	5778	5592

Note: *The dimensions shown are typical of metallic conduit and tubing. Other figures more accurately representing the actual dimensions of a particular product may be substituted, when known. Dimensions of other circular raceways may be obtained from the approved standard to which they are manufactured.*

Table 10
Dimensions of cable for calculating conduit and tubing fill
(See Rule 12-1014.)

Conductor size, AWG or kcmil	R90XLPE*, RW75XLPE*, RW90XLPE* 600 V		R90XLPE*, RW75XLPE*, RW90XLPE* 1000 V		R90XLPE†, RW75XLPE†, R90EP†, RW75EP†, RW90XLPE†, RW90EP‡ 600 V		TWN75, T90 NYLON		TW, TW75		TWU, TWU75, RWU90XLPE*	
	Dia., mm	Area, mm²	Dia., mm	Area, mm²	Dia., mm	Area, mm²	Dia., mm	Area, mm²	Dia., mm	Area, mm²	Dia., mm	Area, mm²
14	3.36	8.89	4.12	13.36	4.12	13.36	2.80	6.18	3.36	8.89	4.88	18.70
12	3.84	11.61	4.60	16.65	4.60	16.75	3.28	8.47	3.84	11.61	5.36	22.56
10	4.47	15.67	5.23	21.45	5.23	21.45	4.17	13.63	4.47	15.67	5.97	27.99
8	5.99	28.17	5.99	28.17	6.75	35.77	5.49	23.66	5.99	28.17	7.76	47.29
6	6.95	37.98	7.71	46.73	8.47	56.39	6.45	32.71	7.71	46.73	8.72	59.72
4	8.17	52.46	8.93	62.67	9.69	73.79	8.23	53.23	8.93	62.67	9.95	77.76
3	8.88	61.99	9.64	73.05	10.40	85.01	8.94	62.83	9.64	73.05	10.67	89.42
2	9.70	73.85	10.46	85.88	11.22	98.82	9.76	74.77	10.46	85.88	11.48	103.5
1	11.23	99.10	12.49	122.6	13.51	143.4	11.33	100.9	12.49	122.6	13.25	137.9
1/0	12.27	118.3	13.53	143.9	14.55	166.4	12.37	120.3	13.53	143.9	14.28	160.2
2/0	13.44	141.9	14.70	169.8	15.72	194.2	13.54	144.0	14.70	169.8	15.45	187.5
3/0	14.74	170.6	16.00	201.0	17.02	227.5	14.84	172.9	16.00	201.0	16.76	220.6
4/0	16.21	206.4	17.47	239.7	18.49	268.5	16.31	209.0	17.47	239.7	18.28	262.4
250	17.90	251.8	19.17	288.5	21.21	353.2	18.04	255.7	19.43	296.4	20.20	320.5
300	19.30	292.6	20.56	332.1	22.60	401.2	19.44	296.9	20.82	340.5	21.54	364.4
350	20.53	331.0	21.79	372.9	23.83	446.0	20.67	335.6	22.05	381.9	22.81	408.6
400	21.79	373.0	23.05	417.3	25.09	494.5	21.93	377.8	23.31	426.8	24.07	455.0
450	22.91	412.2	24.17	458.8	26.21	539.5	23.05	417.3	24.43	468.7	25.19	498.4
500	23.95	450.5	25.21	499.2	27.25	583.2	24.09	455.8	25.47	509.5	26.24	540.8
600	26.74	561.7	27.24	582.9	30.04	708.8	—	—	28.26	627.3	29.02	661.4
700	28.55	640.0	29.05	662.6	31.85	796.5	—	—	30.07	710.0	30.82	746.0
750	29.41	679.3	29.91	702.6	32.71	840.3	—	—	30.93	751.3	31.69	788.7
800	30.25	718.7	30.75	742.6	33.55	884.0	—	—	31.77	792.7	32.53	831.1
900	31.85	796.6	32.35	821.8	35.15	970.2	—	—	33.37	874.5	34.13	914.9
1000	33.32	872.0	33.82	898.4	36.62	1053	—	—	34.84	953.4	35.60	995.4
1250	37.56	1108	38.32	1153	42.38	1411	—	—	39.08	1200	39.08	1199
1500	40.68	1300	41.44	1349	45.50	1626	—	—	42.20	1399	42.96	1449
1750	43.58	1492	44.34	1544	48.40	1840	—	—	45.10	1598	45.86	1652
2000	46.27	1681	47.03	1737	51.09	2050	—	—	47.79	1794	48.55	1851

*Unjacketed.
†Jacketed.
‡Includes EPCV.
Note: *Dimensions for aluminum conductors are subject to the range of sizes for which they are certified.*

Table 11
Conditions of use, voltage, and temperature ratings of flexible cords, heater cords, tinsel cords, equipment wires, Christmas-tree cords, portable power cables, elevator cables, stage lighting, and festoon cables

(See Rules 4-010, 4-018, 4-038, 12-010, 16-112, 16-210, 22-108, 30-310, 30-410, 32-100, 38-011, 38-021, 44-350, 44-354, 50-014, 60-308, 70-108, 76-002, 76-010, 78-058, and 78-104.)

	Use	Kind	CSA type designation	Voltage rating, V	Temperature rating, °C	Reference Notes
Dry locations only	Not for hard usage	Heat-resistant equipment wire	GTF	600	125	4
		Equipment wire	TXF	125	60	—
		Indoor Christmas-tree cord	PXT	125	60	7
Damp (or dry) locations	Not for hard usage	Flexible cord	SV	300	60, 75, 90	—
			SVO, SVOO	300	60, 75, 90	3
			SVT	300	60, 75, 90, 105	—
			SVTO, SVTOO	300	60, 75, 90, 105	3
			SPT-1, NISPT-1	300	60, 75, 90, 105	7
			SPT-2, NISPT-2	300	60, 75, 90, 105	—
			SPT-3	300	105	—
		Heater cord	HPN	300	90, 105	3
	Not for hard usage	Tinsel cord	TPT	300	60	—
			TST	300	60	—
		Equipment wire	TEW	600	105	1, 3, 4, 11
			TBS	600	90	4, 11
			SIS	600	90	4, 11
			REW	300	105	1, 3, 6, 11
			REW	600	105	1, 3, 4, 6, 11
			SEWF-1	300	150	1, 5, 11
			SEW-1	300	200	1, 5, 11
			SEWF-2	600	150	4, 5, 11
			SEW-2	600	200	4, 5, 11
			TEWN	600	105	1, 3, 11
	For hard usage	Flexible cord	SJ	300	60, 75, 90, 105	10
			SJO, SJOO	300	60, 75, 90, 105	3, 10
			SJT	300	60, 75, 90, 105	10
			SJTO, SJTOO	300	60, 75, 90, 105	3, 10
	For hard usage	Heater cord	HSJ, HSJO, HSJOO	300	90, 105	3, 8, 9, 10
	For extra-hard usage	Flexible cord	S	600	60, 75, 90, 105	10
			SO, SOO	600	60, 75, 90, 105	3, 10
			ST,	600	60, 75, 90, 105	10
			STO, STOO	600	60, 75, 90, 105	3, 10
		Dryer and range cable	DRT	300	60	2
	Elevator travelling cable		E	300	60	11
			E	600	60	11
			ETT, ETP	300	60	11
			ETT, ETP	600	60	11
			EO	300	60	3, 11
			EO	600	60	3, 11

(Continued)

Table 11 (Continued)

Use		Kind	CSA type designation	Voltage rating, V	Temperature rating, °C	Reference Notes
Wet (or damp or dry) locations	Not for hard usage	Outdoor Christmas-tree cord	CXWT	300	60	—
			CXWT	600	60	—
			PXWT	300	60	—
		Outdoor equipment wire	TXFW	300	60	—
	For hard usage	Outdoor flexible cord	SJOW, SJOOW	300	60, 75, 90, 105	3
			SJTW	300	60, 75, 90, 105	—
			SJTOW, SJTOOW	300	60, 75, 90, 105	3
	For extra-hard usage	Outdoor flexible cord	SOOW	600	60, 75, 90, 105	3
			SOW	600	60, 75, 90, 105	3
			STW	600	60, 75, 90, 105	—
			STOW, STOOW	600	60, 75, 90, 105	—
		Portable power cables	W	2000	90	12, 13
			G	2000	90	12, 13
			G-GC	2000	90	12
			G-BGC	2000	90	12
			SHC-GC	2000	90	12
			SH,	(2000,	90	12
			SHD,	5000,	90	12
			SHD-GC,	8000,	90	12
			SHD-BGC	15 000, or 25 000)	90	12
	For hard usage	Stage lighting	PPC	600	60, 75, 90, 105	13
		Festoon	Festoon	600	60, 105	14
			Festoon-outdoors	600	60, 105	14

Notes:

(1) Types REW, SEW, SEWF, TEW, and TEWN shall be permitted in raceways for Class 1 circuits in accordance with Rule 16-112(2).

(2) Dryer and range cables are for use only in household dryer and range power-supply cords. These cables are not for sale to the public for general use.

(3) When exposed to oil, the temperature rating of the jacket of Types SVO, SVTO, HSJO, EO, SJO, SJTO, STO, SO, SOW, SJOW, and SJTOW; and the insulation and jacket of Types SVOO, SVTOO, SJOO, SJTOO, STOO, SOO, SOOW, SJOOW, and SJTOOW; and the insulation of Type HPN heater cord and TEWN, REW, and TEW equipment wire is limited to 60 °C regardless of the temperature rating of the product.

(4) Types GTF, REW, TEW, TBS, SIS, SEWF-2, and SEW-2 may be used in raceways in accordance with Rule 30-310(2)(c)(ii).

(5) Types SEWF-1 and SEWF-2 with a nickel or a nickel-coated copper conductor have a temperature rating of 200 °C. Types SEW-1, SEWF-1, SEW-2, and SEWF-2 with a nickel or a nickel-coated copper conductor may also have a temperature rating of 250 °C.

(6) Types having cross-linked PVC insulation are surface marked with the type designation followed by (XLPVC) and types having cross-linked chlorinated polyethylene are surface marked with the type designation followed by (XLCPE).

(7) Types PXT and SPT-1 (Size No. 20 AWG) are not for sale to the public or for general use. They are for decorative lighting and electric clock use respectively.

(8) When Type HSJO heater cord is provided with 90 °C polychloroprene insulation (no asbestos insulation), the type designation "HSJO" is surface printed on this cord.

(9) When Type HSJO heater cords are provided with 90 °C ethylene propylene rubber insulation (no asbestos insulation), the type designation "HSJO" is surface printed on this cord.

(10) Types HSJO, SJ, SJO, SJOO, SJT, SJTO, SJTOO, S, SO, SOO, ST, STO, and STOO flexible cords are now recognized only as components of equipment.

(Continued)

Table 11 (Concluded)

(11) Suitable for use under Rule 38-011(2) when flame-retardant and moisture-resistant.

(12) Natural rubber jackets are not suitable for use in oily environments.

(13) Type PPC is single-conductor cable intended for use in temporary installations, such as portable stage lighting and outdoor functions. If multi-conductor is needed for the same applications, Types W or G may be used if ampacities are higher than flexible cords specified in Section 44.

(14) Festoon cable is of flat configuration incorporating power and/or control conductors and supported by a messenger. It supplies power and control wiring for cranes and hoists, indoors or outdoors. See Rule 40-004.

Table 12
Allowable ampacity of flexible cord and equipment wire (based on ambient temperature of 30 °C)

(See Rules 4-014 and 4-018.)

| | Allowable ampacity | | | | | | | | |
| | Flexible cord | | | | | | | Equipment wire | |
Size, AWG	Tinsel cords — Types TPT, TST	Christmas-tree cord — Type CXWT	Christmas-tree cord — Type PXT	Elevator cable — Types E, EO, ETT, ETP	Types NISPT-1, NISPT-2, PXWT, SV§, SVO§, SVOO§, SJ‡§, SJO‡§, SJOO‡§, SJOW§, SJOOW§, S‡, SO‡, SOO‡, SOW, SOOW, SPT-1, SPT-2, SPT-3, SVT§, SVTO§, SVTOO§, SJT‡§, SJTO‡§, SJTOO‡§, ST‡, STO‡, STOO‡, SJTW§, SJTOW§, SJTOOW§, STW, STOW, STOOW — 2 current-carrying conductors	— 3 current-carrying conductors	Types HSJ‡, HSJO‡, HSJOO, HPN**, DRT	Types TXF, TXFW	Types GTF*, TEW*, SEW*, REW*, TEWN*, SEWF*, TBS*, SIS*
27	0.5	—	—	—	—	—	—	—	1
26	—	—	—	—	—	—	—	—	2
24	—	—	—	—	—	—	—	—	3
22	—	—	—	—	—	—	—	—	4
20	—	—	2	—	2	—	2	2	—
18	—	5	—	5	10	7	10	5	6
16	—	7	—	7	13	10	15	7	8
14	—	15	—	15	18	15	20	—	17
12	—	20	—	20	25	20	25	—	23
10	—	—	—	25	30	25	30†	—	28
8	—	—	—	35	40	35	40†	—	40
6	—	—	—	45	55	45	50†	—	55
4	—	—	—	60	70	60	60†	—	70
3	—	—	—	—	—	—	—	—	80
2	—	—	—	80	95	80	—	—	95
1	—	—	—	—	—	—	—	—	110

(Continued)

Table 12 (Concluded)

Size AWG	Allowable ampacity							Equipment wire	
	Flexible cord								
	Tinsel cords	Christmas-tree cord		Elevator cable	Types NISPT-1, NISPT-2, PXWT, SV§, SVO§, SVOO§, SJ‡§, SJO‡§, SJOO‡§, SJOW§, SJOOW§, S‡, SO‡, SOO‡, SOW, SOOW, SPT-1, SPT-2, SPT-3, SVT§, SVTO§, SVTOO§, SJT‡§, SJTO‡§, SJTOO‡§, ST‡, STO‡, STOO‡, STW, STOW, STOOW		Types HSJ‡, HSJO‡, HSJOO, HPN**, DRT	Types TXF, TXFW	Types GTF*, TEW*, SEW*, REW*, TEWN*, SEWF*, TBS*, SIS*
	Types TPT, TST	Type CXWT	Type PXT	Types E, EO, ETT, ETP	2 current-carrying conductors	3 current-carrying conductors*			
1/0	—	—	—	—	—	—	—	—	125
2/0	—	—	—	—	—	—	—	—	145
3/0	—	—	—	—	—	—	—	—	165
4/0	—	—	—	—	—	—	—	—	195

*The derating factors of Rule 4-014(1)(b), (c), (d), and (e) are to be applied to these values for the types listed in this Column.

†These current ratings are for Type DRT household dryer and range cables only.

‡Types HSJ, HSJO, SJ, SJO, SJOO, SJT, SJTO, S, SO, SOO, ST, STO, and STOO flexible cords are now recognized only as components of equipment.

§Types SJ, SJO, SJOO, SJOW, SJOOW, SVT, SVTO, SVTOO, SV, SVO, SVOO, SJT, SJTO, SJTOO, SJTW, SJTOW, and SJTOOW No. 17 AWG are recognized with an ampacity of 12 A as a component of vacuum cleaners with retractable power supply cords.

**Type HPN No. 17 AWG is recognized with an ampacity of 13 A.

Notes:

(1) It is intended that this Table be used in conjunction with applicable end-use product standards to ensure selection of the proper size and type.

(2) TXF is recognized in size No. 20 AWG only. TXFW is recognized in sizes No. 16 and 18 AWG.

Table 12A

Allowable ampacities for portable power cables (amperes per conductor)

(See Rule 4-040.)

Power conductor size, AWG or kcmil	Single conductor				Two conductor	Three conductor				Four conductor	Five conductor	Six conductor
	≤ 2 kV non-shielded	5 and 8 kV shielded	15 kV shielded	25 kV shielded	2 kV	2 kV non-shielded	2, 5, and 8 kV shielded	15 kV shielded	25 kV shielded	2 kV	2 kV	2 kV
12	—	—	—	—	42	35	—	—	—	31	30	28
10	—	—	—	—	59	49	49	—	—	44	42	39
8	80	—	—	—	74	65	65	—	—	59	55	53
6	105	—	—	—	99	87	102	—	—	79	75	70
4	140	—	—	—	130	114	134	—	—	102	97	91
2	190	—	—	—	174	152	175	180	—	134	128	121
1	220	248	248	244	202	177	202	206	210	157	150	142
1/0	260	286	285	280	234	205	232	236	240	182	—	—
2/0	300	329	328	322	271	237	267	271	274	211	—	—
3/0	350	380	377	371	313	274	307	311	315	243	—	—
4/0	405	440	437	428	361	316	353	358	360	280	—	—
250	455	488	484	473	402	352	390	395	396	308	—	—
300	505	546	540	528	449	393	438	441	441	341	—	—
350	570	604	597	582	495	433	478	482	482	368	—	—
400	615	656	649	629	535	468	517	517	517	392	—	—
500	700	757	746	725	613	536	590	590	590	434	—	—
750	885	—	—	—	—	—	—	—	—	—	—	—
1000	1055	—	—	—	—	—	—	—	—	—	—	—

Notes:

(1) The ampacity values are based on a single isolated cable in air at an ambient temperature of 30 °C. For cables operating at a different ambient air temperature, the ampacity shall be obtained by multiplying the appropriate value from Table 12A by the correction factor for that other ambient temperature, as contained in Table 12B.

(2) The ampacities are based on a conductor temperature of 90 °C. For conductor temperatures other than 90 °C the ampacity of a given conductor type shall be obtained by multiplying the appropriate value from Table 12A by the correction factor for that conductor temperature as contained in Table 12C.

(3) When cables are used with one or more layers wound on a drum, the actual internal temperature of the cable may exceed the 90 °C rating. Thermal overheating may shorten the run life of the cable. The appropriate ampacity correction factors are shown in Table 12D.

(4) For single-conductor cables with metallic shields, the ampacity values are for cables operated with an open circuit shield.

Table 12B
Temperature correction factor

(See Table 12A.)

Ambient air temperature, °C	Correction factor
10	1.14
20	1.07
30	1.00
40	0.91
50	0.82

Table 12C
Conductor rating correction factor

(See Table 12A.)

Conductor temperature rating, °C	Correction factor
60	0.75
75	0.885
90	1.00

Table 12D
Layering correction factor

(See Table 12A.)

Number of layers of cable on drum	Correction factor
1	0.85
2	0.65
3	0.45
4	0.35

Table 13
Rating or setting of overcurrent devices protecting conductors*

(See Rules 14-104 and 28-204.)

Ampacity of conductor, A	Rating or setting permitted, A	Ampacity of conductor, A	Rating or setting permitted, A
0–15	15	126–150	150
16–20	20	151–175	175
21–25	25	176–200	200
26–30	30	201–225	225
31–35	35	226–250	250
36–40	40	251–275	300
41–45	45	276–300	300
46–50	50	301–325	350
51–60	60	326–350	350
61–70	70	351–400	400
71–80	80	401–450	450
81–90	90	451–500	500
91–100	100	501–525	600
101–110	110	526–550	600
111–125	125	551–600	600

*For general use where not otherwise specifically provided for.

Table 14
Watts per square metre and demand factors for services and feeders for various types of occupancy

(See Rule 8-210.)

Type of occupancy	Watts per square metre	Demand factor, %	
		Service conductors	Feeders
Store, restaurant	30	100	100
Office			
First 930 m^2	50	90	100
All in excess of 930 m^2	50	70	90
Industrial and commercial	25	100	100
Church	10	100	100
Garage	10	100	100
Storage warehouse	5	70	90
Theatre	30	75	95
Armouries and auditoriums	10	80	100
Banks	50	100	100
Barbershops and beauty parlours	30	90	100
Clubs	20	80	100
Courthouses	20	100	100
Lodges	15	80	100

Table 15
Bending radii — High-voltage cable

(See Rules 34-400 and 36-102.)

Type of cable	Cable diameter multiplying factor (see Note)		
	Up to and including 25 mm diameter	Over 25 mm diameter and up to and including 50 mm diameter	Over 50 mm diameter
Lead covered	10	12	12
Corrugated aluminum-sheathed	10	12	12
Smooth aluminum-sheathed	12	15	18
Tape shielded	12	12	12
Flat tape armoured	12	12	12
Wire armoured	12	12	12
Non-shielded	7	7	7
Wire shielded	7	7	7
Portable power cables 5 kV and less	6	6	6
Portable power cables over 5 kV	8	8	8

Note: *The bending radius is that measured at the innermost surface and equals the overall diameter of the cable multiplied by the appropriate number shown in Columns 2, 3, and 4.*

Δ

Table 16
Minimum size conductors for bonding raceways and equipment

(See Rules 10-204, 10-626, 10-814, 10-816, 10-906, 12-1814, 24-104, 24-202, 30-1030, 68-058, and 68-406.)

Ampacity of largest ungrounded conductor in the circuit or equivalent for multiple parallel conductors, A	Size of bonding conductor	
	Copper wire, AWG	Aluminum wire, AWG
20	14	12
*30	12	10
40	10	8
60	10	8
100	8	6
200	6	4
300	4	2
400	3	1
500	2	0
600	1	00
800	0	000
1000	00	0000
1200	000	250 kcmil
1600	0000	350 kcmil
2000	250 kcmil	400 kcmil

(Continued)

Table 16 (Concluded)

Ampacity of largest ungrounded conductor in the circuit or equivalent for multiple parallel conductors, A	Size of bonding conductor	
	Copper wire, AWG	Aluminum wire, AWG
2500	350 kcmil	500 kcmil
3000	400 kcmil	600 kcmil
4000	500 kcmil	800 kcmil
5000	700 kcmil	1000 kcmil
6000	800 kcmil	1250 kcmil

Δ

Table 17
Minimum size of grounding conductor

(See Rules 10-206, 10-700, and 10-812.)

Ampacity of largest service conductor or equivalent for multiple conductors not exceeding — A	Size of copper grounding conductor, AWG
100	8
200	6
400	3
600	1
800	0
Over 800	00

Note: *The ampacity of the largest service conductor, or equivalent if multiple conductors are used, is to be determined from the appropriate Table in the Code, taking into consideration the number of conductors in the raceway or cable and the type of insulation.*

Δ

Table 18
(Note: This Table has been deleted.)

Table 19
Conditions of use and maximum allowable conductor temperature of wires and cables other than flexible cords, portable power cables, and equipment wires

(See Rules 4-006, 6-300, 12-100, 12-302, 12-602, 12-606, 12-902, 12-904, 12-1606, 12-2104, 12-2202, 16-112, 16-210, 22-200, 22-202, 22-204, 22-206, 26-642, 30-310, 30-1004, 30-1102, 32-100, 32-202, 38-011, 54-100, 56-104, 60-304, 62-130, 74-004, 78-104, 80-004, and Tables 1, 2, 3, 4, D1, and D3.)

Conditions of use	Trade designation	CSA type designation	Maximum allowable conductor temperature, °C	Reference Notes
For exposed wiring in dry locations only	Armoured cable	TECK90	90	8, 9
		AC90	90	8, 9
For exposed wiring in dry locations where exposed to corrosive action, if suitable for corrosive conditions encountered	Armoured cable	TECK90	90	8, 9
For exposed wiring in dry locations where not exposed to mechanical injury	Non-metallic-sheathed cable	NMD90	90	20
For exposed wiring in dry locations and in Category 1 and 2 locations, where not exposed to mechanical injury	Non-metallic-sheathed cable	NMW, NMWU	60	20
For exposed wiring in dry or damp locations	Rubber (thermoset-) insulated cable	R90	90	7, 8, 9
	Thermoplastic-insulated cable	TW	60	—
	Nylon jacketed thermoplastic-insulated cable	T90 NYLON	90	11
For exposed wiring in wet locations	Non-metallic-sheathed cable	NMD90	90	15, 26
	Armoured cable	TECK90	90	5, 8, 9
		ACWU90	90	5, 8, 9
	Rubber (thermoset-) insulated cable	RW75	75	5, 8, 9
		RL90, RW90	90	5, 8, 9
	Aluminum-sheathed cable	RA75	75	5
		RA90	90	5, 8, 9
	Mineral-insulated cable	MI, LWMI	90	5, 18
	Thermoplastic-insulated cable	TW, TW75	60	5
		TWN75	75	
	Non-metallic-sheathed cable	NMWU	60	5, 6, 20

(Continued)

304

Table 19 (Continued)

Conditions of use	Trade designation	CSA type designation	Maximum allowable conductor temperature, °C	Reference Notes
For exposed wiring where subjected to the weather	Armoured cable	TECK90	90	8, 9, 30
	Rubber (thermoset-) insulated cable	RW75	75	8, 9, 30
		R90, RW90	90	8, 9, 30
	Thermoplastic-insulated cable	TW, TWU, TW75	60	30
		TWU75	75	30
	Neutral-supported cable	NS-75, NS-90	75	30
			90	30, 31
	Non-metallic-sheathed cable	NMWU	60	6, 20, 30
For concealed wiring in dry locations only	Armoured cable	TECK90	90	8, 9
		AC90	90	8, 9
For concealed wiring in dry and damp locations	Non-metallic-sheathed cable	NMD90	90	15, 20
For concealed wiring in dry locations and in Category 1 and 2 locations where not exposed to mechanical injury	Non-metallic-sheathed cable	NMW, NMWU	60	20
For concealed wiring in wet locations	Armoured cable	TECK90	90	5, 8, 9
		ACWU90	90	5, 8, 9
	Non-metallic-sheathed cable	NMWU	60	5, 6, 20
	Aluminum-sheathed cable	RA75	75	5
		RA90	90	5, 8, 9
	Mineral-insulated cable	MI, LWMI	90	5, 18
For use in raceways, except cable trays, in dry or damp locations	Rubber (thermoset-) insulated cable	R90	90	7, 8, 9, 10
	Thermoplastic-insulated cable	TW	60	—
	Nylon jacketed thermoplastic-insulated cable	T90 NYLON	90	11
For use in raceways, except cable trays, in wet locations	Rubber (thermoset-) insulated cable	RW75, RWU75	75	5, 8, 9
		RW90, RWU90	90	5, 8, 9
	Thermoplastic-insulated cable	TW, TWU	60	4, 5
		TW75, TWN75, TWU75	75	
For use in ventilated, non-ventilated, and ladder-type cable trays in dry locations only	Armoured cable	AC90	90	8, 9
		TECK90	90	8, 9

(Continued)

Table 19 (Continued)

Conditions of use	Trade designation	CSA type designation	Maximum allowable conductor temperature, °C	Reference Notes
For use in ventilated, non-ventilated, and ladder-type cable trays in wet locations	Armoured cable	TECK90	90	5, 8, 9
		ACWU90	90	5, 8, 9
	Aluminum-sheathed cable	RA75	75	5
		RA90	90	5, 8, 9
	Mineral-insulated cable	MI, LWMI	90	5
	Rubber (thermoset-) insulated lead-sheathed cable	RL90	90	5, 8, 9
For use in ventilated and non-ventilated cable trays in vaults and switch rooms	Rubber (thermoset-) insulated cable	RW75	75	8, 9, 10
		RW90	90	8, 9, 10
For direct earth burial (with protection as required by inspection authority)	Armoured cable	ACWU90	90	3, 8, 9
		TECK90	90	3, 8, 9
	Non-metallic-sheathed cable	NMWU	60	3, 20
	Rubber (thermoset-) insulated cable	RWU75	75	3, 8, 9
		RL90, RWU90	90	3, 8, 9
	Aluminum-sheathed cable	RA75	75	3
		RA90	90	3, 7, 8
For direct earth burial (with protection as required by inspection authority)	Mineral-insulated cable	MI, LWMI	90	1, 3, 18
	Thermoplastic-insulated cable	TWU	60	3, 5
		TWU75	75	3
	Airport series lighting cable	ASLC	90	19
	Tray cable	TC	90	25
For service entrance above ground	Armoured cable	AC90	90	16
		ACWU90	90	
		TECK90	90	
	Aluminum-sheathed cable	RA75	75	—
		RA90	90	
	Mineral-insulated cable	MI	90	1, 18
	Neutral supported cable	NS-75	75	30
		NS-90	90	30, 31

(Continued)

Table 19 (Continued)

Conditions of use	Trade designation	CSA type designation	Maximum allowable conductor temperature, °C	Reference Notes
For service entrance below ground	Service-entrance cable	USEI75	75	3, 8, 9
		USEI90	90	3, 8, 9, 12
		USEB90	90	3
	Thermoplastic insulated wire	TWU	60	3
		TWU75	75	3
	Rubber (thermoset-) insulated cable	RWU75	75	3, 8, 9
		RWU90	90	3, 8, 9
	Armoured cable	TECK90	90	—
		ACWU90	90	
	Aluminum-sheathed cable	RA75	75	3
		RA90	90	
For high-voltage wiring in luminous-tube signs	Luminous-tube sign cable	GTO, GTOL	60	
For use in raceways in hoistways	Hoistway cable	—	60	13, 14
For use in Class 2 circuits, in exposed or concealed wiring or use in raceways, in dry or damp locations	Extra-low-voltage control cable	LVT	60	—
For use in Class 2 circuits in dry locations in concealed wiring or exposed wiring where not subject to mechanical injury	Extra-low-voltage cable	ELC	60	17
For use when concealed indoors under-carpet squares, in dry or damp locations	Flat conductor cable	FCC	60	—
For use in communication circuits when exposed, concealed or used in raceways, indoors in dry or damp locations, or in ceiling air handling plenums	Communication cable	CMP, CMR, CMG, CM, CMX, CMH	60	21
For use in fire alarm, signal and voice communication circuits where exposed, concealed or used in raceways, indoors in dry or damp locations	Fire alarm and signal cable	FAS	60	22
		FAS90	90	
		FAS105	105	
		FAS200	200	
		MI		
For use in raceways including ventilated, non-ventilated, and ladder-type cable trays in wet locations and where exposed to weather	Tray cable	TC	—	23

(Continued)

Table 19 (Continued)

Conditions of use	Trade designation	CSA type designation	Maximum allowable conductor temperature, °C	Reference Notes
For use in cable trays in Class I, Division 2 and Class II, Division 2 hazardous locations	Tray cable	TC	—	23
For use in buildings in dry or damp locations, where exposed, concealed or used in raceways including cable trays, or in plenums	Non-conductive optical fiber cable	OFNP, OFNR, OFNG, OFN, OFNH	—	24
For use in buildings in dry or damp locations, where exposed, concealed or used in raceways including cable trays, or in plenums	Conductive optical fiber cable	OFCP, OFCR, OFCG, OFC, OFCH	—	24
For use in buildings in dry or damp locations, where exposed or concealed	Hybrid conductor cable	NMDH90	90	—
For concealed wiring used as non-heating leads on heating panels and panel sets	Thermoset-insulated cable	R90, RW90, RWU90 (XLPE, EP insulation only)	90	26
For use in ventilated, non-ventilated and ladder-type cable trays, direct earth burial, in ceiling air handling plenums, for exposed or concealed wiring in wet (or damp or dry) locations	Control and instrumentation cable (without an armour)	CIC	250 (dry or damp locations) 90 (wet locations)	27, 29
For use in ventilated, non-ventilated, and ladder-type cable trays, direct earth burial, for exposed and concealed wiring in wet (or damp or dry) locations, in ceiling air handling plenums	Armoured control and instrumentation cable (other than steel wire armour)	ACIC	250 (dry or damp locations) 0 (wet locations)	27, 28, 29
For use in ventilated, non-ventilated, and ladder-type cable trays, direct earth burial, for exposed and concealed wiring in wet (or damp or dry) locations, in ceiling air handling plenums	Steel wire armour-control and instrumentation cable	SW-ACIC	250 (dry or damp locations) 0 (wet locations)	27, 28, 29
For use in Class 2 circuits in exposed or concealed wiring or use in raceways, in dry or damp locations	Extra-low-voltage control cable	LVT	60	—
	Communication cable	CMP, CMR, CMG, CM, CMX, CMH	60	21

(Continued)

Table 19 (Continued)

Conditions of use	Trade designation	CSA type designation	Maximum allowable conductor temperature, °C	Reference Notes
For use in Class 2 circuits in dry locations in concealed wiring or exposed wiring where not subject to mechanical injury	Extra-low-voltage cable	ELC	60	17
	Communication cable	CMP, CMR, CMG, CM, CMX, CMH	60	21

Notes:

(1) A maximum sheath temperature of 250 °C is permissible for mineral-insulated cable, provided that the temperature at the terminations does not exceed that specified in Tables 1 and 2. Any protective covering provided shall be suitable for the applicable sheath temperature.

(2) May be used where exposed to heat, grease, or corrosive fumes, if suitable for the corrosive condition.

(3) Conductors or cable assemblies acceptable for direct earth burial may be used for underground services in accordance with Rule 6-300.

(4) Types TW and TWU, when provided with a nylon jacket, are also approved for use where adverse conditions may exist, such as in oil refineries and around gasoline storage or pump areas (e.g., where subjected to alkaline conditions in the presence of petroleum solvents).

(5) Types suitable for use in wet locations may also be used in dry or damp locations.

(6) Type NMWU cable is not suitable for use in aerial spans.

(7) Types having silicone rubber insulation are surface marked with the type designation followed by "silicone".

(8) Types having cross-linked polyethylene insulation are surface marked with the type designation followed by "X-Link" or "XLPE", e.g., R90 (X-Link) or R90XLPE.

(9) Types having ethylene-propylene insulation are surface marked with the type designation followed by "EP", e.g., R90 (EP).

(10) Types RW75 and RW90, when used under Rule 12-2202, are required to be flame tested.

(11) When exposed to oil, Type T90 NYLON is limited to 60 °C.

(12) Type USEB90 shall have a non-metallic jacket over concentric neutral conductor.

(13) Hoistway cables may also be provided with 90 °C insulation.

(14) Except for short runs not exceeding 1.5 m in length, the parallel construction is intended for use in raceways in which the cables are laid.

(15) With thermoplastic jacket in damp locations.

(16) For dry locations only.

(17) Type ELC cable is limited to Class 2 circuit application as per Rule 16-210.

(18) Mineral-insulated cable having a stainless steel sheath requires a separate grounding conductor. (See Rule 10-804(e).)

(19) Type ASLC is for use only in accordance with Section 74.

(20) NMD90, NMW, and NMWU were previously marked NMD-7, NMW-9, and NMW-10 respectively.

(21) The following cable substitution may be used:
   (a) Communication cables marked CMP, CMR, CMG, CM, CMX, CMH, FT6, and FT4 have been found to meet the standard criteria for FT1.
   (b) Communication cables marked CMP, CMR, CMG, CM, and FT6 have been found to meet the standard criteria for FT4.
   (c) Communication cables marked CMP have been found to meet the standard criteria for FT6.

(Continued)

Table 19 (Concluded)

(22) Types FAS, FAS90, FAS105, and FAS200 may be provided with mechanical protection such as interlock armour or an aluminum sheath, with or without overall thermoplastic covering. A thermoplastic covering shall be provided over the interlock armoured cable when installed in a damp location.

(23) The maximum allowable conductor temperature for Type TC cables is dependent on the temperature rating of the cable so marked.

(24) OFNP, OFNR, OFNG, OFN, OFNH, OFCP, OFCR, OFCG, OFC, and OFCH shall have a minimum cable temperature rating of 60 °C. Cables having a temperature rating greater than 60 °C shall be permitted, provided that the temperature rating is surface marked on the cable.

(25) Type TC cable directly buried in the earth shall be marked as suitable for the purpose.

(26) Types R90, RW90, and RWU90, when used under Rule 62-130(3), are required to be marked as suitable for use as non-heating leads.

(27) The maximum allowable conductor temperature(s) for wet and/or dry locations depends upon the material in the cable, and is marked on the outer covering of the cable.

(28) When cables are installed in Category 2 locations (Rule 22-202), they shall be suitable for the corrosive conditions encountered.

(29) Proper additional markings are needed when used on cable trays.

(30) Wires and cables suitable for exposed wiring where exposed to the weather are so marked.

(31) 90 °C rated neutral-supported cables have been added and the Type designations changed to reflect their maximum temperature ratings.

Table 20
Spacings for conductors

(See Rules 12-204 and 12-214.)

Voltage of circuit, V	Minimum distance, mm	
	Between conductors	From adjacent surfaces
0 to 300	65	13
301 to 750	100	25

Table 21
Supporting of conductors in vertical runs of raceways

(See Rule 12-120.)

Conductor sizes, AWG or kcmil	Maximum distance, m	
	Copper	Aluminum
14 to 8	30	30
6 to 0	30	60
00 to 0000	24	55
250 to 350	18	40
Over 350 to 500	15	35
Over 500 to 750	12	30
Over 750	10	25

Table 22
Space for conductors in boxes

(See Rule 12-3034.)

Size of conductor, AWG	Usable space required for each insulated conductor, mL
14	24.6
12	28.7
10	36.9
8	45.1
6	73.7

Table 23
Number of conductors in boxes

(See Rule 12-3034.)

Box dimensions trade size		Capacity, mL (in³)	Maximum number of conductors (per AWG size)				
			14	12	10	8	6
Octagonal	4 × 1-1/2	245 (15)	10	8	6	5	3
	4 × 2-1/8	344 (21)	14	12	9	7	4
Square	4 × 1-1/2	344 (21)	14	12	9	7	4
	4 × 2-1/8	491 (30)	20	17	13	10	6
	4-11/16 × 1-1/2	491 (30)	20	17	13	10	6
	4-11/16 × 2-1/8	688 (42)	28	24	18	15	9
Round	4 × 1/2	81 (5)	3	2	2	1	1
Device	3 × 2 × 1-1/2	131 (8)	5	4	3	2	1
	3 × 2 × 2	163 (10)	6	5	4	3	2
	3 × 2 × 2-1/4	163 (10)	6	5	4	3	2
	3 × 2 × 2-1/2	204 (12.5)	8	7	5	4	2
	3 × 2 × 3	245 (15)	10	8	6	5	3
	4 × 2 × 1-1/2	147 (9)	6	5	4	3	2
	4 × 2-1/8 × 1-7/8	229 (14)	9	8	6	5	3
	4 × 2-3/8 × 1-7/8	262 (16)	10	9	7	5	3
Masonry	3-3/4 × 2 × 2-1/2	229 (14)/gang	9	8	6	5	3
	3-3/4 × 2 × 3-1/2	344 (21)/gang	14	12	9	7	4
	4 × 2-1/4 × 2-3/8	331 (20.25)/gang	13	11	9	7	4
	4 × 2-1/4 × 3-3/8	364 (22.25)/gang	14	12	9	8	4
Through box	3-3/4 × 2	3.8/mm (6/in) depth	4	3	2	2	1
Concrete ring	4	7.7/mm (12/in) depth	8	6	5	4	2
FS	1 Gang	229 (14)	9	8	6	5	3
	1 Gang tandem	557 (34)	22	19	15	12	7
	2 Gang	426 (26)	17	14	11	9	5
	3 Gang	671 (41)	27	23	18	14	9
	4 Gang	917 (56)	37	32	24	20	12
FD	1 Gang	368 (22.5)	15	12	10	8	5
	2 Gang	671 (41)	27	23	18	14	9
	3 Gang	983 (60)	40	34	26	21	13
	4 Gang	1392 (85)	56	48	37	30	18

Table 24
Minimum insulation resistances for installations

(See Rule 70-130.)

Installation, copper or aluminum	Insulation resistance, Ω
For circuits of No. 14 or No. 12 AWG	1 000 000
For circuits of No. 10 AWG or larger:	
25 to 50 A	250 000
51 to 100 A	100 000
101 to 200 A	50 000
201 to 400 A	25 000
401 to 800 A	12 000
Over 800 A	5 000

Note: *Where lampholders, receptacles, fixtures, baseboard heaters, or other appliances are connected to the installation or where excessive humidity exists, lower insulation resistance values may be expected.*

Table 25
Overcurrent trip coils for circuit breakers and overload devices for protecting motors

(See Rules 14-306 and 28-304.)

For circuit protection*		For motor overload protection	
Number and location of overcurrent devices (trip coils)	System	Number and location of overload devices such as trip coils, relays, or thermal cut-outs	Kind of motor
3-trip coils, one in each conductor	3-wire, 3-phase ac, ungrounded or with grounded neutral	3 — one in each phase not to be connected in any neutral conductor	3-phase ac
3-trip coils, one in each phase	4-wire, 3-phase ac		
2-trip coils, one in each phase†	4-wire, 2-phase ac, ungrounded	2 — one in each phase, not to be connected in any neutral or grounded conductor	2-phase ac
2-trip coils, one in each outside conductor	3-wire, 2-phase ac		
4-trip coils, one in each ungrounded conductor	4-wire, 2-phase ac, with grounded neutral		
4-trip coils, one in each ungrounded conductor	5-wire, 2-phase ac		
2-trip coils, one in each outside conductor	3-wire, 1-phase ac or dc	1 — in any conductor except a neutral or grounded conductor	1-phase ac or dc
1-trip coil in each ungrounded conductor	2-wire ac or dc, ungrounded or with one conductor grounded‡		
2-trip coils, one in each ungrounded conductor	3-wire, 1-phase ac or dc, with grounded neutral		

*This will not preclude the use of other arrangements that will provide equivalent protection.
†For services see Section 6.
‡This will not prevent the use of one single-pole circuit breaker in each conductor for the protection of an ungrounded 2-wire circuit.

Table 26
(Note: This Table is now Table D16.)

Table 27
Determining conductor sizes for motors for different requirements of service

(See Rules 28-106, 28-112, and 38-013.)

Classification of service	Percentage of nameplate current rating of motor			
	5-minute rating	15-minute rating	30- and 60-minute rating	Continuous rating
Short-time duty operating valves, raising or lowering rolls, etc.	110	120	150	—
Intermittent duty Freight and passenger elevators, tool heads, pumps, drawbridges, turntables, etc.	85	85	90	140
Periodic duty Rolls, ore- and coal-handling machines, etc.	85	90	95	140
Varying duty	110	120	150	200

Note: *For motor-generator arc welders see Section 42.*

Table 28
Determining conductor sizes in the secondary circuits of motors

(See Rule 28-112.)

Resistor duty classification	Duty cycles	Carrying capacity of conductors in per cent of full-load secondary circuit
Light starting duty	5 s on 75 s off	35
Heavy starting duty	10 s on 70 s off	45
Extra-heavy starting duty	15 s on 75 s off	55
Light intermittent duty	15 s on 45 s off	65
Medium intermittent duty	15 s on 30 s off	75
Heavy intermittent duty	15 s on 15 s off	90
Continuous duty	Continuous duty	110

Table 29
Rating or setting of overcurrent devices for the protection of motor branch circuits

(See Rules 28-200, 28-206, 28-208, and 28-308 and Table D16.)

| Type of motor | Full load current, % | | Maximum setting time-limit type circuit breaker |
| | Maximum fuse rating | | |
	Time-delay* fuses	Non-time-delay	
Alternating current			
Single-phase all types	175	300	250
Squirrel-cage and synchronous:			
Full-voltage, resistor and reactor starting	175	300	250
Auto-transformer and star delta starting:			
Not more than 30 A	175	250	200
More than 30 A	175	200	200
Wound rotor	150	150	150
Direct current	150	150	150

Includes time-delay "D" fuses referred to in Rule 14-200.

Notes:

(1) *Synchronous motors of the low-torque, low-speed type (usually 450 rpm, or lower) such as are used to drive reciprocating compressors, pumps, etc., and which start up unloaded, do not require a fuse rating or circuit-breaker setting in excess of 200% of full-load current.*

(2) *For the use of instantaneous trip (magnetic only) circuit breakers in motor branch circuits see Rule 28-210.*

Table 30
Minimum clearances for bus support and rigid conductors

(See Rule 36-108.)

| Maximum system voltage, kV | Minimum air gap distance, mm | | |
| | From live parts to adjacent surfaces other than insulation and bases of conductor supports* | Between live parts** (not centre-to-centre) | |
	Indoor and outdoor	Indoor	Outdoor
Not exceeding 7.5	190	150	180
15	260	230	305
25	305	330	380
		Indoor and outdoor	
34.5	380	460	
46	460	530	
69	740	790	
120	1200	1350	
161	1560	1830	
230	2300	2670	

For ungrounded systems, the Maximum system voltage is the phase-to-phase voltage, and for grounded systems it is the phase-to-ground voltage.
**For all systems, the Maximum system voltage is the phase-to-phase voltage.*

Table 31
Minimum horizontal separations of line conductors
attached to the same supporting structure

(See Rule 36-108.)

Maximum system voltage*, kV	Minimum separation of conductors for span 50 m or less**, mL
Not exceeding 5.0	300
7.5	325
15	400
25	500
34.5	600
46	700
69	950

*For all systems, the Maximum system voltage is the phase-to-phase voltage.
**For voltages greater than 69 kV and for spans greater than 50 m, the requirements of CAN/CSA-C22.3 No. 1 shall apply.

Table 32
Vertical isolation of unguarded live parts

(See Rule 36-110 and Appendix B.)

Maximum system voltage*, kV	Minimum separation from ground, m			
	Areas accessible to pedestrians only			Areas likely to be travelled by vehicles
	Indoors	Outdoors		
		Light snow area**	Heavy snow area**	
Not exceeding 15	2.9	3.4	4.0	4.7
25	3.2	3.7	4.3	5.2
34.5	3.2	3.7	4.3	5.2
46	3.2	3.7	4.3	5.2
69	3.5	4.0	4.6	5.5
120	4.0	4.5	5.1	6.0
161	4.4	4.9	5.5	6.4
230	4.8	5.3	5.9	6.8

*For ungrounded systems, the Maximum system voltage is the phase-to-phase voltage, and for grounded systems it is the phase-to-ground voltage.
**See Appendix B.
Note: Radial clearances from live parts shall be maintained in accordance with this Table where conductors overhang the edge of a building or structure, including any protuberances.

Table 33
Horizontal clearances from adjacent structures*
(including protuberances)

(See Rules 26-302 and 36-110 and Appendix B.)

Maximum system voltage**, kV	Clearance, m
Not exceeding 46.0	3
69	3.7

*See Appendix B.
**For ungrounded systems, the Maximum system voltage is the phase-to-phase voltage, and for grounded systems it is the phase-to-ground voltage.

Table 34
Vertical ground clearances for open line conductors*

(See Rule 36-110 and Appendix B.)

Maximum system voltage**, kV	Minimum vertical clearances above ground, m
Not exceeding 25.0	6.1
34.5	6.7
46	7
69	7.6

*See Appendix B.
**For ungrounded systems, the Maximum system voltage is the phase-to-phase voltage, and for grounded systems it is the phase-to-ground voltage.

Table 35
Spacing for switches and fuses assembled in the field
(not of the metal enclosed type)

(See Rule 36-212.)

Maximum system voltage*, kV	Minimum phase spacing (centre-to-centre), mm	
	Disconnect switches and fuses other than expulsion type	Horn-gap switches and expulsion fuses
Not exceeding 7.5	460	915
15	610	915
25	760	1220
34.5	915	1520
46	1220	1830
69	1520	2130
120	2130	3050
161	2740	4270
230	3960	5500

*For all systems, the Maximum system voltage is the phase-to-phase voltage.

Δ

Table 36A
Maximum allowable ampacity for aluminum conductor neutral supported cables

(See Rule 4-004(5).)

Phase conductor size, AWG or kcmil	Ampacity, A NS75		Ampacity, A NS90	
	Duplex, triplex	Quadruplex	Duplex, triplex	Quadruplex
6	80	70	95	85
4	105	95	125	110
2	140	125	165	150
1	160	145	190	170
1/0	185	165	220	200
2/0	210	190	255	230
3/0	245	220	290	265
4/0	280	255	335	305
266.8	325	290	390	355
336.4	375	335	450	410
397.5	415	370	500	455
477	460	415	560	510
500	500	450	605	555

Notes:

(1) *The ampacity ratings are based on 30 °C ambient and wind velocity of 0.6 m/sec and sun-radiated heat energy of 1025 W/m^2.*

(2) *For ambient temperatures of 30 °C, 35 °C, and 40 °C, multiply the above values by the corresponding correction factor of 1.0, 0.94, and 0.88, respectively.*

Δ

Table 36B
Maximum allowable ampacity for copper conductor neutral supported cables

(See Rule 4-004(5).)

Phase conductor size, AWG	Ampacity, A NS75		Ampacity, A NS90	
	Duplex, triplex	Quadruplex	Duplex, triplex	Quadruplex
6	100	90	120	110
4	135	120	155	140
2	175	160	210	190
1	205	185	240	220
1/0	235	210	280	255
2/0	270	245	325	295
3/0	310	280	370	335
4/0	360	320	430	390

Notes:

(1) *The ampacity ratings are based on 30 °C ambient and wind velocity of 0.6 m/sec and sun-radiated heat energy of 1025 W/m^2.*

(2) *For ambient temperatures of 30 °C, 35 °C, and 40 °C, multiply the above values by the corresponding correction factor of 1.0, 0.92, and 0.84, respectively.*

Table 37
Motor supply conductor insulation minimum temperature rating, °C (based on ambient temperatures of 30 °C)

(See Rule 28-104.)

Motor enclosure	Insulation class rating			
	A	B	F	H
All except totally enclosed, non-ventilated	75	75	90	110
Totally enclosed, non-ventilated	75	90	110	110

Table 38
(Note: This Table has been deleted.)

Table 39
(Note: This Table has been deleted.)

Table 40
External tapered threads for rigid metal conduit

(See Rule 12-1006.)

Trade size of conduit	Number of threads per 25.4 mm	External threads	
		Length of thread	
		Minimum, mm	Maximum, mm
16	14	16.3	19.8
21	14	16.5	20.1
27	11-1/2	20.6	24.9
35	11-1/2	21.3	25.7
41	11-1/2	21.8	26.2
53	11-1/2	22.6	26.9
63	8	33.5	39.9
78	8	34.5	41.4
91	8	36.3	42.7
103	8	37.6	43.9
129	8	40.4	46.7
155	8	43.2	49.5

Table 41
Minimum size of bonding jumper for service raceways

(See Rules 10-614 and 70-126.)

Ampacity of largest service conductor or equivalent for multiple conductors, A	Size of bonding jumper	
	Copper wire, AWG	Aluminum wire, AWG
100 or less	8	6
200	6	4
400	4	2
600	2	0
800	0	00
1000	00	000
1200	000	0000

Table 42

(Note: This Table has been deleted.)

Table 43
Minimum conductor size for concrete encased electrodes

(See Rule 10-700.)

Ampacity of largest service conductor or equivalent for multiple conductors, A	Size of bare copper conductor, AWG
165 or less	4
166–200	3
201–260	2
261–355	0
356–475	00
Over 475	000

Table 44
Three-phase AC motors

(See Rules 28-010 and 28-704.)

3-phase Motor rating, hp	AC motor full-load current, A [see Notes (1), (2), (3), and (5)]										
	Induction type, squirrel cage and wound rotor, A						Synchronous type, unity power factor [see Note (4)], A				
	115 V	200 V	230 V	460 V	575 V	2300 V	200 V	230 V	460 V	575 V	2300 V
1/2	4	2.3	2	1	0.8	—	—	—	—	—	—
3/4	5.6	3.2	2.8	1.4	1.1	—	—	—	—	—	—
1	7.2	4.1	3.6	1.8	1.4	—	—	—	—	—	—
1-1/2	10.4	6.0	5.2	2.6	2.1	—	—	—	—	—	—
2	13.6	7.8	6.8	3.4	2.7	—	—	—	—	—	—
3	—	11.0	9.6	4.8	3.9	—	—	—	—	—	—
5	—	17.5	15.2	7.6	6.1	—	—	—	—	—	—
7-1/2	—	25.3	22	11	9	—	—	—	—	—	—
10	—	32.2	28	14	11	—	—	—	—	—	—
15	—	48	42	21	17	—	—	—	—	—	—
20	—	62	54	27	22	—	—	—	—	—	—
25	—	78	68	34	27	—	62	54	27	22	—
30	—	92	80	40	32	—	75	65	33	26	—
40	—	120	104	52	41	—	99	86	43	35	—
50	—	150	130	65	52	—	124	108	54	44	—
60	—	177	154	77	62	16	147	128	64	51	12
75	—	221	192	96	77	20	185	161	81	65	15
100	—	285	248	124	99	26	243	211	106	85	20
125	—	359	312	156	125	31	304	264	132	106	25
150	—	414	360	180	144	37	—	—	158	127	30
200	—	552	480	240	192	49	—	—	210	168	40

Notes:
(1) *For full-load currents of 208 V motors, increase the corresponding 230 V motor full-load current by 10%.*
(2) *These values of motor full-load current are to be used as guides only. Where exact values are required (e.g., for motor protection), always use those appearing on the motor nameplate.*

(Continued)

Table 44 (Concluded)

(3) *These values of motor full-load current are for motors running at speeds typical for belted motors and motors with normal torque characteristics. Motors built for especially low speeds or high torques may require more running current, and multi-speed motors will have full-load current varying with speed, in which case the nameplate current rating shall be used.*

(4) *For 90 and 80% power factor, multiply the above figures by 1.1 and 1.25 respectively.*

(5) *The voltages listed are rated motor voltages. Corresponding Nominal System Voltages are 120, 208, 240, 480, and 600 V. Refer to CSA CAN3-C235.*

Table 45
Single-phase AC motors

(See Rules 28-010 and 28-704.)

Single-phase AC motors full-load current, A [see Notes (1) to (4)]		
hp rating	115 V	230 V
1/6	4.4	2.2
1/4	5.8	2.9
1/3	7.2	3.6
1/2	9.8	4.9
3/4	13.8	6.9
1	16	8
1-1/2	20	10
2	24	12
3	34	17
5	56	28
7-1/2	80	40
10	100	50

Notes:

(1) *For full-load currents of 208 and 200 V motors, increase the corresponding 230 V motor full-load current by 10 and 15% respectively.*

(2) *These values of motor full-load current are to be used as guides only. Where exact values are required (e.g., for motor protection), always use those appearing on the motor nameplate.*

(3) *These values of full-load current are for motors running at usual speeds and motors with normal torque characteristics. Motors built for especially low speeds or high torques may have higher full-load currents, and multi-speed motors will have full-load current varying with speed, in which case the nameplate current ratings shall be used.*

(4) *The voltages listed are rated motor voltages. Corresponding Nominal System Voltages are 120 and 240 V. Refer to CSA CAN3-C235.*

Table 46
(Note: This Table is now Diagram 1.)

Table 47
(Note: This Table is now Diagram 2.)

Table 48
Size of conduit for mobile homes

(See Rule 70-104.)

Rating of main overcurrent protection, A	Minimum trade size of conduit	
	Excluding system ground	Including system ground
50	27	35
60	35	35
100	35	41
150	53	53
200	53	63

Note: *These sizes are based on the use of copper conductors.*

Table 49

(Note: This Table is now Diagram 3.)

Table 50
Transformers rated over 750 V having primary and secondary overcurrent protection

(See Rule 26-252.)

Transformer rated impedance	Maximum setting or rating of overcurrent device as a percentage of rated current of transformer				
	Primary side		Secondary side		
	Over 750 V		Over 750 V		750 V or less
	Circuit breaker setting, %	Fuse rating, %	Circuit breaker setting, %	Fuse rating, %	Circuit breaker setting or fuse rating, %
Not more than 7.5%	600	300	300	150	250
More than 7.5% and not more than 10%	400	200	250	125	250

Table 51
Minimum size of bare copper grounding conductor
(See Rules 36-300 and 36-308 and Appendix B.)

Maximum available short-circuit current, A	Maximum fault duration			
	0.5 s		1.0 s	
	With exothermic weld, compression or bolted joint	With brazed joint	With exothermic weld, compression or bolted joint	With brazed joint
5 000	6	5	4	3
10 000	3	2	1	1/0
15 000	1	1/0	1/0	3/0
20 000	1/0	2/0	3/0	4/0
25 000	2/0	3/0	4/0	250*
30 000	3/0	4/0	4/0	300*
35 000	4/0	250*	250*	350*
40 000	4/0	300*	300*	400*
50 000	250*	350*	350*	500*
60 000	300*	400*	500*	600*
70 000	350*	500*	500*	700*
80 000	400*	600*	600*	800*
90 000	500*	600*	700*	900*
100 000	500*	700*	700*	1000*

*Wire size in kcmil, all others in AWG.
Note: *Sizes calculated in accordance with IEEE 80.*

Table 52
Tolerable touch and step voltages
(See Rules 36-304, 36-306, 36-308, 36-310, and 36-312.)

Type of ground	Resistivity	Fault duration 0.5 s		Fault duration 1.0 s	
	Ω•m	Step voltage, V	Touch voltage, V	Step voltage, V	Touch voltage, V
Wet organic soil	10	174	166	123	118
Moist soil	100	263	188	186	133
Dry soil	1 000	1 154	405	816	286
150 mm crushed stone	3 000	3 143	885	2 216	626
Bedrock	10 000	10 065	2 569	7 116	1 816

Notes:
(1) *Table values calculated in accordance with IEEE 80.*
(2) *A typical substation installation is designed for 0.5 s fault duration, and the entire ground surface inside the fence is covered with 150 mm of crushed stone having a resistivity of 3000 Ω •m.*

Table 53
Minimum cover requirements for direct buried conductors, cables, or raceways

(See Rule 12-012.)

Wiring method	Minimum cover, mm			
	Non-vehicular areas		Vehicular areas	
	750 V or less	Over 750 V	750 V or less	Over 750 V
Conductors or cable not having a metal sheath or armour	600	750	900	1000
Conductors or cable having a metal sheath or armour	450	750	600	1000
Raceway	450	750	600	1000

Note: *Minimum cover means the distance between the top surface of the conductor, cable, or raceway and the finished grade.*

Table 54
(Note: This Table is now Diagram 4.)

Table 55
(Note: This Table is now Diagram 5.)

Table 56
Minimum working space around electrical equipment having exposed live parts

(See Rule 2-308.)

Nominal voltage-to-ground	Working space, m
0–750	1.0
751–2 500	1.2
2 501–9 000	1.5
9 001–25 000	1.9
25 001–46 000	2.5
46 001–69 000	3.0
Over 69 000	3.7

Table 57
Allowable ampacities for Class 2 copper conductors
(based on ambient temperature of 30 °C†)

(See Rule 16-210(6) and Table 5A.)

Size, AWG	Single conductors in free air, A	Not more than 3 copper conductors in raceway or cable*, A
26	3	1
24	4	2
22	5	2.5
20	7	3.5
19	8	4
18	9	5
16	13	10
Col. 1	Col. 2	Col. 3

Where more than 3 conductors are in a raceway or cable, apply the following derating factors to Column 3:

Conductors in raceway or cable	Derating factor
4– 6	0.8
7–24	0.7
25–42	0.6
43–50	0.5

†*For ambient temperatures over 30 °C for Columns 2 and 3, apply the correction factors of Table 5A, Column 2.*

Table 58
Ampacities of up to four insulated copper conductors in raceway or cable for short time rated crane and hoist motors (based on ambient temperature of 30 °C)

(See Rule 40-002.)

Maximum operating temperature	75 °C		90 °C		110 °C	
Size, AWG or kcmil	60 minutes	30 minutes	60 minutes	30 minutes	60 minutes	30 minutes
16	10	12	—	—	—	—
14	25	26	31	32	38	40
12	30	33	36	40	45	50
10	40	43	49	52	60	65
8	55	60	63	69	73	80
6	76	86	83	94	93	105
5	85	95	95	106	109	121
4	100	117	111	130	126	147
3	120	141	131	153	145	168
2	137	160	148	173	163	190

(Continued)

Table 58 (Concluded)

Maximum operating temperature	75 °C		90 °C		110 °C	
Size, AWG or kcmil	60 minutes	30 minutes	60 minutes	30 minutes	60 minutes	30 minutes
1	143	175	158	192	177	215
0	190	233	211	259	239	294
00	222	267	245	294	275	331
000	280	341	305	372	339	413
0000	300	369	319	399	352	440
250	364	420	400	461	447	516
300	455	582	497	636	554	707
350	486	646	542	716	616	809
400	538	688	593	760	666	856
450	600	765	660	836	740	930
500	660	847	726	914	815	1004

Notes:
(1) Allowable ampacities of copper conductors used with 15-minute motors shall be the 30-minute ratings increased by 12%.
(2) For 5 or more simultaneously energized power conductors in raceway or cable, the ampacity of each shall be reduced to 80% of that shown in the Table.
(3) For conductors subject to ambient temperatures in excess of 30 °C, the derating factors in Table 5A shall apply to the ampacities shown in this Table.

Table 59
Minimum size of protector grounding conductor for communications systems

(See Rule 60-704.)

Size, AWG	Maximum number of protected circuits	
	Fuseless	Fused
No. 14	1	3
No. 12	2	6
No. 10	6	7
No. 6	7 or more	8 or more

Note: The grounding conductor between protectors shall be of at least the minimum size as required in this Table for the maximum number of protected circuits.

Table 60
(Note: This Table has been deleted.)

Table 61
Minimum conductor separation from pools
(See Rule 68-056.)

Type of installation	Minimum separation, m	
	Conductors buried directly in earth	Conductors in underground raceways
Communication and CATV conductors	1.5	0.75
Electrical conductors		
0–750 V	1.5	0.75
751–15 000 V	3.0	1.5
15 001–25 000 V	4.0	2.0

Table 62
Feeder demand factors for elevators
(See Rules 38-013(2) and 38-014.)

Number of elevators on a single feeder	Demand factors (DF)
1	1.00
2	0.95
3	0.90
4	0.85
5	0.82
6	0.79
7	0.77
8	0.75
9	0.73
10 or more	0.72

Note: *Demand factors (DF) are based on 50% duty (i.e., half time load, half time no load).*

Table 63
Hazardous areas for propane dispensing, container filling, and storage
(See Rule 20-034.)

Part	Location	Extent of hazardous locations*	Zone of Class I, Group IIA hazardous location
A	Storage containers other than TC/CTC/DOT cylinders and ASME vertical containers of less than 454 kg water capacity	Within 4.5 m in all directions from connections, except connections otherwise covered in the Table	Zone 2
B	Tank vehicle and tank car loading and unloading†	Within 3 m in all directions from connections regularly made or disconnected from product transfer	Zone 1
		Beyond 3 m but within 7.5 m in all directions from a point where connections are regularly made or disconnected and within the cylindrical volume between the horizontal equator of the sphere and grade (see Diagram 7)	Zone 2
C	Gauge vent openings other than those on TC/CTC/DOT cylinders and ASME vertical containers of less than 454 kg water capacity	Within 1.5 m in all directions from point of discharge	Zone 1
		Beyond 1.5 m but within 4.5 m in all directions from point of discharge	Zone 2
D	Relief device discharge other than those on CTC/DOT cylinders and ASME vertical containers of less than 454 kg water capacity	Within direct path of discharge‡	Zone 1
		Within 1.5 m in all directions from point of discharge	Zone 1
		Beyond 1.5 m but within 4.5 m in all directions from point of discharge except within the direct path of discharge	Zone 2
E	Pumps, vapour compressors, gas-air mixers, and vaporizers (other than direct-fired or indirect-fired with an attached or adjacent gas-fired heat source)	—	—
	Indoors without ventilation	Entire room and any adjacent room not separated by a gas-tight partition	Zone 1
		Within 4.5 m of the exterior side of any exterior wall or roof that is not vapour-tight or within 4.5 m of any exterior opening	Zone 2
	Indoors with adequate ventilation	Entire room and any adjacent room not separated by a gas-tight partition	Zone 2
	Outdoors in open air at or above grade	Within 4.5 m in all directions from this equipment and within the cylindrical volume between the horizontal equator of the sphere and grade (see Diagram 8)	Zone 2

(Continued)

Table 63 (Continued)

Part	Location	Extent of hazardous locations*	Zone of Class I, Group IIA hazardous location
F	Service station dispensing units	Entire space within dispenser's enclosure, or up to a solid partition within the enclosure at any height above the base. The space within 450 mm horizontally from the dispenser enclosure up to 1.2 m above the base or to the height of a solid partition within the enclosure. Entire pit or open space beneath the dispenser	Zone 1
		The space above a solid partition within the dispenser enclosure. The space up to 450 mm above grade within 6 m horizontally from any edge of the dispenser enclosure§	Zone 2
G	Pits or trenches containing or located beneath propane gas valves, pumps, vapour compressors, regulators, and similar equipment	—	—
	Without mechanical ventilation	Entire pit or trench	Zone 1
		Entire room and any adjacent room not separated by a gas-tight partition	Zone 2
		Within 4.5 m in all directions from pit or trench when located outdoors	Zone 2
	With adequate mechanical ventilation	Entire pit or trench	Zone 2
		Entire room and any adjacent room not separated by a gas-tight partition	Zone 2
		Within 4.5 m in all directions from pit or trench when located outdoors	Zone 2
H	Special buildings or rooms for storage of portable containers	Entire room	Zone 2
I	Pipelines and connections containing operational bleeds, drips, vents, or drains	Within 1.5 m in all directions from point of discharge	Zone 1
		Beyond 1.5 m from point of discharge, same as Part E of this Table	—
J	Container filling	—	—
	Indoors with adequate ventilation	Within 1.5 m in all directions from the dispensing hose inlet connections for product transfer	Zone 1
		Beyond 1.5 m and entire room	Zone 2
	Outdoors in open air	Within 1.5 m in all directions from the dispensing hose inlet connections for product transfer	Zone 1
		Beyond 1.5 but within 4.5 m in all directions from the dispensing hose inlet connections and within the cylindrical volume between the horizontal equator of the sphere and grade (see Diagram 9)	Zone 2

(Continued)

Table 63 (Concluded)

Part	Location	Extent of hazardous locations*	Zone of Class I, Group IIA hazardous location
K	Outdoor storage area for portable cylinders or containers	—	—
	Aggregate storage up to and including 454 kg water capacity	Within 1.5 m in all directions from connections	Zone 2
	Aggregate storage over 454 kg water capacity	Within 4.5 m in all directions from connections	Zone 2

The classified area shall not extend beyond an unpierced wall, roof, or solid vapour-tight partition.
†When classifying extent of hazardous area, consideration shall be given to possible variations in the locating of tank cars and tank vehicles at the unloading points and to the effect these variations of location may have on the point of connection.
‡Fixed electrical equipment should not be installed in this space.
§For pits within this area, see Part G of this Table.

Table 64
Class I, Zone 1 space surrounding compressed natural gas (NGV) storage facilities

(See Rule 20-064.)

Storage volume water capacity, L	Distance measured from containers, m*
Up to and including 4 000	2.5
Over 4 000, up to and including 10 000	4
Over 10 000	10

When a wall with a 4-hour fire resistance rating is located within these distances, the distances shall be measured either around the end of or over the wall, but not through it. This wall shall not be located closer than 1 m from a fuel container up to 10 000 L in storage volume, and 1.5 m from a fuel container with a storage volume greater than 10 000 L.

Where the wall of an adjacent building other than a compressor enclosure is within the specified distance and serves as the 4-hour fire resistance rated wall, it shall have no doors, windows, or openings in it unless the building is also classified as a Class I, Zone 1 location.

Table 65
Enclosure selection table for non-hazardous locations

(See Rules 2-400 and 2-402.)

Provides a degree of protection against the following environmental conditions	Indoor use						Indoor/outdoor use					Submersible	
	1	2	5	12*	12K†	13	3	3R	3S	4	4X	6	6P
Accidental contact with live parts	X	X	X	X	X	X	X	X	X	X	X	X	X
Falling dirt	X	X	X	X	X	X	X	X	X	X	X	X	X
Dripping and light splashing of noncorrosive liquids	—	X	X	X	X	X	X	X	X	X	X	X	X
Circulating dust, lint, fibres, and flyings	—	—	—	X	X	X	X	—	X	X	X	X	X
Settling dust, lint, fibres, and flyings	—	—	X	X	X	X	X	—	X	X	X	X	X
Hosedown and splashing water	—	—	—	—	—	—	—	—	—	X	X	X	X
Corrosion	—	—	—	—	—	—	—	—	—	—	X	—	X
Occasional temporary submersion	—	—	—	—	—	—	—	—	—	—	—	X	X
Occasional prolonged submersion	—	—	—	—	—	—	—	—	—	—	—	—	X
Oil and coolant seepage, spraying and splashing	—	—	—	—	—	X	—	—	—	—	—	—	—
Rain, snow, and external formation of ice‡	—	—	—	—	—	—	X	X	X	X	X	X	X
External formation of ice§	—	—	—	—	—	—	—	—	X	—	—	—	—
Wind-blown dust	—	—	—	—	—	—	X	—	X	X	X	X	X

*Without knockouts.
†With knockouts.
‡External operating mechanism(s) are not required to operate when the enclosure is ice covered.
§External operating mechanisms(s) must be operable when the enclosure is ice covered.

Diagrams

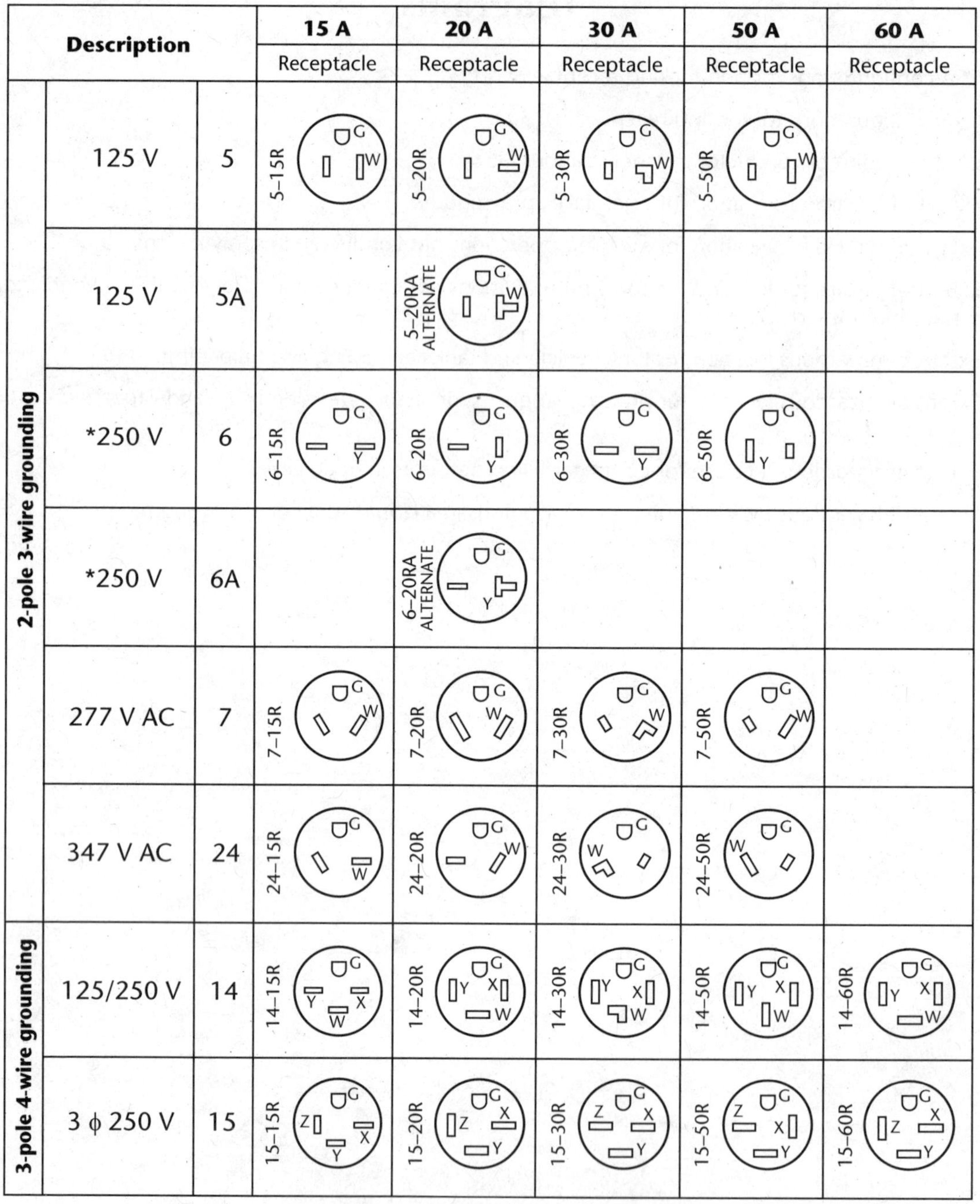

Description		15 A Receptacle	20 A Receptacle	30 A Receptacle	50 A Receptacle	60 A Receptacle
125 V	5	5-15R	5-20R	5-30R	5-50R	
125 V	5A		5-20RA ALTERNATE			
*250 V	6	6-15R	6-20R	6-30R	6-50R	
*250 V	6A		6-20RA ALTERNATE			
277 V AC	7	7-15R	7-20R	7-30R	7-50R	
347 V AC	24	24-15R	24-20R	24-30R	24-50R	
125/250 V	14	14-15R	14-20R	14-30R	14-50R	14-60R
3 φ 250 V	15	15-15R	15-20R	15-30R	15-50R	15-60R

(Left side row-group labels: "2-pole 3-wire grounding" spans rows 5 through 24; "3-pole 4-wire grounding" spans rows 14 and 15.)

*For configurations 6-15R, 6-20R, 6-20RA, 6-30R, and 6-50R, Y denotes the identified terminal when used on circuits derived from 3-phase, 4-wire 416 V circuits.

Note: Except as noted above, in Diagrams 1 and 2,
(a) G represents the terminal for bonding to ground;
(b) W represents the identified terminal; and
(c) X, Y, and Z represent the terminals for ungrounded conductors.

Diagram 1
CSA configurations for non-locking receptacles
(See Rules 26-700, 26-710, 26-744, 38-023, 78-052, 78-102, and 82-014,
Diagram 2, and Appendix B.)

Description			15 A Receptacle	20 A Receptacle	30 A Receptacle	50 A Receptacle	60 A Receptacle
2-pole 3-wire grounding	125 V	L5	L5-15R	L5-20R	L5-30R	L5-50R	L5-60R
	250 V	L6	L6-15R	L6-20R	L6-30R	L6-50R	L6-60R
	277 V AC	L7	L7-15R	L7-20R	L7-30R	L7-50R	L7-60R
	480 V AC	L8		L8-20R	L8-30R	L8-50R	L8-60R
	600 V AC	L9		L9-20R	L9-30R	L9-50R	L9-60R
3-pole 4-wire grounding	125/250 V	L14		L14-20R	L14-30R	L14-50R	L14-60R
	3 φ 250 V	L15		L15-20R	L15-30R	L15-50R	L15-60R
	3 φ 480 V	L16		L16-20R	L16-30R	L16-50R	L16-60R
	3 φ 600 V	L17			L17-30R	L17-50R	L17-60R
4-pole 5-wire grounding	3 φ 208 Y / 120 V	L21		L21-20R	L21-30R	L21-50R	L21-60R
	3 φ 480 Y / 277 V	L22		L22-20R	L22-30R	L22-50R	L22-60R
	3 φ 600 Y / 347 V	L23		L23-20R	L23-30R	L23-50R	L23-60R

Note: *Except as noted above, in Diagrams 1 and 2,*

(a)   *G represents the terminal for bonding to ground;*

(b)   *W represents the identified terminal; and*

(c)   *X, Y, and Z represent the terminals for ungrounded conductors.*

Diagram 2

CSA configurations for locking receptacles

(See Rules 12-020, 26-700, 78-052, 78-102, and 82-014, Diagram 1, and Appendix B.)

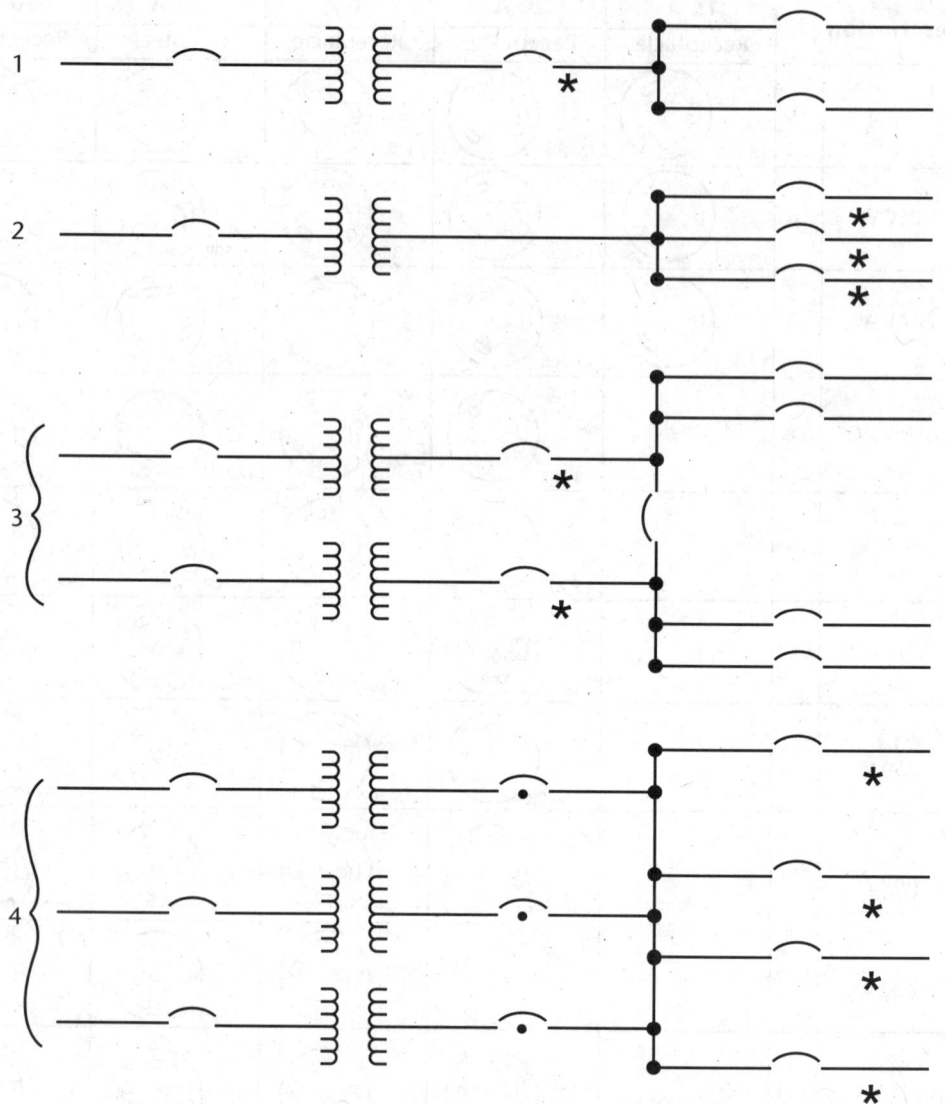

Notes:

(1) The symbol ‿⁀‿ represents a circuit breaker, a combination of circuit breaker and fuses, or a fused switch.

(2) The symbol ‿•‿ represents a network protector that protects against reverse current.

(3) An asterisk (*) indicates the ultimate point beyond which the downstream ungrounded circuit conductors must be de-energized in the event of a ground fault in the circuit fed by such conductors.

Diagram 3
Ultimate point of conductor de-energization
(See Rule 14-102 and Appendix B.)

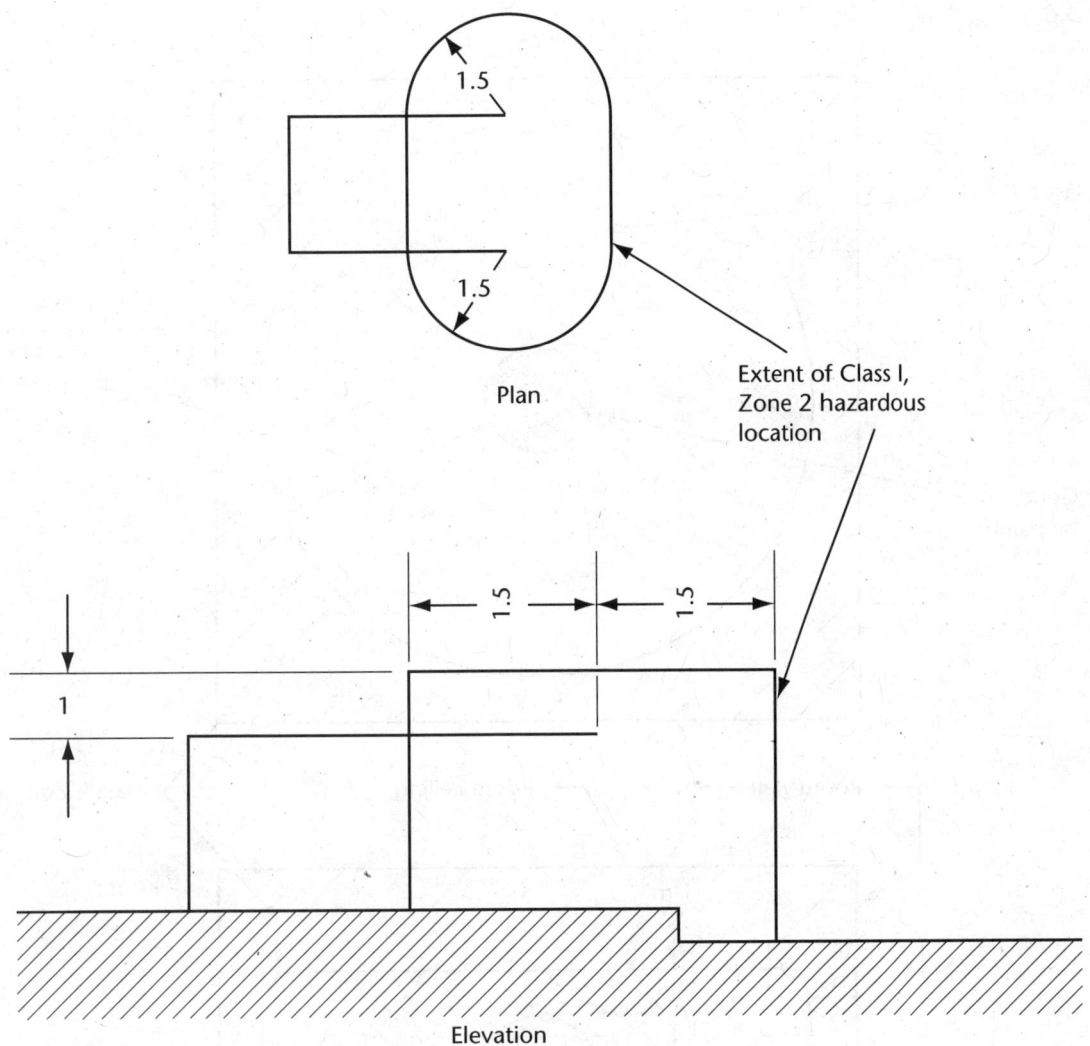

Plan

Extent of Class I,
Zone 2 hazardous
location

Elevation

Note: *All dimensions given are in metres.*

Diagram 4
Extent of hazardous location for open face spray booths
(See Rule 20-402(2).)

PLAN

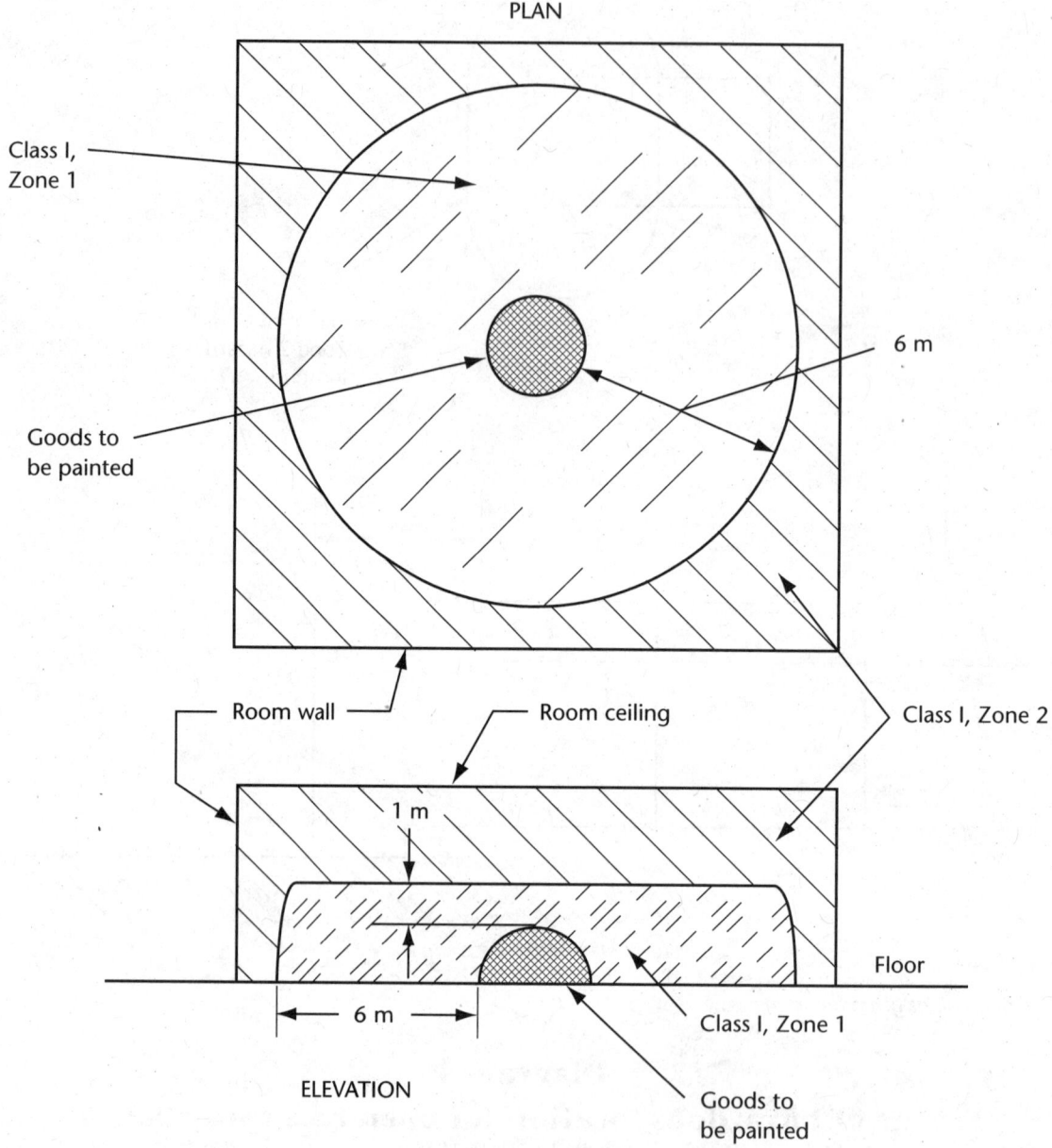

Class I, Zone 1

Goods to be painted

6 m

Room wall

Room ceiling

Class I, Zone 2

1 m

Floor

6 m

Class I, Zone 1

ELEVATION

Goods to be painted

Diagram 5
Extent of hazardous location for spraying operations not conducted in spray booths
(See Rule 20-402.)

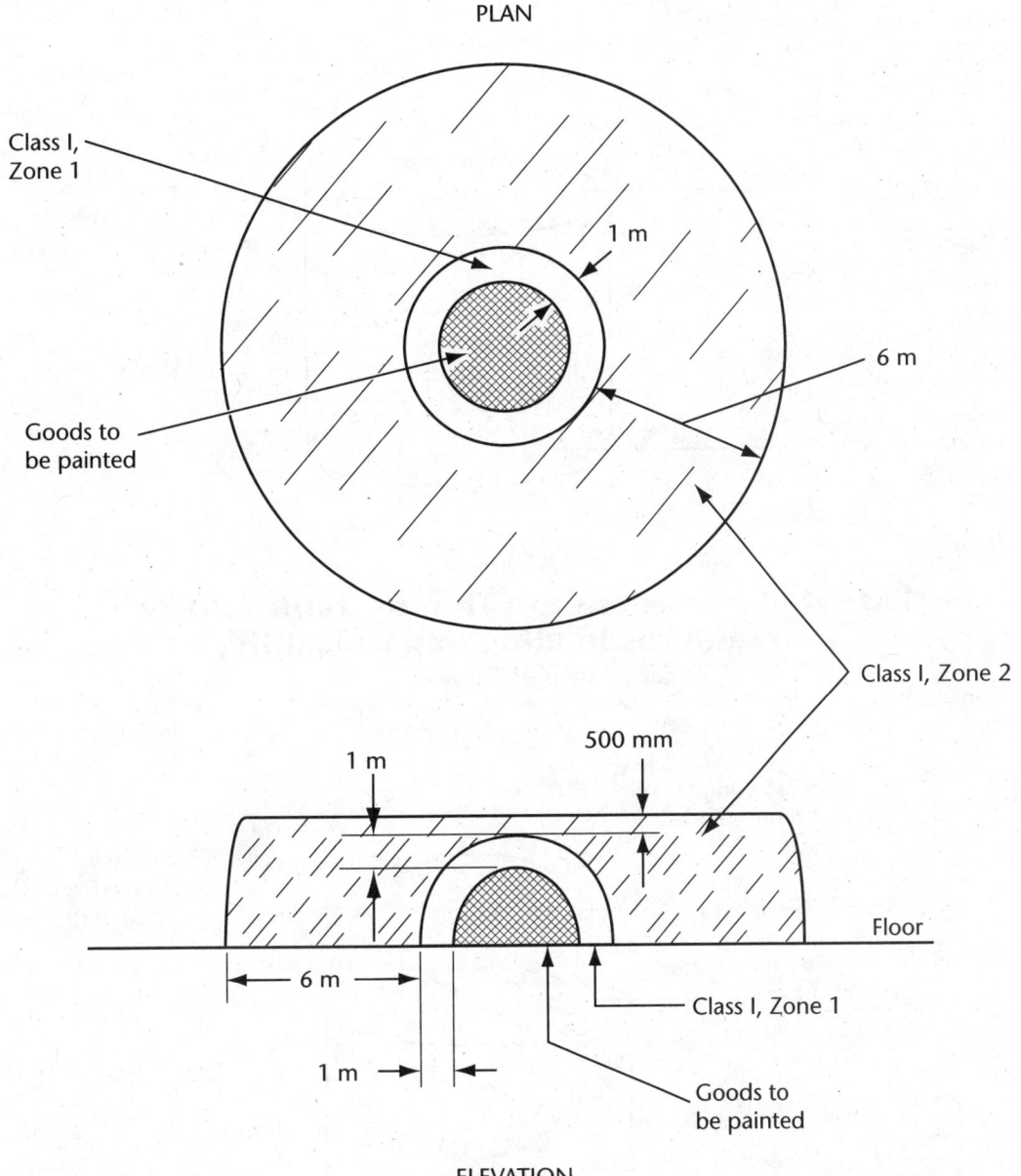

PLAN

Class I,
Zone 1

1 m

Goods to
be painted

6 m

Class I, Zone 2

1 m

500 mm

Floor

6 m

Class I, Zone 1

1 m

Goods to
be painted

ELEVATION

Diagram 6
Extent of hazardous location for spraying operations not conducted in spray booths — Ventilation system interlocked
(See Rule 20-402(7).)

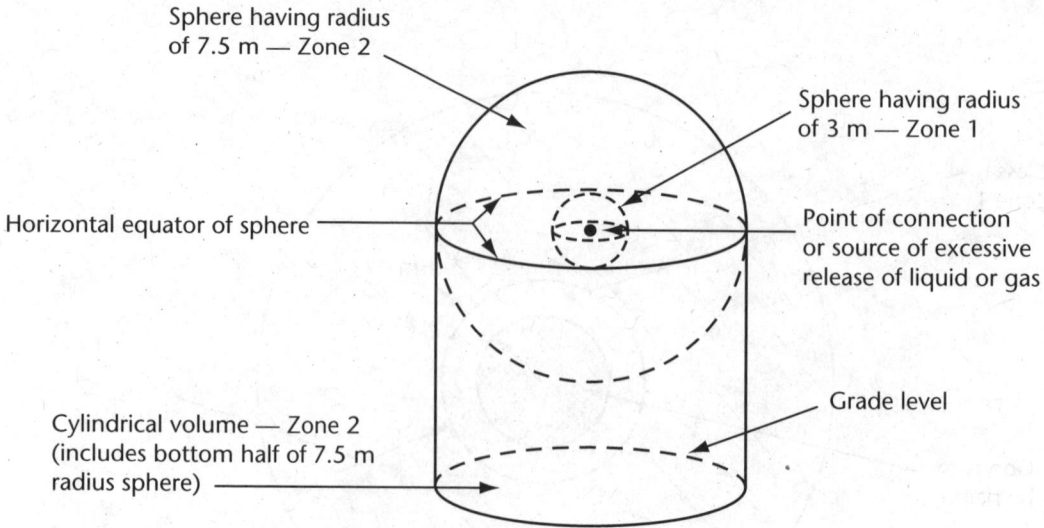

Sphere having radius of 7.5 m — Zone 2

Sphere having radius of 3 m — Zone 1

Horizontal equator of sphere

Point of connection or source of excessive release of liquid or gas

Grade level

Cylindrical volume — Zone 2 (includes bottom half of 7.5 m radius sphere)

Diagram 7
Extent of hazardous location for tank vehicle and tank car loading and unloading
(See Part B of Table 63.)

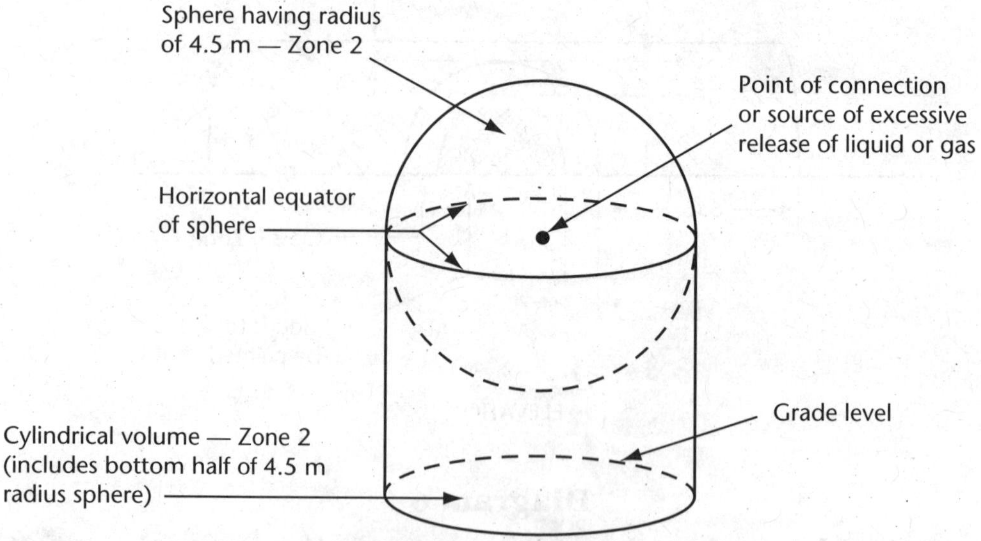

Sphere having radius of 4.5 m — Zone 2

Point of connection or source of excessive release of liquid or gas

Horizontal equator of sphere

Grade level

Cylindrical volume — Zone 2 (includes bottom half of 4.5 m radius sphere)

Diagram 8
Extent of hazardous location for pumps, vapour compressors, gas air mixers, and vaporizers outdoors in open air
(See Part E of Table 63.)

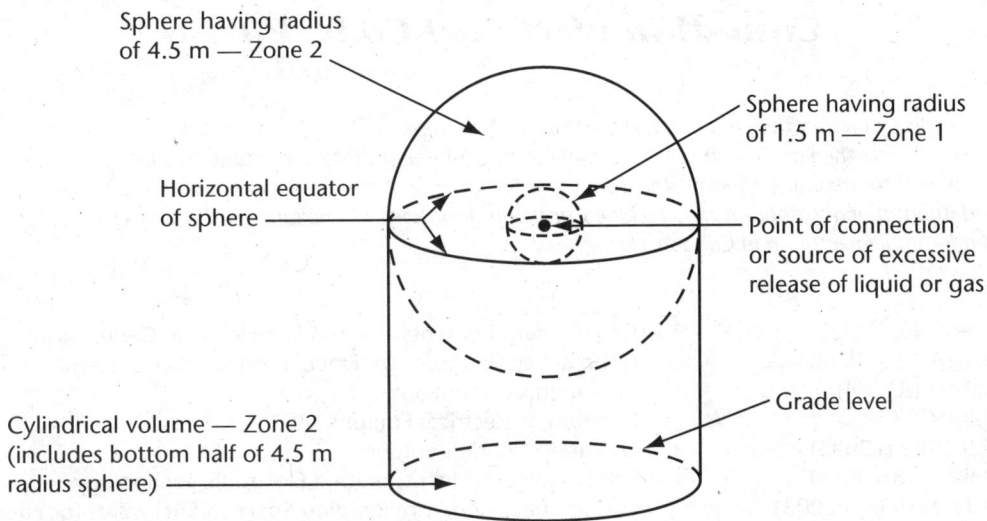

Diagram 9
Extent of hazardous location for container filling outdoors in open air
(See Part J of Table 63.)

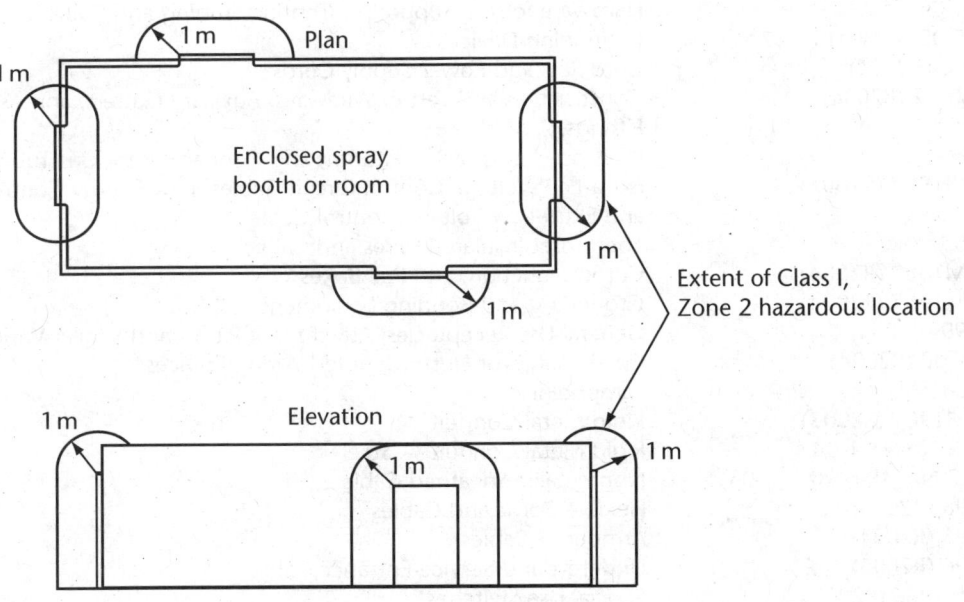

Diagram 10
Extent of hazardous location adjacent to openings in a closed spray booth or room
(See Rule 20-402(3).)

341

Appendix A — Safety standards for electrical equipment, *Canadian Electrical Code, Part II*

Notes:

(1) *This Appendix is a normative (mandatory) part of this Standard.*

(2) *This Appendix lists the Part II Standards available as of September 2005. The standards listed include harmonized binational and trinational standards and adopted international standards as well as standards developed by CSA.*

(3) *Adopted international standards may include Canadian deviations. Compliance with these Canadian deviations is required for implementation in Canada.*

General

CAN/CSA-C22.2 No. 0-M91 (R2001)	General Requirements — Canadian Electrical Code, Part II
C22.2 No. 0.1-M1985 (R2003)	General Requirements for Double-Insulated Equipment
C22.2 No. 0.2-93 (R2004)	Insulation Coordination
C22.2 No. 0.4-04	Bonding of Electrical Equipment
C22.2 No. 0.5-1982 (R2003)	Threaded Conduit Entries
C22.2 No. 0.8-M1986 (R2003)	Safety Functions Incorporating Electronic Technology
C22.2 No. 0.12-M1985 (R2003)	Wiring Space and Wire Bending Space in Enclosures for Equipment Rated 750 V or Less
C22.2 No. 0.15-01	Adhesive Labels
CAN/CSA-C22.2 No. 0.16-M92 (R2001)	Measurement of Harmonic Currents
CAN/CSA-C22.2 No. 0.17-00 (R2004)	Evaluation of Properties of Polymeric Materials

Wiring Products

C22.2 No. 0.3-01	Test Methods for Electrical Wires and Cables
CAN/CSA-C22.2 No. 18-98 (R2003)	Outlet Boxes, Conduit Boxes, Fittings, and Associated Hardware
C22.2 No. 18.1-04	Metallic Outlet Boxes
C22.2 No. 18.3-04	Conduit, Tubing and Cable Fittings
C22.2 No. 18.4-04	Hardware for the Support of Conduit, Tubing and Cable
C22.2 No. 18.5-02 (R2004)	Positioning Devices
C22.2 No. 21-95 (R2004)	Cord Sets and Power Supply Cords
C22.2 No. 26-1952 (R2004)	Construction and Test of Wireways, Auxiliary Gutters, and Associated Fittings
C22.2 No. 34-M1987 (R2004)	Electrode Receptacles, Fittings, and Connectors for Gas Tubes
C22.2 No. 35-M1987 (R2004)	Extra-Low-Voltage Control Circuit Cables, Low-Energy Control Cable, and Extra-Low-Voltage Control Cable
C22.2 No. 38-05	Thermoset Insulated Wires and Cables
C22.2 No. 40-M1989 (R2004)	Cutout, Junction, and Pull Boxes
C22.2 No. 41-M1987 (R2004)	Grounding and Bonding Equipment
C22.2 No. 42-99 (R2004)	General Use Receptacles, Attachment Plugs, and Similar Wiring Devices
C22.2 No. 42.1-00 (R2004)	Cover Plates for Flush-Mounted Wiring Devices
C22.2 No. 43-04	Lampholders
C22.2 No. 45-M1981 (R2003)	Rigid Metal Conduit
CAN/CSA-C22.2 No. 45.1-04	Rigid Metal Conduit — Steel
CAN/CSA-C22.2 No. 48-M90 (R2005)	Nonmetallic Sheathed Cable
C22.2 No. 49-98 (R2003)	Flexible Cords and Cables
C22.2 No. 51-95 (R2004)	Armoured Cables
C22.2 No. 52-96 (R2005)	Underground Service-Entrance Cables
C22.2 No. 55-M1986 (R2003)	Special Use Switches
C22.2 No. 56-04	Flexible Metal Conduit and Liquid-Tight Flexible Metal Conduit
C22.2 No. 57-M1985 (R2005)	Appliance Plugs for Heater Cord Sets
C22.2 No. 59.1-M1987 (R2001)	Fuses (Both Plug and Cartridge-Enclosed Types)
C22.2 No. 62-93 (R2003)	Surface Raceway Systems
C22.2 No. 62.1-03	Nonmetallic Surface Raceways and Fittings
C22.2 No. 65-03	Wire Connectors
C22.2 No. 75-03	Thermoplastic-Insulated Wires and Cables
C22.2 No. 79-1978 (R2003)	Cellular Metal and Cellular Concrete Floor Raceways and Fittings
C22.2 No. 80-1978 (R2003)	Underfloor Raceways and Fittings

C22.2 No. 82-1969 (R2004)	Tubular Support Members and Associated Fittings for Domestic and Commercial Service Masts
C22.2 No. 83-M1985 (R2003)	Electrical Metallic Tubing
C22.2 No. 83.1-04	Electrical Metallic Tubing — Steel
CAN/CSA-C22.2 No. 85-M89 (R2001)	Rigid PVC Boxes and Fittings
C22.2 No. 96-03	Portable Power Cables
C22.2 No. 96.1-04	Mine Power Feeder Cables
C22.2 No. 111-00	General-Use Snap Switches
C22.2 No. 123-96 (R2005)	Aluminum Sheathed Cables
C22.2 No. 124-04	Mineral-Insulated Cable
CAN/CSA-C22.2 No. 126.1-02	Metal Cable Tray Systems
CAN/CSA-C22.2 No. 126.2-02	Nonmetallic Cable Tray Systems
C22.2 No. 127-99 (R2004)	Equipment and Lead Wires
C22.2 No. 129-05	Neutral Supported Cable
CAN/CSA-C22.2 No. 130-03	Requirements for Electrical Resistance Heating Cables and Heating Device Sets
CAN/CSA-C22.2 No. 130.1-M90 (R2001) (withdrawn)	Heat-Tracing Cable Systems for Use in Industrial Locations
CAN/CSA-C22.2 No. 131-M89 (R2004)	Type TECK 90 Cable
C22.2 No. 153-M1981 (R2003)	Quick-Connect Terminals
C22.2 No. 159-M1987 (R2004)	Attachment Plugs, Receptacles, and Similar Wiring Devices for Use in Hazardous Locations: Class I, Groups A, B, C, and D; Class II, Group G, in Coal or Coke Dust, and in Gaseous Mines
C22.2 No. 174-M1984 (R2003)	Cables and Cable Glands for Use in Hazardous Locations
C22.2 No. 179-00	Airport Series Lighting Cables
C22.2 No. 182.1-02	Plugs, Receptacles, and Cable Connectors of the Pin and Sleeve Type
C22.2 No. 182.2-M1987 (R2004)	Industrial Locking Type, Special Use Attachment Plugs, Receptacles, and Connectors
C22.2 No. 182.3-M1987 (R2004)	Special Use Attachment Plugs, Receptacles, and Connectors
CAN/CSA-C22.2 No. 182.4-M90 (R2001)	Plugs, Receptacles, and Connectors for Communication Systems
C22.2 No. 184-M1988 (R2004)	Solid-State Lighting Controls
C22.2 No. 184.1-96 (R2001)	Solid-State Dimming Controls
C22.2 No. 188-04	Splicing Wire Connectors
C22.2 No. 197-M1983 (R2003)	PVC Insulating Tape
C22.2 No. 198.1-99 (R2004)	Extruded Insulating Tubing
C22.2 No. 198.2-M1986 (R2003)	Underground Cable Splicing Kits
C22.2 No. 198.3-95 (R2005)	Coated Electrical Sleeving
CAN/CSA-C22.2 No. 203-M91 (R2001)	Modular Wiring Systems for Office Furniture
C22.2 No. 203.1-94 (R2004)	Manufactured Wiring Systems
C22.2 No. 208-03	Fire Alarm and Signal Cable
C22.2 No. 210-05	Appliance Wiring Material Products
C22.2 No. 211.0-03	General Requirements and Methods of Testing for Nonmetallic Conduit
C22.2 No. 211.1-M1984 (R2003)	Rigid Types EB1 and DB2/ES2 PVC Conduit
C22.2 No. 211.2-M1984 (R2003)	Rigid PVC (Unplasticized) Conduit
C22.2 No. 211.3-96 (R2002)	Reinforced Thermosetting Resin Conduit (RTRC) and Fittings
C22.2 No. 214-02	Communications Cables
C22.2 No. 222-M1986 (R2003)	Type FCC Under-Carpet Wiring System
CAN/CSA-C22.2 No. 227.1-97 (R2002)	Electrical Nonmetallic Tubing
C22.2 No. 227.2-93 (R2003)	Flexible Liquid-Tight Nonmetallic Conduit
C22.2 No. 227.2.1-04	Liquid-Tight Flexible Nonmetallic Conduit
C22.2 No. 227.3-05	Nonmetallic Mechanical Protection Tubing (NMPT)
C22.2 No. 230-M1988 (R2004)	Tray Cables
C22.2 No. 232-M1988 (R2004)	Optical Fiber Cables
CAN/CSA-C22.2 No. 233-M89 (R2004)	Cords and Cord Sets for Communication Systems
CAN/CSA-C22.2 No. 239-97 (R2001)	Control and Instrumentation Cables
C22.2 No. 241-97 (R2000)	IEEE Standard for Cable Joints for Use with Extruded Dielectric Cable Rated 5000–138 000 V and Cable Joints for Use with Laminated Dielectric Cable Rated 2500–500 000 V

CAN/CSA-C22.2 No. 242-92 (R2001)	IEEE Standard Test Procedures and Requirements for High-Voltage Alternating-Current Cable Terminations
C22.2 No. 245-95 (R2004)	Marine Shipboard Cable
C22.2 No. 249-96 (R2000)	Standard Tests for Determining Compatibility of Cable-Pulling Lubricants with Wire and Cable
CAN/CSA-C22.2 No. 262-04	Optical Fibre Cable and Communication Cable Raceway Systems

Industrial Products

C22.2 No. 4-04	Enclosed and Dead-Front Switches
CAN/CSA-C22.2 No. 5-02	Molded-Case Circuit Breakers, Molded-Case Switches and Circuit-Breaker Enclosures
C22.2 No. 13-1962 (R2001)	Transformers for Luminous-Tube Signs, Oil- or Gas-Burner Ignition Equipment, Cold-Cathode Interior Lighting
C22.2 No. 14-05	Industrial Control Equipment
C22.2 No. 22-M1986 (R2004)	Electrical Equipment for Flammable and Combustible Fuel Dispensers
C22.2 No. 25-1966 (R2004)	Enclosures for Use in Class II Groups E, F, and G Hazardous Locations
C22.2 No. 27-00	Busways
C22.2 No. 29-M1989 (R2004)	Panelboards and Enclosed Panelboards
C22.2 No. 30-M1986 (R2003)	Explosion-Proof Enclosures for Use in Class I Hazardous Locations
C22.2 No. 31-04	Switchgear Assemblies
C22.2 No. 33-M1984 (R2004)	Construction and Test of Electric Cranes and Hoists
C22.2 No. 39-M1987 (R2003)	Fuseholder Assemblies
CAN/CSA-C22.2 No. 47-M90 (R2001)	Air-Cooled Transformers (Dry-Type)
C22.2 No. 58-M1989 (R2005)	High-Voltage Isolating Switches
C22.2 No. 60-M1990 (R2001)	Arc Welding Equipment
C22.2 No. 66-M1988 (R2001)	Specialty Transformers
C22.2 No. 73-1953 (R2004)	Construction and Test of Electrically Equipped Machine Tools
CAN/CSA-C22.2 No. 76-M92 (R2002)	Splitters
C22.2 No. 77-95 (R2004)	Motors with Inherent Overheating Protection
C22.2 No. 88-1958 (R2004)	Construction and Test of Industrial Heating Equipment
CAN/CSA-C22.2 No. 94-M91 (R2001)	Special Purpose Enclosures
C22.2 No. 100-04	Motors and Generators
C22.2 No. 102-1958 (R2004)	Brooders and Incubators
C22.2 No. 105-1953 (R2004)	Electrical Equipment for Woodworking Machinery
CAN/CSA-C22.2 No. 106-M92 (R2001)	HRC Fuses
C22.2 No. 107.1-01	General Use Power Supplies
CAN/CSA-C22.2 No. 107.2-01	Battery Chargers
CAN/CSA-C22.2 No. 107.3-03	Uninterruptible Power Systems
CAN/CSA-C22.2 No. 108-01	Liquid Pumps
C22.2 No. 115-M1989 (R2005)	Meter-Mounting Devices
C22.2 No. 137-M1981 (R2004)	Electric Luminaires for Use in Hazardous Locations
C22.2 No. 139-1982 (R2005)	Electrically Operated Valves
C22.2 No. 142-M1987 (R2004)	Process Control Equipment
CAN/CSA-C22.2 No. 144-M91 (R2001)	Ground Fault Circuit Interrupters
C22.2 No. 145-M1986 (R2004)	Motors and Generators for Use in Hazardous Locations
C22.2 No. 152-M1984 (R2001)	Combustible Gas Detection Instruments
C22.2 No. 155-M1986 (R2004)	Electric Duct Heaters
C22.2 No. 156-M1987 (R2004)	Solid-State Speed Controls
CAN/CSA-C22.2 No. 157-92 (R2002)	Intrinsically Safe and Non-Incendive Equipment for Use in Hazardous Locations
C22.2 No. 158-M1987 (R2004)	Terminal Blocks
C22.2 No. 160-M1985 (R2003)	Voltage and Polarity Testers
CAN/CSA-C22.2 No. 165-92 (R2001)	Electric Boilers
C22.2 No. 173-M1983 (R2003)	Transformers for Toy and Hobby Use
CAN/CSA-C22.2 No. 177-92 (R2002)	Clock-Operated Switches
C22.2 No. 178-1978 (R2001)	Automatic Transfer Switches
C22.2 No. 178.2-04	Requirements for Manually Operated Generator Transfer Panels
C22.2 No. 180-M1983 (R2004)	Series Isolating Transformers for Airport Lighting

C22.2 No. 183.1-M1982 (R2004)	Alternating-Current (AC) Electrical Installations on Boats
C22.2 No. 183.2-M1983 (R2004)	DC Electrical Installations on Boats
C22.2 No. 190-M1985 (R2004)	Capacitors for Power Factor Correction
C22.2 No. 193-M1983 (R2004)	High Voltage Full-Load Interrupter Switches
C22.2 No. 201-M1984 (R2004)	Metal-Enclosed High Voltage Busways
C22.2 No. 204-M1984 (R2004)	Line Isolation Monitors
C22.2 No. 213-M1987 (R2004)	Non-incendive Electrical Equipment for Use in Class I, Division 2 Hazardous Locations
CAN/CSA-C22.2 No. 223-M91 (R2003)	Power Supplies with Extra-Low-Voltage Class 2 Outputs
C22.2 No. 229-M1988 (R2004)	Switching and Metering Centres
C22.2 No. 235-04	Supplementary Protectors
CAN/CSA-C22.2 No. 248.1-00 (R2005)	Low Voltage Fuses — Part 1: General Requirements
CAN/CSA-C22.2 No. 248.2-00 (R2005)	Low Voltage Fuses — Part 2: Class C Fuses
CAN/CSA-C22.2 No. 248.3-00 (R2005)	Low Voltage Fuses — Part 3: Class CA and CB Fuses
CAN/CSA-C22.2 No. 248.4-00 (R2005)	Low Voltage Fuses — Part 4: Class CC Fuses
CAN/CSA-C22.2 No. 248.5-00 (R2005)	Low Voltage Fuses — Part 5: Class G Fuses
CAN/CSA-C22.2 No. 248.6-00 (R2005)	Low Voltage Fuses — Part 6: Class H Non-Renewable Fuses
CAN/CSA-C22.2 No. 248.7-00 (R2005)	Low Voltage Fuses — Part 7: Class H Renewable Fuses
CAN/CSA-C22.2 No. 248.8-00 (R2005)	Low Voltage Fuses — Part 8: Class J Fuses
CAN/CSA-C22.2 No. 248.9-00 (R2005)	Low Voltage Fuses — Part 9: Class K Fuses
CAN/CSA-C22.2 No. 248.10-00 (R2005)	Low Voltage Fuses — Part 10: Class L Fuses
CAN/CSA-C22.2 No. 248.11-00 (R2005)	Low Voltage Fuses — Part 11: Plug Fuses
CAN/CSA-C22.2 No. 248.12-00 (R2005)	Low Voltage Fuses — Part 12: Class R Fuses
CAN/CSA-C22.2 No. 248.13-00 (R2005)	Low Voltage Fuses — Part 13: Semiconductor Fuses
CAN/CSA-C22.2 No. 248.14-00 (R2005)	Low Voltage Fuses — Part 14: Supplemental Fuses
CAN/CSA-C22.2 No. 248.15-00 (R2005)	Low Voltage Fuses — Part 15: Class T Fuses
CAN/CSA-C22.2 No. 248.16-00 (R2005)	Low Voltage Fuses — Part 16: Test Limiters
CAN/CSA-E691-94 (R2004)	Thermal-links — Requirements and Application Guide
CAN/CSA-E691A-94 (R2002)	Amendment 1:2002 to CAN/CSA-E691-94, Thermal-Links — Requirements and Application Guide
CAN/CSA-E691B-94 (R2002)	Amendment 2:2002 to CAN/CSA-E691-94, Thermal-Links — Requirements and Application Guide
CAN/CSA-E60079-0-02	Electrical Apparatus for Explosive Gas Atmospheres — Part 0: General Requirements
CAN/CSA-E60079-1-02	Electrical Apparatus for Explosive Gas Atmospheres — Part 1: Flameproof Enclosures "d"
CAN/CSA-E60079-2-02	Electrical Apparatus for Explosive Gas Atmospheres — Part 2: Electrical Apparatus — Type of Protection "p"
CAN/CSA-E60079-5-02	Electrical Apparatus for Explosive Gas Atmospheres — Part 5: Powder Filling "q"
CAN/CSA-E60079-6-02	Electrical Apparatus for Explosive Gas Atmospheres — Part 6: Oil-Immersion "o"
CAN/CSA-E60079-7-03	Electrical Apparatus for Explosive Gas Atmospheres — Part 7: Increased Safety "e"
CAN/CSA-E60079-11-02	Electrical Apparatus for Explosive Gas Atmospheres — Part 11: Intrinsic Safety "i"
CAN/CSA-E60079-15-02	Electrical Apparatus for Explosive Gas Atmospheres — Part 15: Electrical Apparatus with Type of Protection "n"
CAN/CSA-E79-18-95 (R2004)	Electrical Apparatus for Explosive Gas Atmospheres — Part 18: Encapsulation "m"
CAN/CSA-E742-94 (R2004)	Isolating Transformers and Safety Isolating Transformers — Requirements
CAN/CSA-E947-1-99 (R2004)	Low-Voltage Switchgear and Controlgear — Part 1: General Rules
CAN/CSA-E1008-1-98 (R2003)	Residual Current Operated Circuit-Breakers Without Integral Overcurrent Protection for Household and Similar Uses (RCCBs) — Part 1: General Rules
CAN/CSA-E60127-6-03	Miniature Fuses — Part 6: Fuse-Holders for Miniature Cartridge Fuse-Links

CAN/CSA-E61241-1-1-02   Electrical Apparatus for Use in the Presence of Combustible Dust — Part 1: Electrical Apparatus Protected by Enclosures — Section 1: Specification for Apparatus

CAN/CSA-E61496-1-04   Safety of Machinery — Electro-Sensitive Protective Equipment — Part 1: General Requirements and Tests

CAN/CSA-E61496-2-04   Safety of Machinery — Electro-Sensitive Protective Equipment — Part 2: Particular Requirements for Equipment Using Active Opto-electronic Protective Devices (AOPDs)

Consumer and Commercial Products

C22.2 No. 1-04	Audio, Video, and Similar Electronic Equipment
C22.2 No. 3-M1988 (R2004)	Electrical Features of Fuel-Burning Equipment
C22.2 No. 8-M1986 (R2004)	Electromagnetic Interference (EMI) Filters
C22.2 No. 9.0-96 (R2001)	General Requirements for Luminaires
C22.2 No. 9.0S1-97 (R2004)	Supplement No. 1 to C22.2 No. 9.0-96
C22.2 No. 10-1965 (R2004)	Electric Floor Surfacing and Cleaning Machines
C22.2 No. 12-1982 (R2004)	Portable Luminaires
CAN/CSA-C22.2 No. 15-M91 (R2001)	Electrically Heated Warming Pads
C22.2 No. 23.1-M1986 (R2004)	Electric Furnaces in Combination with Solid Fuel Fired Furnaces
C22.2 No. 24-93 (R2003)	Temperature-Indicating and -Regulating Equipment
C22.2 No. 36-M1989 (R2004)	Hairdressing Equipment
C22.2 No. 37-M1989 (R2004)	Christmas Tree and Other Decorative Lighting Outfits
C22.2 No. 46-M1988 (R2001)	Electric Air-Heaters
C22.2 No. 53-1968 (R2004)	Electric Washing Machines
CAN/CSA-C22.2 No. 61-M89 (R2004)	Household Cooking Ranges
C22.2 No. 63-93 (R2004)	Household Refrigerators and Freezers
CAN/CSA-C22.2 No. 64-M91 (R2003)	Household Cooking and Liquid-Heating Appliances
CAN/CSA-C22.2 No. 68-92 (R2004)	Motor-Operated Appliances (Household and Commercial)
CAN/CSA-C22.2 No. 71.1-M89 (R2004)	Portable Electric Tools
CAN/CSA-C22.2 No. 71.2-M89 (R2001)	Electric Bench Tools
C22.2 No. 72-M1984 (R2004)	Heater Elements
CAN/CSA-C22.2 No. 74-96 (R2000)	Equipment for Use With Electric Discharge Lamps
CAN/CSA-C22.2 No. 81-M90 (R2001)	Electric Irons
C22.2 No. 84-05	Incandescent Lamps
C22.2 No. 89-1976 (R2004)	Swimming Pool Luminaires, Submersible Luminaires and Accessories
C22.2 No. 92-1971 (R2004)	Dehumidifiers
C22.2 No. 98-1954 (R2002)	Construction and Test of Power-Operated Radio Transmitters
C22.2 No. 99-1954 (R2004)	Construction and Test of Domestic Electric Ironing Machines
C22.2 No. 101-M1984 (R2004)	Electrically Heated Bedding Appliances for Household Use
CAN/CSA-C22.2 No. 103-M92 (R2001)	Electric Fence Controllers
C22.2 No. 104-01	Humidifiers
C22.2 No. 109-M1981 (R2004)	Commercial Cooking Appliances
CAN/CSA-C22.2 No. 110-94 (R2004)	Construction and Test of Electric Storage-Tank Water Heaters
CAN/CSA-C22.2 No. 112-97 (R2002)	Electric Clothes Dryers
C22.2 No. 113-M1984 (R2004)	Fans and Ventilators
C22.2 No. 117-1970 (R2002)	Room Air-Conditioners
C22.2 No. 118-1959 (R2004)	Construction and Test of Picture Machines and Appliances
CAN/CSA-C22.2 No. 120-M91 (R2004)	Refrigeration Equipment
C22.2 No. 122-M1989 (R2004)	Hand-Held Electrically Heated Tools
C22.2 No. 128-95 (R2004)	Vending Machines
C22.2 No. 140.2-96 (R2001)	Hermetic Refrigerant Motor-Compressors
C22.2 No. 140.3-M1987 (R2004)	Refrigerant-Containing Components for Use in Electrical Equipment
C22.2 No. 141-02	Unit Equipment for Emergency Lighting
CAN/CSA-C22.2 No. 147-M90 (R2004)	Motor-Operated Gardening Appliances
C22.2 No. 149-1972 (R2004)	Electrically Operated Toys
CAN/CSA-C22.2 No. 150-M89 (R2004)	Microwave Ovens
CAN/CSA-C22.2 No. 164-M91 (R2002)	Electric Sauna Heating Equipment
C22.2 No. 166-M1983 (R2004)	Stage and Studio Luminaires

C22.2 No. 167-97 (R2002)	Household Dishwashers
C22.2 No. 168-M1981 (R2004)	Commercial Dishwashing Equipment
C22.2 No. 169-97 (R2002)	Electric Clothes Washing Machines and Extractors
C22.2 No. 187-M1986 (R2003)	Electrostatic Air Cleaners
CAN/CSA-C22.2 No. 189-M89 (R2004)	High-Voltage Insect Killers
CAN/CSA-C22.2 No. 191-M89 (R2004)	Engine Heaters and Battery Warmers
C22.2 No. 195-M1987 (R2004)	Motor Operated Food Processing Appliances (Household and Commercial)
CAN/CSA-C22.2 No. 199-M89 (R2004)	Combustion Safety Controls and Solid-Stage Igniters for Gas- and Oil-Burning Equipment
C22.2 No. 200-M1985 (R2004)	Electric Waterbed Heater Systems
C22.2 No. 205-M1983 (R2004)	Signal Equipment
C22.2 No. 206-M1987 (R2004)	Lighting Poles
CAN/CSA-C22.2 No. 207-M89 (R2004)	Portable and Stationary Electric Signs and Displays
C22.2 No. 209-M1985 (R2004)	Thermal Cut-Offs
CAN/CSA-C22.2 No. 218.1-M89 (R2001)	Spas, Hot Tubs, and Associated Equipment
C22.2 No. 218.2-93 (R2004)	Hydromassage Bathtub Appliances
C22.2 No. 221-M1986 (R2004)	Electrically Heated Hobby and Educational Type Kilns
CAN/CSA-C22.2 No. 224-M89 (R2004)	Radiant Heaters and Infrared and Ultraviolet Lamp Assemblies for Cosmetic or Hygienic Purposes in Nonmedical Applications
CAN/CSA-C22.2 No. 226-92 (R2001)	Protectors in Telecommunication Networks
CAN/CSA-C22.2 No. 231 Series-M89 (R2001)	CSA Safety Requirements for Electrical and Electronic Measuring and Test Equipment
C22.2 No. 231SP-M1990 (R2000)	Audit Checklist and Test Pack for Use with CAN/CSA-C22.2 No. 231 Series-M89, CSA Safety Requirements for Electrical and Electronic Measuring and Test Equipment
C22.2 No. 236-05	Heating and Cooling Equipment
C22.2 No. 243-01	Vacuum Cleaners, Blower Cleaners, and Household Floor Finishing Machines
CAN/CSA-C22.2 No. 247-92 (R2004)	Operators and Systems of Doors, Grates, Draperies, and Louvres
C22.2 No. 250.0-04	Luminaires
C22.2 No. 255-04	Neon Transformers and Power Supplies
CAN/CSA-C22.2 No. 745-2-30-95 (R2004)	Safety of Portable Electric Tools — Part 2: Particular Requirements for Staplers
CAN/CSA-C22.2 No. 745-2-31-95 (R2004)	Safety of Portable Electric Tools — Part 2: Particular Requirements for Diamond Core Drills
CAN/CSA-C22.2 No. 745-2-32-95 (R2004)	Safety of Portable Electric Tools — Part 2: Particular Requirements for Magnetic Drill Presses
CAN/CSA-C22.2 No. 745-2-36-95 (R2004)	Safety of Portable Electric Tools — Part 2: Particular Requirements for Hand Motor Tools
CAN/CSA-C22.2 No. 745-2-37-95 (R2004)	Safety of Portable Electric Tools — Part 2: Particular Requirements for Plate Jointers
CAN/CSA-C22.2 No. 745-4-3-95 (R2004)	Safety of Portable Battery-Operated Tools — Part 4: Particular Requirements for Grinders, Polishers and Disk-Type Sanders
CAN/CSA-C22.2 No. 745-4-36-95 (R2004)	Safety of Portable Battery-Operated Tools — Part 4: Particular Requirements for Hand Motor Tools
CAN/CSA-C22.2 No. 60745-1-04	Hand-held Motor Operated Electric Tools — Safety — Part 1: General Requirements
CAN/CSA-C22.2 No. 60745-2-1-04	Hand-held Motor Operated Electric Tools — Safety — Part 2: Particular Requirements for Drills and Impact Drills
CAN/CSA-C22.2 No. 60745-2-2-04	Hand-held Motor Operated Electric Tools — Safety — Part 2: Particular Requirements for Screwdrivers and Impact Wrenches
CAN/CSA-C22.2 No. 745-2-3-95 (R2004)	Safety of Portable Electric Tools — Part 2: Particular Requirements for Grinders, Polishers and Disk-Type Sanders
CAN/CSA-C22.2 No. 60745-2-4-04	Hand-held Motor Operated Electric Tools — Safety — Part 2: Particular Requirements for Sanders and Polishers Other Than Disk-Type
CAN/CSA-C22.2 No. 60745-2-5-04	Hand-held Motor Operated Electric Tools — Safety — Part 2: Particular Requirements for Circular Saws

CAN/CSA-C22.2 No. 60745-2-6-04	Hand-held Motor Operated Electric Tools — Safety — Part 2: Particular Requirements for Hammers
CAN/CSA-C22.2 No. 60745-2-8-04	Hand-held Motor Operated Electric Tools — Safety — Part 2: Particular Requirements for Shears and Nibblers
CAN/CSA-C22.2 No. 60745-2-9-04	Hand-held Motor Operated Electric Tools — Safety — Part 2: Particular Requirements for Tappers
CAN/CSA-C22.2 No. 60745-2-11-04	Hand-held Motor Operated Electric Tools — Safety — Part 2: Particular Requirements for Reciprocating Saws
CAN/CSA-C22.2 No. 60745-2-12-05	Hand-held Motor-Operated Electric Tools — Safety — Part 2-12: Particular Requirements for Concrete Vibrators
CAN/CSA-C22.2 No. 60745-2-14-04	Hand-held Motor Operated Electric Tools — Safety — Part 2: Particular Requirements for Planers
CAN/CSA-C22.2 No. 60745-2-17-04	Hand-held Motor Operated Electric Tools — Safety — Part 2: Particular Requirements for Routers and Trimmers
CAN/CSA-C22.2 No. 60745-2-18-05	Hand-held Motor-Operated Electric Tools — Safety — Part 2-18: Particular Requirements for Strapping Tools
CAN/CSA-C22.2 No. 60745-2-20-05	Hand-held Motor-Operated Electric Tools — Safety — Part 2-20: Particular Requirements for Band Saws
CAN/CSA-C22.2 No. 60745-2-21-05	Hand-held Motor-Operated Electric Tools — Safety — Part 2-21: Particular Requirements for Drain Cleaners
CAN/CSA-C22.2 No. 61010-1-04	Safety Requirements for Electrical Equipment for Measurement, Control, and Laboratory Use, Part 1: General Requirements
CAN/CSA-C22.2 No. 61010-2-010-04	Safety Requirements for Electrical Equipment for Measurement, Control, and Laboratory Use — Part 2-010: Particular Requirements for Laboratory Equipment for the Heating of Materials
CAN/CSA-C22.2 No. 1010.2.020-94 (R2004)	Safety Requirements for Electrical Equipment for Measurement, Control, and Laboratory Use — Part 2-020: Particular Requirements for Laboratory Centrifuges
CAN/CSA-C22.2 No. 1010.2.020A-97	Amendment 1 to CAN/CSA-C22.2 No. 1010.2.020-94, Safety Requirements for Electrical Equipment for Measurement, Control, and Laboratory Use — Part 2-020: Particular Requirements for Laboratory Centrifuges
CAN/CSA-C22.2 No. 1010.2.031-94 (R2004)	Safety Requirements for Electrical Equipment for Measurement, Control, and Laboratory Use — Part 2-031: Particular Requirements for Hand-Held Probe Assemblies for Electrical Measurement and Test
CAN/CSA-C22.2 No. 61010-2-032-04	Safety Requirements for Electrical Equipment for Measurement, Control, and Laboratory Use — Part 2-032: Particular Requirements for Hand-Held and Hand Manipulated Current Sensors for Electrical Measurement and Test
CAN/CSA-C22.2 No. 1010.2.041-96 (R2004)	Safety Requirements for Electrical Equipment for Measurement, Control, and Laboratory Use — Part 2-041: Particular Requirements for Autoclaves using Steam for the Treatment of Medical Materials, and for Laboratory Processes
C22.2 No. 1335.1-93 (R2004)	Portable Electrical Motor-Operated and Heating Appliances: General Requirements
C22.2 No. 1335.2.9-93 (R2004)	Portable Electrical Motor-Operated and Heating Appliances: Particular Requirements for Portable Electric Cooking Appliances
C22.2 No. 1335.2.14-93 (R2004)	Portable Electrical Motor-Operated and Heating Appliances: Particular Requirements for Motor-Operated Kitchen Machines
CAN/CSA-C22.2 No. 1335.2.15-93 (R2004)	Portable Electrical Motor-Operated and Heating Appliances: Particular Requirements for Liquid-Heating Appliances
CAN/CSA-C22.2 No. 60065-03	Audio, Video and Similar Electronic Apparatus — Safety Requirements
CAN/CSA-C22.2 No. 60529:05	Degrees of Protection Provided by Enclosures (IP Code)
CAN/CSA-C22.2 No. 60950-1-03	Information Technology Equipment — Safety — Part 1: General Requirements
CAN/CSA-C22.2 No. 60950-21-03	Information Technology Equipment — Safety — Part 21: Remote Power Feeding

CAN/CSA-C22.2 No. 61010-2-042-98 (R2003)	Safety Requirements for Electrical Equipment for Measurement, Control, and Laboratory Use — Part 2-042: Particular Requirements for Autoclaves and Sterilizers Using Toxic Gas for the Treatment of Medical Materials, and for Laboratory Processes
CAN/CSA-C22.2 No. 61010-2-043-98 (R2003)	Safety Requirements for Electrical Equipment for Measurement, Control, and Laboratory Use — Part 2-043: Particular Requirements for Dry Heat Sterilizers Using Either Hot Air or Hot Inert Gas for the Treatment of Medical Materials, and for Laboratory Processes
CAN/CSA-C22.2 No. 61010-2-045-04	Safety Requirements for Electrical Equipment for Measurement, Control, and Laboratory Use — Part 2-045: Particular Requirements for Washer Disinfectors Used in Medical, Pharmaceutical, Veterinary and Laboratory Fields
CAN/CSA-C22.2 No. 61010-2-051-04	Safety Requirements for Electrical Equipment for Measurement, Control, and Laboratory Use — Part 2-051: Particular Requirements for Laboratory Equipment for Mixing and Stirring
CAN/CSA-C22.2 No. 61010-2-061-04	Safety Requirements for Electrical Equipment for Measurement, Control, and Laboratory Use — Part 2-061: Particular Requirements for Laboratory Atomic Spectrometers with Thermal Atomization and Ionization
CAN/CSA-C22.2 No. 61010-2-081-04	Safety Requirements for Electrical Equipment for Measurement, Control, and Laboratory Use — Part 2-081: Particular Requirements for Automatic and Semi-Automatic Laboratory Equipment for Analysis and Other Purposes
CAN/CSA-C22.2 No. 61010-2-101-04	Safety Requirements for Electrical Equipment for Measurement, Control, and Laboratory Use — Part 2-101: Particular Requirements for In Vitro Diagnostic (IVD) Medical Equipment
CAN/CSA-E155-98 (R2003)	Glow-Starters for Fluorescent Lamps
CAN/CSA-E335-1/2E-94 (R2004)	Safety of Household and Similar Electrical Appliances — Part 1: General Requirements (Second edition)
CAN/CSA-E335-1/3E-94 (R2004)	Safety of Household and Similar Electrical Appliances — Part 1: General Requirements (Third edition)
CAN/CSA-E60335-1/4E-03	Household and Similar Electrical Appliances — Safety — Part 1: General Requirements (Fourth edition)
CAN/CSA-E335-2-2-94 (R2004)	Safety of Household and Similar Electrical Appliances — Part 2: Particular Requirements for Vacuum Cleaners and Water Suction Cleaning Appliances
CAN/CSA-E335-2-3-94 (R2004)	Safety of Household and Similar Electrical Appliances — Part 2: Particular Requirements for Electric Irons
CAN/CSA-E335-2-4-94 (R2000)	Safety of Household and Similar Electrical Appliances — Part 2: Particular Requirements for Spin Extractors
CAN/CSA-E335-2-5-94 (R2004)	Safety of Household and Similar Electrical Appliances — Part 2: Particular Requirements for Dishwashers
CAN/CSA-E60335-2-6-01	Safety of Household and Similar Electrical Appliances — Part 2: Particular Requirements for Stationary Cooking Ranges, Hobs, Ovens and Similar Appliances
CAN/CSA-E60335-2-7-01	Safety of Household and Similar Electrical Appliances — Part 2-7: Particular Requirements for Washing Machines
CAN/CSA-E335-2-8-94 (R2004)	Safety of Household and Similar Electrical Appliances — Part 2: Particular Requirements for Shavers, Hair Clippers and Similar Appliances
CAN/CSA-E335-2-9-94 (R2004)	Safety of Household and Similar Electrical Appliances — Part 2: Particular Requirements for Toasters, Grills, Roasters and Similar Appliances
CAN/CSA-E335-2-10-94 (R2004)	Safety of Household and Similar Electrical Appliances — Part 2: Particular Requirements for Floor Treatment Machines and Wet Scrubbing Machines
CAN/CSA-E60335-2-11-01	Safety of Household and Similar Electrical Appliances — Part 2-11: Particular Requirements for Tumble Dryers
CAN/CSA-E335-2-12-94 (R2004)	Safety of Household and Similar Electrical Appliances — Part 2: Particular Requirements for Warming Plates and Similar Appliances
CAN/CSA-E335-2-13-94 (R2004)	Safety of Household and Similar Electrical Appliances — Part 2: Particular Requirements for Frying Pans, Deep Fat Fryers and Similar Appliances

CAN/CSA-E60335-2-14-05	Safety of Household and Similar Electrical Appliances — Part 2: Particular Requirements for Kitchen Machines
CAN/CSA-E60335-2-15-05	Safety of Household and Similar Electrical Appliances — Part 2: Particular Requirements for Appliances for Heating Liquids
CAN/CSA-E60335-2-16-01	Safety of Household and Similar Electrical Appliances — Part 2: Particular Requirements for Food Waste Disposers
CAN/CSA-E335-2-18-94 (R2001)	Safety of Household and Similar Electrical Appliances — Part 2: Guide for Preparing Safety Requirements for Battery-Powered Motor-Operated Appliances and Their Charging and Battery Assemblies
CAN/CSA-E335-2-19-94 (R2001)	Safety of Household and Similar Electrical Appliances — Part 2: Particular Requirements for Battery-Powered Shavers, Hair Clippers and Similar Appliances and Their Charging and Battery Assemblies
CAN/CSA-E335-2-20-94 (R2001)	Safety of Household and Similar Electrical Appliances — Part 2: Particular Requirements for Battery-Powered Tooth-Brushes and Their Charging and Battery Assemblies
CAN/CSA-E60335-2-21-01	Safety of Household and Similar Electrical Appliances — Part 2: Particular Requirements for Storage Water Heaters
CAN/CSA-E60335-2-23-05	Safety of Household and Similar Electrical Appliances — Part 2: Particular Requirements for Appliances for Skin or Hair Care
CAN/CSA-E335-2-24-94 (R2004)	Safety of Household and Similar Electrical Appliances — Part 2: Particular Requirements for Refrigerators, Food-Freezers and Ice-Makers
CAN/CSA-E60335-2-25-01	Safety of Household and Similar Electrical Appliances — Part 2: Particular Requirements for Microwave Ovens
CAN/CSA-E60335-2-26-01	Safety of Household and Similar Electrical Appliances — Part 2: Particular Requirements for Clocks
CAN/CSA-E60335-2-27-01	Safety of Household and Similar Electrical Appliances — Part 2: Particular Requirements for Appliances for Skin Exposure to Ultraviolet and Infrared Radiation
CAN/CSA-E60335-2-28-01	Safety of Household and Similar Electrical Appliances — Part 2: Particular Requirements for Sewing Machines
CAN/CSA-E60335-2-29-01	Safety of Household and Similar Electrical Appliances — Part 2: Particular Requirements for Battery Chargers
CAN/CSA-E60335-2-30-01	Safety of Household and Similar Electrical Appliances — Part 2: Particular Requirements for Room Heaters
CAN/CSA-E60335-2-31-05	Safety of Household and Similar Electrical Appliances — Part 2: Particular Requirements for Range Hoods
CAN/CSA-E60335-2-32-01	Safety of Household and Similar Electrical Appliances — Part 2: Particular Requirements for Massage Appliances
CAN/CSA-E335-2-33-94 (R2004)	Safety of Household and Similar Electrical Appliances — Part 2: Particular Requirements for Coffee Mills and Coffee Grinders
CAN/CSA-E60335-2-34-01	Safety of Household and Similar Electrical Appliances — Part 2: Particular Requirements for Motor-Compressors
CAN/CSA-E60335-2-35-01	Safety of Household and Similar Electrical Appliances — Part 2-35: Particular Requirements for Instantaneous Water Heaters
CAN/CSA-E60335-2-36-01	Safety of Household and Similar Electrical Appliances — Part 2-36: Particular Requirements for Commercial Electric Cooking Ranges, Ovens, Hobs and Hob Elements
CAN/CSA-E60335-2-37-01	Safety of Household and Similar Electrical Appliances — Part 2-37: Particular Requirements for Commercial Electric Deep Fat Fryers
CAN/CSA-E60335-2-38-01	Safety of Household and Similar Electrical Appliances — Part 2-38: Particular Requirements for Commercial Electric Griddles and Griddle Grills
CAN/CSA-E60335-2-39-01	Safety of Household and Similar Electrical Appliances — Part 2-39: Particular Requirements for Commercial Electric Multi-Purpose Cooking Pans
CAN/CSA-E60335-2-40-01	Safety of Household and Similar Electrical Appliances — Part 2: Particular Requirements for Electrical Heat Pumps, Air-Conditioners and Dehumidifiers

CAN/CSA-E60335-2-41-01 — Safety of Household and Similar Electrical Appliances — Part 2: Particular Requirements for Liquids Having a Temperature Not Exceeding 35 Degrees Celsius

CAN/CSA-E60335-2-42-01 — Safety of Household and Similar Electrical Appliances — Part 2-42: Particular Requirements for Commercial Electric Forced Convection Ovens, Steam Cookers and Steam-Convection Ovens

CAN/CSA-E60335-2-43-01 — Safety of Household and Similar Electrical Appliances — Part 2: Particular Requirements for Clothes Dryers and Towel Rails

CAN/CSA-E60335-2-44-01 — Safety of Household and Similar Electrical Appliances — Part 2: Particular Requirements for Ironers

CAN/CSA-E60335-2-45-01 — Safety of Household and Similar Electrical Appliances — Part 2: Particular Requirements for Portable Electric Heating Tools and Similar Appliances

CAN/CSA-E60335-2-47-01 — Safety of Household and Similar Electrical Appliances — Part 2-47: Particular Requirements for Commercial Electric Boiling Pans

CAN/CSA-E60335-2-48-01 — Safety of Household and Similar Electrical Appliances — Part 2-48: Particular Requirements for Commercial Electric Grillers and Toasters

CAN/CSA-E60335-2-49-01 — Safety of Household and Similar Electrical Appliances — Part 2-49: Particular Requirements for Commercial Electric Hot Cupboards

CAN/CSA-E60335-2-50-01 — Safety of Household and Similar Electrical Appliances — Part 2-50: Particular Requirements for Commercial Electric Bains-Marie

CAN/CSA-E60335-2-51-01 — Safety of Household and Similar Electrical Appliances — Part 2: Particular Requirements for Stationary Circulation Pumps for Heating and Service Water Installations

CAN/CSA-E60335-2-52-01 — Safety of Household and Similar Electrical Appliances — Part 2: Particular Requirements for Oral Hygiene Appliances

CAN/CSA-E60335-2-53-05 — Safety of Household and Similar Electrical Appliances — Part 2: Particular Requirements for Sauna Heating Appliances

CAN/CSA-E60335-2-54-01 — Safety of Household and Similar Electrical Appliances — Part 2: Particular Requirements for Surface-Cleaning Appliances Employing Liquids

CAN/CSA-E60335-2-55-05 — Safety of Household and Similar Electrical Appliances — Part 2: Particular Requirements for Electrical Appliances for Use with Aquariums and Garden Ponds

CAN/CSA-E60335-2-56-01 — Safety of Household and Similar Electrical Appliances — Part 2: Particular Requirements for Projectors and Similar Appliances

CAN/CSA-E335-2-57-94 (R2004) — Safety of Household and Similar Electrical Appliances — Part 2: Particular Requirements for Ice-Cream Appliances with Incorporated Motor-Compressors

CAN/CSA-E60335-2-58-01 — Safety of Household and Similar Electrical Appliances — Part 2: Particular Requirements for Commercial Electric Dishwashing Machines

CAN/CSA-E335-2-59-94 (R2004) — Safety of Household and Similar Electrical Appliances — Part 2: Particular Requirements for Insect Killers

CAN/CSA-E335-2-61-95 (R2000) — Safety of Household and Similar Electrical Appliances — Part 2: Particular Requirements for Thermal Storage Room Heaters

CAN/CSA-E60335-2-62-01 — Safety of Household and Similar Electrical Appliances — Part 2: Particular Requirements for Commercial Electric Rinsing Sinks

CAN/CSA-E335-2-63-94 (R2004) — Safety of Household and Similar Electrical Appliances — Part 2: Particular Requirements for Commercial Electric Water Boilers and Liquid Heaters

CAN/CSA-E60335-2-64-01 — Safety of Household and Similar Electrical Appliances — Part 2: Particular Requirements for Commercial Electric Kitchen Machines

CAN/CSA-E335-2-65-95 (R2000) — Safety of Household and Similar Electrical Appliances — Part 2: Particular Requirements for Air-Cleaning Appliances

CAN/CSA-E335-2-66-95 (R2000) — Safety of Household and Similar Electrical Appliances — Part 2: Particular Requirements for Water-Bed Heaters

CAN/CSA-E60335-2-67-01 — Safety of Household and Similar Electrical Appliances — Part 2: Particular Requirements for Floor Treatment and Floor Cleaning Machines, for Industrial and Commercial Use

CAN/CSA-E60335-2-68-01	Safety of Household and Similar Electrical Appliances — Part 2: Particular Requirements for Spray Extraction Appliances, for Industrial and Commercial Use
CAN/CSA-E60335-2-69-01	Safety of Household and Similar Electrical Appliances — Part 2: Particular Requirements for Wet and Dry Vacuum Cleaners, Including Power Brush, for Industrial and Commercial Use
CAN/CSA-E335-2-70-95 (R2004)	Safety of Household and Similar Electrical Appliances — Part 2: Particular Requirements for Milking Machines
CAN/CSA-E335-2-71-95 (R2004)	Safety of Household and Similar Electrical Appliances — Part 2: Particular Requirements for Electrical Heating Appliances for Breeding and Rearing Animals
CAN/CSA-E60335-2-76-05	Safety of Household and Similar Electrical Appliances — Part 2: Particular Requirements for Electric Fence Energizers
CAN/CSA-E60335-2-82-01	Safety of Household and Similar Electrical Appliances — Part 2: Particular Requirements for Service Machines and Amusement Machines
CAN/CSA-E342-1-95 (R2001)	Safety Requirements for Electric Fans and Regulators — Part 1: Fans and Regulators for Household and Similar Purposes
CAN/CSA-E60384-1-03	Fixed Capacitors for Use in Electronic Equipment — Part 1: Generic Specification
CAN/CSA-E384-14-95 (R2004)	Fixed Capacitors for Use in Electronic Equipment — Part 14: Sectional Specification: Fixed Capacitors for Electromagnetic Interference Suppression and Connection to the Supply Mains
CAN/CSA-E384-14-1-95 (R2004)	Fixed Capacitors for Use in Electronic Equipment — Part 14: Blank Detail Specification: Fixed Capacitors for Electromagnetic Interference Suppression and Connection to the Supply Mains — Assessment Level D
CAN/CSA-E432-1-98 (R2003)	Safety Specifications for Incandescent Lamps — Part 1: Tungsten Filament Lamps for Domestic and Similar General Lighting Purposes
CAN/CSA-E432-2-98 (R2003)	Safety Specifications for Incandescent Lamps — Part 2: Tungsten Halogen Lamps for Domestic and Similar General Lighting Purposes
CAN/CSA-E491-94 (R2003)	Safety Requirements for Electronic Flash Apparatus for Photographic Purposes
CAN/CSA-E570-98 (R2002)	Electrical Supply Track Systems for Luminaires
CAN/CSA-E60598-1-02	Luminaires — Part 1: General Requirements and Tests
CAN/CSA-E598-2-1-98 (R2002)	Luminaires — Part 2: Section 1: Fixed General Purpose Luminaires
CAN/CSA-E598-2-2-98 (R2002)	Luminaires — Part 2: Section 2: Recessed Luminaires
CAN/CSA-E60598-2-3-98 (R2003)	Luminaires — Part 2: Section 3: Luminaires for Road and Street Lighting
CAN/CSA-E60598-2-4-98 (R2003)	Luminaires — Part 2: Section 4: Portable General Purpose Luminaires
CAN/CSA-E60598-2-5-02	Luminaires — Part 2: Section 5: Floodlights
CAN/CSA-E598-2-6-98 (R2002)	Luminaires — Part 2: Section 6: Luminaires with Built-in Transformers for Filament Lamps
CAN/CSA-E598-2-7-98 (R2002)	Luminaires — Part 2: Section 7: Portable Luminaires for Garden Use
CAN/CSA-E598-2-8-98 (R2002)	Luminaires — Part 2: Section 8: Handlamps
CAN/CSA-E598-2-9-98 (R2003)	Luminaires — Part 2: Section 9: Photo and Film Luminaires (non-professional)
CAN/CSA-E598-2-10-98 (R2002)	Luminaires — Part 2: Section 10: Portable Child-Appealing Luminaires
CAN/CSA-E598-2-17-98 (R2002)	Luminaires — Part 2: Section 17: Luminaires for Stage Lighting, Television and Film Studios (Outdoor and Indoor)
CAN/CSA-E598-2-18-98 (R2002)	Luminaires — Part 2: Section 18: Luminaires for Swimming Pools and Similar Applications
CAN/CSA-E598-2-19-98 (R2002)	Luminaires — Part 2: Section 19: Air-Handling Luminaires (Safety Requirements)
CAN/CSA-E60730-1-02	Automatic Electrical Controls for Household and Similar Use — Part 1: General Requirements
CAN/CSA-E730-2-1-94 (R2004)	Automatic Electrical Controls for Household and Similar Use — Part 2: Particular Requirements for Electrical Controls for Electrical Household Appliances
CAN/CSA-E730-2-2-94 (R2004)	Automatic Electrical Controls for Household and Similar Use — Part 2: Particular Requirements for Thermal Motor Protectors

CAN/CSA-E730-2-3-94 (R2004)	Automatic Electrical Controls for Household and Similar Use — Part 2: Particular Requirements for Thermal Protectors for Ballasts for Tubular Fluorescent Lamps
CAN/CSA-E730-2-4-94 (R2004)	Automatic Electrical Controls for Household and Similar Use — Part 2: Particular Requirements for Thermal Motor Protectors for Motor-Compressors of Hermetic and Semi-Hermetic Type
CAN/CSA-E730-2-5-94 (R2004)	Automatic Electrical Controls for Household and Similar Use — Part 2: Particular Requirements for Automatic Electrical Burner Control Systems
CAN/CSA-E730-2-6-94 (R2004)	Automatic Electrical Controls for Household and Similar Use — Part 2: Particular Requirements for Automatic Electrical Pressure Sensing Controls Including Mechanical Requirements
CAN/CSA-E730-2-7-94 (R2004)	Automatic Electrical Controls for Household and Similar Use — Part 2: Particular Requirements for Timers and Time Switches
CAN/CSA-E60730-2-8-01	Automatic Electrical Controls for Household and Similar Use — Part 2-8: Particular Requirements for Electrically Operated Water Valves, Including Mechanical Requirements
CAN/CSA-E60730-2-9-01	Automatic Electrical Controls for Household and Similar Use — Part 2: Particular Requirements for Temperature Sensing Controls
CAN/CSA-E730-2-10-94 (R2004)	Automatic Electrical Controls for Household and Similar Use — Part 2: Particular Requirements for Electrically Operated Motor Starting Relays
CAN/CSA-E730-2-11-94 (R2004)	Automatic Electrical Controls for Household and Similar Use — Part 2: Particular Requirements for Energy Regulators
CAN/CSA-E730-2-12-94 (R2004)	Automatic Electrical Controls for Household and Similar Use — Part 2: Particular Requirements for Electrically Operated Door Locks
CAN/CSA-E920-98 (R2003)	Ballasts for Tubular Fluorescent Lamps — General and Safety Requirements
CAN/CSA-E922-98 (R2003)	Ballasts for Discharge Lamps (Excluding Tubular Fluorescent Lamps) — General and Safety Requirements
CAN/CSA-E926-98 (R2003)	Auxiliaries for Lamps — Starting Devices (Other than Glow Starters) — General and Safety Requirements
CAN/CSA-E928-98 (R2003)	Auxiliaries for Lamps — A.C. Supplied Electronic Ballasts for Tubular Fluorescent Lamps — General and Safety Requirements
CAN/CSA-E967-94 (R2001)	Safety of Electrically Heated Blankets, Pads and Similar Flexible Heating Appliances for Household Use
CAN/CSA-E968-99 (R2003)	Self-Ballasted Lamps for General Lighting Services — Safety Requirements
CAN/CSA-E60974-1-00	Arc Welding Equipment — Part 1: Welding Power Sources
CAN/CSA-E60974-1A-00	Amendment 1:2002 to CAN/CSA-E60974-1-00, Arc Welding Equipment — Part 1: Welding Power Sources
CAN/CSA-E60974-5-03	Arc Welding Equipment — Part 5: Wire Feeders
CAN/CSA-E60974-7-02	Arc Welding Equipment — Part 7: Torches
CAN/CSA-E1029-1-94 (R2004)	Safety of Transportable Motor-Operated Electric Tools — Part 1: General Requirements
CAN/CSA-E1029-2-1-94 (R2004)	Safety of Transportable Motor-Operated Electric Tools — Part 2: Particular Requirements for Circular Saws
CAN/CSA-E1029-2-2-94 (R2004)	Safety of Transportable Motor-Operated Electric Tools — Part 2: Particular Requirements for Radial Arm Saws
CAN/CSA-E1029-2-3-94 (R2004)	Safety of Transportable Motor-Operated Electric Tools — Part 2: Particular Requirements for Planers and Thicknessers
CAN/CSA-E1029-2-4-94 (R2004)	Safety of Transportable Motor-Operated Electric Tools — Part 2: Particular Requirements for Bench Grinders
CAN/CSA-E1029-2-5-94 (R2004)	Safety of Transportable Motor-Operated Electric Tools — Part 2: Particular Requirements for Band Saws
CAN/CSA-E1029-2-6-94 (R2004)	Safety of Transportable Motor-Operated Electric Tools — Part 2: Particular Requirements for Diamond Drills with Water Supply
CAN/CSA-E1029-2-7-94 (R2004)	Safety of Transportable Motor-Operated Electric Tools — Part 2: Particular Requirements for Diamond Saws with Water Supply
CAN/CSA-E1048-98 (R2003)	Capacitor for Use in Tubular Fluorescent and Other Discharge Lamp Circuits — General and Safety Requirements

CAN/CSA-E60825-1-03	Safety of Laser Products — Part 1: Equipment Classification, Requirements and User's Guide
CAN/CSA-E61347-1-03	Lamp Controlgear — Part 1: General and Safety Requirements
CAN/CSA-E61347-2-3-03	Lamp Controlgear — Part 2-3: Particular Requirements for A.C. Supplied Electronic Ballasts for Fluorescent Lamps
CAN/CSA-E61558-1-03	Safety of Power Transformers, Power Supply Units and Similar — Part 1: General Requirements and Tests
CAN/CSA-E61558-2-1-03	Safety of Power Transformers, Power Supply Units and Similar — Part 2: Particular Requirements for Separating Transformers for General Use
CAN/CSA-E61558-2-2-03	Safety of Power Transformers, Power Supply Units and Similar — Part 2-2: Particular Requirements for Control Transformers
CAN/CSA-E61558-2-4-03	Safety of Power Transformers, Power Supply Units and Similar — Part 2: Particular Requirements for Isolating Transformers for General Use
CAN/CSA-E61558-2-5-03	Safety of Power Transformers, Power Supply Units and Similar — Part 2-5: Particular Requirements for Shaver Transformers and Shaver Supply Units
CAN/CSA-E61558-2-6-03	Safety of Power Transformers, Power Supply Units and Similar — Part 2: Particular Requirements for Safety Isolating Transformers for General Use
CAN/CSA-E61558-2-13-03	Safety of Power Transformers, Power Supply Units and Similar Devices — Part 2-13: Particular Requirements for Auto-Transformers for General Use
CAN/CSA-E61965-04	Mechanical Safety of Cathode Ray Tubes

Health Care Products

CAN/CSA-C22.2 No. 114-M90 (R2005)	Diagnostic Imaging and Radiation Therapy Equipment
C22.2 No. 125-M1984 (R2004)	Electromedical Equipment
C22.2 No. 151-M1986 (R2004)	Laboratory Equipment
CAN/CSA-C22.2 No. 601.1-M90 (R2001)	Medical Electrical Equipment — Part 1: General Requirements for Safety
CAN/CSA-C22.2 No. 601.1S1-94 (R1999)	Supplement No. 1-94 to CAN/CSA-C22.2 No. 601.1-M90
CAN/CSA-C22.2 No. 601.1B-90 (R2002)	Amendment 2 to CAN/CSA-C22.2 No. 601.1-M90
CAN/CSA-C22.2 No. 60601-1-1-02	Medical Electrical Equipment — Part 1: General Requirements for Safety. Collateral Standard: Safety Requirements for Medical Electrical Systems
CAN/CSA-C22.2 No. 60601-1-2-03	Medical Electrical Equipment — Part 1: General Requirements for Safety. Collateral Standard: Electromagnetic Compatibility — Requirements and Tests
CAN/CSA-C22.2 No. 601.1.3-98 (R2002)	Medical Electrical Equipment — Part 1: General Requirements for Safety — Part 3: Collateral Standard: General Requirements for Radiation Protection in Diagnostic X-Ray Equipment
CAN/CSA-C22.2 No. 60601-1-4-02	Medical Electrical Equipment — Part 1-4: General Requirements for Safety — Collateral Standard: Programmable Electrical Medical Systems
CAN/CSA-C22.2 No. 60601-2-1-01	Medical Electrical Equipment — Part 2-1: Particular Requirements for the Safety of Electron Accelerators in the Range 1 MeV to 50 MeV
CAN/CSA-C22.2 No. 60601-2-2-01	Medical Electrical Equipment — Part 2-2: Particular Requirements for the Safety of High Frequency Surgical Equipment
CAN/CSA-C22.2 No. 601.2.3-92 (R2001)	Medical Electrical Equipment, Part 2: Particular Requirements for the Safety of Short-Wave Therapy Equipment
CAN/CSA-C22.2 No. 60601-2-4-04	Medical Electrical Equipment — Part 2: Particular Requirements for the Safety of Cardiac Defibrillators
CAN/CSA-C22.2 No. 60601-2-5-02	Medical Electrical Equipment — Part 2: Particular Requirements for the Safety of Ultrasonic Physiotherapy Equipment
CAN/CSA-C22.2 No. 601.2.6-92 (R2001)	Medical Electrical Equipment, Part 2: Particular Requirements for the Safety of Microwave Therapy Equipment
CAN/CSA-C22.2 No. 60601-2-7-01	Medical Electrical Equipment — Part 2-7: Particular Requirements for the Safety of High-Voltage Generators of Diagnostic X-Ray Generators
CAN/CSA-C22.2 No. 60601-2-8-01	Medical Electrical Equipment — Part 2-8: Particular Requirements for the Safety of Therapeutic X-Ray Equipment Operating in the Range 10 kV to 1 MV
CAN/CSA-C22.2 No. 60601-2-9-01	Medical Electrical Equipment — Part 2: Particular Requirements for the Safety of Patient Contact Dosemeters Used in Radiotherapy with Electrically Connected Radiation Detectors

CAN/CSA-C22.2 No. 601.2.10-92 (R2001)	Medical Electrical Equipment — Part 2: Particular Requirements for the Safety of Nerve and Muscle Stimulators
CAN/CSA-C22.2 No. 60601-2-11-01	Medical Electrical Equipment — Part 2: Particular Requirements for the Safety of Gamma Beam Therapy Equipment
CAN/CSA-C22.2 No. 60601-2-12-03	Medical Electrical Equipment — Part 2: Particular Requirements for the Safety of Lung Ventilators — Critical Care Ventilators
CAN/CSA-C22.2 No. 60601-2-13-02	Medical Electrical Equipment — Part 2: Particular Requirements for the Safety of Anaesthetic Workstations
CAN/CSA-C22.2 No. 60601-2-16-01	Medical Electrical Equipment — Part 2-16: Particular Requirements for the Safety of Haemodialysis, Haemodialfiltration and Haemofiltration Equipment
CAN/CSA-C22.2 No. 60601-2-17-04	Medical Electrical Equipment — Part 2: Particular Requirements for the Safety of Automatically-Controlled Brachytherapy Afterloading Equipment
CAN/CSA-C22.2 No. 60601-2-18-01	Medical Electrical Equipment — Part 2: Particular Requirements for the Safety of Endoscopic Equipment
CAN/CSA-C22.2 No. 601.2.19-92 (R2001)	Medical Electrical Equipment — Part 2: Particular Requirements for the Safety of Baby Incubators
CAN/CSA-C22.2 No. 601.2.20-92 (R2001)	Medical Electrical Equipment — Part 2: Particular Requirements for the Safety of Transport Incubators
CAN/CSA-C22.2 No. 601.2.21-98 (R2002)	Medical Electrical Equipment — Part 2: Particular Requirements for the Safety of Infant Radiant Warmers
CAN/CSA-C22.2 No. 60601-2-22-01	Medical Electrical Equipment — Part 2: Particular Requirements for the Safety of Diagnostic and Therapeutic Laser Equipment
CAN/CSA-C22.2 No. 60601-2-23-02	Medical Electrical Equipment — Part 2: Particular Requirements for the Safety, Including Essential Performance, of Transcutaneous Partial Pressure Monitoring Equipment
CAN/CSA-C22.2 No. 60601-2-24-01	Medical Electrical Equipment — Part 2-24: Particular Requirements for the Safety of Infusion Pumps and Controllers
CAN/CSA-C22.2 No. 601.2.25-94 (R2003)	Medical Electrical Equipment, Part 2: Particular Requirements for the Safety of Electrocardiographs
CAN/CSA-C22.2 No. 60601-2-26-04	Medical Electrical Equipment — Part 2: Particular Requirements for the Safety of Electroencephalographs
CAN/CSA-C22.2 No. 601.2.27-98 (R2002)	Medical Electrical Equipment — Part 2: Particular Requirements for the Safety of Electrocardiographic Monitoring Equipment
CAN/CSA-C22.2 No. 601.2.28-94 (R2003)	Medical Electrical Equipment, Part 2: Particular Requirements for the Safety of X-Ray Source Assemblies and X-Ray Tube Assemblies for Medical Diagnosis
CAN/CSA-C22.2 No. 60601-2-29-02	Medical Electrical Equipment, Part 2: Particular Requirements for the Safety of Radiotherapy Simulators
CAN/CSA-C22.2 No. 60601-2-30-02	Medical Electrical Equipment — Part 2: Particular Requirements for the Safety, Including Essential Performance, of Automatic Cycling Non-invasive Blood Pressure Monitoring Equipment
CAN/CSA-C22.2 No. 601.2.31-98 (R2002)	Medical Electrical Equipment — Part 2: Particular Requirements for the Safety of External Cardiac Pacemakers with Internal Power Source
CAN/CSA-C22.2 No. 601.2.32-98 (R2002)	Medical Electrical Equipment — Part 2: Particular Requirements for the Safety of Associated Equipment of X-Ray Equipment
CAN/CSA-C22.2 No. 60601-2-33-04	Medical Electrical Equipment — Part 2: Particular Requirements for the Safety of Magnetic Resonance Equipment for Medical Diagnosis
CAN/CSA-C22.2 No. 60601-2-34-02	Medical Electrical Equipment — Part 2: Particular Requirements for the Safety, Including Essential Performance, of Invasive Blood-Pressure Monitoring Equipment
CAN/CSA-C22.2 No. 60601-2-36-98 (R2003)	Medical Electrical Equipment — Part 2: Particular Requirements for the Safety of Equipment for Extracorporeally Induced Lithotripsy
CAN/CSA-C22.2 No. 60601-2-37-03	Medical Electrical Equipment — Part 2-37: Particular Requirements for the Safety of Ultrasonic Medical Diagnostic and Monitoring Equipment
CAN/CSA-C22.2 No. 60601-2-38-03	Medical Electrical Equipment — Part 2: Particular Requirements for the Safety of Electrically Operated Hospital Beds

CAN/CSA-C22.2 No. 60601-2-39-02   Medical Electrical Equipment — Part 2-39: Particular Requirements for the Safety of Peritoneal Dialysis Equipment

CAN/CSA-C22.2 No. 60601-2-40-01   Medical Electrical Equipment — Part 2-40: Particular Requirements for the Safety of Electromyographs and Evoked Response Equipment

CAN/CSA-C22.2 No. 60601-2-41-02   Medical Electrical Equipment — Part 2-41: Particular Requirements for the Safety of Surgical Luminaires and Luminaires for Diagnosis

CAN/CSA-C22.2 No. 60601-2-43-03   Medical Electrical Equipment — Part 2-43: Particular Requirements for the Safety of X-Ray Equipment for Interventional Procedures

CAN/CSA-C22.2 No. 60601-2-44-03   Medical Electrical Equipment — Part 2-44: Particular Requirements for the Safety of X-Ray Equipment for Computed Tomography

CAN/CSA-C22.2 No. 60601-2-45-02   Medical Electrical Equipment — Part 2-45: Particular Requirements for the Safety of Mammographic X-Ray Equipment and Mammographic Stereotactic Devices

CAN/CSA-C22.2 No. 60601-2-46-01   Medical Electrical Equipment — Part 2-46: Particular Requirements for the Safety of Operating Tables

CAN/CSA-C22.2 No. 60601-2-47-03   Medical Electrical Equipment — Part 2-47: Particular Requirements for the Safety, Including Essential Performance, of Ambulatory Electrocardiographic Systems

CAN/CSA-C22.2 No. 60601-2-49-04   Medical Electrical Equipment — Part 2-49: Particular Requirements for the Safety of Multifunction Patient Monitoring Equipment

CAN/CSA-C22.2 No. 60601-2-50-03   Medical Electrical Equipment — Part 2-50: Particular Requirements for the Safety of Infant Phototherapy Equipment

CAN/CSA-C22.2 No. 60601-2-51-04   Medical Electrical Equipment — Part 2-51: Particular Requirements for the Safety, Including Essential Performance of, Recording and Analysis, Single Channel and Multichannel Electrocardiographs

Appendix B — Notes on Rules

Note: *This Appendix is an informative (non-mandatory) part of this Standard.*

The notes and diagrams in this Appendix are for information and clarification purposes only and apply to the following Rules:

Section 0

Δ **Object**

The safety provisions of this Code are not intended to limit installation methods to those specifically described by the rules in this Code. The safety objectives of this Code may be met by utilizing alternate installation methods based on the fundamental safety principles of IEC 60364-1.

Such alternate methods are intended only for industrial and similar installations where objective-based installation criteria are addressed under the provisions of safety management systems or equivalent programs developed between users (industrial plants, independent power producers, etc.) and the authorities adopting and enforcing this Code.

Chapter 13 of IEC 60364-1 offers the fundamental safety principles and is included in Appendix K.

Circuit

For the purposes of this Code, a circuit is generally considered to mean that portion of a wiring installation which is connected to the load side terminals of an ac or dc system and forms a complete path or paths through which electrical current is normally intended to flow, including utilization equipment. For example, a cable connecting the load side terminals of a three-phase circuit breaker up to and including utilization equipment is generally considered to be a circuit, as would be a similar three-wire single-phase installation.

Ground fault circuit interrupter

Class A, when applied to a ground fault circuit interrupter (GFCI), is an interrupter that will interrupt the circuit to the load when the ground fault current is 6 mA or more but not when the ground fault current is 4 mA or less (when the ambient air temperature is less than –5 °C or more than 40 °C, the minimum tripping current may be 3.5 mA instead of 4) in a time

(a) not greater than that given by the equation

$$T = \left(\frac{20}{I}\right)^{1.43}$$

where
T is in seconds; and
I is the ground fault current in rms milliamperes for fault currents between 4 mA and 260 mA; and

(b) not greater than 25 ms for ground fault currents over 260 mA.

In addition, a Class A GFCI is to be capable of interrupting the circuit to the load, in keeping with the above requirements if the identified circuit conductor (neutral) becomes inadvertently grounded between the interrupter and the load.

The prime function of a Class A GFCI is to provide protection against hazardous electric shocks from leakage current flowing to ground from defective circuits or equipment. It does not provide protection against shock if a person makes contact with two of the circuit conductors on the load side of the GFCI.

Class A GFCIs are marked "GROUND FAULT CIRCUIT INTERRUPTER CLASS A" or with an abbreviated form such as "GFCI CL A", "GFCI A", or "CL A" where the area available for marking makes the complete text impracticable.

Ladder cable tray

Although a single-rail cable tray by definition is not a ladder cable tray, it is subject to the same performance requirements as a ladder-type cable tray in CSA C22.2 No. 126.1 (NEMA VE 1).

Mobile home

The following is the complete definition of mobile home as defined in CAN/CSA-Z240.0.1:

Mobile home — a transportable, single- or multiple-section single family dwelling conforming to the CAN/CSA-Z240 MH Series at time of manufacture. It is ready for occupancy upon completion of set-up in accordance with required factory-recommended installation instructions.

Mobile industrial or commercial structure

Such structures are built specifically for commercial or industrial use, such as construction offices, bunk houses, wash houses, kitchen and dining units, libraries, TV units, industrial display units, laboratory units, and medical clinics.

Neutral

By definition, a neutral conductor of a circuit requires at least three conductors in that circuit. However, in the trade, the term "neutral conductor" is commonly applied to that conductor of a 2-wire circuit that is connected to a conductor grounded at the supply end. Care should therefore be used in the use of this term when applying the Code.

Non-combustible construction

The specific details for buildings of non-combustible construction are located in Part 3 of the *National Building Code of Canada*.

Outdoor location

Locations that are sheltered from the weather are not considered outdoor locations.

Park model trailer

The following is the complete definition of park model trailer as defined in CAN/CSA-Z241:

Park model trailer — a recreational vehicle unit that meets the following criteria:
(a)   it is built on a single chassis mounted on wheels;
(b)   it is designed to facilitate relocation from time to time;
(c)   it is designed as living quarters for seasonal camping and may be connected to those utilities necessary for operation of installed fixtures and appliances; and
(d)   it has a gross floor area, including lofts, not exceeding 50 m² when in the set-up mode, and having a width greater than 2.6 m in the transit mode.

Recreational vehicle

The following is the complete definition of recreational vehicle as defined in CAN/CSA-Z240.0.2 in the CAN/CSA-Z240 RV Series of Standards:

Recreational vehicle — a structure designed to provide temporary living accommodation for travel, vacation, or recreational use and to be driven, towed, or transported. Except for fifth-wheel trailers, it has an overall length not exceeding 12.5 m and an overall width not exceeding 2.6 m, where the width is the sum of the distances from the vehicle centreline to the outermost projections on each side. These dimensions exclude safety-related equipment such as side safety and warning lights and entry and exit handholds. Also excluded are water spray suppression attachments, load-induced tire bulge, and equipment used to secure cargo on a vehicle. These excluded items are allowed to extend not more than 100 mm when the vehicle is folded or stowed for transit. For a fifth-wheel trailer the maximum overall length is 11.3 m taken from the rear extremity to the front of the main body, measured at the floorline. Such structures include travel trailers, slide-in campers, camping trailers, motor homes, and fifth-wheel trailers:

> **Camping trailer** — a recreational vehicle built on its own chassis, having a rigid or canvas top and side walls, that may be folded or otherwise stowed for transit and that is designed to be towed behind a motor vehicle.

> **Fifth-wheel trailer** — a recreational vehicle designed to be coupled to the towing vehicle by a fifth-wheel-type coupler, through which a substantial part of the trailer weight is supported by the towing vehicle.

> **Motor home** — a recreational vehicle that is self-propelled. For purposes of applying the CAN/CSA-Z240 RV Series, this includes a van conversion containing at least one
> (a)   plumbing fixture;
> (b)   fuel-burning appliance; or
> (c)   120 V electrical component.

Slide-in camper — a recreational vehicle designed to be loaded onto, and unloaded from, the bed of a pick-up truck.

Travel trailer — a recreational vehicle designed to be towed behind a motor vehicle by means of a bumper or frame hitch.

System

Within this Code, the word "system" is used in many different contexts. However, within the context of electrical power distribution, a system is intended to mean an electrical installation in which the energy provided by that installation to utilization equipment is derived from a single energy source. For example, an electrical installation supplied from a transformer or bank of transformers can be considered a system; an installation supplied from a different transformer would be considered a different system.

Ventilated cable tray

Ventilated cable tray includes cable tray of the wire mesh type.

Section 2

Rule 2-026

As a condition of approval of certain types of electrical equipment, the manufacturer supplies instructions pertaining to its installation. It is of the utmost importance that the installer closely follow installation instructions supplied by the manufacturer to fulfill the terms of the approval agreement.

Rule 2-100

Evidence of approval may consist of either of the following:
(a)   the certification mark of the certification agency, usually in the form of a monogram or seal of that agency; or
(b)   the special inspection label or document of the authority having jurisdiction.

Δ   **Rule 2-102**

The intent of this Rule is that any molded case circuit breaker and molded switches subject to any undesirable conditions such as smoke, water or large fault currents greater than that for which the breaker is rated, shall be removed from service and destroyed. For the purpose of this Rule the replacement of interchangeable trip units or addition of contacts does not apply.

Rule 2-124

Specific requirements pertaining to penetration of fire separations in buildings can be found in Subsections 3.1.7 and 3.1.9 of the *National Building Code of Canada* or in the appropriate provincial/territorial legislation.

Rule 2-126

The flame spread requirements for wiring and cables in buildings are located in the *National Building Code of Canada* as follows:
(a)   combustible building construction — Article 3.1.4.3;
(b)   non-combustible building construction — Article 3.1.5.17; and
(c)   plenum spaces in buildings — Article 3.6.4.3.

The markings for wires and cables meeting the flame spread requirements of the *National Building Code of Canada* (without additional fire protection) are
(a)   FT1* — wires and cables that are suitable for installation in buildings of combustible construction; and
(b)   FT4** — wires and cables that are suitable for installation in
(i)   buildings of non-combustible and combustible construction; and
(ii)   spaces between a ceiling and floor, or ceiling and roof, that may be used as a plenum in buildings of combustible or non-combustible construction.

Communication and optical fiber cables marked CMP, CMR, CMG, CM, CMX, CMH, OFNP, OFCP, OFNR, OFCR, OFNG, OFCG, OFN, OFC, OFNH, OFCH, and communications and optical fiber cables marked FT4 have been found to meet the standard criteria for FT1.

**Communication and optical fiber cables marked CMP, CMR, CMG, OFNP, OFCP, OFNR, OFCR, OFNG, and OFCG have been found to meet the standard criteria for FT4.*

Wires and cables with combustible insulation, jackets, or sheaths that do not meet the above classifications but are located in
(a)   totally enclosed non-combustible raceways;
(b)   masonry walls;
(c)   concrete slabs;
(d)   a service room separated from the remainder of the building by a fire separation having not less than 1 h fire-resistance rating; or
(e)   totally enclosed non-metallic raceways conforming to Rule 2-128

may be considered to be in compliance with the *National Building Code of Canada* requirements relating to flame spread.

Δ **Rule 2-128**

The flame spread requirement for totally enclosed non-metallic raceways is located in Article 3.1.5.19 of the *National Building Code of Canada*. The *National Building Code of Canada* permits the use of totally enclosed non-metallic raceways for wires and cables, provided that the totally enclosed non-metallic raceways do not exceed 175 mm in outside diameter, or equivalent cross-sectional area, and conform to the vertical flame test requirement specified in Clause 6.16 of CSA C22.2 No. 211.0. The marking on totally enclosed non-metallic raceways to indicate compliance with the above flame test is "FT4".

Δ **Rule 2-130**

It is the intent of this Rule to protect totally enclosed non-metallic raceways and insulation of conductors against adverse effects from direct exposure to rays of the sun. Electrical conductors and non-metallic raceways marked for such application are suitable for installation and use for direct exposure to rays of the sun.

Rule 2-132

When insulation resistance or dielectric strength tests are performed, precautions should be taken to ensure that voltage-sensitive devices such as ground fault circuit interrupters are not subjected to voltages that will damage the device.

Δ **Rule 2-306**

NFPA 70E-2004, *Electrical Safety in the Workplace*, provides assistance in determining severity of potential exposure, planning safe work practices, and selecting personal protective equipment to protect against shock and arc flash hazards.

ANSI Z535.4-2002, *Product Safety Signs and Labels*, provides guidelines for the design of safety signs and labels for application to products.

IEEE 1584-2002, *Guide for Performing Arc-Flash Hazard Calculations*, provides assistance in determining the arc flash hazard distance and incident energy that workers may be exposed to from electrical equipment.

Rule 2-310

In order to obtain the path of travel required in Subrule (2), a second exit may be required.

Rule 2-318

Equipment considered as having relatively high losses will, in general, consist of generators, motors, transformers, and similar apparatus. Approximately 3.5 to 4.3 m³ of air per minute for each kilowatt of loss is normally required for ventilating 40 °C rise equipment. A value of 2.8 m³ of air per minute will have a temperature rise of approximately 18 °C when absorbing 1 kW loss.

The temperature rise of all such equipment is based on a 40 °C ambient temperature.

Δ **Rule 2-400**

Enclosures are not intended to protect against conditions such as condensation, icing, corrosion, or contamination that may occur within the enclosure or enter via the conduit or unsealed openings.

Ingress protection (IP) describes the degree of protection an enclosure provides from ingress of water or foreign bodies. (See Ingress Protection table in Notes to Rule 18-106(5)). For further information on the "IP" designations, refer to CAN/CSA-C22.2 No. 60529. The "IP" designations are supplemental to the Enclosure Types specified in Rule 2-400 and Table 65. There is no direct correlation between the "IP" designations and Enclosure

Types, and furthermore CAN/CSA-C22.2 No. 60529 does not cover environmental considerations such as resistance to corrosion or the effects of ultraviolet.

Section 4

Rule 4-004

The allowable ampacities of Tables 1, 2, 3, and 4 are based on temperature alone and do not take voltage drop into consideration (see Table D3).

The ampacities shown in Tables D8A to D11B inclusive and D12A to D15B inclusive have been determined using the calculation method in IEEE 835, for the cable arrangements shown in Diagrams B4-1, B4-2, B4-3, and B4-4.

For "stacked" arrangements of two single conductors per phase in parallel (one row located vertically over another row), it is recommended that they be obtained from Detail 5 of Table D8A or D8B (copper) or Table D9A or D9B (aluminum) for direct buried cables, or from Detail 2 of Table D10A or D10B (copper) or Table D11A or D11B (aluminum) for cables in underground raceways.*

For single-conductor metal-armoured and metal-sheathed cables in which the sheath, armour, or bonding conductors are bonded at more than one point, the derating factors of Rule 4-008 apply, unless the ampacity has been determined by detailed calculation according to the method outlined in Items (1)(d) and (2)(d).

It is recommended that ampacities for three single conductors per phase and for five single conductors per phase with spacings, directly buried in the earth, be selected from Table D8A or D8B or Table D9A or D9B for the installation configurations of Diagrams B4-1, Detail 5 and Detail 7 respectively. It is recommended that ampacities for three single conductors per phase and for five single conductors per phase installed in separate underground conduits in a single bank be selected from Table D10A or D10B or D11A or D11B for the installation configurations of Diagram B4-2, Detail 3 and Detail 4 respectively.*

It is recommended that ampacities for three-conductor cables directly buried in the earth be selected from Table D12A or D12B (copper) or D13A or D13B (aluminum) for installation configurations of Diagram B4-3, and those for three-conductor cables in separate underground raceways be selected from Table D14A or D14B (copper) or D15A or D15B (aluminum) for installation configurations of Diagram B4-4.

It is recommended that the ampacities of groups of conductors in twos, and two-conductor cables, be obtained from the ampacity Tables D12A to D15B inclusive for groups of three conductors, and three-conductor cables, for the appropriate spacings between groups and numbers of conductors in parallel.* The neutral conductor of a three-phase, four-wire system need not be counted in the determination of ampacities.

Underground ampacities for conductor temperatures of 75 °C and 60 °C respectively may be obtained by multiplying the appropriate ampacity at 90 °C conductor temperature from Tables D8A to D15B by the derating factor 0.886 (for 75 °C) or 0.756 (for 60 °C).*

Ampacities for underground installations at ambient earth temperatures other than the assumed values of 20 °C may be obtained by multiplying the appropriate underground ampacity obtained from Tables D8A to D15B by the following factor:

$$\sqrt{\left[(90 - T_{ae})/70\right]}$$

where
T_{ae} = the new ambient temperature.*

Where precisely calculated values are not available.

The ampacities of underground cables in conductor sizes smaller than 1/0 AWG should be selected in accordance with Subrules 4-004(1)(b) and (2)(b) for copper and aluminum conductors respectively.

Ampacities of underground installations based on conditions of use not as set out in the foregoing notes or the defined assumptions preceding them should either be justified by precise calculation according to the method of Subrule (1)(d) or (2)(d), or derived in accordance with Subrule (1)(b) or (2)(b).

Rules 4-004(1) and (2)

IEEE 835, *Standard Power Cable Ampacity Tables.*

As an alternative to Rule 4-004(1) and (2) and related Tables 1, 2, 3, and 4, IEEE tables may be used to compute the ampacities of copper and aluminum conductors if the electrical inspection department is agreeable and the data submitted as stipulated below is satisfactory to the department.

The following data should be supplied, in writing, where using IEEE tables is desired:
(a) **Cable:**
    (i)   type of cable and voltage rating;
    (ii)   maximum conductor temperature rating;
    (iii)   maximum continuous current rating; and
    (iv)   short-circuit rating;
(b) **Equipment:**
    (i)   voltage and current rating;
    (ii)   short-circuit rating; and
    (iii)   type of overcurrent devices with their characteristics;
(c) **Operating conditions:**
    (i)   nominal voltage of the system;
    (ii)   highest voltage of the system;
    (iii)   system frequency;
    (iv)   type of grounding, and where the neutral is not effectively grounded, the maximum permitted duration of ground fault conditions on any one occasion;
    (v)   rated current for continuous operation;
    (vi)   load factor;
    (vii)   maximum currents that may flow during short-circuits between phases and to ground; and
    (viii) maximum time for which short-circuit current may flow;
    Note: *The conductor heating during short-circuit is governed by the following:*

$$For\ copper: \frac{I^2}{A^2}\ t = 0.0297\ log_{10}\ \frac{(T_2 + 234)}{(T_1 + 234)}$$

$$For\ aluminum: \frac{I^2}{A^2}\ t = 0.0125\ log_{10}\ \frac{(T_2 + 228)}{(T_1 + 228)}$$

where
I   = short-circuit current in amperes
A   = conductor area in circular mils
t   = time of short-circuit in seconds
T_1 = initial conductor temperature in degrees Celsius. (The maximum conductor temperature rating is to be used.)
T_2 = final conductor temperature in degrees Celsius. (This temperature must not exceed the maximum short-circuit temperature rating of the cable.)

(d) **Underground raceway installation data:**
    (i)   length and profile of route including location of maintenance holes;
    (ii)   depth of duct bank;
    (iii)   number and geometric arrangement of the ducts if there are more than one;
    (iv)   spacing of the ducts;
    (v)   duct material including encasements;
    (vi)   thermal resistivity and kinds of soil along the route and whether data are based on measurement and inspection or only assumed;
    (vii)   maximum ground temperature if in excess of 20 °C at the duct depth; and
    (viii) proximity of other current-carrying cables or other heat sources with details;
(e) **Direct burial data:**
    (i)   length and profile of route;
    (ii)   depth of burial;
    (iii)   spacing of cables;

        (iv)   thermal resistivity and kinds of soil along the route and whether data are based on measurement and inspection or only assumed;

        (v)   maximum ground temperature if in excess of 20 °C at burial depth; and

        (vi)   proximity of other current-carrying cables or other heat sources with details;

(f)   **Conduit installation data:**

        (i)   length and profile of route;

        (ii)   number and geometric arrangement of conduits if there are more than one;

        (iii)   spacing of the conduits;

        (iv)   whether exposed to direct sunlight;

        (v)   details of ventilation for cables indoors or in tunnels;

        (vi)   proximity of other current-carrying cables or other heat sources with details; and

        (vii)   maximum air ambient temperature;

(g)   **Cables in free air:**

        (i)   length and profile of route;

        (ii)   cable spacing and if spacing is maintained;

        (iii)   whether exposed to direct sunlight;

        (iv)   details of ventilation for cables indoors or in tunnels;

        (v)   proximity of other current-carrying cables or other heat sources with details; and

        (vi)   maximum air ambient temperature.

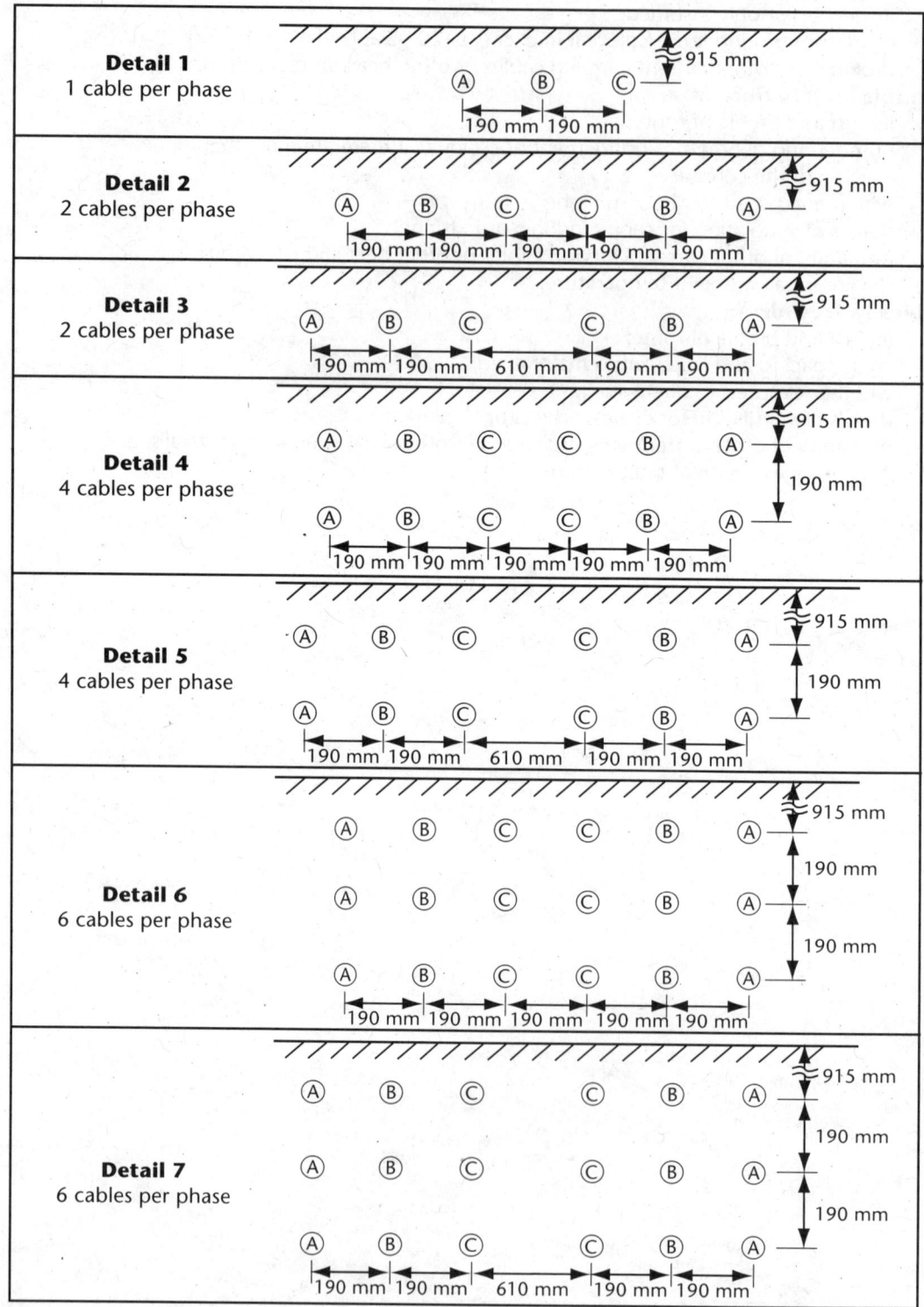

Diagram B4-1

Installation configurations — Direct buried

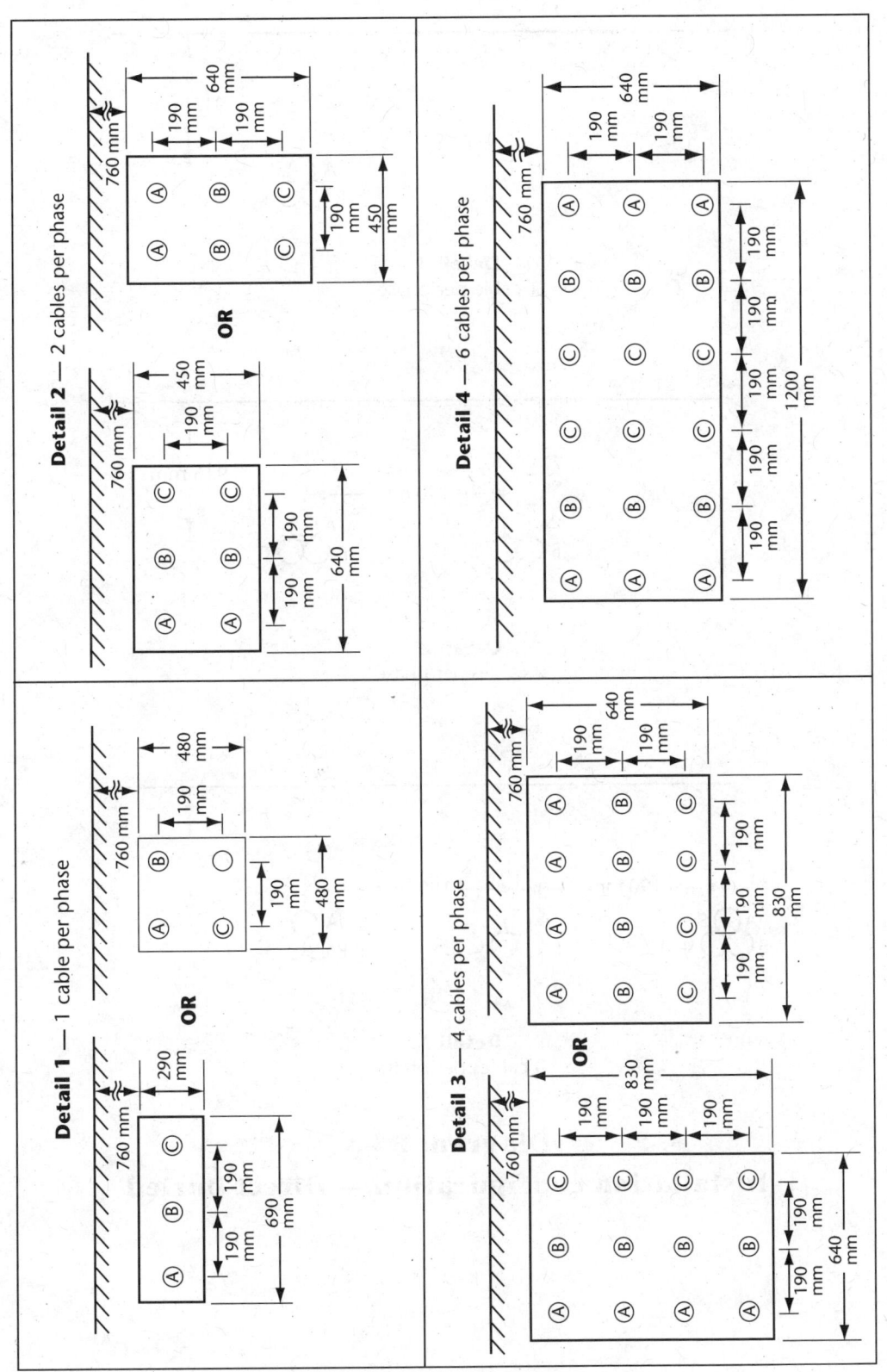

Diagram B4-2
Installation configurations — Conduit

Detail 1
1 cable per phase

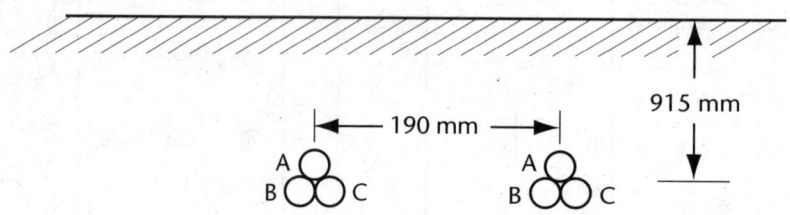

Detail 2
2 cables per phase

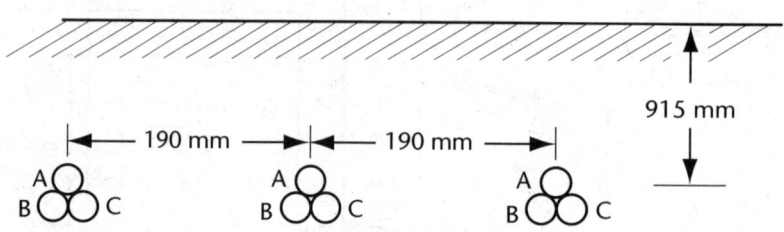

Detail 3
3 cables per phase

Diagram B4-3
Installation configuration — Direct buried

(Continued)

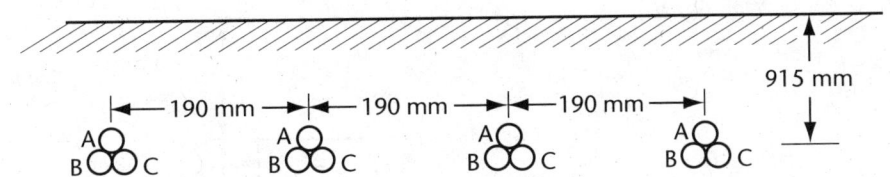

Detail 4
4 cables per phase

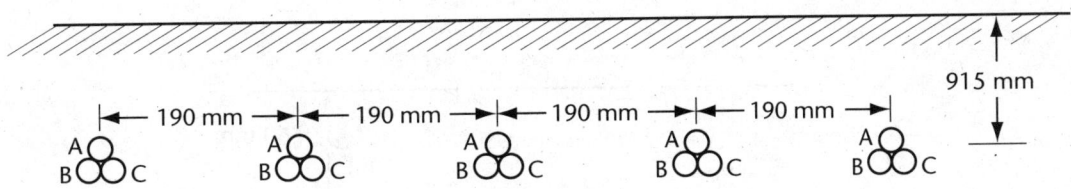

Detail 5
5 cables per phase

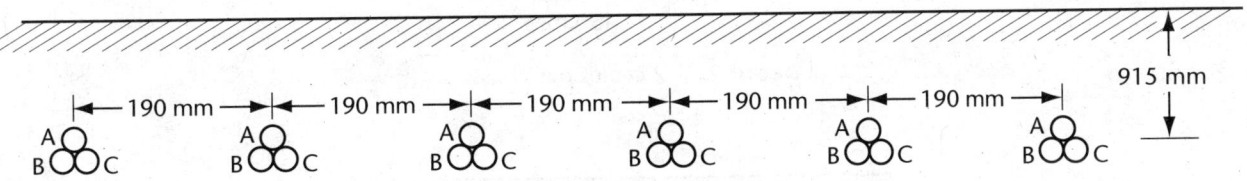

Detail 6
6 cables per phase

Diagram B4-3 (Concluded)

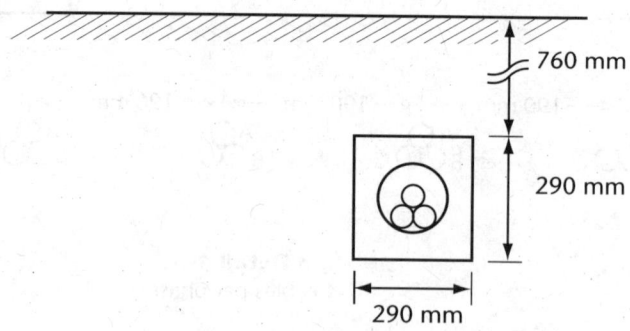

Detail 1 — 1 cable per phase

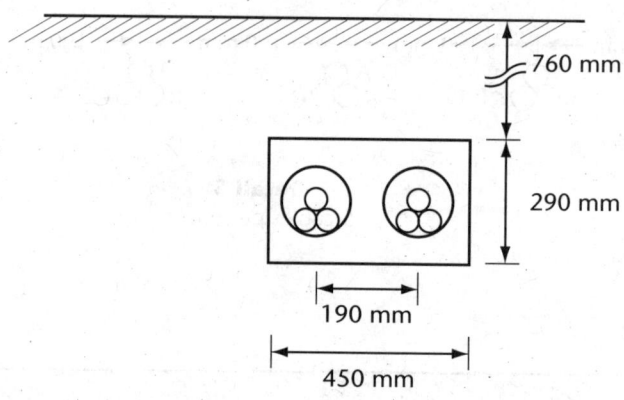

Detail 2 — 2 cables per phase

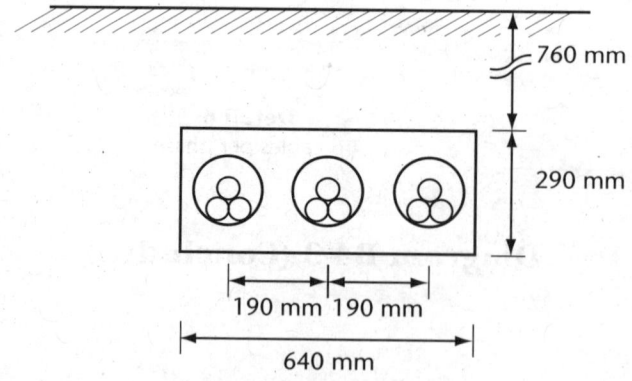

Detail 3 — 3 cables per phase

Diagram B4-4
Installation configuration — Raceway

(Continued)

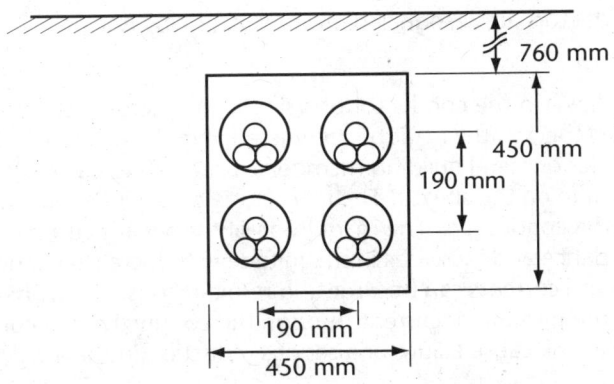

Detail 4 — 4 cables per phase

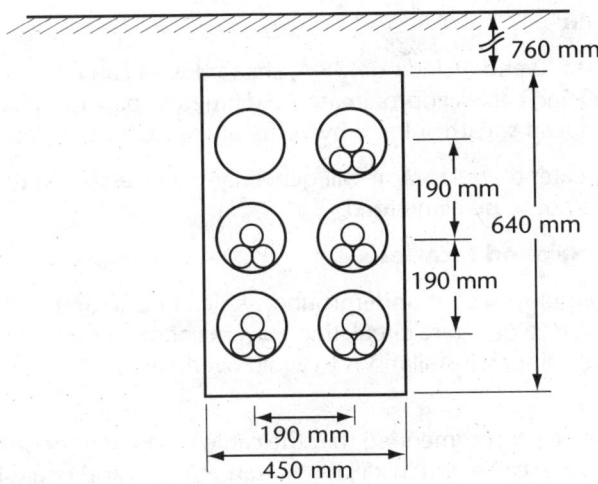

Detail 5 — 5 cables per phase

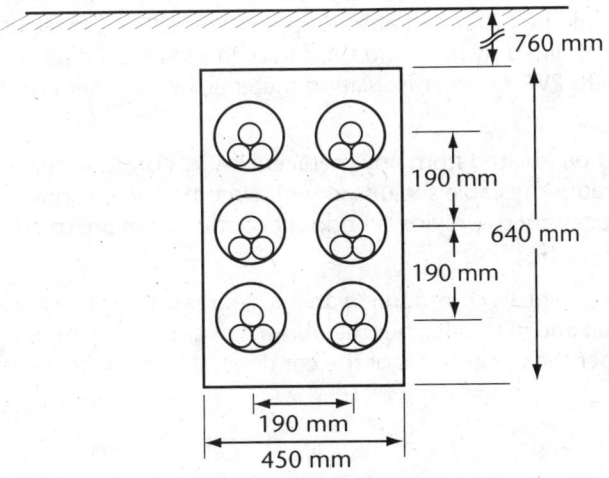

Detail 6 — 6 cables per phase

Diagram B4-4 (Concluded)

△ **Rule 4-004(17)**

Nickel and nickel-plated conductors are recognized in CSA C22.2 No. 124.

Rule 4-008

When an alternating current flows in the conductor of a single-conductor metal-sheathed cable, a voltage is induced on the sheath. Should the sheath circuit be completed, e.g., by grounding both ends or by connections to the sheaths of adjacent cables or metal building members, a circulating current called sheath current will flow as a result of the induced voltage on the sheath. The magnitude of the induced voltage is relative to the magnitude of the current in the conductors. The magnitude of the sheath current is a function of the induced voltage and the sheath impedance and increases in magnitude with increased conductor spacing within the range of typical spacings. Sheath currents can be large and result in considerable heating of the sheath. Coupled with the heat resulting from the passage of current through the conductor, the conductor insulation will be subjected to temperatures that will cause failure or a serious reduction in the life expectancy of the cable.

Cables carrying currents 200 A and less, with typical spacings in air, do not constitute a problem because induced voltages and sheath resistances are of the order to minimize sheath losses.

Single conductors in free air

With cables carrying currents up to and including 425 A, sheath losses can be reduced to tolerable levels, i.e., with no need to derate, by spacing cables approximately a diameter apart to minimize the effects of mutual heating while reducing the induced sheath voltage by virtue of the field cancellation effect at close spacing.

For cables carrying currents greater than 425 A, it will generally be necessary to derate to avoid overheating of the cable if sheath currents are not to be eliminated.

Single conductors in underground locations

The wider spacings generally employed with underground single-conductor installations, in comparison with cables in free air, can be expected to generate circulating currents of greater magnitude in the metallic sheath or armour. It is necessary to derate all such installations to avoid overheating of the cable if sheath currents are not to be eliminated.

If derating is the desired course, it is recommended that the cable manufacturer be consulted because this factor will depend on the type and size of cable and installation arrangements and could be more favourable than the factor given in Subrule (1)(a).

To prevent the flow of sheath currents, it is necessary to ensure that all paths by which they may circulate are kept open. Cable sheaths should be grounded at the supply end only and thereafter isolated from ground and each other. Isolation may be attained by installing the cables in individual ducts of insulating material, by employing cables jacketed with PVC or other insulating material, or by mounting the cables on insulated supports.

The sheath or sheaths should be isolated from any metal enclosures or other terminations at the load end that might bridge them. Because the cable sheaths in such circumstances cannot be used for bonding the electrical system, it will be necessary to provide a bonding conductor of adequate size for this purpose (see Rule 10-304(2)).

The phenomenon of sheath currents is common in varying degrees to single conductors enclosed in ferrous metals as in galvanized conduit and in non-ferrous metals such as copper, aluminum, and lead employed as cable sheaths and will occur whether the enclosure is of the continuous tube or the spiral-armoured type.

Section 6

Rule 6-102

The supply authority should be consulted on the number and location of the supply service(s).

Rule 6-112

Clearances for overhead conductors in this Code apply under the conditions existing at the time of installation rather than at maximum sag and are therefore greater than, but consistent with, the clearances specified in CAN/CSA-C22.3 No. 1.

Rule 6-112(4)

Components that meet the requirements of CSA C22.2 No. 82 include
(a)   tubular support members;
(b)   support clamps;
(c)   roof plate;
(d)   supply service attachments, e.g., racks;
(e)   service-entrance head; and
(f)   support member termination (i.e., at lower end).

An assembly considered to be installed in an acceptable manner is one where
(a)   the consumer's service does not exceed 200 A or 750 V;
(b)   the supply service does not exceed 30 m;
(c)   the maximum unguyed projection of the support member does not exceed 1.5 m and the cantilever load does not exceed 270 kg;
(d)   three support clamps are used, the upper one being located at the roof line and the two others being installed on the wall of the building;
(e)   a roof plate is provided to prevent entrance of moisture;
(f)   the supply conductors are attached on the mast to comply with Rule 6-116; and
(g)   a clearance of not less than 915 mm is provided between the roof and the supply service attachment, except that the clearance may be reduced to 600 mm to bottom of drip loops.

Rule 6-206(2)

The local regulatory authority may forbid the locking of service boxes in the ON position, for example, on construction sites.

Where the operating means of a service switch or circuit breaker is rendered inaccessible, it is recommended that a notice be displayed on the outside of the service box enclosure, room, or building advising of the location of the key to gain access to the service box operating means.

Rule 6-402(2)

The supply authority should be consulted regarding the acceptability of installing the metering equipment on the supply side of the service box.

Section 8

Rules 8-104, 62-114

Service boxes, fusible switches, circuit breakers, and panelboards not marked as suitable for continuous operation at either 80% or 100% of the rating of their overcurrent devices are considered to be suitable for continuous operation at 80%.

Rules 8-200, 8-202

If more than one electric range is involved, the initial range will be provided for according to Rule 8-200(1)(a)(iv) or Rule 8-202(1)(a)(v), and any subsequent ranges will be provided for by Rule 8-200(1)(a)(vi) or Rule 8-202(1)(a)(vi).

Rule 8-208

For the purpose of this Rule, a motel unit with cooking facilities may be considered as an apartment.

Section 10

Rule 10-106(2)

When a ground detection device is installed, it is intended that it will be identified as to its purpose and will be visible to those persons responsible for knowing the status of the system.

Δ Rules 10-204, 10-812, 10-814

The following types of system grounding are examples of grounded AC systems commonly used in Canada:
(a)   single-phase, 3-wire solidly grounded systems (see Figure 1);
(b)   3-phase, 4-wire solidly grounded systems (see Figures 2 and 3);
(c)   3-phase, 4-wire impedance grounded system (see Figure 4).

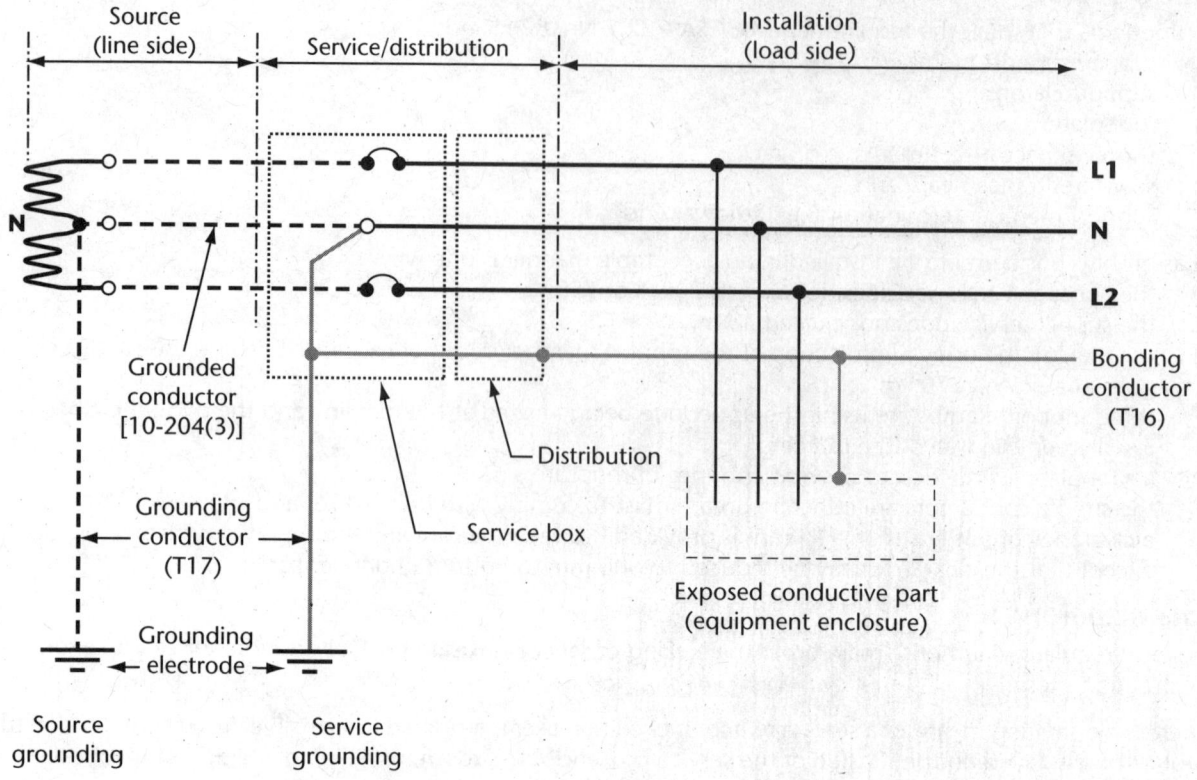

Notes:

(1) *Neutral and bonding conductor functions are combined in a single conductor (system grounded conductor) on the line side of the service [Rule 10-624(2)].*

(2) *Neutral (grounded circuit conductor) and bonding conductor function are separate on the load side of the service [Rule 10-624(1)].*

Figure 1
Single-phase 3-wire solidly grounded system (midpoint grounded)

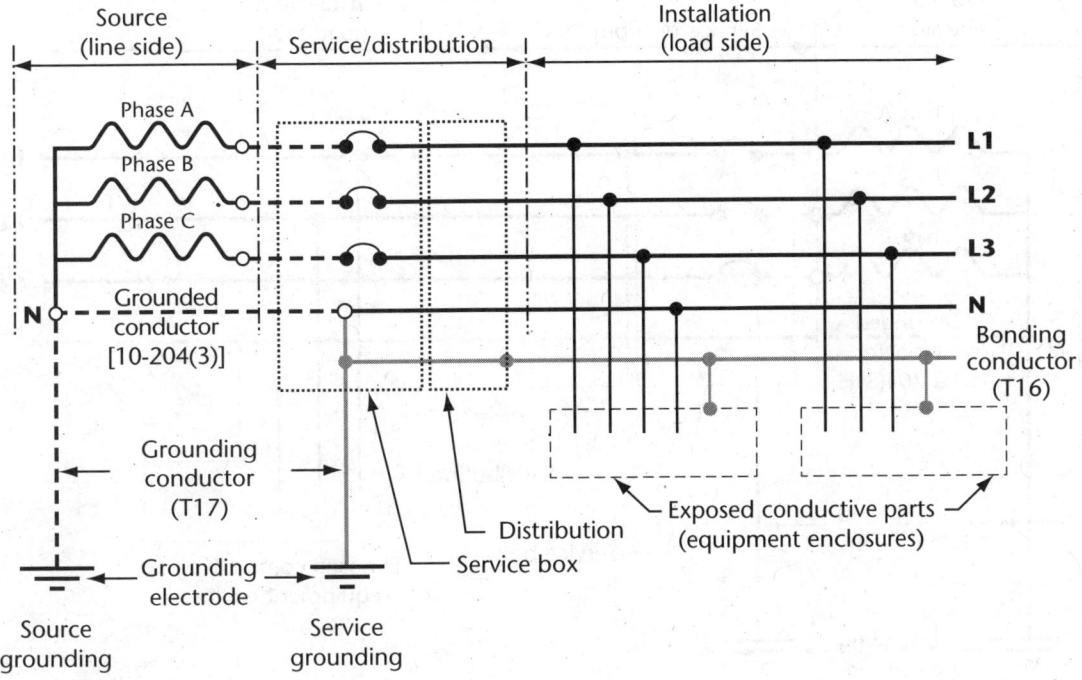

Notes:

(1) *Neutral and bonding conductor functions are combined in a single conductor (system grounded conductor) on the line side of the service [Rule 10-624(2)].*

(2) *Neutral (grounded circuit conductor) and bonding conductor function are separate on the load side of the service [Rule 10-624(1)].*

Figure 2
Three-phase 4-wire solidly grounded system (midpoint grounded)

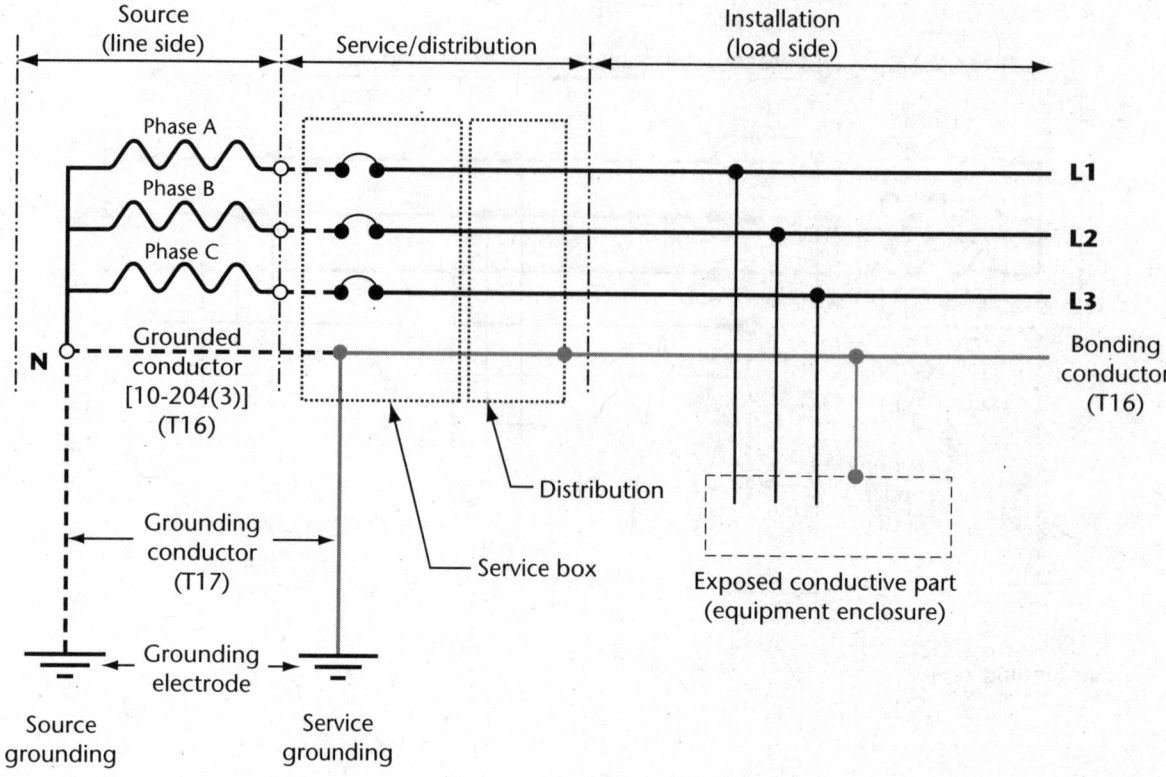

Notes:

(1) *The grounded conductor on the load side of the service functions as a bonding conductor with no distributed neutral throughout the system.*

(2) *The grounded conductor on the line side of the service (system grounded conductor) with no neutral currents is sized as specified for bonding conductors (Table 16).*

Figure 3
Three-phase 4-wire solidly grounded system with no neutral load (3-wire on load side) (midpoint grounded)

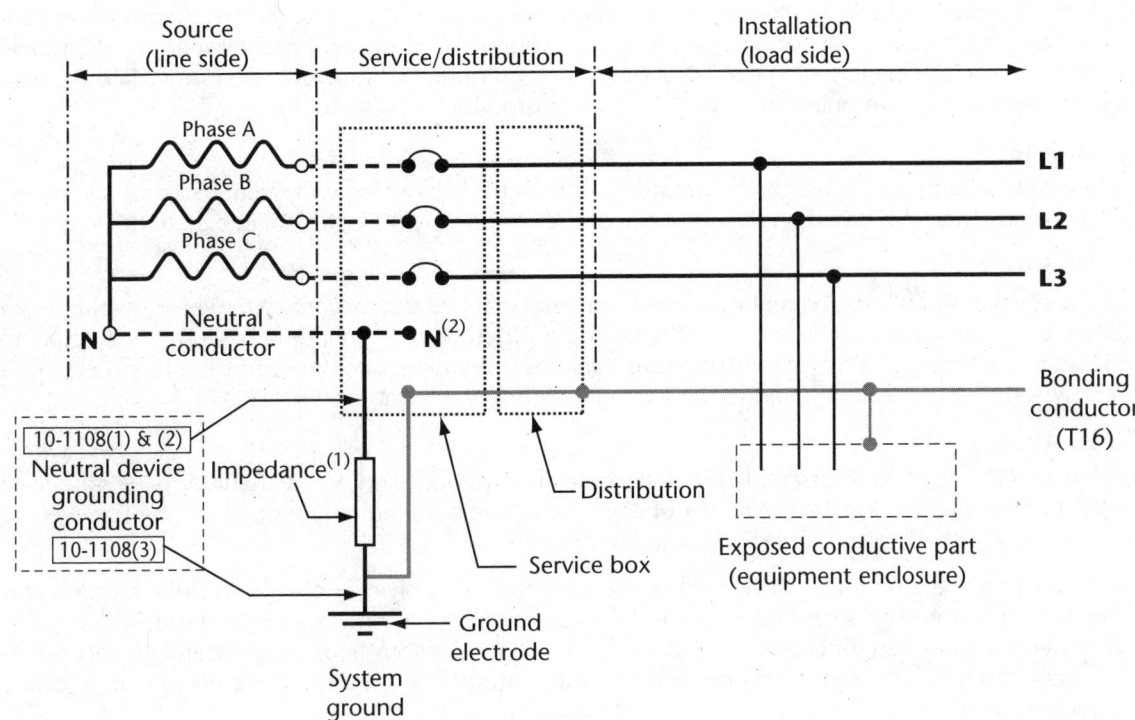

Notes:
(1) *System connected to ground via sufficiently high impedance.*
(2) *The neutral may or may not be distributed.*
(3) *See Subsection "Installation of Neutral Grounding Devices" (1100 series) of Section 10.*

Figure 4
Three-phase 4-wire impedance grounded system (midpoint grounded)

Rule 10-204(1)

The supply side of the disconnecting means is deemed to include all of the interior of the service box.

Δ **Rule 10-204(3)**

The term "grounded conductor" is a commonly used term in the Code that may serve different functions.

On the load side of the service disconnecting means, it serves only as the identified conductor intended to carry the unbalanced load (neutral currents) and can be referred to as the "grounded circuit conductor" (see Rules 10-204(1)(d) and 10-624(1)).

On the supply side of the service disconnecting means, it serves as a neutral conductor only in a single- or multi-phase system requiring a grounded circuit (neutral) conductor (see Rule 10-210). Where the grounded circuit conductor (neutral) is not intended to be used in a wiring system, the grounded conductor serves as a bonding conductor to carry fault currents to source. On the supply side of the service, the grounded conductor is also referred to as the "grounded service conductor" (see Rules 10-624(2)).

For the purposes of Rule 10-204, the term "grounded conductor" refers to the grounded conductor (or "system grounded conductor") on the supply side of the service that serves as a bonding conductor to carry fault currents and that may also serve as a grounded circuit conductor to carry neutral currents.

Like the bonding conductor, the system grounded conductor's primary function is to provide a low impedance path capable of withstanding any fault currents that may be imposed on it. In addition, the system grounded conductor may have to carry neutral or harmonic currents. Therefore, in addition to sizing the grounded conductor as specified in Table 16, further consideration for increasing the size of the system grounded conductor is required where line-to-neutral loads are present as prescribed in Rule 4-022. Consideration should also be given to the possibility of further increasing the size of the system grounded conductor to accommodate any non-linear loads that may impose harmonic currents on the system grounded conductor.

Rule 10-206

The intent of this Rule is to ensure that a grounding conductor is provided when required and is sized in accordance with Table 17, based on the ampacity of the conductors from the isolated system.

Rule 10-406

Exposed metal piping systems and metal building structural material may become energized by voltage gradients resulting from relatively high fault currents. The bonding together and grounding of all such equipment will provide additional safety. To eliminate dangerous voltages from appearing between the ground electrodes of different systems, under fault conditions, these should be bonded together.

Δ Rule 10-406(6)

Where the metallic floor panel assembly is not electrically continuous and where the floor panels are electrically connected to metallic supports at the corner of each panel, the bonding requirement can be achieved by ensuring every fourth metallic support is bonded to ground.

Where the floor panel assembly is electrically continuous, the assembly may only need to be bonded at one location. However, for large areas and to meet the requirement of an "effective equipotential plane", the floor area may need to be tested for continuity. For an electrically continuous floor panel assembly, ensuring that at least every 300 m^2 is bonded to ground in accordance with the Rule is deemed to provide an effective equipotential plane.

Rule 10-408(3)

Tools and appliances approved with a protective system of double insulation, or its equivalent, are marked with the words "double insulated" or the symbol ▣ as specified in CAN/CSA-C22.2 No. 0.

Rule 10-500

To have an impedance sufficiently low to
(a) facilitate the operation of the overcurrent devices in the circuit on the occurrence of a fault of negligible impedance from an energized or phase conductor to exposed metal; and
(b) limit the duration of the voltage above ground on this exposed metal
the complete fault path (line ground loop) of the grounding and bonding arrangement of the consumer's installation would normally have to be such that a current of not less than 5 times the rating of the overcurrent device protecting the circuit will flow on the occurrence of a fault of negligible impedance.

Δ Rules 10-700(1)(a), 10-700(4)

Manufactured grounding electrodes are those manufactured and certified to CSA C22.2 No. 41.

It is important that in-situ grounding electrodes provide an equivalent surface area contact with earth as do manufactured electrodes (see CSA C22.2 No. 41). Consideration should also be given to the effects that corrosion may have on the in-situ ground electrode impacting durability and life-expectancy. For example, an underground metal water piping system located at least 600 mm below finished grade and extending at least 3 m has traditionally been recognized as a suitable grounding electrode. Similarly, the metallic reinforcement of a concrete slab, concrete piling, or concrete foundation and iron pilings in significant contact with earth at 600 mm or more below finished grade have also been found to be suitable in-situ electrodes.

Any metallic material encapsulated with a non-conductive compound to protect it from corrosion would not meet the criteria for use as in-situ ground electrode.

Rule 10-706

Recommended practices for the installation of a lightning protection system including lightning rods, interconnecting conductors, and ground electrodes are given in the CAN/CSA-B72. Other national and international industry recognized standards on the subject of lightning protection may also be available.

Δ **Rule 10-806(4)**

When currents are imposed on the grounding conductor, the magnetic field encircling the conductor is increased by magnetic material surrounding the conductor unless the magnetic material is bonded at both ends. The increased magnetic field correspondingly increases the inductive voltage drop thus increasing conductor impedance.

Non-magnetic material such as PVC, aluminum, and some metal alloys are not affected by the magnetic field and accordingly need not be bonded at both ends.

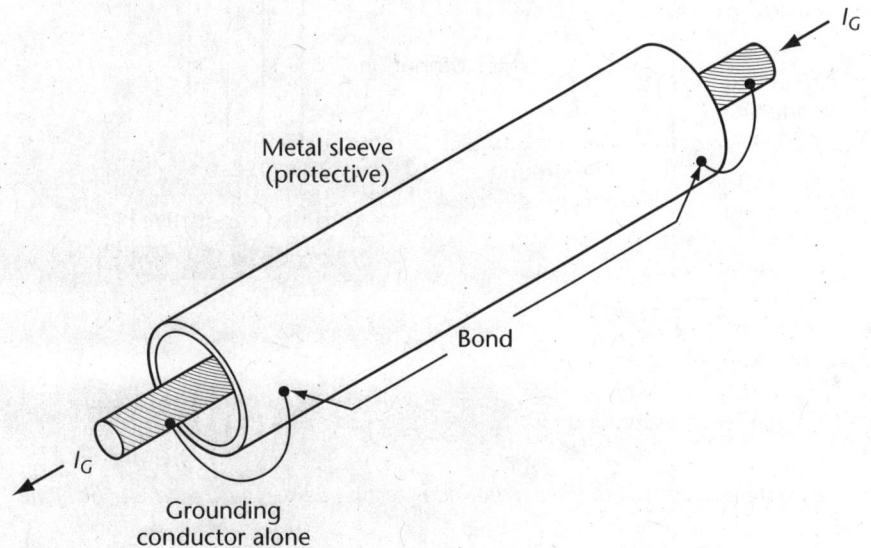

Δ **Rule 10-812(c)**

The following diagram depicts an AC system that is not grounded.

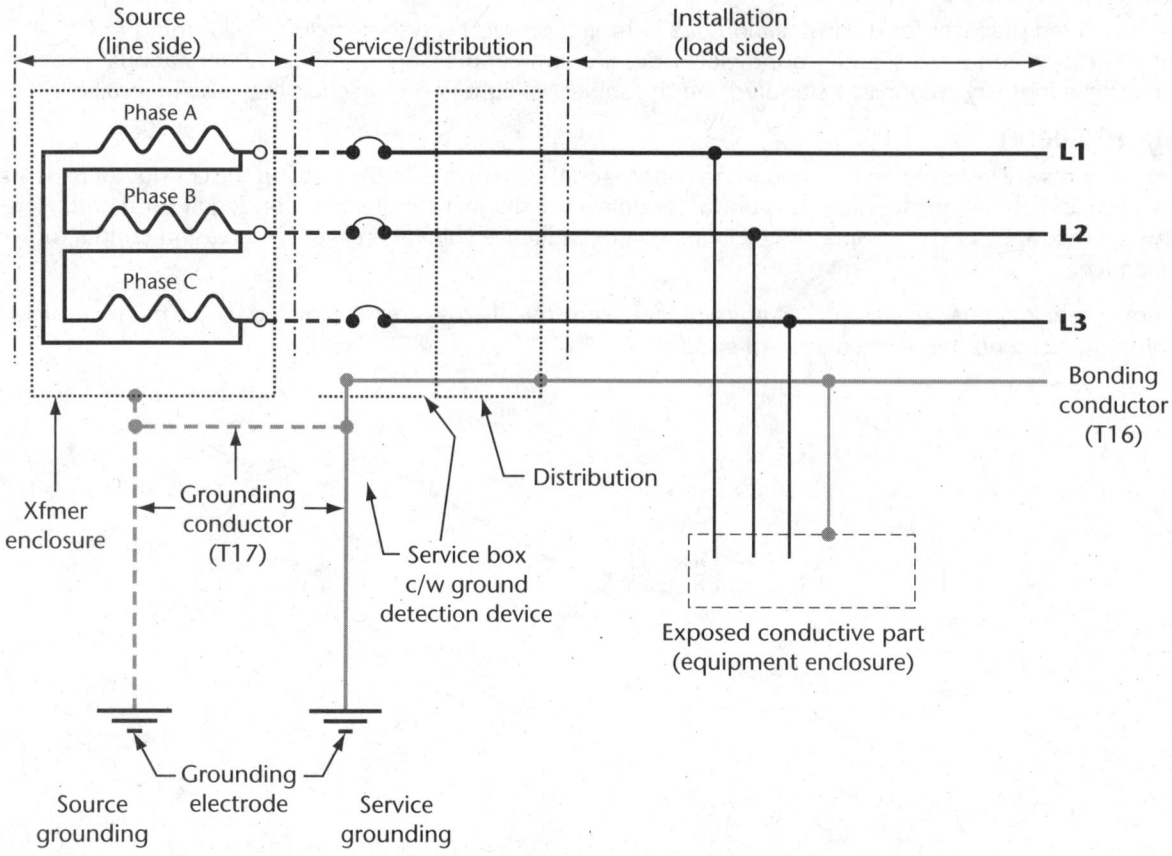

Note: *An ungrounded system is required to have a ground detection device in accordance with Rule 10-106(2).*

Rule 10-814(1)

When a raceway or cable sheath enclosing the circuit conductors is permitted to be used as a bonding conductor for the equipment being supplied, it is deemed to be of adequate size for the purposes of this Rule.

Rule 10-814(3)

An application of this Rule is as follows:

If the circuit conductors are paralleled in 3 separate raceways and are protected by an overcurrent device rated or set at 500 A, the bonding conductor size is determined by dividing 500 A by 3 and selecting from Table 16 the conductor adequate for 167 A.

For this application, use 200 A. Therefore the conductor size is No. 6 AWG copper or No. 4 AWG aluminum.

Rule 10-904

In making the point of attachment accessible, care should be taken to avoid exposing the grounding conductor and ground clamp to mechanical damage thereby jeopardizing assurance of a permanent ground. Ensuring that the grounding conductor and ground clamp are not exposed to mechanical damage by installing them below grade may be considered an acceptable method of "protection by location". Maintaining reasonable access where practicable can be achieved by having the connection just slightly below finished grade.

Rule 10-1100

Neutral grounding devices for impedance grounding an alternating current system include grounding resistors, grounding transformers, ground-fault neutralizers, reactors, capacitors, or combinations of these.

Rule 10-1102

Unless the installation is exceptional in some way, the ground grid system for the supply authority should always be interconnected with the consumer's ground grid system as outlined in Rule 10-204(3) for a grounded system. Where neutral grounding devices are employed, a grounded service conductor will not be available for this purpose and it may be necessary to install a separate conductor to interconnect the two ground grid systems.

Section 12

Rule 12-000

Reference should be made to the *National Building Code of Canada* or to appropriate sections of the provincial/territorial building codes regarding the installation and use of combustible electrical equipment such as raceways, boxes, and conductors.

Rule 12-012

Wooden planks, when buried in the ground, should be treated with a solution of pentachlorophenol or other suitable material as recommended by a manufacturer of wood preservatives. The use of creosote as a wood preservative in such installations is not recommended since it is known to damage rubber and thermoplastic insulations and will act as a catalyst in the corrosion of lead.

If polyethylene water pipe is used for mechanical protection for conductors and cables for direct earth burial under Subrule (3)(e), pipe in conformance with CAN/CSA-B137.1 is considered acceptable.

Rule 12-100

Table 19 indicates the maximum allowable conductor temperature for various types of building wires and cables. Where the surface temperature and/or the temperature on the insulation of conductors, cable assemblies, or raceway systems exceeds 90 °C, such assemblies are a potential fire hazard if installed adjacent to combustible material, and in such cases the assemblies should be relocated or supported in a manner to remove this potential hazard.

The low temperature marking on conductors indicates compliance with a test at that temperature, as specified in the product Standard, but does not guarantee safe installation at that temperature.

Care should be taken with the installation of cables at low temperatures. Measures that should be considered include pre-conditioning at higher temperatures prior to installation, and avoidance of mechanical shock from dropping the cable, unreeling the cable too quickly, or bending sharply or quickly at bends. Manufacturers should be consulted when further information is desired.

Rule 12-108

The following configurations are acceptable for conductors in parallel to minimize the difference in inductive reactance and the unequal division of current. Additional conductors in parallel should be arranged in repetitive configurations of those illustrated. Factors other than reactance that affect load sharing and should be considered are as outlined in Subrule (1).

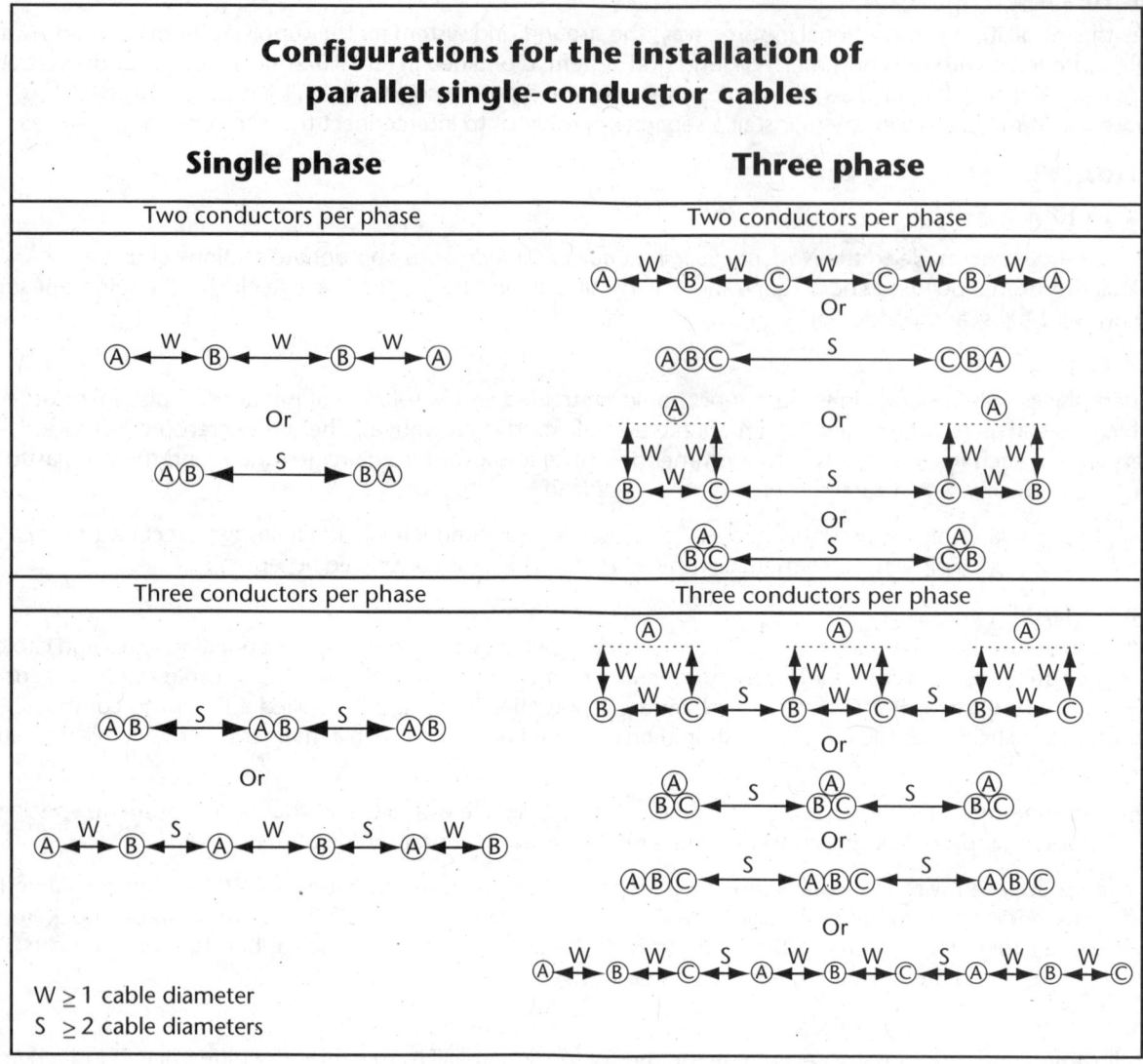

Configurations for the installation of parallel single-conductor cables

Single phase · **Three phase**

Notes:
(1) *Neutral may be located outside of the above groups in the most convenient location.*
(2) *All above configurations may not result in equal ampacity on division of current for all cable systems (see also Rule 4-008).*

Rule 12-116

Solderless wire connectors or their cartons are identified as follows:
(a) "solid" or equivalent for conductor sizes Nos. 18, 16, and 8 AWG and larger;
(b) "stranded" or equivalent for conductor sizes Nos. 14 to 10 AWG; and
(c) if not so marked, they are suitable for connecting stranded conductors in sizes Nos. 18, 16, and 8 AWG and larger, and solid conductors in sizes Nos. 14 to 10 AWG.

A wire connector marked as indicated in Item (a) or (b) is suitable for use with the marked type of construction only.

Rule 12-504

The specific details for buildings of non-combustible construction are located in Part 3 of the *National Building Code of Canada* or in the appropriate provincial/territorial legislation.

Rule 12-506

Specific requirements pertaining to materials suitable as thermal barriers can be found in Article 3.1.5.11 of the *National Building Code of Canada*.

Rule 12-602

The steel wire armour (SWA) used in cables features inherently different mechanical characteristics than conventional interlocked armour. Steel wire armour, due to its physical structure, provides high tensile strength but may provide less mechanical protection than interlocked armour, for example, when subjected to puncture. The user should consider the particular application when selecting SWA for cables.

Rule 12-714

Mineral-insulated cable has either a copper, aluminum, or stainless steel sheath. Box connectors suitable for use with the particular sheath material should therefore be used.

Rule 12-802

The bottom shield may or may not be incorporated as an integral part of the whole. The top shield and the metal tape may be two separate components or may be a single integral component of the Type FCC system.

Rule 12-814

Tapes having a conductive surface in intimate electrical contact with metal shields throughout the Type FCC system are considered to be bonded when approved for the purpose.

Rule 12-934

Where raceways pass across structural expansion or control joints, possible relative movement that could damage the raceway should be accommodated in the design.

Rules 12-942, 12-3000(3)

The intent of this Rule is to protect metal raceways and boxes against corrosion in concrete slabs of unheated parkades and similar structures, where permeation of salt represents a corrosion hazard. Users of this Code should be aware that CSA S413 restricts the use of metal raceways and boxes embedded in concrete slabs where they are subject to corrosion.

Rule 12-1006

When field threading rigid metal conduit, see ANSI/ASME B1.20.1.

Rules 12-1012, 12-1118, 12-1214

The following provides information on the linear expansion of materials where extreme temperature changes are encountered:

Coefficient of linear expansion (mm per m per °C)	
Wood	0.0050
Brick	0.0047 to 0.0090
Steel conduit	0.0114
Rigid RE conduit	0.0108 to 0.0135
Concrete	0.0144
Aluminum conduit	0.0220
PVC conduit	0.0520
HFT conduit	0.0700

Example: The change in length, in millimetres, of a run of rigid PVC conduit, due to the maximum expected variation in temperature, is found by multiplying the length of the run, in metres, by the maximum expected temperature change, in degrees Celsius, and by the coefficient of linear expansion.

For a 20 m run of rigid PVC conduit when the minimum expected temperature is –40 °C and the maximum expected temperature is 30 °C, the change in length is

$20 \times (40 + 30) \times 0.0520 = 73$ mm

Rule 12-1014

The maximum permitted number of conductors in a raceway is based on the actual measured dimensions of the raceways and the wires and cables. Where calculations of the maximum number of permitted conductors in a given raceway are performed, based on supplied dimensions or from standards for the products, they should be validated by measurement of products concerned before installation.

Calculated maximum permissible number of conductors of the same size in a given raceway provides an alternative basis for determination of the maximum number of permitted conductors, provided that the dimensions of such raceways and conductors are derived from Tables 9 and 10. The maximum numbers of single conductors of one size permitted in a conduit or raceway as determined from Table 6 are based on the aforementioned tables, with no allowance for bare bonding or grounding conductors.

Dimensions of bare bonding or grounding conductors, such as bonding conductors that are required for some raceway installations under applicable Code Rules, may be obtained from Table D5. They should be verified by measurement before installation.

Rules 12-1104, 12-1154, 12-1508

Tests show that 90 °C conductors, continuously loaded, under conditions of 50% fill and 30 °C ambient, do not result in a temperature exceeding 75 °C. Conductors having insulation ratings in excess of 90 °C may be used in PVC conduit, provided that the ampacity is derated to 90 °C.

Rule 12-1108

When bending PVC conduit, an open flame should not be used.

Rule 12-1118

Refer to the Note in this Appendix for Rule 12-1012. In the example given, the change in length is 73 mm and therefore one or more expansion joints would be required depending upon the maximum range that a particular joint is capable of handling.

Rule 12-1150

Precautions should be taken to ensure that no pour of concrete and/or its reinforcement exert a load on the conduit that will render the conduit unsuitable for use. Too high a pour may cause failure through collapse on overloading when the concrete is in a wet (uncured) state.

Rule 12-1154

See the Note to Rule 12-1104.

Rule 12-1156

In general, a thermostatically controlled heat gun may be used for field bends on trade sizes up to 53 trade size. For sizes 27 to 53 trade sizes, springs or equivalent devices should be used in conjunction with the heat gun to prevent reduction of the internal diameter. For sizes larger than 53 trade size, special jigs, moulds, springs, and heating arrangements are required.

Rule 12-1156

See the Note to Rule 12-1108.

Rule 12-1158

When connecting rigid Types EB1 and DB2/ES2 PVC conduit to conduit made of materials other than PVC through the use of a taper threaded connection, it is preferable that a female threaded adapter be made from the same PVC material as the PVC conduit and the male threaded adapter be made from materials suited to the other than PVC material.

Rule 12-1204

Specific requirements pertaining to use of combustible conduits and tubing in buildings can be found in Articles 3.1.4.3, 3.1.5.15, 3.1.5.17, and 3.1.5.19 of the *National Building Code of Canada* or in the appropriate provincial/territorial legislation.

Rule 12-1214

See the Note to Rule 12-1012.

Rule 12-1508

See the Note to Rule 12-1104.

Rule 12-1602

See Rules 2-124 and 2-128.

Rule 12-1606

In applying this Rule, the "minimum available cross-sectional area" is the minimum cross-sectional area of the surface raceway minus the maximum cross-sectional area of any device installed in the surface raceway that projects into the surface raceway.

Rule 12-2200

Recommended installation requirements are available from the manufacturer of cable trays. Additional points to consider include the following:
(a)   The ideal support point for cable tray is at the one-quarter span point.
(b)   There should not be more than one joint between support points.
(c)   Allowance for wind and snow loading should be included within the maximum design load.
(d)   Some fittings (particularly horizontal elbows) may require additional support, depending on the loading.

Rule 12-2202

Particular concerns that should be addressed in the installation of cable in cable tray are
(a)   protection by elevation or other means at traffic areas;
(b)   protection from falling objects through the use of covers, shields, or other means;
(c)   protection from radiant heat by heat shields, insulation, or other means;
(d)   protection from people walking on cables in trays by covers, guards, location, or other means; and
(e)   protection from movable objects, stored goods, or other material by elevation, covers, guards, or other means.

Rule 12-3000

Sealing around outlet boxes and wires and cables to provide an air barrier may be required. Requirements for air and vapour barriers are in the *National Building Code of Canada*, Subsections 9.25.3 and 9.25.4.

Rule 12-3000(3)

See the Note to Rule 12-942.

Rule 12-3022(7)

In addition to the use of non-ferrous box connectors, locknuts, and bushings, inductive heating caused by single-conductor cables entering separate openings in ferrous metal may be minimized by
(a)   cutting a slot in the metal between the individual holes through which the individual conductors pass; or
(b)   passing all the conductors in the circuit through separate openings in a plate of non-ferrous metal or insulating material that are sufficiently large for all conductors of the circuit.

Section 14

Rule 14-012

It is the intent of this Rule to ensure that the overcurrent protective and control devices, the total impedance, and other characteristics of the circuit to be protected are selected and coordinated so that the circuit protective devices will clear a fault without extensively damaging the electrical components of the circuit.

Interrupting ratings of overcurrent devices (circuit breakers or fuses) are 5000 A symmetrical maximum for circuit breakers rated 100 A or less and 250 V or less, and 10 000 A symmetrical maximum for circuit breakers rated above 100 A or above 250 V unless otherwise marked. The interrupting rating of fuses is 10 000 A symmetrical maximum unless otherwise marked.

Rule 14-014

A series rated system is one in which either a circuit breaker or a fuse is in series with a downstream circuit breaker that has an interrupting rating less than the fault current available at the line terminals of the upstream overcurrent device. The upstream device always has an interrupting rating at least equal to the available fault current.

This series combination is tested and approved at the higher rating in accordance with special requirements for series rated devices in the CSA Standards for the equipment involved (e.g., circuit breakers, panelboards, or metering equipment).

The tests verify that the combination acts together to safely clear a fault up to the maximum rating of the line side overcurrent device.

The downstream equipment is marked, as a part of its electrical rating, with its series rating and with the specific upstream overcurrent device required to achieve the series rating.

Where motors are connected in the system between the series connected devices, any significant motor contribution to the fault current should be considered. It is generally agreed that the contribution of asynchronous motors to the short-circuit current may be neglected if the sum of the rated currents of motors connected directly to the point between the series connected devices is 1% or less of the interrupting rating of the downstream circuit breaker. (See IEC 60781.)

Δ **Rule 14-100**

This Rule applies only to conductors interconnecting electrical equipment. It does not apply to overcurrent protection of electrical equipment as required by other Rules of the Code.

Rule 14-102, Diagram 3

It is recognized that ground fault protection may be desired for circuits other than those described in this Rule.

Ground fault protective equipment at the supply will make it necessary to review the overall system for proper coordination with other overcurrent protection. Additional ground fault protective equipment may be needed on feeders and branch circuits where maximum continuity of electrical supply to the remainder of the system is required.

It should be noted that with disconnecting devices located as shown in Diagram 3, no protection is given for faults between the transformer and the disconnecting device. If this protection is required, the primary disconnecting device must be tripped.

In any ground fault protective scheme, the protective equipment should be applied so that availability of the external tripping power (if required) is assured whenever the circuit being protected is energized.

It should be noted that ground fault relays are usually factory-set at the lowest current and shortest time settings available to ensure against unnecessary equipment damage during early stages of construction. These settings should be adjusted to the intended settings prior to final commissioning of the equipment.

Multi-fed installations with ties and interconnected neutrals may be grounded at more than two locations. In such cases, it will be the responsibility of the system designer to ensure that tripping occurs only on the main breaker associated with the supply affected by the ground fault (see Items 3 and 4 of Diagram 3). It is also recognized that for purposes of selectivity and continuity of service, many installations, in addition to utilizing multi-feeds for systems, also use multi-stage ground fault protection as described in Subrule (8). This results in complex schemes. In both of these special cases, the designer should be required to submit data to the inspection department showing that such considerations have been made in the design of the system involved.

Rule 14-104

Although Item (a) permits overcurrent devices larger than the conductor ampacity, they are restricted in rating or setting to the upper limits of Table 13. The conductor size should be determined and then Table 13 consulted only when an overcurrent device of the exact rating is not available. In other words, Table 13 is the last resort, although the use of time-delay fuses or fuses referred to in Rule 14-212(b) will assist in selecting a fuse of proper rating due to their availability in a greater range of values.

Δ **Rule 14-114**

Supplementary overcurrent protectors used as components of some appliances and equipment are not suitable for the protection of branch circuit conductors.

Rule 14-204

There are two types of non-interchangeable plug fuses currently available, i.e., Types C and S as described in CSA C22.2 No. 59.1.

Rule 14-212

The fuse "Classes" are cross-referenced to the former "Form" and "HRC" fuse designations in the following table:

CAN/CSA-C22.2 No. 248.1 to 248.15, published in 1994, 1996, and 2000	CSA C22.2 No. 106-M1985, CAN/CSA-C22.2 No. 106-M92, and C22.2 No. 59.1-M1987	CSA C22.2 No. 106-1953 and Electrical Bulletin No. 832-1971
Class J	HRCI-J	HRC Form I
Class R	HRCI-R	HRC Form I
Class T	HRCI-T, HRC-T	HRC Form I
Class CA	HRCI-CA	HRC Form I
Class CB	HRCI-CB	HRC Form I
Class CC	HRCI-CC	HRC Form I
—	HRCI-Misc***	HRC Form I
Class G*	—	—
Class K*	—	—
Class H**	—	—
Class L	HRC-L	Class L
Class C	HRCII-C	HRC Form II
—	HRCII-Misc***	HRC Form II

Class G and Class K fuses did not formerly have a fuse classification.
**Class H fuses were formerly referred to as "standard fuses", but are now marked as "Class H" in accordance with CAN/CSA-C22.2 No. 248.6 and CAN/CSA-C22.2 No. 248.7.*
***"Misc" (miscellaneous) fuse designations will remain in use for fuses that meet the requirements of CAN/CSA-C22.2 No. 106 and that do not have a "Class" designation.*

Rule 14-302

The voltage ratings referred to in Item (b) covering single-pole circuit breakers suitable for use with handle ties appear on each breaker, e.g., 120/240 V.

Rule 14-508

The requirements for general use ac/dc switches are given in CSA C22.2 No. 111.

Rule 14-510

The requirements for manually operated general use ac switches are given in CSA C22.2 No. 111.

Rule 14-512

The requirements for manually operated general use 347 V ac switches are given in CSA C22.2 No. 111.

Section 16

Rule 16-200

A primary battery for Class 2 power sources consists of one or more cells electrically connected under the same cover with each cell producing an electrical current by an electro-chemical reaction that is not reversible, that is, non-rechargeable.

Rule 16-210

Both LVT and ELC are approved cables for Class 2 circuit applications under the conditions outlined in Table 19. Type ELC conductors do not have an overall protective jacket and are further limited in use under this Rule to certain Class 2 circuits operating at 30 V or less, such as doorbells, intrusion devices, etc., in dwelling units in buildings of combustible construction. Type ELC is not permitted for the wiring of circuits related to fire safety, such as fire alarms or smoke alarm devices.

Rule 16-222(1)(a)

With respect to the acceptance of equipment for connection to Class 2 circuits operating at not more than 42.4 V peak or dc, consideration should be given to the fact that while Class 2 circuits limit the power that can be dissipated in the circuit continuously, this power is more than sufficient to be a fire hazard if dissipated in a fault within improperly designed equipment, e.g., shorted turns in a coil.

Section 18

Rule 18-000

Through the exercise of ingenuity in the layout of electrical installations for hazardous locations, it is frequently possible to locate much of the equipment in less hazardous or in non-hazardous locations and thus to reduce the amount of special equipment required. It is recommended that the authority enforcing this Code be consulted before such layouts are prepared.

To assist users in the proper design and selection of equipment for electrical installations in hazardous locations, numerous reference documents are available. The following are 3 tables that list documents most commonly referenced.

Table A
Documents generally applicable to all classes of hazardous locations

Publishing organization	Reference publication
CSA	CAN/CSA-C22.2 No. 130, *Requirements for Electrical Resistance Heating Cables and Heating Device Sets* C22.2 No. 137, *Electric Luminaires for Use in Hazardous Locations* C22.2 No. 145, *Motors and Generators for Use in Hazardous Locations* CAN/CSA-C22.2 No. 157, *Intrinsically Safe and Non-Incendive Equipment for Use in Hazardous Locations* C22.2 No. 159, *Attachment Plugs, Receptacles, and Similar Wiring Devices for Use in Hazardous Locations: Class I, Groups A, B, C, and D; Class II, Group G, in Coal or Coke Dust, and in Gaseous Mines* C22.2 No. 174, *Cables and Cable Glands for Use in Hazardous Locations* (These standards are also listed in Appendix A.)
IEC CSA (adopted IEC)	C22.2 No. 60529, *Degrees of protection provided by enclosures (IP Code)*
ISA	RP 12.6, *Wiring Practices for Hazardous (Classified) Locations — Instrumentation — Part 1: Intrinsic Safety*
NFPA	No. 70, *National Electrical Code* No. 91, *Standard for Exhaust Systems for Air Conveying of Vapors, Gases, Mists, and Noncombustible Particulate Solids* No. 496, *Standard for Purged and Pressurized Enclosures for Electrical Equipment* No. 505, *Fire Safety Standard for Powered Industrial Trucks Including Type Designations, Areas of Use, Maintenance, and Operation*

Table B
Documents applicable specifically to Class I hazardous locations

Publishing organization	Reference publication
CSA	C22.2 No. 22, *Electrical Equipment for Flammable and Combustible Fuel Dispensers* C22.2 No. 30, *Explosion-Proof Enclosures for Use in Class I Hazardous Locations* C22.2 No. 152, *Combustible Gas Detection Instruments* C22.2 No. 213, *Non-incendive Electrical Equipment for Use in Class I, Division 2 Hazardous Locations* (These standards are also listed in Appendix A.)
CSA (adopted IEC)	CAN/CSA-E60079-0, *Electrical Apparatus for Explosive Gas Atmospheres — Part 0: General Requirements* CAN/CSA-E60079-1, *Electrical Apparatus for Explosive Gas Atmospheres — Part 1: Flameproof Enclosures "d"* CAN/CSA-E60079-2, *Electrical Apparatus for Explosive Gas Atmospheres — Part 2: Electrical Apparatus — Type of Protection "p"* CAN/CSA-E60079-5, *Electrical Apparatus for Explosive Gas Atmospheres — Part 5: Powder Filling "q"* CAN/CSA-E60079-6, *Electrical Apparatus for Explosive Gas Atmospheres — Part 6: Oil-Immersion "o"* CAN/CSA-E60079-7, *Electrical Apparatus for Explosive Gas Atmospheres — Part 7: Increased Safety "e"* CAN/CSA-E60079-11, *Electrical Apparatus for Explosive Gas Atmospheres — Part 11: Intrinsic Safety "i"* CAN/CSA-E60079-15, *Electrical Apparatus for Explosive Gas Atmospheres — Part 15: Electrical Apparatus with Type of Protection "n"* CAN/CSA-E79-18, *Electrical Apparatus for Explosive Gas Atmospheres — Part 18: Encapsulation "m"* (These standards are also listed in Appendix A.)
API	API 500, *Recommended Practice for Classification of Locations for Electrical Installations at Petroleum Facilities Classified as Class I, Division 1 and Division 2* API 505, *Recommended Practice for Classification of Locations for Electrical Installations at Petroleum Facilities Classified as Class I, Zone 0, Zone 1, and Zone 2*
IEC	60079-1A, *Electrical apparatus for explosive gas atmospheres — Part 1: Flameproof enclosures "d"* 60079-4, *Electrical apparatus for explosive gas atmospheres — Part 4: Method of test for ignition temperature* 60079-4A, *Electrical apparatus for explosive gas atmospheres — Part 4: Method of test for ignition temperature — First supplement* 60079-10, *Electrical apparatus for explosive gas atmospheres — Part 10: Classification of hazardous areas* 60079-12, *Electrical apparatus for explosive gas atmospheres — Part 12: Classification of mixtures of gases of vapours with air according to their maximum experimental safe gaps and minimum igniting currents* 60079-13, *Electrical apparatus for explosive gas atmospheres — Part 13: Construction and use of rooms or buildings protected by pressurization* 60079-14, *Electrical apparatus for explosive gas atmospheres — Part 14: Electrical installations in hazardous areas (other than mines)* 60079-16, *Electrical apparatus for explosive gas atmospheres — Part 16: Artificial ventilation for the protection of analyser(s) houses* 60079-17, *Electrical apparatus for explosive gas atmospheres — Part 17: Inspection and maintenance of electrical installations in hazardous areas (other than mines)* 60079-19, *Electrical apparatus for explosive gas atmospheres — Part 19: Repair and overhaul for apparatus used in explosive atmospheres (other than mines or explosives)* 60079-20, *Electrical apparatus for explosive gas atmospheres — Part 20: Data for flammable gases and vapours, relating to the use of electrical apparatus* 60079-25, *Electrical apparatus for explosive gas atmospheres — Part 25: Intrinsically safe systems* 60079-26, *Electrical apparatus for explosive gas atmospheres — Part 26: Construction, test and marking of Group II Zone 0 electrical apparatus*

Publishing organization	Reference publication
NFPA	No. 51A, *Standard for Acetylene Cylinder Charging Plants* No. 497, *Classification of Flammable Liquids, Gases, or Vapors and of Hazardous (Classified) Locations for Electrical Installations in Chemical Process Areas* No. 655, *Standard for the Prevention of Sulfur Fires and Explosions*

Table C
Documents applicable specifically to Class II hazardous locations

Publishing organization	Reference publication
CSA	C22.2 No. 25, *Enclosures for Use in Class II Groups E, F, and G Hazardous Locations* (These standards are also listed in Appendix A.)
CSA (adopted IEC)	CAN/CSA-E61241-1-1, *Electrical apparatus for use in the presence of combustible dust — Part 1-1: Electrical apparatus protected by enclosures and surface temperature limitation — Specification for apparatus* (These standards are also listed in Appendix A.)
IEC	61241-1-1, *Electrical apparatus for use in the presence of combustible dust — Part 1-1: Electrical apparatus protected by enclosures and surface temperature limitation — Specification for apparatus* 61241-1-2, *Electrical apparatus for use in the presence of combustible dust — Part 1-2: Electrical apparatus protected by enclosures and surface temperature limitation — Selection, installation and maintenance* 61241-2-1, *Electrical apparatus for use in the presence of combustible dust — Part 2 — Test methods — Section 1: Methods for determining the minimum ignition temperatures of dust* 61241-2-2, *Electrical apparatus for use in the presence of combustible dust — Part 2 — Test methods — Section 2: Method for determining the electrical resistivity of dust in layers* 61241-2-3, *Electrical apparatus for use in the presence of combustible dust — Part 2 — Test methods — Section 3: Method for determining minimum ignition energy of dust/air mixtures* 61241-3, *Electrical apparatus for use in the presence of combustible dust — Part 3: Classification of areas where combustible dust are or may be present* 61241-4, *Electrical apparatus for use in the presence of combustible dust — Part 4: Type of protection "pD"*

Δ **Rules 18-000, 18-006**

The Zone and Division systems of area classification are deemed to provide equivalent levels of safety; however, the Code has been written to give preference to the Zone system of area classification. It is important to understand that while the Code gives preference to the Zone system of area classification, it does not give preference to the IEC type of equipment. Equipment approved as Class I, or Class I, Division 1 will be acceptable in Zone 1 and Zone 2, and equipment marked Class I, Division 2 will be acceptable only in Zone 2. See Rules 18-100, 18-150, and the table in Appendix J, Section J1.2.

The Scope of this Section recognizes that there are cases where renovations or additions will occur on existing installations employing the Class/Division system of classification. It is expected that such installations will comply with the requirements for Class I installations as found in Appendix J.

Δ **Rule 18-002**

Cable seals — seals that are designed to prevent the escape of flames from an explosion-proof enclosure. Because cables are not designed to withstand the pressures of an explosion, transmission of an explosion into a cable could result in ignition of gases or vapours in the area outside the enclosure.

Conduit seals — seals that are designed to prevent the passage of flames from one portion of the electrical installation to another through the conduit system and to minimize the passage of gases or vapours at atmospheric pressure. Unless specifically designed for the purpose, conduit seals are not intended to prevent the passage of fluids at a continuous pressure differential across the seal. Even at differences in pressure across the seal equivalent to a few centimetres of water, there may be passage of gas or vapour through the seal and/or through the conductors passing through the seal. Where conduit seals are exposed to continuous pressure, there may be a danger of transmission of flammable fluids to "safe areas" resulting in fire or explosions.

Primary seals — seals that are typically a part of electrical devices such as pressure, temperature or flow measuring devices and devices (such as canned pumps) where the electrical connections are immersed in the process fluids.

Secondary seals — seals that are designed to prevent flammable process fluids entering the electrical wiring system upon failure of a primary seal. These devices typically prevent passage of fluids at process pressure by a combination of sealing and pressure relief.

Δ **Rules 18-004, 18-006**

Reference material for area classification can be found in the following documents:
(a) IEC 60079-10, *Area Classification*;
(b) Institute of Petroleum (British), *Model Code of Safe Practice — Part 15: Area Classification Code for Petroleum Installations*;
(c) ANSI/API 505, *Recommended Practice for Classification of Locations for Electrical Installations at Petroleum Facilities Classified as Class I, Zone 0, Zone 1, and Zone 2*;
(d) API 500, *Recommended Practice for Classification of Locations for Electrical Installations at Petroleum Facilities Classified as Class I, Division 1 and Division 2*;
(e) NFPA 497, *Classification of Flammable Liquids, Gases, or Vapors and of Hazardous (Classified) Locations for Electrical Installations in Chemical Process Areas*; and
(f) see also the Note to Rule 18-064 in this Appendix.

Δ **Rule 18-006**

Typical situations leading to a Zone 0 area classification are
(a) the interiors of storage tanks that are vented to atmosphere and that contain flammable liquids stored above their flash point;
(b) enclosed sumps containing flammable liquids stored above their flash point continuously or for long periods; and
(c) the area immediately around atmospheric vents that are venting from a Zone 0 hazardous area.

Typical situations requiring a Class I, Zone 1 hazardous locations are
(a) inadequately ventilated buildings or enclosures;
(b) adequately ventilated buildings or enclosures, such as remote unattended and unmonitored facilities, that have insufficient means of limiting the duration of explosive gas atmospheres when they do occur; and
(c) enclosed sumps containing flammable liquids stored above their flash point during normal operation.

Typical situations leading to a Zone 2 area classification are
(a) areas where flammable volatile liquids, flammable gases, or vapours are handled, processed, or used, but in which liquids, gases, or vapours are normally confined within closed containers or closed systems from which they can escape only as a result of accidental rupture or breakdown of the containers or systems or the abnormal operation of the equipment by which the liquids or gases are handled, processed, or used;
(b) adequately ventilated buildings that have means of ensuring that the length of time where abnormal operation resulting in the occurrence of explosive gas atmospheres exist will be limited to a "short time"; and
(c) most outdoor areas except those around open vents, or open vessels or sumps containing flammable liquids.

ANSI/API 505 defines "adequate ventilation" as "Ventilation (natural or artificial) that is sufficient to prevent the accumulation of significant quantities of vapour-air or gas-air mixtures in concentrations above 25% of their lower flammable (explosive) limit, LFL, (LEL)". Appendix B of ANSI/API 505 outlines a method for calculating the ventilation requirements for enclosed areas based on fugitive emissions.

Industry documents such as ANSI/API 505 provide guidance on how industry interprets a "short time".

Rule 18-008

Class II, Division 1 locations usually include the working areas of grain-handling and storage plants; rooms containing grinders or pulverizers, cleaners, graders, scalpers, open conveyors or spouts, open bins or hoppers, mixers or blenders, automatic or hopper scales, packing machinery, elevator heads and boots, stock distributors, dust and stock collectors (except all-metal collectors vented to the outside), and all similar dust-producing machinery and equipment in grain processing plants, starch plants, sugar pulverizing plants, malting plants, hay grinding plants, and other occupancies of similar nature; coal pulverizing plants (except where the pulverizing equipment is essentially dust-tight); all working areas where metal dusts and powders are produced, processed, handled, packed, or stored (except in tight containers); and all other similar locations where combustible dust may, under normal operating conditions, be present in the air in quantities sufficient to produce explosive or ignitable mixtures.

Combustible dusts that are electrically non-conducting will include dusts produced in the handling and processing of grain and grain products, pulverized sugar and cocoa, dried egg and milk powders, pulverized spices, starch and pastes, potato and wood flour, oil meal from beans and seed, dried hay, and other organic materials that may produce combustible dusts when processed or handled. Only Group E dusts are considered to be electrically conductive for the purposes of classification. Metallic dusts of magnesium, aluminum, and aluminum bronze are particularly hazardous, and every precaution should be taken to avoid ignition and explosion.

Class II, Division 2 locations include those in which dangerous concentrations of suspended dust are not likely, but where dust accumulation might form on, in, or in the vicinity of electrical equipment, and include rooms and areas containing only closed spouting and conveyors, closed bins or hoppers, or machines and equipment from which appreciable quantities of dust might escape only under abnormal conditions; rooms or areas adjacent to Class II, Division 1 locations and into which explosive or ignitable concentrations of suspended dust might be communicated only under abnormal operating conditions; rooms or areas where the formulation of explosive or ignitable concentrations of suspended dust is prevented by the operation of effective dust control equipment; warehouses and shipping rooms where dust-producing materials are stored or handled only in bags or containers; and other similar locations.

There are many dusts, such as fine sulphur dust, that cannot be equated specifically to dusts mentioned above, and in a number of cases further information may be obtained by reference to Standards included in the NFPA National Fire Codes, e.g., NFPA No. 655 gives information on prevention of sulphur fires and explosions and makes reference to electrical wiring and equipment.

Rule 18-010

Class III, Division 1 locations include parts of rayon, cotton, and other textile mills; combustible fibre manufacturing and processing plants; cotton gins and cotton-seed mills; flax processing plants; clothing manufacturing plants; woodworking plants; and establishments and industries involving similar hazardous processes or conditions.

Readily ignitable fibres and flyings include rayon, cotton (including cotton linters and cotton waste), sisal or henequen, istle, jute, hemp, tow, cocoa fibre, oakum, baled waste, kapok, Spanish moss, excelsior, and other materials of similar nature.

Δ ## Rule 18-050

At the present time, the marking requirements of the IEC-based Standards and the North American based Standards differ concerning the gas groups for which apparatus is approved.

With the E60079 Series of Standards (IEC-based), apparatus marked IIB is also suitable for applications requiring Group IIA apparatus. Similarly, apparatus that is marked IIC is also suitable for applications requiring Group IIB or IIA apparatus.

With the C22.2 Series of Standards (North American based), apparatus bears the mark of each Group for which it is certified, i.e., apparatus that is approved for Groups B, C, and D is marked to indicate such by including all three Groups.

Rule 18-050

NFPA No. 505 recognizes the use of electric trucks, Types EE and EX, in Class III hazardous locations.

Rules 18-050, 18-066

It should be noted that battery-operated and self-generating equipment is not excluded from the Rules of Section 18, regardless of the voltage involved. Examples of such equipment are flashlights, transceivers, paging receivers, tape recorders, combustible gas detectors, vibration monitors, tachometers, battery- or voice-powered telephones, and portable test equipment that may be carried into or located within a hazardous area. Such equipment may be eligible for approval under CAN/CSA-C22.2 No. 157 or CAN/CSA-E60079-11.

Where general-purpose enclosures are used for such equipment and the Rules of this Section require the equipment to be specifically approved for the hazardous location, the electrical equipment is required to be approved for the location as intrinsically safe in accordance with Rule 18-066 and marked in accordance with Rule 18-052.

In cases where the Rules of this Section permit general-purpose enclosures with the qualification that acceptable non-incendive circuits are incorporated, the electrical equipment should be approved as such and marked in accordance with Rule 18-052.

Rules 18-050, 18-066

The users of this Code should recognize that the Class/Zone system of classification uses a method to identify gas groups that is different from that used by the Class/Division system.

The following table illustrates the correspondence between the two systems.

Temperature and gas groups

Atmosphere Typical North American name	Name in IEC 60079-20 (if different) (see Note (1))	CAS reference number	Minimum ignition temperature, °C	Gas group North America	Gas group IEC
acetylene		74-86-2	305	A	IIC
butadiene	buta-1,3-diene	106-99-0	430	B	IIB
hydrogen		1333-74-0	560	B	IIC
manufactured gases containing more than 30% hydrogen (by volume)			500 (Note (2))	B	
propylene oxide	(Note (3))	75-56-9	430	B	IIB
acetaldehyde		75-07-0	204	C	IIA
cyclopropane		75-19-4	498	C	IIA
diethyl ether		60-29-7	160	C	IIB
ethylene		74-85-1	425	C	IIB
hydrogen sulphide		7783-06-4	270	C	IIB
unsymmetrical dimethyl hydrazine (UDMH 1,1-dimethyl hydrazine)	N,N-Dimethylhydrazine	57-14-7	240	C	IIB
acetone		67-64-1	535	D	IIA
acrylonitrile		107-13-1	480	D	IIB
alcohol (see ethyl alcohol)				D	
ammonia		7664-41-7	630	D	IIA
benzene		71-43-2	560	D	IIA
benzine (see petroleum naphtha)					
benzol (see benzene)					
butane		106-97-8	372	D	IIA
1-butanol (butyl alcohol)	butan-1-ol	71-36-3	359	D	IIA
2-butanol (secondary butyl alcohol)	butan-2-ol	78-92-2	405 (Note (2))	D	IIA
butyl acetate		123-86-4	370	D	IIA
isobutyl acetate		110-19-0	421 (Note (2))	D	IIA
ethane		74-84-0	515	D	IIA
ethanol (ethyl alcohol)		64-17-5	363	D	IIA
ethyl acetate		141-78-6	460	D	IIA

(Continued)

Atmosphere Typical North American name	Name in IEC 60079-20 (if different) (see Note 1)	CAS reference number	Minimum ignition temperature, °C	Gas group North America	Gas group IEC
ethylene dichloride	1,2-Dichloroethane	107-06-2	438	D	IIA
gasoline	petroleum	86290-81-5	560	D	IIA
heptanes		142-82-5	215	D	IIA
hexanes		110-54-3	233	D	IIA
isoprene		78-79-5	395 (Note (2))	D	IIA
methane		74-82-8	537	D	IIA
methanol (methyl alcohol)		67-56-1	386	D	IIA
3-methyl-1-butanol (isoamyl alcohol)		123-51-3	350 (Note (2))	D	IIA
methyl ethyl ketone	butanone	78-93-3	404	D	IIA
methyl isobutyl ketone	4-methylpentan-2-one	108-10-1	475	D	IIA
2-methyl-1-propanol (isobutyl alcohol)		78-83-1	415 (Note (2))	D	IIA
2-methyl-2-propanol (tertiary butyl alcohol)		75-65-0	478 (Note (2))	D	IIA
naphtha (see petroleum naphtha)					
natural gas			482 (Note (2))		
petroleum naphtha	naphtha	64742-95-6	290	D	IIA
octanes	octane	111-65-9	206	D	IIA
pentanes (mixed isomers)	pentanes	109-66-0	258	D	IIA
1-pentanol (amyl alcohol)	pentan-1-ol	71-41-0	298	D	IIA
propane		74-98-6	470	D	IIA
1-propanol (propyl alcohol)	propan-1-ol	71-23-8	405	D	IIA
2-propanol (isopropyl alcohol)	propan-2-ol	67-63-0	425	D	IIA
propylene propene		115-07-1	455	D	IIA
styrene		100-42-5	490	D	IIA
toluene		108-88-3	535	D	IIA
vinyl acetate		108-05-4	425	D	IIA
vinyl chloride	chloroethylene	75-01-4	415	D	IIA
xylenes		106-42-3	464	D	IIA

Notes:

(1) *Most of the values in this table have been obtained from IEC 60079-20, Electrical apparatus for explosive gas atmospheres — Part 20: Data for flammable gases and vapours, relating to the use of electrical apparatus, 1st edition (1996-10). In many cases, the name used in the IEC standard differs from the name typically used in North America for the same substance. In fact, chemicals may have several different names. The CAS number provided above is a well-known method of uniquely identifying chemicals and is a required feature of MSDS documentation. Further information on the CAS numbering system may be found at http://www.cas.org/faq.html.*

(2) *This substance is not listed in IEC 60079-20.*

(3) *The name is incorrectly stated in IEC 60079-20 as "1,2-epoxypropene". Propylene oxide is also known as 1,2-epoxypropane.*

Rule 18-050(3)(a)

Information on classification of areas (as well as ventilation requirements) in plants engaged in the generation and compression of acetylene and in the charging of acetylene cylinders may be found in NFPA No. 51A.

Rule 18-050(3)(d)

Information on classification of areas for refrigeration systems utilizing flammable gases (including ammonia) may be found in CSA B52.

Rule 18-052

The markings described in Subrule (1)(a) to (d) appear on equipment. A typical example of the marking is as follows:

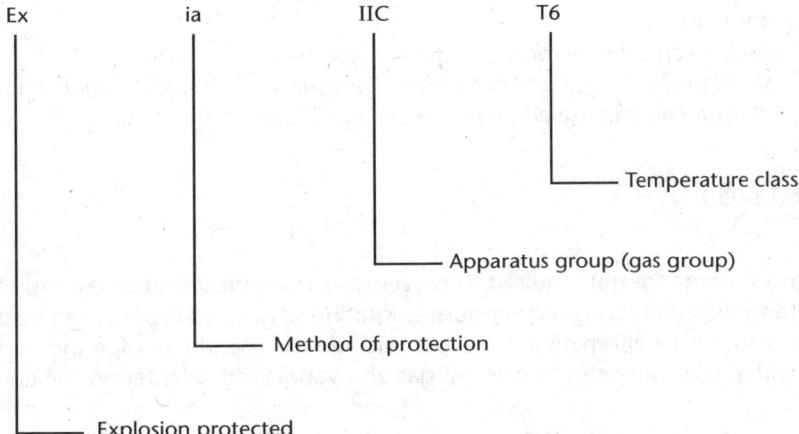

Marking on equipment may also show an "EEx" in place of the normal "Ex" to signify that the equipment has also been approved to a European standard.

The users of this Code are reminded that, in all cases, the identifying mark of an accredited certification organization is required as well as the marking requirements of this Section.

Rule 18-052

Equipment marked for Class I but not marked with a Division is suitable for both Zones 1 and 2.

Rule 18-052(1)

Some equipment permitted for use in Zone 2 hazardous locations is not marked to indicate the class and group because it is not specifically required to be approved for the location (e.g., motors and generators for Class I, Zone 2 that do not incorporate arcing, sparking, or heat-producing components — see Rule 18-168(2)).

Rules 18-054, 18-168

Equipment of the heat-producing type is currently required by product standards to have a temperature code (T-Code) marking if its temperature exceeds 100 °C. However, for equipment manufactured prior to the T-Code requirement and motors applied in accordance with Rule 18-168, there may be no such marking. Therefore, the suitability of older hazardous locations equipment of the heat-producing type and motors applied in accordance with Rule 18-168 should be reviewed prior to being installed in a hazardous location to ensure compliance with Rule 18-054. For the purpose of this Rule, equipment such as boxes, terminals, fittings, and RTDs are not considered to be heat-producing devices.

Rule 18-062

For the purposes of this Rule, metal-covered cable includes a cable with a metal sheath or with a metal armour of the interlocking type, the wire type, or the flat-tape type, or with metal shielding.

Rule 18-062(1)

Suitable lightning protective devices should include primary devices and also secondary devices if overhead secondary lines exceed 90 m in length or if the secondary is ungrounded. Interconnection of all grounds should include grounds for primary and secondary lightning protective devices, secondary system grounds, if any, and grounds of conduit and equipment of the interior wiring system.

Rule 18-062(2)(b)

Where single-conductor metal-covered or armoured cables with jackets are used in hazardous locations, the armour must be grounded in the hazardous location only to prevent circulating currents. As a result, there will be a standing voltage on the metal covering in the non-hazardous location area. There is, therefore, a need to properly isolate the armour in the non-hazardous area to ensure that circulating currents will not occur.

Rule 18-064

To meet the intent of the Rule for effectively maintaining a protective gas pressure, the following references for pressurization are recommended:

(a)   CAN/CSA-E60079-2, *Electrical apparatus — Type of protection "p"*;
(b)   NFPA No. 496, *Standard for Purged and Pressurized Enclosures for Electrical Equipment*; and
(c)   IEC 60079-13, *Construction and use of rooms or buildings protected by pressurization*.

Rule 18-066

See the Note to Rule 18-050.

Rule 18-066(3)

Intrinsically safe wiring systems are not required to prevent the transmission of an explosion and therefore the only concern is the transmission of gases and vapours. Migration of gas and vapours can be prevented by the use of conduit and cable seals. Other alternatives for cables include the use of a compound such as silicone rubber applied around the end of the connector to prevent gas and vapours from entering the end of the cable.

Rule 18-070

It is intended that this Rule be used only where suitable equipment, certified for use in the hazardous location, is not available. For example, Class I, Division 1 ignition systems for internal combustion engines are not available; only Class I, Division 2 ignition systems are available. Therefore, ignition systems rated for Class I, Division 2 are currently the only hazardous location ignition systems available and could possibly be used in Class I, Zone 1 locations.

In many situations, proper area classification will eliminate the need to use this Rule. Rule 18-070 should not be used to compensate for improper area classification.

When this Rule is used, the gas detection system should consist of an adequate number of sensors to ensure the sensing of flammable gases or vapours in all areas where they may accumulate.

Electrical equipment that is suitable for non-hazardous locations and that has unprotected arcing, sparking, or heat-producing components must not be installed in a Zone 2 location. Arcing, sparking, or heat-producing components may be protected by encapsulating, hermetically sealing, or sealing by other means such as restricted breathing.

Before applying this Rule, the user should fully understand the risks associated with such an installation. When applying this Rule, it remains the responsibility of the owner of the facility, or agents of the owner, to ensure that the resulting installation is safe. Simply complying with the requirements of this Rule may not ensure a safe installation in all situations.

Rules 18-090, 18-100, 18-150

Equipment certified for hazardous locations is marked with the area classification where the equipment can be installed. For example, a piece of equipment marked with Class I may be installed in a Class I location. This Code recognizes the IEC system of marking equipment as providing a specific method of protection. This equipment may only be installed in locations where this method of protection is recognized. For example, Rule 18-100 recognizes that equipment marked with the method of protection "d" is acceptable for installation in a Class I, Zone 1 location. Additional information related to methods of protection may be found in the CAN/CSA-E60079 Series of Standards listed in Appendix A.

The following table is provided to illustrate some of the equipment and methods of protection permitted in the three Zones. This table is not intended to be comprehensive. Other equipment may be permitted by specific Rules (e.g., fuses in Rule 18-164).

Class I, Zone 0	Class I, Zone 1	Class I, Zone 2
i, ia*	Class I, Division 1	Class I, Division 1
	i, ia, and ib*	Class I, Division 2
	d	i, ia, and ib*
	e	d
	o	e
	p	o
	q	p
	m	q
		m
		n
		Non-incendive
		Non-arcing, non-sparking, and non-heat-producing equipment.

*Equipment may be certified to CAN/CSA-E60079-11.

Δ **Rules 18-092(1), 18-108(1), 18-158(1)**

Where devices such as pressure switches, flow devices, etc., are connected to a process containing flammable fluids, failure of the seal (primary) in these devices could release the flammable fluids into the wiring system where they may migrate to a safe area where electrical or other devices are not constructed to prevent explosions. Because conduit and cable seals are not designed to seal against continuous pressure and will allow slow passage of flammable fluids, secondary seals designed to prevent the passage of flammable fluids into the wiring system are required.

Δ **Rules 18-092(2), 18-108(2), 18-158(2)**

Various methods can be used to detect the failure of a primary seal. Monitoring the secondary seal vent with flow detection or gas detection are two possible means. It is also possible to connect the vents of a number of secondary seals and monitor the common discharge. Any method used should not restrict the venting of the seal to atmosphere.

Rule 18-100

Equipment marked for Class I, Division 1 is suitable for use in Zone 1 and Zone 2.

Rules 18-100, 18-150

The following briefly explains the various methods of protection used in the Zone system. For further information see the applicable Part II Standards.

(a) **Intrinsically safe (intrinsic safety) (i, ia, or ib):** a method of protection based on the limitation of electrical energy to levels where any open spark or thermal effect occurring in equipment or interconnecting wiring that may occur in normal use, or under fault conditions likely to occur in practice, is incapable of causing an ignition. The use of intrinsically safe equipment in a hazardous location also requires that associated wiring and equipment, which is not necessarily located in a hazardous area, be assessed as part of any intrinsically safe system. The primary difference between equipment marked Ex ia and Ex ib is that equipment marked Ex ia must continue to provide explosion protection after two faults have been applied, whereas equipment marked Ex ib must continue to provide explosion protection after one such fault has been applied (see CAN/CSA-E60079-11).

(b) **Flame-proof (d):** a method of protection of electrical apparatus in which the enclosure will withstand an internal explosion of a flammable mixture that has penetrated into the interior, without suffering damage and without causing ignition, through any joints or structural openings in the enclosure of an external explosive atmosphere consisting of one or more of the gases or vapours for which it is designed (see CAN/CSA-E60079-1).

(c) **Increased safety (e):** a method of protection by which additional measures are applied to an electrical apparatus to give increased security against the possibility of excessive temperatures and of the occurrence of arcs and sparks during the service life of the apparatus. It applies only to an electrical apparatus, no parts of which produce arcs or sparks or exceed the limiting temperature in normal service (see CAN/CSA-E60079-7).

(d) **Oil immersed (o):** a method of protection where electrical apparatus is made safe by oil immersion in the sense that an explosive atmosphere above the oil or outside the enclosure will not be ignited (see CAN/CSA-E60079-6).

(e) **Pressurized (p):** a method of protection using the pressure of a protective gas to prevent the ingress of an explosive atmosphere to a space that may contain a source of ignition and, where necessary, using continuous dilution of an atmosphere within a space that contains a source of emission of gas, which may form an explosive atmosphere (see CAN/CSA-E60079-2).

(f) **Powder-filled (sand-filled) (q):** a method of protection where the enclosure of electrical apparatus is filled with a mass of granular material such that, if an arc occurs, the arc will not be liable to ignite the outer flammable atmosphere (see CAN/CSA-E60079-5).

(g) **Encapsulation (m):** a method of protection in which parts that could ignite an explosive atmosphere by either sparking or heating are enclosed in a compound in such a way that this explosive atmosphere cannot be ignited (see CAN/CSA-E79-18).

(h) **Non-sparking, restricted breathing, etc. (n):** a method of protection applied to an electrical apparatus such that, in normal operation, it is not capable of igniting a surrounding explosive atmosphere, and a fault capable of causing ignition is not likely to occur (see CAN/CSA-E60079-15).

Note: *Method of Protection "n" includes a number of means of providing protection. In addition to non-sparking, component parts of apparatus that in normal operation arc, spark, or produce surface temperatures of 85 °C or greater may be protected by one of the following:*

(a)   enclosed break device;

(b)   non-incendive component;

(c)   hermetically sealed device;

(d)   sealed device;

(e)   energy-limited apparatus and circuits; or

(f)   restricted breathing enclosure.

Some of these methods are similar to methods previously allowed in Class I, Division 2 locations. To better understand the various methods allowed under Method of Protection "n", refer to CAN/CSA-E60079-15.

Rules 18-106, 18-156(1)(b), 18-202, 18-252(1)(b), 18-302

Cables approved for hazardous locations are suitable for all locations but the termination fittings must be approved for the particular hazardous location.

Rules 18-106(3)(a), 18-156(3)(a)

Where tapered threads are used, the requirement to have five fully engaged threads (i.e., threads done up tight) is critical for the following reasons:

(a) When the threads are not fully engaged, the flame path is compromised making it possible for an explosion occurring within the conduit system to be transmitted to the area outside the conduit.

(b) If there are not five fully engaged threads, the flame path may be too short to cool the gases resulting from an internal explosion to a temperature below that which could ignite gas in the surrounding area.

(c) As the conduit forms a bonding path to ground, not making the conduit tight will introduce resistance into the flame path and if a fault occurs arcing at the interface may result.

While it may not always be possible to install certain fittings without backing off, it is important to ensure the connection is as tight as possible. Properly made conduit connections are critical to the safety of hazardous location wiring systems.

Rule 18-106(4)

It is recognized that electrical equipment that has been certified to IEC-based Canadian Standards (IEC Standards adapted for Canadian Standards use) may have conduit or cable entries with threads that are either tapered or straight complying with ISO (International Organization for Standardization) Standards. Subrule 18-106(4) requires that approved adapters be used to ensure an effective connection to this equipment where threadforms differ between the equipment and the wiring method.

Two references for understanding ISO threads are ISO 965, Parts 1 and 3.

Rule 18-106(5)

The method of protection "increased safety" incorporates protection from ingress of water or foreign bodies. CAN/CSA-E60079-7 requires that enclosures containing bare conductive parts provide at least degree of protection IP54 (see following table) and that enclosures containing only insulated conductive parts provide at least degree of protection IP44. It is important that conduit and cable entries maintain at least the degree of protection provided by the enclosure by the use of devices such as gaskets and sealing lock nuts. Increased safety enclosures should restrict easy entry of gases or vapours in order to minimize the entry of gases or vapours from short-term releases.

Ingress protection (IP) describes the degree of protection an enclosure provides. The first number of the IP designation describes the degree of protection against physical contact (i.e., fingers, tools, dust, etc.) with internal parts; the second number designates the IP against liquids. For example, an IP54 rating will require an enclosure to be dust-protected and protected against water splashing from any direction. While the minimum requirement for an increased safety enclosure is IP44 or IP54 as stated above, typically most increased safety enclosures meet IP65 or IP66 rating.

For further information on the IEC "IP" designations, refer to IEC 60529.

Ingress Protection

Protection against contact and solid objects		Protection against liquids	
Number	Description	Number	Description
0	No protection	0	No protection
1	Objects greater than 50 mm	1	Vertically dripping water
2	Objects greater than 12 mm	2	Dripping water when tilted up to 15°
3	Objects greater than 2.5 mm	3	Spraying water at an angle up to 60°
4	Objects greater than 1 mm	4	Splashing water from any direction
5	Dust-protected	5	Low-pressure water jets
6	Dust-tight	6	Strong jets of water
		7	The effects of immersion to a depth of 1 m
		8	Submersion

Rules 18-108, 18-158

Seals are provided in conduit or cable systems to prevent the passage of gases, vapours, or flames from one portion of the electrical installation to another through the system.

Passage of gases, vapours, or flames through mineral-insulated cable is inherently prevented by construction of the cable, but sealing compound is used in cable glands to exclude moisture and other fluids from the cable insulation, and is required to be of a type approved for the conditions of use.

Cables and flexible cords are not tested to determine their ability to resist internal explosions. Therefore, regardless of size, each cable must be sealed at the point of entry into any enclosure that is required to be explosion-proof.

Some designs of cable glands incorporate an integral seal, and these are marked "SL" to indicate that the seal is provided by the cable gland. Cable glands of this type are identified with the class designation. Designs requiring a field- or factory-installed sealing fitting have the group designation marked on this component.

As the appropriate sealing characteristics may be achieved by different means, the manufacturer's instructions should be followed.

Sealing of conductors in the conduit, or in most cables, requires that the sealing compound completely surround each individual insulated conductor to ensure that the seal performs its intended function. In certain constructions of cables, specifically those containing bundles of shielded pairs, triads, or quads, removal of the shielding or overall covering from the bundles negates the purpose for which the shielding was provided. Testing of this type of cable now includes testing for flame propagation along the length of the individual subassemblies of the cable.

The letters A, B, C, or D, or a combination thereof, may be added to signify the group(s) for which the cable has been tested, for example,
(a) "HL-CD" indicates the cable has been tested for flame propagation for gas groups C and D; and
(b) "TC-BCD" indicates the cable has been tested for flame propagation for gas groups B, C, and D.

See also the Table in the Note to Rule 18-050.

Δ Rule 18-108(3)(d)

Seals are required on conduit systems where a conduit enters an enclosure not required to be explosion-proof or flame-proof (typically a type "e" enclosure) because the conduit system is required to be maintained as an explosion-proof or flame-proof wiring system in a Class I, Zone 1 hazardous location. The conduit entry into a type "e" enclosure must also meet the ingress protection rating of the enclosure.

Δ Rule 18-108(4)

Reducers may have one side larger than the trade size of the conduit where the entry to the explosion-proof or flame-proof enclosure is larger than the trade size of the conduit.

Δ Rules 18-108(4), 18-158(4)

Conduit fittings approved for Class I locations and similar to the "L", "T", or "Cross" type would not usually be classed as enclosures when not larger than the trade size of the conduit.

Δ Rule 18-108(6)

It is important that the manufacturer's instructions are followed closely or seals will not function properly to prevent the transmission of an explosion or to prevent the transmission of flammable fluids to non-hazardous areas where they will be exposed to unprotected ignition sources. Improper sealing has been the primary factor in a number of explosions resulting in loss of life and/or major equipment damage. Users are reminded that only the sealing compound outlined with the instructions may be used in a seal. Use of other manufacturer's compounds in a seal may compromise the integrity of the installation.

Rule 18-108(6)(a)

All motors and generators approved under the applicable Part II Standards for Class I locations are required to have a seal provided by the manufacturer between the main motor or generator enclosure and the enclosure for the conduit entry (connection box). A marking regarding the seal being provided is therefore not necessary on this particular class of product.

For cables, compliance with Subrule (2)(a)(i) and (ii) can be accomplished by
(a) a cable gland approved for Class I hazardous locations for appropriate cable type(s) and a field-installed sealing fitting;
(b) a cable gland approved for Class I hazardous locations for appropriate cable types with an integral seal; or
(c) a cable gland for approved cable types used with an approved enclosure provided with sealing as specified in Subrule (2)(a)(ii).

Cable glands with integral seals are marked "SL".

Rule 18-108(6)(a)(i)

The term "accessible" as used in this Rule is in accordance with the Code definition in Section 0 for "Accessible as applied to wiring methods".

Rules 18-114, 28-314

Users are cautioned that combining a variable frequency drive (VFD) with a motor may increase the operating temperature of the motor as a result of the harmonics produced by the drive. This may cause the motor temperature to exceed its temperature code rating. This is of particular concern where the operating temperature of the motor is close to the ignition temperature of hazardous materials that may be in the area. Because of the generally lower ignition temperatures associated with Class II materials, it will be of particular concern in Class II areas. It remains the responsibility of the user to ensure that the operating temperature of the motor, in combination with the drive, is below the minimum ignition temperature of the hazardous material in the area. The motor manufacturer should be consulted where necessary. Some references that may assist the user in determining the suitability of an installation are

(a)   CSA Technical Information Letter E-22, "Motors and Generators For Use in Class I, Division 2 and Class II, Division 2 Locations";

(b)   API RP 2216, *Ignition Risk of Hydrocarbon Vapors by Hot Surfaces in the Open Air;* and

(c)   IEEE Paper No. PCIC-97-04, "Flammable Vapor Ignition Initiated by Hot Rotor Surfaces Within an Induction Motor — Reality or Not?".

Rule 18-116

It should be recognized that gas turbines in hazardous locations also need safeguards against potential hazards from other than electrical ignition systems, such as exhaust and fuel systems. The complete engine assembly should, therefore, be investigated for its suitability in Class I, Zone 1 hazardous locations.

Rule 18-150

Equipment marked Class I, Division 2 is suitable only for Zone 2. See the Note to Rule 18-100.

Rule 18-156

See the Note to Rule 18-106.

Rule 18-156(8)

Cable glands should be compatible with the degree of ingress protection and explosion protection provided by the enclosure on which they are installed.

For example, to maintain the protection of an enclosure required to be explosion-proof, a sealing-type gland approved for the location should be used. Where unarmoured cables must enter an enclosure required to be explosion-proof, a combination of an approved sealing fitting and a non-sealing cable gland may be used.

Where equipment normally considered suitable for use in ordinary locations is acceptable in Zone 2 locations, such as terminal boxes and motors, ordinary location cable glands that maintain the degree of protection of the enclosure may be used. Similarly where purged enclosures are used in Zone 1 and Zone 2, ordinary location cable glands that maintain the degree of protection of the enclosure may be used.

Where equipment is specifically designed for use in Zone 2 locations, such as "Ex n<u>X</u>", ordinary location cable glands that maintain the degree of protection of the enclosure may be used. One means of achieving equivalent protection would be to use a cable gland with the same or better IP rating as the enclosure. (See the Ingress Protection table in the Note to Rule 18-106(5)). If the gland does not have an IP rating, other ratings, such as weatherproof, may be matched to the enclosure rating.

Rule 18-158

See the Note to Rule 18-108.

Δ   Rule 18-158(3)(c)

This Rule allows the seal at the boundary between an outdoor Class I, Zone 2 location and an outdoor non-hazardous location, to be located further than 300 mm from the boundary of the Class I, Zone 2 location provided that it is located on the conduit prior to its entering an enclosure or a building. Because gas is present in Class I, Zone 2 locations only for short periods, it is unlikely that gas or vapour could be released through conduit couplings at sufficiently high rates to form an explosive mixture in outdoor areas. However, the seal must be located on the conduit before it enters an enclosure or a building because, depending on the ventilation rate, gas transmitted through the conduit may build up to flammable concentrations.

Rule 18-158(4)

See the Note to Rule 18-108(4).

Rule 18-160

This Rule includes service and branch circuit switches and circuit breakers; motor controllers including push buttons, pilot switches, relays, and motor-overload protective devices; and switches and circuit breakers for the control of lighting and appliance circuits. Oil-immersed circuit breakers and controllers of the ordinary general-use type may not confine completely the arc produced in the interruption of heavy overloads, and specific approval for locations of this Class and Division is therefore necessary.

Rule 18-166

A group of three fuses protecting an ungrounded 3-phase circuit, and a single fuse protecting the ungrounded conductor of an identified 2-wire single-phase circuit, would each be considered a set of fuses.

Rules 18-168, 18-210, 18-260, 18-308, 18-358

See the Note to Rule 18-114.

Rule 18-170

It should be recognized that internal combustion engines in hazardous locations also need safeguards against potential hazards from other than electrical ignition systems such as exhaust and fuel systems. The complete engine assembly should therefore be investigated for its suitability in Class I, Zone 2 hazardous locations.

Rule 18-202

See the Note to Rule 18-106.

Rules 18-210, 18-212, 18-260, 18-262, 18-308, 18-310, 18-358, 18-360

As overheated windings of large pipe-ventilated motors (or fire in these motors) are not readily detected by odour or smoke, it is advisable, especially in the case of buildings not provided with automatic sprinklers, that the following precautions be taken:

(a) if ventilation air is supplied from a separate source, an air-pressure-operated switch should supervise the supply of air, and be arranged to shut down the pipe-ventilated motor in case of air failure;

(b) an automatic fire detector should be placed at the air discharge end of the pipe-ventilated motor and be arranged to shut down the motor on the occurrence of overheating or fire;

(c) a port with a self-closing shutter should be provided at the motor air intake end to facilitate discharge into the motor frame of a fire extinction medium;

(d) to complement Item (c), fire dampers fitted with fusible links should be provided for the air intake and discharge ends of the motor to confine fire and the fire extinction medium to the motor frame;

(e) intake and discharge ducts should be carefully installed with respect to combustible construction or storage, and should not pierce firewalls, fire partitions, floors, or ceilings unless provided with automatic fire shutters or dampers where they pierce the fire section or division of the building (see NFPA No. 91); and

(f) intake and discharge ducts should be kept clear of accumulation of combustible lint or dust.

Rule 18-252

See the Note to Rule 18-106.

Rule 18-260

See the Note to Rule 18-210.

Rule 18-260(2)

It is the responsibility of the owner of the facility to demonstrate to the authority having jurisdiction that the conditions outlined in the Rule will exist.

Accumulations of the dust can be considered to be moderate if the colour of a surface is visible through the dust layer.

Rule 18-262

See the Note to Rule 18-210.

Rule 18-302

See the Note to Rule 18-106.

Rules 18-308, 18-310

See the Note to Rule 18-210.

Rules 18-358, 18-360

See the Note to Rule 18-210.

Section 20

Rule 20-004

For the purposes of Subrules (6) and (7), buildings such as kiosks in which electrical equipment such as cash registers and/or self-service console controls are located are considered to be buildings not suitably cut off.

Rule 20-030

Information on non-electrical aspects for propane tank systems, refill centres, and filling plants may be found in CAN/CSA-B149.2.

Rule 20-060

Information on non-electrical aspects of compressed natural gas (NGV) refuelling stations and NGV storage facilities may be found in CAN/CSA-B149.1.

Section 22

Rule 22-002

Examples of some, but not all, of the occupancies in which Category 1 or Category 2 locations may be encountered are as follows:

Category 1	
Basements (other than in residential occupancies)	Dairies (commercial and farm)
	Dye works
Bathhouses	Ice cream plants
Bottling works	Ice plants
Breweries	Laundries (commercial)
Canneries	Stables for cattle only
Cold storage plants	Stables for horses in rural farm areas

Category 2	
Abattoirs	Metal refineries
Casing rooms	Potato storage facilities
Chemical works (some)	Pulp mills
Fertilizer rooms	Railway round houses
Glue houses	Stables for horses
Hide cellars	Sugar mills
Meat-packing plants	Tanneries

Rule 22-100

Circuit breakers located in a Category 2 location have experienced nuisance tripping due to internal corrosion and may not operate as designed in all cases if located in a corrosive environment.

Rule 22-204

Farm buildings that are of weathertight construction and that are not Category 1 or Category 2 locations can be considered dry locations (e.g., machinery warehouses).

Additional information on the ventilation of buildings housing livestock and poultry can be found in the *National Farm Building Code of Canada*.

Rule 22-204(5)

The mechanical requirements of this Rule are met by rigid conduit and electrical metallic tubing.

Δ Rule 22-704

Sewage lift and treatment plants

Sewage lift and treatment plants produce a combination of conditions that may require specialized attention to the electrical installation. Abnormal hazardous conditions can occur due to the buildup of methane gas and the spills of chemical, gasoline or other volatile liquids into the sewer system. Reference material for hazardous area classification can be found in NFPA No. 820. Wet well areas normally contain an atmosphere of high humidity and corrosive hydrogen sulphide vapours.

An extreme hazard to personnel working in wet wells exists because of the presence of sewer gas (hydrogen sulphide). This gas is treacherous because the ability to sense it by smell is quickly lost. If workers ignore the first notice of the gas, their senses will give them no further warning. If the concentration is high enough, loss of consciousness and death can result.

Before work in wet well locations begins, the air in the wet well area should be purged, and ventilation with fresh air should be maintained while work continues in the area.

Section 24

Rule 24-000

This Section consolidates, in one section of this Code, requirements that arise from safety considerations that are unique to specific areas in health care facilities. It was compiled by the Technical Committee on Applications of Electricity in Health Care and the Committee on the *Canadian Electrical Code, Part I*, working in partnership.

This Section no longer includes the original requirements affecting the use of flammable anaesthetics, since such anaesthetics are no longer used, but it still incorporates Rules designed to minimize the risks inherent in the use of electricity in patient care areas, as well as risks inherent in the interruption of a supply of electricity.

The content of this Section has been modified to reflect the changing nature of health care. Procedures once reserved for hospitals are now performed in medical clinics. As such, this Section has been modified to apply to patient care areas of health care facilities and its requirements are based on the care area (e.g., basic, intermediate, or critical). This approach is consistent with that taken in CAN/CSA-Z32.

Rule 24-002

Hazard index

The hazard index with one isolated conductor connected to ground is not necessarily the same as the hazard index with the other isolated conductor connected to ground; of the two, the greater hazard index governs.

Patient care environment

The patient care environment is a zone fixed to the patient bed, table, or other supporting mechanism and does not move with the patient as the patient moves through the health care facility or room.

Rule 24-100

Users of this Code should also consult CAN/CSA-Z32, as it recommends additional precautions that should be taken in the design, construction, use, and maintenance of electrical systems in such areas.

It is highly desirable that the intended use of all patient care areas be designated by the health care facility's administration in a manner that may be readily understood by the facility's staff.

Rule 24-102, 24-106

Basic care areas should not, even in long-term facilities, be considered residential occupancies that are governed by General Rules such as 26-710 and 26-724. Users of this Code should be aware that the need for circuits and receptacles in patient care areas is frequently greater than in most other locations. Users are directed to CAN/CSA-Z32 for recommendations regarding the minimum number of receptacles and circuits normally required in the various patient care areas.

Rule 24-102(3)

This Subrule is not intended to restrict the number of patient care environments served by a branch circuit. Users of this Code should consider the nature of the care area (e.g., basic, intermediate, or critical), Table 6 of CAN/CSA-Z32, and the voltage drop requirements of Rule 8-102. It is the intent that receptacles supplied by the branch circuit supply only medical electrical equipment. The actual use of the receptacle by health care facility staff within the patient care environment is beyond the scope of this Subrule. The word "load" within this Subrule is intended to refer exclusively to permanently connected medical electrical equipment or receptacles intended for medical electrical equipment.

402

Rules 24-104(1), 24-112

The object of this Rule is to limit the voltage difference in the vicinity of the patient and thus to minimize the risk of electric shock. The adequacy of the installation may be verified by test in accordance with the procedure outlined in CAN/CSA-Z32. It is important to note the specifications of the measurement instrument.

Bonding to ground in patient care areas must accomplish two functions:
(a)   limit the voltage that occurs on exposed metal parts in the event of a fault in the electrical insulation of the wiring system or of a utilization device; and
(b)   eliminate small but potentially hazardous voltage differences that might otherwise exist between grounded points in the vicinity of the patient.

In order that the integrity of the bonding conductor can be checked, the conductor should not be permitted to make intermediate contact with grounded metal, such as would be the case in metal conduit and in some armoured cables.

Δ Rule 24-104(1)(b)

The uninsulated bonding conductor of a Type AC90 cable is not considered suitable for the purpose.

Rules 24-104(2), 24-112

It is intended that the bonding methods specified by this Subrule may be mixed, i.e., some bonding conductors for an area may be terminated at a grounding bus, and others may be terminated at the panelboard. In some situations, "daisy chaining" of bonding conductors from outlet to outlet may prove to be more effective than installing a separate conductor from each outlet to a common point.

Rule 24-104(5)

This Subrule is not intended to isolate the 3-phase equipment bonding conductor from other equipment within the patient care environment.

Δ Rule 24-106(3)

Rule 26-700(11) requires that receptacles installed within 1.5 m of wash-basins, bathtubs, or shower stalls be protected by a ground fault circuit interrupter of the Class A type.

Rules 24-106(3), 24-114

The intent of the Subrule is to provide protection against electric shock hazard when using personal grooming appliances. The Note on Rule 24-102 also applies to Rule 24-106.

Δ Rule 24-106(6)

It is not intended by this Subrule to mandate installation of hospital grade receptacles in bathrooms or washrooms contained within a patient care area.

Extent of a patient care area is outlined in the CAN/CSA-Z32.

Rule 24-108

CSA has embarked on a harmonization of CAN/CSA-C22.2 No. 114 and CSA C22.2 No. 125 with the requirements of the IEC 601 Standards. To that effect, Canada has adopted the IEC 601 Series of Standards as National Standards of Canada. The harmonization process will be complete when CAN/CSA-C22.2 No. 114 and CSA C22.2 No. 125 are phased out. Currently, CAN/CSA-C22.2 No. 114, CSA C22.2 No. 125, and the CAN/CSA-C22.2 No. 601 Series are accepted as equivalent, and manufacturers are able to certify to either Series of Standards. As of December 31, 2002, new products will only be certified to the CAN/CSA-C22.2 No. 601 series. After January 1, 2005, all equipment previously manufactured and certified to CAN/CSA-C22.2 No. 114 or CSA C22.2 No. 125 is required to comply with the applicable Standard within the CAN/CSA-C22.2 No. 601 Series of Standards.

Rule 24-110

In intermediate and critical care areas, either grounded or isolated systems may be used. Users of this Code should consult CAN/CSA-Z32 regarding the relative merits of each system.

Rule 24-112

See Notes for Rule 24-104.

Rule 24-114

See Notes for Rule 24-106.

Δ **Rule 24-116**

Areas subject to standing fluids on the floor or drenching of the work area can create a condition where the patient or staff can become a path for ground fault current under fault conditions.

Routine housekeeping procedures and incidental spillage of liquids are not intended to be considered for the purpose of this Rule.

Use of receptacles protected by a ground fault circuit interrupter of a Class A type is intended for those wet locations within a patient care area where interruption of power to the receptacles by actuation of a GFCI is deemed to be acceptable in accordance with provisions of the CAN/CSA-Z32.

These receptacles are intended to be supplied by an isolated system where such power interruption to the receptacles is not acceptable in accordance with CAN/CSA-Z32.

Rules 24-200, 24-202(2), 24-204(7)

Users of this Code should recognize that while fixed lighting fixtures and medical electrical equipment, permanently connected or otherwise, identified in these Rules may physically fall outside of the patient care environment, they nonetheless serve the patient care environment.

Rule 24-204(6)

Users of this Code should refer to CAN/CSA-Z32 regarding methods for verifying the impedance to ground of an isolated system.

Rule 24-206

It is imperative that the impedance to ground of individually isolated branch circuits be tested at regular intervals, and that maintenance procedures be instituted for the system and the equipment connected to it as necessary to limit the hazard index to 2 mA.

Rule 24-208

Three-phase isolated systems should be subjected to a periodic test of the impedance to ground of the system together with any connected load, unless an approved isolation sensing device (e.g., isolation monitor) is used.

Rule 24-300

Users of this Code are directed to CAN/CSA-Z32, which makes further recommendations regarding the design, installation, use, and maintenance of these systems.

Rule 24-302(1)

CAN/CSA-Z32 provides advice as to what loads should be supplied by the vital, delayed vital, or conditional branch of an essential electrical system.

Rule 24-304

The intent of the requirement that transfer switches be mechanically held is to ensure that, once the essential system has been connected to the emergency supply, it will not be disconnected until the normal supply has been restored.

Section 26

Rule 26-004

Electrical equipment certified after September 30, 1986, is acceptable for mounting directly over combustible surfaces without additional protection unless it bears a cautionary marking requiring additional protection. Equipment certified prior to that date is not required to carry the cautionary marking.

Rule 26-008

The intent of the Rule is to protect electrical apparatus within ventilated enclosures from the direct spray from sprinkler heads. The intent of the Rule is considered to be met when
(a)   water following a direct line of sight path from the sprinkler head cannot strike live parts within the enclosure through ventilation openings in the sides and tops of electrical equipment; and

404

(b)   water accumulating on the top of the equipment, cannot flow into the interior through significant openings. Examples of significant openings are ventilation openings, openings around bus duct, and dry-type armoured cable connectors. Bolts and seams are not considered to be significant openings.

The intent of the Rule can also be met through use of weatherproof equipment.

Rule 26-012

Dielectric liquid-filled circuit breakers or switches should have their vents piped directly to an outside area in accordance with the manufacturer's instructions or recommendations.

Construction criteria for service rooms are provided in Section 3.5 of the *National Building Code of Canada*. Where a service room is required by this Code, it must be separated from the remainder of the building by a fire separation having a 1 h fire-resistance rating, unless the service room is sprinklered.

Rule 26-012(2)

Separation of liquid-filled equipment, indoors, exclusive of installations in electrical equipment vaults

Total amount of liquid at location, L	Radius, m		
	Open floor location, R_a	Against wall location, R_b	Corner location, R_c
5	0.36	0.51	0.72
15	0.63	0.88	1.25
30	0.88	1.25	1.77
45	1.08	1.53	2.17
60	1.25	1.77	2.50
75	1.40	1.98	2.80

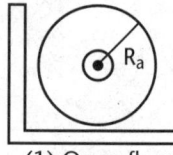 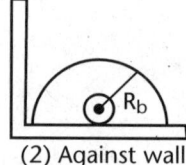 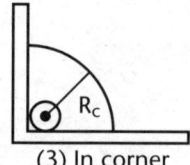

(1) Open floor   (2) Against wall   (3) In corner

Radius is calculated from formula $R_a = \sqrt{\dfrac{area}{\pi}} = \sqrt{\dfrac{0.06184 \times litres}{\pi}}$

Similarly $R_b = \sqrt{\dfrac{0.16388 \times litres}{\pi}}$ and $R_c = \sqrt{\dfrac{0.32738 \times litres}{\pi}}$

Note: *Radii to be measured from centre of liquid-filled container.*

Example: Two pieces of equipment each containing 75 L of liquid, one installed in a corner and the other along the wall as shown, must have the centre points of the containers at least 2.80 + 1.98 = 4.78 m apart.

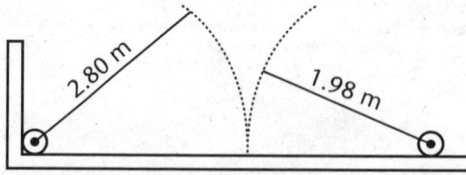

Rule 26-014(3)

The normal enclosure for the equipment is not to be considered as the barrier referred to in this Subrule.

Rule 26-210

In addition to the circuit overcurrent protection provided by this Rule, overcurrent protection should be provided for the capacitors to protect them against bursting should a unit become defective.

Where capacitors for power factor correction are assembled in the field to form banks or groups of banks, the manufacturer's instructions with respect to proper application and connection should be obtained in order to ensure that such overcurrent protection is properly provided.

406

Generally, individual capacitor fusing or the single fusing of a capacitor bank is used. It becomes an application engineering problem and involves the coordination of the time-current (i.e., blowing) characteristics of the fuse with those of the container with respect to bursting caused by the generation of gas pressure under fault conditions. The selection of the fuse also requires consideration to be given to the available fault current of the circuit and to proper connection of the capacitors in the circuit (i.e., whether parallel, series-parallel, Y-connected with a floating neutral, etc.). Improper capacitor connections can also cause overvoltage on adjacent units upon failure of a unit.

Rules 26-244, 26-246

For the purposes of these Rules, a non-propagating liquid is one that when subjected to a source of ignition may burn, but the flame will not spread from the source of ignition. The flash point of a liquid is the minimum temperature at which the liquid gives off sufficient vapour to form an ignitable mixture with air near the surface of the liquid or within the test vessel used.

Rule 26-256

Selection of overcurrent devices with too low a rating for the primary of a dry-type transformer can result in unintended operation when the transformer is being energized (such as might occur after a power outage). To avoid such operation, the overcurrent device should be able to carry
(a)   12 times the transformer rated primary full load current for 0.1 second; and
(b)   25 times the transformer rated primary full load current for 0.01 second.

Rule 26-260(1)

The purpose of installing primary fuses between the power lines and instrument voltage transformers is to protect the power system from possible destructive power arc-over due to breakdown of the major insulation of the transformers. Such fuses must have adequate interrupting capacity for the power system to which they are connected, either self-contained or in conjunction with suitable current-limiting resistors.

Rule 26-260(3)

The reference in Subrule (3)(c) to primary terminals outside the enclosure being common to both voltage transformers and current transformers includes the "centre" (common) phases primary terminals of "Open Delta" connected voltage transformers and the primary grounded neutral terminal.

Rule 26-266

The neutral current to a zero sequence filter is three times the phase current.

Installation of a zero sequence filter can increase the single phase-to-ground fault current to 1.5 times the available phase-to-phase fault current.

Rule 26-354

Construction requirements for electrical equipment vaults are found in Article 3.6.2.7 of the *National Building Code of Canada*.

It is recommended, wherever practicable, that vaults be located where they can be ventilated directly from and to an outside area without the use of flues or ducts. In order to minimize the possible explosion hazard from gases that might seep into a vault, vaults should be located remote from points where gas, sewer, water, and other pipelines and conduits enter the building. It is also recommended that they not be located adjacent to, or close to, vertical openings such as elevator shafts.

Rule 26-400

Where in dwelling units, the branch circuit breakers are equipped with ground fault circuit interrupters of the Class A type, panelboards containing these circuit breakers should be provided with a self-adhesive label indicating the test procedure for the GFCI and a chart for recording the tests.

Rule 26-520

Low-voltage surge protective devices or transient voltage surge suppressors are surge suppression products designed for repeated suppression of transient voltage surges on 50 and 60 Hz power circuits not exceeding 750 V. IEEE C62.41 describes three general categories of operating environment for surge protective devices. The three system exposure environments are called Category A, Category B, and Category C. Surge rating for Category A is intended for outlets and long branch circuits that exceed 10 m in length from Category B or 20 m in length from Category C. Surge rating for Category B is intended for feeders and short branch circuits, distribution panels, heavy appliance outlets with short connections to service entrance, and lighting systems in large buildings. Surge rating for Category C is intended for outdoor and service entrance installations, service drop from a pole to a building, and underground line to a well pump. Surge suppression voltage ratings are also listed in accordance with UL 1449.

Rule 26-546

Sufficient ventilation should be provided to prevent the hydrogen gas from building up to a level of 2% by volume in the room air at any time.

When batteries are operated in constant-voltage-float service and the float voltage is maintained at appropriate levels, generation of gas is very slight.

The rate of ventilation required to maintain the volume of hydrogen gas below the 2% level in a battery room may be calculated in accordance with IEEE 484.

As an example, the volume of hydrogen gas generated daily by a 60 cell, 840 ampere hour lead calcium grid battery charging at 2.2 V per cell is determined as follows:

Total m^3/min of hydrogen gas = number of cells × gas generation rate of battery type in m^3/min × float current in amperes × minutes/day.

$$\text{Volume of gas production} = 60 \text{ cells} \times 7.6 \times 10^{-6} \frac{m^3}{min.} \times \frac{0.006 \text{ A}}{100 \text{ A.H.}} \times 840 \text{ A.H.} \times \frac{60 \text{ min.}}{h} \times \frac{24 \text{ h}}{day}$$

$$= \frac{0.03309 \text{ } m^3 \text{ gas}}{day}$$

For a room volume of 30 m^3, the total volume of gas that should be allowed to accumulate in this room is 30 m^3 × 2% = 0.6 m^3.

Therefore, to meet this 2% maximum level, one air change is required for each

$$\frac{0.6 \text{ } m^3}{0.03309 \text{ } m^3 \frac{gas}{day}} = 18 \text{ days}$$

However, a minimum of 1 to 4 air changes per hour in the battery room is recommended to prevent pockets of hydrogen gas from accumulating and to ensure the comfort of the maintenance personnel.

Rule 26-546(2)

The freezing point of the electrolyte used in a lead-acid battery is –15 °C for a specific gravity of 1.150, –20 °C for a specific gravity of 1.175, and –27 °C for a specific gravity of 1.200. The freezing point will be higher if the battery is completely discharged. Therefore, batteries should not be located in areas where the temperature is likely to fall below –7 °C.

Rule 26-700, Diagrams 1 and 2

The configurations of Diagrams 1 and 2 are all of the grounding type. With two exceptions (5-20R and 6-20R), these configurations are identical to those similarly designated in the United States, dimensional details of which are given in NEMA WD 6.

The CSA 5-20R and 6-20R configurations differ from those in the United States in that they will not accept a 15 A attachment plug of the same voltage rating. Receptacles of the CSA 5-20RA and 6-20RA configurations are intended to accommodate both 15 A and 20 A rated attachment plugs.

Δ ### Rule 26-700(11)

The term "sink" is intended to include kitchen sinks, bar sinks, laundry sinks, utility room sinks, wash basins, etc., connected to a plumbing drain pipe. It is not intended to include portable wash basins.

It is not intended that the 1.5 m dimension be extended through a wall opening that is fitted with a door. Where a room combines a wash basin or shower/bathing facilities with an area serving another purpose such as an ensuite bathroom in a bedroom, requirements for receptacles located in such rooms or areas should be considered as similar to the requirements for receptacles located in bathrooms and washrooms.

GFCI protection is not intended to apply to receptacles supplying specific-use appliances located behind such appliances as washers, dryers, fridges, ranges, built-in microwaves, and other similar appliances provided that those receptacles, by virtue of their location, are rendered essentially inaccessible for use by other portable appliances.

Δ Rule 26-710(h)

It is intended by this Rule to specify the location of the dedicated receptacle for a cord-connected hydromassage bathtub. The intent of this Rule is to render this receptacle inaccessible to a person in the hydromassage bathtub by installing this receptacle behind an access panel, a skirt, an apron or other suitable barrier described in the manufacturer's instructions.

Rule 26-710(m)

For the purposes of this Subrule, the installation of the duct should be considered as being completed when enough duct has been installed to allow identification of the central vacuum unit location.

Δ Rule 26-710(o)

It is the intent of this Rule that all receptacles of residential occupancies installed outdoors and within 2.5 m of finished grade be protected by a GFCI. This includes receptacles located on buildings or structures associated with the residential occupancy such as garages, carports, sheds, and receptacles on posts or fences, etc.

Without precluding its use in such applications, this Rule is not intended to apply to receptacles located in parking lots of apartments installed solely for use as automobile heater receptacles.

Rule 26-712(a)

In laying out the location of receptacle outlets in residential premises, consideration should be given to the placement of electric baseboard heaters, hot air registers, and hot water or steam registers, with a view of eliminating cords having to pass over hot or conductive surfaces wherever possible.

Δ Rule 26-712(d)(v)

A peninsular countertop is measured from the connecting edge.

Rule 26-714(a)

For single dwellings of the detached, semi-detached, and row housing types, it is recommended that outdoor receptacles be placed at both the front and rear of the house and that these be controlled by a switch located inside the house.

Rule 26-804

The following are excerpts from CSA C22.2 No. 3:

4.8.5 The nominal supply voltage of a safety control circuit shall not exceed 120 V.

4.8.6 A safety control circuit intended to be supplied by a nominal 120 V branch circuit shall comply with the following:
(a)   the circuit shall not be grounded within the equipment;
(b)   the ungrounded conductor shall have an overcurrent protection device rated at not more than 125% of the current drawn by the circuit, except that this value may be increased because of inrush currents and ambient temperatures. These requirements shall apply only where the maximum current to the appliance exceeds 12 A and the safety controls are in series with the total load they control.

4.8.7 A safety control circuit supplied other than as specified in Clause 4.8.6, such as one supplied by a battery or a transformer, shall comply with the following:
(a)   it shall be a 2-wire circuit not exceeding 120 V;
(b)   one side of the circuit shall be grounded;
(c)   except for the condition specified in Item (d), the ungrounded conductor shall have an overcurrent protection device rated at not more than 125% of the current drawn by the circuit, except that for circuits drawing currents up to and including 2 A the protection shall be rated at not more than 200%. These values may be increased because of inrush currents and ambient temperatures; and
(d)   a safety control circuit supplied by a Class 2 transformer shall not require overcurrent protection.

4.8.8 A safety control shall interrupt the current in the ungrounded conductor of the circuit between the overcurrent protection and the load.

4.8.9 Except for multiphase loads and circuits in which the load to be controlled exceeds the contact rating of the safety control*, safety controls which open an electrical circuit to the burner or to the shut-off device shall directly open the circuit regardless of whether the switching mechanism is integral with, or remote from, the sensing element.

In these instances, the safety control may interrupt the coil circuit of a magnetic relay or contactor, which in turn directly opens the circuit to the burner or shut-off device.

Note: *The purpose of this requirement is to minimize the interposing of other controls in the safety control circuit, the failure of which might create an unsafe condition that the safety control is intended to prevent.*

Rule 26-806(1)

Subrule (1) is intended to apply only to central heating equipment that does not use electricity as the source of heat. It is not intended to apply to electrical components of non-electric heating equipment such as water heaters, fireplace inserts, room heaters, or other similar auxiliary heating equipment provided with electric auto-ignition, controls, or blower motors rated not more than 1/8 hp.

Section 28

Example to determine motor conductors and protection

The following is a sample calculation for determining copper conductor size, overcurrent and overload protection for 1 to 100 hp, 1 to 30 hp, and 2 to 7-1/2 hp motors 575 V 3-phase, at full-voltage start.

Conductors

It is necessary to determine the motor full-load currents. Obtain preferably from the motor nameplate or from Table 44. Conductor sizes for the individual motors (see Rule 28-106 and Table 2) are as follows:

	100 hp	30 hp	7-1/2 hp
FLC (from Table 44)	99 A	32 A	9 A
125% calculation	124 A	40 A	11 A
From Table 2			
75 °C conductors	No. 1	No. 8	No. 14
90 °C conductors	No. 1	No. 8	No. 14

Feeder conductor ampacity [see Rule 28-108(1)(a)]

Conductor ampacity would be 125% of 99 A plus 32 A plus 2 times 9 A equals 174 A for the 4 motors. Conductor size from Table 2 is No. 4/0 for 60 °C conductors and No. 2/0 for 75 or 90 °C conductors.

Protection

Overload protection

The maximum allowable setting of overload devices is determined from Rule 28-306. Assuming a 1.15 service factor, the ratings are 123.8 A for the 100 hp, 40 A for the 30 hp, and 11.3 A for the 7-1/2 hp motors.

Overcurrent protection

Branch circuit overcurrent protection for each motor is determined by using Rule 28-200. For the purpose of the described motors, use Rule 28-200(a) and Table 29. Listed below are actual currents with the standard size protector shown in brackets.

	100 hp	30 hp	7-1/2 hp
Time-delay fuse	99 × 175% = 174 A (150)	32 × 175% = 56 A (50)	9 × 175% = 15.8 A (15)
Non-time-delay fuse	99 × 300% = 297 A (250)	32 × 300% = 96 A (90)	9 × 300% = 27.0 A (25)
Time-limit circuit breaker	99 × 250% = 248 A (225)	32 × 250% = 80 A (80)	9 × 250% = 22.5 A (20)

Note that Table D16 may be used to select the size of overcurrent devices in accordance with Rule 28-200 where the full-load current rating of the motor is shown in the Table.

Feeder overcurrent protection

The maximum allowable feeder overcurrent protection for motors is determined from Rule 28-204.

Using Rule 28-204(1), the ratings are as shown below with standard sizes shown in parentheses:

Time-delay fuse	174 + 32 + 9 + 9 = 224 A (200)
Non-time-delay fuse	297 + 32 + 9 + 9 = 347 A (300)
Time-limit circuit breaker	248 + 32 + 9 + 9 = 298 A (250)

Diagram of circuits, control, and protective devices for motors

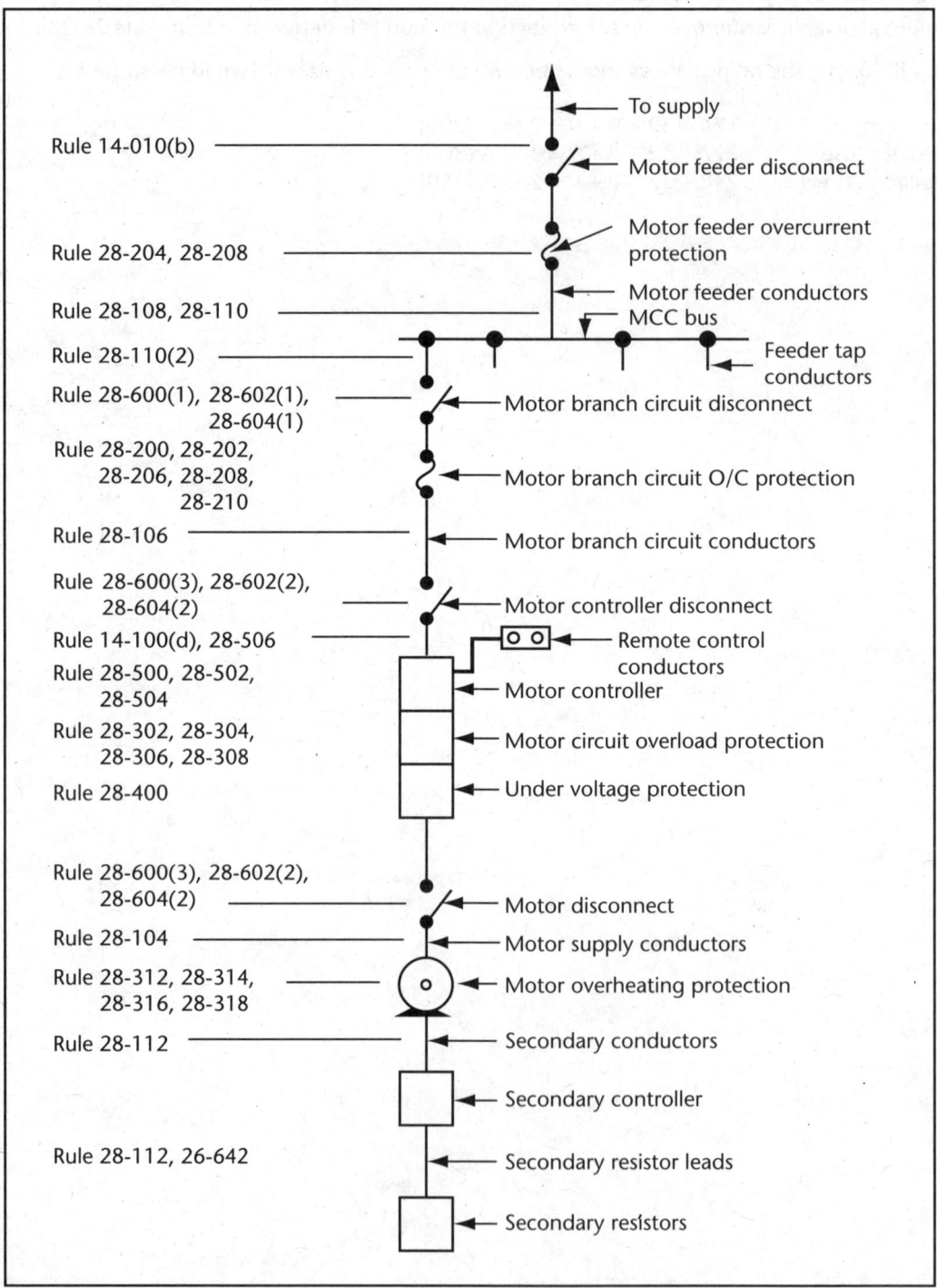

Rule 14-010(b) ——————————— To supply

Motor feeder disconnect

Rule 28-204, 28-208 ——————————— Motor feeder overcurrent protection

Rule 28-108, 28-110 ——————————— Motor feeder conductors
MCC bus

Rule 28-110(2) ——————————— Feeder tap conductors

Rule 28-600(1), 28-602(1), 28-604(1) ——————— Motor branch circuit disconnect

Rule 28-200, 28-202, 28-206, 28-208, 28-210 ——————— Motor branch circuit O/C protection

Rule 28-106 ——————————— Motor branch circuit conductors

Rule 28-600(3), 28-602(2), 28-604(2) ——————— Motor controller disconnect

Rule 14-100(d), 28-506 ——————— Remote control conductors

Rule 28-500, 28-502, 28-504 ——————— Motor controller

Rule 28-302, 28-304, 28-306, 28-308 ——————— Motor circuit overload protection

Rule 28-400 ——————————— Under voltage protection

Rule 28-600(3), 28-602(2), 28-604(2) ——————— Motor disconnect

Rule 28-104 ——————————— Motor supply conductors

Rule 28-312, 28-314, 28-316, 28-318 ——————— Motor overheating protection

Rule 28-112 ——————————— Secondary conductors

Secondary controller

Rule 28-112, 26-642 ——————— Secondary resistor leads

Secondary resistors

Rule 28-100

Where a motor is supplied by rigid conduit and is provided with noise or vibration damping, a flexible fitting installed between the motor terminal enclosure and the conduit will prevent damage to the conduit system due to vibration.

Rule 28-210

Instantaneous-trip circuit breakers are a magnetic-only trip device without time delay that may be provided with a dampening means to accommodate the transient motor inrush current.

Rule 28-210(b)

The intent of this Rule is to allow an increase in the trip setting above 1300% for motors with high locked rotor currents that will trip on the asymmetrical inrush at the 1300% rating. For example, a motor with 800% locked rotor current would result in a trip setting of up to 1720% of full load current. Higher locked rotor currents are common in energy-efficient motors.

Rule 28-304

There are several conditions that may create current imbalance in 3-phase motor circuits of sufficient magnitude to overload one or more phase conductors of the circuit. When the overload occurs in only one of the three conductors, a current-sensitive element is necessary in each conductor of the circuit to protect the motor against burnout, since two such elements will not protect it if the overload is in the third or unprotected conductor. Among the conditions that may create this situation are

(a)   single phasing in the primary of wye-delta or delta-wye connected transformers feeding the motors;
(b)   a single-phase load taken from the same circuit that feeds the motor, where one of the phases feeding the single-phase load is open;
(c)   a single-phase load fed from the same feeder that serves the motor, where the line drop is not negligible. No single-phase condition is necessary in this case to create current imbalance that may overload one phase of the motor; and
(d)   a large and a small motor fed from the same feeder, in which single phasing occurs. The small motor may be damaged, since it attempts to act as a phase converter to maintain current balance in the larger motor.

Rule 28-306

The manufacturer's instructions should be consulted to determine how to match the trip setting or rating information to the motor full-load current rating.

Rule 28-314

See Note to Rule 18-114.

Rule 28-316

The abbreviations "TP" and "ZP" may be used for marking "Thermally Protected" and "Impedance Protected" respectively on motors having less than 100 W input.

Rule 28-400

Upon the inspection of an installation, if it is the opinion of an inspector that automatic restarting of such motor-operated machinery as saws, routers, millers, wood and metal turning lathes, conveyors, or other moving machinery would create a hazard on return of voltage after stopping due to failure of voltage, the motor control device will be required to provide low-voltage protection.

Rule 28-602

A motor branch circuit disconnecting means is required to be located in close proximity to the branch circuit overcurrent device(s); therefore the use of fused motor circuit switches or circuit breakers is obvious. Unfused motor circuit switches, moulded case switches, and instantaneous trip circuit interrupters, etc., are often used in certain switchgear and control gear along with a separate overcurrent or overload device(s) to meet this requirement.

Rule 28-602(1)(d)

An approved combination motor controller, including a self-protected control device, is a type of equivalent device that is suitable for use as the motor branch circuit disconnecting means.

Rule 28-602(3)(b)

The use of a manually operated across-the-line type of motor controller that serves as both a starter and disconnecting means is limited by Rule 28-602(3) to the following:

(a)   the manual across-the-line controller is part of an approved combination controller, motor control unit, or controller that also includes the overcurrent protection as required by Rules 28-200 and 28-202; or
(b)   the manual across-the-line controller is used on the load side of the branch circuit over-current protection.

Section 30

Rule 30-002

Cabinet lighting system

The intent of this definition is to describe a complete, extra-low-voltage cabinet lighting system that is packaged by the manufacturer and intended for installation in accordance with the manufacturer's instructions.

Convertible luminaire

Convertible luminaire — a recessed luminaire that can be converted by the installer from a Type Non-IC to a Type IC or from a Type IC to a Type Non-IC recessed luminaire.

Undercabinet lighting system

The intent of this definition is to describe a complete, extra-low-voltage undercabinet lighting system that is packaged by the manufacturer and intended for installation in accordance with the manufacturer's instructions.

Rule 30-104

In the application of this Rule, mogul base includes Edison screw, end prong, extended end prong, side prong, bipost, and prefocus.

Rules 30-710, 30-804

Manually operated general-use 347 V ac switches can be identified by their 347 V rating and the marking "AC ONLY". Those intended for use in a box are not interchangeable in their mounting centres with switches of other types. Boxes having mounting centres spaced 89.7 mm are required.

Δ Rule 30-710(4)(a)

Suitable markings meeting the intent of this Rule are "Electric Discharge Lamp Control", where contactor rating is expressed in amperes, or "HP", where contactor rating is expressed in horsepower.

Where contactor rating is expressed in horsepower, conversion to amperes can be effected by referring to the appropriate table contained into C22.2 No. 14, the product standard for contactors.

Rule 30-910(1)

Examples of conductors other than branch circuit conductors are
(a) through conductors that only pass directly through the supply junction box to feed other luminaires down the line; and
(b) remote ballast secondary circuit conductors.

Rule 30-1000

Clause 4.3.1.3 of CSA C22.3 No. 7 addresses the grounding of poles under the authority of electric and communication utilities.

Section 32

Rule 32-000

For further information pertaining to the installation of fire alarm systems, reference should be made to CAN/ULC-S524.

It is essential that fire alarm systems be maintained in an operating condition at all times. The inspection, maintenance, and testing procedures are detailed in the *National Fire Code of Canada*, or the appropriate provincial fire code.

For further information pertaining to the installation of fire pumps, reference should be made to NFPA No. 20.

Δ Rule 32-110

It is intended by this Rule that 120 V smoke alarms conforming to CAN/ULC S531 be installed on a branch circuit that supplies lighting or a mix of lighting and receptacles in each dwelling unit and in each sleeping room not within a dwelling unit.

Rule 32-110(a)

It is not intended by this Rule to allow smoke alarms to be installed in a branch circuit supplying receptacles only.

Δ **Rule 32-200**

The intent of this Rule is to protect the feeder conductors between a fire pump and an emergency power source from fire damage.

The *National Building Code of Canada* requires that conductors supplying a life and fire safety equipment be protected against exposure to fire to ensure continued operation of this equipment for a period not less than 1 h.

NFPA 20 also mandates protection of circuits feeding fire pumps against possible damage by fire.

The following examples illustrate acceptable methods for achieving this protection:
(a)   using mineral-insulated cables conforming to fire rating requirements as specified in Clause 5.3 of the CSA C22.2 No. 124;
(b)   embedding the raceway containing fire pump feeder conductors in not less than 50 mm of concrete; and
(c)   installing the raceway containing fire pump feeder conductors in a shaft enclosure or service space of at least 1 h fire resistance construction.

Specific requirements pertaining to the fire resistance rating of a material or an assembly of materials can be found in subsection 3.1.7 of the *National Building Code of Canada* or in the appropriate provincial/territorial legislation.

Rule 32-202

Consideration should be given to the location, routing, and design of wiring to minimize hazards that might cause failure due to explosions, floods, fires, icing, vandalism, and other adverse external conditions that might impair the function of a fire pump.

Cables supplying power to regular service fire pumps should be located underground where possible.

Δ **Rule 32-206**

The intent of this Rule is to allow only a circuit breaker specifically approved for a fire pump service to be installed upstream from the fire pump controller in a normal power supply circuit, or upstream from the fire pump transfer switch in an emergency power supply circuit. It is also intended by this Rule that this circuit breaker potentially be used in the fire pump service box described in Rule 32-204. When this circuit breaker is installed in the emergency power supply circuit, upstream from the fire pump transfer switch, then the circuit breaker overcurrent protection provided by requirements of Subrule (4) should be able to allow the fire pump to operate up to locked rotor current condition. This will allow an emergency generator to provide necessary power to the required fire pumps while supplying all other loads connected to the generator. It is intended that compliance with Rule 28-200 potentially be met by selecting overcurrent protection in conformance with Table 29.

The circuit breaker installed in the normal power supply circuit, upstream from the fire pump controller, should have a rating/setting that is coordinated with the integral overcurrent protection of the fire pump controller in such a manner that the upstream overcurrent device does not disconnect the circuit prior to the operation of the fire pump controller overcurrent protection.

Note: *Clause 7-4.3.3 of NFPA No. 20 (1999) requires that the controller have an instantaneous trip setting of not more than 20 times the full load current. Clause 7-4.4 of NFPA No. 20 requires that a fire pump controller carry locked rotor current for a period of 8 to 20 s.*

Δ **Rule 32-212**

For the purpose of this Rule, a fire pump circuit is defined as the circuit supplied from the emergency power source referred to in Rule 32-200, or the circuit supplied from a separate service box in accordance with Rule 32-204 to a fire pump equipment.

Section 34

Rule 34-200

The enclosure should be constructed to prevent the emission of flames or any burning or ignited material. Openings for ventilation should be arranged to be at least 100 mm from live parts.

Metal sign enclosures should be not less than 0.68 mm thick (22 MSG). At the point where it is intended that the supply connections be made, the sign enclosure should be not less than 1.34 mm thick (16 MSG). Each enclosure housing a neon supply, transformer, or other components should be marked in accordance with the requirements of Section 2.

For neon supplies, the enclosure volume should be three times the volume of the transformer and/or internal box.

Section 36

Rule 36-000(4)

Gas-filled high-voltage switchgear and control gear enclosures may not be subject to regulation or inspection by local boiler and pressure vessel authorities. Equipment owners and electrical inspectors should be cognizant of such circumstances and take the necessary precautions to ensure the installation of safe and reliable equipment. Compliance with the following manufacturing standards for the design, construction, testing, inspection, and certification of enclosures is recommended:

CAN/CSA-C50052, *Cast Aluminium Alloy Enclosures for Gas-Filled High-Voltage Switchgear and Controlgear* (adopted EN 50052 (1986) with Canadian deviations);

CAN/CSA-C50064, *Wrought Aluminium and Aluminium Alloy Enclosures for Gas-Filled High-Voltage Switchgear and Controlgear* (adopted EN 50064 (1989) with Canadian deviations);

CAN/CSA-C50068, *Wrought Steel Enclosures for Gas-Filled High-Voltage Switchgear and Controlgear* (adopted EN 50068 (1991) with Canadian deviations);

CAN/CSA-C50069, *Welded Composite Enclosures of Cast and Wrought Aluminium Alloys for Gas-Filled High-Voltage Switchgear and Controlgear* (adopted EN 50069 (1991) with Canadian deviations);

CAN/CSA-C50089, *Cast Resin Partitions for Metal-Enclosed Gas-Filled High-Voltage Switchgear and Controlgear* (adopted EN 50089 (1992) with Canadian deviations);

CAN/CSA-C1264, *Ceramic Pressurized Hollow Insulators for High-Voltage Switchgear and Controlgear* (adopted CEI/IEC 1264 (1994) with Canadian deviations).

Δ Rule 36-006(1)

For a small access gate intended for foot traffic, a warning notice should typically be mounted on the gate. For a large vehicle access gate that may remain open for periods of time during construction work, consideration should be given to mounting the warning notice on the station fence adjacent to the gate lock for improved visibility under all situations.

Rule 36-100(4)

The marking must be designed to draw attention to the location and nature of the embedded equipment, and it must be indelible and easily legible, using materials such as metal markers and dye markings.

Rule 36-104

Any fabric tape, semi-conducting or otherwise, over the insulation should be removed completely with the metal shielding and the surface of the insulation thoroughly cleaned to remove any current-carrying residue. At all terminations and joints, stress cones should be made and adequate leakage distance provided from bare live parts. Electrical continuity of the metal shielding should be maintained across insulated joints.

Grounding should be effected at several convenient points if possible. The manufacturer's instructions and kits, if necessary, should be made available with each order of shielded conductor to ensure proper installation.

Rules 36-300, 36-308, Table 51

The conductor sizes shown in Table 51 are the minimum required to prevent conductor damage due to heating of the conductor.

Precautions should be taken where other factors are to be considered, particularly the intended application and class of use. Reference to IEEE 80, IEEE 837, and CSA C22.2 No. 41 may be necessary in the selection of appropriate devices and material. Special attention should be paid to downleads, as they may be subjected to the total fault current passed into the grid.

Rule 36-302(2)

The design of any station grounding system other than explicitly approved by the Rules should be documented and signed by an engineer in addition to being subject to acceptance in accordance with Rule 2-030.

Rule 36-302(6)

ANSI/IEEE 80 should be consulted for conductor sizing to prevent thermal damage to the rebar during fault conditions.

Rule 36-304(3)

The procedure required by Rule 36-304(3), Station Ground Electrode Design, may be found in CEA Report 249 D541.

Rule 36-308

See the Note to Rule 36-300.

Section 38

Rule 38-001

For further information see:
(a)   CSA B44, *Safety Code for Elevators and Escalators*;
(b)   CAN/CSA-B44.1/ASME-A17.5, *Elevator and Escalator Electrical Equipment*;
(c)   CSA B355, *Lifts for Persons With Physical Disabilities*; and
(d)   CSA B613, *Private Residence Lifts for Persons With Physical Disabilities*.

Rule 38-002

The control system hardware is permitted to be located in a single enclosure or a combination of enclosures, including the separate functions of motor control, motion control, and operational control.

Rule 38-003

See CAN/CSA-B44.1/ASME-A17.5 for voltage limitations within equipment.

Rule 38-013

The heating of conductors depends on rms current values, which, with generator field control, are reflected by the nameplate current rating of the motor-generator driving motor rather than by the rating of the driving machine motor, which represents actual but short-time and intermittent full-load current values.

Rule 38-013(2)(a)

Driving machine motor currents, or those of similar functions, shall be permitted to exceed the nameplate value, but since they are inherently intermittent duty, and the heating of the motor and conductors is dependent on the rms current value, conductors are sized for duty cycle service as required by Rule 28-106 and Table 27.

Rule 38-013(2)(b)

Motor controller nameplate current rating shall be permitted to be derived based on the rms value of the motor current using an intermittent duty cycle and other control system loads, if applicable.

Rule 38-013(2)(c)

The nameplate current rating of a power transformer supplying a motor controller reflects the nameplate current rating of the motor controller at line voltage (transformer primary).

Rule 38-021(1)(a)(ii)

Only electrical protective devices as required by CSA B44 and CAN/CSA-B355 are recognized as devices that can introduce a direct life hazard. (See Rule 16-010.)

If a life hazard can occur, Class 2 cables are not permitted. Cables suitable for Class 1 extra-low-voltage circuits are permitted, such as communication cable, fire alarm and signal cable, multi-conductor jacketed thermoplastic-insulated cable, and hard-usage and extra-hard-usage cable (see Table 19).

Rule 38-021(2)(a)

See the Note to Rule 38-021(1)(a)(ii).

Rule 38-021(3)(a)

See the Note to Rule 38-021(1)(a)(ii).

Rule 38-023

See CSA B44 and CAN/CSA-B355 for illumination levels.

It is not intended that devices that have machine or control spaces incorporated within the equipment, e.g., stair chairlifts that have the machine built into the chair housing, need to have separate lighting and receptacles within that space (housing).

Rule 38-024

See CSA B44 for illumination levels.

Rule 38-041

Unsupported length for the hoistway suspension means is the length of cable measured from the point of suspension in the hoistway to the bottom of the loop, with the elevator car located at the bottom landing. Unsupported length for the car suspension means is that length of cable as measured from the point of suspension on the car to the bottom of the loop, with the elevator car located at the top landing.

Rule 38-051

In order to completely isolate all control conductors in the circuits to a machine operating as one of a group, it is necessary to disconnect all the selector or programming circuitry, thereby taking all the other cars out of service. Not only is it impractical to shut down all cars in order to service one, it is often necessary to locate troubles by checking performance of components with the controller energized. Where undue shock hazards might exist, the enforcing authority shall be permitted to require the provision of cautions or warning labels pertaining thereto.

Rule 38-091(1)

The elevator must operate on such emergency power in accordance with the emergency power system requirements of CSA B44. For additional information, see CSA B44, Section 2.27.2.

Section 40

Rule 40-024(1)

In the case of equipment supplied by contact conductors, metal-to-metal contact between wheels and tracks may constitute effective grounding when a low impedance grounding path is ensured. Where local conditions, such as paint or other insulating material, prevent reliable metal-to-metal contact between wheels and tracks, a separate bonding conductor must be provided for bonding to ground.

Section 46

Rule 46-108

Reference should be made to the *National Building Code of Canada* or the appropriate sections of the provincial building codes for additional requirements for fire protection of electrical conductors used in conjunction with emergency equipment (see Section 3.2.6 of the *National Building Code of Canada*).

Rule 46-200

For additional requirements regarding the location and fire protection of emergency power supplies, reference should be made to Sections 3.2.7 and 3.6.2 of the *National Building Code of Canada*, or to the appropriate provincial/territorial legislation.

Rule 46-300

CSA C22.2 No. 141 defines unit equipment for emergency lighting as follows:

Unit equipment for emergency lighting means equipment that
(a)   is intended to provide automatically, in response to a failure of the power supply to which it is connected, a specified light output and a specified amount of power for illumination purposes, for a specified period of time, but in any case not less than 30 min;
(b)   consists of, in a unit construction, a storage battery; charging means to maintain the battery in a charged condition automatically; lamps or output terminals to which specifically listed lamps may be connected;

means to energize the lamps when the normal power supply fails and to de-energize the lamps when the normal power supply is restored; and means to indicate and test the operating condition of the equipment; and

(c)   is designed for use in applications in which the provision of emergency illumination is required by a governmental or other agency having jurisdiction.

Unit equipment certified to CSA C22.2 No. 141 is required to be marked in accordance with Clause 5.11 of that Standard as follows:

UNIT EQUIPMENT — CERTIFIED TO CSA C22.2 NO. 141, AND RECOGNIZED BY SECTION 46, *CANADIAN ELECTRICAL CODE, PART I.*

and

APPAREIL AUTONOME — CERTIFIÉ EN VERTU DE LA NORME ACNOR C22.2 NO. 141, ET CONFORME À LA SECTION 46 DU *CODE CANADIEN DE L'ÉLECTRICITÉ, PREMIÈRE PARTIE.*

Unit equipment is not to be confused with other emergency lighting equipment certified by CSA as emergency lighting units.

Such emergency lighting units do not bear the above marking but are marked "ELECTRICAL ONLY" and "CERTIFIÉ DU POINT DE VUE ÉLECTRIQUE SEULEMENT", adjacent to the CSA mark.

Rule 46-306(1)

Where approved unit equipment includes a list of lamps suitable for remote installations, the requirements of CSA C22.2 No. 141 take into account a voltage drop of 5% in the remote lamps circuit unless the list specifies that certain lamps conform to the requirements with a greater voltage drop.

Table D4 can be used to determine approximate permissible circuit lengths.

Rule 46-306(2)

The requirements of CSA C22.2 No. 141 are designed to ensure that any lamps forming part of the equipment, or specified in a list provided with the equipment as suitable for remote connection, will not exhibit an undue diminution of light intensity during the emergency period.

Rule 46-400

This Rule applies only to exit signs connected to an electrical circuit. Other requirements for exit signs, including those not connected to an electrical circuit, may be found in Section 3.4.5 of the *National Building Code of Canada.*

Rule 46-400(2)

The circuit supplying emergency lighting could be ac or dc. One requirement of the *National Building Code of Canada* is that exit signs be illuminated continuously while the building is occupied. Caution should be taken to ensure that a circuit supplying both emergency lighting and exit signs is not controlled by a switch, time clock, or other means.

Section 48

Rule 48-022

As nitrocellulose film is known to be subject to exothermic decomposition, which may result in the generation of combustible and usually poisonous gases, these requirements are necessary to retard the possible migration of such gases, as well as to prevent the spread of fire or heated gases to other vaults through raceways.

Section 50

Rule 50-000

It is intended that solar photovoltaic systems installed in hazardous locations will be installed in accordance with Section 18 requirements regardless of the current and voltage ratings.

Rule 50-002

The following diagrams of typical solar photovoltaic systems show the various terms and circuits referenced in this Section.

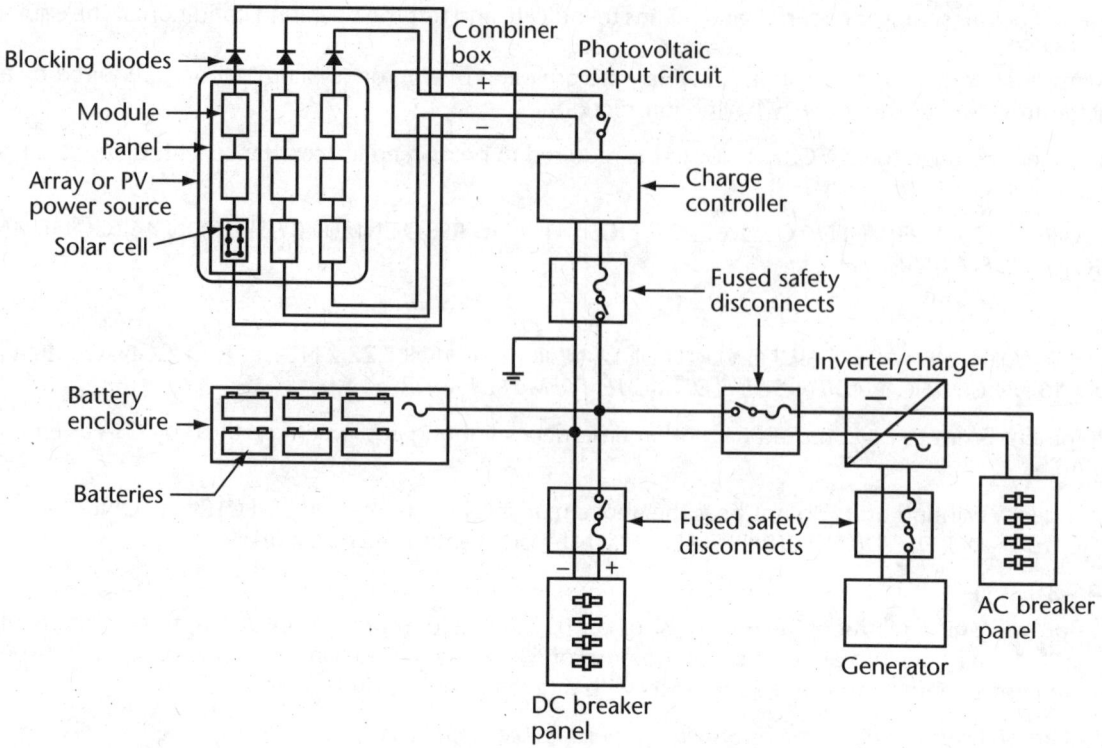

(1) Stand-alone system

Different inverters may require different photovoltaic array and wiring configurations. These configurations may be divided in two groups:

(a) a floating array (as shown in Diagram 2) that requires a two-pole disconnect switch; and

(b) a grounded array that requires a single-pole disconnect switch, except for a 3-wire neutral-grounded array that requires a two-pole disconnect switch to interrupt both ungrounded wires.

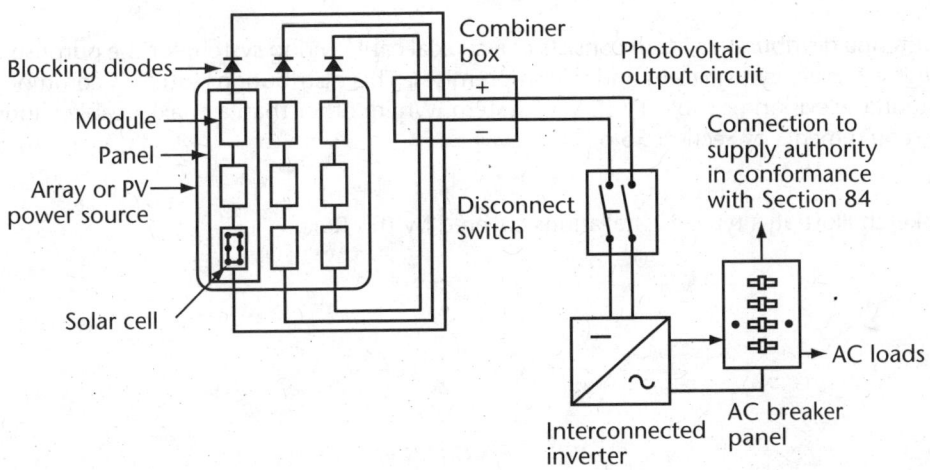

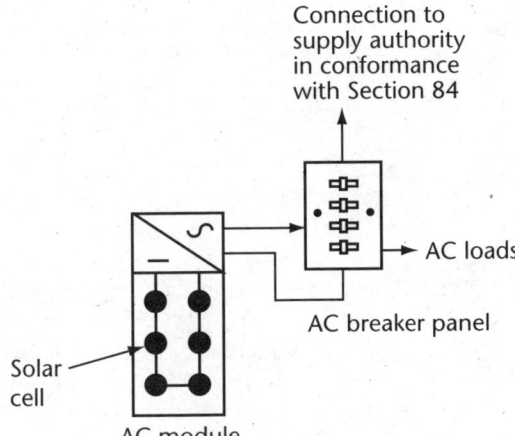

(2) Interconnected system

AC module

AC modules do not provide access to the photovoltaic output circuit which is internally connected to the power conditioning unit. The output of an ac module is then referenced as the power conditioning unit output.

Rule 50-004

When a reflecting system is used for irradiance enhancement, increased levels of output power may result. Marking of equipment should indicate the increased levels when such equipment is used.

Rule 50-012

Because photovoltaic modules are energized while exposed to light, the installation, replacement, or servicing of array components while the modules are irradiated may expose persons to the hazard of electric shock.

It is intended that means will be provided to isolate and disable portions of an array or panel that may require servicing. An opaque covering is an acceptable means of disabling the array.

Rule 50-016

When connectors and attachment plugs are used to interconnect modules or arrays, they should be used in such a manner that the blades of the attachment plug are not energized when withdrawn.

Section 54

Rule 54-004

A community antenna distribution system consists of a coaxial cable wiring system for the purpose of distributing radio and television frequency signals to and within premises. This distribution system is commonly known as a community antenna television or cable TV (CATV) system. Where other than a coaxial cable wiring system is employed, the requirements of Section 56 or 60 prevail.

Rule 54-102

The following sketch illustrates typical installations covered by this Rule.

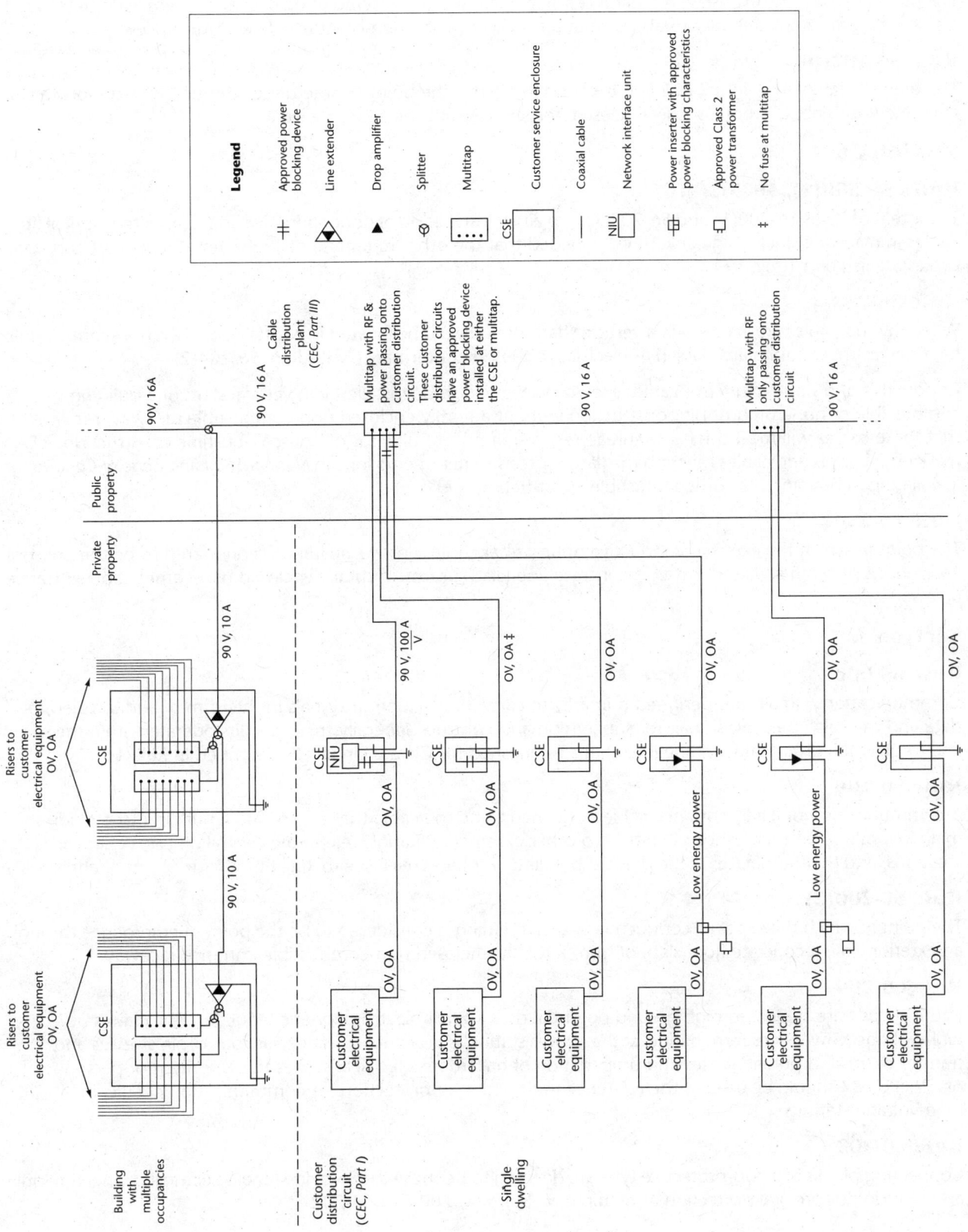

Rule 54-200(1)

The point at which the exposed conductors enter a building is considered to be the point of emergence through an exterior wall, a concrete floor slab, or from a totally enclosed non-combustible entrance raceway.

Rule 54-202(b)

The point of cable entry into a building is considered to be the point of emergence through an exterior wall, a concrete floor slab, or from a totally enclosed non-combustible entrance raceway.

Section 56

Rules 56-200(1), 56-202(1)

The intent of Rules 56-200(1) and 56-202(1) is to allow installation of nonconductive and conductive optical fiber cables in raceways, including cable trays, provided that the other requirements in Section 56 are met and such cables are listed in Table 19.

Rule 56-208

Where hybrid cables are installed in a vertical shaft, they should be located in a totally enclosed non-combustible raceway, as these cables are classed as electrical cables in conformance with Rule 56-204(2).

Conductive and non-conductive cables should be allowed to be installed in a vertical shaft of a building of combustible or non-combustible construction without a totally enclosed non-combustible raceway, provided that these cables will meet the flame spread requirements for buildings of non-combustible construction. CSA marking for wires and cables meeting the flame spread requirements for the *National Building Code of Canada* for installation in buildings of non-combustible construction is FT4.

Rule 56-214

The point at which the exposed conductive optical fiber cables enter a building is considered to be the point of emergence through an exterior wall, a concrete floor slab, or from a totally enclosed non-combustible entrance raceway.

Section 60

Rule 60-000

Communications circuits are designed primarily to carry information or signals in the form of audio, video, or data and may also transmit signals for supervision and control. Generally these circuits operate within the line-to-ground current and voltage limitations established for Class 2 circuits as described in Rule 16-200.

Rule 60-104

A communication building entrance cable is deemed to be the cable that enters into a building to provide the main incoming communication circuits from other external buildings, telephone central offices, or similar locations, and that terminates at the point of building entrance (see Note to Rule 60-700 for "point of entrance").

Rule 60-200(2)

The point at which the exposed conductors enter a building is considered to be the point of emergence through an exterior wall, a concrete floor slab, or from a totally enclosed non-combustible entrance raceway.

Rule 60-204

These circuits are subject to high ground potential rises and/or electromagnetic induction from faults of high-voltage power lines terminating at the power station. This could cause extraneous cable stresses and transfer of ground potential rises and propagation of hazardous electrical surges to equipment or personnel. ANSI/IEEE 487 should be used as the reference in provision of protection for communication circuits in high-voltage stations.

Rule 60-402

Connecting blocks of a non-protective type are deemed to be those that provide for an electrical connection only and that do not provide protection as required by Rules 60-200 and 60-202.

Rule 60-700

The point of cable entry into a building is considered to be the point of emergence through an exterior wall, a concrete floor slab, or from a totally enclosed non-combustible entrance raceway.

Rule 60-702

It is the intent of this Rule that where a conductor is installed to exclusively bond to ground the entrance cable sheath, the maximum required size is No. 6 AWG copper. However, where the conductor will also function as a common bonding conductor for other equipment, a larger gauge size may be required.

Δ ## Rule 60-704(c)

The wave front of a lightning surge has a rise time in the order of microseconds and is approximated by equivalent frequencies between 25 and 250 kHz. As a result, the self-impedance of the grounding conductor to the wave front of a lightning surge is very significant. For this reason it is paramount to keep the length of the grounding conductor as short as practicable to guarantee the effectiveness of the protector. As a guide, it is suggested to make efforts to limit the length of the grounding conductor to 6 m maximum.

Rule 60-706

It is intended that the grounding electrode for power circuits will be the common point of any multi-bonding with other grounding electrodes.

Section 62

Rule 62-114

See Note to Rule 8-104.

Rule 62-124

Splicing, modification, or assembly of series heat tracing cables in the field should be done only by trained personnel. Design of series heat tracing cables should be done by persons qualified to do so. This Rule recognizes the ability of a qualified person, using the manufacturer's design tools, to do similar design to that normally done by the manufacturer. Owners of industrial establishments are responsible for ensuring that persons undertaking work outlined in this Rule are properly qualified.

Rule 62-126(1)

Concrete floors exposed to periodic hosing down or rain (such as might occur at loading docks) should be considered wet locations.

Rule 62-206

Heating cables embedded in concrete or in plaster attached to the concrete have about the same effect on temperatures within the concrete; therefore Item (b) will apply in either case.

Rule 62-222

Formula to determine spacing between heating cable runs on or in a ceiling

Cable spacing between heating cable runs on or in a ceiling may be determined by the following formula, except that in no case may the spacing be less than 50 mm for cable sets when in contact with gypsum board or plaster lath, or when embedded in plaster or sand that is in contact with gypsum board or plaster lath (Rule 62-208(1)), or less than 25 mm for cable sets when embedded in concrete or poured masonry (Rule 62-224).

$$S = \frac{(L - 300) \times (W - 300) - A}{U}$$

where
S = cable spacing between runs in millimetres
L = length of room in millimetres
W = width of room in millimetres
A = total area in square millimetres of unusable ceiling area such as that required for recessed lighting fixtures, closets, cabinets, and any outlet to which a lighting fixture or any other heat-producing equipment is liable to be connected [Rule 62-214(9)]
U = length of cable in millimetres

Rule 62-300

Heating cable sets with type designations 5A and 5B meet the requirements of CAN/CSA-C22.2 No. 130.1 (withdrawn) and are suitable for use in industrial applications such as petrochemical, chemical, smelting, and similar plants in non-hazardous locations.

Rule 62-300(3)

This Rule is intended to apply to industrial applications of heating panel sets intended for industrial use. This Rule is not intended to adapt non-industrial heating panel sets to industrial applications.

Rule 62-400

Usage marking of heating devices and cross-references to Table 60 of the *Canadian Electrical Code, Part I* (18th edition, 1998) — Heating cable set type designations and applications

Usage marking and usage		Weather resistance	Wet location applications	Wet location under pressure	Previous type designations and applications
No usage mark or G	General use	No	No	No	1A — Ceiling 1B — Floor embedded in concrete (if dry location) 3A, 3C, 5A — Dry and damp location external pipe or vessel tracing
S	With weather resistance	Yes	No	No	All applications where a degree of weather resistance is required including damp locations. (No previous equivalent.)
W	With wet rating	No	Yes	No	2A — Soil heating 2B — Snow melting and floor embedded in concrete (wet location) 2C — Animal pens 2D — Pool decks 3B, 3D, 5B — Wet location pipe and vessel surface heating 4A, 4B — Pipe interior heating systems (unpressurized applications)
P	With wet/pressure test	No	Yes	Yes	4A, 4B — Pipe interior heating systems
WS	With wet test and weather resistance	Yes	Yes	No	2E — Roof de-icing
PS	With wet/pressure test and weather resistance	Yes	Yes	Yes	4A, 4B — Pipe interior heating systems
X*	Specially investigated heating device or heating device set	—	—	—	e.g., 2C — Animal pens 4A, 4B — Pipe interior heating in potable water Applications where mechanical abuse is very unlikely, allowing for lower mechanical test levels.

*X indicates additional requirements and/or exemptions for specific applications. The manufacturer's instructions must include a complete explanation of these additional requirements or exemptions.

Rule 62-500

Sauna heaters should be secured in place ensuring that the minimum clearances specified in the nameplate are not reduced. If the heater is provided with legs, they should not be removed in favour of other supports. Covering combustible surfaces with non-combustible material, such as metal tile or asbestos board, does not ensure safety from fire.

Sauna heaters marked "FOR INSTALLATION ON CONCRETE FLOORS ONLY" shall not be installed on combustible floors even if the floor is covered with ceramic tile, asbestos board, or other non-combustible material.

Equipment or material of other than an electrical nature should not be installed or placed so close to electrical equipment as to create a dangerous condition. Benches, shelves, guard rails, other structures, or obstructions should not be placed closer to the heating unit than is permitted for the clearances specified on the nameplate.

To properly control the maximum temperature in the room, the heat sensor for the temperature control should be located near the heater. A timer should be installed to turn off the heater after a predetermined or preset time.

Section 68

Rule 68-054

The following sketch illustrates the minimum clearances for conductors over swimming pools. No conductors would be permitted under any circumstances in the area under Line 1. In the area above Line 1, insulated communication conductors and neutral supported cables operating at 600 V or less might be permitted (see Subrules (2) and (3)). Any other conductors operating at not more than 50 kV might be permitted above the area outlined by Line 2 (see Subrules (2) and (4)).

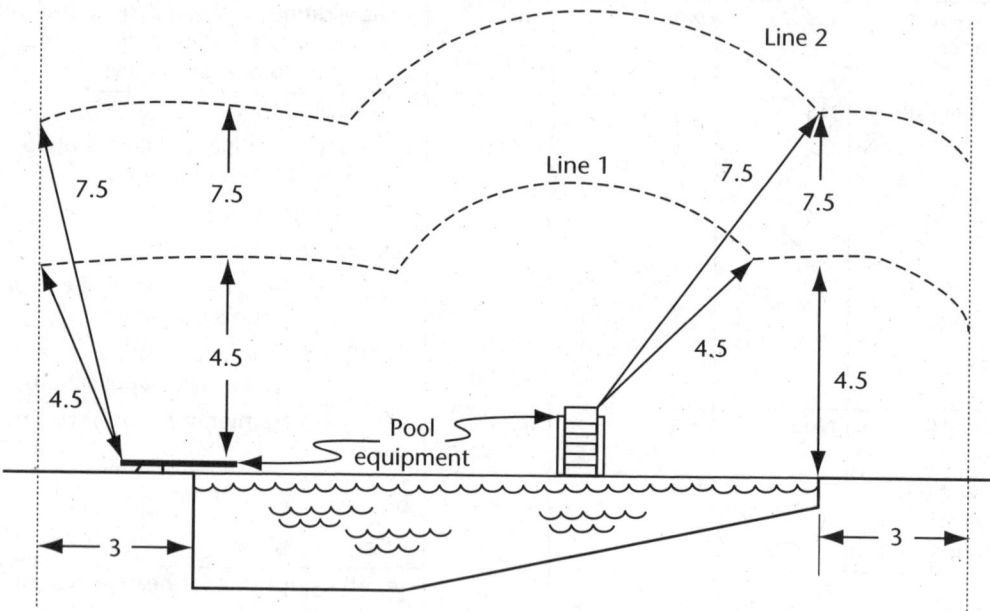

Note: *All dimensions given are in metres.*

Rule 68-058(3)

Where reinforcing steel encapsulated with a non-conductive compound (epoxy coated rebar) is used in the construction of a pool, an alternative means to eliminate voltage gradients could be a loop around the pool of a minimum No. 6 copper conductor installed below normal water level.

Rule 68-058(8)

Even though no electrical equipment is located within 3 m of a pool, it is recommended that metal parts of an in-ground pool be interconnected with a minimum No. 6 AWG copper conductor. This guards against the effects of stray currents in the ground and reduces the cost of meeting *Canadian Electrical Code* requirements, should electrical equipment be added later.

Rule 68-060(6)

The deck in the vicinity of the deck box may be sloped up to the top of the deck box from the normal deck level.

If a deck box is located so that the top of the box is above the finished level of the pool deck, the box should not be located in a walkway unless afforded additional protection such as by location under diving boards, adjacent to fixed structures, and similar areas.

Rule 68-062(3)

Audio isolation transformers should

(a)   have either the primary and secondary windings wound on separate bobbins on the core legs or a grounded metal shield between the primary and secondary windings; and

(b)   withstand a 60 Hz test voltage of 2500 V applied between the primary and secondary windings for a period of 1 min without breakdown.

Microphones used in the vicinity of pools and baptismal fonts must have audio isolation transformers and cables with ungrounded conductors installed between them and any mixer, pre-amplifier, amplifier, or like equipment.

Rule 68-068

The 1.5 m separation is intended to prevent the occupant of the pool from resetting the ground fault circuit interrupter. This separation should be the shortest unobstructed distance, which need not follow a straight line.

Δ   Rule 68-306

The warning label to be affixed to the receptacle for the cord-connected hydromassage bathtub is intended to warn against and prevent the connection of any other equipment to the receptacle.

Δ   Rule 68-404(4)

The intent of the Subrule is to provide protection against entrapment hazards associated with spas and hot tubs located in any occupancy except for a dwelling unit, by installing an emergency shutoff switch within sight and 15 m of the spa or hot tub. It is also intended that such a switch be located at least 1 m horizontally from the spa or hot tub, as specified in Subrule (1). The shutoff switch used for this purpose may be an inline-operated device or a remote-control circuit that causes the pump circuit to open.

Rule 68-408(3)

An inverted U-shaped pipe installed in the air pipe so that the bottom of the top loop of the pipe is not less than 300 mm above the tub rim is considered an acceptable means to prevent water from contacting blower live parts.

Section 70

Rule 70-106

The conduit service facility that must be provided for the main communication entrance service into a mobile home is usually located in the living room, main hallway, or kitchen area. An additional conduit service facility should be provided wherever an extension communication outlet is to be installed such as in the master bedroom. If the conduit service facility is to be used for both telephone and cable television services, the minimum conduit size should be 21 trade size inside diameter.

Rule 70-122(1)

In applying Rule 26-712(a) for this Rule, it is not necessary to include in the linear measurements the space occupied by standard door openings, closets, or cupboards that have been designed to render the wall space unusable for electrical equipment.

Δ   Rule 70-122(4)

The intent of this Rule is to ensure that receptacles that are mounted on the underside of trailers to supply power to heating cable sets for freeze protection of plumbing pipes on trailers are provided with Class A ground fault protection and that they are properly identified for that use. Heating cable sets (particularly the self-limiting type) without ground fault protection have caused numerous fires on trailers.

Rule 70-130

When insulation resistance or ac dielectric strength tests are performed, precautions should be taken to ensure that voltage-sensitive devices such as ground fault circuit interrupters are not subjected to voltages that will damage the device.

Section 72

Rule 72-110

The 30 A, 125 V, 2-pole, 3-wire NEMA WD 6, Figure TT receptacle referred to in Subrule (1)(b) is:

The 5-15R and 14-50R receptacle configurations referred to in Subrules (1)(a) and (1)(c) are shown in Diagram 1.

Section 74

Δ **Rule 74-004(4)**

Because there are many variables in the structural design of airport runways and aprons depending on the required design strength of the surface material and in order to ensure that the installation of a conduit within the surface material will not result in damage to the runway or apron, this Rule intends that such an installation be designed by a civil engineer in accordance with good engineering practice and that the design be acceptable to the airport authorities.

Section 78

Rule 78-108

To minimize deterioration due to marine environmental conditions, the following materials have been found to be generally acceptable:
(a)   copper-free aluminum with aluminum or stainless steel hardware;
(b)   fibreglass with stainless steel hardware;
(c)   epoxy-coated rigid steel threaded conduit;
(d)   PVC-coated rigid steel threaded conduit;
(e)   19 mm plywood, either penta-treated or painted with two coats of marine grade paint and used with galvanized or stainless steel hardware;
(f)   rigid PVC boxes and enclosures with stainless steel hardware; or
(g)   hot-dipped galvanized structural steel.

Section 80

Rule 80-000

NACE International Standards are recommended as guides to the design, materials specifications, installation, and operation of cathodic protection systems.

Rule 80-000

This Section has the following objectives:
(a)   to recognize that cathodic protection systems have to be installed using wiring methods that may not be consistent with those of other sections of the Code;
(b)   to address the electrical safety of the cathodic protection systems and not their efficacy.

Rule 80-006

Care should be taken to select clamps that maintain a secure electrical connection and that will be anodic to the material being protected when in the presence of an electrolyte, so that the clamp will not itself corrode the material if the connection becomes wet.

When welding to oil or gas piping, reference should be made to CAN/CSA-Z662.

Section 84

Δ **Rule 84-000**

Where power generating equipment such as photovoltaic arrays, fuel cells, micro-turbines, etc., supply power through an approved inverter, the output of the inverter is considered to be the "electric power production source".

Δ **Rule 84-002**

The consumer electric power generator owner should consult with the supply authority before planning the interconnection.

The interconnection arrangements should not adversely affect the safety of the supply authority system.

The output of the electric power production source, when interconnected with a supply authority electric system, should not adversely affect the voltage, frequency, or wave shape of the system to which it is connected.

Δ **Rule 84-008**

Where the utility loses one phase of a 3-phase system, some transformer configurations allow a voltage to continue to be present on all phases, and the voltage drop is often not high enough to cause the electric power production source to shut down. Because the electric power production source continues to detect a voltage within tolerance on all phases, it is not expected that the electric power production source shut down.

The words "disconnected" and "disconnection" in the context of this Rule do not necessarily mean "disconnecting means" as used elsewhere in the Code.

Δ **Rule 84-022**

The supply authority disconnecting means is intended to allow the supply authority a single point of access to simultaneously isolate one or more electric power production sources on a premise. The main service box, or the equivalent, is normally used to provide this function.

Δ **Rule 84-024**

In some circumstances, the supply authority may use provisions of Rule 84-002 to require the disconnecting means to have "contact operation verifiable by direct visible means". This is a common worker safety feature used by supply authority workers to provide added assurance that the circuit is open before work is initiated.

Where inverters approved for interconnection are used, the anti-islanding feature automatically isolates the generation equipment from the supply authority upon loss of supply authority voltage so that having "contact operation verifiable by visible means" may not be required. CSA C22.2 No. 107.1-01 Section 15 applies to utility-interconnected inverters and requires the inverter to automatically cease to deliver ac power to the utility in accordance with an anti-islanding test — within the time in Table 3.1 and after the output V and frequency of the utility source are adjusted to each condition in Table 3.1. Utility abandonment of the interface disconnect switch would require the utility to rely 100% on the inverter to be fail-safe under normal operation and component fault mode re-energizing a dead utility bus. A small generator can magnetize a 1-phase distribution transformer when the transformer is disconnected from the primary conductor.

Δ **Rule 84-028**

The isolating transformer referred to in Subrule (2) can be remote from or integral to the inverter.

Δ **Rule 84-030**

The single line diagram should identify related components of the interconnected system including switching arrangements, interlocks, isolation points, and their relative locations. See Diagram (2) in the Appendix B Note to Rule 50-002 for diagrams of an interconnected photovoltaic system.

Section 86

Rule 86-102

Installation of electric vehicle charging equipment is not intended to be considered a hazardous location unless the location is specified as hazardous in accordance with Section 18 or 20, or where ventilation is required and not provided.

Rule 86-400

It is the intent of this Rule to provide ventilation with electric vehicle charging equipment unless the equipment is marked for use with electric vehicles not requiring ventilation, or where the manufacturer's installation instructions specify that ventilation is not required. When designing ventilation for indoor charging sites where vented storage batteries are used, both supply and mechanical exhaust equipment should be installed and located to intake from, and vent directly to, the outdoors.

See the Note to Rule 26-546 for similar considerations.

Tables

Table 32

A light snow area is considered to be an area in which the mean annual recorded depth of snow is 500 mm or less. This information for any area in Canada can be obtained from the following:

(a)   Meteorological Service of Canada, Environment Canada; and

(b)   *Atlas of Canada*, published by Natural Resources Canada.

Tables 33 and 34

The spacings and clearances shown in these tables differ intentionally from those found in CAN/CSA-C22.3 No. 1, as explained in Clause 4.2.1 of that Standard:

4.2.1 Construction and day-to-day clearances

The stated clearances for wires and conductors are minimum values related to maximum specified loads and service conditions; that is, those stated clearances represent design limits rather than clearances for construction or day-to-day operation. Clearances provided at the time of construction, under ambient conditions then prevailing, must by design be sufficiently greater than the stated minimum clearances to ensure that the actual clearances that will result under the maximum specified loads and service conditions will not be less than the stated minimum clearances. Consequently, clearances under day-to-day conditions will be greater than the stated minimum clearances when loads and service conditions are less severe than the specified maximum conditions.

Note: *Clearances specified in the* Canadian Electrical Code, Part I, *apply at the time of installation rather than under specified maximum conditions and therefore are larger than those specified in the* Canadian Electrical Code, Part III, *for the reasons explained in Clause 4.2.1 above.*

Table 51

See the Note to Rule 36-300.

Diagrams

Diagrams 1 and 2

See the Note to Rule 26-700.

Diagram 3

See the Note to Rule 14-102.

Appendix C
The Technical Committee on the Canadian Electrical Code,
Part I — *Organization and rules of procedure*

Appendix C — The Technical Committee on the *Canadian Electrical Code, Part I* — Organization and rules of procedure

Note: *This Appendix is a normative (mandatory) part of this Standard.*

C1. General

C1.1
The Technical Committee on the *Canadian Electrical Code, Part I* (hereinafter called the Committee on Part I) shall operate under the authority of the Strategic Steering Committee on Requirements for Electrical Safety and in accordance with *CSA Policy governing standardization — Code of good practice for standardization* and *CSA Directives and guidelines governing standardization.*

C1.2
The Committee on Part I shall be responsible for the development of the *Canadian Electrical Code, Part I* (hereinafter called the *CE Code, Part I*), which shall consist of safety standards for the installation and maintenance of electrical equipment.

C2. Committee on Part I

C2.1 Terms of reference
The Committee on Part I shall be responsible for
(a) establishing Committees and Subcommittees, appointing a Chair and Vice-Chair, and establishing the terms of reference for them;
(b) planning, programming, coordinating, and monitoring the activities of Committees and Subcommittees;
(c) recommending adoption of amendments to the *CE Code, Part I*;
(d) determining the form and arrangement of the *CE Code, Part I*;
(e) interpreting the *CE Code, Part I*;
(f) all policy matters related to the *CE Code, Part I*;
(g) setting up procedures that will facilitate feedback to the Committee on Part I from regulatory authorities, the Canadian Standards Association, industry, users, and others; and
(h) establishing and maintaining liaison with the Canadian Advisory Council on Electrical Safety, the Technical Committee on the *CE Code, Part II*, other Strategic Steering Committees, and national and international organizations responsible for safety standards for the installation and maintenance of electrical equipment.

C2.2 Structure

C2.2.1
The Committee on Part I shall consist of
(a) members as specified in Clause C2.3;
(b) a Chair and Vice-Chair appointed from the members, each of whom shall serve, subject to the approval of the Strategic Steering Committee on Requirements for Electrical Safety, a term of 3 years and shall be eligible for reappointment;
(c) an Executive Committee;
(d) Subcommittees; and
(e) a Project Manager (nonvoting) appointed by CSA.

C2.2.2
Chairs of Subcommittees, if they are not voting members of the Committee on Part I, shall be recorded as ex officio, nonvoting members of the Committee on Part I.

C2.3 Members

C2.3.1 Matrix

C2.3.1.1 General
The Committee on Part I shall be composed of not more than 41 voting members, representing the following interests:

	Range	
	Minimum	Maximum
Regulatory authority	11	16
Owner/Operator/Producer	9	14
General interest	9	14

The Committee shall also include associate, liaison, and ex officio members (nonvoting) as required.

C2.3.1.2 Regulatory authorities
The regulatory authorities shall be selected from the various provincial, territorial, and municipal electrical inspection authorities.

C2.3.1.3 Owners/Operators/Producers
The Owner/Operator/Producer representatives shall be selected from groups with national stature representing the viewpoints of
(a)   electrical equipment manufacturers;
(b)   electrical installation designers and installers; and
(c)   electrical installation users.

C2.3.1.4 General interest representatives
The general interest representatives shall be selected from groups with national stature representing the viewpoints of
(a)   fire chiefs;
(b)   electric utilities;
(c)   committees responsible for related electrical codes and standards;
(d)   fire insurers;
(e)   labour;
(f)   issuers of building codes; and
(g)   educators.

Δ   **C2.3.2**
Members shall be nominated by the interest or organization that they represent, and their appointment shall be subject to the approval of the Executive Committee or the Chair, in concurrence with the Project Manager.

C2.3.3
Members shall participate actively in the work of the Committee on Part I, attend meetings, accept the Chair of Section Subcommittees, and participate in Subcommittee work.

Δ   **C2.3.4 Termination of membership**
In consultation with the Project Manager and after enquiry, the Chair of the Committee on Part I (on behalf of the Executive Committee) should recommend that a TC member be removed from the Committee if the member has failed to
(a)   attend three consecutive meetings;
(b)   respond to three consecutive letter ballots; or
(c)   be actively and effectively involved in the work and responsibilities of the Committee.
    Notice of pending termination should be sent to that member by CSA staff. Subsequent failure to comply with the requirements should result in termination as directed by the Vice-President or Program Director, Standards Development.

C2.3.5
The Executive Committee shall recommend the removal of a member after consultation with the nominating interest and the Project Manager.

C2.4 Meetings

C2.4.1
The Committee on Part I shall meet at least once a year.

Appendix C
The Technical Committee on the Canadian Electrical Code,
Part I — *Organization and rules of procedure*

C2.4.2

Notices and agendas of meetings shall be sent at least 4 weeks in advance of the meeting date.

C2.4.3

One half of the total membership shall constitute a quorum. Proxies shall not be included. Alternates shall be included.

C2.4.4

Voting by proxy shall be permitted for the Committee on Part I, provided that notice of the proxy is filed with the Chair prior to the meeting.

C2.4.5

An absent member may, with the approval of the Chair, be represented at a meeting by an alternate who may vote in that member's stead.

C2.4.6

In the event of a lack of quorum, or if desired by those at a meeting where a quorum is present, the vote shall be taken by a letter ballot at a later date. To be valid, a ballot shall be returned within 30 days.

C3. Regulatory authority committee

C3.1 Terms of reference

The Regulatory Authority Committee shall be responsible for advising the Committee on Part I when the language of an amendment is deemed unacceptable from an enforcement or legal standpoint.

C3.2

The Regulatory Authority Committee shall consist of
(a)   the regulatory authorities' representatives who are members of the Committee on Part I;
(b)   a Chair and Vice-Chair appointed from the members of the Regulatory Authority Committee; and
(c)   a Project Manager (nonvoting) provided by CSA.

C3.3

The voting members of the Regulatory Authority Committee shall consist of the provincial and territorial inspection authority representatives who are members of the Committee on Part I.

C3.4

The Regulatory Authority Committee shall have the authority, within its terms of reference, to agree or disagree with the proposed amendments to the *CE Code, Part I*. Its terms of reference shall not give it the authority to amend the *CE Code, Part I*. (See Clause C7.2.1.3.)

C3.5 Legal amendments

C3.5.1

A legal amendment changes
(a)   the words but not the intent of Rules in one or more Sections of the Code; or
(b)   the administrative Rules in Section 2 or the Scope of Section 0 affecting regulatory implementation of the Code.
A legal amendment is initiated by a member of the Regulatory Authority Committee.

C3.5.2

The Regulatory Authority Committee may act as the Section Subcommittee for legal amendments.

C3.5.3

The Chair of the Regulatory Authority Committee may appoint a member of the Regulatory Authority Committee to act as the Chair of the Subcommittee for purposes of achieving consensus, preparing a Subcommittee report, or resolving negative Part I ballots.

C3.5.4

While the Regulatory Authority Committee is acting as a Subcommittee for a legal amendment, the voting process in Clause C7.2 does not necessarily apply.

C3.5.5

After a report has been submitted to the Committee on Part I, it will be sent to each affected Part I Subcommittee for a minimum of one month and a maximum of two months to identify and report on those Rules where the proposed legal amendment may have changed the intent.

C3.5.6

The report, as revised by the Regulatory Authority Committee (in its role as a Subcommittee) due to comments from the affected *CE Code, Part I* Subcommittees, will then be processed as any other Subcommittee report and submitted for Part I letter ballot in accordance with Clause C7.1.2.

C3.6 Editorial changes

C3.6.1

An editorial change is one that revises the words of a Rule or a portion of an Appendix to improve clarity of expression without changing the intent of the original wording or affecting the safety or cost of an installation.

C3.6.2

The Regulatory Authority Committee may act as the Section Subcommittee for editorial amendments.

C3.6.3

The Regulatory Authority Committee may appoint a working group(s) to review particular Rules and prepare recommendations.

C3.6.4

Editorial changes shall be voted on following the process described in Clauses C7.2.2 and C7.2.3.

C3.6.5

Chairs of the Part I Section Subcommittee responsible for the Rules under discussion will be advised when there is to be a Regulatory Authority Committee ballot. Any negative comment shall be treated as a negative vote by a Regulatory Authority Committee member.

C4. Executive Committee

C4.1 Terms of reference

The Executive Committee shall

(a)   act in an advisory capacity to the Committee on Part I on administrative matters;

(b)   assist the Chair in monitoring the rate of progress of the Subcommittees and be ready to offer administrative assistance in the event of delays;

(c)   assist the Chair in the appointment or replacement of Subcommittee Chairs;

(d)   recommend to the Committee on Part I any proposed changes that it deems necessary in procedures, operation, or policy of the Committee on Part I;

(e)   work with CSA staff to implement any changes in procedures or operations that have been approved by the Committee on Part I;

(f)   make recommendations to the Committee on Part I on

(i)   requests for membership on the Committee on Part I; and

(ii)   replacements of members of the Committee on Part I; and

(g)   periodically review the matrix of the Committee on Part I and make recommendations to that Committee.

C4.2 Membership

The members of the Executive Committee shall be members of the Committee on Part I and shall consist of the following:

(a)   the Chair and Vice-Chair of the Committee on Part I (who shall be Chair and Vice-Chair, respectively, of the Executive Committee);

(b)   two representatives from each of the regulatory authority, owner/operator/producer, and general interest categories who shall be elected by the Committee on Part I; and

(c)   a Project Manager (nonvoting) provided by CSA.

C5. Section Subcommittees

C5.1 Terms of reference

Subcommittees shall be responsible for the preparation, amendment, and interpretation of the Sections assigned to them by the Committee on Part I.

Appendix C
The Technical Committee on the Canadian Electrical Code,
Part I — *Organization and rules of procedure*

C5.2 Structure

Subcommittees shall have a Chair appointed by the Committee on Part I. The Chair of a Subcommittee shall be a member of the Committee on Part I and may be a voting or nonvoting member (see Clause C2.2.2).

C5.3 Members

C5.3.1

The Chair of a Subcommittee shall appoint the Subcommittee members. Requests for membership on Subcommittees shall be directed through the Project Manager of the Committee on Part I, who shall coordinate such requests with the Subcommittee Chair.

C5.3.2

The Chair of the Committee on Part I or the Subcommittee Chair may appoint a member of the Subcommittee to act as Vice-Chair. The Vice-Chair shall act as Chair in the absence of the Chair.

C5.3.3

It is recommended that representation on Section Subcommittees be chosen from among the following categories, in accordance with the major interests of the Subcommittee:

(a)   inspection authorities;
(b)   manufacturers of electrical equipment;
(c)   employers;
(d)   employees;
(e)   consultants;
(f)   utilities;
(g)   testing laboratories, underwriters, or fire marshals;
(h)   primary and secondary industries;
(i)   respective code-making panels of the *National Electrical Code*; and
(j)   users.

C5.3.4

It is further recommended that

(a)   Subcommittees for the General Sections (Sections 0 to 16 and 26) be composed of not more than 12 members;
(b)   Subcommittees for the other Sections (Sections 18 to 24 and 28 to 86) be composed of not more than 8 members;
(c)   at the discretion of the Subcommittee Chair, the number of members be increased if further representation is required;
(d)   at least one member of a Subcommittee in addition to the Chair be from the Committee on Part I;
(e)   75% of a Subcommittee membership be non-Part I members with not more than one-third of the membership from any one category;
(f)   if practicable, the Subcommittee membership be balanced in representation from the various geographical areas of Canada; and
(g)   requests for representation from categories such as manufacturers, electrical contractors, consultants, and utilities be directed to the organization if such exists.

C5.3.5

Subcommittees should consult with individuals or organizations outside the membership of the Subcommittee or the Committee on Part I when specific data or information may be required. Experts on specific subjects may be asked to attend meetings of the Subcommittee, or to submit special data or information to the Subcommittee for its use.

C5.3.6

The Subcommittee Chair may set up task groups to study and report on specific problems. Task groups may include individuals with expertise not available within the Subcommittee.

C5.3.7

Members of a Subcommittee shall be provided with the names and addresses of the other members of the Subcommittee.

C5.3.8

The Subcommittee Chair should review periodically the performance of each member of the Subcommittee and decide on any changes to the Subcommittee membership. Consideration should be given to the calibre of responses to correspondence, promptness in responding to requests for comment, and attendance at meetings.

C5.3.9

Members of Section Subcommittees who are not members of the Committee on Part I should be advised by the Project Manager about the action taken by the Committee on Part I.

C5.3.10

Members of the Subcommittee should review, on a continuing basis, the Section of the Code for which they are responsible and should propose amendments where necessary.

C5.4 Subcommittee operation

C5.4.1

After receiving a proposal from the Project Manager of the Committee on Part I, the Subcommittee Chair shall review the proposal and shall submit it to the Subcommittee members.

C5.4.2

Meetings shall be held as necessary.

C5.4.3

Decisions at meetings or through correspondence shall be based on the consensus principle.

Δ **Note:** *As defined in CSA Directives, CSA-SDP-2.1, consensus in standardization practice is achieved when substantial agreement is achieved. Consensus implies much more than a simple majority, but not necessarily unanimity.*

C5.4.4

The Subcommittee Chair shall report the Subcommittee's recommendation on the proposal to the Project Manager of the Committee on Part I.

C5.4.5

A Subcommittee recommendation shall be submitted to the Project Manager of the Committee on Part I within a period not exceeding 42 months from the date that the original proposal was received, in accordance with Clause C5.4.1.

C5.4.6

Subcommittee reports should be presented in the standard format (see Annex A) and should include, in addition to the proposal, the name and affiliation of the submitter, the reason for the proposal, a summary of the Subcommittee's deliberations, and the Subcommittee's recommendation.

C5.4.7

The summary of Subcommittee's deliberations should include the comments of any members who may not be in agreement with the Subcommittee's recommendation.

C5.4.8

If the proposed amendment could affect a product, the Subcommittee report to the Committee on Part I shall include a recommendation that the Technical Committee on the pertinent standard(s) of the *Canadian Electrical Code, Part II*, be advised.

C5.4.9

If the Subcommittee report recommends removal of a product design requirement, the Subcommittee report shall state whether or not the particular requirement is included in applicable equipment standards.

C5.4.10

The Project Manager shall submit the Subcommittee report to the Committee on Part I for letter ballot in accordance with Clause C7.1.2.

C5.4.11

The submitter of a proposal that has been sent to the Subcommittee in accordance with Clause C5.4.1 has the right to withdraw the proposal any time before the Subcommittee's recommendation is sent to the Committee on Part I for approval.

Appendix C
The Technical Committee on the Canadian Electrical Code,
Part I — Organization and rules of procedure

C5.4.12

If the submitter of a proposal requests its withdrawal after the Subcommittee recommendation has been sent to the Committee on Part I for approval, the withdrawal is subject to the approval of the Subcommittee.

Where the Subcommittee agrees not to take sponsorship of a proposal withdrawn by the original submitter, the subject shall be closed.

Notes:
(1) *Amendments that affect Section 24 are also balloted by the Technical Committee on Electrical Installations in Health Care Facilities and by the Strategic Steering Committee on Health Care Technology.*
(2) *Amendments that affect Section 52 are also balloted by the Technical Committee on Equipment for Radiology and Nuclear Medicine and by the Strategic Steering Committee on Health Care Technology.*

C6. Requests for amendments to the *CE Code, Part I* — General

C6.1

A request for an amendment to the *CE Code, Part I*, may be submitted to the Project Manager of the Committee on Part I by any person, organization, or committee (see Annex B).

C6.2

A request for an amendment to the Code shall include a specifically worded proposal, reasons for the proposal, and supporting data. The wording to be added, changed, or deleted shall be submitted in such a way that the intent is clear. An unclear proposal may be returned to the submitter by the Project Manager after consultation with the Section Chair and the Chair of the Committee on Part I.

C6.3

The Project Manager shall assign the request a subject number and submit it to the Chair of the appropriate Section Subcommittee for the preparation of a report and recommendation by the Subcommittee (see Clause C5.4).

C6.4

If the report on the assigned subject is not completed by the Section Subcommittee in accordance with Clause C5.4.5, the subject may be closed on a recommendation made by the Chair of the Committee on Part I.

Δ **C6.5**

If the proposed change affects new products, the Project Manager shall request that the appropriate Subcommittee Chair give priority to these proposed amendments. As soon as the Subcommittee report is received from the Section Subcommittee it shall be forwarded by the Project Manager to the Committee on Part I for 30-day ballot.

C7. Approval of amendments to the *CE Code, Part I*

C7.1 Approval by the Committee on Part I

C7.1.1 General

C7.1.1.1

The Chair and Vice-Chair shall be entitled to vote.

C7.1.1.2

If the recommendation is approved and would result in a change to the Code, it shall be submitted to the Regulatory Authority Committee for review.

C7.1.1.3

If the recommendation is approved and the resulting Rule amendment affects a Standard under the *CE Code, Part II*, the Project Manager of the Committee on Part I shall inform the relevant Part II Technical Committee Chair and Project Manager that a modification of the Standard is required, emphasizing that the Committee on Part I requires an answer within 8 months regarding the action to be taken.

C7.1.1.4

In reviewing amendments to the *CE Code, Part I*, the Committee on Part I members shall act without regard to the individual viewpoint of the interest that they represent. In approving amendments, the Committee attests that

(a)   the amendment satisfies the intent;

(b)   the amendment has been subjected to proper procedures; and

(c)   as far as it is aware, the amendment does not conflict with other amendments, with published CSA Standards, or with National Standards of Canada.

C7.1.1.5

In addition to the criteria given in Clause C7.1.1.4, members may vote on the technical adequacy of an amendment. Any points raised by the members shall be dealt with by the appropriate Section Subcommittee.

C7.1.2 Approval by letter ballot

C7.1.2.1

The Subcommittee report and recommendation shall be submitted to the Committee on Part I for letter ballot approval, unless otherwise authorized by the Chair of the Committee on Part I. To be valid, a ballot shall be returned within 30 days.

C7.1.2.2

If there are no negative votes, the recommendation shall be considered approved, provided that at least 50% of the total voting membership voted affirmative.

C7.1.2.3

If there are any negative votes that cannot be resolved by the action outlined in Clause C7.3, the subject shall be included in the agenda of the next meeting of the Committee on Part I for the purpose of resolution.

Δ ### C7.1.2.4

If a member of the Regulatory Authority Committee submits a negative vote for regulatory reasons, it shall be indicated as such and shall be accompanied by a revised amendment carrying the same intent regarding safety and technical requirements.

C7.1.2.5

If the recommendation is not approved, the subject shall be included in the agenda of the next meeting when it shall be either returned to the Subcommittee or closed.

C7.1.3 Approval at a meeting

C7.1.3.1

Meetings shall be conducted in accordance with the procedures in CSA Directives, CSA-SDP-2.1-99, Clause 9.6.9, "Rules of procedure for conducting a meeting".

C7.1.3.2

The Subcommittee's recommendations on motions or on an amendment or interpretation shall be considered approved, provided that at least 50% of the total voting membership voted affirmative and that at least two-thirds of the votes cast are affirmative.

C7.1.3.3

When a subject is placed on the floor, the Chair should allow a general discussion of it prior to a motion being made.

Δ ### C7.1.3.4

If the Section Subcommittee's recommendation is to reject the submitter's proposal and the Committee on Part I rejects the Section Subcommittee's recommendation, the Committee on Part I shall return the proposal to the Section Subcommittee for further review of the reasons for rejection provided by the Committee on Part I.

Δ ### C7.1.3.5

If the Section Subcommittee's recommendation is either to accept the proposal as submitted or to accept the proposal with amendment and the Committee on Part I rejects that recommendation, the Committee on Part I shall

(a)   accept a motion to close the subject; or

(b)   accept a motion to return the subject to the Subcommittee for further review.

C7.1.3.6

Amendments to a motion cannot be accepted if they have the effect of defeating the main motion.

C7.1.3.7

When a motion to close the subject is passed by the Committee on Part I, the submitter shall be informed by the Project Manager of the Committee on Part I of the proposal's rejection, with the reasons for rejection, unless the

Appendix C
The Technical Committee on the Canadian Electrical Code,
Part I — *Organization and rules of procedure*

submitter is also a member of the Committee on Part I. The matter may be resubmitted after a period of 6 months.

C7.2 Approval of subjects by the regulatory authority committee

C7.2.1 General

C7.2.1.1

The Chair and Vice-Chair shall be entitled to vote.

C7.2.1.2

If at the Committee on Part I stage no voting member of the Regulatory Authority Committee has voted negative as a Part I member for reasons concerning the suitability of the amendment for use in a regulation and the subject is approved by that Committee, no further vote is necessary in the Regulatory Authority Committee.

Δ C7.2.1.3

If the Regulatory Authority Committee disagrees with the proposed amendment accepted by the Committee on Part I, the Regulatory Authority Committee shall submit a revised amendment carrying the same intent in terms of safety and technical requirements to the Section Subcommittee for further consideration.

Δ C7.2.1.4

A revised amendment from the Section Subcommittee shall be proposed to the Committee on Part I by letter ballot or be a recorded vote at a meeting.

C7.2.2 Approval by letter ballot

C7.2.2.1

Proposals for amendment as submitted by the Committee on Part I shall be distributed for letter ballot approval. To be valid, a ballot shall be returned within 30 days.

C7.2.2.2

If there are no negative votes, the recommendation shall be considered approved, provided that at least 50% of the total voting membership voted affirmative.

C7.2.2.3

If there are any negative votes that cannot be resolved by the Chair of the Regulatory Authority Committee, the subject shall be included in the agenda of the next meeting of the Regulatory Authority Committee for the purpose of resolution.

C7.2.3 Approval at a meeting

C7.2.3.1

Letter ballots referred to a meeting shall be reconsidered and an open vote taken. The amendment shall be considered approved, provided that at least 50% of the total voting membership voted affirmative and that at least two-thirds of the votes cast are affirmative.

C7.2.3.2

Subjects that have not been submitted for letter ballot may be considered at a meeting, provided that all voting members present are agreeable or that the subject has been placed on the agenda and no objection is registered.

C7.3 Consideration of negative votes

The Chair of the Committee on Part I shall consult with the Chair of the Section Subcommittee and CSA staff, and one or more of the following courses of action shall be taken, as appropriate:

(a) an attempt shall be made to resolve each negative vote by editorial changes or explanation and thereby have the negative vote withdrawn;

(b) a negative vote may be ruled non-germane if

  (i) the negative vote is not accompanied by supporting comments;

  (ii) the negative vote and supporting comments are not considered to conform to the criteria outlined in Clauses C7.1.1.4 and C7.1.1.5; or

  (iii) the negative vote and supporting reasons are not considered relevant to the items being balloted upon;

(c) a negative vote may be ruled non-persuasive if the particular reasons for the vote have been previously discussed and rejected by the Section Subcommittee*; or

  **In such instances, this decision should be supported by Subcommittee records.*

(d) a negative vote shall be referred to the next meeting of the Committee on Part I (see Clause C7.1.3) if
 (i) the particular reasons for the vote are considered to be of a technical nature not previously discussed by the Section Subcommittee, except when the reasons for the vote are referred to the Section Subcommittee, in which case the Subcommittee report shall be presented at the next meeting of the Committee on Part I (see Clause C7.2.3.2); or
 (ii) the vote has not been disposed of under Items (a), (b), or (c).

Δ
C8. Approval of other subjects

Voting on other motions shall comply with Clause C7.1.3.2.

C9. Interpretation of the *CE Code, Part I*

C9.1
Interpretation of the *CE Code, Part I*, shall be a function of the Committee on Part I in accordance with Clauses C9.2 to C9.9.

C9.2
Requests for interpretation shall be submitted in writing to the Project Manager of the Committee on Part I in the form of a question that can be answered by a categoric "yes" or "no".

C9.3
The request shall make specific reference to the relevant Rule or Rules and shall provide an explanation of circumstances surrounding the actual field situation.

C9.4
Requests for interpretation shall not be accepted for
(a) the degree and extent of a hazardous location area;
(b) the suitability of isolation or guarding; or
(c) items that involve an intimate knowledge of the installation rather than the meaning of the Rule.

C9.5
The request for an interpretation shall be referred to the appropriate Section Subcommittee.

C9.6
Interpretations shall be based on the literal text and not on the intent.

C9.7
The Section Subcommittee shall present a recommended interpretation to the Committee on Part I for vote. Voting shall be in accordance with Clause C7.

C9.8
The results of the letter ballot shall be made known by the Project Manager to the submitter and to the Committee on Part I.

C9.9
Interpretations shall be published in "CSA Info Update".

C9.10
When an interpretation of a Rule has been adopted by the Committee on Part I in accordance with Clause 9.7
(a) the responsible Subcommittee Chair shall ensure that an appropriate proposal for a new subject is made promptly to the Subcommittee of a new subject in accordance with Clause C6.1 to reword the subject Rule in a way that removes ambiguity of meaning; and
(b) the interpretation shall be published in Appendix I and the Rule shall reference the Appendix if the rewording of the Rule has not been approved by the cut-off date for that edition.

C10. Appeals

C10.1
Any Committee member or individual who believes that this CSA Standard is being prepared under procedures that do not conform to the *CSA Policy governing standardization — Code of good practice for standardization,* the *CSA Directives and guidelines governing standardization,* and these Rules of Procedures may appeal to the Strategic Steering Committee on Requirements for Electrical Safety for a review of the project.

Appendix C
The Technical Committee on the Canadian Electrical Code,
Part I — Organization and rules of procedure

C10.2

Appeals shall be based on procedural matters and not on technical considerations.

C10.3

Application for appeal shall not necessarily be considered cause for delaying the development or publication of this Standard.

C10.4

All appeals shall be submitted in writing to the Standards Policy Board Secretary.

C10.5

The Standards Policy Board Secretary shall notify the appellant of the decision of the Standards Policy Board and shall refer the decision to the appropriate CSA staff for implementation.

C11. Code format and Rule terminology

C11.1

Because the *Canadian Electrical Code, Part I,* may be adopted as a regulatory document, it is important that all Rules, Subrules, and Items be stated in mandatory language in accordance with CSA guidelines. In this respect, the verb form "shall" must be used rather than "is to be", "are to be", "will be", "should be", etc., or as "shall not" if the negative is required. Requirements shall be stated in the positive rather than the negative.

C11.2

The term "may" shall not be used in a permissive sense, because it may indicate to the user or the enforcer of the Code that the permissive idea may or may not be acceptable to the enforcer. The term "shall be permitted" shall be used because it indicates definitely that the enforcer has no alternative but to allow the easement.

C11.3

Recommendations or explanatory notes shall not be included in the text of the Code, but they may be included in an Appendix and shall be written in a permissive, not mandatory, manner.

Δ **C11.4**

Each Section should be assigned an even number, the odd numbers being reserved for new Sections pending their inclusion in the next edition of the *CE Code, Part I.* The title shall be descriptive of the contents of the Section.

Δ **C11.5**

Rules and Subrules should occur in a logical sequence. Where the Section is not a General Section, the first Rule should contain a statement that the Section supplements and/or amends the general requirements of the Code.

Δ **C11.6**

Where reference is made to a Subrule or Item in the same Rule, only the Subrule number and/or Item letter and the word "Subrule" and/or "Item" need be mentioned. If the reference is to another Rule or Section, then the Rule number and the word "Rule" shall be stated.

Δ **C11.7**

Each Rule of a Section shall be provided with a title or caption following the Rule number that indicates the contents of the Rule.

C11.8

Each Rule should be assigned an even number, the odd numbers being reserved for new Rules pending their inclusion in the next edition of the *CE Code, Part I.*

C11.9

References to other Codes or Standards shall be to a specific edition of the referenced Code or Standard rather than to the latest edition, except for Standards forming part of the *Canadian Electrical Code, Part II.*

C11.10

The term "ampacity", as defined in Section 0, applies to the current-carrying capacity of conductors only and shall not be used in relation to switches, panels, motors, etc.

C11.11

Maximum and minimum limits shall be expressed in the following ways as appropriate:
(a)   "...shall not exceed...volts-to-ground...";
(b)   "...shall have a clearance of not less than...between conductors...";

(c)   "...shall be supported at intervals not exceeding...".

C11.12

The term "voltage" shall be used instead of the term "potential".

C11.13

The term "approved" as defined in Section 0 and described in Rule 2-024 is applicable to all electrical equipment. Use of the term outside Section 0 and Rule 2-024 is deemed redundant unless the equipment is required to be marked for a specific use.

C11.14

The term "acceptable" should be used to describe equipment that is not required to be approved.

C11.15

Use of the term "acceptable" in the dictionary sense is permissible, and the attributes that are important in deciding on acceptance should be included in the wording (e.g., "acceptable in terms of clearance, ruggedness, separation from..., colour, legibility, location, etc.").

C11.16

In the technical sections of the Code, the phrase "acceptable to (x)" shall not be used (where x may stand for the electrical inspector, the building inspector, or the supply authority).

C11.17

Where one wishes to indicate that the supply authority has to agree, the phrase "in accordance with the requirements of the supply authority" shall be used.

C11.18

Standard terms have been established by usage or practice and shall be used in preference to similar terms not having such wide or established recognition. Some examples are
(a)   "authority having jurisdiction";
(b)   "disconnecting means" (not "disconnection means");
(c)   "ducts" (only for air-handling purposes; not as raceways);
(d)   "electric" (as applied to equipment);
(e)   "electrical" (as applied to requirements, standards, or codes);
(f)   "equipment" (both singular and plural);
(g)   "metal" (not "metallic", in general);
(h)   "metallic" (only where directly related to material using this term, e.g., electrical metallic tubing or non-metallic sheathed cable);
(i)   "not exceeding" instead of "not more than";
(j)   "not less than"; and
(k)   "provided with mechanical protection" (instead of "protection against mechanical damage").

C11.19

Terms such as "adequate", "adjacent", "reasonable", "near to", "large", "small", "high", "low", etc., shall be replaced by more definitive terms.

C11.20

The following terms shall not be used:
(a)   "fire-resistant";
(b)   "fireproof";
(c)   "flame-retarding"; and
(d)   "that will (not) burn in air".

C11.21

Numbers shall be used to express values and shall be expressed as numerals instead of as words, except at the beginning of a sentence or where two numbers in sequence would be confusing. Words shall be used to express a quantity of items.

C11.22

There shall be only one sentence per Rule, per Subrule, or per Item.

Appendix C
The Technical Committee on the Canadian Electrical Code,
Part I — *Organization and rules of procedure*

C11.23

The following is a list of standardized comments that Subcommittees may wish to use in reports:

(a) Accept:
  (i) Acceptance is unconditional; or
  (ii) Acceptance is conditional on acceptance by the Subcommittee or Committee on Section _____ or Standard _____ ;

(b) Hold pending:
  (i) Submission of further supporting data;
  (ii) Further study;
  (iii) Receipt of fact-finding report; or
  (iv) Receipt of the findings of a Task Group;

(c) Reject:
  (i) The supporting comment does not justify the proposed amendment or addition;
  (ii) See the Subcommittee or Committee on Part I action on Subject No. _____ ;
  (iii) The Subcommittee agrees with the intent of the proposal. However, please note the intent of the action taken on Subject No. _____ ;
  (iv) No additional clarification would be achieved by this proposal;
  (v) The present wording adequately represents the intent;
  (vi) The supporting data are not consistent with the proposal;
  (vii) The supporting comment is not persuasive as to the necessity;
  (viii) The supporting data are not adequate;
  (ix) The Subcommittee disagrees with the supporting data;
  (x) The proposal is already adequately covered by ...;
  (xi) The extension of the coverage as proposed is not appropriate at this time;
  (xii) Safety is not enhanced by the proposal;
  (xiii) The intent of the proposal is not specific or definite;
  (xiv) The proposal is primarily a design consideration and adds nothing to the safety of the product involved or the method;
  (xv) The proposal covers a method or practice that is not prohibited by the present Code, and thus is not necessary;
  (xvi) The proposal is beyond the scope of the Code; or
  (xvii) Deletion of the present requirement as suggested by the proposal is not desirable because ....

C11.24

Rules dealing with flammability limits for wiring systems in a building shall be contained in the *National Building Code of Canada* in liaison with the Committee on Part I, and Rules accomplishing these limits shall be contained in the *CE Code, Part I*, in liaison with the *National Building Code of Canada* Committees.

C11.25

Where *National Building Code of Canada* requirements are referenced in the *CE Code, Part I*, informational notes concerning those requirements should be contained in Appendix B, and the *National Building Code of Canada* article or sentence number should be listed in Appendix G.

C11.26

Where reference is made to the Building Code, the phrase "the *National Building Code of Canada*" shall be used (e.g., "in accordance with the *National Building Code of Canada*").

Δ C11.27

Except for the last Subrule or Item, each Subrule or Item in a listing shall end with a semi-colon. The penultimate Subrule or Item shall end with either an "and" or an "or".

C11.28

Where it is necessary to include messages or warnings in the Code, the content of those messages or warnings shall be included in the Code but specific wordings shall be avoided.

Annex A — Standard format for subcommittee reports

<table>
<tr><td colspan="3" align="center">**Canadian Standards Association**
Toronto, Ontario

Section Subcommittee Report</td></tr>
<tr><td>**SUBJECT NO.**</td><td>**TITLE:**</td><td>**CHAIR:**

DATE:</td></tr>
<tr><td colspan="2">Submitted by:

Affiliation:

Request or proposal:

Reason for request or proposal:

Supporting information:

Summary of Subcommittee deliberations:

Subcommittee recommendation:</td><td>Date:</td></tr>
</table>

Appendix C
The Technical Committee on the Canadian Electrical Code,
Part I — *Organization and rules of procedure*

Annex B — Request for an amendment to the *Canadian Electrical Code, Part I*

Note: *This Annex contains the suggested form to be used when requesting a change to the* CE Code, Part I. *See Clause C6.*

TO: The Project Manager of the Committee on Part I

FROM:

AFFILIATION:

DATE:

RE: Request for an Amendment to Rule(s):

Request (specifically worded):

Reasons for request:

Supporting information:

Part I Subject No. assigned: _____

Completed form to be sent by CSA to: Section Chair
                                       Part I Chair
                                       Submitter

Appendix D — Tabulated general information

Note: *This Appendix is an informative (non-mandatory) part of this Standard.*

Table D1
Type designations, voltage ratings, and construction of wires and cables other than flexible cords

Note: *These data are subject to frequent revision and in cases where any doubt exists, the latest edition of the appropriate Standard of the CE Code, Part II, or appropriate laboratory requirements should be consulted.*

Trade designation	CSA type designation	Maximum voltage rating	Number of circuit conductors	Size-range AWG or kcmil	Kind of insulation	Covering on each insulated conductor	Outer covering	Reference Notes
Armoured cable	AC90	600	1 or more	14 to 2000	Cross-linked polyethylene	None or thermoset or thermoplastic	Interlocking metal armour	1, 8, 10, 25
	ACWU90						Interlocking metal armour and flame-tested thermoplastic	
	TECK90	5000	1	6 to 2000	Cross-linked polyethylene		Thermoset or thermoplastic jacket and interlocking metal armour with or without thermoplastic or thermoset covering overall	8, 10, 14
					Ethylene-propylene rubber			8, 14, 17
			2 or more	14 to 2000	Cross-linked polyethylene			1, 8, 10
					Ethylene-propylene rubber			1, 8, 17
Non-metallic sheathed cable	NMD90	300	2, 3, or 4	14 to 2	90 °C heat-resistant thermoplastic	Nylon sheath	Thermoplastic	1
					Cross-linked polyethylene			28
	NMW				Moisture-resistant thermoplastic	None		31
	NMWU							1

(Continued)

Table D1 (Continued)

Trade designation	CSA type designation	Maximum voltage rating	Number of circuit conductors	Size-range AWG or kcmil	Kind of insulation	Covering on each insulated conductor	Outer covering	Reference Notes
Thermoset insulated wires and cables	RL90	600 and 1000	1	14 to 2000	Cross-linked polyethylene or EPCV	None or thermoset jacket	Thermoplastic jacket over lead, or lead	1, 8, 10, 37
			2 or more					1, 8, 10, 11, 37
			1		Ethylene-propylene rubber	Thermoset jacket		1, 8, 17
			2 or more					1, 8, 11, 17
		5000	1	8 to 2000	Cross-linked polyethylene or EPCV	None or thermoset jacket		5, 8, 10, 37
			2 or more					5, 8, 10, 11, 37
			1		Ethylene-propylene rubber	Thermoset jacket		5, 8, 17
			2 or more					5, 8, 11, 17
	RW75, R90, RW90	600 and 1000	1	14 to 2000	Cross-linked polyethylene	None or thermoplastic or thermoset	None	1, 10
			2 or more				Thermoplastic or thermoset	1, 8, 10, 11
			1		Ethylene-propylene rubber	Thermoplastic or Thermoset	Thermoplastic or thermoset	1, 8, 17
			2 or more				Thermoplastic or thermoset	1, 8, 11, 17
			1		EPCV	None or thermoplastic or thermoset	Thermoplastic or thermoset	1, 37
			2 or more				Thermoplastic or thermoset	1, 8, 11, 37
		5000	1	8 to 2000	Cross-linked polyethylene	None or thermoplastic or thermoset	None	5, 10
			2 or more				Thermoplastic or thermoset	5, 8, 10, 11
			1		Ethylene-propylene rubber	Thermoplastic or thermoset	Thermoplastic or thermoset	5, 8, 17
			2 or more				Thermoplastic or thermoset	5, 8, 11, 17
			1		EPCV	None or thermoplastic or thermoset	Thermoplastic or thermoset	5, 37
			2 or more					5, 8, 37

(Continued)

Table D1 (Continued)

Trade designation	CSA type designation	Maximum voltage rating	Number of circuit conductors	Size-range AWG or kcmil	Kind of insulation	Covering on each insulated conductor	Outer covering	Reference Notes
Thermoset insulated wires and cables	RWU75 and RW90	1000	1	14 to 2000	Cross-linked polyethylene	None or thermoplastic or thermoset	None	1, 8, 10
					Ethylene-propylene rubber	Thermoplastic or thermoset		1, 8, 17
					EPCV	Thermoplastic or thermoset		1, 8, 37
Aluminum sheathed cables	RA75, RA90	600 and 1000	1	14 to 2000	Cross-Linked Polyethylene	None or thermoplastic or thermoset	Aluminum or aluminum with thermoplastic covering	1, 8
			2 or more					1, 8, 10, 11
			1		Ethylene-propylene rubber	Thermoplastic or thermoset		1, 8, 17
			2 or more					1, 8, 11, 17
		5000	1	8 to 2000	Cross-linked polyethylene	None or thermoplastic or thermoset		5, 8
			2 or more					5, 8, 10, 11
			1		Ethylene-propylene rubber	Thermoplastic or thermoset		5, 8, 17
			2 or more					5, 8, 11, 17
Mineral-insulated cable	MI	600	7	16 to 10	Magnesium oxide or silicon dioxide	None	Copper or stainless steel	—
			4	16 to 6				
			2 or 3	16 to 4				
			1	16 to 250				
			2, 3, 4, or 7	18 to 10				
	LWMI	300	2 or 3	16 to 10			Copper	22

(Continued)

Table D1 (Continued)

Trade designation	CSA type designation	Maximum voltage rating	Number of circuit conductors	Size-range AWG or kcmil	Kind of insulation	Covering on each insulated conductor	Outer covering	Reference Notes
Thermoplastic cable	TW	600	1	14 to 2000	Moisture-resistant flame-tested thermoplastic	None or nylon	None	1, 12
	TWU							
	TWN75			14 to 1000	Heat- and moisture-resistant flame-tested thermoplastic	Nylon	None	1
	TW75					None		
	TWU75			14 to 4/0				
	T90 NYLON			14 to 5000	Heat-resistant and flame-tested thermoplastic	Nylon		1
Neutral supported cable	NS 75	600	2, 3, 4, or 5	8 to 4/0 Cu, 6 to 500 kcmil Al (Minimum size of neutral 8 Cu, 6 Al)	Polyethylene or cross-linked polyethylene, with bare or insulated neutral	None or flame-tested polyvinyl chloride		16
	NS 90							
Service-entrance cable	USEI75	600	2, 3, or 4	6 to 500 kcmil	Polyethylene	Polyvinyl chloride	(See Note 3)	2, 10
					Cross-linked polyethylene			
	USEI90				Ethylene propylene rubber			2, 17
	USEB90				Cross-linked polyethylene	None	Polyvinyl chloride jacket	2, 10
					Ethylene propylene			2, 17
Switchboard wire	TBS	600	1	14 to 4/0	Thermoplastic	Cotton or rayon flame-tested	None	8, 21

(Continued)

Table D1 (Continued)

Trade designation	CSA type designation	Maximum voltage rating	Number of circuit conductors	Size-range AWG or kcmil	Kind of insulation	Covering on each insulated conductor	Outer covering	Reference Notes
Luminous tube sign and oil burner ignition cable	GTO	15 000	1	14 to 10	Rubber or polyethylene	Cotton, flame-tested and moisture-resistant or polychloroprene or thermoplastic	None	3
	GTOL						Lead	
Ignition cable	ICS	10 000	1	16 nickel-plated copper or 20 stainless steel	Silicone rubber	Moisture-resistant braid and silicone rubber	None	7
Extra-low-voltage control cable	LVT	30	2 or more	22 to 16	Thermoplastic	None	Flame-tested thermoplastic	9, 32
Low-energy control cable	Low-energy control cable	30	2 or 3	18	Flame-tested thermoplastic		None	13
Extra-low-voltage control cable	ELC	30	1 or more	26 to 16	Thermoplastic			—
Thermoset insulated equipment wire	REW	300	1	26 to 10	Flame-tested cross-linked PVC or flame-tested cross-linked chlorinated polyethylene			7, 19
		600		24 to 4/0				4, 7, 19
		300	2 or more	26 to 10	Flame-tested cross-linked PVC	None or shield	Shield and flame-tested cross-linked PVC insulating covering	7, 19, 20
		600		24 to 4/0				
	SIS	600	1	14-4/0	Cross-linked polyethylene	None	None	7
	TEWN			18 and 16	Flame-tested thermoplastic	Extruded nylon		4, 7

(Continued)

Table D1 (Continued)

Trade designation	CSA type designation	Maximum voltage rating	Number of circuit conductors	Size-range AWG or kcmil	Kind of insulation	Covering on each insulated conductor	Outer covering	Reference Notes
Thermoplastic insulated equipment wire	TEW	600	1	26 to 4/0	Flame-tested thermoplastic	None	None	4, 7
				24 to 10		Wire shield	Flame-tested thermoplastic insulating covering	7
			2 or more	26 to 4/0		None or shield	None or shield and flame-tested thermoplastic insulating covering	7, 20
	TXF	125	1	20		None		7
	TXFW	300		18 and 16				7
	GTF	600		18 to 10		Lacquered glass braid		4, 7
Silicone rubber insulated equipment wire	SEW-1	300	1	22 to 16	Silicone rubber	Glass-braid-treated	None	7
	SEWF-1		2 or more					
	SEW-1							
	SEW-2	600	1	22 to 4/0				
			2 or more					
	SEWF-2	600	1	22 to 6	Silicone rubber	Glass-braid-treated		
			2 or more					
Insulated conductors for power-operated electronic devices	RR-64	600	1	28 to 14	Cross-linked PVC	None	None or shield with or without thermoplastic insulating covering	19, 26, 33
			2 or more					
	RR-32	1400	1	28 to 10				
			2 or more					
	RR-64	600	2 up to 7	28 to 10		None or shield	Shield and cross-linked PVC insulating covering	20, 26, 33
	RR-32	1400		24 to 10				

(Continued)

Table D1 (Continued)

Trade designation	CSA type designation	Maximum voltage rating	Number of circuit conductors	Size-range AWG or kcmil	Kind of insulation	Covering on each insulated conductor	Outer covering	Reference Notes
Insulated conductors for power-operated electronic devices	RR-64	600	1	28 to 14	Cross-linked chlorinated polyethylene	None	None	19, 26, 33
	RR-32	1400		24 to 10				
	TRSR-64	600		28 to 14	Semi-rigid PVC	None or extruded nylon	None or shield with or without nylon	26, 33
	TR-64	600	2 or more		Thermoplastic	None or shield	thermoplastic insulating covering shield and thermoplastic insulating covering	20, 26, 33
	TR-32	1400	1	24 to 10				
			2 or more					
			2 up to 7					
	Twin Lead	—	2	24 to 20	Flame-tested polyethylene	None	None or thermoplastic covering	29
					Polyvinyl chloride			30
	TTR	600	1	26 to 14	Thermoplastic	Cotton, or rayon braid treated, or nylon	None	7, 26, 33
	TV-6	6000 (dc)	1	24 minimum	Flame-tested polyethylene, cross-linked low-density polyethylene, cross-linked high-density polyethylene, cross-linked polyvinyl chloride, silicone rubber, fluorinated ethylene propylene	None or polyvinyl chloride, cross-linked low-density polyethylene, cross-linked high-density polyethylene	None or shield with 2 kV dc PVC insulating covering	7, 34
	TV-10	10 000 (dc)						
	TV-15	15 000 (dc)						
	TV-20	20 000 (dc)						
	TV-30	30 000 (dc)						
	TV-40	40 000 (dc)						
	TV-50	50 000 (dc)						

(Continued)

Table D1 (Continued)

Trade designation	CSA type designation	Maximum voltage rating	Number of circuit conductors	Size-range AWG or kcmil	Kind of insulation	Covering on each insulated conductor	Outer covering	Reference Notes
Arc-welding cable	Arc-welding cable	(See Note 7)	1	8 to 300	Rubber or flame-tested polychloroprene covering	None	None or shield with 2 kV dc PVC insulating covering	5
Coil-lead wire	CL1251	600		22 to 500	Cross-linked polyethylene	None	None	4, 7, 27
	CL902	300			Polychloroprene	None or cotton braid lacquered		4, 7
	CL903	600						
	CL1052	300						
	CL1053	600			Chloro-sulfonyl polyethylene	None		
	CL1151	300						
	CL1152	600						
	CL904	600			Thermoplastic	None or extruded nylon		
	CL1501, CL2001				Silicone rubber	Glass-braid-treated		
	CL1051	300		22 to 16	Thermoplastic	Flame-tested treated acrylic fibre		
	CL1054			22 to 14	Cross-linked ethylene copolymer	None		
	CL905	600		22 to 4/0	Ethylene propylene rubber			
	CL1253			22 to 500	Styrene-ethylene butylene copolymer			
	CL1056	300		22 to 14	Styrene-ethylene butylene copolymer			4

(Continued)

Table D1 (Continued)

Trade designation	CSA type designation	Maximum voltage rating	Number of circuit conductors	Size-range AWG or kcmil	Kind of insulation	Covering on each insulated conductor	Outer covering	Reference Notes
Coil-lead wire	CL1254	600	1	22 to 4/0	Ethylene propylene rubber	None	None	4, 7
	CL1055			22 to 4/0	Ethylene propylene rubber			
	CL1255			22 to 12				
	CL910	300		22 to 500	NBR/PVC			
	CL1252	300		22 to 10	Cross-linked polyethylene			
	CL906	600		22 to 500	Inner layer of ethylene propylene rubber, outer layer of chloro-sulfonyl polyethylene			
	CL907							
	CL1502				Inner polyamide tape and outer wall of cross-linked ethylene copolymer			
	CL908			22 to 8	Chlorinated polyethylene (CPE)			4, 7
	CL909	300						
Pendant weatherproof lampholder lead wire	TLW	600		14	Flame-tested thermoplastic			7

(Continued)

Table D1 (Continued)

Trade designation	CSA type designation	Maximum voltage rating	Number of circuit conductors	Size-range AWG or kcmil	Kind of insulation	Covering on each insulated conductor	Outer covering	Reference Notes
Hoistway cable	—	600	Parallel construction 2 to 4	18	Flame-tested thermoplastic	None	None	—
			Twisted construction 2 to 75	18 to 14			None or PVC jacket	
Airport series lighting cable	ASLC	5000	1	8, 6, or 4	Cross-linked polyethylene		None	35
Inside wiring cable	IWC	150	4 or more	22 or 24	Thermoplastic	None	Thermoplastic	—
Station wire	ZSW	150	2 to 6	22 or 24	Thermoplastic	None	Thermoplastic	—
Communication flat cable	CFC	150	40 to 50	22, 24, 26, 28, 30	Thermoplastic	None	Thermoplastic	36
Communication building cable	CBC	150	20 or more	22, 24, 26	Thermoplastic	None	Thermoplastic	—
Premise communication cable	PCC	150	2 or more	22, 24, 26	Thermoplastic	None	Thermoplastic	—
Fire alarm and signal cable	FAS	300	1 or more	26 to 12	Thermoplastic	None	None or thermoplastic or thermoset jacket, or interlocking metal armour or aluminum sheath with or without overall thermoplastic covering	—
	FAS 90				Thermoplastic or thermoset			
	FAS 105				Thermoplastic			
	FAS 200				Thermoplastic or thermoset			

(Continued)

Table D1 (Continued)

Notes:

(1) No. 14 AWG for copper conductors; No. 12 AWG for aluminum conductors.

(2) USEI90 — PVC jacketed individual conductors are twisted together without overall covering.

(3) For Type GTO cable, the maximum voltage rating shall be designated as follows:

(a) GTO-5 and GTOL-5 — for use at not more than 5 000 V;

(b) GTO-10 and GTOL-10 — for use at not more than 10 000 V; and

(c) GTO-15 and GTOL-15 — for use at not more than 15 000 V.

(4) When used in applications where the current is limited or controlled, or both, by means of ballast, resistor, transformer, etc., the following types of wire may be operated at the voltages shown below:

(a) Types GTF, REW rated at 600 V; TEW; TEWN, CL904, CL1251, CL1501, CL1502, and CL2001 — 1000 V;

(b) Types CL752, CL901, CL903, CL905, CL907, CL908, CL911, CL1053, and CL1152 — 750 V; and

(c) Types CL751, CL902, CL906, CL909, CL910, CL1051, CL1052, CL1054, CL1151, and CL1252 — 600 V.

(5) No. 8 AWG minimum for copper conductors; No. 6 AWG minimum for aluminum conductors.

(6) Arc-welding cables are intended only for use with electric welders having an open-circuit secondary voltage of 100 V or less.

(7) Maximum allowable conductor temperatures will be found in Table 19 for all wire types except the following:

(a) Types TXF, TXFW, TLW, and Arc-Welding Cable — 60 °C;

(b) Types VHR-64, VHR-32, CL751, and CL752 — 75 °C;

(c) Types TV-6, TV-10, TV-15, TV-20, TV-30, TV-40, and TV-50

— with flame-tested polyethylene — 80, 90, or 105 °C

— with cross-linked low-density polyethylene — 80, 90, 105, or 125 °C

— with cross-linked high-density polyethylene — 90 or 105 °C

— with cross-linked polyvinyl chloride — 80, 90, or 105 °C

— with silicone rubber — 150 or 200 °C

— with fluorinated ethylene propylene — 150 °C;

(d) Types RR-64, RR-32, TR-64, TR-32, TRB-64, TRB-32, TTR, CL901, CL902, CL903, CL904, CL905, CL906, CL907, CL908, CL909, CL910, and CL911 — 90 °C, except Type TR-64 may also have a maximum allowable temperature of 105 °C;

(e) Types REW, TEW, TEWN, CL1051, CL1052, CL1053, CL1054, and Low-Energy Control Cable — 105 °C;

(f) Type SIS — 90 °C;

(g) Types CL1151 and CL1152 — 115 °C;

(h) Types CL1251, GTF, and CL1252 — 125 °C;

(i) Types CL1501, ICS, SEWF-1, and SEWF-2 — 150 °C, except Types SEWF-1 and SEWF-2 with a nickel-coated copper or nickel conductor — 200 °C;

(j) Types CL2001, SEW-1, and SEW-2 — 200 °C;

(k) Types CL1501, CL1502, ICS, SEWF-1, and SEWF-2 — 150 °C, except Types SEWF-1 and SEWF-2 with a nickel-coated copper conductor — 200 °C;

(l) Type SEWF-1 with a nickel conductor — 200 °C or 250 °C and Type SEWF-2 with a nickel conductor — 200 °C;

(m) Types CL2001 and SEW-2 — 200 °C; and

(n) Type SEW-1 with a copper or nickel-coated copper conductor — 200 °C, and with a nickel conductor — 200 °C or 250 °C.

Thermoset coverings include polychloroprene and chloro-sulfonyl polyethylene where applicable.

(8) Type LVT may be provided with an overall armour consisting of a single layer of closely wound, D-shaped, soft aluminum wire.

(9)

(Continued)

Table D1 (Concluded)

(10) Conductors having cross-linked polyethylene insulation are surface marked with the type designation followed by "XLPE".

(11) For 2-conductor parallel construction, the maximum size is No. 6 AWG.

(12) Types TW and TWU when provided with a nylon jacket are also approved for use where adverse conditions may exist, such as in oil refineries and around gasoline storage or pump areas (e.g., where subjected to alkaline conditions in the presence of petroleum solvents), and are limited to sizes No. 14 to 1000 kcmil in copper only.

(13) For operation at 30 V or less, low-energy control cable is suitable for Class 1 remote control, signal, and extra-low-voltage power circuits, and Class 2 remote control, signal, and low-energy power circuits in accordance with Section 16.

(14) No. 6 AWG minimum for copper conductors; No. 4 AWG minimum for aluminum conductors.

(15) A nickel-plated iron conductor may be used as an alternative to copper or nickel conductors.

(16) In 5-conductor neutral supported cable, the fifth conductor is to control or supply power to an auxiliary device, e.g., a water heater, street light, etc. The fifth conductor is No. 10 AWG minimum for copper, or No. 8 AWG minimum for aluminum.

(17) Conductors having ethylene propylene rubber insulation are surface marked with the type designation followed by "EP".

(18) Conductors having silicone rubber insulation are surface marked with the type designation followed by "Silicone".

(19) Cross-linked PVC insulation is surface marked (XLPVC) and cross-linked chlorinated polyethylene is surface marked (XLCPE).

(20) The 2-conductor type may be of parallel or twisted construction.

(21) Nickel-coated copper or nickel alloy conductors may be used as an alternative to copper or nickel conductors. Where nickel alloy is employed for the conductor, the suffix letters "NA" shall be added to the type designation.

(22) The voltage rating is ink printed on the surface of the copper sheath of 300 V Type LWMI cables.

(23) Conductors are of silver coated copper wires stranded.

(24) Conductors are of nickel coated copper wires stranded.

(25) Single-conductor armoured cables in sizes No. 4 AWG and smaller, and single-conductor armoured cables without concentric grounding conductor in sizes larger than No. 4 AWG are intended for use as grounding conductors only, and the covering over the insulation or the insulation, where a covering is not provided, is coloured green. Single-conductor armoured cables with a concentric grounding conductor are not intended for use as grounding conductors, and the insulation and the covering over the insulation or the insulation, where a covering is not provided, are not coloured green.

(26) These voltages are peak values.

(27) When Type CL1251 wire is provided with gasoline vapour-resistant insulation, it is surface marked "Gasoline Vapour-Resistant".

(28) When Type NMD90 is provided with nylon sheaths over the insulated conductors, "NMD90 NYLON" is marked on the surface of the jacket.

(29) Polyethylene insulated non-jacketed twin lead with 15 mil average thickness over the conductors is surface marked "TWIN LEAD-32 80C PE", and polyethylene insulated jacketed twin lead is surfaced marked "TWIN LEAD 80C".

(30) PVC insulated twin lead with 15 mil average thickness over the conductors is surface marked "TWIN LEAD-64 90C PVC" and PVC insulated twin lead with 30 mil average thickness over the conductors is surface marked "TWIN LEAD-32 90C PVC".

(31) Type NMD90 with cross-linked polyethylene insulation is surface marked on the jacket "NMD90 XLPE".

(32) Type LVT is surface marked on the jacket "LVT".

(33) Peak voltage rating is as assigned by CSA C22.2 No. 1.

(34) The shield and PVC insulating covering are only recognized over Types TV-20, TV-30, and TV-40 with cross-linked high-density polyethylene insulation and PVC jacket over Types TV-6, TV-10, TV-15, TV-20, TV-30, TV-40, and TV-50 with cross-linked PVC insulation.

(35) Airport series lighting cable is surface marked "ASLC 5000 V".

(36) CFC conductors that are used to electrically connect communications equipment to a telecommunications network shall not be smaller than No. 26 AWG copper. Conductors of No. 28 and No. 30 AWG copper shall be permitted for other types of communication applications.

(37) Conductors having EPCV insulation are surface marked with the type designation, followed by "EPCV".

460

Table D2
Direct current motors

Motor rating, hp	DC full-load current rating, A (see Notes (1) and (2))		
	120 V	240 V	500 V
1/4	2.9	1.5	—
1/3	3.6	1.8	—
1/2	5.2	2.6	—
3/4	7.4	3.7	1.8
1	9.4	4.7	2.3
1-1/2	13.2	6.6	3.2
2	17	8.5	4.1
3	25	12.5	6.0
5	40	20	9.7
7-1/2	58	29	14
10	76	38	18
15	110	55	26
20	145	72	35
25	179	89	43
30	212	106	51
40	280	140	68
50	349	174	84
60	418	209	101
75	518	259	124
100	—	343	165
125	—	426	205
150	—	507	243
200	—	675	324

Notes:

(1) *These values of full-load current are for motors running at moderate base speeds usual for belted motors and motors with normal torque characteristics. Motors built for especially low speeds may require more running current, in which case the nameplate current rating should be used.*

(2) *These values of full-load current are to be used as guides only. When exact values are required (e.g., for motor protection), always use those appearing on the motor nameplate.*

Table D3
Distance to centre of distribution for a 1% drop in voltage on nominal 120 V, 2-conductor copper circuits

(See Appendix B Note to Rule 4-004.)

Current, A	Copper conductor size, AWG														
	18	16	14	12	10	8	6	4	3	2	1	1/0	2/0	3/0	4/0
	Distance to centre of distribution measured along the conductor run, m (calculated for conductor temperature of 60 °C)														
1.00	24.2	38.5	61.4												
1.25	19.4	30.8	49.1												
1.6	15.1	24.1	38.4	61.0											
2.0	12.1	19.3	30.7	48.8											
2.5	9.7	15.4	24.6	39.0	62										
3.2	7.6	12.0	19.2	30.5	48.5										
4.0	6.1	9.6	15.3	24.4	38.8	61.7									
5.0	4.8	7.7	12.3	19.5	31.0	49.3									
6.3	3.8	6.1	9.7	15.5	24.6	39.1	62.2								
8.0	3.0	4.8	7.7	12.2	19.4	30.8	49.0								
10.0	2.4	3.9	6.1	9.8	15.5	24.7	39.2	62.4							
12.5		3.1	4.9	7.8	12.4	19.7	31.4	49.9	62.9						
16		2.4	3.8	6.1	9.7	15.4	24.5	39.0	49.1	62.0					
20			3.1	4.9	7.8	12.3	19.6	31.2	39.3	49.6	62.5				
25				3.9	6.2	9.9	15.7	24.9	31.4	39.7	50.0	63.1			
32					4.8	7.7	12.2	19.6	24.6	31.0	39.1	49.3	62.1		
40					3.9	6.2	9.8	15.6	19.7	24.8	31.3	39.4	49.7	62.7	
50						4.9	7.8	12.5	15.7	19.8	25.0	31.5	39.8	50.1	63.2
63						3.9	6.2	9.9	12.5	15.7	19.8	25.0	31.6	39.8	50.2
80						3.1	4.9	7.8	9.8	12.4	15.6	19.7	24.8	31.3	39.5
100							3.9	6.2	7.9	9.9	12.5	15.8	19.9	25.1	31.6
125								5.0	6.3	7.9	10.0	12.6	15.9	20.1	25.3
160									4.9	6.2	7.8	9.9	12.4	15.7	19.8
200										5.0	6.3	7.9	9.9	12.5	15.8
250												6.3	8.0	10.0	12.6
320													6.2	7.8	9.9

Notes:

(1) Table D3 is calculated for copper wire sizes No. 18 AWG to No. 4/0 AWG and, for each size specified, gives the approximate distance in metres to the centre of distribution measured along the conductor run for a 1% drop in voltage at a given current, with the conductor at a temperature of 60 °C. Inductive reactance has not been included because it is a function of conductor size and spacing.

(2) The distances for a 3% or 5% voltage drop are 3 or 5 times those for a 1% voltage drop.

(Continued)

Table D3 (Concluded)

Δ   **(3)**   *Because the distances in Table D3 are based on conductor resistances at 60 °C, these distances must be multiplied by the correction factors below according to the temperature rating of the conductor used and the percentage load with respect to the allowable ampacity. Where the calculation of allowable ampacity falls between two columns, the factor in the higher percentage column shall be used.*

Rated conductor temperature	Distance correction factor						
	Per cent of allowable ampacity						
	100	90	80	70	60	50	40
60 °C	1.00	1.02	1.04	1.06	1.07	1.09	1.10
75 °C	0.96	1.00	1.00	1.03	1.06	1.07	1.09
85–90 °C	0.91	0.95	1.00	1.00	1.04	1.06	1.08
110 °C	0.85	0.90	0.95	1.00	1.02	1.05	1.07
125 °C	0.82	0.87	0.92	0.97	1.00	1.04	1.07
200 °C	0.68	0.76	0.83	0.90	0.96	1.00	1.04

(4)   *For other nominal voltages, multiply the distances in metres by the other nominal voltage (in volts) and divide by 120.*

(5)   *Aluminum conductors have equivalent resistance per unit length to copper conductors that are smaller in area by two AWG sizes. Table D3 may be used for aluminum conductors because of this relationship, i.e., for No. 6 AWG aluminum use the distances listed for No. 8 AWG copper in Table D3. Similarly, for No. 2/0 AWG aluminum use the distances for No. 1 AWG copper.*

(6)   *The distances and currents listed in Table D3 follow a pattern. When the current, for any conductor size, is increased by a factor of 10, the corresponding distance decreases by a factor of 10. This relationship can be used when no value is shown in the table. In that case, look at a current 10 times larger. The distance to the centre of distribution is then 10 times larger than the listed value.*

(7)   *For multi-conductor cables, ensure wire size obtained from this table is suitable for ampacity from Table 2 or 4, and Rule 4-004.*

(8)   *For currents intermediate to listed values use the next higher current value.*

Δ   **(9)**   *Example on use of table:*
Consider a two-conductor circuit of No. 12 AWG copper NMD90 carrying 16 A at nominal 240 V under maximum ambient of 30 °C.
The maximum run distance from the centre of distribution to the load without exceeding a 3% voltage drop is:
Maximum run length for No. 12 AWG, 16 A, 1% voltage drop at nominal 120 V from Table is: 6.1 m
Distance correction factor to be used is:
From Table 2, allowable ampacity for 2-conductor No. 12 AWG NMD90 (90 °C rating per Table 19) is 20 A. The given current is 16 A or 80% (16/20) of the allowable ampacity.
The distance correction factor to be used, from Note (3), 90 °C row, 80% column, is 1.00.
The maximum run length is

$$6.1\,\text{m} \times 3(\%) \times 1.00 \times \frac{240\,\text{V}}{120\,\text{V}} = 37\,\text{m}$$

Beyond this distance a larger size of conductor is required, e.g., No. 10 AWG (30 A allowable ampacity) beyond 37 m up to and including 62 m.

$$9.7\,\text{m} \times 3(\%) \times 1.06 \times \frac{240\,\text{V}}{120\,\text{V}} = 62\,\text{m}$$

If the distance is between 37 and 60.5 m, a larger size of conductor is required, e.g., No. 10 AWG (30 A allowable ampacity)

$$9.7\,\text{m} \times 3(\%) \times 1.04 \times 240\,\text{V}/120\,\text{V} = 60.5\,\text{m}.$$

Table D4
Copper conductor sizes for 5% drop in voltage
on 6 V — Two conductors

(See Appendix B Note to Rule 46-306(1).)

Current, A	One way distance from power source measured along the conductor, m (calculated for conductor temperature of 20 °C)				
	No. 12 AWG	No. 10 AWG	No. 8 AWG	No. 6 AWG	No. 4 AWG
2-1/2	11.5	18.3	28.5	49.4	72.2
4-1/4	6.8	10.8	16.8	26.7	42.5
7	4.1	6.5	10.2	16.2	25.8
10	2.9	4.6	7.1	11.3	18.0
12	2.4	3.8	5.9	9.5	15.0
15	1.9	3.1	4.8	7.6	12.0
20	1.4	2.3	3.6	5.7	9.0

Note: *Acceptable one way distance in metres (L) for a selected wire size may be calculated using the following formula where one or all of the actual current (I), the actual voltage (V), and the actual voltage drop permitted (P) differ from Table D4:*

$$L = \frac{V}{6} \times \frac{P}{5} \times \frac{I_t}{I} \times L_t$$

where

I_t = *current shown in Table D4 closest to the actual current (I)*
L_t = *one way distance in metres shown in Table D4 corresponding to the wire size used for the value I_t selected*
Example: System characteristics
V = *12 V*
P = *7%*
I = *3 A*
Wire size = *No. 12 AWG*
Values of I_t and L_t to be used in the calculation are
I_t = *2.5 A*
L_t = *11.5 m*

$$L = \frac{12}{6} \times \frac{7}{5} \times \frac{2.5}{3} \times 11.5 = 26.8 \text{ m}$$

Table D5
Strandings for building wires and cables
(See Appendix B Note to Rule 12-1014.)

Nominal Conductor size, AWG or kcmil	Conductor area, mm²	Standard* Number of wires†	Diameter, mm	Occupied area‡, mm²	Flexible Number of wires	Diameter, mm	Occupied area‡, mm²	Extra flexible Number of wires	Diameter, mm	Occupied area‡, mm²
14	2.08	7	1.84	2.74	19	1.87	2.74	37	1.87	2.74
12	3.31	7	2.32	4.34	19	2.35	4.34	37	2.35	4.34
10	5.26	7	2.95	6.94	19	2.97	6.94	37	2.97	6.94
8	8.37	7	3.71	11.1	19	3.76	11.1	37	3.76	11.1
6	13.3	7	4.67	17.5	19	4.72	17.5	37	4.72	17.5
4	21.2	7	5.89	28.0	19	5.97	28.0	37	5.99	28.2
3	26.7	7	6.60	35.0	19	6.68	35.0	37	6.71	35.3
2	33.6	7	7.42	44.4	19	7.52	44.4	37	7.54	44.7
1	42.4	19	8.43	56.2	37	8.46	56.2	61	8.46	56.2
1/0	53.5	19	9.47	70.9	37	9.50	70.9	61	9.53	71.3
2/0	67.4	19	10.6	89.4	37	10.7	89.4	61	10.7	89.8
3/0	85.0	19	11.9	112	37	12.0	112	61	12.0	113
4/0	107	19	13.4	142	37	13.4	142	61	13.5	142
250	127	37	14.6	168	61	14.6	168	91	14.7	169
300	152	37	16.0	202	61	16.0	202	91	16.1	202
350	177	37	17.3	236	61	17.3	236	91	17.3	236
400	203	37	18.5	269	61	18.5	269	91	18.5	270
450	228	37	19.6	280	61	19.6	280	91	19.7	304
500	253	37	20.7	337	61	20.7	337	91	20.7	337
550	279	61	21.7	370	91	21.7	370	127	21.7	371
600	304	61	22.7	405	91	22.7	405	127	22.7	405
650	329	61	23.6	438	91	23.6	438	127	23.6	438
700	355	61	24.5	472	91	24.5	472	127	24.5	472
750	380	61	25.3	506	91	25.4	506	127	25.4	506
800	405	61	26.2	540	91	26.2	540	127	26.2	541
900	456	61	27.8	606	91	27.8	606	127	27.8	608

(Continued)

Table D5 (Concluded)

Nominal Conductor size, AWG or kcmil	Standard*			Flexible			Extra flexible		
	Conductor area, mm²	Number of wires†	Diameter, mm	Occupied area‡, mm²	Number of wires	Diameter, mm	Occupied area‡, mm²	Number of wires	Diameter, mm	Occupied area‡, mm²

Nominal Conductor size, AWG or kcmil	Conductor area, mm²	Number of wires†	Diameter, mm	Occupied area‡, mm²	Number of wires	Diameter, mm	Occupied area‡, mm²	Number of wires	Diameter, mm	Occupied area‡, mm²
1000	507	61	29.3	674	91	29.3	674	127	29.3	675
1250	633	91	32.7	843	127	32.8	843	169	32.8	843
1500	760	91	35.9	1010	127	35.9	1010	169	35.9	1010
1750	887	127	38.8	1180	169	38.8	1180	217	38.8	1180
2000	1010	127	41.5	1350	169	41.5	1350	217	41.5	1350

*Compact conductor diameters of equivalent cross-sectional area are reduced by up to 10% of the dimension indicated. Compressed conductor diameters of equivalent cross-sectional area are reduced by 2% of the dimension indicated.
†The number of wires indicated may be reduced by one in each layer.
‡Area of circumscribing circle; use for conduit space calculations.

Table D6
Recommended* tightening torques for wire-binding screws, connectors with slotted screws, and connectors for external drive wrench

(See Table D7.)

Type of connection	Wire size, AWG or kcmil	Tightening torques, Newton•metres
Wire-binding screws	14 to 10	1.4
Connectors with slotted screws	14, 12, and 10	2.3
	8	3.4
	6 and 4	4.0
	3 to 4/0 inclusive	4.5
Connectors for external drive wrench	1/0	19.8
	2/0	19.8
	3/0	28.3
	4/0	28.3
	250	39.5
	300	39.5
	350	39.5
	400	39.5
	500	45.2
	600	45.2
	700	45.2
	750	45.2
	800	50.8
	900	50.8
	1000	50.8
	1250	67.8
	1500	67.8
	1750	67.8
	2000	67.8

*For proper termination of conductors it is very important that field connections be properly tightened. In the absence of manufacturer's instructions on the equipment, the torque values given in Tables D6 and D7 are recommended.

Because it is normal for some relaxation to occur in service, checking torque values sometime after installation is not a reliable means of determining the values of torque applied at installation.

Table D7
Recommended* tightening torques for connectors with hexagonal socket screws

(See Table D6.)

Socket size (across flats), in	Tightening torque, Newton•metres
5/32	11.3
3/16	13.6
7/32	17.0
1/4	19.8
5/16	28.3
3/8	39.5
1/2	50.8
9/16	67.8

*For proper termination of conductors it is very important that field connections be properly tightened. In the absence of manufacturer's instructions on the equipment, the torque values given in Tables D6 and D7 are recommended.

Because it is normal for some relaxation to occur in service, checking torque values sometime after installation is not a reliable means of determining the values of torque applied at installation.

Table D8A
Allowable copper conductor ampacities for the installation configuration of Diagram B4-1

(See Appendix B Note to Rule 4-004.)

Size, AWG or kcmil	1/Phase Detail 1	2/Phase Detail 2	2/Phase Detail 3	4/Phase Detail 4	4/Phase Detail 5	6/Phase Detail 6	6/Phase Detail 7
1/0	245	245	245	203	220	165	179
2/0	285	285	285	229	248	186	202
3/0	330	330	330	258	280	210	228
4/0	385	385	385	291	315	236	256
250	425	421	425	317	343	256	278
350	530	500	520	375	408	304	331
500	660	605	630	452	489	365	396
600	740	659	682	491	534	397	433
750	845	745	775	554	596	447	482
1000	980	846	890	627	683	505	551
1250	1083	935	985	691	753	556	607
1500	1176	1011	1068	746	813	600	655
1750	1257	1078	1140	793	865	637	696
2000	1325	1133	1200	832	909	669	730

Notes:

(1) *This Table gives the allowable current for 90 °C rated single copper conductors with spacings directly buried in earth, subject to Rule 4-004(13) and (14), where*

  (a)   for any load, the cable terminates at equipment of any type other than a service box, fusible switch, circuit breaker, or panelboard; or

  (b)   the load is noncontinuous and either end of the cable terminates at a service box, fusible switch, circuit breaker, or panelboard.

(2) *The ampacities provided in the Table are the lesser of*

  (a)   the value obtained in accordance with Rule 4-004(1)(d); or

  (b)   the value obtained in accordance with Rule 8-104(7).

Table D8B
Allowable copper conductor ampacities for the installation configuration of Diagram B4-1

(See Appendix B Note to Rule 4-004.)

Size, AWG or kcmil	1/Phase Detail 1		2/Phase Detail 2		2/Phase Detail 3		4/Phase Detail 4		4/Phase Detail 5		6/Phase Detail 6	6/Phase Detail 7	
	100%	80%	100%	80%	100%	80%	100%	80%	100%	80%	—	100%	80%
1/0	208	172	208	172	208	172	203	172	208	172	165	179	172
2/0	242	200	242	200	242	200	229	200	242	200	186	202	200
3/0	280	231	280	231	280	231	258	231	280	231	210	228	
4/0	327	270	327	270	327	270	291	270	315	270	236	256	
250	361	298	361	298	361	298	317	298	343	298	256	278	
350	450	371	450	371	450	371	375	371	408	371	304	331	
500	561	462	561	462	561	462		452	489	462	365	396	
600	629	518	629	518	629	518		491	534	518	397	433	
750	718	592	718	592	718	592		554	596	592	447	482	
1000	850	700	846	700	846	700	627			683	505	551	
1250	960	791	935	791	935	791	691			753	556	607	
1500	1071	882	1011	882	1011	882	746			813	600	655	
1750	1165	959	1078	959	1078	959	793			865	637	696	
2000	1250	1029	1133	1029	1133	1029	832			909	669	730	

Notes:

(1) This Table gives the allowable current for 90 °C rated single copper conductors with spacings directly buried in earth, subject to Rule 4-004(13) and (14), where
   (a)   the load is continuous; and
   (b)   either end of the cable terminates at a service box, fusible switch, circuit breaker, or panelboard.

(2) The columns with the heading "80%" indicate that the equipment identified in (1)(b) above is not marked as certified to carry its ampere rating continuously.

(3) The columns with the heading "100%" indicate that the equipment identified in (1)(b) above is marked as certified to carry its ampere rating continuously.

(4) The ampacities provided in the Table are the lesser of
   (a)   the value obtained in accordance with Rule 4-004(1)(d); or
   (b)   the value obtained in accordance with Rule 8-104(7).

Table D9A
Allowable aluminum conductor ampacities for the installation configuration of Diagram B4-1

(See Appendix B Note to Rule 4-004.)

Size, AWG or kcmil	1/Phase Detail 1	2/Phase Detail 2	2/Phase Detail 3	4/Phase Detail 4	4/Phase Detail 5	6/Phase Detail 6	6/Phase Detail 7
1/0	190	190	190	158	171	129	140
2/0	220	220	220	178	193	145	157
3/0	255	255	255	201	218	163	178
4/0	300	300	300	227	246	183	200
250	330	328	330	247	267	200	217
350	415	390	410	292	318	237	258
500	515	471	495	352	383	284	309
600	585	513	541	382	419	308	340
750	665	580	610	431	469	348	379
1000	780	659	710	488	542	393	437
1250	868	750	790	554	604	446	487
1500	952	821	865	605	660	487	531
1750	1027	880	932	647	706	520	568
2000	1094	934	991	686	749	552	602

Notes:

(1) *This Table gives the allowable current for 90 °C rated single aluminum conductors with spacings directly buried in earth, subject to Rule 4-004(13) and (14), where*

    (a)   for any load, the cable terminates at equipment of any type other than a service box, fusible switch, circuit breaker, or panelboard; or

    (b)   the load is noncontinuous and either end of the cable terminates at a service box, fusible switch, circuit breaker, or panelboard.

(2) *The ampacities provided in this Table are the lesser of*

    (a)   the value obtained in accordance with Rule 4-004(2)(d); or

    (b)   the value obtained in accordance with Rule 8-104(7).

Table D9B
Allowable aluminum conductor ampacities for the installation configuration of Diagram B4-1

(See Appendix B Note to Rule 4-004.)

Size, AWG or kcmil	1/Phase Detail 1		2/Phase Detail 2		2/Phase Detail 3		4/Phase Detail 4		4/Phase Detail 5		6/Phase Detail 6	6/Phase Detail 7	
	100%	80%	100%	80%	100%	80%	100%	80%	100%	80%	—	100%	80%
1/0	162	133	162	133	162	133	158	133	162	133	129	140	133
2/0	187	154	187	154	187	154	178	154	187	154	145	157	154
3/0	217	179	217	179	217	179	201	179	217	179	163	178	
4/0	255	210	255	210	255	210	227	210	246	210	183	200	
250	281	231	281	231	281	231	247	231	267	231	200	217	
350	353	291	353	291	353	291	292	291	318	291	237	258	
500	438	361	438	361	438	361	352		383	361	284	309	
600	498	410	498	410	498	410	382		419	410	308	340	
750	570	469	570	469	570	469	431		469		348	379	
1000	680	560	659	560	680	560	488		542		393	437	
1250	770	634	750	634	770	634	554		604		446	487	
1500	867	714	821	714	865	714	605		660		487	531	
1750	956	788	880	788	932	788	647		706		520	568	
2000	1037	854	934	854	991	854	686		749		552	602	

Notes:

(1) This Table gives the allowable current for 90 °C rated single aluminum conductors with spacings directly buried in earth, subject to Rule 4-004(13) and (14), where

   (a)   the load is continuous; and

   (b)   either end of the cable terminates at a service box, fusible switch, circuit breaker, or panelboard.

(2) The columns with the heading "80%" indicate that the equipment identified in (1)(b) above is not marked as certified to carry its ampere rating continuously.

(3) The columns with the heading "100%" indicate that the equipment identified in (1)(b) above is marked as certified to carry its ampere rating continuously.

(4) The ampacities provided in this Table are the lesser of

   (a)   the value obtained in accordance with Rule 4-004(2)(d); or

   (b)   the value obtained in accordance with Rule 8-104(7).

Table D10A
Allowable copper conductor ampacities for the installation configuration of Diagram B4-2

(See Appendix B Note to Rule 4-004.)

Size, AWG or kcmil	1/Phase Detail 1	2/Phase Detail 2	4/Phase Detail 3	6/Phase Detail 4
1/0	231	201	159	146
2/0	264	228	180	164
3/0	301	260	204	186
4/0	345	296	231	211
250	379	325	252	230
350	461	391	303	275
500	564	475	364	330
600	621	521	404	365
750	706	589	448	406
1000	823	682	526	474
1250	920	759	571	515
1500	1004	824	618	556
1750	1077	880	659	592
2000	1139	928	692	622

Notes:

(1)   *This Table gives the allowable current for 90 °C rated single copper conductors with spacings installed in non-metallic underground raceway, subject to Rule 4-004(13) and (14), where*

    (a)   for any load, the cable terminates at equipment of any type other than a service box, fusible switch, circuit breaker, or panelboard; or

    (b)   the load is noncontinuous and either end of the cable terminates at a service box, fusible switch, circuit breaker, or panelboard.

(2)   *The ampacities provided in this Table are the lesser of*

    (a)   the value obtained in accordance with Rule 4-004(1)(d); or

    (b)   the value obtained in accordance with Rule 8-104(7).

Table D10B
Allowable copper conductor ampacities for the installation configuration of Diagram B4-2

(See Appendix B Note to Rule 4-004.)

Size, AWG or kcmil	1/Phase Detail 1		2/Phase Detail 2		4/Phase Detail 3	6/Phase Detail 4
	100%	80%	100%	80%	—	—
1/0	208	172	201	172	159	146
2/0	242	200	228	200	180	164
3/0	280	231	260	231	204	186
4/0	327	270	296	270	231	211
250	361	298	325	298	252	230
350	450	371	391	371	303	275
500	561	462	475	462	364	330
600	621	518	521	518	404	365
750	706	592	589		448	406
1000	823	700	682		526	474
1250	920	791	759		571	515
1500	1004	882	824		618	556
1750	1077	959	880		659	592
2000	1139	1029	928		692	622

Notes:

(1) *This Table gives the allowable current for 90 °C rated single copper conductors with spacings installed in non-metallic underground raceways, subject to Rule 4-004(13) and (14), where*
  (a)   the load is continuous; and
  (b)   either end of the cable terminates at a service box, fusible switch, circuit breaker, or panelboard.

(2) *The columns with the heading "80%" indicate that the equipment identified in (1)(b) above is not marked as certified to carry its ampere rating continuously.*

(3) *The columns with the heading "100%" indicate that the equipment identified in (1)(b) above is marked as certified to carry its ampere rating continuously.*

(4) *The ampacities provided in this Table are the lesser of*
  (a)   the value obtained in accordance with Rule 4-004(1)(d); or
  (b)   the value obtained in accordance with Rule 8-104(7).

Table D11A
Allowable aluminum conductor ampacities for the installation configuration of Diagram B4-2

(See Appendix B Note to Rule 4-004.)

Size, AWG or kcmil	1/Phase Detail 1	2/Phase Detail 2	4/Phase Detail 3	6/Phase Detail 4
1/0	180	157	123	114
2/0	205	178	140	128
3/0	235	203	158	145
4/0	269	231	180	164
250	296	253	197	179
350	360	306	236	213
500	442	372	283	257
600	488	409	314	284
750	556	464	349	315
1000	653	541	409	370
1250	738	608	457	413
1500	813	667	501	452
1750	880	719	538	484
2000	940	766	571	513

Notes:

(1) *This Table gives the allowable current for 90 °C rated single aluminum conductors with spacings installed in non-metallic underground raceway, subject to Rule 4-004(13) and (14), where*

    (a)   for any load, the cable terminates at equipment of any type other than a service box, fusible switch, circuit breaker, or panelboard; or

    (b)   the load is noncontinuous and either end of the cable terminates at a service box, fusible switch, circuit breaker, or panelboard.

(2) *The ampacities provided in this Table are the lesser of*

    (a)   the value obtained in accordance with Rule 4-004(2)(d); or

    (b)   the value obtained in accordance with Rule 8-104(7).

Table D11B
Allowable aluminum conductor ampacities for the installation configuration of Diagram B4-2

(See Appendix B Note to Rule 4-004.)

Size, AWG or kcmil	1/Phase Detail 1		2/Phase Detail 2		4/Phase Detail 3	6/Phase Detail 4
	100%	80%	100%	80%	—	—
1/0	162	133	157	133	123	114
2/0	187	154	178	154	140	128
3/0	217	179	203	179	158	145
4/0	255	210	231	210	180	164
250	281	231	253	231	197	179
350	353	291	306	291	236	213
500	438	361	372	361	283	257
600	488	410		409	314	284
750	556	469		464	349	315
1000	653	560		541	409	370
1250	738	634		608	457	413
1500	813	714		667	501	452
1750	880	788		719	538	484
2000	940	854		766	571	513

Notes:

(1) *This Table gives the allowable current for 90 °C rated single aluminum conductors with spacings installed in non-metallic underground raceways, subject to Rule 4-004(13) and (14), where*
  (a)   the load is continuous; and
  (b)   either end of the cable terminates at a service box, fusible switch, circuit breaker, or panelboard.

(2) *The columns with the heading "80%" indicate that the equipment identified in (1)(b) above is not marked as certified to carry its ampere rating continuously.*

(3) *The columns with the heading "100%" indicate that the equipment identified in (1)(b) above is marked as certified to carry its ampere rating continuously.*

(4) *The ampacities provided in this Table are the lesser of*
  (a)   the value obtained in accordance with Rule 4-004(2)(d); or
  (b)   the value obtained in accordance with Rule 8-104(7).

Table D12A
Allowable copper conductor ampacities for the installation configuration of Diagram B4-3

(See Appendix B Note to Rule 4-004.)

Size, AWG or kcmil	1/Phase Detail 1	2/Phase Detail 2	3/Phase Detail 3	4/Phase Detail 4	5/Phase Detail 5	6/Phase Detail 6
1/0	243	209	186	174	164	157
2/0	274	235	209	195	184	176
3/0	311	266	236	220	207	198
4/0	360	306	271	253	237	227
250	383	326	288	268	252	242
350	470	397	350	326	306	293
500	548	460	404	375	352	337
600	600	502	440	408	383	366
750	667	556	486	450	421	403
1000	758	628	548	508	475	454
1250	831	682	593	547	511	488
1500	889	727	630	581	542	517
1750	927	755	653	602	561	535
2000	962	781	674	621	578	552

Notes:

(1) *This Table gives the allowable current for 90 °C rated multiple copper conductor cables, or single conductors in contact, or multiplexed single copper conductors, directly buried in earth, subject to Rule 4-004(13) and (14), where*

(a) for any load, the cable terminates at equipment of any type other than a service box, fusible switch, circuit breaker, or panelboard; or

(b) the load is noncontinuous and either end of the cable terminates at a service box, fusible switch, circuit breaker, or panelboard.

(2) *The ampacities provided in this Table are the lesser of*

(a) the value obtained in accordance with Rule 4-004(1)(d); or

(b) the value obtained in accordance with Rule 8-104(7).

Table D12B
Allowable copper conductor ampacities for the installation configuration of Diagram B4-3

(See Appendix B Note to Rule 4-004.)

Size, AWG or kcmil	1/Phase Detail 1		2/Phase Detail 2		3/Phase Detail 3		4/Phase Detail 4		5/Phase Detail 5	6/Phase Detail 6
	100%	80%	100%	80%	100%	80%	100%	80%	—	—
1/0	208	172	208	172	186	172	174	172	164	157
2/0	242	200	235	200	209	200	195		184	176
3/0	281	231	266	231	236	231	220		207	198
4/0	327	270	306	270	271	270	253		237	227
250	361	298	326	298	288		268		252	242
350	451	371	397	371	350		326		306	293
500	561	462	460		404		375		352	337
600	629	518	502		440		408		383	366
750	667	592	556		486		450		421	403
1000	758	700	628		548		508		475	454
1250	831	791	682		593		547		511	488
1500	889	882	727		630		581		542	517
1750	927		755		653		602		561	535
2000	962		781		674		621		578	552

Notes:

(1) This Table gives the allowable current for 90 °C rated multiple copper conductor cables, or single copper conductors in contact, or multiplexed single copper conductors, directly buried in earth, subject to Rule 4-004(13) and (14), where

(a) the load is continuous; and

(b) either end of the cable terminates at a service box, fusible switch, circuit breaker, or panelboard.

(2) The columns with the heading "80%" indicate that the equipment identified in (1)(b) above is not marked as certified to carry its ampere rating continuously.

(3) The columns with the heading "100%" indicate that the equipment identified in (1)(b) above is marked as certified to carry its ampere rating continuously.

(4) The ampacities provided in this Table are the lesser of

(a) the value obtained in accordance with Rule 4-004(1)(d); or

(b) the value obtained in accordance with Rule 8-104(7).

Table D13A
Allowable aluminum conductor ampacities for the installation configuration of Diagram B4-3

(See Appendix B Note to Rule 4-004.)

Size, AWG or kcmil	1/Phase Detail 1	2/Phase Detail 2	3/Phase Detail 3	4/Phase Detail 4	5/Phase Detail 5	6/Phase Detail 6
1/0	190	164	146	137	129	124
2/0	217	186	166	155	146	140
3/0	242	207	184	171	161	154
4/0	280	238	211	197	185	177
250	304	258	229	213	200	192
350	366	309	273	254	238	228
500	440	370	325	302	283	271
600	486	406	356	330	309	296
750	540	450	393	364	341	326
1000	613	508	444	411	384	367
1250	684	562	488	451	421	402
1500	734	600	520	480	448	427
1750	774	631	545	503	469	447
2000	809	657	567	522	487	464

Notes:

(1) *This Table gives the allowable current for 90 °C rated multiple aluminum conductor cables, or single aluminum conductors in contact, or multiplexed single aluminum conductors, directly buried in earth, subject to Rule 4-004(13) and (14), where*

    (a) *for any load, the cable terminates at equipment of any type other than a service box, fusible switch, circuit breaker, or panelboard; or*

    (b) *the load is noncontinuous and either end of the cable terminates at a service box, fusible switch, circuit breaker, or panelboard.*

(2) *The ampacities provided in this Table are the lesser of*

    (a) *the value obtained in accordance with Rule 4-004(2)(d); or*

    (b) *the value obtained in accordance with Rule 8-104(7).*

Table D13B
Allowable aluminum conductor ampacities for the installation configuration of Diagram B4-3

(See Appendix B Note to Rule 4-004.)

Size, AWG or kcmil	1/Phase Detail 1		2/Phase Detail 2		3/Phase Detail 3		4/Phase Detail 4		5/Phase Detail 5	6/Phase Detail 6
	100%	80%	100%	80%	100%	80%	100%	80%	—	—
1/0	162	133	162	133	146	133	137	133	129	124
2/0	187	154	186	154	166	154	155	154	146	140
3/0	217	179	207	179	184	179	171		161	154
4/0	255	210	238	210	211	210	197		185	177
250	281	231	258	231	229		213		200	192
350	353	291	309	291	273		254		238	228
500	438	361	370	361	325		302		283	271
600	486	410	406		356		330		309	296
750	540	469	450		393		364		341	326
1000	613	560	508		444		411		384	367
1250	684	634	562		488		451		421	402
1500	734	714	600		520		480		448	427
1750	774		631		545		503		469	447
2000	809		657		567		522		487	464

Notes:

(1) *This Table gives the allowable current for 90 °C rated multiple aluminum conductor cables, or single aluminum conductors in contact, or multiplexed single aluminum conductors, directly buried in earth, subject to Rule 4-004(13) and (14), where*

  (a) the load is continuous; and

  (b) either end of the cable terminates at a service box, fusible switch, circuit breaker, or panelboard.

(2) *The columns with the heading "80%" indicate that the equipment identified in (1)(b) above is not marked as certified to carry its ampere rating continuously.*

(3) *The columns with the heading "100%" indicate that the equipment identified in (1)(b) above is marked as certified to carry its ampere rating continuously.*

(4) *The ampacities provided in this Table are the lesser of*

  (a) the value obtained in accordance with Rule 4-004(2)(d); or

  (b) the value obtained in accordance with Rule 8-104(7).

Table D14A
Allowable copper conductor ampacities for the installation configuration of Diagram B4-4

(See Appendix B Note to Rule 4-004.)

Size, AWG or kcmil	1/Phase Detail 1	2/Phase Detail 2	3/Phase Detail 3	4/Phase Detail 4	5/Phase Detail 5	6/Phase Detail 6
1/0	180	164	152	141	131	125
2/0	206	187	172	160	149	142
3/0	235	213	196	181	168	160
4/0	269	242	223	205	190	181
250	298	267	244	225	208	198
350	361	321	293	268	248	235
500	437	386	350	319	294	279
600	480	423	383	349	321	303
750	538	471	425	386	355	335
1000	620	540	485	439	400	380
1250	676	583	521	472	433	408
1500	724	623	555	502	459	434
1750	756	648	576	521	476	449
2000	785	671	596	538	491	463

Notes:

Δ

(1) *This Table gives the allowable current for 90 °C rated single copper conductor cables, or single copper conductors in contact, or multiplexed single copper conductors, installed in underground raceway, subject to Rule 4-004(13) and (14), where*

   (a) for any load, the cable terminates at equipment of any type other than a service box, fusible switch, circuit breaker, or panelboard; or

   (b) the load is noncontinuous and either end of the cable terminates at a service box, fusible switch, circuit breaker, or panelboard.

(2) *The ampacities provided in this Table are the lesser of*

   (a) the value obtained in accordance with Rule 4-004(1)(d); or

   (b) the value obtained in accordance with Rule 8-104(7).

Table D14B
Allowable copper conductor ampacities for the installation configuration of Diagram B4-4

(See Appendix B Note to Rule 4-004.)

Size, AWG or kcmil	1/Phase Detail 1		2/Phase Detail 2	3/Phase Detail 3	4/Phase Detail 4	5/Phase Detail 5	6/Phase Detail 6
	100%	80%	—	—	—	—	—
1/0	180	172	164	152	141	131	125
2/0	206	200	187	172	160	149	142
3/0	235	232	213	196	181	168	160
4/0		269	242	223	205	190	181
250		298	267	244	225	208	198
350		361	321	293	268	248	235
500		437	386	350	319	294	279
600		480	423	383	349	321	303
750		538	471	425	386	355	335
1000		620	540	485	439	400	380
1250		676	583	521	472	433	408
1500		724	623	555	502	459	434
1750		756	648	576	521	476	449
2000		785	671	596	538	491	463

Notes:

Δ

(1) *This Table gives the allowable current for 90 °C rated multiple copper conductor cables, or single copper conductors in contact, or multiplexed single copper conductors, installed in underground raceway, subject to Rule 4-004(13) and (14), where*
(a) the load is continuous; and
(b) either end of the cable terminates at a service box, fusible switch, circuit breaker, or panelboard.

(2) *The column with the heading "80%" indicates that the equipment identified in (1)(b) above is not marked as certified to carry its ampere rating continuously.*

(3) *The column with the heading "100%" indicates that the equipment identified in (1)(b) above is marked as certified to carry its ampere rating continuously.*

(4) *The ampacities provided in this Table are the lesser of*
(a) the value obtained in accordance with Rule 4-004(1)(d); or
(b) the value obtained in accordance with Rule 8-104(7).

Table D15A
Allowable aluminum conductor ampacities for the installation configuration of Diagram B4-4

(See Appendix B Note to Rule 4-004.)

Size, AWG or kcmil	1/Phase Detail 1	2/Phase Detail 2	3/Phase Detail 3	4/Phase Detail 4	5/Phase Detail 5	6/Phase Detail 6
1/0	142	129	119	111	103	99
2/0	163	148	136	126	117	112
3/0	186	168	155	143	132	126
4/0	214	192	177	163	151	143
250	236	212	194	178	165	157
350	288	256	233	214	198	187
500	351	310	281	257	237	224
600	388	341	309	281	259	245
750	435	381	344	313	287	271
1000	502	437	392	355	326	307
1250	556	480	429	389	356	336
1500	589	514	458	415	379	358
1750	632	541	481	435	397	375
2000	660	564	501	452	413	389

Notes:

Δ

(1) *This Table gives the allowable current for 90 °C rated multiple aluminum conductor cables, or single aluminum conductors, installed in underground raceway, subject to Rule 4-004(13) and (14), where*

    (a)   for any load, the cable terminates at equipment of any type other than a service box, fusible switch, circuit breaker, or panelboard; or

    (b)   the load is noncontinuous and either end of the cable terminates at a service box, fusible switch, circuit breaker, or panelboard.

(2) *The ampacities provided in this Table are the lesser of*

    (a)   the value obtained in accordance with Rule 4-004(2)(d); or

    (b)   the value obtained in accordance with Rule 8-104(7).

Table D15B
Allowable aluminum conductor ampacities for the installation configuration of Diagram B4-4

(See Appendix B Note to Rule 4-004.)

Size, AWG or kcmil	1/Phase Detail 1		2/Phase Detail 2	3/Phase Detail 3	4/Phase Detail 4	5/Phase Detail 5	6/Phase Detail 6
	100%	80%	—	—	—	—	—
1/0	142	133	129	119	111	103	99
2/0	163	154	148	136	126	117	112
3/0	186	179	168	155	143	132	126
4/0	214	210	192	177	163	151	143
250	236	231	212	194	178	165	157
350		288	256	233	214	198	187
500		351	310	281	257	237	224
600		388	341	309	281	259	245
750		435	381	344	313	287	271
1000		502	437	392	355	326	307
1250		556	480	429	389	356	336
1500		589	514	458	415	379	358
1750		632	541	481	435	397	375
2000		660	564	501	452	413	389

Notes:

Δ

(1) *This Table gives the allowable current for 90 °C rated multiple aluminum conductor cables, or single aluminum conductors in contact, or multiplexed single aluminum conductors, installed in underground raceway, subject to Rule 4-004(13) and (14), where*
  (a)   the load is continuous; and
  (b)   either end of the cable terminates at a service box, fusible switch, circuit breaker, or panelboard.

(2) *The column with the heading "80%" indicates that the equipment identified in (1)(b) above is not marked as certified to carry its ampere rating continuously.*

(3) *The column with the heading "100%" indicates that the equipment identified in (1)(b) above is marked as certified to carry its ampere rating continuously.*

(4) *The ampacities provided in this Table are the lesser of*
  (a)   the value obtained in accordance with Rule 4-004(2)(d); or
  (b)   the value obtained in accordance with Rule 8-104(7).

Table D16
Sizes of conductors, fuse ratings, and circuit breaker settings for motor overload protection and motor circuit overcurrent protection

(See Appendix B Note to Section 28.)

Note: *This Table is based on Table 29 and a room temperature of 30 °C.*

Full-load current rating of motor, A	Minimum allowable ampacity of conductor, A	Overcurrent protection maximum allowable rating of fuses and maximum allowable setting for circuit breakers of the time-limit type for motor circuits								
		Single-phase — All types and squirrel cage and synchronous (full voltage, resistor and reactor starting)			Squirrel cage and synchronous (autotransformer and star-delta starting)			DC or wound rotor AC		
		Non-time-delay fuses, A	Time-delay* fuses, A	Circuit breaker, A	Non-time-delay fuses, A	Time-delay* fuses, A	Circuit breaker, A	Non-time-delay fuses, A	Time-delay* fuses, A	Circuit breaker, A
1	1.25	15	15	15	15	15	15	15	15	15
2	2.50	15	15	15	15	15	15	15	15	15
3	3.80	15	15	15	15	15	15	15	15	15
4	5.00	15	15	15	15	15	15	15	15	15
5	6.25	15	15	15	15	15	15	15	15	15
6	7.50	15	15	15	15	15	15	15	15	15
7	8.75	20	15	15	15	15	15	15	15	15
8	10.00	20	15	20	20	15	15	15	15	15
9	11.25	25	15	20	20	15	15	15	15	15
10	12.50	30	15	25	25	15	20	15	15	15
11	13.75	30	15	25	25	15	20	15	15	15
12	15.00	35	20	30	30	20	20	15	15	15
13	16.25	35	20	30	30	25	20	15	15	15
14	17.50	40	20	35	35	25	30	20	20	20
15	18.75	45	25	35	35	25	30	20	20	20
16	20.00	45	25	40	40	25	30	20	20	20
17	21.3	50	25	40	40	35	30	25	25	25
18	22.5	50	30	45	45	35	30	25	25	25
19	23.8	50	30	45	45	35	30	25	25	25
20	25.0	60	35	50	50	35	40	30	30	30

(Continued)

Table D16 (Continued)

Overcurrent protection maximum allowable rating of fuses and maximum allowable setting for circuit breakers of the time-limit type for motor circuits

Full-load current rating of motor, A	Minimum allowable ampacity of conductor, A	Single-phase squirrel cage and synchronous (full voltage, resistor and reactor starting)			Squirrel cage and synchronous (autotransformer and star-delta starting)			DC or wound rotor AC		
		Non-time-delay fuses, A	Time-delay* fuses, A	Circuit breaker, A	Non-time-delay fuses, A	Time-delay* fuses, A	Circuit breaker, A	Non-time-delay fuses, A	Time-delay* fuses, A	Circuit breaker, A
22	27.5	60	35	50	50	35	40	30	30	30
24	30.0	70	40	60	60	45	40	35	35	35
26	32.5	70	45	60	60	45	50	35	35	35
28	35.0	80	45	70	70	45	50	40	40	40
30	37.5	90	50	70	70	50	60	45	45	45
32	40.0	90	50	80	60	50	60	45	45	45
34	42.5	100	50	80	60	50	60	50	50	50
36	45.0	100	60	90	70	60	70	50	50	50
38	47.5	110	60	90	70	60	70	50	50	50
40	50.0	110	70	100	80	80	70	60	60	60
42	52.5	125	70	100	80	80	70	60	60	60
44	55.0	125	70	110	80	80	70	60	60	60
46	57.5	125	80	110	90	90	70	60	60	60
48	60.0	125	80	110	90	90	70	70	70	70
50	62.5	150	80	125	100	90	100	70	70	70
52	65.0	150	90	125	100	90	100	70	70	70
54	67.5	150	90	125	100	90	100	80	80	80
56	70.0	150	90	125	110	110	100	80	80	80
58	72.5	150	100	125	110	110	100	80	80	80
60	75.0	175	100	150	110	110	100	90	90	90
62	77.5	175	100	150	110	110	100	90	90	90
64	80.0	175	110	150	125	110	125	90	90	90
66	82.5	175	110	150	125	110	125	90	90	90

(Continued)

Table D16 (Continued)

Full-load current rating of motor, A	Minimum allowable ampacity of conductor, A	Overcurrent protection maximum allowable rating of fuses and maximum allowable setting for circuit breakers of the time-limit type for motor circuits										
		Single-phase — All types and squirrel cage and synchronous (full voltage, resistor and reactor starting)			Squirrel cage and synchronous (autotransformer and star-delta starting)			DC or wound rotor AC			
		Non-time-delay fuses, A	Time-delay* fuses, A	Circuit breaker, A	Non-time-delay fuses, A	Time-delay* fuses, A	Circuit breaker, A	Non-time-delay fuses, A	Time-delay* fuses, A	Circuit breaker, A		
68	85.0	200	110	150	125	110	125	100	100	100		
70	87.5	200	110	175	125	110	125	100	100	100		
72	90.0	200	125	175	125	125	125	100	100	100		
74	92.5	200	125	175	125	125	125	110	110	110		
76	95.0	225	125	175	150	125	150	110	110	110		
78	97.5	225	125	175	150	125	150	110	110	110		
80	100.0	225	125	200	150	125	150	110	110	110		
82	102.5	225	125	200	150	125	150	110	110	110		
84	105.0	250	125	200	150	125	150	125	125	125		
86	107.5	250	150	200	150	150	150	125	125	125		
88	110.0	250	150	200	175	150	175	125	125	125		
90	112.5	250	150	225	175	150	175	125	125	125		
92	115.0	250	150	225	175	150	175	125	125	125		
94	117.5	250	150	225	175	150	175	125	125	125		
96	120.0	250	150	225	175	150	175	125	125	125		
98	122.5	250	150	225	175	150	175	125	125	125		
100	125.0	300	175	250	200	175	200	150	150	150		
105	131.3	300	175	250	200	175	200	150	150	150		
110	137.5	300	175	250	200	175	200	150	150	150		
115	143.8	300	200	250	225	200	225	150	150	150		
120	150.0	350	200	300	225	200	225	175	175	175		
125	156.3	350	200	300	250	200	250	175	175	175		
130	162.5	350	225	300	250	225	250	175	175	175		
135	168.8	400	225	300	250	225	250	200	200	200		

(Continued)

Table D16 (Continued)

Overcurrent protection maximum allowable rating of fuses and maximum allowable setting for circuit breakers of the time-limit type for motor circuits

Full-load current rating of motor, A	Minimum allowable ampacity of conductor, A	Single-phase — All types and squirrel cage and synchronous (full voltage, resistor and reactor starting)			Squirrel cage and synchronous (autotransformer and star-delta starting)			DC or wound rotor AC		
		Non-time-delay fuses, A	Time-delay* fuses, A	Circuit breaker, A	Non-time-delay fuses, A	Time-delay* fuses, A	Circuit breaker, A	Non-time-delay fuses, A	Time-delay* fuses, A	Circuit breaker, A
140	175.0	400	225	350	250	225	250	200	200	200
145	181.3	400	250	350	250	250	250	200	200	200
150	187.5	450	250	350	300	250	300	225	225	225
155	193.8	450	250	350	300	250	300	225	225	225
160	200.0	450	250	400	300	250	300	225	225	225
165	206.3	450	250	400	300	250	300	225	225	225
170	212.5	500	250	400	300	250	300	250	250	250
175	218.8	500	300	400	350	300	350	250	250	250
180	225.0	500	300	450	350	300	350	250	250	250
185	231.3	500	300	450	350	300	350	250	250	250
190	237.5	500	300	450	350	300	350	250	250	250
195	243.8	500	300	450	350	300	350	250	250	250
200	250.0	600	350	500	400	350	400	300	300	300
210	262.5	600	350	500	400	350	400	300	300	300
220	275.0	600	350	500	400	350	400	300	300	300
230	287.5	600	400	500	450	400	450	300	300	300
240	300.0		400		450	400	450	350	350	350
250	312.5		400		500	400	500	350	350	350
260	325.0		450	600	500	450	500	350	350	350
270	337.5		450	600	500	450	500	400	400	400
280	350.0		450		500	450	500	400	400	400
290	362.5		500		500	500	500	400	400	400
300	375.0		500		600	500	600	450	450	450
320	400		500		600	500	600	450	450	450

(Continued)

Table D16 (Concluded)

Overcurrent protection maximum allowable rating of fuses and maximum allowable setting for circuit breakers of the time-limit type for motor circuits

Full-load current rating of motor, A	Minimum allowable ampacity of conductor, A	Single-phase — All types and squirrel cage and synchronous (full voltage, resistor and reactor starting)			Squirrel cage and synchronous (autotransformer and star-delta starting)			DC or wound rotor AC		
		Non-time-delay fuses, A	Time-delay* fuses, A	Circuit breaker, A	Non-time-delay fuses, A	Time-delay* fuses, A	Circuit breaker, A	Non-time-delay fuses, A	Time-delay* fuses, A	Circuit breaker, A
340	425		500		600	500	600	500	500	500
360	450		600			600		500	500	500
380	475					600		500	500	500
400	500							600	600	600
420	525							600	600	600
440	550							600	600	600
460	575							600		600
480	600									
500	625									

*Includes time-delay "D" fuses referred to in Rule 14-200.
Note: For overload protection, refer to Rule 28-306 and the motor nameplate data.

Appendix E — Dust-free rooms
(See Rules in Section 18 for Class II and Class III hazardous locations)

Note: *This Appendix is an informative (non-mandatory) part of this Standard.*

E1. Introduction

This Appendix covers recommended practices for the housing of electrical equipment. With respect to actual constructional requirements, it is not practical to cover all materials and methods that are or may become available and, therefore, any specific details mentioned in this Appendix are to be considered examples rather than specific requirements. The *National Building Code of Canada* or any other building code that may be in force in any particular locality should take precedence over anything stated here and should be consulted in this respect.

E2. Scope

This Appendix covers the construction of dust-free rooms built adjacent to or as part of buildings that, by the nature of their use or occupancy, are subject to accumulations of dusts that may create a fire or explosion hazard or that may be detrimental to the proper operation of electrical equipment not provided with dust-tight enclosures.

E3. Definition

The following definition applies in this Appendix:

Dust-free room — a room, building, or other area, of a size that permits the entrance of persons for operation and maintenance purposes and constructed so that the quantity of dust that can enter will not create a hazardous condition.

E4. Use

E4.1
Dust-free rooms are intended to be used to house electrical equipment except equipment required by this Code to be installed in vaults.

E4.2
Dust-free rooms should not be used for any manufacturing, processing, maintenance, storage, or other purposes except those that may be essential for the proper operation and maintenance of the electrical equipment that they house.

E5. Enclosing of electrical equipment

E5.1
Electrical equipment in dust-free rooms need not be of types approved for Class II locations.

E5.2
Where access to the dust-free room by unauthorized persons is permitted, the electrical equipment should be enclosed, guarded, protected, etc., as required by this Code for ordinary locations.

E5.3
Where access is to authorized persons only, enclosures may be omitted as provided for by this Code (e.g., Section 26).

E6. Materials and methods of construction

Materials used for the construction of dust-free rooms and the method of construction should fulfill the following conditions:
(a)   the enclosure of the room should be as impervious to the passage of dust from outside the room as is practicable;

(b)   the various components, walls, floors, ceilings, etc., should be capable of safely supporting the live and dead loads (including impact) to which they are liable to be subjected;

(c)   there should be no liability of dust passages being created due to shrinkage, breaking, or cracking;

(d)   the completed structure should have a fire resistance rating of 1 h or better;

(e)   in buildings constructed of non-combustible materials, the rooms should also be constructed of non-combustible material, but in other buildings, the rooms may be constructed of combustible materials with non-combustible facing on at least the inner surfaces; and

(f)   if it is necessary that shafts or other rotating or sliding members be used to connect equipment outside the room with that inside, suitable means such as seals, gaskets, baffles, etc., should be provided to prevent passage of dust through the necessary openings.

E7. Floors

Some acceptable floor constructions are as follows:

(a)   solid concrete slab of minimum thickness 75 mm and reinforced as necessary; or

(b)   steel joists with welded plate metal floor or concrete slabs of minimum thickness 50 mm over the joists and 22 mm Portland cement plaster on metal lath under them.

Wherever necessary, floors should be surfaced with insulating material to prevent shock hazard.

E8. Walls

Some acceptable wall constructions are as follows, and it is to be noted that lath and plaster on both sides must have other construction built into the wall to ensure continuance of the dust-free features, and plywood joints must be backed by studding:

(a)   monolithic concrete of minimum thickness 100 mm reinforced as necessary;

(b)   built-up masonry consisting of
  (i)   solid bricks of minimum thickness 95 mm;
  (ii)   hollow tile of minimum thickness 75 mm if plastered both sides, and 150 mm if not plastered; or
  (iii)   hollow concrete or cinder concrete block of minimum thickness 125 mm if plastered both sides, and 200 mm if not plastered; or

(c)   stud construction of 50 mm if of metal or 100 mm if of wood, faced on the outside with metal or perforated gypsum lath with 19 mm gypsum or 22 mm Portland cement plaster, faced on the inside with
  (i)   metal-faced laminated wood;
  (ii)   laminated wood with 19 mm gypsum or 22 mm Portland cement plaster on metal or perforated gypsum lath;
  (iii)   laminated wood with fire-resisting non-metallic facing; or
  (iv)   sheet metal equivalent to 0.0667 inch (No. 14 MSG) steel with welded or riveted locked-seam joints secured to metal studs by welding or by self-locking screws.

E9. Ceilings

Ceilings, if load-bearing, should have the same construction as floors, but if non-load-bearing, may be constructed similarly to walls with wood joists, if used, of greater depth in accordance with the span.

E10. Cubicle construction

E10.1
Where the room is constructed as a free-standing cubicle with walls and ceiling not forming a part of and spaced away from the structure of the building proper, the walls and ceiling may be of sheet steel not less than 0.0667 inch (No. 14 MSG) thick, suitably joined and reinforced as may be necessary.

E10.2
If the floor is elevated above the floor of the building proper, it may be constructed in accordance with Clause E7 and the plaster may be omitted.

E11. Doors

E11.1
Doors giving access to the room from dusty locations should be
(a) either metal-clad or hollow metal and weatherstripped or otherwise arranged to prevent dust leakage at the edges of frames and sills; and
(b) equipped with self-closers.

E11.2
If operation of the electrical equipment requires entry to the room from dusty locations, two hinged doors with a 1.5 m vestibule between them should be provided, but where entry is necessary for maintenance only, single doors may be used.

E11.3
Doors giving access to the room from a dust-free atmosphere may be of ordinary types.

E11.4
If more than one point of access to a room is provided, all but the principal one should be either securely locked or other adequate means should be used to prevent unauthorized traffic through the room.

E11.5
All doors should be provided with means whereby they can be readily unlocked and opened from the inside without the use of a removable key.

E12. Windows

E12.1
Windows facing dusty locations should have a fixed metal sash and wired glass.

E12.2
Windows in exterior walls may be arranged for opening if it is reasonably certain that the surrounding exterior area will remain sufficiently dust free.

E13. Ventilation

E13.1
Ventilation by clean air should be adequate for the dissipation of heat from the electrical apparatus installed.

E13.2
If ventilation is by means of forced air circulation, the air should be forced into the room rather than exhausted out of it.

E13.3
It is recommended that the air in the room be kept at a pressure slightly above atmospheric, which will tend to blow any dust out of the room rather than have it sucked in.

E13.4
The amount of ventilating air required is difficult to specify in an empirical value because it depends on
(a) the size of the room;
(b) the dissipating ability of walls and ceilings;
(c) the amount and nature of the electrical equipment; and
(d) the temperature of the incoming air.

E13.5
The cooling effect may be obtained by radiation alone (if the area is sufficient), by ventilation, or by a combination of both.

Appendix F — Recommended installation practice for intrinsically safe and non-incendive electrical equipment and wiring
(See Rules 18-000, 18-066, 18-090, 18-100, and 18-150)

Note: *This Appendix is an informative (non-mandatory) part of this Standard.*

F1. Introduction

F1.1
Approved intrinsically safe and non-incendive electrical equipment is designed and tested to ensure a high level of safety for use in hazardous locations when properly installed and wired and used in the intended manner. Intrinsically safe equipment, type i or ia, is approved for use in Class I, Zone 0, 1, or 2, Class I or Class II, Division 1 or 2 hazardous locations. Intrinsically safe equipment, type ib, is approved for use in Class I, Zone 1 or 2 and Class I, Division 2 hazardous locations. Non-incendive equipment is approved for use only in Class I, Zone 2 or Class I, Division 2 hazardous locations. The safety provided by the design of such equipment can be seriously jeopardized or defeated if it is misapplied or improperly installed. The guidelines given in this Appendix are intended to point out certain necessary precautions which, if observed, should result in a satisfactory installation and continued safe operation. Additional guidelines may be found in ISA RP 12.6.

F1.2
Intrinsically safe equipment capable of tolerating up to two countable faults while still providing protection is marked i or ia. Intrinsically safe equipment capable of tolerating only one countable fault while still providing protection is marked ib.

F1.3
The term "control room", as used in this Appendix, refers to those areas where the signal voltages are received and utilized.

F1.4
The recommendations in Clauses F2 to F7 apply to intrinsically safe equipment, those in Clauses F8 to F13 apply to non-incendive equipment, and those in Clause F14 apply to both intrinsically safe and non-incendive equipment.

F2. General

F2.1
Intrinsically safe circuits may consist of a simple signal or control circuit or a more complex loop comprising a specific controller, monitor, barrier device, or similar control room equipment serving as the energy source, connected by means of control room wiring and interconnecting field wiring to a specific signal transmitter, transducer, positioner, or similar field-mounted equipment located in the remote hazardous area. Alternatively, the intrinsically safe circuit may originate in a powered field-mounted device suitable for and located in the hazardous area, and be connected by means of interconnecting field wiring and control room wiring to a specific receiver input located either in the control room safe area or elsewhere in the hazardous area. Care must be taken in the interpretation of the guidelines in this Appendix to appreciate the significance of the possible directions of energy flow in order to ensure that the proper precautions are taken.

F2.2
An intrinsically safe circuit should not contain any additional items of equipment that are not specifically included in the description of the approved circuit as marked on the principal components of the approved circuit or indicated on the loop circuit diagram drawing referred to in the marking on such components.

F2.3
Control room equipment connected to the control room side of approved barrier devices need not be approved for connection to intrinsically safe circuits but the equipment should comply with the requirements of Clause F3.1.3 regarding supply voltage and internally generated voltage.

F2.4
When intrinsically safe circuits operate at voltage and current levels sufficient to constitute a shock hazard,

the same precautions against shock hazard are necessary when installing or servicing such circuits as with non-intrinsically safe circuits.

F2.5

In order to prevent the development of possible hazardous ground loop currents due to differences in ground voltage between field-mounted equipment and control room equipment during normal operation or under fault conditions, an intrinsically safe circuit should be grounded only as provided for in the approved description of the intrinsically safe circuit, the source equipment, or the barrier device itself.

F3. Control room equipment — Intrinsic safety

F3.1 General

F3.1.1

Control room equipment associated with intrinsically safe circuits should be approved for the specific type of location for which the control room itself is classified.

F3.1.2

If the control room is classified as a hazardous location, the equipment and wiring in that area should comply with the requirements of Section 18.

F3.1.3

Control room equipment, including non-intrinsically safe equipment connected to barrier devices or associated apparatus, should not be powered by more than 250 V rms ac or dc line-to-line or line-to-ground and should not generate internally any voltage in excess of these values, unless either the barrier device or the associated apparatus is specifically approved and marked for a higher voltage or the equipment itself is approved and marked to permit connection.

F3.1.4

Where it is necessary to derive the control room equipment voltage from a source in excess of 250 V rms ac, a suitable overcurrent protected isolating transformer should be used.

F3.2 Control room equipment — Barrier devices

F3.2.1

Approved barrier devices should be installed and used in strict accordance with the manufacturer's published instructions.

F3.2.2

Proper and reliable grounding of shunt diode barrier devices is essential to the intrinsic safety that they provide. To ensure proper and reliable grounding, it is recommended that
(a) duplicate grounding conductors be used to connect the barrier device to a designated ground electrode;
(b) the barrier grounding system be insulated from ground at all places except at the point of connection to a designated ground electrode;
(c) each grounding conductor be capable of carrying the maximum system fault current in accordance with Section 10;
(d) grounding path connections be secure, permanent, visible, and accessible;
(e) the resistance from the furthest barrier to the designated ground electrode not exceed 1 Ω; and
(f) the grounding conductors have mechanical strength or physical protection to ensure that they are not likely to be broken. For unprotected conductors, at least a No. 12 AWG conductor is recommended.

F3.2.3

Aluminum should not be used for the busbar for a barrier device or for its connection to ground unless precautions are taken against electrolytic corrosion.

F4. Control room wiring (safe area) — Intrinsic safety

F4.1

All intrinsically safe field wiring should be guarded against intrusion from non-intrinsically safe voltages by separation, shielding by metal braid, or by other means.

F4.2

Intrinsically safe control room wiring should be routed and restrained in such a way that any wire that comes

Appendix F
Recommended installation practice for intrinsically safe
and non-incendive electrical equipment and wiring

loose from an intrinsically safe terminal is not likely to touch a non-intrinsically safe terminal. Conversely, non-intrinsically safe wiring should be routed and restrained in such a way that any wire that comes loose from a non-intrinsically safe terminal is not likely to touch an intrinsically safe terminal. A separation of 50 mm or a grounded metal partition or a non-metallic partition between intrinsically safe and non-intrinsically safe terminals is considered to conform to these requirements.

F4.3
Where intrinsically safe circuits in the safe area are connected by single or multiple plug and socket connectors not approved as part of the intrinsically safe apparatus, intrinsically safe and non-intrinsically safe circuits should not be mixed in the same connector.

F4.4
Plug and socket connectors used to connect intrinsically safe circuits in the safe area should be non-interchangeable with any other plugs or sockets connected to non-intrinsically safe circuits in the same location or should be identified so that interchangeability is unlikely and uniformity of non-interchangeable connector types should be maintained through the entire installation area wherever practicable.

F5. Field-mounted equipment (hazardous area) — Intrinsic safety

F5.1
Field-mounted equipment connected to intrinsically safe circuits may be provided with a general-purpose (non-explosion-proof or non-flame-proof) enclosure. Field-mounted intrinsically safe equipment should be installed only in clean, dry, indoor, or protected locations unless its operation would not be adversely affected by weather, excessive moisture, dust, or other contaminants.

F5.2
In the case of a switch contact in a hazardous area intended for connection to an intrinsically safe circuit, the switch contact should
(a)   be in the form of a discrete device that does not contain any non-intrinsically safe circuits (e.g., a pressure switch or a limit switch);
(b)   be enclosed in a separate compartment; or
(c)   be positively segregated from any non-intrinsically safe circuit by means of
     (i)   50 mm spacing;
     (ii)   a grounded metal partition; or
     (iii) a non-metallic partition.
   If more than one intrinsically safe circuit is controlled by adjacent or ganged switches, the design of the switches should be such as to positively prevent inadvertent parallel or series interconnection of the different intrinsically safe circuits.

F5.3
In case of intrinsically safe circuits intended for connection to a fixed resistance temperature detector or adjustable resistive device in the hazardous area, the device should
(a)   be in the form of a discrete device that does not contain any non-intrinsically safe circuits (e.g., a resistance-type strain gauge, a load cell, a potentiometer);
(b)   be enclosed in a separate compartment;
(c)   be positively separated from any non-intrinsically safe circuit by means of
     (i)   50 mm spacing;
     (ii)   a grounded metal partition; or
     (iii) a non-metallic partition;
(d)   not incorporate any energy-storing components such as capacitors, inductors, or inductive winding.

F5.4
A selector switch used to select one from a number of thermocouples and to connect its output to the control room equipment via an intrinsically safe circuit located in the hazardous area should be of the non-shorting type to prevent series or parallel connection of thermocouple circuits.

F5.5
In the case of monitoring equipment that has multiple intrinsically safe input circuits, special care should be taken to ensure that the intrinsic safety of each and every such input circuit is maintained so as not to jeopardize the intrinsic safety of the remaining input circuits. The same multiple input monitor should not be connected to a

mixture of intrinsically safe and non-intrinsically safe signal sources unless the description of the approved loop circuits specifically permits such use.

F6. Interconnecting field wiring (hazardous area) — Intrinsic safety

F6.1

The interconnecting field wiring for intrinsically safe circuits may be wired in the same manner as comparable circuits for use in ordinary locations, for example Class 1 or Class 2 safety circuits. However, they should be mechanically protected since the energy in an intrinsically safe circuit is inherently limited by the source equipment and no additional overcurrent protection is usually required in such circuits.

F6.2

Different intrinsically safe systems are defined as, for example, those
(a)   operating at different voltage levels or polarities;
(b)   having different signal ground reference points; or
(c)   approved for different hazardous location groups.
    An intrinsically safe system may include more than one intrinsically safe circuit.

F6.3

Different intrinsically safe systems should not be run in the same multi-conductor cable. Different intrinsically safe circuits of the same intrinsically safe system should not be run in the same cable unless at least 0.25 mm thick insulation is used on each conductor or unless no hazard can result from interconnection.

F6.4

Different intrinsically safe systems approved for different hazardous location groups may be considered a single intrinsically safe system if approved for use in the most severe location in which any part of the system is used and the system otherwise complies with Clause F6.2.

F6.5

Where more than one intrinsically safe circuit occupies the same raceway, compartment, outlet, junction box, or similar fitting, all such circuits should be clearly identified to help ensure that the intended field-mounted equipment is connected to the proper control room equipment.

F6.6

Where intrinsically safe circuits are contained in multi-conductor cable, and flammable mixtures can be transmitted through the cable core, the cable core should be sealed and/or vented so that the cable will not transmit the flammable atmosphere from one division or zone of a hazardous location to another or from a hazardous to a non-hazardous location.

F6.7

Where conduit or other raceway is used to enclose intrinsically safe wiring, the conduit or raceway is to be sealed and/or vented so that the conduit or raceway will not transmit the flammable atmosphere from one division or zone of a hazardous location to another or from a hazardous to a non-hazardous location. Raceways containing intrinsically safe wiring should be identified and used for this purpose only.

F6.8

The intrinsic safety of a circuit can be affected by the cumulative energy-storing effect of distributed inductance and capacitance in the field wiring and the unprotected terminal inductance and capacitance of the hazardous location device. Cable selection should be based on the difference between the maximum allowable parameters of the control room equipment as stated in the manufacturer's instructions and the unprotected terminal inductance and capacitance of the field-mounted device.

F6.9

In general, the inductive effect does not depend on the length of such cables, provided that the ratio of inductance to resistance of the cable (L/R ratio) is held to a sufficiently low value, where the manufacturer's literature is not explicit. In general, the inductive effect of a cable will not be significant if the cable L/R ratio is less than the following [see Rule 18-050(2) and (3)]:
(a)   Class I, Group IIC or Group A — 25 $\mu H/\Omega$;
(b)   Class I, Group IIC or Group B — 25 $\mu H/\Omega$;
(c)   Class I, Group IIB or Group C — 60 $\mu H/\Omega$; and
(d)   Class I, Group IIA or Group D — 200 $\mu H/\Omega$.

Appendix F
Recommended installation practice for intrinsically safe
and non-incendive electrical equipment and wiring

F6.10

Similarly, the cumulative effect of distributed capacitance depends on the length of the loop wiring, the type of wiring to the hazardous location device, and the voltage of the circuit. The capacitive effect in general will not seriously affect the intrinsic safety of the circuit, provided that the following total cumulative capacitance values of the system are not exceeded:

Maximum total loop capacitance
(Microfarads area classification)

Circuit voltage, V	Class I Group IIC (A and B)	Class I Group IIB (C)	Class I Group IIA (D)
0–15	0.8	2.4	6.4
Over 15–30	0.11	0.33	0.88
Over 30–60	0.028	0.084	0.224
Over 60–100	0.013	0.039	0.104
Over 100–200	—	—	—
Over 200–300	—	—	—

F6.11

Information on the distributed inductance and capacitance of signal cables is generally available in literature published by the cable manufacturer, from which the L/R ratio and total capacitance of a circuit loop may be calculated. Alternatively, the inductance, resistance, and capacitance for a particular loop may be determined by actual measurement using appropriate instrumentation (see Clause F7.1 for precautions needed to carry out such tests).

F7. Testing, servicing, and maintenance — Intrinsic safety

F7.1

Testing of control room wiring or wiring of equipment associated with intrinsic safety may be carried out with the equipment connected unless the test equipment utilizes voltages in excess of the marked barrier voltage.

F7.2

An annual check is recommended and should include examination for signs of corrosion, tightness of terminal connections and, for barrier device installations in particular, measurement of ground path resistance (resistance to ground rod is not to exceed 1 Ω).

F7.3

When wiring within the safe area is subjected to high-voltage insulation tests, all barrier devices and field wiring should be disconnected at the safe area termination to prevent damage to the energy-limiting devices and to prevent transmission of the high voltage into the hazardous area.

F7.4

Although there is no explosion hazard in short-circuiting intrinsically safe circuits, care should be taken to avoid short-circuiting of intrinsically safe circuits derived from barrier devices, as this may blow the barrier fuse and require replacement of the barrier.

F8. General — Non-incendive

F8.1

Non-incendive circuits may consist of a simple signal or control circuit or a more complex loop comprising a specific controller, monitor, barrier device, or similar control room equipment serving as the energy source, connected by means of control room wiring and interconnecting field wiring to a specific signal transmitter, transducer, positioner, or similar field-mounted equipment located in the remote hazardous area. Alternatively, the non-incendive circuit may originate in a powered field-mounted device suitable for and located in the hazardous area, and can be connected by means of interconnecting field wiring and control room wiring to a specific receiver input located either in the control room safe area or elsewhere in the hazardous area. Care should be taken in the interpretation of the guidelines in this Appendix to appreciate the significance of the possible directions of energy flow in order to ensure that the proper precautions are taken.

F8.2

When non-incendive circuits operate at voltage and current levels sufficient to constitute a shock hazard, the same precautions against shock hazard are recommended when installing or servicing such circuits as with other electric circuits.

F9. Control room equipment — Non-incendive

F9.1

Control room equipment associated with non-incendive circuits should be approved for the specific type of location for which the control room itself is classified.

F9.2

If the control room is classified as a hazardous location, the equipment and wiring in that area should comply with the requirements of Section 18.

F10. Control room wiring — Non-incendive

F10.1

All non-incendive field wiring should be guarded against intrusion from incendive voltages by separation, shielding by metal braid, or by other means.

F10.2

Where plug and socket connectors are used to connect non-incendive circuits, they should not be interchangeable with plugs and sockets of other wiring or they should be identified such that interchange is unlikely. The uniformity of non-interchangeable connector types or markings should be maintained throughout the entire installation when practicable.

F11. Field-mounted equipment — Non-incendive

F11.1

Field-mounted equipment connected to non-incendive circuits may be provided with general-purpose (non-explosion-proof) enclosures but the enclosure should be approved for the expected environmental conditions.

F11.2

A switch contact intended for connection to a non-incendive circuit should be

(a)   a stand-alone device that does not contain any other circuits (e.g., a pressure switch or limit switch);
(b)   a stand-alone device enclosed in a separate compartment and not electrically connected to other circuits; or
(c)   approved for connection to non-incendive circuits or intrinsically safe circuits.

   If more than one non-incendive circuit is controlled by adjacent or ganged switches, the design should be such that parallel or series connection of different circuits is not possible.

F11.3

A resistance temperature detector (RTD) or variable resistive device should be

(a)   a stand-alone device that does not contain any other circuit (e.g., slide wire feedback);
(b)   a stand-alone device enclosed in a separate compartment and not electrically connected to any other circuits; or
(c)   approved for connection to non-incendive circuits or intrinsically safe circuits.

F11.4

When a selector switch is used on the thermocouples to connect them to a non-incendive circuit, the design of the switch should be such that parallel or series connection of different circuits is not possible.

F12. Interconnecting field wiring — Non-incendive

F12.1

The interconnecting field wiring for non-incendive circuits may be wired in the same manner as comparable circuits for use in ordinary locations (e.g., Class 1 or Class 2 safety circuits). The wires should be mechanically protected but overcurrent protection is not usually required.

Appendix F
*Recommended installation practice for intrinsically safe
and non-incendive electrical equipment and wiring*

F12.2

Where more than one non-incendive circuit occupies the same junction box or similar fitting where joints or connections are made, all such circuits should be identified to ensure proper connection.

F12.3

Where non-incendive circuits are contained in multi-conductor cable, and flammable mixtures can be transmitted through the cable core, the cable core should be sealed so that the cable will not transmit the flammable atmosphere from one division or zone of a hazardous location to another or from a hazardous to a non-hazardous location.

F12.4

Where conduit or other raceway is used to enclose non-incendive wiring, the conduit or raceway should be sealed so that the conduit or raceway will not transmit the flammable atmosphere from one division or zone of a hazardous location to another or from a hazardous to a non-hazardous location.

F12.5

The non-incendive characteristics of a circuit can be affected by the cumulative energy-storing effect of distributed inductance and capacitance in the field wiring and the unprotected terminal inductance and capacitance of the hazardous location device. Cable selection should be based on the difference between the maximum allowable parameters of the control room equipment as stated in the manufacturer's instructions and the unprotected terminal inductance and capacitance of the field-mounted device.

F12.6

In general, the inductive effect does not depend on the length of such cables, provided that the ratio of inductance to resistance of the cable (L/R ratio) is held to a sufficiently low value where the manufacturer's literature is not explicit. In general, the inductive effect of a cable will not be significant if the cable L/R ratio is less than the following [see Rule 18-050(2)]:

(a)   Class I, Group IIC or Group A — 25 $\mu H/\Omega$;
(b)   Class I, Group IIC or Group B — 25 $\mu H/\Omega$;
(c)   Class I, Group IIB or Group C — 60 $\mu H/\Omega$; and
(d)   Class I, Group IIA or Group D — 200 $\mu H/\Omega$.
    Field device inductance and capacitance are marked L_i and C_i on the device.

F12.7

Similarly, the cumulative effect of distributed capacitance depends on the length of the loop wiring, the type of wiring to the hazardous location device, and the voltage of the circuit. The capacitive effect in general will not seriously affect the non-incendive characteristics of the circuit provided that the following total cumulative capacitance values of the system are not exceeded:

Maximum total loop capacitance
(Microfarads area classification)

Circuit voltage, V	Class I Group IIC (A and B)	Class I Group IIB (C)	Class I Group IIA (D)
0–15	2.79	8.37	22.32
Over 15–30	0.27	0.81	2.16
Over 30–60	0.052	0.156	0.416
Over 60–100	0.0222	0.066	0.176

Note: *The above numbers were obtained from the ignition curve for hydrogen and a factor of 3 for Group C and 8 for Group D and include a 10% factor of safety to allow for tolerances in actual voltages and currents.*

F12.8

Information on the distributed inductance and capacitance of signal cables is generally available in literature published by the cable manufacturer, from which the L/R ratio and total capacitance of a circuit loop may be calculated. Alternatively, the inductance, resistance, and capacitance for a particular loop may be determined by actual measurement using appropriate instrumentation (see Clause F13.1 for precautions to be taken in carrying out such tests).

F13. Testing, servicing, and maintenance — Non-incendive

F13.1

No test equipment should be connected to a non-incendive circuit unless all the areas where the circuit is contained are proven to be non-hazardous.

F13.2

When wiring within the safe area is subjected to high-voltage insulation tests, all barrier devices and field wiring should be disconnected at the safe area termination to prevent damage to the energy-limiting devices and to prevent transmission of the high voltage into the hazardous area.

F13.3

Although there is no explosion hazard in short-circuiting non-incendive circuits, care should be taken to avoid damaging protective components.

F14. Allowable equipment loops

F14.1

Equipment approved as intrinsically safe, type i or ia, and connected as outlined in the approved loop description or as marked on the equipment is suitable for use in Class I, Zone 0, 1, or 2; Class I, Division 1 or 2; and Class II, Division 1 or 2 hazardous locations.

F14.2

Equipment approved as intrinsically safe, type ib, and connected as outlined in the approved loop description or as marked on the equipment is suitable for use in Class I, Zone 1 or 2 and Class I, Division 2 hazardous locations.

F14.3

Intrinsically safe and non-incendive equipment should be used only in the groups for which it is approved (e.g., Class I, Groups IIB and IIA (C and D) equipment should not be used in Class I, Groups IIC (A or B) or in Class II).

F14.4

The manufacturer's instruction manual may contain a list of equipment or combination of equipment that may be classified as non-incendive. The non-incendive loops specified in the manual should not be changed by adding other equipment to the non-incendive circuit or by combining one or more approved loops in the same circuit.

F14.5

As an alternative to the markings discussed in Clause F14.1, the manufacturer may mark the following on equipment having intrinsically safe field circuit connections:
(a) V_{oc} — the maximum open-circuit voltage;
(b) I_{sc} — the maximum short-circuit current;
(c) C_a — the maximum allowable connected capacitance; and
(d) L_a — the maximum allowable connected inductance.

F14.6

As an alternative to the markings discussed in Clause F14.1, the manufacturer may mark the following on equipment receiving power from the intrinsically safe circuits:
(a) V_{max} — the maximum voltage the equipment can receive;
(b) I_{max} — the maximum current the equipment can receive;
(c) C_i — the maximum unprotected internal capacitance; and
(d) L_i — the maximum unprotected internal inductance.

F14.7

Field equipment may be connected to equipment having intrinsically safe field wiring terminals provided that the marked value of
(a) $V_{oc} \leq V_{max}$;
(b) $I_{sc} \leq I_{max}$;
(c) C_i plus cable capacitance $\leq C_a$; and
(d) L_i plus cable inductance $\leq L_a$.

Appendix G — Electrical installations of fire protection systems

Note: *This Appendix is an informative (non-mandatory) part of this Standard.*

G1. Introduction

G1.1
This Appendix lists requirements related to electrical installations that are not governed by Rules of the *Canadian Electrical Code, Part I*, but are required by the *National Building Code of Canada*.

G1.2
References listed in this Appendix are associated with electrical installations that are a part of the fire protection requirements contained in the *National Building Code of Canada*.

G2. Application

G2.1
The intent of this Appendix is to advise *Canadian Electrical Code, Part I*, users of performance requirements for electrically connected fire-protective equipment required by the *National Building Code of Canada*.

G2.2
Special fire protection requirements, such as use of thermal insulation, fire spread, flame spread requirements for electrical wiring and cables, flame-spread requirements for combustible raceways, and construction of electrical equipment vaults are covered by this Code (e.g., Rules 2-122, 2-124, 2-126, 2-128, 26-354, etc.).

G2.3
Provincial and municipal building codes may alter the *National Building Code of Canada,* and users of this list should also check those codes.

G3. *Canadian Electrical Code* reference to the *National Building Code of Canada* — 2005 Edition

CEC Sections and Rules	NBCC Sections, Articles, and Sentences (unless otherwise stated, references refer to Division B of the *NBCC*)
Section 0	1.4.1.2 of Division A Definitions of words and phrases
	1.5.1 of Division A Jurisdiction of Building Code over referenced documents
	1.3.1 and Table 1.3.1.2, Effective date and edition of referenced documents (e.g., NFPA No. 96, CAN/ULC-S524, etc.)
Rule 2-118	3.8.1.5, Mounting height of electrical controls in barrier-free areas
Rule 2-124	3.1.9.1.(1) and (2), Fire stopping of service penetrations through fire-rated assemblies or fire separations
Rule 2-124	3.1.9.3, Penetration of fire-rated assemblies or fire separations by wires, cables, boxes, and raceways
Rule 2-124	3.1.13.4, Flame-spread rating for combustible light diffusers and lenses
Rule 2-124	9.10.9.6, Electrical wiring and boxes and penetrating a fire separation
Rule 2-126	3.1.4.3.(1), Wires and cables in combustible buildings
Rule 2-126	3.1.5.18.(1), Wires and cables in noncombustible buildings
Rule 2-126	3.4.4.4.(1)(b), Restrictions on wiring penetrating an exit enclosure
Rule 2-126	3.6.4.3.(1), Equipment and wiring within plenums
Rule 2-126	3.6.1.2; 9.34.1, Electrical facilities — general

Section 32	9.10.13.11, Hold-open devices on doors in fire separations
Section 32	9.10.18.1, Continuity of fire alarm system
Section 32	9.10.18.2, Where fire alarm required
Section 32	9.10.18.3, Heat and smoke detectors
Section 32	9.10.18.4, Duct-type smoke detectors
Section 32	9.10.18.5, Fire alarm system for portions of buildings separated from remainder of building
Section 32	9.10.18.6, Fire alarm system, design, and installation
Section 32	9.10.18.7, Shutdown of central vacuum on fire alarm system activation
Section 32	9.10.18.8, Fire alarm system in storage garages
Section 32	9.32.3.8 and 9.32.3.9, Interconnection of smoke alarms and carbon monoxide detectors (where required)
Section 38	3.2.6.5.(6), Protection of electrical conductors for fire fighters' elevators
Section 38	3.3.1.7.(1), Protection of conductors for elevator used in barrier-free floor area
Section 38	3.2.4.14, Alternate floor recall for elevators
Section 38	3.2.6.4, Emergency operation of elevators
Section 38	3.5.1, Vertical transportation
Section 38	3.8.3.5, Elevators in barrier-free areas
Rule 46-108	3.2.6.9, Protection of electrical conductors in high buildings
Rule 46-202(3)	3.2.6.6, Use of building exhaust system to vent smoke from sprinklered floor areas
Rule 46-400	3.4.5, Exit signs
Rule 46-400	9.9.10, Exit signs
Section 46	3.2.7.3, Emergency lighting, where required
Section 46	3.2.7.4, Power supply for emergency lighting, time (duration) requirements
Section 46	3.2.7.5, Emergency power supply installation, general
Section 46	3.2.7.6, Emergency power supply installation for hospitals and nursing homes
Section 46	3.2.7.9, Emergency power supply for building services (elevators, fire pumps, fans, etc.)
Section 46	3.3.1.13.(6), Electrical operation of locked egress doors serving contained-use area or impeded egress zone
Section 46	3.3.1.24, Exit signs in service spaces
Section 46	3.3.3.7.(4)(c), Electrical release of doors in contained-use area
Section 46	3.4.6.13.(2), Electrical release of a sliding door used as an exit door
Section 46	9.9.11.3, Emergency lighting
Section 46	9.34.3, Emergency lighting
Section 76	5.6.1.10.(2) of the *National Fire Code of Canada*, 2005, Temporary electrical installations

Appendix H
Combustible gas detection instruments
for use in Class I hazardous locations

Appendix H — Combustible gas detection instruments for use in Class I hazardous locations

(See Rule 18-070)

Note: *This Appendix is an informative (non-mandatory) part of this Standard.*

H1. Introduction

H1.1
All combustible gases and vapours have a lower explosive limit (LEL) below which they will not ignite in air. In most Class I hazardous locations, the concentration of combustible gases and vapours is below the LEL most of the time.

H1.2
Combustible gas detection instruments are approved in accordance with CSA C22.2 No. 152, which covers the performance of these instruments in addition to their safety in hazardous locations because their failure to perform may give a false indication that an area is safe. Many permanently installed detection instruments not only indicate the level of combustible gas present but also have low and high alarms with contacts capable of initiating corrective action. For this reason, combustible gas detection instruments may be used in special circumstances as a method of protection for equipment that is not specifically approved for a Class I location.

H1.3
Combustible gas detection instruments should never be used as a substitute for safe electrical design. Because it is possible to defeat such devices either by poor maintenance or by deliberate tampering, they should be used only as protective devices where it is impractical to use another form of protection (e.g., explosion-proof or intrinsic safety) and where it is likely that the maintenance and the training of the personnel involved is adequate.

H1.4
The following guidelines provide suggestions for the use of combustible gas detection instruments as supplementary protection against explosions when certain equipment installed in the area is not approved for the area classification. Essential services such as lighting, instruments essential to the safe operation of the process, and the gas detection instrument itself must be approved for the Class, Group, and Zone (Division) of the area. Only equipment that can be disconnected from the supply without warning should be protected by gas detection instruments.

H2. General

H2.1
Combustible gas detection instruments used for the protection of equipment in Class I hazardous locations must be the stationary type permanently installed in a fixed location and must be certified to the requirements of CSA C22.2 No. 152 in their entirety, including the control unit.

H2.2
The gas detector must be certified for the highest classification to be encountered and calibrated specifically for the gas that is the basis for the classification. If more than one gas may be present, the gas detector must be adjusted to detect all gases that may be encountered with a direct reading for the gas giving the lowest response.

H2.3
Ignition temperature will be a significant factor in locations where heated surfaces are possible by normal or abnormal operation of apparatus. Where such surfaces exceed 80% of the auto-ignition temperature in degrees Celsius of the gas involved, the equipment must be protected by some other means.

H2.4
Combustible gas detection instrumentation is certified in accordance with CSA C22.2 No. 152 for its ability to function satisfactorily within certain environmental parameters involving temperature, humidity, air velocity, and vibration. Reference should be made to the Standard to determine if the expected environmental operating conditions of the gas detection application at hand fall within the parameters so specified. Instrument applications involving environmental conditions outside the parameters specified above should be given special consideration. In addition, the continued satisfactory performance of gas detection instrumentation under various environmental operating conditions will be achieved through consideration of environmental factors such

as, but not limited to, the following:

(a) **Temperature** — Gas detection instrumentation may be expected to operate satisfactorily within a wide temperature range, and interpretation of instrument indication under such conditions should be related to the physical properties of the particular gas or vapour involved. For example, at low temperatures, the lower explosive limit of a particular hydrocarbon vapour may be a value above the saturation limit of that substance at such temperatures. It should be recognized that any quantities of such hydrocarbon released to the environment at such temperatures may be present and detectable only to the saturated concentration, which may be well below the lower explosive limit.

(b) **Airborne particles** — Any airborne particles, such as dust, fibres, and aerosols, that could potentially prevent diffusion of the atmosphere to be monitored to the sensing element of a combustible gas detection instrument must be adequately guarded against through maintenance based on operating experience in such conditions, the provision of contaminant-excluding hardware, and the orientation and location of sensing elements and/or sampling points to minimize such effects. Similar consideration should be given to the effects of rain, ice, and snow.

(c) **Contaminants** — The gas sensing element of combustible gas detection instrumentation may be susceptible to desensitization by certain airborne compounds, such as silicone, silanes, halogenated compounds, etc., as listed in the instruction manual associated with the instrument. The effect of such exposure must be guarded against through maintenance based on operating experience in such conditions, the orientation and location of sensing heads, and surveys for potential sources of such materials in each individual application.

(d) **Corrosive compounds** — The presence of corrosive compounds in the combustible gas detection instrument environment must be considered for satisfactory operation both for material compatibility and for compatibility with any gases generated as a result of chemical reactions involving such corrosive compounds.

H3. Application recommendations

H3.1

Equipment for ordinary (non-hazardous) locations must be assumed to be ignition capable. In addition, if it contains heated parts, it may operate at a temperature above the auto-ignition temperature of the gas or vapour in the location. If this is the case, it may be above the ignition temperature for some time after it has been de-energized and therefore such equipment is not considered suitable for this means of protection. (See Clause H2.3.)

Equipment suitable for non-hazardous locations may be used in Class I, Zone 2 (Division 2) hazardous locations under the following conditions:

(a) the equipment can be switched off at any time, without warning, without causing any hazards;

(b) an audible and visible alarm is actuated when the combustible gas concentration reaches 20% of the LEL; and

(c) the equipment is automatically disconnected from the electrical supply when the combustible gas concentration reaches 40% of the LEL.

If the location is such that the gas concentration can be reduced by forced ventilation, the 20% alarm contact may also be used to switch on fans or other such devices to reduce the possibility of a shutdown. However, the alarms must not be capable of being reset until the concentration drops below the 20% LEL level.

H3.2

Equipment suitable for Class I, Zone 2 (Division 2) hazardous locations contains no normal sources of ignition. It can become ignition capable only in the case of an electrical fault within the equipment. Such equipment may be used in a Class I, Zone 1 (Division 1) hazardous location under the following conditions:

(a) the equipment can be switched off at any time, with a 30-minute warning, without causing any hazards;

(b) an audible or visible alarm is activated when the combustible gas concentration reaches 20% of the LEL; and

(c) a timer is started when the combustible gas concentration reaches 40% of the LEL, that will disconnect the electrical supply to the equipment after the high gas has persisted for 30 min. Any other actions that must be taken within this time to prevent other hazards are to be automatically completed within this time.

Appendix H
Combustible gas detection instruments
for use in Class I hazardous locations

H4. Installation recommendations

H4.1

It is recommended that consideration be given to the following factors when locating remote detector heads:

(a) density of the gases or vapours to be detected (relating to air);

(b) locations of the potential gas or vapour sources and the hazardous division for which the equipment is approved;

(c) provision for extra (i.e., redundant) detector heads;

(d) effects of ventilation systems on the flow of hazardous gases or vapours from the hazardous location and the possibility of gas or vapour concentration gradients in the hazardous location;

(e) adverse environments at detector locations; and

(f) accessibility for calibration and maintenance.

H4.2

Audible and/or visual alarms distinguishable from any other alarms should be installed and repeated at the central control locations to warn those in the area protected by the detectors and those approaching the area that a potential hazard exists.

H4.3

The system should be arranged so that it is fail-safe (i.e., the equipment being protected is automatically de-energized in the event of a failure in the gas detection instrument).

Note: *Spacing and location of alarms are dependent on many factors and will vary from site to site. In the absence of any specific requirements in this Appendix, NFPA No. 72,* National Fire Alarm Code, *may be used as a guide.*

H5. Maintenance

It is recommended that the installation be calibrated in accordance with the manufacturer's instructions or Paragraph 4.8.2(f) of API 500.

Δ

Appendix I — Interpretations approved since publication of the 2002 Code
(See Clause C9.10(b).)

Notes:
(1) *This Annex is an informative (non-mandatory) part of this Standard.*
(2) *Committee interpretations of the* Canadian Electrical Code, Part I, *are published in CSA's* Info Update. *These interpretations are reproduced here for information.*

Scope
Question: Do street-lighting installations, installed and operated by a municipality, installed on rights of way and public thoroughfares, fall within the Scope of the *Canadian Electrical Code, Part I*?

Answer: Yes.

Definitions — Service box
Question: For equipment such as a combination panelboard, switchgear or other similar equipment marked "Suitable for use as service entrance equipment", does a service disconnecting means such as a fused switch or circuit breaker within the equipment have to be located in a separate compartment in order for the compartment to be deemed a "Service box" as defined in Section 0 of the *Canadian Electrical Code, Part I*?

Answer: Yes.

Rule 2-308(2)
Question: Is draw-out type equipment required to be completely drawn out for the purpose of fulfilling the criteria of Rule 2-308(2) of being in a "fully disconnected position"?

Answer: No.

Rule 6-202
Question: Is a Central Distribution Panelboard (CDP) an acceptable piece of equipment for use to subdivide the main consumer's service? Example: Supplying metered feeders to individual bays in warehouses and similar types of occupancies.

Answer: No.

Rule 12-906
Question: Does an EMT connector without an insulated throat or bushing meet the requirements of Subrule 12-906(1)?

Answer: Yes.

Rule 36-308(6)
Question: Is the requirement of Rule 36-308(6) applicable for a transformer connected to a grounded neutral system, where such a grounded neutral system represents the high voltage (primary side)?

Answer: Yes.

Rule 46-304(2)
Question: In circumstances where Rule 46-304(2) is not exceeded, is the installation permitted in Subrule (3), in any way related to the objectives of the *Canadian Electrical Code, Part I*, preferable to a permanent connection?

Answer: No.

Appendix J
Rules and Notes to Rules for installations using the
Division system of classification for Class I locations

Δ Appendix J — Rules and Notes to Rules for installations using the Division system of classification for Class I locations

Note: *This informative (non-mandatory) Appendix has been written in normative (mandatory) language to facilitate its use when referenced from Rule 18-000(4) or 20-000(3), or where regulatory authorities wish to adopt it formally as additional requirements to this Standard.*

Table of contents

J1. Introduction

J1.1
As indicated in Section 18, new installations in Class I hazardous locations will now employ the Zone system of classification. For existing installations, it is left to the discretion of the user/owner whether to reclassify these facilities to the Zone system or continue the use of the Division system of classification. For installations that continue the use of the Division system of classification, the Rules contained in Annexes J18 and J20 shall apply.

J1.2
The following overview table is given to show either the type of equipment permitted or the methods of protection acceptable for use in the Zone and Division systems.

Acceptable equipment comparison for Class I locations

Zone system		Division system	
Intrinsically safe, ia	**Zone 0**		
Equipment acceptable in Zone 0, Class I, Division 1			
Powder-filled q			Class I, Division 1
Flame-proof d	**Zone 1**	**Division 1**	
Pressurized p			Intrinsically safe i, ia
Oil immersed o			
Increased safety e			
Intrinsically safe ib			
Encapsulation m			
Equipment acceptable in Zone 0			Class I, Division 1
Equipment acceptable in Zone 1, Class I, Division 2			Class I, Division 2 Flame-proof d
Method of protection n	**Zone 2**		Pressurized p
Non-incendive			Intrinsically safe ib
Other electrical apparatus*		**Division 2**	Oil immersed o
			Increased safety e
			Powder-filled q
			Method of protection n
			Encapsulation m
			Non-incendive
			Other electrical apparatus*

*"Other electrical apparatus" means electrical apparatus complying with the requirements of a recognized standard for industrial electrical apparatus that does not in normal service have ignition-capable hot surfaces and does not in normal service produce incendive arcs or sparks.

"Other electrical apparatus" also makes reference to equipment or systems currently acceptable as alternative means of protection, (see Rules J18-064 and J18-070).

Because Division 1 encompasses the equivalent of Zone 0, methods of protection designed for Zone 1 would not be allowed in Division 1; e.g., an increased safety device would not be allowed in a Class I, Division 1 location.

Appendix J
Rules and Notes to Rules for installations using the
Division system of classification for Class I locations

Annex J18
Class I hazardous locations classified as Division 1 or Division 2
Scope and introduction

J18-000 Scope (see Annex JB)

(1) This Annex applies to Class I, Division 1 and Division 2 hazardous locations in which electrical equipment and wiring are subject to the conditions outlined in this Annex.

(2) This Annex supplements or amends the general requirements of this Code.

J18-002 Special terminology (see Annex JB)

In this Section, the following definitions apply:

Cable gland — a device or combination of devices intended to provide a means of entry of a cable or flexible cord into an enclosure situated in a hazardous location and that also provides strain relief and shall be permitted to provide sealing characteristics where required, either by an integral means or when combined with a separate sealing fitting.

Cable seal — a seal that is installed at a cable termination to prevent the release of an explosion from an explosion-proof enclosure and that minimizes the passage of gases or vapours at atmospheric pressure.

Conduit seal — a seal that is installed in a conduit to prevent the passage of an explosion from one portion of the conduit system to another and that minimizes the passages of gases or vapours at atmospheric pressure.

Degree of protection — the measures applied to the enclosures of electrical apparatus to ensure

(a) the protection of persons against contact with live or moving parts inside the enclosure and protection of apparatus against the ingress of solid foreign bodies; and

(b) the protection of apparatus against ingress of liquids.

Explosive gas atmosphere — a mixture with air, under atmospheric conditions, of flammable substances in the form of gas, vapour, or mist in which, after ignition, combustion spreads throughout the unconsumed mixture.

Explosive limits — the lower and upper percentage by volume of concentration of gas in a gas-air mixture that will form an ignitable mixture.

    LEL — lower explosive limit.

    UEL — upper explosive limit.

Methods of protection — defined methods to reduce the risk of ignition of explosive gas atmospheres.

Non-incendive circuit — a circuit in which any spark or thermal effect that may occur under normal operating conditions or due to opening, shorting, or grounding of field wiring is incapable of causing an ignition of the prescribed flammable gas or vapour.

Normal operation — the situation when the plant or equipment is operating within its design parameters.

Primary seal — a seal that isolates process fluids from an electrical system and has one side of the seal in contact with the process fluid.

Protective gas — the gas used to maintain pressurization or to dilute a flammable gas or vapour.

Secondary seal — a seal that is designed to prevent the passage of process fluids at the pressure it will be subjected to upon failure of the primary seal.

J18-004 Classification (see Annex JB)

Class I locations are those in which flammable gases or vapours are or may be present in the air in quantities sufficient to produce explosive or ignitable mixtures.

J18-006 Division of Class I locations (see Annex JB)

Class I locations shall be further divided into two Divisions based upon frequency of occurrence and duration of an explosive gas atmosphere as follows:

(a) Division 1, consisting of Class I locations in which explosive gas atmospheres are likely to be present continuously, intermittently or periodically during normal operation;

(b) Division 2, consisting of Class I locations in which
- (i) explosive gas atmospheres are not likely to occur in normal operation and if they do occur, they will exist for a short time only; or
- (ii) the location is adjacent to a Class I, Division 1 location from which explosive gas atmospheres could be communicated, unless such communication is prevented by adequate positive-pressure ventilation from a source of clean air, and effective safeguards against ventilation failure are provided.

General

J18-050 Electrical equipment (see Annex JB)
(1) Where electrical equipment is required by this Section to be approved for the class of location, it shall also be approved for the specific gas or vapour that will be present.
(2) For equipment approved with a method of protection permitted in a Class I location, such approval shall be permitted to be indicated by one or more of the following atmospheric group designations:
- (a) **Group II C,** consisting of atmospheres containing acetylene, carbon disulphide, or hydrogen, or other gases or vapours of equivalent hazard;
- (b) **Group II B,** consisting of atmospheres containing acrylonitrile, butadiene, diethyl ether, ethylene, ethylene oxide, hydrogen sulphide, propylene oxide, or unsymmetrical dimethyl hydrazine (UDMH), or other gases or vapours of equivalent hazard;
- (c) **Group II A,** consisting of atmospheres containing acetaldehyde, acetone, cyclopropane, alcohol, ammonia, benzine, benzol, butane, ethylene dichloride, gasoline, hexane, isoprene, lacquer solvent vapours, naphtha, natural gas, propane, propylene, styrene, vinyl acetate, vinyl chloride, xylenes, or other gases or vapours of equivalent hazard;
- (d) **Group II,** consisting of all Group II gases; or
- (e) **Group IIXXXXX,** where XXXXX is a chemical formula or chemical name suitable for that specific gas only.
(3) For equipment approved for Class I, Division 1 or 2, the specific gas shall be permitted to be indicated by one or more of the following atmospheric group designations:
- (a) Group A, consisting of atmospheres containing acetylene;
- (b) Group B, consisting of atmospheres containing butadiene, ethylene oxide, hydrogen (or gases or vapours equivalent in hazard to hydrogen, such as manufactured gas), or propylene oxide;
- (c) Group C, consisting of atmospheres containing acetaldehyde, cyclopropane, diethyl ether, ethylene, hydrogen sulphide, or unsymmetrical dimethyl hydrazine (UDMH), or other gases or vapours of equivalent hazard; or
- (d) Group D, consisting of atmospheres containing acetone, acrylonitrile, alcohol, ammonia, benzine, benzol, butane, ethylene dichloride, gasoline, hexane, isoprene, lacquer solvent vapours, naphtha, natural gas, propane, propylene, styrene, vinyl acetate, vinyl chloride, xylenes, or other gases or vapours of equivalent hazard.
(4) Notwithstanding Subrule (2)(b), where the atmosphere contains
- (a) butadiene, Group D equipment shall be permitted to be used if such equipment is isolated in accordance with Rule J18-108(1) by sealing all conduit 16 trade size or larger; or
- (b) ethylene oxide or propylene oxide, Group C equipment shall be permitted to be used if equipment is isolated in accordance with Rule J18-108(3) by sealing all conduit 16 trade size or larger.

J18-052 Marking (see Annex JB)
(1) Electrical equipment intended for use in Class I hazardous locations shall be permitted to be marked with the following:
- (a) the letters "Ex" or "EEx";
- (b) the symbol(s) to indicate method(s) of protection used;
- (c) the gas group as specified in Rule J18-050(2); and
- (d) the temperature rating in accordance with Subrule (4) for equipment of the heat-producing type.
(2) Notwithstanding Subrule (1), electrical equipment approved for
- (a) Class I or Class I, Division 1 or 2 locations shall be permitted to be marked with the class and the group described in Rule J18-050(3), or the specific gas or vapour for which it has been approved; and
- (b) Class I, Division 2 only, shall be permitted to be so marked.
(3) Electrical equipment approved for use in Class I hazardous locations shall be permitted to be marked with
- (a) the maximum external temperature; or

Appendix J
Rules and Notes to Rules for installations using the
Division system of classification for Class I locations

(b)   one of the following temperature codes to indicate the maximum external temperature:

Temperature code	Maximum external temperature
T1	450 °C
T2	300 °C
T2A	280 °C
T2B	260 °C
T2C	230 °C
T2D	215 °C
T3	200 °C
T3A	180 °C
T3B	165 °C
T3C	160 °C
T4	135 °C
T4A	120 °C
T5	100 °C
T6	85 °C

(4)   If no maximum surface temperature is shown on Class I equipment approved for the Class and Group, the equipment, if of the heat-producing type, shall be considered as having a maximum surface temperature of 100 °C or less for the purpose of compliance with Rule J18-054.

(5)   Electrical equipment approved for operation at ambient temperatures exceeding 40 °C shall, in addition to the marking specified in Rule J18-052(2), be marked with the maximum ambient temperature for which the equipment is approved, and the maximum external temperature of the equipment at that ambient temperature.

J18-054 Temperature (see Annex JB)

(1)   In Class I hazardous locations, equipment shall not be installed in an area where vapours or gases are present that have an ignition temperature less than the maximum external temperature of the equipment as referred to in Rule J18-052(3) and (4).

(2)   Where the equipment is not required to be approved for hazardous locations, the maximum external temperature referred to in Subrule (1) shall be the surface temperature at any point, internal or external, of the equipment.

J18-056 Non-essential electrical equipment

(1)   No electrical equipment shall be used in a hazardous location, unless it is essential to the processes being carried on in that location.

(2)   Service equipment, panelboards, switchboards, and similar electrical equipment shall, where practicable, be located in rooms or sections of the building in which hazardous conditions do not exist.

J18-058 Rooms, sections, or areas

Each room, section, or area, including motor- and generator-rooms and rooms for the enclosure of control equipment, shall be considered as a separate location for the purpose of determining the classification of the hazard.

J18-060 Equipment rooms

(1)   Where walls, partitions, floors, or ceilings are used to form hazard-free rooms or sections, they shall be
(a)   of substantial construction;
(b)   built of or lined with non-combustible material; and
(c)   such as to ensure that the rooms or sections will remain free from hazards.

(2)   Where a non-hazardous location within a building communicates with a Class I, Division 2 location, the locations shall be separated by close-fitting, self-closing, approved fire doors.

(3)   For communication from a Class I, Division 1 location, the provisions of Rule J18-006(b)(ii) shall apply.

J18-062 Metal-covered cable (see Annex JB)

(1)   Where exposed overhead conductors supply mineral-insulated cable in a hazardous location, surge arresters shall be installed to limit the surge voltage level to 5 kV on the cable.

(2)   Where single-conductor metal-covered cable is used in hazardous locations, it shall be installed in such a manner as to prevent sparking between cable sheaths or between cable sheaths and metal bonded to ground, and cables in the circuit shall

(a)   be clipped or strapped together in a manner that will ensure good electrical contact between metal coverings, at intervals of not more than 1.8 m, and the metal coverings shall be bonded to ground; or

(b)   have the metal coverings continuously covered with insulating material, and the metal coverings shall be bonded to ground at the point of termination in the hazardous location only.

J18-064 Pressurized equipment or rooms (see Annex JB)

Electrical equipment and associated wiring in Class I locations shall be permitted to be located in enclosures or rooms constructed and arranged so that a protective gas pressure is effectively maintained, in which case the provisions of Rules J18-100 to J18-178 of this Section need not apply.

J18-066 Intrinsically safe and non-incendive electrical equipment and wiring (see Annex JB and Appendix F)

(1)   Where electrical equipment is approved as intrinsically safe and associated circuits are designed and installed as intrinsically safe for the intended hazardous location, they shall be permitted and the provisions of Rules J18-100 to J18-178 of this Section need not apply.

(2)   Where electrical equipment is approved as non-incendive and associated circuits are designed and installed as non-incendive, they shall be permitted in Class I, Division 2 locations and the provisions of Rules J18-152 to J18-158 need not apply.

(3)   Raceways or cable for intrinsically safe and non-incendive wiring and equipment in Class I locations shall be properly sealed to prevent migration of gas or vapour into enclosures or raceways required to be explosion-proof, as well as to other locations.

(4)   The conductors in an intrinsically safe and non-incendive circuit shall not be placed in any raceway, compartment, outlet, junction box, or similar fitting with the conductors of any other system unless the conductors of the two systems are separated by a suitable mechanical barrier.

J18-068 Cable trays

Cable trays shall not be used to support cables in hazardous locations except where

(a)   the type of cable is approved in Rules of this Section for use in the particular hazardous location;

(b)   the type of cable is approved for use in cable trays in accordance with Rule 12-2202; and

(c)   there can be no hazardous accumulation of combustible process dust or fibre in or upon the cable, the cable tray, or the supports.

J18-070 Combustible gas detection (see Annex JB and Appendix H)

Electrical equipment suitable for non-hazardous locations shall be permitted to be installed in a Class I, Division 2 hazardous location, and electrical equipment suitable for Class I, Division 2 hazardous locations shall be permitted to be installed in a Class I, Division 1 hazardous location provided that

(a)   no specific equipment suitable for the purpose is available;

(b)   the equipment, during its normal operation, does not produce arcs, sparks, or hot surfaces capable of igniting an explosive gas atmosphere; and

(c)   the location is continuously monitored by a combustible gas detection system that

(i)   will activate an alarm when the gas concentration reaches 20% of the lower explosive limit;

(ii)   will activate ventilating equipment or other means designed to prevent the concentration of gas from reaching the lower explosive limit when the gas concentration reaches 20% of the lower explosive limit, where such ventilating equipment or other means is provided;

(iii)   will automatically de-energize the electrical equipment being protected when the gas concentration reaches 40% of the lower explosive limit, where the ventilating equipment or other means referred to in Item (c)(ii) is provided;

(iv)   will automatically de-energize the electrical equipment being protected when the gas concentration reaches 20% of the lower explosive limit, where the ventilating equipment or other means referred to in Item (c)(ii) cannot be provided; and

(v)   will automatically de-energize the electrical equipment being protected upon failure of the gas detection instrument.

J18-072 Explosive fluid seals

Electrical equipment containing an explosive fluid seal intended to prevent explosive fluids from reaching the electrical housing or conduit system shall not be used at pressures in excess of the marked maximum working pressure (MWP).

514

Appendix J
Rules and Notes to Rules for installations using the
Division system of classification for Class I locations

Installations in Class I, Division 1 locations

J18-100 Equipment in Class I, Division 1 locations (see Annex JB)
Where required by other Rules of this Code, electrical equipment installed in a Class I, Division 1 location shall be approved
(a)   for Class I, or Class I, Division 1 locations; or
(b)   as intrinsically safe, type i or ia.

J18-102 Transformers and capacitors, Class I, Division 1
(1)   Transformers and electrical capacitors that contain a liquid that will burn shall be installed in electrical equipment vaults in accordance with Rules 26-350 to 26-356, and the following shall apply:
(a)   there shall be no door or other connecting opening between the vault and the hazardous area;
(b)   the vault shall be ventilated to ensure the continuous removal of hazardous gases or vapours;
(c)   vent openings or vent ducts shall lead to a safe location outside the building containing the vault;
(d)   vent openings and vent ducts shall be of sufficient area to relieve pressure caused by explosions within the vault;
(e)   every portion of a vent duct within the building shall be constructed of reinforced concrete.
(2)   Transformers and electrical capacitors that do not contain a liquid that will burn shall be
(a)   installed in electrical equipment vaults conforming to Subrule (1); or
(b)   in compliance with Rule J18-100.

J18-104 Meters, instruments, and relays, Class I, Division 1
(1)   Where practicable, meters, instruments, and relays, including kilowatt-hour meters, instrument transformers, and resistors, rectifiers, and thermionic tubes, shall be located outside the hazardous location.
(2)   Where it is not practicable to install meters, instruments, and relays outside Class I, Division 1 locations, they shall comply with the requirements of Rule J18-100.

J18-106 Wiring methods, Class I, Division 1 (see Annex JB)
(1)   The wiring method shall be threaded rigid metal conduit or cables approved for hazardous locations with associated cable glands that comply with the requirements of Rule J18-100.
(2)   All boxes, fittings, and joints shall be threaded for connection to conduit or cable glands, and shall be explosion-proof with boxes and fittings that comply with the requirements of Rule J18-100.
(3)   Threaded joints shall have at least five full threads fully engaged, and running threads shall not be used.
(4)   Cables shall be installed and supported in a manner to avoid tensile stress at the cable glands.
(5)   Where it is necessary to use flexible connections at motor terminals and similar places, flexible fittings of the explosion-proof type approved for the location shall be used.

J18-108 Sealing, Class I, Division 1 (see Annex JB)
(1)   Secondary seals shall be provided, between devices containing a primary seal and conduit or cable seals, where failure of a single component in the device containing the primary seal could allow passage of process fluids.
(2)   Where secondary seals are installed, drains, vents or other devices intended to make primary seal leakage obvious shall be installed.
(3)   Conduit seals shall be provided in conduit systems where
(a)   the conduit enters an explosion-proof or flame-proof enclosure containing devices that may produce arcs, sparks or high temperatures, and shall be located as close as practicable to the enclosure, or as marked on the enclosure, but not further than 450 mm from the enclosure;
(b)   the conduit is 53 trade size or larger and enters an explosion-proof or flame-proof enclosure housing terminals, splices, or taps, and shall be located no further than 450 mm from the enclosure;
(c)   the conduit leaves the Class I, Division 1 location with no box, coupling, or fitting in the conduit run between the seal and the point at which the conduit leaves the location, except that a rigid unbroken conduit that passes completely through a Class I, Division 1 area with no fittings less than 300 mm beyond each boundary, provided that the termination points of the unbroken conduit are in non-hazardous areas, need not be sealed; or
(d)   the conduit enters an enclosure that is not required to be explosion-proof or flame-proof, except that a seal is not required where an unbroken and continuous run of conduit connects two enclosures that are not required to be explosion-proof or flame-proof.
(4)   Only explosion-proof unions, couplings, reducers, and elbows that are not larger than the trade size of the conduit shall be permitted between the sealing fitting and an explosion-proof or flame-proof enclosure.

(5) Cable seals shall be provided in a cable system where
    (a) the cable enters an enclosure required to be explosion-proof or flame-proof; or
    (b) the cable enters an enclosure not required to be explosion-proof or flame-proof, and the following apply:
        (i) the cable leaves the Division 1 area and is less than 10 m in length; or
        (ii) the other end of the cable terminates in a Division 2 or non-hazardous location in which a negative atmospheric pressure greater than 0.2 kPa exists.

(6) Where secondary seals, cable seals or conduit seals are required, they shall conform to the following:
    (a) the seal shall be made
        (i) in a field-installed sealing fitting or cable gland that shall be accessible and shall comply with the requirements of Rule J18-100; or
        (ii) in a sealing fitting provided as part of an enclosure approved for the area and where the seal is factory-made, the enclosure shall be marked to indicate that such a seal is provided;
    (b) splices and taps shall not be made in fittings intended only for sealing with compound, nor shall other fittings in which splices or taps are made be filled with compound;
    (c) where there is a probability that liquid or other condensed vapour may be trapped within enclosures for control equipment or at any point in the raceway system, approved means shall be provided to prevent accumulation or to permit periodic draining of such liquid or condensed vapour; and
    (d) where there is a probability that liquid or condensed vapour may accumulate within motors or generators, joints and conduit systems shall be arranged to minimize entrance of liquid, but if means to prevent accumulation or to permit periodic draining are judged necessary, such means shall be provided at the time of manufacture and shall be deemed an integral part of the machine.

(7) Runs of cables, each having a continuous sheath, either metal or non-metal, shall be permitted to pass through a Class I, Division 1 location without seals.

(8) Cables that do not have a continuous sheath, either metal or non-metal, shall be sealed at the boundary of the Division 1 location.

J18-110 Switches, motor controllers, circuit breakers, and fuses, Class I, Division 1

Switches, motor controllers, circuit breakers, and fuses, including push buttons, relays, and similar devices, shall be provided with enclosures, and the enclosure in each case together with the enclosed apparatus shall be approved as a complete assembly and shall comply with the requirements of Rule J18-100.

J18-112 Control transformers and resistors, Class I, Division 1

Transformers, impedance coils, and resistors used as or in conjunction with control equipment for motors, generators, and appliances and the switching mechanism, if any, associated with them shall comply with the requirements of Rule J18-100.

J18-114 Motors and generators, Class I, Division 1 (see Annex JB)

Motors, generators, and other rotating electrical machines shall comply with the requirements of Rule J18-100.

J18-116 Ignition systems for gas turbines, Class I, Division 1 (see Annex JB)

Ignition systems for gas turbines shall comply with the requirements of Rule J18-100.

J18-118 Lighting fixtures, Class I, Division 1

(1) Fixtures for fixed and portable lighting shall be approved as complete assemblies for Class I locations and shall be clearly marked to indicate the maximum wattage of lamps for which they are approved.

(2) Fixtures intended for portable use shall be specifically approved as complete assemblies for that use.

(3) Each fixture shall be protected against physical damage by a suitable guard or by location.

(4) Pendant fixtures shall be
    (a) suspended by and supplied through threaded rigid conduit stems, and threaded joints shall be provided with set screws or other effective means to prevent loosening; and
    (b) for stems longer than 300 mm, provided with permanent and effective bracing against lateral displacement at a level not more than 300 mm above the lower end of the stem, or provided with flexibility in the form of a fitting or flexible connector approved for the purpose and for the location not more than 300 mm from the point of attachment to the supporting box or fitting.

(5) Boxes, box assemblies, or fittings used for the support of lighting fixtures shall be approved for the purpose and for Class I locations.

Appendix J
Rules and Notes to Rules for installations using the
Division system of classification for Class I locations

J18-120 Utilization equipment, fixed and portable, Class I, Division 1

(1)   Utilization equipment, fixed and portable, including electrically heated and motor-driven equipment, shall comply with the requirements of Rule J18-100.

(2)   Ground fault protection shall be provided to de-energize all normally ungrounded conductors of an electric heat tracing cable set with the ground fault trip setting adjusted to allow normal operation of the heater.

J18-122 Flexible cords, Class I, Division 1

Flexible cords shall be permitted to be used only for connection between a portable lamp or other portable utilization equipment and the fixed portion of its supply circuit and, where used, shall

(a)   be of a type approved for extra-hard usage;

(b)   contain, in addition to the conductors of the circuit, a bonding conductor; and

(c)   be provided with glands approved for the class and group where the flexible cord enters a box, fitting, or enclosure of the explosion-proof type.

J18-124 Receptacles and attachment plugs, Class I, Division 1

Receptacles and attachment plugs shall be of the type providing for connection to the bonding conductor of the flexible cord, and shall comply with the requirements of Rule J18-100.

J18-126 Conductor insulation, Class I, Division 1

Where condensed vapours or liquids may collect on or come in contact with the insulation on conductors, such insulation shall be of a type approved for use under such conditions or the insulation shall be protected by a sheath of lead or by other approved means.

J18-128 Signal, alarm, remote-control, and communication systems, Class I, Division 1

Signal, alarm, remote-control, and communication systems shall conform to the following:

(a)   all apparatus and equipment shall comply with the requirements of Rule J18-100; and

(b)   all wiring shall comply with Rules J18-104 and J18-106.

J18-130 Live parts, Class I, Division 1

No live parts of electrical equipment or of an electrical installation shall be exposed.

J18-132 Bonding, Class I, Division 1

(1)   Exposed non-current-carrying metal parts of electrical equipment, including the frames or metal exteriors of motors, fixed or portable lamps or other utilization equipment, lighting fixtures, cabinets, cases, and conduit, shall be bonded to ground in accordance with Section 10.

(2)   The bonding path continuity and adequacy in a hazardous location and in a non-hazardous location from which the hazardous location is supplied shall be ensured by the use of threaded connections, bonding jumpers with proper fittings, or other approved means, meeting the requirements of Rules 10-606(1)(a), (1)(c), and (1)(d) and 10-606(2).

Installations in Class I, Division 2 locations

J18-150 Equipment in Class I, Division 2 locations (see Annex JB)

Where required by other Rules of this Code, electrical equipment installed in a Class I, Division 2 location shall be approved:

(a)   for Class I or Class I, Division 1 locations;

(b)   for Class I, Division 2 locations;

(c)   as non-incendive; or

(d)   as providing one or more of the following protective methods:

(i)   intrinsically safe i, ia, or ib;

(ii)   flame-proof d;

(iii)   increased safety e;

(iv)   oil immersed o;

(v)   pressurized p;

(vi)   powder-filled q;

(vii)   encapsulation m;

(viii) method of protection n; or

(ix)   equipment specifically allowed in Rules J18-152 through J18-178.

J18-152 Process instrumentation, communication, and remote-control equipment, Class I, Division 2

Except for transformers, solenoids, and other windings that do not incorporate sliding or make-and-break contacts or heat-producing resistance devices, process instrumentation, communication, and remote-control equipment shall comply with the requirements of Rule J18-150.

J18-154 Transformers and capacitors, Class I, Division 2

Installation of transformers and capacitors shall be permitted, provided that they do not contain arcing or spark-producing components.

J18-156 Wiring methods, Class I, Division 2 (see Annex JB)

(1)   The wiring method shall be
    (a)   threaded metal conduit;
    (b)   cables approved for hazardous locations;
    (c)   Type TC cable, installed in cable tray in accordance with Rule 12-2202;
    (d)   Type ACWU cable;
    (e)   control and instrument cables with an interlocking metallic armour and a continuous jacket in control circuits (Type ACIC); or
    (f)   Type CIC cable (non-armoured control and instrumentation cable) installed in cable tray in accordance with the installation requirements of Rule 12-2202(2), where
        (i)   the voltage rating of the cable is not less than 300 V;
        (ii)   the circuit voltage is 150 V or less; and
        (iii)   the circuit current is 5 A or less.

(2)   Explosion-proof or flame-proof boxes, fittings, and joints shall be threaded for connection to conduit and cable glands.

(3)   Threaded joints that are required to be explosion-proof or flame-proof shall be permitted to be either tapered or straight, and shall comply with the following:
    (a)   tapered threads shall have at least five fully engaged threads, and running threads shall not be used;
    (b)   where straight threads are used in Groups IIA and IIB atmospheres, they shall have at least five fully engaged threads; and
    (c)   where straight threads are used in Group IIC atmospheres, they shall have at least eight fully engaged threads.

(4)   Where thread forms differ between the equipment and the wiring system, approved adapters shall be used.

(5)   Cables shall be installed and supported in a manner to avoid tensile stress at the cable glands.

(6)   Where it is necessary to use flexible connections at motor terminals and similar places, flexible metal conduit shall be permitted to be used.

(7)   Boxes, fittings, and joints need not be explosion-proof or flame-proof except as required by the Rules in this Section.

(8)   Cable glands shall be compatible with the degree of ingress protection and explosion protection provided by the enclosure that the cable enters, where the area classification and environmental conditions require these degrees of protection.

J18-158 Sealing, Class I, Division 2 (see Annex JB)

(1)   Secondary seals shall be provided, between devices containing a primary seal and conduit or cable seals, where failure of a single component in the device containing the primary seal could allow passage of process fluids.

(2)   Where secondary seals are installed, drains, vents or other devices intended to make primary seal leakage obvious shall be installed.

(3)   Conduit seals shall be provided in a conduit system where
    (a)   the conduit enters an enclosure that is required to be explosion-proof or flame-proof and shall be located as close as practicable to the enclosure, or as marked on the enclosure, but not further than 450 mm from the enclosure;
    (b)   the conduit leaves the Class I, Division 2 location with no box, coupling, or fitting in the conduit run between the seal and the point at which the conduit leaves the location, except that a rigid unbroken conduit that passes completely through a Class I, Zone 2 area with no fittings less than 300 mm beyond each boundary, provided that the termination points of the unbroken conduit are in non-hazardous areas, need not be sealed; or

Appendix J
Rules and Notes to Rules for installations using the
Division system of classification for Class I locations

(c)   the conduit leaves a Class I, Division 2 location outdoors, the seal may be located more than 300 mm beyond the Class I, Division 2 boundary provided that it is located on the conduit prior to entering an enclosure or building.

(4)   Only explosion-proof unions, couplings, reducers and elbows that are not larger than the trade size of the conduit shall be permitted between the sealing fitting and an explosion-proof or flame-proof enclosure.

(5)   Cable seals shall be provided in a cable system where
    (a)   the cable enters an enclosure required to be explosion-proof or flame-proof; or
    (b)   the cable enters an enclosure not required to be explosion-proof or flame-proof, and the following apply:
        (i)   the cable leaves the Division 2 area and is less than 10 m in length; or
        (ii)   the other end of the cable terminates in a non-hazardous location in which a negative atmospheric pressure greater than 0.2 kPa exists.

(6)   Where a run of conduit enters an enclosure that is required to be explosion-proof or flame-proof, every part of the conduit from the seal to that enclosure shall comply with Rule J18-106.

(7)   Runs of cables, each having a continuous sheath, either metal or non-metal, shall be permitted to pass through a Class I, Division 2 location without seals.

(8)   Cables that do not have a continuous sheath, either metal or non-metal, shall be sealed at the boundary of the Division 2 location.

(9)   Where seals are required, Rule J18-108(6) shall apply.

J18-160 Switches, controllers, and circuit breakers, Class I, Division 2 (see Annex JB)

(1)   Switches, controllers, and circuit breakers shall be provided with enclosures that comply with the requirements of Rule J18-150.

(2)   Notwithstanding Subrule (1), switches, controllers, and circuit breakers that comply with the requirements of Rule J18-150 shall be permitted to be provided with general-purpose enclosures.

J18-162 Isolating switches, Class I, Division 2

Isolating switches shall conform to the following:
(a)   they shall be interlocked with their associated current-interrupting devices so that they cannot be opened under load; and
(b)   they shall be permitted to have enclosures of the general-purpose type, provided that they are unfused.

J18-164 Fuses for motors, appliances, and portable lamps, Class I, Division 2

Where fuses are used in Class I, Division 2 locations for the protection of motors, appliances, and portable lamps:
(a)   a standard plug or cartridge fuse shall be permitted to be used if placed within an explosion-proof or flame-proof enclosure; or
(b)   a fuse of a type in which the operating element is immersed in oil or other approved liquid, or is enclosed within a chamber hermetically sealed against the entrance of gases and vapours, shall be permitted to be used if approved for the purpose and placed within a general-purpose enclosure.

J18-166 Sets of fuses or circuit breakers for fixed lighting, Class I, Division 2 (see Annex JB)

(1)   In this Rule, "sets of fuses" means a group containing as many fuses as are required to perform a single protective function in a circuit, but excluding fuses conforming to Rule J18-164.

(2)   Where not more than
    (a)   ten sets of approved enclosed fuses; or
    (b)   ten circuit breakers that are not used as switches for the normal operation of the lamps
    are installed in Class I, Division 2 locations for the protection of a branch circuit or a feeder circuit that supplies only lamps in a fixed position, the enclosures for the fuses or circuit breakers shall be permitted to be of the general-purpose type.

J18-168 Motors and generators, Class I, Division 2 (see Annex JB)

(1)   Motors, generators, and other rotating electrical machines that incorporate arcing, sparking, or heat-producing resistance components shall be approved for the location unless these components are connected to intrinsically safe or non-incendive circuits or are provided with enclosures approved for Class I locations.

(2)   Motors, generators, and other rotating electrical machines that do not incorporate arcing, sparking, or heat-producing components or that contain such components that are connected to intrinsically safe or non-incendive circuits shall be permitted to be of the open or non-explosion-proof type.

J18-170 Ignition systems for stationary internal combustion engines, Class I, Division 2
(see Annex JB)

Ignition systems for stationary internal combustion engines shall comply with the requirements of Rule J18-150.

J18-172 Lighting fixtures, Class I, Division 2
(1)　Lighting fixtures shall conform to the following:
　　(a)　portable lamps shall conform to Rules J18-118(1) and (2); and
　　(b)　fixed lighting shall be
　　　　(i)　protected from physical damage by suitable guards or by location; and
　　　　(ii)　approved as complete assemblies in compliance with the requirements of Rule J18-150 and shall be clearly marked to indicate the maximum wattage, voltage, and specific type designation of the lamps for which they are approved.
(2)　Pendant fixtures shall be
　　(a)　suspended by threaded rigid conduit stems or by other approved means; and
　　(b)　for stems longer than 300 mm, provided with permanent and effective bracing against lateral displacement at a level not more than 300 mm above the lower end of the stem, or flexibility in the form of a fitting or flexible connector approved for the purpose shall be provided not more than 300 mm from the point of attachment to the supporting box or fitting.
(3)　Boxes, box assemblies, or fittings used for the support of lighting fixtures shall be approved for the purpose.
(4)　Switches that are part of an assembled fixture or of an individual lampholder shall conform to Rule J18-160.
(5)　Starting and control equipment for electric-discharge lighting equipment shall be provided with enclosures that comply with the requirements of Rule J18-150.

J18-174 Utilization equipment, fixed and portable, Class I, Division 2
(1)　Electrically heated utilization equipment, whether fixed or portable, shall comply with the requirements of Rule J18-150.
(2)　Motors of motor-driven utilization equipment shall conform to Rule J18-168.
(3)　Switches, circuit breakers, and fuses forming part of or used in connection with utilization equipment shall conform to Rules J18-160 to J18-164.

J18-176 Flexible cords, Class I, Division 2
Flexible cords shall be permitted to be used only for connection between permanently mounted lighting fixtures, portable lamps, or other portable utilization equipment and the fixed portion of supply circuits, and where used shall
(a)　be of a type approved for extra-hard usage;
(b)　contain, in addition to the circuit conductors, a bonding conductor; and
(c)　be provided with glands approved for the class and group where the flexible cord enters a box, fitting, or enclosure of the explosion-proof type.

J18-178 Receptacles and attachment plugs, Class I, Division 2
Receptacles and attachment plugs shall comply with the requirements of Rule J18-150.

J18-180 Live parts, Class I, Division 2
No live parts of electrical equipment or of an electrical installation shall be exposed.

J18-182 Bonding, Class I, Division 2
Electrical equipment shall be bonded to ground in the manner required by Rule J18-132.

Appendix J
Rules and Notes to Rules for installations using the
Division system of classification for Class I locations

Annex J20
Flammable liquid and gas dispensing and service stations, garages, bulk storage plants, finishing processes, and aircraft hangars

J20-000 Scope
This Annex supplements or amends the general requirements of this Code and applies to installation as follows:
(a)   gasoline dispensing and service stations — Rules J20-002 to J20-014;
(b)   propane dispensing, container filling, and storage — Rules J20-030 to J20-042;
(c)   compressed natural gas refuelling stations and compressor and storage facilities — Rules J20-060 to J20-072;
(d)   commercial garages, repairs, and storage — Rules J20-100 to J20-114;
(e)   residential storage garages — Rules J20-200 to J20-206;
(f)   bulk storage plants — Rules J20-300 to J20-312;
(g)   finishing processes — Rules J20-400 to J20-414; and
(h)   aircraft hangars — Rules J20-500 to J20-522.

Gasoline dispensing and service stations

J20-002 General
(1)   Rules J20-004 to J20-014 apply to electrical apparatus and wiring installed in gasoline dispensing and service stations, and other locations where gasoline or other similar volatile flammable liquids are dispensed or transferred to the fuel tanks of self-propelled vehicles.
(2)   Other areas used as lubritoriums, service rooms, and repair rooms, and offices, salesrooms, compressor rooms, and similar locations shall conform to Rules J20-100 to J20-114 with respect to electrical wiring and equipment.

J20-004 Hazardous areas (see Annex JB)
(1)   Except as provided for in Subrule (3), the space within a dispenser enclosure up to 1.2 m vertically above its base, including the space below the dispenser that may contain electrical wiring and equipment, shall be considered to be a Class I, Division 1 location.
(2)   The space within a nozzle boot of a dispenser shall be considered a Class I, Division 1 location.
(3)   The space within a dispenser enclosure above the Class I, Division 1 location as specified in Subrule (1) or spaces within a dispenser enclosure isolated from the Division 1 location by a solid vapour-tight partition or by a solid nozzle boot but not completely surrounded by a Division 1 location shall be considered a Class I, Division 2 location.
(4)   The space within 450 mm horizontally from the Division 1 location within the dispenser enclosure as specified in Subrule (1) shall be considered a Class I, Division 1 location.
(5)   The space outside the dispenser within 450 mm horizontally from the opening of a solid nozzle boot located above the vapour-tight partition shall be considered a Class I, Division 2 location, except that the classified area need not be extended beyond the plane in which the boot is located.
(6)   In an outside location, any area beyond the Class I, Division 1 area (and in buildings not suitably cut off) within 6 m horizontally from the exterior enclosure of any dispenser shall be considered a Class I, Division 2 location that extends to a level 450 mm above driveway or ground level.
(7)   In an outside location, any area beyond the Class I, Division 1 location (and in buildings not suitably cut off) within 3 m horizontally from any tank fill-pipe shall be considered a Class I, Division 2 location extending upward to a level 450 mm above driveway or ground level.
(8)   Electrical wiring and equipment, any portion of which is below the surface of areas defined as Class I, Division 1 or Division 2 in Subrule (1), (4), (6), or (7) shall be considered to be within a Class I, Division 1 location that extends at least to the point of emergence above grade.
(9)   Areas within the vicinity of tank vent-pipes shall be classified as follows:
(a)   the spherical volume within a 900 mm radius from point of discharge of any tank vent-pipe shall be considered a Class I, Division 1 location and the volume between the 900 mm to 1.5 m radius from point of discharge of a vent shall be considered a Class I, Division 2 location;
(b)   for any vent that does not discharge upward, the cylindrical volume below both the Division 1 and Division 2 locations extending to the ground shall be considered a Class I, Division 2 location;

    (c)   the hazardous area shall not be considered to extend beyond an unpierced wall.

(10)   Areas within lubrication rooms shall be classified as follows:

    (a)   the area within any pit or space below grade or floor level in a lubrication room shall be considered a Class I, Division 1 location, unless the pit or space below grade is beyond the hazardous areas specified in Subrules (6), (7), and (9), in which case the pit or space below grade shall be considered a Class I, Division 2 location;

    (b)   notwithstanding Item (a), for each floor below grade that is located beyond the hazardous area specified in Subrules (6), (7), and (9) and where adequate mechanical ventilation is provided, a Class I, Division 2 location shall extend up to a level of only 50 mm above each such floor; and

    (c)   the area within the entire lubrication room up to 50 mm above the floor or grade, whichever is the higher, and the area within 900 mm measured in any direction from the dispensing point of a hand-operated unit dispensing volatile flammable liquids shall be considered a Class I, Division 2 location.

J20-006 Wiring and equipment within hazardous areas

(1)   Electrical wiring and equipment within the hazardous areas defined in Rule J20-004 shall conform to Annex J18 requirements.

(2)   Where dispensers are supplied by rigid metal conduit, a union and a flexible fitting shall be installed between the conduit and the dispenser junction box in addition to any sealing fittings required by Annex J18.

(3)   The flexible metal fitting required by Subrule (2) shall be installed in a manner that allows for relative movement of the conduit and the dispenser.

(4)   Where dispensers are supplied by a cable approved for hazardous locations, provisions shall be made to separate the cable from the dispenser junction box without rendering ineffective the explosion-proof cable seal.

J20-008 Wiring and equipment above hazardous areas

Wiring and equipment above hazardous areas shall conform to Rules J20-106 and J20-110.

J20-010 Circuit disconnects

Each circuit leading to or through a dispensing pump shall be provided with a switching means that will simultaneously disconnect all ungrounded conductors of the circuit from the source of supply.

J20-012 Sealing

(1)   Seals as required by Annex J18 shall be provided in each conduit run entering or leaving a dispenser or any cavities or enclosures in direct communication with a dispenser.

(2)   Additional seals shall be provided in conformance with Rules J18-108 and J18-158, and the requirements of Rules J18-108(3)(c) and J18-158(3)(b) shall include horizontal and vertical boundaries.

J20-014 Bonding

All non-current-carrying metal parts of dispensing pumps, metal raceways, and other electrical equipment shall be bonded to ground in accordance with Section 10.

Propane dispensing, container filling, and storage

J20-030 Scope (see Annex JB)

Rules J20-032 to J20-042 apply to locations where propane is dispensed or transferred to the fuel tanks of self-propelled vehicles or to portable containers and to locations where propane is stored or transferred from rail cars or tanker vehicles to storage containers.

J20-032 Special terminology

In this Subsection, the following definitions apply:

Container refill centre — a facility such as a propane service station that is open to the public and where propane is dispensed into containers or the fuel tanks of motor vehicles and that consists of propane storage containers, piping, and pertinent equipment including pumps and dispensing devices.

Filling plant — a facility such as a bulk propane plant, the primary purpose of which is the distribution of propane, that receives propane in tank car or truck transport for storage and/or distribution in portable containers or tank trucks, that has bulk storage, and that usually has container-filling and truck-loading facilities on the premises.

Appendix J
Rules and Notes to Rules for installations using the
Division system of classification for Class I locations

Propane — any material that is composed predominantly of the following hydrocarbons either by themselves or as mixtures: propane, propylene, butane (normal butane or iso-butane), and butylene.

J20-034 Hazardous areas

In container refill centres and in filling plants, the hazardous areas shall be classified as listed in Table JT-63.

J20-036 Wiring and equipment in hazardous areas

(1)　All electrical wiring and equipment in the hazardous areas referred to in Rule J20-034 shall conform to the requirements of Annex J18.

(2)　Where dispensing devices are supplied by rigid metal conduit, the requirements of Rule J20-006(2) and (3) shall be met.

J20-038 Sealing

(1)　Seals shall be installed as required by Annex J18, and the requirements shall be applied to horizontal as well as vertical boundaries of the defined hazardous locations.

(2)　Seals for dispensing devices shall be provided as required by Rule J20-012.

J20-040 Circuit disconnects

Each circuit leading to or through a propane dispensing device or pump shall be provided with a switching means that will disconnect simultaneously from the source of supply all ungrounded conductors of the circuit.

J20-042 Bonding

All non-current-carrying metal parts of equipment and raceways shall be bonded to ground in accordance with Section 10.

Compressed natural gas refuelling stations and compressor and storage facilities

J20-060 Scope (see Annex JB)

(1)　Rules J20-062 to J20-072 apply to locations where compressed natural gas is dispensed to the fuel tanks of self-propelled vehicles and to associated compressor and storage facilities.

(2)　The Rules in this Section do not apply to vehicle refuelling appliances installed in accordance with CAN/CSA-B149.1 without storage facilities.

J20-062 Hazardous areas

(1)　The areas surrounding compressors shall be classified as follows:

(a)　in an outdoor location, the space within 1.5 m in all directions from the compressor shall be considered a Class I, Division 1 location;

(b)　in an outdoor location, the space between 1.5 and 4.5 m in all directions from the compressor shall be considered a Class I, Division 2 location;

(c)　if the compressor is enclosed, the space within the compressor enclosure shall be considered a Class I, Division 1 location;

(d)　if the compressor is enclosed, the space within 3 m in all directions from non-gas-tight, non-welded seams and openings in the enclosure shall be considered a Class I, Division 2 location;

(e)　a compressor shall be regarded as enclosed when it is sheltered by a building or enclosure having four sides, a roof, and limited ventilation;

(f)　the space within a compressor enclosure shall be classified as Class I, Division 2 when the enclosure is provided with an exhaust fan interlocked with an approved gas detection system that functions to shut down the compressor and actuate the exhaust fan when the concentration of gas within the enclosure reaches 20% of the lower explosive limit; and

(g)　when a gas-tight wall is located within the distances specified in Items (a), (b), and (d), the distances shall be measured around the end of the wall, over the wall, or through any doors, windows, or openings in the wall.

(2)　The areas surrounding a natural gas dispensing point located outdoors shall be classified as follows:

(a)　for fast fill dispensing, the space within 3 m in all directions from the dispensing point shall be considered a Class I, Division 2 location; and

(b)　for slow fill dispensing, the space within 1.5 m in all directions from the dispensing point shall be considered a Class I, Division 2 location; and

(c)　the distances specified in Items (a) and (b) shall be measured from the breakaway coupling at the transition point between rigid piping and the refuelling hose.

(3)   For dispensing devices, the entire space within the dispenser enclosure and the space below the dispenser shall be considered a Class I, Division 1 location.

J20-064 Hazardous areas surrounding gas storage facilities
The electrical classification of areas surrounding gas storage facilities shall be as indicated in Table JT-64.

J20-066 Wiring and equipment in hazardous areas
(1)   All electrical wiring and equipment in the hazardous areas defined in Rules J20-062 and J20-064 shall comply with the requirements of Annex J18.
(2)   Where dispensing devices are supplied with rigid metal conduit, the requirements of Rule J20-006(2) and (3) shall be complied with.

J20-068 Sealing
(1)   Seals shall be installed as required by Annex J18, and the requirements shall be applied to horizontal as well as vertical boundaries of the defined hazardous locations.
(2)   Seals for dispensing devices shall be provided as required by Rule J20-012.

J20-070 Circuit disconnects
Each circuit leading to a compressor or a dispensing device shall be provided with a switching means that will disconnect simultaneously from the source of supply all ungrounded conductors of the circuit.

J20-072 Bonding
All non-current-carrying metal parts of equipment and raceways shall be bonded to ground in accordance with Section 10.

Commercial garages, repairs, and storage

J20-100 Scope
Rules J20-102 to J20-114 apply to locations used for service and repair operations in connection with self-propelled vehicles in which volatile flammable liquids or flammable gases are used for fuel or power, and locations in which more than three such vehicles are, or may be, stored at one time.

J20-102 Hazardous areas
(1)   For each floor at or above grade, the entire area up to a level 50 mm above the floor shall be considered a Class I, Division 2 location.
(2)   For each floor below grade, the entire area up to a level 50 mm above the bottom of outside doors or other openings that are at, or above, grade level shall be considered a Class I, Division 2 location; except that where adequate mechanical ventilation is provided, the hazardous location shall extend up to a level of only 50 mm above each such floor.
(3)   Notwithstanding Subrule (2), in storage garages, only the area up to a level of 50 mm above each floor that is below grade shall be considered a Class I, Division 2 location.
(4)   Any pit or depression below floor level shall be considered a Class I, Division 2 location that extends up to the floor level.
(5)   Adjacent areas in which hazardous vapours are not likely to be released, such as stockrooms, switchboard rooms, and other similar locations having floors elevated at least 50 mm above adjacent garage floor, or separated from the garage floor by tight-fitting barriers such as curbs, ramps, or partitions at least 50 mm high, shall not be classed as hazardous.

J20-104 Wiring and equipment in hazardous area
Within hazardous areas as defined in Rule J20-102, wiring and equipment shall conform to the applicable requirements of Annex J18.

J20-106 Wiring above hazardous areas
(1)   All fixed wiring above hazardous areas shall be in accordance with Section 12 and suitable for the type of building and occupancy.
(2)   For pendants, flexible cord of the hard-usage type shall be used.
(3)   For connection of portable lamps, portable motors, or other portable utilization equipment, flexible cord of the hard-usage type shall be used.

J20-108 Sealing
(1)   Seals shall be installed as required by Annex J18, and the requirements of Rule J18-158(3)(b) shall include horizontal and vertical boundaries.

524

Appendix J
Rules and Notes to Rules for installations using the
Division system of classification for Class I locations

(2) Raceways embedded in a masonry floor or buried beneath a floor shall be considered to be within the hazardous area above the floor if any connections or extensions lead into or through such an area.

J20-110 Equipment above hazardous areas

(1) Fixed equipment that is less than 3.6 m above the floor level and that may produce arcs, sparks, or particles of hot metal, such as cut-outs, switches, charging panels, generators, motors, or other equipment (excluding receptacles, lamps, and lampholders) having make-and-break or sliding contacts, shall be the totally enclosed type or constructed to prevent escape of sparks or hot metal particles.

(2) Lamps and lampholders for fixed lighting that are located over lanes through which vehicles are commonly driven or that may otherwise be exposed to physical damage shall be located not less than 3.6 m above floor level unless they are a totally enclosed type or constructed to prevent escape of sparks or hot metal particles.

(3) Portable lamps shall comply with the following:
    (a) they shall be of the totally enclosed gasketted type, equipped with handle, lampholder, hook, and substantial guard attached to the lampholder or handle, and all exterior surfaces that may come in contact with battery terminals, wiring terminals, or other objects shall be of non-conducting materials or shall be effectively protected with an insulating jacket;
    (b) the lampholders shall be of unswitched type; and
    (c) they shall not be provided with receptacles for attachment plugs.

J20-112 Battery-charging equipment

Battery chargers and their control equipment, and batteries being charged, shall not be located within the hazardous areas classified in Rule J20-102.

J20-114 Electric vehicle charging

(1) Flexible cords used for charging shall be the extra-hard-usage type.

(2) Connectors shall have a rating not less than the ampacity of the cord and in no case less than 50 A.

(3) Connectors shall be designed and installed so that they will break apart readily at any position of the charging cable, and live parts shall be guarded from accidental contact.

(4) No connector shall be located within the hazardous area defined in Rule J20-102.

(5) Where plugs are provided for direct connection to vehicles, the point of connection shall not be within a hazardous area as defined in Rule J20-102.

(6) Where a cord is suspended from overhead, it shall be arranged so that the lowest point of sag is at least 150 mm above the floor.

(7) Where the vehicle is equipped with a plug that will readily pull apart, and where an automatic arrangement is provided to pull both cord and plug beyond the range of mechanical damage, no additional connector shall be required in the cable or outlet.

Residential storage garages

J20-200 Scope

Rules J20-202 to J20-206 apply to a building or part of a building in which not more than three vehicles of the type described in Rule J20-100 are, or may be, stored, but that will not normally be used for service or repair operations on stored vehicles.

J20-202 Non-hazardous location

Where the lowest floor is at or above adjacent grade or driveway level, and where there is at least one outside door at or below floor level, the garage area shall not be classed as a hazardous location.

J20-204 Hazardous location

Where the lowest floor is below adjacent grade or driveway level, the following shall apply:
    (a) the entire area of the garage or of any enclosed space that includes the garage shall be classified as a Class I, Division 2 location up to a level 50 mm above the garage floor; and
    (b) adjacent areas in which hazardous vapours or gases are not likely to be released, and where floors are elevated at least 50 mm above the garage floor or separated from the garage floor by tight curbs or partitions at least 50 mm high, shall not be classed as hazardous.

J20-206 Wiring

(1) Wiring above the hazardous locations shall conform to Section 12.

(2) Wiring in the hazardous locations shall conform to Annex J18.

Bulk storage plants

J20-300 Scope
Rules J20-302 to J20-312 apply to locations where gasoline or other similar volatile flammable liquids are stored in tanks having an aggregate capacity of one carload or more, and from which such products are distributed (usually by tank truck).

J20-302 Hazardous areas
(1) Areas containing pumps, bleeders, withdrawal fittings, meters, and similar devices that are located in pipelines handling flammable liquids under pressure shall be classified as follows and comply with the following requirements:

    (a) ventilated indoor areas shall be considered Class I, Division 2 locations within a 1.5 m distance extending in all directions from the exterior surface of such devices as well as 7.5 m horizontally from any surface of these devices and extending upwards to 900 mm above floor or grade level, provided that

        (i) the design of the ventilation systems takes into account the relatively high relative density of the vapours;

        (ii) where openings are used in outside walls, they are of adequate size and located at floor level unobstructed except by louvres or coarse screens; and

        (iii) where natural ventilation is inadequate, mechanical ventilation is provided;

    (b) indoor areas not ventilated in accordance with Subrule (1)(a) shall be considered Class I, Division 1 locations within a 1.5 m distance extending in all directions from the exterior surface of such devices as well as 7.5 m horizontally from any surface of the device and extending upward 900 mm above floor or grade level; and

    (c) outdoor areas shall be considered Class I, Division 2 locations within a 900 mm distance extending in all directions from the exterior surfaces of such devices as well as up to 450 mm above grade level within 3 m horizontally from any surface of the devices.

(2) Areas where flammable liquids are transferred shall be classified as follows:

    (a) in outdoor areas or where positive and reliable mechanical ventilation is provided in indoor areas in which flammable liquids are transferred to individual containers, such areas shall be considered Class I, Division 1 locations within 900 mm of the vent or fill opening extending in all directions and Class I, Division 2 locations within the area extending between a 900 mm and 1.5 m radius from the vent or fill opening extending in all directions, and including the area within a horizontal radius of 3 m from the vent or fill opening and extending to a height of 450 mm above floor or grade levels; or

    (b) where positive and reliable mechanical ventilation is not provided in indoor areas in which flammable liquids are transferred to individual containers, such areas shall be considered Class I, Division 1 locations.

(3) Areas in outside locations where loading and unloading of tank vehicles and tank cars takes place shall be classified as follows:

    (a) the area extending 900 mm in all directions from the dome when loading through an open dome or from the vent when loading through a closed dome with atmospheric venting shall be considered a Class I, Division 1 location;

    (b) the area extending between a 900 mm and a 1.5 m radius from the dome when loading an open dome or from the vent when loading through a closed dome with atmospheric venting shall be considered a Class I, Division 2 location; and

    (c) the area extending within 900 mm in all directions from a fixed connection used in bottom loading or unloading, loading through a closed dome with atmospheric venting, or loading through a closed dome with a vapour recovery system shall be considered a Class I, Division 2 location, except that in the case of bottom loading or unloading this classification shall also be applied to the area within a 3 m radius from point of connection and extending 450 mm above grade.

(4) Areas within the vicinity of above-ground tanks shall be classified as follows:

    (a) the area above the roof and within the shell of a floating roof type tank shall be considered a Class I, Division 1 location;

    (b) for all types of above-ground tanks

        (i) the area within 3 m from the shell, ends, and roof of other than a floating roof shall be considered a Class I, Division 2 location; and

Appendix J
Rules and Notes to Rules for installations using the
Division system of classification for Class I locations

      (ii)   where dikes are provided, the area inside the dike and extending upwards to the top of the dike shall be considered a Class I, Division 2 location;

(c)   the area within 1.5 m of a vent opening and extending in all directions shall be considered a Class I, Division 1 location; and

(d)   the area between 1.5 m and 3 m of a vent opening and extending in all directions shall be considered a Class I, Division 2 location.

(5)   Pits and depressions shall be classified as follows:

    (a)   any pit or depression, any part of which lies within a Division 1 or Division 2 location, unless provided with positive and reliable mechanical ventilation, shall be considered a Class I, Division 1 location;

    (b)   any such areas, when provided with positive and reliable mechanical ventilation, shall be considered Class I, Division 2 locations; and

    (c)   any pit or depression not within a Division 1 or Division 2 location as defined in this Section but that contains piping, valves, or fittings, shall be considered a Class I, Division 2 location.

(6)   Garages in which tank vehicles are stored or repaired shall be considered to be Class I, Division 2 locations up to 450 mm above floor or grade level unless conditions warrant more severe classification or a greater extent of the hazardous area.

(7)   Buildings such as office buildings, boiler rooms etc., that are outside the limits of hazardous areas as defined in this Section and that are not used for handling or storage of volatile flammable liquids or containers for such liquids shall not be considered to be hazardous locations.

J20-304 Wiring and equipment in hazardous areas

All electrical wiring and equipment in hazardous areas defined in Rule J20-302 shall conform to the requirements of Annex J18.

J20-306 Wiring and equipment above hazardous areas

(1)   Wiring installed above a hazardous location shall conform to the requirements of Section 12 and be suitable for the type of building and the occupancy.

(2)   Fixed equipment that may produce arcs, sparks, or particles of hot metal, such as lamps and lampholders, cut-outs, switches, receptacles, motors, or other equipment having make-and-break or sliding contacts, shall be of the totally enclosed type or constructed to prevent the escape of sparks or hot metal particles.

(3)   Portable lamps or utilization equipment and the flexible cords supplying them shall conform to the requirements of Annex J18 for the class of location above which they are connected or used.

J20-308 Sealing

(1)   Seals shall be installed in accordance with Annex J18 and shall be applied to horizontal as well as vertical boundaries of the defined hazardous locations.

(2)   Buried raceways under defined hazardous areas shall be considered to be within such areas.

J20-310 Gasoline dispensing

Where gasoline dispensing is carried on in conjunction with bulk station operations, the applicable provisions of Rules J20-002 to J20-014 inclusive shall apply.

J20-312 Bonding

All non-current-carrying metal parts of equipment and raceways shall be bonded to ground in accordance with Section 10.

Finishing processes

J20-400 Scope

Rules J20-402 to J20-414 apply where paints, lacquers, or other flammable finishes are regularly or frequently applied by spraying, dipping, brushing, or other means, and where volatile flammable solvents or thinners are used, or where readily ignitable deposits or residues from such paints, lacquers, or finishes may occur.

J20-402 Hazardous locations

(1)   The following areas shall be considered Class I, Division 1 locations:

    (a)   the interiors of spray booths and their exhaust ducts;

    (b)   all space within 6 m horizontally in any direction, extending to a height of 1 m above the goods to be painted, from spraying operations more extensive than touch-up spraying and not conducted within the spray booth and as otherwise shown in Diagram JD-5;

(c)   all space within 6 m horizontally in any direction from dip tanks and their drain boards, such space extending to a height of 1 m above the dip tank and drain board; and

(d)   all other spaces where hazardous concentrations of flammable vapours are likely to occur.

(2)   For spraying operations within an open-faced spray booth, the extent of the Class I, Division 2 location shall extend not less than 1.5 m from the open face of the spray booth, and as otherwise shown in Diagram JD-4.

(3)   For spraying operations confined within a closed spray booth or room, or for rooms where hazardous concentrations of flammable vapours are likely to occur, such as paint mixing rooms, the space within 1 m in all directions from any openings in the booth or room shall be considered a Class I, Division 2 location, and as otherwise shown in Diagram JD-10.

(4)   All space within the room but beyond the limits for Class I, Division 1 as classified in Subrule (1) for extensive open spraying, and as otherwise shown in Diagram JD-5, for dip tanks and drain boards, and for other hazardous operations, shall be considered Class I, Division 2 locations.

(5)   Adjacent areas that are cut off from the defined hazardous area by tight partitions without communicating openings, and within which hazardous vapours are not likely to be released, shall be permitted to be classed as non-hazardous.

(6)   Drying and baking areas provided with positive mechanical ventilation to prevent formation of flammable concentrations of vapours and provided with effective interlocks to de-energize all electrical equipment not approved for Class I locations in case the ventilating equipment is inoperative shall be permitted to be classed as non-hazardous.

(7)   Notwithstanding the requirements of Subrule (1)(b), where adequate mechanical ventilation with effective interlocks is provided at floor level and as otherwise shown in Diagram JD-6,

(a)   the space within 1 m horizontally in any direction from the goods to be painted and such space extending to a height of 1 m above the goods to be painted shall be considered a Class I, Division 1 location; and

(b)   all space between a 1 m and a 1.5 m distance above the goods to be painted and all space within 6 m horizontally in any direction beyond the limits for a Class I, Division 1 location shall be considered a Class I, Division 2 location.

(8)   Notwithstanding the requirements of Subrule (2), where a baffle of sheet metal of not less than No. 18 MSG is installed vertically above the front face of an open-face spray booth to a height of 1 m or to the ceiling, whichever is less, and extending back on the side edges for a distance of 1.5 m, the space behind this baffle shall be considered a non-hazardous location.

(9)   Notwithstanding the requirements of Subrule (3), where a baffle of sheet metal of not less than No. 18 MSG is installed vertically above an opening in a closed spray booth or room to a height of 1 m or to the ceiling, whichever is less, and extends horizontally a distance of 1 m beyond each side of the opening, the space behind the baffle shall be considered a non-hazardous location.

J20-404 Ventilation and spraying equipment interlock

The spraying equipment for a spray booth shall be interlocked with the spray booth ventilation system so that the spraying equipment is made inoperable when the ventilation system is not in operation.

J20-406 Wiring and equipment in hazardous areas

(1)   All electrical wiring and equipment within the hazardous areas as defined in Rule J20-402 shall conform to the requirements of Annex J18.

(2)   Unless specifically approved for both readily ignitable deposits and the flammable vapour location, no electrical equipment shall be installed or used where it may be subject to a hazardous accumulation of readily ignitable deposits or residue.

(3)   Illumination of readily ignitable areas through panels of glass or other transparent or translucent materials is permissible only where

(a)   fixed lighting units are used as the source of illumination;

(b)   the panel is non-combustible and effectively isolates the hazardous area from the area in which the lighting unit is located;

(c)   the panel is of such a material or is protected so that breakage will be unlikely; and

(d)   the arrangement is such that normal accumulations of hazardous residue on the surface of the panel will not be raised to a dangerous temperature by radiation or conduction from the source of illumination.

(4)   Portable electric lamps or other utilization equipment shall

(a)   not be used within a hazardous area during operation of the finishing process;

Appendix J
Rules and Notes to Rules for installations using the
Division system of classification for Class I locations

(b)   be a type specifically approved for Class I locations when used during cleaning or repairing operations.

(5)   Notwithstanding Subrule (2),

(a)   totally enclosed and gasketted lighting shall be permitted to be used on the ceiling of a spray room where adequate and positive mechanical ventilation is provided; and

(b)   infrared paint drying units shall be permitted to be utilized in a spray room if the controls are interlocked with those of the spraying equipment so that both operations cannot be performed simultaneously, and if portable, the paint drying unit shall not be brought into the spray room until spraying operations have ceased.

J20-408 Fixed electrostatic equipment

Electrostatic spraying and detearing equipment shall conform to the following:

(a)   no transformers, power packs, control apparatus, or other electrical portions of the equipment except high-voltage grids and their connections shall be installed in any of the hazardous areas defined in Rule J20-402, unless of a type specifically approved for the location;

(b)   high-voltage grids or electrodes shall be

(i)   located in suitable non-combustible booths or enclosures provided with adequate mechanical ventilation;

(ii)   rigidly supported and of substantial construction; and

(iii)   effectively insulated from ground by means of nonporous, non-combustible insulators;

(c)   high-voltage leads shall be

(i)   effectively and permanently supported on suitable insulators;

(ii)   effectively guarded against accidental contact or grounding; and

(iii)   provided with automatic means for discharging any residual charge to ground when the supply voltage is interrupted;

(d)   where goods are being processed

(i)   they shall be supported on conveyors in such a manner that minimum clearance between goods and high-voltage grids or conductors cannot be less than twice the sparking distance; and

(ii)   a conspicuous sign indicating the sparking distance shall be permanently posted near the equipment;

(e)   automatic controls shall be provided that will operate without time delay to disconnect the power supply and to signal the operator in case of

(i)   stoppage of ventilating fans;

(ii)   failure of ventilating equipment;

(iii)   stoppage of the conveyor carrying goods through the high-voltage field;

(iv)   occurrence of a ground or of an imminent ground at any point on the high-voltage system; or

(v)   reduction of clearance below that specified in Item (d); and

(f)   adequate fencing, railings, or guards that are electrically conducting and effectively bonded to ground shall be provided for safe isolation of the process, and signs shall be permanently posted designating the process as dangerous because of high voltage.

J20-410 Electrostatic hand spraying equipment

Electrostatic hand spray apparatus and devices used with them shall conform to the following:

(a)   the high-voltage circuits shall be intrinsically safe and not produce a spark of sufficient intensity to ignite any vapour-air mixtures, nor result in appreciable shock hazard upon coming in contact with a grounded object;

(b)   the electrostatically charged exposed elements of the hand gun shall be capable of being energized only by a switch that also controls the paint supply;

(c)   transformers, power packs, control apparatus, and all other electrical portions of the equipment, with the exception of the hand gun itself and its connections to the power supply, shall be located outside the hazardous area;

(d)   the handle of the spray gun shall be bonded to ground by a metallic connection and be constructed so that the operator in normal operating position is in intimate electrical contact with the handle in order to prevent buildup of a static charge on the operator's body;

(e)   all electrically conductive objects in the spraying area shall be bonded to ground and the equipment shall carry a prominent, permanently installed warning regarding the necessity for this bonding feature;

(f)   precautions shall be taken to ensure that objects being painted are maintained in metallic contact with the conveyor or other grounded support and shall include the following:

(i)   hooks shall be regularly cleaned;

(ii)   areas of contact shall be sharp points or knife edges; and

(iii)   points of support of the object shall be concealed from random spray where feasible, and where the objects being sprayed are supported from a conveyor, the point of attachment to the conveyor shall be located so as to not collect spray material during normal operation; and

(g)   the spraying operation shall take place within a spray area that is adequately ventilated to remove solvent vapours released from the operation and the electrical equipment shall be interlocked with the ventilation of the spraying area so that the equipment cannot be operated unless the ventilation system is in operation.

J20-412 Wiring and equipment above hazardous areas

(1)   All fixed wiring above hazardous areas shall conform to Section 12.

(2)   Equipment that may produce arcs, sparks, or particles of hot metal, such as lamps and lampholders for fixed lighting, cut-outs, switches, receptacles, motors, or other equipment having make-and-break or sliding contacts, where installed above a hazardous area or above an area where freshly finished goods are handled, shall be of the totally enclosed type or constructed to prevent the escape of sparks or hot metal particles.

J20-414 Bonding

All metal raceways and all non-current-carrying metal portions of fixed or portable equipment, regardless of voltage, shall be bonded to ground in accordance with Section 10.

Aircraft hangars

J20-500 Scope

Rules J20-502 to J20-522 apply to locations used for storage or servicing of aircraft in which gasoline, jet fuels, or other volatile flammable liquids, or flammable gases, are used but shall not include those locations used exclusively for aircraft that have never contained such liquids or gases, or that have been drained and properly purged.

J20-502 Hazardous areas

(1)   Any pit or depression below the level of the hangar floor shall be considered a Class I, Division 1 location which shall extend up to the floor level.

(2)   The entire area of the hangar including any adjacent and communicating areas not suitably cut off from the hangar shall be considered a Class I, Division 2 location up to a level 450 mm above the floor.

(3)   The area within 1.5 m horizontally from aircraft power plants, aircraft fuel tanks, or aircraft structures containing fuel shall be considered a Class I, Division 2 location that extends upward from the floor to a level 1.5 m above the upper surface of wings and of engine enclosures.

(4)   Adjacent areas in which hazardous vapours are not likely to be released such as stock rooms, electrical control rooms, and other similar locations shall be permitted to be classed as non-hazardous when adequately ventilated and when effectively cut off from the hangar itself in accordance with Rule J18-060.

J20-504 Wiring and equipment in hazardous areas

(1)   All fixed and portable wiring and equipment that is or may be installed or operated within any of the hazardous locations defined in Rule J20-502 shall conform to the requirements of Annex J18.

(2)   All wiring installed in or under the hangar floor shall conform to the requirements for Class I, Division 1 locations.

(3)   Wiring systems installed in pits or other spaces in or under the hangar floor shall be provided with adequate drainage and shall not be placed in the same compartment with any other service except piped compressed air.

(4)   Attachment plugs and receptacles in hazardous locations shall be explosion-proof or shall be designed so that they cannot be energized while the connections are being made or broken.

J20-506 Wiring not within hazardous areas

(1)   All fixed wiring in a hangar not within a hazardous area as defined in Rule J20-502 shall be installed in metal raceways or shall be armoured cable, type MI cable, or aluminum-sheathed cable, except that wiring in a non-hazardous location as set out in Rule J20-502(4) shall be permitted to be of any type recognized in Section 12 as suitable for the type of building and the occupancy.

(2)   For pendants, flexible cord of the hard-usage type and containing a separate bonding conductor shall be used.

Appendix J
Rules and Notes to Rules for installations using the
Division system of classification for Class I locations

(3)   For portable utilization equipment and lamps, flexible cord approved for hard usage and containing a separate bonding conductor shall be used.

(4)   Suitable means shall be provided for maintaining continuity and adequacy of the bonding between the fixed wiring system and the non-current-carrying metal portions of pendant fixtures, portable lamps, and other portable utilization equipment.

J20-508 Equipment not within hazardous areas

(1)   In locations other than those described in Rule J20-502, equipment that is less than 3 m above wings and engine enclosures of aircraft and that may produce arcs, sparks, or particles of hot metal, such as lamps and lampholders or fixed lighting, cut-outs, switches, receptacles, charging panels, generators, motors, or other equipment having make-and-break or sliding contacts, shall be of a totally enclosed type or constructed to prevent the escape of sparks or hot metal particles, except that equipment in areas described in Rule J20-502(4) shall be permitted to be the general-purpose type.

(2)   Lampholders of metal shell, fibre-lined types shall not be used for fixed lighting.

(3)   Portable lamps that are used within a hangar shall comply with Rule J18-118.

(4)   Portable utilization equipment that is, or may be, used within a hangar shall be a type suitable for use in Class I, Division 2 locations.

J20-510 Stanchions, rostrums, and docks

(1)   Electric wiring, outlets, and equipment including lamps on, or attached to, stanchions, rostrums, or docks that are located, or likely to be located, in a hazardous area as defined in Rule J20-502(3) shall conform to the requirements for Class I, Division 2 locations.

(2)   Where stanchions, rostrums, and docks are not located, or are not likely to be located, in a hazardous area as defined in Rule J20-502(3), wiring and equipment shall conform to Rules J20-506 and J20-508, except for the following:

(a)   receptacles and attachment plugs shall be the locking type that will not break apart readily; and

(b)   wiring and equipment not more than 450 mm above the floor in any position shall conform to Subrule (1).

(3)   Mobile stanchions with electrical equipment conforming to Subrule (2) shall carry at least one permanently affixed warning sign stating that the stanchions be kept 1.5 m clear of aircraft engines and fuel tank areas.

J20-512 Sealing

(1)   Seals shall be installed in accordance with Annex J18 and shall apply to horizontal as well as to vertical boundaries of the defined hazardous areas.

(2)   Raceways embedded in a masonry floor or buried beneath a floor shall be considered within the hazardous area above the floor when any connections or extensions lead into or through the hazardous area.

J20-514 Aircraft electrical systems

Aircraft electrical systems shall be de-energized when the aircraft is stored in a hangar and, whenever possible, while the aircraft is undergoing maintenance.

J20-516 Aircraft battery-charging and equipment

(1)   Aircraft batteries shall not be charged when installed in an aircraft located inside or partially inside a hangar.

(2)   Battery chargers and their control equipment shall not be located or operated within any of the hazardous areas defined in Rule J20-502 but shall be permitted to be located or operated in a separate building or in an area complying with Rule J20-502(4).

(3)   Mobile chargers shall carry at least one permanently affixed warning sign stating that the chargers must be kept 1.5 m clear of aircraft engines and fuel tank areas.

(4)   Tables, racks, trays, and wiring shall not be located within a hazardous area and shall conform to the provisions of Section 26 pertaining to storage batteries.

J20-518 External power sources for energizing aircraft

(1)   Aircraft energizers shall be designed and mounted so that all electrical equipment and fixed wiring will be at least 450 mm above floor level and shall not be operated in a hazardous area as defined in Rule J20-502(3).

(2)   Mobile energizers shall carry at least one permanently affixed sign stating that the energizer be kept 1.5 m clear of aircraft engines and fuel tank areas.

(3)   Aircraft energizers shall be equipped with polarized external power plugs and with automatic controls to isolate the ground power unit electrically from the aircraft in case excessive voltage is generated by the ground power unit.

(4)   Flexible cords for aircraft energizers and ground support equipment shall be of the extra-hard-usage type and shall include a bonding conductor.

J20-520 Mobile servicing equipment with electrical components

(1)   Mobile servicing equipment such as vacuum cleaners, air compressors, and air movers, etc., having electrical wiring and equipment not suitable for Class I, Division 2 locations shall

    (a)   be designed and mounted so that all such wiring and equipment will be at least 450 mm above the floor;

    (b)   not be operated within the hazardous areas defined in Rule J20-502(3); and

    (c)   carry at least one permanently affixed warning sign stating that the equipment be kept 1.5 m clear of aircraft engines and fuel tank areas.

(2)   Flexible cords used for mobile equipment shall be of the extra-hard-usage type and shall include a bonding conductor.

(3)   Attachment plugs and receptacles shall provide for the connection of the bonding conductor to the raceway system.

(4)   Equipment shall not be operated in areas where maintenance operations likely to release hazardous vapours are in progress, unless the equipment is at least suitable for use in a Class I, Division 2 location.

J20-522 Bonding

All metal raceways, and all non-current-carrying metal portions of fixed or portable equipment, regardless of voltage, shall be bonded to ground in accordance with Section 10.

Annex JB
Notes to Rules for Annexes J18 and J20

Rule JB18-000

Through the application of ingenuity in the layout of electrical installations for hazardous locations, it is frequently possible to locate much of the equipment in less hazardous or in non-hazardous areas and thus reduce the amount of special equipment required. It is recommended that the authority enforcing this Code be consulted before such layouts are prepared.

Rule JB18-002

Cable seals — seals that are designed to prevent the escape of flames from an explosion-proof enclosure. Because cables are not designed to withstand the pressures of an explosion, transmission of an explosion into a cable could result in ignition of gases or vapours in the area outside the enclosure.

Conduit seals — seals that are designed to prevent the passage of flames from one portion of the electrical installation to another through the conduit system and to minimize the passage of gases or vapours at atmospheric pressure. Unless specifically designed for the purpose, conduit seals are not intended to prevent the passage of fluids at a continuous pressure differential across the seal. Even at differences in pressure across the seal equivalent to a few inches of water, there may be passage of gas or vapour through the seal and/or through the conductors passing through the seal. Where conduit seals are exposed to continuous pressure, there may be a danger of transmission of flammable fluids to "safe areas" resulting in fire or explosions.

Primary seals — seals that are typically a part of electrical devices such as pressure, temperature or flow measuring devices and devices (such as canned pumps) where the electrical connections are immersed in the process fluids.

Secondary seals — seals that are designed to prevent flammable process fluids entering the electrical wiring system upon failure of a primary seal. These devices typically prevent passage of fluids at process pressure by a combination of sealing and pressure relief.

Rules JB18-004, JB18-006

Reference material for area classification can be found in the following documents:

(a)   IEC 60079-10, *Classification of Hazardous Areas;*

(b)   Institute of Petroleum (British), *Model Code of Safe Practice — Part 15: Area Classification Code for Petroleum Installations;*

Appendix J
Rules and Notes to Rules for installations using the
Division system of classification for Class I locations

(c)   ANSI/API 505, *Recommended Practice for Classification of Locations for Electrical Installations at Petroleum Facilities Classified as Class I, Zone 0, Zone 1, and Zone 2;*

(d)   API 500, *Recommended Practice for Classification of Locations for Electrical Installations at Petroleum Facilities Classified as Class I, Division 1 and Division 2;*

(e)   NFPA 497, *Classification of Flammable Liquids, Gases, or Vapors and of Hazardous (Classified) Locations for Electrical Installations in Chemical Process Areas;* and

(f)   see also the Note to Rule J18-064 in this Annex.

Rule JB18-006

Typical situations leading to a Division 1 area classification are

(a)   the interiors of storage tanks that are vented to atmosphere and that contain flammable liquids stored above their flash point;

(b)   enclosed sumps containing flammable liquids stored above their flash point during normal operation;

(c)   the area immediately around atmospheric vents;

(d)   inadequately ventilated buildings or enclosures; and

(e)   adequately ventilated buildings or enclosures, such as remote unattended and unmonitored facilities, that have insufficient means of limiting the duration of explosive gas atmospheres when they do occur.

Typical situations leading to a Division 2 area classification are

(a)   areas where flammable volatile liquids, flammable gases, or vapours are handled, processed, or used, but in which liquids, gases, or vapours are normally confined within closed containers or closed systems from which they can escape only as a result of accidental rupture or breakdown of the containers or systems or the abnormal operation of the equipment by which the liquids or gases are handled, processed, or used;

(b)   adequately ventilated buildings that have means of ensuring the length of time abnormal operation resulting in the occurrence of explosive gas atmospheres exist will be limited to a "short time"; and

(c)   most outdoor areas except those around open vents, or open vessels or sumps containing flammable liquids.

API RP500 defines "adequate ventilation" as "Ventilation (natural or artificial) that is sufficient to prevent the accumulation of significant quantities of vapour-air or gas-air mixtures in concentrations above 25% of their lower flammable (explosive) limit, LFL, (LEL)". Appendix B of API RP500 outlines a method for calculating the ventilation requirements for enclosed areas based on fugitive emissions.

Industry documents such as API 505 provide guidance on how industry interprets a "short time".

Rules JB18-050, JB18-066

It should be noted that battery-operated and self-generating equipment is not excluded from the Rules of Annex J18, regardless of the voltage involved. Examples of such equipment are flashlights, transceivers, paging receivers, tape recorders, combustion gas detectors, vibration monitors, tachometers, battery- or voice-powered telephones, and portable test equipment that may be carried into or located within a hazardous area. Such equipment may be eligible for approval under CAN/CSA-C22.2 No. 157.

Where general-purpose enclosures are used for such equipment and the Rules of this Section require the equipment to be specifically approved for the hazardous location, the electrical equipment is required to be approved for the location as intrinsically safe in accordance with Rule J18-066 and marked in accordance with Rule J18-052.

In cases where the Rules of this Section permit general-purpose enclosures with the qualification that acceptable non-incendive circuits are incorporated, the electrical equipment should be approved as such and marked in accordance with Rule J18-052.

Rule JB18-050(3)(a)

Information on classification of areas (as well as ventilation requirements) in plants engaged in the generation and compression of acetylene and in the charging of acetylene cylinders may be found in NFPA No. 51A.

Rule JB18-050(3)(d)

Information on classification of areas for refrigeration systems utilizing flammable gases (including ammonia) may be found in CSA B52.

Rule JB18-052

Equipment marked for Class I but not marked with a Division is suitable for both Divisions 1 and 2.

Where parts of equipment are approved to the E60079 series of standards they will be marked with the "Method of Protection", as indicated in the following table:

Approval	Symbol
Intrinsic safety	i, ia, ib
Flame-proof	d
Increased safety	e
Oil immersed	o
Pressurized	p
Powder-filled	q
Encapsulation	m
Method of protection	n

The mark will include the letters "Ex" or "EEx" to indicate that the equipment is "explosion protected" followed by one or more of the symbols.

Where equipment is approved to several requirements, each type of approval is shown. For example, equipment with an explosion-proof or flame-proof enclosure for the supply circuit and intrinsic safety for the output terminals is marked with either "explosion-proof housing with intrinsically safe output" or "Ex dia".

Rules JB18-052(2), JB18-054

Some equipment permitted for use in Division 2 hazardous locations is not marked to indicate the class and group because it is not specifically required to be approved for the location; for example, motors and generators for Class I, Division 2 that do not incorporate spark-producing components or integral resistance devices. See Rule J18-168(2).

Atmosphere Typical North American name	Name in IEC 60079-20 (if different) (see Note 1)	CAS reference number	Minimum ignition temperature, °C	Gas group North America	Gas group IEC
acetylene		74-86-2	305	A	IIC
butadiene	buta-1,3-diene	106-99-0	430	B	IIB
hydrogen		1333-74-0	560	B	IIC
manufactured gases containing more than 30% hydrogen (by volume)			500 (Note (2))	B	
propylene oxide	(Note (3))	75-56-9	430	B	IIB
acetaldehyde		75-07-0	204	C	IIA
cyclopropane		75-19-4	498	C	IIA
diethyl ether		60-29-7	160	C	IIB
ethylene		74-85-1	425	C	IIB
hydrogen sulphide		7783-06-4	270	C	IIB
unsymmetrical dimethyl hydrazine (UDMH 1,1-dimethyl hydrazine)	N,N-Dimethylhydrazine	57-14-7	240	C	IIB
acetone		67-64-1	535	D	IIA
acrylonitrile		107-13-1	480	D	IIB
alcohol (see ethyl alcohol)				D	
ammonia		7664-41-7	630	D	IIA
benzene		71-43-2	560	D	IIA

(Continued)

Appendix J
Rules and Notes to Rules for installations using the
Division system of classification for Class I locations

Atmosphere — Typical North American name	Name in IEC 60079-20 (if different) (see Note 1)	CAS reference number	Minimum ignition temperature, °C	Gas group North America	Gas group IEC
benzine (see petroleum naphtha)					
benzol (see benzene)					
butane		106-97-8	372	D	IIA
1-butanol (butyl alcohol)	butan-1-ol	71-36-3	359	D	IIA
2-butanol (secondary butyl alcohol)	butan-2-ol	78-92-2	405 (Note (2))	D	IIA
butyl acetate		123-86-4	370	D	IIA
isobutyl acetate		110-19-0	421 (Note (2))	D	IIA
ethane		74-84-0	515	D	IIA
ethanol (ethyl alcohol)		64-17-5	363	D	IIA
ethyl acetate		141-78-6	460	D	IIA
ethylene dichloride	1,2-Dichloroethane	107-06-2	438	D	IIA
gasoline	petroleum	86290-81-5	560	D	IIA
heptanes		142-82-5	215	D	IIA
hexanes		110-54-3	233	D	IIA
isoprene		78-79-5	395 (Note (2))	D	IIA
methane		74-82-8	537	D	IIA
methanol (methyl alcohol)		67-56-1	386	D	IIA
3-methyl-1-butanol (isoamyl alcohol)		123-51-3	350 (Note (2))	D	IIA
methyl ethyl ketone	butanone	78-93-3	404	D	IIA
methyl isobutyl ketone	4-methylpentan-2-one	108-10-1	475	D	IIA
2-methyl-1-propanol (isobutyl alcohol)		78-83-1	415 (Note (2))	D	IIA
2-methyl-2-propanol (tertiary butyl alcohol)		75-65-0	478 (Note (2))	D	IIA
naphtha (see petroleum naphtha)					
natural gas			482 (Note (2))		
petroleum naphtha	naphtha	64742-95-6	290	D	IIA
octanes	octane	111-65-9	206	D	IIA
pentanes	pentanes (mixed isomers)	109-66-0	258	D	IIA
1-pentanol (amyl alcohol)	pentan-1-ol	71-41-0	298	D	IIA
propane		74-98-6	470	D	IIA
1-propanol (propyl alcohol)	propan-1-ol	71-23-8	405	D	IIA
2-propanol (isopropyl alcohol)	propan-2-ol	67-63-0	425	D	IIA
propylene	propene	115-07-1	455	D	IIA
styrene		100-42-5	490	D	IIA
toluene		108-88-3	535	D	IIA
vinyl acetate		108-05-4	425	D	IIA
vinyl chloride	chloroethylene	75-01-4	415	D	IIA
xylenes		106-42-3	464	D	IIA

Notes:

(1) *Most of the values in this table have been obtained from IEC 60079-20, Electrical apparatus for explosive gas atmospheres — Part 20: Data for flammable gases and vapours, relating to the use of electrical apparatus, 1st edition (1996-10). In many cases, the name used in the IEC standard differs from the name typically used in North America for the same substance. In fact, chemicals may have several different names. The CAS number provided above is a well-known method of uniquely identifying chemicals and is a required feature of MSDS documentation. Further information on the CAS numbering system may be found at http://www.cas.org/faq.html.*

(2)   *This substance is not listed in IEC 60079-20.*

(3)   *The name is incorrectly stated in IEC 60079-20 as "1,2-epoxypropene". Propylene oxide is also known as 1,2-epoxypropane.*

Rules JB18-054, JB18-168

Equipment of the heat-producing type is currently required by product standards to have a temperature code (T-Code) marking if its temperature exceeds 100 °C. However, for equipment manufactured prior to the T-Code requirement and motors applied in accordance with Rule J18-168, there may be no such marking. Therefore, the suitability of older hazardous locations equipment of the heat-producing type and motors applied in accordance with Rule J18-168 should be reviewed prior to being installed in a hazardous location to ensure compliance with Rule J18-054. For the purpose of this Rule, equipment such as boxes, terminals, fittings, and RTDs are not considered to be heat-producing devices.

Rule JB18-062

For the purposes of this Rule, metal-covered cable includes a cable with a metal sheath or with a metal armour of the interlocking type, the wire type, or the flat-tape type or with metal shielding.

Rule JB18-062(1)

Suitable lightning protective devices should include primary devices and also secondary devices if overhead secondary lines exceed 90 m in length or if the secondary is ungrounded.

Interconnection of all grounds should include grounds for primary and secondary lightning protective devices, secondary system grounds, if any, and grounds of conduit and equipment of the interior wiring system.

Rule JB18-062(2)(b)

Where single-conductor metal-covered or armoured cables with jackets are used in hazardous locations, the armour must be grounded in only the hazardous location to prevent circulating currents. As a result, there will be a standing voltage on the metal covering in the non-hazardous location area. Therefore, there is a need to properly isolate the armour in the non-hazardous area to ensure that circulating currents will not occur.

Rule JB18-064

To meet the intent of the Rule for effectively maintaining a protective gas pressure, the following references for pressurization are recommended:

(a)   CAN/CSA-E60079-2, *Electrical apparatus — Type of protection "p";*

(b)   NFPA No. 496, *Standard for Purged and Pressurized Enclosures for Electrical Equipment;* and

(c)   IEC 60079-13, *Construction and use of rooms or buildings protected by pressurization.*

Rule JB18-066

See the Note to Rule J18-050.

Rule JB18-066(3)

Intrinsically safe wiring systems are not required to prevent the transmission of an explosion and therefore the only concern is the transmission of gases and vapours. Migration of gas and vapours can be prevented by the use of conduit and cable seals. Other alternatives for cables include the use of a compound such as silicone rubber applied around the end of the connector to prevent gas and vapours from entering the end of the cable.

Rule JB18-070

It is intended that this Rule be used only where suitable equipment, certified for use in the hazardous location, is not available. For example, Class I, Division 1 ignition systems for internal combustion engines are not available; only Class I, Division 2 ignition systems are available. Therefore ignition systems rated for Class I, Division 2 are currently the only hazardous location ignition systems available and could possibly be used in Class I, Division 1 locations.

In many situations, proper area classification will eliminate the need to use this Rule. Rule J18-070 should not be used to compensate for improper area classification. When this Rule is used, the gas detection system should consist of an adequate number of sensors to ensure the sensing of flammable gases or vapours in all areas where they may accumulate.

Electrical equipment suitable for non-hazardous locations and having unprotected arcing, sparking, or heat-producing components must not be installed in a Division 2 location. Arcing, sparking, or heat-producing

Appendix J
Rules and Notes to Rules for installations using the
Division system of classification for Class I locations

components may be protected by encapsulating, hermetically sealing, or sealing by other means such as restricted breathing.

Before applying this Rule, the user should fully understand the risks associated with such an installation. When applying this Rule, it remains the responsibility of the owner of the facility, or his or her agents, to ensure that the resulting installation is safe. Simply complying with the requirements of Rule J18-070 may not ensure a safe installation in all situations.

Rule JB18-100

Electrical equipment marked with a method of protection, (e.g., "d" "Flame-proof") is not approved for Class I Division 1 locations unless also marked with the "Class" of location.

Rules JB18-106, JB18-156

Cables approved for hazardous locations are suitable for all locations, but the termination fittings must be approved for the particular hazardous location.

Rule JB18-106(3)

Where tapered threads are used, the requirement to have five fully engaged threads (i.e., threads done up tight) is critical for three reasons:

(a) when the threads are not fully engaged, the flame path is compromised making it possible for an explosion occurring within the conduit system to be transmitted to the area outside the conduit;

(b) if there are not five fully engaged threads, the flame path may be too short to cool the gases resulting from an internal explosion to a temperature below that which could ignite gas in the surrounding area; and

(c) as the conduit forms a bonding path to ground, not making the conduit tight will introduce resistance into the flame path and if a fault occurs arcing may result at the interface.

While it may not always be possible to install certain fittings without backing off, it is important to ensure the connection is as tight as possible. Properly made conduit connections are critical to the safety of hazardous location wiring systems.

Rules JB18-108, JB18-158

Seals are provided in conduit or cable systems to prevent the passage of gases, vapours, or flames from one portion of the electrical installation to another through the system.

Passage of gases, vapours, or flames through mineral-insulated cable is prevented by the construction of the cable, but sealing compound is used in cable glands to exclude moisture and other fluids from the cable insulation and is required to be of a type approved for the conditions of use.

Sealing of conductors in the conduit, or in most cables, requires the sealing compound to completely surround each individual insulated conductor to ensure that the seal performs its intended function. In certain constructions of cables, specifically those that contain bundles of shielded pairs, triads, or quads, removal of the shielding or overall covering from the bundles negates the purpose for which the shielding was provided. Testing of this type of cable now includes testing for flame propagation along the length of the individual subassemblies of the cable.

The letters A, B, C, or D, or a combination thereof, may be added to signify the group(s) for which the cable has been tested, for example,

(a) "HL-CD" indicates the cable has been tested for flame propagation for gas groups C and D; and

(b) "TC-BCD" indicates the cable has been tested for flame propagation for gas groups B, C, and D.

See also the Table in the Note to Rule J18-052(2).

Rules JB18-108(1), JB18-158(1)

Where devices such as pressure switches, flow devices, etc., are connected to a process containing flammable fluids, failure of the seal (primary) in these devices could release the flammable fluids into the wiring system where they may migrate to a safe area where electrical or other devices are not constructed to prevent explosions. Because conduit and cable seals are not designed to seal against continuous pressure and will allow slow passage of flammable fluids, secondary seals designed to prevent the passage of flammable fluids into the wiring system are required.

Rules JB18-108(2), JB18-158(2)

Various methods could be used to detect the failure of a primary seal. Monitoring the secondary seal vent with flow detection or gas detection are two possible means. It would also be possible to connect the vents of a number of secondary seals and monitor the common discharge. Other methods are also possible. Any method used should not restrict the venting of the seal to atmosphere.

Rules JB18-108(4), JB18-158(4)

Conduit fittings approved for Class I locations and similar to the "L", "T", or "Cross" type would not usually be classed as enclosures when not larger than the trade size of the conduit.

Reducers may have one side larger than the trade size of the conduit where the entry to the explosion-proof or flame-proof enclosure is larger than the trade size of the conduit.

Rules JB18-108(5)(a), JB18-158(5)(a)

Cables and flexible cords are not tested to determine their ability to resist internal explosions. Therefore, regardless of size, each cable must be sealed at the point of entry into any enclosure that is required to be explosion-proof.

Some designs of cable glands incorporate an integral seal and these are marked "SL" to indicate that the seal is provided by the cable gland. Cable glands of this type are identified with the class designation. Designs requiring a field- or factory-installed sealing fitting have the group designation marked on this component.

Because the appropriate sealing characteristics may be achieved by different means, the manufacturer's instructions should be followed.

Rule JB18-108(6)

It is important that the manufacturer's instructions are adhered to closely or seals will not function properly to prevent the transmission of an explosion or to prevent the transmission of flammable fluids to non-hazardous areas where they will be exposed to unprotected ignition sources. Improper sealing has been the primary factor in a number of explosions, resulting in loss of life and/or major equipment damage. Users are reminded that only the sealing compound outlined with the instructions may be used in a seal. Use of other manufacturer's compounds in a seal may compromise the integrity of the installation.

Rule JB18-108(6)(a)(ii)

All motors and generators approved under the applicable Part II Standards for Class I locations are required to have a seal provided by the manufacturer between the main motor or generator enclosure and the enclosure for the conduit entry (connection box). Therefore a marking regarding the seal being provided is not necessary on this particular class of product.

For cables, compliance with Rule J18-108(6) can be accomplished by
(a) a cable gland approved for Class I hazardous locations for appropriate cable type(s) and a field-installed sealing fitting;
(b) a cable gland approved for Class I hazardous locations for appropriate cable types with an integral seal; or
(c) a cable gland for approved cable types used with an approved enclosure provided with sealing as specified in Rule J18-108(6).
Cable glands with integral seals are marked "SL".

Rule JB18-114, JB-168

Users are cautioned that combining a variable frequency drive (VFD) with a motor may increase the operating temperature of the motor as a result of the harmonics produced by the drive. This may cause the motor temperature to exceed its temperature code rating. This is of particular concern where the operating temperature of the motor is close to the ignition temperature of hazardous materials that may be in the area. Because of the generally lower ignition temperatures associated with Class II materials, it will be of particular concern in Class II areas. It remains the responsibility of the user to ensure that the operating temperature of the motor, in combination with the drive, is below the minimum ignition temperature of the hazardous material in the area. The motor manufacturer should be consulted where necessary. Some references that may assist the user in determining the suitability of an installation are:
(a) CSA Technical Information Letter E-22, "Motors and Generators For Use in Class I, Division 2 and Class II, Division 2 Locations";

Appendix J
Rules and Notes to Rules for installations using the
Division system of classification for Class I locations

(b) API RP 2216, *Ignition Risk of Hydrocarbon Vapors by Hot Surfaces in the Open Air*; and

(c) IEEE Paper No. PCIC-97-04, "Flammable Vapor Ignition Initiated By Hot Rotor Surfaces Within an Induction Motor — Reality or Not?".

Rule JB18-116

Gas turbines in hazardous locations also need safeguards against potential hazards from other than electrical ignition systems, such as exhaust and fuel systems. The complete engine assembly should be investigated for its suitability in Class I, Division 1 hazardous locations.

Rule JB18-150

Equipment marked Class I, Division 2 is suitable only for Division 2.

Rule JB18-156

See the Note to Rule J18-106.

Rule J18-156(8)

Cable glands should be compatible with the degree of ingress protection and explosion protection provided by the enclosure on which they are installed.

For example, to maintain the protection of an enclosure required to be explosion-proof, a sealing-type gland approved for the location should be used. Where unarmoured cables must enter an enclosure required to be explosion-proof, a combination of an approved sealing fitting and a non-sealing cable gland may be used.

Where equipment normally considered suitable for use in ordinary locations is acceptable in Division 2 locations, such as terminal boxes and motors, ordinary location cable glands that maintain the degree of protection of the enclosure may be used. Similarly, where purged enclosures are used in Division 1 and Division 2 locations, ordinary location cable glands that maintain the degree of protection of the enclosure may be used.

Where equipment is specifically designed for use in Division 2 locations, ordinary location cable glands that maintain the degree of protection of the enclosure may be used. One means of achieving equivalent protection would be the use of a cable gland with the same or better IP rating as the enclosure. (See Ingress Protection table in Appendix B Note to Rule 18-106(5).) If the gland does not have an IP rating, other ratings, such as weatherproof, may be matched to the enclosure rating.

Rule JB18-158(1)

See the Note to Rule J18-108(1).

Rule JB18-158(3)(c)

This Rule allows the seal at the boundary between an outdoor Class I, Division 2 location and an outdoor non-hazardous location, to be located further than 300 mm from the boundary of the Class I, Division 2 location provided that it is located on the conduit prior to it entering an enclosure or a building. As gas is present only in Class I, Division 2 locations for short periods, it is unlikely that gas or vapour could be released through conduit couplings at sufficiently high rates to form an explosive mixture in outdoor areas. However, the seal must be located before the conduit enters an enclosure or a building as, depending on the ventilation rate, gas transmitted through the conduit may build up to flammable concentrations.

Rule JB18-158(4)

See the Note to Rule J18-108(4).

Rule JB18-158(5)(a)

See the Note to Rule J18-108(5)(a).

Rule JB18-160

This Rule includes service and branch circuit switches and circuit breakers; motor controllers including push buttons, pilot switches, relays, and motor-overload protective devices; and switches and circuit breakers for the control of lighting and appliance circuits. Oil-immersed circuit breakers and controllers of ordinary general-use type may not confine completely the arc produced in the interruption of heavy overloads, and specific approval for locations of this class and division is therefore necessary.

Rule JB18-166

A group of three fuses protecting an ungrounded 3-phase circuit, and a single fuse protecting the ungrounded conductor of an identified 2-wire single-phase circuit, would each be considered as a set of fuses.

Rule JB18-168

See the Note to Rule J18-114.

Rule JB18-170

It should be recognized that internal combustion engines in hazardous locations also need safeguards against potential hazards from other than electrical ignition systems, such as exhaust and fuel systems. The complete engine assembly should therefore be investigated for its suitability in Class I, Division 2 hazardous locations.

Rule JB20-004

For purposes of Subrules (6) and (7), buildings such as kiosks in which electrical equipment such as cash registers and/or self-service console controls are located are considered to be buildings not suitably cut off.

Rule JB20-030

Information on non-electrical aspects for propane tank systems, refill centres, and filling plants may be found in CAN/CSA-B149.2.

Rule JB20-060

Information on non-electrical aspects of compressed natural gas (NGV) refuelling stations and NGV storage facilities may be found in CAN/CSA-B149.1.

Appendix J
Rules and Notes to Rules for installations using the
Division system of classification for Class I locations

Annex JD
Diagrams for Annex J20

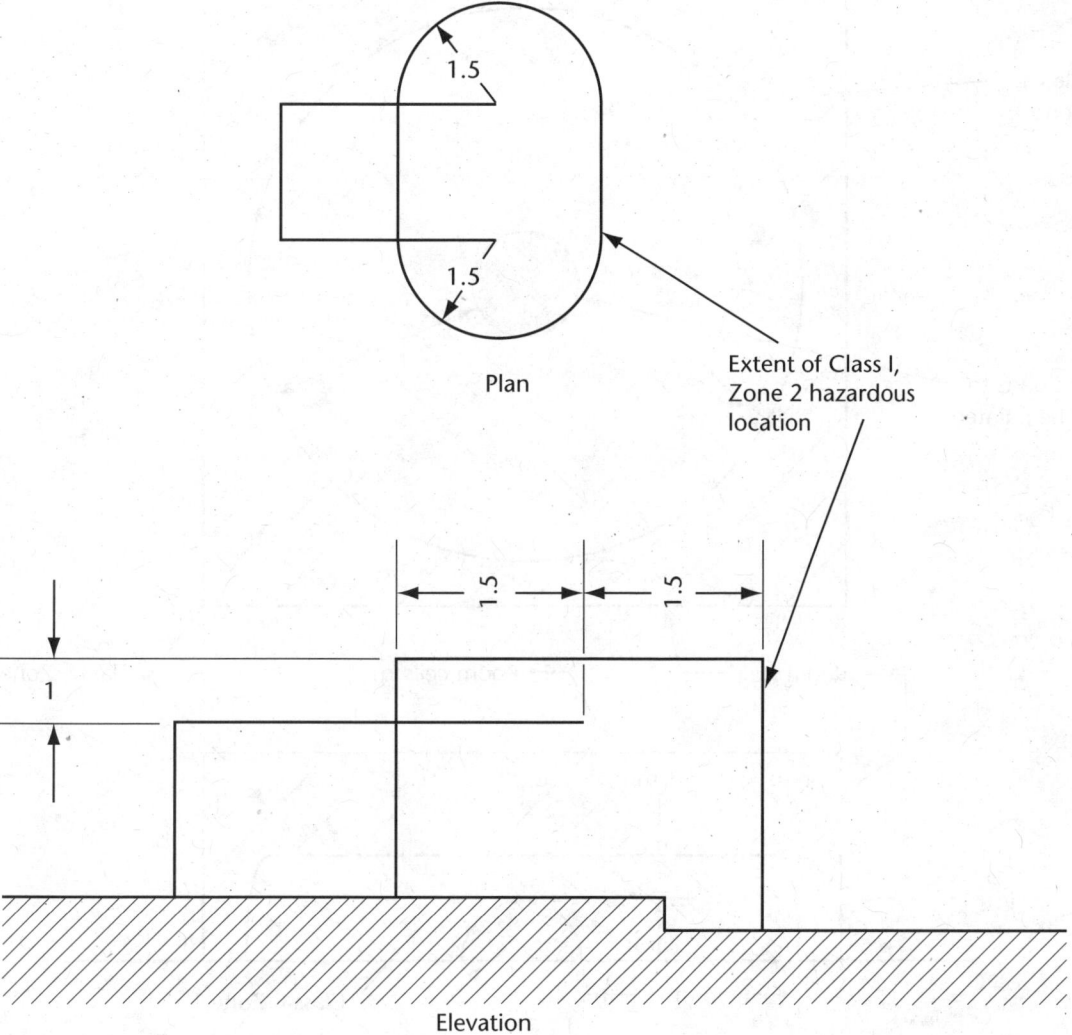

Plan

Extent of Class I,
Zone 2 hazardous
location

Elevation

Note: *All dimensions given are in metres.*

Diagram JD-4
Extent of hazardous location for open-face spray booths
(See Rule J20-402(2).)

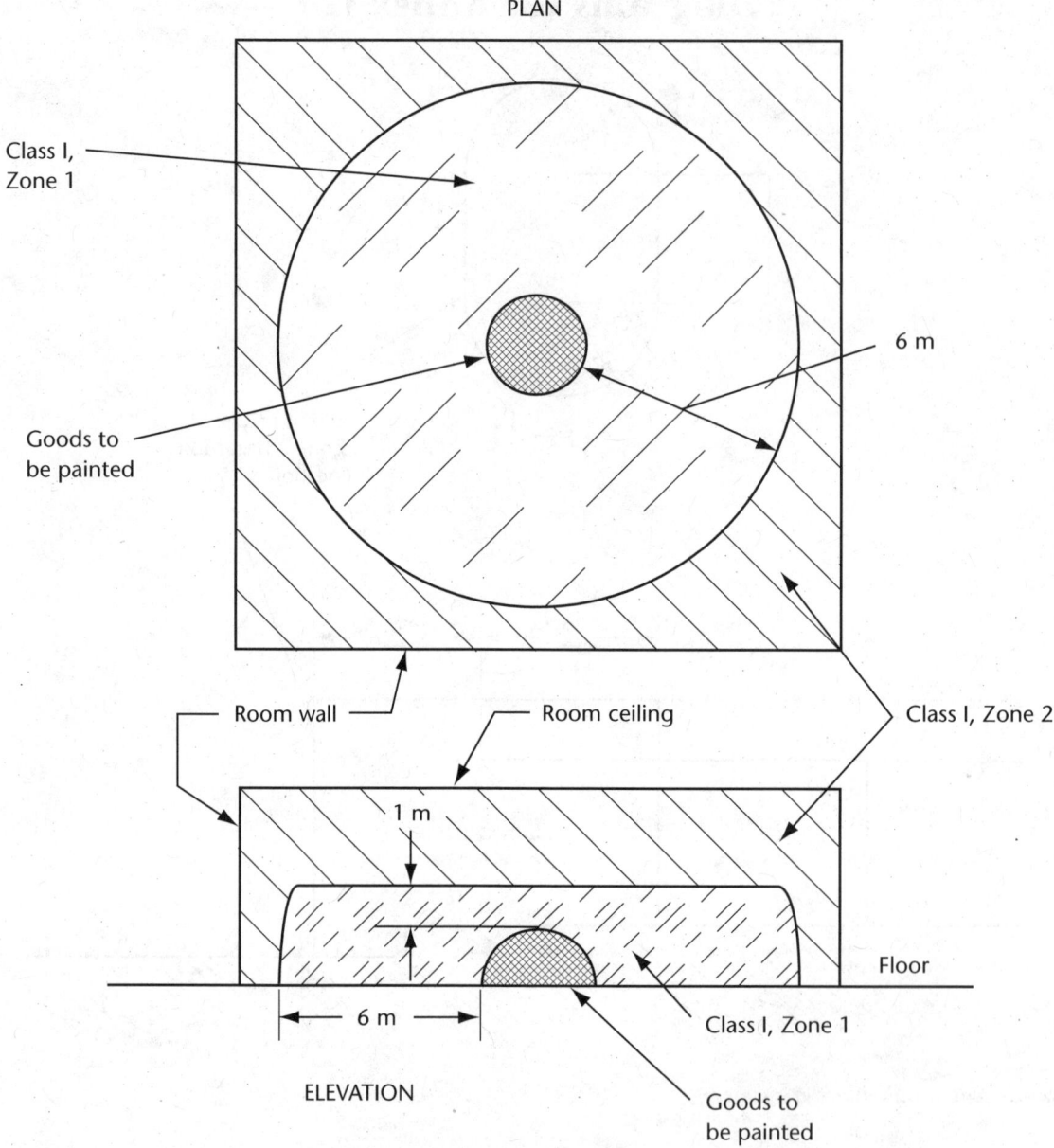

PLAN

Class I,
Zone 1

Goods to
be painted

6 m

Room wall — — Room ceiling

Class I, Zone 2

1 m

Floor

6 m

Class I, Zone 1

ELEVATION

Goods to
be painted

Diagram JD-5
Extent of hazardous location for spraying
operations not conducted in spray booths
(See Rules J20-402(1)(b) and J20-402(4).)

Appendix J
Rules and Notes to Rules for installations using the
Division system of classification for Class I locations

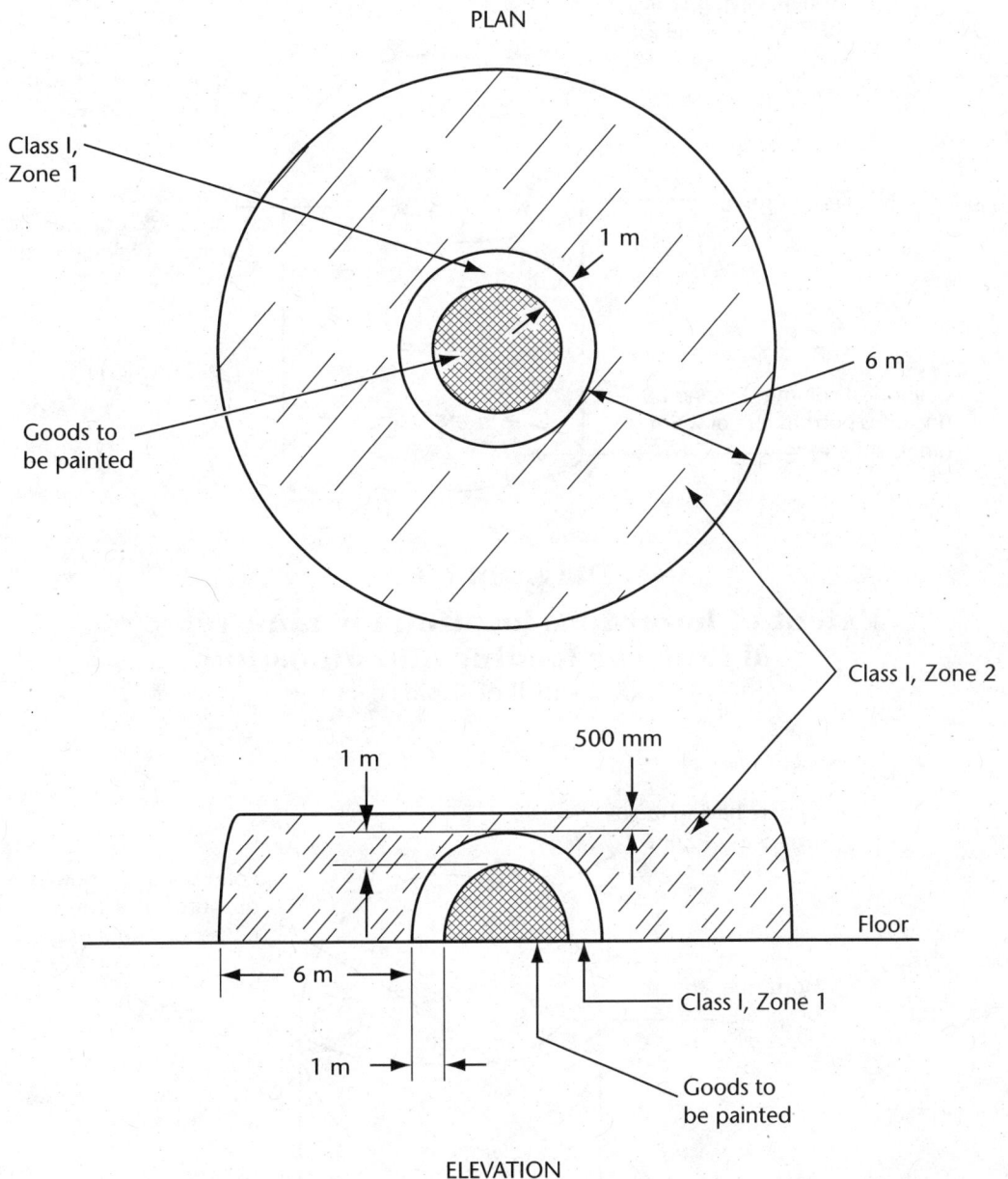

PLAN

Class I,
Zone 1

1 m

6 m

Goods to
be painted

Class I, Zone 2

500 mm

1 m

Floor

6 m

Class I, Zone 1

1 m

Goods to
be painted

ELEVATION

Diagram JD-6
Extent of hazardous location for spraying operations not conducted
in spray booths — Ventilation system interlocked
(See Rule J20-402(7).)

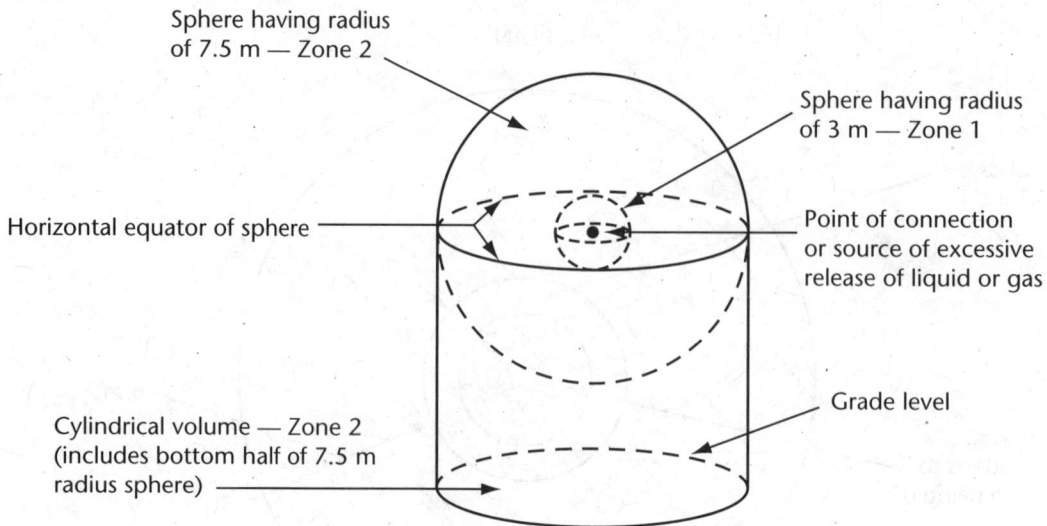

Diagram JD-7

**Extent of hazardous location for tank vehicle
and tank car loading and unloading**

(See Part B of Table JT-63.)

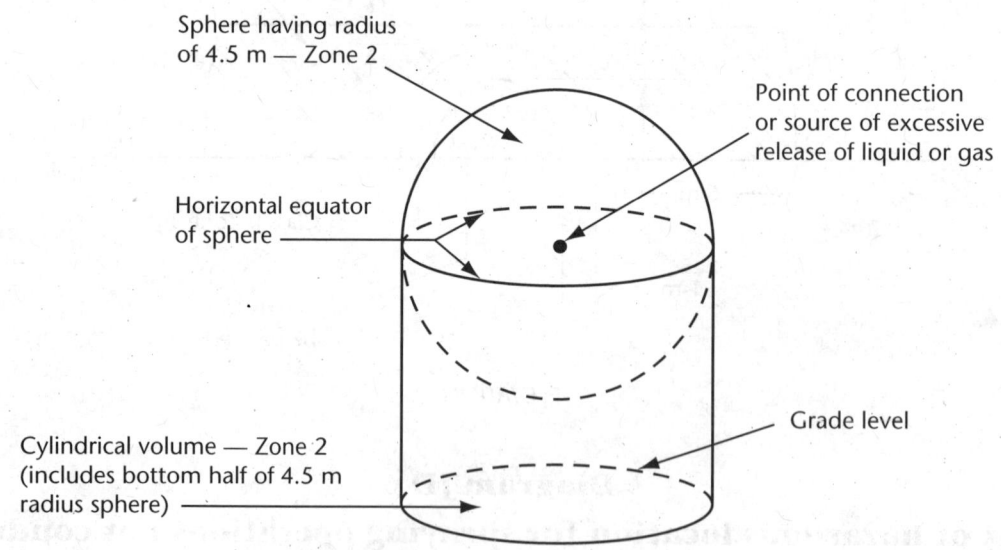

Diagram JD-8

**Extent of hazardous location for pumps, vapour compressors,
gas-air mixers, and vaporizers outdoors in open air**

(See Part E of Table JT-63.)

Appendix J
Rules and Notes to Rules for installations using the
Division system of classification for Class I locations

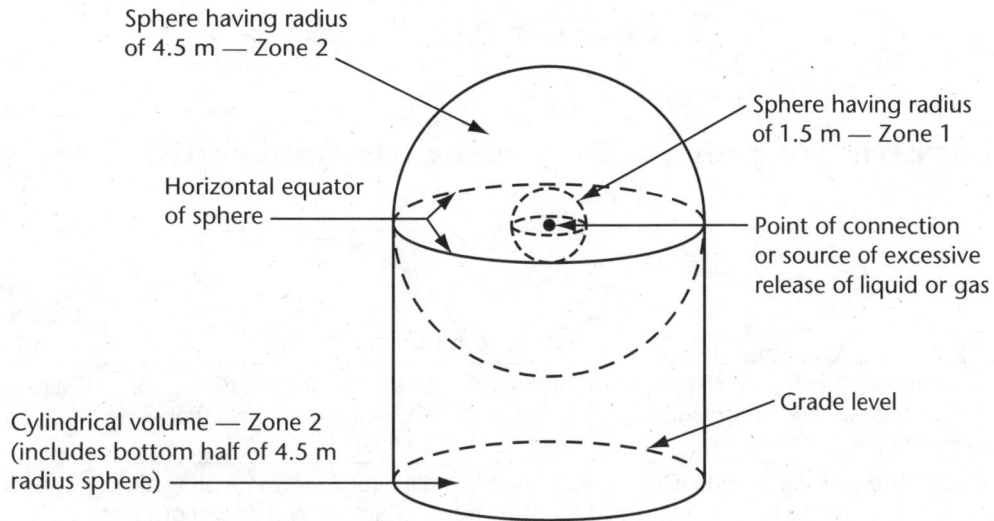

Diagram JD-9
Extent of hazardous location for container filling outdoors in open air
(See Part J of Table JT-63.)

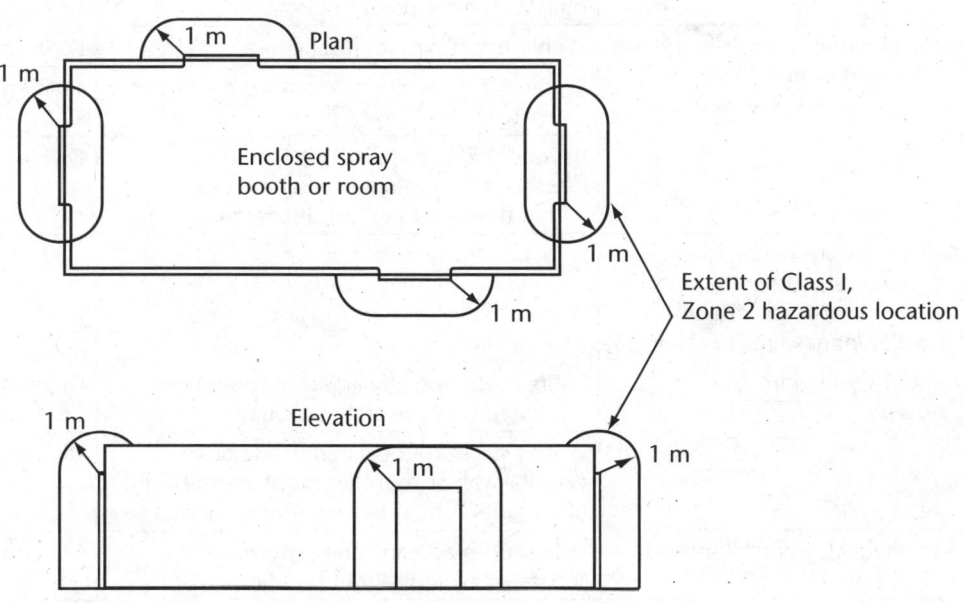

Diagram JD-10
Extent of hazardous location adjacent to openings in a closed spray booth or room
(See Rule J20-402(3).)

Annex JT
Tables for Annex J20

Table JT-63
Hazardous areas for propane dispensing, container filling, and storage
(See Rule J20-034.)

Part	Location	Extent of hazardous locations*	Division of Class I, Group D hazardous location
A	Storage containers other than CTC/DOT cylinders and ASME vertical containers of less than 454 kg water capacity	Within 4.5 m in all directions from connections, except connections otherwise covered in this Table	Division 2
B	Tank vehicle and tank car loading and unloading†	Within 3 m in all directions from connections regularly made or disconnected from product transfer	Division 1
		Beyond 3 m but within 7.5 m in all directions from a point where connections are regularly made or disconnected and within the cylindrical volume between the horizontal equator of the sphere and grade (see Diagram JD-7)	Division 2
C	Gauge vent openings other than those on CTC/DOT cylinders and ASME vertical containers of less than 454 kg water capacity	Within 1.5 m in all directions from point of discharge	Division 1
		Beyond 1.5 m but within 4.5 m in all directions from point of discharge	Division 2
D	Relief device discharge other than those on CTC/DOT cylinders and ASME vertical containers of less than 454 kg water capacity	Within direct path of discharge‡	Division 1
		Within 1.5 m in all directions from point of discharge	Division 1
		Beyond 1.5 m but within 4.5 m in all directions from point of discharge except within the direct path of discharge	Division 2
E	Pumps, vapour compressors, gas-air mixers, and vaporizers (other than direct-fired or indirect-fired with an attached or adjacent gas-fired heat source)		
	Indoors without ventilation	Entire room and any adjacent room not separated by a gas-tight partition	Division 1
		Within 4.5 m of the exterior side of any exterior wall or roof that is not vapour-tight or within 4.5 m of any exterior opening	Division 2
	Indoors with adequate ventilation	Entire room and any adjacent room not separated by a gas-tight partition	Division 2
	Outdoors in open air at or above grade	Within 4.5 m in all directions from this equipment and within the cylindrical volume between the horizontal equator of the sphere and grade (see Diagram JD-8)	Division 2

(Continued)

Appendix J
Rules and Notes to Rules for installations using the
Division system of classification for Class I locations

Table JT-63 (Continued)

Part	Location	Extent of hazardous locations*	Division of Class I, Group D hazardous location
F	Service station dispensing units	Entire space within dispenser's enclosure, or up to a solid partition within the enclosure at any height above the base. The space within 450 mm horizontally from the dispenser enclosure up to 1.2 m above the base or to the height of a solid partition within the enclosure. Entire pit or open space beneath the dispenser	Division 1
		The space above a solid partition within the dispenser enclosure. The space up to 450 mm above grade within 6 m horizontally from any edge of the dispenser enclosure§	Division 2
G	Pits or trenches containing or located beneath propane gas valves, pumps, vapour compressors, regulators, and similar equipment	Entire pit or trench	Division 1
	Without mechanical ventilation	Entire room and any adjacent room not separated by a gas-tight partition	Division 2
		Within 4.5 m in all directions from pit or trench when located outdoors	Division 2
	With adequate mechanical ventilation	Entire pit or trench	Division 2
		Entire room and any adjacent room not separated by a gas-tight partition	Division 2
		Within 4.5 m in all directions from pit or trench when located outdoors	Division 2
H	Special buildings or rooms for storage of portable containers	Entire room	Division 2
I	Pipelines and connections containing operational bleeds, drips, vents, or drains	Within 1.5 m in all directions from point of discharge	Division 1
		Beyond 1.5 m from point of discharge, same as Part E of this Table	—
J	Container filling: Indoors with adequate ventilation	Within 1.5 m in all directions from the dispensing hose inlet connections for product transfer	Division 1
		Beyond 1.5 m and entire room	Division 2
	Outdoors in open air	Within 1.5 m in all directions from the dispensing hose inlet connections for product transfer	Division 1
		Beyond 1.5 m but within 4.5 m in all directions from the dispensing hose inlet connections and within the cylindrical volume between the horizontal equator of the sphere and grade (see Diagram JD-9)	Division 2

(Continued)

Table JT-63 (Concluded)

Part	Location	Extent of hazardous locations*	Division of Class I, Group D hazardous location
K	Outdoor storage area for portable cylinders or containers Aggregate storage up to and including 454 kg water capacity	Within 1.5 m in all directions from connections	Division 2
	Aggregate storage over 454 kg water capacity	Within 4.5 m in all directions from connections	Division 2

The classified area shall not extend beyond an unpierced wall, roof, or solid vapour-tight partition.

†*When classifying the extent of a hazardous area, consideration shall be given to possible variations in the locating of tank cars and tank vehicles at the unloading points and the effect these variations in location may have on the point of connection.*

‡*Fixed electrical equipment should preferably not be installed in this space.*

§*For pits within this area, see Part G of this Table.*

Table JT-64
Class I, Division 1 space surrounding compressed natural gas (NGV) storage facilities

(See Rule J20-064.)

Storage volume water capacity, L	Distance measured from containers, m*
Up to and including 4000	2.5
Over 4000, up to and including 10 000	4
Over 10 000	10

When a wall with a 4-hour fire resistance rating is located within these distances, the distances shall be measured either around the end of or over the wall but not through it. This wall shall not be located closer than 1 m from a fuel container with up to 10 000 L in storage volume, and 1.5 m from a fuel container with a storage volume greater than 10 000 L.

Where the wall of an adjacent building other than a compressor enclosure is within the specified distance and serves as the 4-hour fire resistance rated wall, it shall have no doors, windows, or openings in it unless the building is also classified as a Class I, Division 1 location.

Δ

Appendix K — Extract from IEC 60364-1
Chapter 13

Notes:
(1) *This Appendix is an informative (non-mandatory) part of this Standard.*
(2) *This extract is reprinted from IEC 60634-1 with permission.*

13 Fundamental principles

Note: *Where countries not yet having national regulations for electrical installations deem it necessary to establish legal requirements for this purpose, it is recommended that such requirements be limited to fundamental principles which are not subject to frequent modification on account of technical development. The contents of clause 13 may be used as a basis for such legislation.*

131 Protection for safety

131.1 General

The requirements stated in this subclause are intended to ensure the safety of persons, livestock and property against dangers and damage which may arise in the reasonable use of electrical installations.

Note: *In electrical installations, two major types of risk exist:*
— shock currents;
— excessive temperatures likely to cause burns, fires and other injurious effects.

131.2 Protection against electric shock

131.2.1 Protection against direct contact

Persons and livestock shall be protected against dangers that may arise from contact with live parts of the installation.

This protection can be achieved by one of the following methods:
— preventing a current from passing through the body of any person or any livestock;
— limiting the current which can pass through a body to a value lower than the shock current.

131.2.2 Protection against indirect contact

Persons and livestock shall be protected against dangers that may arise from contact with exposed-conductive-parts in case of a fault.

This protection can be achieved by one of the following methods:
— preventing a fault current from passing through the body of any person or any livestock;
— limiting the fault current which can pass through a body to a value lower than the shock current;
— automatic disconnection of the supply in a determined time on the occurrence of a fault likely to cause a current to flow through a body in contact with exposed-conductive-parts, where the value of that current is equal to or greater than the shock current.

Note: *In connection with the protection against indirect contact, the application of the method of equipotential bonding is one of the important principles for safety.*

131.3 Protection against thermal effects

The electrical installation shall be so arranged that there is no risk of ignition of flammable materials due to high temperature or electric arc. In addition, during normal operation of the electrical equipment, there shall be no risk of persons or livestock suffering burns.

131.4 Protection against overcurrent

Persons or livestock shall be protected against injury and property shall be protected against damage due to excessive temperatures or electromechanical stresses caused by any overcurrents likely to arise in live conductors.

This protection can be achieved by one of the following methods:
— automatic disconnection on the occurrence of an overcurrent before this overcurrent attains a dangerous value taking into account its duration;
— limiting the maximum overcurrent to a safe value and duration.

131.5 Protection against fault currents

Conductors, other than live conductors, and any other parts intended to carry a fault current shall be capable of carrying that current without attaining an excessive temperature.

NOTE 1 Particular attention should be given to earth fault currents and leakage current.

NOTE 2 For live conductors, compliance with 131.4 assures their protection against overcurrents caused by faults.

131.6 Protection against overvoltage

131.6.1 Persons or livestock shall be protected against injury and property shall be protected against any harmful effects as a consequence of a fault between live parts of circuits supplied at different voltages.

131.6.2 Persons or livestock shall be protected against injury and property shall be protected against damage as a consequence of any excessive voltages likely to arise due to other causes (e.g. atmospheric phenomena or switching overvoltages).

132 Design

132.1 General

For the design of the electrical installation, the following factors shall be taken into account to provide:
— the protection of persons, livestock and property in accordance with clause 131;
— the proper functioning of the electrical installation for the use intended;

The information required as a basis for design is listed in 132.2 to 132.5. The requirements with which the design should comply are stated in 132.6 to 132.12.

132.2 Characteristics of available supply or supplies

132.2.1 Nature of current: a.c. and/or d.c.

132.2.2 Nature and number of conductors:
— For a.c.: phase conductor(s);
  neutral conductor;
  protective conductor.
— For d.c.: conductors equivalent to those listed above.

132.2.3 Values and tolerances:
— voltage and voltage tolerances;
— frequency and frequency tolerances;
— maximum current allowable;
— prospective short-circuit current.

132.2.4 Protective measures inherent in the supply, e.g. earthed (grounded) neutral or mid-wire

132.2.5 Particular requirements of the supply undertaking

132.3 Nature of demand

The number and type of circuits required for lighting, heating, power, control, signalling, telecommunication, etc. are to be determined by:
— location of points of power demand;
— loads to be expected on the various circuits;
— daily and yearly variation of demand;
— any special conditions;
— requirements for control, signalling, telecommunication, etc.

132.4 Emergency supply or supplies

— Source of supply (nature, characteristics).
— Circuits to be supplied by the emergency source.

132.5 Environmental conditions

See IEC 60364-5-51 and IEC 60721.

132.6 Cross-section of conductors

The cross-section of conductors shall be determined according to:
(a) their admissible maximum temperature;
(b) the admissible voltage drop;
(c) the electromechanical stresses likely to occur due to short-circuits;
(d) other mechanical stresses to which the conductors can be exposed;
(e) the maximum impedance with respect to the functioning of the short-circuit protection.

Note: *The above-listed items concern primarily the safety of electrical installations. Cross-sectional areas greater than those required for safety may be desirable for economic operation.*

132.7 Type of wiring and methods of installation

The choice of the type of wiring and the methods of installation depend on:
— the nature of the locations;
— the nature of the walls or other parts of the building supporting the wiring;
— accessibility of wiring to persons and livestock;
— voltage;
— the electromechanical stresses likely to occur due to short-circuits;
— other stresses to which the wiring can be exposed during the erection of the electrical installation or in service.

132.8 Protective equipment

The characteristics of protective equipment shall be determined with respect to their function which may be, e.g., protection against the effects of:
— overcurrent (overload, short-circuit);
— earth-fault current;
— overvoltage;
— undervoltage and no-voltage.

The protective devices shall operate at values of current, voltage and time which are suitably related to the characteristics of the circuits and to the possibilities of danger.

132.9 Emergency control

Where, in case of danger, there is necessity for immediate interruption of supply, an interrupting device shall be installed in such a way that it can be easily recognized and effectively and rapidly operated.

132.10 Disconnecting devices

Disconnecting devices shall be provided so as to permit disconnection of the electrical installation, circuits or individual items of apparatus as required for maintenance, testing, fault detection or repair.

132.11 Prevention of mutual influence

The electrical installation shall be arranged in such a way that no mutual detrimental influence will occur between the electrical installation and non-electrical installations of the building.

132.12 Accessibility of electrical equipment

The electrical equipment shall be arranged so as to afford as may be necessary:
— sufficient space for the initial installation and later replacement of individual items of electrical equipment;
— accessibility for operation, testing, inspection, maintenance and repair.

133 Selection of electrical equipment

133.1 General

Every item of electrical equipment used in electrical installations shall comply with such IEC standards as are appropriate.

133.2 Characteristics

Every item of electrical equipment selected shall have suitable characteristics appropriate to the values and conditions on which the design of the electrical installation (see clause 132) is based and shall, in particular, fulfil the following requirements.

133.2.1 Voltage

Electrical equipment shall be suitable with respect to the maximum steady voltage (r.m.s. value for a.c.) likely to be applied, as well as overvoltages likely to occur.

Note: *For certain equipment, it may be necessary to take account of the lowest voltage likely to occur.*

133.2.2 Current

All electrical equipment shall be selected with respect to the maximum steady current (r.m.s. value for a.c.) which it has to carry in normal service, and with respect to the current likely to be carried in abnormal conditions and the period (e.g. operating time of protective devices, if any) during which it may be expected to flow.

133.2.3 Frequency

If frequency has an influence on the characteristics of electrical equipment, the rated frequency of the equipment shall correspond to the frequency likely to occur in the circuit.

133.2.4 Power

All electrical equipment, which is selected on the basis of its power characteristics, shall be suitable for the duty demanded of the equipment, taking into account the load factor and the normal service conditions.

133.3 Conditions of installation

All electrical equipment shall be selected so as to withstand safely the stresses and the environmental conditions (see 132.5) characteristic of its location and to which it may be exposed. If, however, an item of equipment does not have by design the properties corresponding to its location, it may be used on condition that adequate additional protection is provided as part of the completed electrical installation.

133.4 Prevention of harmful effects

All electrical equipment shall be selected so that it will not cause harmful effects on other equipment or impair the supply during normal service including switching operations. In this context, the factors which can have an influence include
— power factor;
— inrush current;
— asymmetrical load;
— harmonics.

134 Erection and initial verification of electrical installations

134.1 Erection

134.1.1 For the erection of the electrical installation, good workmanship by suitably qualified personnel and the use of proper materials shall be provided for.

134.1.2 The characteristics of the electrical equipment, as determined in accordance with clause 133, shall not be impaired in the process of erection.

134.1.3 Conductors shall be identified in accordance with IEC 60446.

134.1.4 Connections between conductors and between conductors and other electrical equipment shall be made in such a way that safe and reliable contact is ensured.

134.1.5 All electrical equipment shall be installed in such a manner that the designed cooling conditions are not impaired.

134.1.6 All electrical equipment likely to cause high temperatures or electric arcs shall be placed or guarded so as to eliminate the risk of ignition of flammable materials. Where the temperature of any exposed parts of electrical equipment is likely to cause injury to persons, those parts shall be so located or guarded as to prevent accidental contact therewith.

134.2 Initial verification

Electrical installations shall be tested and inspected before being placed in service and after any important modification to verify proper execution of the work in accordance with this standard.

Index

A

B

D

E

O

P

S

W

X

Y